ANALYSIS OF LANDFORMS

ANALYSIS OF LANDFORMS

C. R. Twidale

Department of Geography
University of Adelaide

John Wiley & Sons Australasia Pty Ltd

SYDNEY   NEW YORK   LONDON   TORONTO

Printed in Singapore by Toppan Printing Co (S) Pte Ltd.

10 9 8 7 6 5 4 3 2 1

National Library of Australia Cataloguing in Publication data:

Twidale, Charles Rowland
  Analysis of landforms.

  Index.
  Bibliography.
  ISBN 0 471 89465 6

  1. Landforms.   I. Title.

551.4

Contents

Preface vi

PART 1 Introduction   I

1   Geomorphology—Scope, Problems and Methods 3

PART 2 Structural Geomorphology 19

2   Major Features of the Planet Earth 21
3   Fractures in the Crust: Joints 42
4   Fractures in the Crust: Faults 81
5   Landforms Developed on Sedimentary Sequences 101
6   Volcanoes and Volcanic Landforms 125
7   Pseudostructural Landforms 159

PART 3 Geomorphological Processes and Climatic Geomorphology 171

8   The Concept of Morphogenetic Regions 173
9   Weathering Processes 179
10   Geomorphological Significance of Weathering 195
11   Mass Movement of Debris 217
12   Running Water and Some Related Forms 229
13   Valleys and Valley Side Slopes 256
14   Rivers and Running Water in the Tropics 273
15   Wind Action on Land Surfaces 282
16   Snow and Ice 317
17   Glaciers and Their Work 333
18   Coastal Processes and Landforms 366

PART 4 Changes in Time—Historical Geomorphology 401

19   Models of Landscape Evolution 403
20   Drainage Patterns 431
21   Climatic Change and Direct Effects of Late Cainozoic Glaciation 453
22   Indirect Effects of Late Cainozoic Climatic Changes 480
23   Older Inherited Forms 505
24   Landforms Related to the Activities of Organisms, Including Man 511

PART 5 529

25   Factors in the Analysis of Landforms 531

Appendix—Geological Background 535

Index 549

Preface

This book is intended for use as a text for advanced secondary students and for tertiary students taking courses in geography, geology, earth sciences and environmental sciences. It was hoped to provide a general text with an emphasis on Australia, yet with an adequate coverage of world examples. In the belief that photographs and diagrams, in many instances, are better able than words to convey both the essentials and the complexities of landforms, the book has been lavishly illustrated. As its title suggests, *Analysis of Landforms* is intended to illustrate how evidence is used on geomorphological analysis. Unfortunately, it falls short of this high and important aim, if only because to have carried out an exhaustive analysis of each and every landform would have expanded the book to impossible lengths; in the event, relatively few landforms have been subjected to close examination. However, the methods of analysis are outlined and exemplified.

The book has been a long time in the making and many have from time to time contributed to its eventual completion. Some have assisted by providing unpublished information or photographs; their help is acknowledged in the text. Others, notably Kerry Nicol, Pam Metcalf and especially Jennifer Bourne, have acted as guinea pigs on whom various chapters have been tried out. Jennie has also contributed greatly with proof reading and the preparation of the index. Finally, I must express my thanks to a patient and understanding family, and an equally indulgent publisher.

Adelaide,
October, 1975

C. R. TWIDALE

Part 1

INTRODUCTION

"*While man has to learn, mankind must have different opinions. It is the prerogative of man to form opinions; these indeed are often, commonly I may say, erroneous; but they are commonly corrected and it is thus that truth in general is made to appear.*"
—James Hutton, 1788.

CHAPTER 1

Geomorphology — Scope, Problems and Methods

A. THE SCOPE AND AIMS OF GEOMORPHOLOGY

More than ever before are we prone to travel, and this together with the penetration of our minds (and homes) by the various communications media — television, films, books, magazines — has been instrumental in bringing to our attention the different landscapes and landforms in various parts of the earth. We see Peter O'Toole as Lawrence of Arabia staggering across the searing desert of sand and salt in the Sinai Peninsula, or John Mills as Captain Scott perishing in the icy wastes of another desert — Antarctica. We can contrast the magnificent mountain scenery of the opening sequence of *The Sound of Music* with the featureless Fens of eastern England which form the setting of another of Lord Peter Wimsey's triumphs of detection in Dorothy Sayers' *The Nine Tailors*, with the stark gritstone tors standing above the bleak Yorkshire Moors in *Wuthering Heights,* or with the bleak stony deserts in central Australia encountered in imagination by Patrick White[1] and in harsh reality by Sturt and by Burke and Wills. We can compare the rolling chalk Downs which formed the setting of Hardy's *Far From the Madding Crowd* with the huge rounded boulders of granite which provide cover (for goodies and baddies alike) during the remarkably ineffectual gun-fights depicted in so many Western films. Are these variations in the form of the land surface capricious, or are they susceptible to rational analysis?

The scientific discipline which is concerned with the explanation of the varied morphology of the form of the land surface is *geomorphology,* and its basic tenets are that morphological variations are not accidental or haphazard; but that patterns and processes can be identified; that it is possible to understand and account for the various landform assemblages which have been described from various parts of the world. It is true that many landforms have long and complex histories. Some features may always defy complete comprehension because of the obliteration or the ambiguous character of the evidence. Many landforms are explicable in terms of more than one hypothesis, and only rarely is it possible to state unequivocally that a particular landform evolved in a particular way at a specific time or period.

Uncertainty is a characteristic of most geomorphological work. Lack of precision and finality are frustrating to some, though in most of the natural sciences any feelings of finality are probably presumptuous, illusory and premature. Indeed, far from really clarifying problems, many investigations either bring to light additional difficulties, which is good, or merely confuse the issue, which is not. At times, and particularly after the introduction of extraneous and irrelevant arguments, one can only agree with the cynical (or realistic?) remark attributed to Mark Twain, that "The researches of many commentators have already thrown much darkness on this subject, and it is probable that, if they continue, we shall soon know nothing at all about it." But though applicable in some instances, this is not typical, and numerous investigations and the application of new techniques over the past half century have brought new understanding of the earth's surface. It is confidently anticipated that, with the aid of new instruments and techniques, similar or even greater advances will be made during the next fifty years.

Geomorphology is literally a discourse on the shape of the earth, but in practice the discipline is nowadays much less ambitious, comprehensive and all-embracing in its coverage. The literal shape of the earth is the concern of the geodesist, and speculations as to its origin are the province of the astronomer. Problems relating to the earth's major relief are properly the realm of the geophysicist. Yet each of these topics has, quite apart from its inherent interest for the earth scientist, facets which are of direct importance to the geomorphologist and brief reference is therefore made to them in Chapter 2. Geomorphology has also been principally concerned not with the whole of the earth's surface but only with a comparatively small part of it. The oceanic floors have been neglected, mainly for practical reasons. Attention has centred on the continents and their shallow submarine extensions. Yet almost 71 per cent of the globe is occupied by the oceans and seas, and the morphological diversity of the seabed is well known. Furthermore, as with problems of the earth's origin and major relief, several aspects of submarine morphology are relevant, indeed vital, to an understanding of widespread geomorphological features, and the geomorphologist ignores the oceanic areas at his peril. He is guilty of keeping his feet too firmly on the ground and many of his problems may be solved only when the oceans and their floors are examined with the same intensity as the land.

Even within the continental areas the geomorphologist is mainly concerned with features that are minute on a global scale. The vast majority of landforms and landform assemblages considered by the geomorphologist are dwarfed by such features as Mount Everest (8882 m above sealevel), and the great ocean deeps such as the Kermadec, Philippine and Mariana trenches. Yet consider the following. The finest discernible line which can be drawn with a pencil is about 0·25 mm or 0·01 in thick. On a globe 30 cm (12 in) in diameter outlined by such a fine pencil line, most of the ocean deeps at this scale are represented by a line less than 0·12 mm thick and therefore encompassed twice over by the thickness of this fine pencil line. Only the greatest of them necessitates a line of 0·25 mm. Most of the land areas are included in a hypothetical line 0·025 mm, and Mount Everest is included in a line only 0·25 mm thick.

Thus geomorphology is in the main concerned with the detailed features of the continents and their adjacent submarine extensions. The purpose of geomorphology is not only to describe the morphology of the land surface but to explain its variations. The approach is genetic. The explanatory part of geomorphology has come under criticism, particularly from some geographers, but the effort involved in the investigation of the origin of landforms can be justified on a number of counts. The ultimate justification for geomorphology, as for any intellectual discipline, is that it satisfies and stimulates a curiosity. It may be of no particular utilitarian significance to know why Ayers Rock stands in isolation above the sand plains of central Australia, or why there are shorelines of former lakes in areas like Scotland, southern New Zealand, Utah and around Lake Chad, or why blocks of rock which originated in Norway are found in eastern England, or why the Great Lakes exist, or why some sand dunes change in colour along their length. But from time to time people have wondered why. These are the sorts of problems geomorphology endeavours to resolve to the intellectual satisfaction of the investigator; and to the pleasure and benefit of the increasing number of laymen who are asking questions about the nature of the landscapes they see in their journeys, and for whom even a partial answer lends a new dimension to their travels and leisure.

On a more professional plane, equally good reasons can be offered to justify genetic geomorphology. Landforms are a tool widely, though often bluntly and inadequately, used by geologists in their interpretations of subsurface structures. Moreover, studies of modern processes materially aid in the understanding of former environments. For instance, the presence of boulders was at one time taken axiomatically as implying glaciation: they were interpreted as glacial erratics and glacial periods were inserted into local stratigraphic histories on this sort of evidence.[3] It is now appreciated[4] that such boulders can develop as a result of the differential subsurface weathering of joint blocks and the subsequent exposure of the corestones so formed. Other examples are cited in the following chapters.

For the geographer the earth's surface is the setting, on which biota, including man, have developed, proliferated and become organised. The land surface is one of many interrelated variables which together form what is called the environment. Many geographers prefer to neglect the findings of geomorphology, or at most to take only perfunctory and superficial cognisance of the physical environment. As many examples at various scales have shown, geographers would do well to examine the subtleties of the physical setting.[5] Others[6] assert that so far as geography is concerned, it is enough for the geomorphologist to describe the land surface, preferably in quantitative terms. The genetic approach is not only more satisfying intellectually but it is also necessary for any understanding of the environmental complex. Any change in one facet of the complex, particularly one so basic as the setting, has an impact on and is reflected in other variables. Moreover, a genetic approach is vital to methods of geographical survey and to extrapolation from observed data. It is again especially important in a geomorphological context because the land surface is one element of the environment which can be readily observed either in the field or on aerial photographs, so that associations established in the field can, within limits and with checks at frequent intervals, be extrapolated with confidence, but only if they are soundly based and understood.[7]

Moreover, because they are integral parts of the environmental complex, the dynamics and history of the land surface are significant to the understanding and appreciation of many pollution and conservation problems.

Thus without wishing to spoil the argument by protesting too much, there are good professional reasons for taking a genetic approach in geomorphology. The final justification, however, is educational. Almost anyone is capable of describing and measuring the forms of the land surface. The process is to some extent mechanical and indeed the task is being accomplished to an ever increasing degree by machines which make maps from air photographs. But the use made of the data derived from such observations calls for skill of a high order. Geomorphology is above all else controversial, and the weighing and marshalling of evidence provides discipline for the mind, no matter to what purpose the training is later put. Geomorphology of course has no monopoly on this educational function of inculcating habits of observation, of reasoning from evidence, and of the formulation of hypotheses, but it possesses this most useful attribute to a high degree. The ability to select and to reason will prove far more important in the long term than elaborate and esoteric techniques, masses of detailed information or extraordinarily refined maps.[8] Quality of thought is an asset of inestimable value in any walk of life.

B. METHODS AND METHODOLOGICAL PROBLEMS

It has been said that geomorphology is one of the more difficult of the physical sciences.[9] These difficulties stem in part from the complexity of the many factors which contribute to the shaping of the land surface. Some idea of the range of knowledge required to investigate and understand the evolution of landforms satisfactorily is brought out in the following chapters. Ideally the geomorphologist should have a working knowledge, and be aware, of the techniques and possibilities of such varied disciplines as chemistry, physics, biology, pedology, meteorology and climatology, hydraulics, the theory of structures and the strength of materials, agricultural practice and the history of settlement, as well as the geological sciences. In addition he should have a working knowledge of statistical methods, not only in respect of the long-established techniques used in sedimentary petrology, but also for the closer description and comparison of land surfaces, which in turn define and in some instances identify problems (though as yet statistical methods have not shown the way to any significant new laws or concepts).

All this, combined with the necessity of mastering such field techniques as topographic survey, air photo interpretation, and so on, plus the desirability of being able to exist and function in the field, means that the ideal or complete geomorphologist (like the complete botanist, historian, chemist, psychologist, mathematician, or geologist) does not exist. All geomorphologists are more competent in some aspects of their discipline than in others (and this too is brought out, although involuntarily, in the following chapters). Yet each facet of the discipline is to a greater or lesser extent relevant to all others; a team approach to problems, or at least maximum communication between the practitioners, is obviously not only desirable but necessary, and any impedence to such exchange of ideas should be ruthlessly eliminated.

Another difficulty is that much of the evidence on which the interpretation of landforms rests lies beneath the surface. Buried landforms or the sequence of sediments may be vital to interpretation, and very often the history of an area is more clearly reflected in its sedimentary deposits than in the ambiguous nuances of the land surface. And though bores and geophysical data help, adequate sections which display subsurface structure and stratigraphy are all too rarely available.

In addition to these practical difficulties, there are several insuperable and very involved methodological problems. The fundamental difficulty facing all investigators is that the origin of landforms has to be deduced from what can be observed of modern processes, though the landforms themselves developed in the past — possibly yesterday, or last year, but more commonly a few thousand or several million years ago. It is a problem of detection on a vast scale, a problem in which the geomorphologist has to deduce the nature of past events, the sequence of those events and their timing, from the evidence which has survived to the present day.

Attempts to overcome this problem take several forms. First, modern processes at work on and near the surface are studied in detail. Both the nature and rates of activity of internal *(endogenetic)* forces like volcanicity, folding and faulting, as well as of external *(exogenetic)* agencies like rivers and glaciers, are investigated in the belief that this will lead to a sounder and more reliable interpretation of the forms and deposits due to these agents in the past. In theory this is reasonable, though many geomorphologists will echo the wry comment of G.W. Lamplugh who for several years worked on the glacial deposits of eastern Yorkshire, in England, and who in 1886 journeyed to the Muir Glacier, in Alaska. He remarked: "This is the first glacier I have visited and I brought away the impression that on the whole it was easier to give explanations of glacial phenomena before I had seen ice."[10]   Modern geomorphological complexes and agents are so intricate and varied that far from illuminating past activities their close observation often, initially at any rate, renders interpretation of their past effects more difficult. But eventually such observations will surely lead to conclusions which are closer to the truth.

One of the basic assumptions underlying this aspect of geomorphological work is that "The present is the key to the past", that the processes observed at present are the same as those operating through geological time. This is the principle of *uniformitarianism*. It is argued that by a mental process of association or comparison, and by using modern processes as a yardstick, the results of past processes (as evidenced by the land surface or by the rocks) and the course of events in the geological past can be reconstructed. But the blind and unrefined application of the law of uniformitarianism can be misleading. It is increasingly clear that the present is far from being typical of geological time. Though many geologists consider that rhythmic changes are a characteristic feature of the geological record,[11] not all variations need be repetitious. The present is either the aftermath of a major glacial era or it may be an interglacial phase. Glaciations account for only 3 per cent of geological time, and the climate characteristic of geological time seems to have been more of what is now called a savannah type. Moreover the continents, until recently — until the explosion in human population and settlement during the last few centuries — were more densely clothed in vegetation than ever before with important repercussions for denudational rates in various parts of the world. Although the laws of physics have not changed (water has never flowed uphill, for example) the spread of vegetation has had important effects on geomorphological development and it is, moreover, difficult to establish standards whereby modern processes and particularly their former rates of activity may be evaluated.

Another, though related, difficulty stems in part from possible ignorance of the modern world and in part from evolutionary processes. It may be that the processes observed at present are the same as those operating through geological time, but were there other processes at work in the past which are now not active? Silcrete is an important geomorphological feature in many parts of Australia, but as far as can be deduced from the literature it is not forming anywhere at the present time. Perhaps it has not been detected; or possibly in some areas biological and/or climatic conditions prevailed in the past which do not exist anywhere now. Possibly many of the plants which clothed the land surface during the mid Tertiary (when most of the massive and extensive silcretes formed) had, to a marked degree, the ability to accumulate and concentrate silica in their tissues, as indeed some still do.[12]

Consider an absurd and hypothetical example of the former activity of a process which is now obsolete or not represented on earth. Many depressions in the plains of the American West have been attributed to buffaloes rolling around in the mud and transporting away with them the soil which stuck to their coats, thus lowering the land surface in localised areas: buffalo wallows.[13] Imagine that in the geological record there are preserved enormous hollows for which no structural or climatic explanation is forthcoming. They are so huge that they cannot be attributed to even the largest herds of buffaloes and let us say for the sake of argument that they predate the evolution of these animals. Could they have their origin in the mudbaths of herds (or whatever the correct collective term is) of playful dinosaurs? Such features could have been formed only so long as dinosaurs existed, and could not be formed today.

Nevertheless, applied with caution and imagination, as a working tool and not as an infallible law, uniformitarianism is a reasonable and perhaps necessary basis for geomorphological investigations.

A second difficulty confronting geomorphological work is that many factors influence the shaping of the land surface. There are many interdependent variables. Hence complexity is the rule and simplicity of explanation the exception in geomorphology. Yet many have taken the line that the most simple explanation is the most likely. For instance P.W. Bridgman, a noted physicist and scientist, states that Occam's Razor, or the so-called Law of Parsimony "appeals to me as a cardinal intellectual principle ..." though his intellectual honesty shines through and he goes on to concede that "I do not know what logical justification be offered for the principle".[14] The complexity of the explanation adopted is surely dictated by the available data; but it is naive to take a simplistic approach as a matter of course and still worse as a matter of principle. There can surely be no justification for automatically adopting the Law of Parsimony in scientific research.[15] It is more in keeping with the realities of geomorphology to cite part of Paul Valery's aphorism that "*Tout ce qui est simple est faux*" (everything simple is wrong), (though the second part of the quotation, that "*Tout ce qui est complexe est inutilisable*" (everything complex is useless) is no more encouraging).

Although not all the variables in the system are independent, there are nevertheless so many plausible combinations of possible effects that, though statistically one result is more likely than others, there always is and probably always will be a strong element of uncertainty or of indeterminacy in geomorphological interpretation.[16] To pretend otherwise is delusion. In addition, the possibilities of convergent evolution, or equifinality—the development of morphologically similar forms by quite different processes—further cloud the picture. Thus deliberately to opt for and to seek the simplest explanation tends to discourage the search for evidence of, for example, climatic effects which have left only subtle impressions on the landscape. For instance, deep chemical weathering of rocks has been reported from many modern arid areas. Is this to be related to former periods of higher rainfall, or is it to be interpreted as due to the infrequent rains of the present climatic regime? To take the latter course because it removes the assumption of an earlier pluvial would tend to retard the search, overt or subconscious, for evidence of climatic change. To adopt the former course on the other hand would blur the appreciation of the realities of present desert climates. It is surely better to keep an open mind and to entertain all the opinions? As with the concept of uniformitarianism, the more hypotheses for testing the better. Acceptance of the Law of Parsimony as of uniformitarianism, "leads to poverty, where riches are desired".[17]

The laboratory is a third avenue explored by many geomorphologists. Attempts have been made to make landforms in the laboratory, essentially isolating one or at most a few factors and thus simplifying natural conditions, in an endeavour to determine what factors are necessary and what agents are responsible for certain features, either erosional or depositional. Some experiments in test tubes, wave tanks, flumes and wind tunnels, for instance those by Griggs on insolational weathering,[18] on river meanders by Friedkin,[19] on dunes by Bagnold[20] and on chemical weathering by Pedro,[21] Henin and Pedro,[22] and Raussell-Colom *et al.*,[23] have been of great value in suggesting how these features may have formed in nature. But it is not possible to reproduce the complexity of nature in the laboratory. It is not possible to reproduce in the laboratory the immensity of geological time, which is the only context in which to see landforms. It is not possible to reproduce exactly the environment in which a given landform evolved, since this is not known in many cases. Thus, though the results of the best laboratory work cannot be neglected and have indeed been of decisive importance in some fields, they at best indicate how landforms *may* have developed, not how they in fact did so.

C. MULTIPLE WORKING HYPOTHESES

In practice the geomorphologist uses a variety of research methods, which vary according to the nature of the problem under investigation, in an attempt to overcome or to reduce the dangers of misrepresentation inherent in these difficulties. Fundamentally, they involve the application of the associative or comparative method combined with the rigorous testing of a number of working hypotheses which are formulated as possible explanations of the feature under consideration. This is called the method of multiple working hypotheses, and its complexity varies with the problem in hand. Some associations are reasonably certain and obvious, others are more obscure, but as Gilbert pointed out, "hypotheses are always suggested through analogy".[24]

Consider first two comparatively straightforward examples of association, where the origin of features developed in the past is deduced from comparison with modern or essentially modern processes.

In many parts of Australia, ranging from the north coast to the southern island state of Tasmania, from Western Australia to the east coast, plateaux and uplands capped by a ferruginous duricrust called laterite are prominent features of the landscape. Laterite is a weathering or soil profile which includes an horizon rich in iron oxides and oxides of aluminium. Laterite is forming today in the intertropical regions of South America, Africa, India and Southeast Asia, in each case in humid tropical conditions, though in many areas there is a dry, or less wet, season.[25] High temperatures, heavy rainfall and abundant vegetation appear to be conducive to laterite formation which, though many processes are apparently at work, essentially involves the deep leaching of an horizon with the redeposition of iron and aluminium oxides in preferred horizons. The precise chemistry of the soil reactions involved and the various types of laterite-producing reaction complexes are far from clear.[26] Yet, as the Australian laterites are relict forms (though there are from time to time claims or hints that laterites are forming there now) and are indeed being destroyed at the present time, it seems reasonable to suppose that when the laterites were evolved, at some time or times in the Mesozoic or Tertiary, humid tropical conditions prevailed. Some confirmation for this comes from studies of marine fossils, particularly foraminifera, which show that the seas off southern Australia in the Tertiary period were tropical.[27] Corroboration, perhaps in the form of terrestrial fossils directly associated with the laterite soil is certainly desirable, but the inference stated concerning the environment of the Australian Tertiary laterites appears reasonable *pro tempore*.

The example of laterite involves a direct analogy. Modern laterites can be observed. Their transformation from the soft, easily cut material of most laterites to the indurated tough cappings known in Australia and Africa, for instance, can be witnessed. In southern India, laterite is quarried with spade and trowel and is cut into blocks of a shape and size suitable for building purposes; but after a brief exposure to the atmosphere the rock dries and hardens (the iron oxides apparently becoming irreversibly crystalline) and it forms an extremely durable brick.

In some cases direct observation of modern processes is impossible for physical reasons, but a combination of association of features and obvious feasibility render some interpretations of features found in the geological record so credible as to be irrefutable. For instance, the movement of glaciers and their contained bed loads over rock outcrops causes the latter to be polished, scratched, grooved and pitted. Polished surfaces, striae and crescentic friction cracks are characteristic features of glaciated pavements. Scratched or striated rock surfaces are of course also formed in ways other than by glacial scouring. They form in diapirs (see Chapter 5) and under mud-flows and avalanches.[28] They are formed by the rush of volcanic *nuées ardentes*.[29] Striae in the form of slickensides are a characteristic feature of faulting. Yet in all these instances the general setting and the general association of features demonstrates that the scratches are not of glacial origin. Far more difficult is the case of scratches formed on quite hard rocks by sliding masses of snow, [30] which also accomplish the transportation of large blocks of rock. But again, the overall setting most often makes clear the likely origin of the features. For example, embedded within the Precambrian sequences of the Kimberleys in Western Australia are several pavements which display not only striations, but are also polished, show crescentic friction cracks (Pl. 1.1), and are overlain unconformably by a glacially-deposited sediment, the Walsh Tillite.[31] This brings out the point that though one scratch does not imply a glaciation, the assemblage of evidence usually resolves any ambiguities or arguments arising from single features. Thus, though the actual erosion of glacial striae has not been observed beneath modern glaciers, these pavements in Western Australia which are about 1500 million years old can be identified as of glacial origin because of the association of several features which all point to the former activity of glaciers. Similar evidence from India, Africa, Europe and North America confirms the occurrence of at least one glaciation during the later Precambrian era (see Chapter 25).

Let us now turn to a rather more complex analysis concerned with a minor but fairly widely distributed landform: the flared slopes found on many residual hills in southern and central Australia, and in other areas.

Elegant flared slopes developed in granite have been observed by the writer near the Devils Postpile and in the Alabama Hills, both in the Sierra Nevada of California, on Paarl Mountain and in Namaqualand, both in Cape Province, Republic of South Africa, and in Rhodesia (Fig. 1.1a-c). They are widespread in semiarid and arid Australia, especially around the lower margins of granite inselbergs in South Australia and Western Australia, though they have also been observed in the Northern Territory, in New England and in the Kosciusko area in N.S.W. and near Mt Buffalo in Victoria (Fig. 1.1d). They occur on sandstone at Ayers Rock, Northern Territory, and in the Flinders Ranges of South Australia; on conglomerate at the Olgas, also in the Northern Territory; on volcanic porphyry in the Gawler Ranges and on volcanic rocks also in north western Mexico,* and on limestone, as in the Weka Pass, South Island, New Zealand. These flared slopes characteristically take the form of concave steepenings which stand in strong contrast with the overall smooth convex forms of the granite, sandstone or conglomerate residuals (Pl. 1.2).

---

* J. Fish, personal communication.

Plate 1.1 Polished and striated glacial pavement of Precambrian age exposed in the Kimberley region, in the northwest of Western Australia, in the area between the Hann and the Traine rivers. Note the striae and the crescentic friction cracks, which are concave southwards. The regional evidence suggests ice flow from the north. (*Dept Nat. Dev., Canberra.*)

Fig. 1.1a Flared slopes in granite near the Devils Postpile, high in the Sierra Nevada, California. Note inclination of flared slope.

Fig. 1.1b Flared base of granite slope on the Alabama Hills, Sierra Nevada, California; in the background is the Owens Range, which attains in Mt Whitney an elevation of 4418 m and which stands about 1600 m above the site shown in the foreground.

Fig. 1.1c Flared slope about 2 m high bordering a depression floored with soil and vegetation on Dombashawa, a complex granite dome near Salisbury, Rhodesia.

Fig. 1.1d Flared slopes flanking elongate granite boulders near the crest of Mt Buffalo, Victoria (elevation 1725 m). (*Drawn from photograph supplied by Miss R.M. Thomson*)

Plate 1.2 Pildappa Hill, northwestern Eyre Peninsula, South Australia, from the west, showing the flared lower slopes which stand in strong contrast to the overall convex form of the dome which rises some 24 m above the adjacent plain. (*C.R. Twidale.*)

Plate 1.3a Joint-controlled cleft, 4–5 m deep with flared bounding slopes on Yarwondutta Rocks, northwestern Eyre Peninsula. The slope on the right is overhanging and the joint zone itself is weathered more deeply and is etched out to form a linear depression in the floor of the cleft. (*C.R. Twidale.*)

Plate 1.3b Joint-controlled cleft with overhanging slope on Pearson Island, literally a granite *inselberg* (island mountain) in the Great Australian Bight, South Australia. (*C.R. Twidale.*)

An examination of many such overhanging and flared slopes in several areas which have developed on different types of bedrock suggests that they have the following characteristics.[32]

1. They are developed only on massive rocks, such as granite or sandstone, which possess few open joints.
2. They are best and most commonly developed on the shady sides of residuals, that is, on the southern and eastern sides of hills in the southern hemisphere.
3. They occur in valleys and clefts within the residuals as well as around their bases (Pls 1.3a and b).
4. Multiple flares are displayed in many areas (Pls 1.4 and 1.5).
5. Especially pronounced steepening is developed:
   (a) on the points of spurs (Pls 1.6 and 1.7), and
   (b) in broad embayments within the residuals (Pl. 1.5).
6. The zone of steepening is not horizontal everywhere (Pl. 1.8).
7. Incipient flares are revealed in excavations and by augering to be developed below the natural land surface beneath weathered bedrock (Pl. 1.9). This is also suggested by the steep plunge of bedrock surfaces beneath weathered material at the margins of inselbergs (see Pl. 1.8).
8. Flared slopes are developed on boulders as well as on major residuals (Pls 1.7 and 1.10, Fig. 1.1b). This was noted by Hellstrom[33] on Mt Buffalo.
9. Though difficult to date precisely, the steepening is characteristically of recent geological age.

Several possible modes of origin can be suggested. These working hypotheses are examined and tested against the field evidence and characteristics.

Plate 1.4 Doubly-flared slope on the southern side of Pildappa Hill. (*C.R. Twidale.*)

Plate 1.5 Overhanging wall known as Wave Rock, on the northern side of Hyden Rock, Western Australia. The wall, which is some 11 m high, displays a minor double concavity, easily discerned by tracing the black streaks of desiccated organic slime down the curve of the feature. (*C.R. Twidale.*)

Plate 1.6 Prominent overhang developed on the point of a spur on Ucontitchie Hill, northwestern Eyre Peninsula. The point of the spur stands 5–6 m above the rock platform below. (*C.R. Twidale.*)

Plate. 1.7 Overhanging slopes developed on a spur of granite (left) and on a large boulder, 3 m maximum diameter, at northern margin of Ucontitchie Hill. (*C.R. Twidale.*)

Plate 1.8 Sloping zone of basal steepening on the northern side of Chilpuddie Hill, northwestern Eyre. Peninsula. The flared zone falls away from the observer. Note the steep plunge of the bedrock surface at the soil junction. (*C.R. Twidale.*)

Plate 1.9 Exposure of the weathering front in an almost dry reservoir at the northern end of Yarwondutta Rocks, north-western Eyre Peninsula. Note the flared slope in the background; the similarly flared shape of the weathering front from which the weathered granite has been removed and which is exposed in the floor of the reservoir; and the concrete retaining wall and pillars, the latter intended to support a corrugated iron roof which reduced evaporation loss and contamination of the water supply. (*C.R. Twidale.*)

Plate 1.10   Flared granite boulder near Smiggins Holes, northwest of Jindabyne, New South Wales. (*C.R. Twidale.*)

1. That the Basal Steepening is due to Wave Attack

High Pleistocene stands of the sea have been postulated,[34] and it is possible that these steepened slopes are, in part at least, of marine origin: that they are sea cliffs since abandoned by the recession of the ocean. Against this argument is the fact that the smooth curvilinear form is not, by comparison with modern sea cliffs, consistent with such an origin. Most sea cliffs are faceted and where visors and nips (or undercuts) have been formed, the latter are angular. It is true that where groundwaters seep out at the coast, these features can become rounded as for instance, at Shag Point, some 9–10 km north of Palmerston, South Island, New Zealand, but such cases are rare.

Despite arguments to the contrary, there is no evidence — such as marine sediments and fossils — to indicate extensive marine transgressions during the recent geological past, even in the present near-coastal regions like the southwest of Western Australia and northwestern Eyre Peninsula. The argument is even more difficult to sustain in the case of the Olgas and Ayers Rock (approximately 350 m above sealevel),

the Sierra Nevada of California and the high plateaux of southern Africa. Turning to more specific points, the wave attack hypothesis can account neither for the inclined zones of steepening, for the preferential development of flares on shady aspects in embayments (where wave attack should be dissipated because of refraction) nor for their incipient development beneath the present land surface.

2. That the Basal Steepening is due to Aeolian Activity (Sand Blasting)

Several, though not all, of the flared slopes occur in areas that either are — or recently were — arid. Sand blasting extends to a height of only a metre or so above the plain surface and is known to polish surfaces, so that such activity could be deemed responsible for the undercutting, steepening and polishing of the bases of hills and boulders (but see Chapter 14).

Again, however, both general argument and specific evidence seemingly deny this possibility. On northwestern Eyre Peninsula the longitudinal sand dunes, which formerly extended across the area but which are now fixed by vegetation, migrated into the area from the northwest. They are a manifestation of a huge counter-clockwise circulation which in expanded or extended form controlled the orientation of former as well as of the present dune fields.[35] In detail the dunes extend from the northwest onto the salt-encrusted beds of desiccated lakes. In some few instances, dune sand encroaches on granite residuals, also from the northwest. Hence if the dominant winds came from the northwest and it was these winds which, armed with sand, shaped the lower margins of the granite hills, those slopes which face northwest should be most markedly steepened. Yet the converse is true; on Eyre Peninsula the highest and steepest flares are developed on the southern and eastern aspects in the vast majority of cases.

The occurrence of flared slopes in the sides of valleys and clefts within the inselbergs which are located well above plain level is difficult to understand in terms of aeolian sand blasting, as is the subsurface development of bedrock flares, which are covered with weathered bedrock *in situ* and not by transported material. Finally, the observed common occurrence of quite thick vegetation, which takes advantage of the greater amounts of moisture available in the scarpfoot zone around the bases of the residual hills, tends to mitigate against effective wind erosion in these areas.

3. That the Basal Steepening is due to Running Water

The rate of run-off from the bare rock surfaces of the inselbergs is high. As the volume of run-off increases downslope it may be suggested that erosion should increase downslope too, thus producing a steepening. It could be argued that such an increase in erosion is not necessarily gradual but that it may increase rapidly beyond critical limits, thus accounting for the sudden change from convex slope to concave. Finally, it could be suggested that the rush of water down the bare rock slopes could produce a sheet-like jet which, drawing in air — the Coanda effect[36] — could conceivably produce an ephemeral vacuum between the jet and the rock face.

Collapse of the vacuum could cause cavitation (see Chapter 12).

If these arguments have any validity then areas where run-off converges should be areas where erosion increases and where, therefore, the flared slopes are best developed. At the heads of narrow clefts, for instance, there should be very well-developed flares. But the field evidence shows that it is just in such areas that flared slopes are either absent or poorly developed. It is true that there are well-developed overhanging walls in broad embayments (Pl. 1.5), but on the other hand some of the most pronounced overhangs are displayed on spurs (Pl. 1.6). A further problem facing the run-off hypothesis is that flares are well-formed on boulders (Pl. 1.10) which can scarcely be deemed to generate enough run-off to accomplish much erosion. The hypothesis also fails to explain the subsurface flares in bedrock. It is unlikely even during the heaviest periods of run-off that sheets of water would be projected above the convex rock surface, and still less that they would be sufficiently continuous for the evacuation of air from the area between them and the rock face to occur. In any case there is no evidence of localised shattering of the rock surface, such as might be expected to result from collapse and cavitation.

4. That the Basal Steepening is due to Subsurface Weathering

Thus all of these possible explanations have been found wanting in some vital respects. They are both inherently unlikely and fail to explain the local field evidence. The only hypothesis so far conceived which is supported by the field evidence and in terms of which not only the observed characteristics but also the unusual cases can be understood, involves essentially subsurface weathering. [37] It may be suggested that the flared slopes evolve in two distinct, though not necessarily temporally separate phases (see Fig. 1.2a-c).

1. A phase of subsurface weathering of the bedrock around the base of the residuals by water which has flowed from the bare rock of the slopes above and which, under the prevailing arid or semiarid conditions, has soaked into the subsurface rather than run far out to the plains and ultimately to the sea.

2. A phase of baselevel lowering during which the plains are lowered, the previously weathered rock is evacuated and the limit of weathering, or weathering front, is exposed as the flared slope.*

This hypothesis not only takes account of the salient characteristics of the forms but is also consistent with their general structural and climatic setting. For it will have been apparent that, in order to have a steepened slope, there must be a reason for the slope being there in the first place. In other words there must be a reason for the inselberg being

---

* It is of interest that Hellstrom,[39] though he had no clear-cut ideas as to the origin of the flared boulders he observed at Mt Buffalo, Victoria, tentatively suggested that the marginal granite was weathered by heating and cooling, and evacuated by wind and water. Though the processes involved are not those responsible for flared slopes, a two-phase mechanism was clearly and correctly suggested.

there. There is now sufficient evidence (see Chapter 3) to suggest that inselbergs survive because they are essentially monolithic—they have fewer joints which can be penetrated by water—on this account they are more resistant to weathering than the closely-jointed compartments of rock all around; and they therefore resist erosion and survive as upstanding masses of rock. [38] The massiveness of the rocks which underlie the residuals is also important because only in such a setting would weathering proceed so evenly as to produce a continuously smooth front.

Climate is also important. It is in arid or semiarid conditions (including areas of winter cold, but with a marked

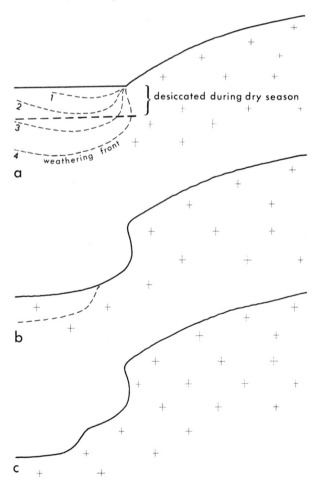

a

b

c

Fig. 1.2 Stages in the development of flared and overhanging slopes by scarpfoot weathering and subsequent differential erosion: (a) subsurface infiltration of moisture in the scarpfoot area, and lowering of the weathering front; (b) lowering of baselevel and stripping of weathered debris resulting in exposure of the weathering front as a flared slope; (c) repetition of process and development of double flare.

dry season) that the hillslopes tend to be devoid of soil and vegetation. Hence high rates of run-off are achieved. It is also in arid areas that the subsoil tends to be dry so that run-off readily infiltrates beneath the surface to accomplish weathering close to the base of the hillslopes.

Subsurface scarpfoot weathering also explains the following field characteristics.

1. The flared slopes are best and most commonly developed on the shady sides of the hills because it is there that soil moisture, which is held to be responsible for the weathering, is longest preserved.

2. The junction between the upper normal convex slope and the basal concavity marks the former limit of the soil cover, that is, the former hill–plain junction. That is, as a matter of observation, not everywhere horizontal, and there is no reason to suppose that it was in the past.

3. Maximum preservation of soil moisture occurs not at the surface, where it is subject to evaporation and transpiration, but some distance beneath the surface. Hence the maximum weathering, possibly because of the longer duration of attack, occurs some distance beneath the land surface and not at it. This is the cause of the concave shape of the weathering front.

4. Pronounced overhangs are developed on the points of spurs where there is attack by soil moisture from two sides, where two weathering fronts merge (Pls 1.6, 1.7 and 1.11).

5. Pronounced overhangs are also developed in embayments, which are collecting areas for run-off converging from the surrounding slopes (Pl. 1.5).

6. Multiple flares are comprehensible in terms of repetitions of the two-phase sequence of events (Fig. 1.3, Pl. 1.9).

7. The flares found in excavations beneath the land surface (Pl. 1.9) are flared slopes in process of development, flares which have not yet proceeded beyond the first of the two-phase development.

This hypothesis has been found to account satisfactorily for most examples of flared slopes examined in the field. It is true that in most areas the precise chronology of events remains obscure, as does the precise nature of the weathering

Plate 1.11 Flared slopes developed on both sides of a rock spur at the eastern end of a low inselberg 3 km northwest of Mt Wudinna, northern Eyre Peninsula. The spur at this point is rather more than 3 m high. Note the high continuous flared slope on the northern (right) side of the residual. (*C.R. Twidale.*)

processes which manifestly have caused the disintegration of the granite or other bedrock. Again, is there any climatic control of the two phases to which the flared slopes are attributed? Did the weathering occur in a climate more moist than today's, and did the lowering of the land surface and the exposure of the weathering front happen in essentially arid conditions? Or did the subsurface weathering take place in arid or semiarid climates and the lowering and exposure in a moister climate when there were some surface streams? The latter seems more likely and is consistent with evidence concerning late Cainozoic climatic changes in arid and semiarid Australia for example, but the matter is not yet settled.

The identification of characteristic features is a useful phase of investigation, but should not be taken to suggest that all the examples of a particular form need necessarily display all characteristics. It merely implies that most do. The value of such a listing of characteristic features lies partly in the isolation of features that any reasonable hypothesis must explain. But it also draws attention to anomalies, or apparent anomalies, which should also be explicable in terms of the successful hypothesis. For instance, the great overhanging wall of Wave Rock, near Hyden in Western Australia (Pl. 1.5) lies on the northern slope of Hyden Rock. This is unusual in an Australian context but the greater exposure to insolation which that wall suffers is compensated by the fact that water from a wide area of the inselberg pours into the joint-controlled residual on which Wave Rock has evolved. [40] Again on a small residual 3 km northwest of Mt Wudinna, on northwestern Eyre Peninsula, the highest and most steeply flared slope also occurs on the northern side of the inselberg. Examination of air photographs (see Fig. 1.3) suggests that this broadly linear northern face of the residual is determined by a major joint, which also finds expression in the nearby Mt Wudinna. Certainly this slope and a major fracture pattern within and partially delineating Mt Wudinna are in alignment. If this is the case then weathering could have taken advantage of the avenue of weakness provided by the joint system, developing a pronounced weathering front and flared slope in association with it (cf. Pl. 1.3a).

Lastly, it may reasonably be asked why these flared slopes are so magnificently developed and preserved in the arid zone of Australia, and, seemingly, not to the same degree elsewhere. For although flared forms occur in southern Africa and in the western United States, there is nothing to compare with Wave Rock (Pl. 1.5) or with the lower slopes of Ucontitchie Hill and Pildappa Hill (Pls 1.2, 1.4, 1.6, 1.7). Why?

The explanation may well be found in the varied stability of the areas under discussion. Optimal development of flared slopes under the two-stage mechanism outlined above requires stability of the land surface so that weathering can penetrate deeply into the margin of the massive rock compartments, and so that vertically and laterally extensive weathering fronts can develop. These are then exposed as flares as a result of erosion of the hillfoot zone. In Australia such conditions are met, for there appear to have been both comparatively long periods of relative landscape stability, during which such deep penetrations of the weathering front could have taken place, and the plain surfaces have been slightly lowered from time to time causing the weathering fronts to be revealed as flared slopes. Furthermore, the prevailing aridity of the regions ensured· the persistence and preservation of the forms once they were exposed.

In the western United States on the other hand, the crust is notoriously and manifestly unstable, with widespread and incontrovertible evidence of faulting and uplift in·the recent past (see Chapter 4). Hence there is a considerable relief amplitude and streams are actively cutting into the uplands. Weathered debris tends to be removed almost as soon as it is formed and consequently there is little time for the development of smooth concave weathering fronts of any extent. In southern Africa there is considerable relief and erosion of the

margins of many inselbergs, for instance in the Valley·of the Thousand Hills (in the core of the Natal Monocline) and in the Paarl region of Cape Province. But elsewhere, in the interior of the vast plateaux, there are inselbergs which have not suffered sudden and pronounced basal erosion. It should be possible to demonstrate the presence or absence of incipient flares marginal to these inselbergs which stand above undissected high plains (as for instance east of Salisbury, Rhodesia) by dint of augering in the weathered granite and establishing the form of the weathering front.

When using the method of multiple working hypotheses, the landforms under investigation are first described and analysed. Their characteristics are ascertained and the sequence of events they represent is established, at least in relative terms and if possible in geological or absolute terms. Possible explanations are formulated to account for the unusual as well as the characteristic features. The deducible consequences of each working hypothesis are then matched against the field evidence either already to hand or sought in consequence of the hypothesis and deduction. It may be that in terms of the field evidence all the hypotheses, either singly or in combination, are found wanting, in which case the solution to the problem awaits the imaginative formulation of more hypotheses for testing. But it may be that one hypothesis, or a combination of two or more hypotheses, survives rigorous testing and apparently explains the field evidence. This may be accepted for .the time being as a possible solution, though of course the emergence of new evidence may either confirm the explanation, cause its modification, or compel its abandonment.

The method is partly deductive, partly inductive. It depends on creative, controlled imagination, working on the basis of established facts, for the development of possible explanations by matching the deducible consequences of each hypothesis with the realities observed in the field. That a discipline purporting to be scientific can proceed in such an apparently haphazard way may surprise some, but not all, and certainly not those who have been actively involved in research. Very rarely, as has been pointed out elsewhere, [41] is the solution of a scientific problem achieved by following a path preordained by inevitable logic. Certainly the interpretation of evidence and the linking of observations and hypotheses are more obvious in some instances than in others; but very frequently the answer to a complex problem comes from inspiration, no doubt generated by the subconscious reflection and mulling over of the evidence, but essentially the result of creative and controlled imagination. Cold logic is used to test and evaluate the results of such inspiration; and it must be emphasised that the rigorous testing of ideas is just as important as their creation.

The successful utilisation of the method of multiple working hypotheses calls for:

(a) a sound theoretical knowledge, so that all possibilities can be entertained in the interpretation of landforms;

(b) field experience, so that landforms and their associated deposits can be recognised for what they are and what they imply, and so that the significance of anomalies and oddities can be appreciated;

(c) the ability to relate remote and seemingly unconnected observations; and

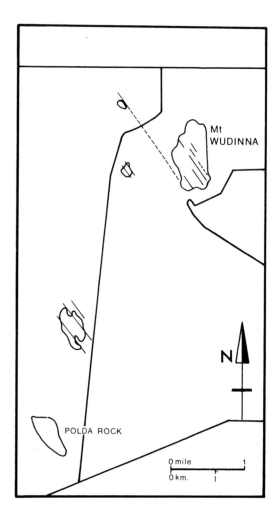

Fig. 1.3 Map of area around Mt Wudinna (Wudinna Hill), northern Eyre Peninsula, South Australia, showing major granite outcrops and major joints. Note the major fracture on the northeastern side of the smaller inselberg to the northwest. (*Drawn from air photographs.*)

(d) imagination, controlled but untrammelled by tradition or orthodoxy, in the formulation of hypotheses.

C. **CONCLUDING STATEMENT**

In the case of the flared slopes just discussed, the basic reason for the development of the forms is the concentration of weathering in the scarpfoot zone. The reason for this is partly climatic, partly structural. Structure also plays a part in that flared forms are preserved only on massive bedrock, while climatic changes may be responsible in some areas for the alternations of weathering and erosion dominance implied in the two-stage development described. Thus three principal factors have contributed to the formation of these spectacular features: the structure of the crust, the processes at work at and near the land surface, and changes in these two factors in the course of time.

These are the three main factors which control the development of the land surface, and to which reference must be made in analysing and interpreting landform assemblages. As the great American geomorphologist W.M. Davis put it, landforms are a function of, or vary with, structure, process and time.[42] The three are commonly interrelated in that most landforms are, in varying degrees, expressions of all three controls, so that it is unrealistic to isolate any one from the other two. But some assemblages are principally manifestations of structural controls. Others result mainly from the activity of a particular climatically-induced process. And yet others bear the clear imprint of more than one set of processes, either exogenetic or endogenetic or both, which have varied in time. Thus it is both rational and organisationally convenient to consider these three factors in turn. Hence the subdivision of this book into structural, climatic and historical geomorphology.

References Cited

1. P. White, *Voss*, Eyre and Spottiswoode, London, 1957.
2. C. Sturt, *Expedition into Central Australia*, Boone, London, 1849, Vol. 1; A. Moorehead, *Coopers Creek*, Hamish Hamilton, London, 1963.
3. See for example C. F. Hartt, *Geology and Physical Geography of Brazil*, Frubner and Co., London, 1870, p. 28; J.C. Branner, "The Supposed Glaciation of Brazil", *J. Geol.*, 1893, **1**, pp. 753–72; J. Romanes, "Geology of a Part of Costa Rica", *Q. J. geol. Soc. London*, 1912, **68**, pp. 133–6.
4. See for instance E. S. Larsen, "Batholith and Associated Rocks of Corona, Elsinore and San Luis Rey Quadranges, Southern California", *Mem. geol. Soc. Am.*, 1948, **29**, pp. 113–6; D. L. Linton, "The Problem of Tors", *Geogr. J.*, 1955, **121**, pp. 470–87; C. R. Twidale, *Structural Landforms*, Australian National University Press, Canberra, 1971, pp. 20 *et seq.*
5. For example, H. C. Darby, *The Drainage of the Fens*, Cambridge University Press, Cambridge, 1940; S. W. Wooldridge, "On Taking the 'Ge-' out of Geography", *Geography*, 1949, **34**, pp. 9–18; S. W. Wooldridge and D. L. Linton, "The Loam Terrains of southeast England and their Relation to Early History", *Antiquity*, 1933, **7**, pp. 297–310; *idem*, "Some Aspects of the Saxon Settlement in southeast England considered in Relation to the Geographical Background", *Geography*, 1935, **20**, pp. 161–75; J. M. Lambert, J. N. Jennings, C. T. Smith, C. Green and J. N. Hutchinson, "The Making of the Broads", *R. Geogr. Soc. Res. Ser.*, 1963, **3**; C. R. Twidale and D. L. Smith, "A 'Perfect Desert' Transformed. The Agricultural Development of

northwestern Eyre Peninsula, South Australia", *Aust. Geogr.*, 1971, **11**, pp. 437–54.
6. For instance, R. J. Russell, "Geographical Geomorphology", *Ass. Am. Geogr. Ann.*, 1949, **39**, pp. 1–11; R. Hartshorne, *Perspective on the Nature of Geography*, Murray, London, 1959, pp. 93–6; G. H. Dury, "Rival Hypotheses — Some Aspects of Geographical Judgement Forming", *Aust. J. Sci.*, 1968, **30**, pp. 357–62.
7. See for example C. S. Christian, J. N. Jennings and C. R. Twidale, "Geomorphology", Chapter 5 in *Guide Books to Research Data for Arid Zone Development*, UNESCO, Paris, 1957, pp. 51–65.
8. P. Haggett, *Locational Analysis in Human Geography*, Methuen, London, 1966, p. 310.
9. D. L. Linton, "The Origin of the Pennine Tors—An Essay in Analysis", *Z. Geomorph.*, 1964, 8NS, pp. 5–24.
10. Cited in L. F. Penny, "Early Discoverers, XXIV, George William Lamplugh (1859–1926)", *J. Glaciol.*, 1966, **6**, pp. 307–9.
11. J. Barrell, "Rhythms and the Measurement of Geologic Time", *Bull. geol. Soc. Am.*, 1917, **28**, pp. 745–904.
12. See T. S. Lovering, "Geologic Significance of Accumulator Plants in Rock-weathering", *Bull. geol. Soc. Am.*, 1959, **70**, pp. 781–800; and several papers by Jones and his collaborators referred to in Chapter 10.
13. W. D. Johnson, "The High Plains and their Utilisation", *Rep. U.S. Geol. Surv. 21st Ann.*, 1899–1900, Washington, pp. 599–768; N. H. Darton *et al.*, "Guidebook of the Western United States. Part C. The Santa Fe

Route", *Bull. U.S. Geol. Surv.*, 1915, **613**, pp. 36–7; S. JUDSON, "Depressions of the Northern Portion of the southern High Plains of eastern New Mexico", *Bull. geol. Soc. Am.*, 1950, **61**, pp. 253–74.

14. P. W. BRIDGMAN, *The Way Things Are*, Harvard University Press, Cambridge, 1959, p. 333.

15. See for example C. D. OLLIER, *Weathering*, Oliver & Boyd, Edinburgh, 1969, p. 119.

16. L. B. LEOPOLD and W. B. LANGBEIN, "Association and Indeterminacy in Geomorphology", in C. C. Albritton, Jr. (Ed.), *The Fabric of Geology*, Freeman, Cooper and Co., Stanford, 1963, pp. 184–92.

17. H. B. BAKER, "Inductive Logic and Lyellian Uniformitarianism", *J. Mich. Acad. Sci.*, 1938, pp. 1–5.

18. D. T. GRIGGS, "The Factor of Fatigue in Rock Exfoliation", *J. Geol.*, 1936, **44**, pp. 625–35.

19. J. F. FRIEDKIN, *A Laboratory Study of the Meandering of Alluvial Rivers*, U.S. Waterways Experimental Station, Vicksburg, Miss., 1945.

20. R. A. BAGNOLD, "*The Physics of Blown Sand and Desert Dunes*", Methuen, London, 1941.

21. G. PEDRO, "An Experimental Study on the Geochemical Weathering of Crystalline Rocks by Water", *Clay Miner. Bull.*, 1961, **26**, pp. 266–81.

22. S. HENIN and G. PEDRO, "The Laboratory Weathering of Rocks", in E. G. Hallsworth and D. V. Crawford (Eds), *Experimental Pedology*, Butterworths, London, 1965, pp. 29–39.

23. J. A. RAUSSELL-COLOM, T. R. SWEATMAN, C. B. WELLS and K. NORRISH, "Studies in the Artificial Weathering of Mica", in E. G. Hallsworth and D. V. Crawford (Eds), *Experimental Pedology*, Butterworths, London, 1965, pp. 40–72.

24. G. K. GILBERT, "The Origin of Hypotheses, illustrated by the Discussion of a Topographic Problem", *Science (N.Y.)*, 1896, **3**, pp. 1–13.

25. R. MAIGNIEN, "Review of Research on Laterites", *Nat. Resour. Res.*, **IV**, UNESCO, Paris, 1966.

26. See R. MAIGNIEN, "Research on Laterites".

27. F. H. DORMAN, "Australian Tertiary Palaeotemperature", *J. Geol.*, 1966, **74**, pp. 49–61; W. K. HARRIS, B. McGOWRAN and J. M. LINDSAY, personal communication.

28. See C. K. WENTWORTH, "Soil Avalanches on Oahu, Hawaii", *Bull. geol. Soc. Am.*, 1943, **54**, pp. 53–64; and Chapter 11.

29. See for example G. A. TAYLOR, "The 1951 Eruption of Mt Lamington, Papua", *Bull. Bur. Miner. Resour. Geol. Geophys. Aust.*, 1958, **38**, pp. 41–2.

30. See for instance J. L. DYSON, "Snowslide Striations", *J. Geol.*, 1937, **45**, pp. 549–77; A. B. COSTIN, J. N. JENNINGS, H. P. BLACK and B. G. THOM, "Snow Action on Mt Twynham, Snowy Mountains, Australia", *J. Glaciol.*, 1966, **5**, pp. 219–28.

31. W. J. PERRY and H. G. ROBERTS, "Late Precambrian Glaciated Pavements in the Kimberley Region, Western Australia", *J. geol. Soc. Aust.*, 1968, **15**, pp. 51–6.

32. C. R. TWIDALE, *Structural Landforms*, pp. 90–96; *idem*, "Steepened Margins of Inselbergs from northwestern Eyre Peninsula, South Australia", *Z. Geomorph*, 1962, **6**NS pp. 51–69; *idem*, "Origin of Wave Rock, Hyden, Western Australia", *Trans. R. Soc. S. Aust.*, 1968, **92**, pp. 115–23.

33. B. HELLSTROM, "Några Iakttagelser över Vitting, Erosion Och Slambildning I Malaya ich Australien", *Geogr. Ann.*, 1941, **23**, pp. 102–24.

34. See for example F. E. ZEUNER, *Dating the Past*, Methuen, London, 1950, 2nd Ed., pp. 127–34; W. T. WARD, "Eustatic and Climatic History of the Adelaide Area, South Australia", *J. Geol.*, 1965, **73**, pp. 592–602; W. T. WARD and R. W. JESSUP, "Changes in Sea Level in Southern Australia", *Nature (Lond.)*, 1965, **205** (4973), pp. 791–2.

35. C. T. MADIGAN, "The Australian Sandridge Deserts", *Geogr. Rev.*, 1936, **26**, pp. 205–27; D. KING, "The Sand-ridge Deserts of South Australia and Related Aeolian Landforms of the Quaternary Arid Cycles", *Trans. R. Soc. S. Aust.*, 1960, **83**, pp. 93–108; J. N. JENNINGS, "A Revised Map of the Desert Dunes of Australia", *Aust. Geogr.*, 1968, **10**, pp. 408–9.

36. I. REBA, "Applications of the Coanda Effect", *Sci. Am.*, 1966, **214** (6), pp. 84–92.

37. C. R. TWIDALE, "Steepened Margins", pp. 51–69; "Origin of Wave Rock", pp. 115–23; *Structural Landforms*, pp. 90–6.

38. B. HELLSTROM, "Nagra Iakttagelser", pp. 102–24.

39. For review, see C. R. TWIDALE, *Structural Landforms*, pp. 50–5.

40. C. R. TWIDALE, "Origin of Wave Rock", pp. 115–23.

41. P. B. MEDAWAR, "Scientific Method", *The Listener*, 1967, **78** (2011), pp. 452–6.

42. W. M. DAVIS, *Geographical Essays*, Dover, Boston, 1909, p. 249.

Additional References

T. C. CHAMBERLIN, "The Method of Multiple Working Hypotheses", *Science (N.Y.)*, 1890, **15**, pp. 92–6.

G. K. GILBERT, "The Inculcation of Scientific Method", *Am. J. Sci.*, 1886, **31**, pp. 248–99.

C. A. M. KING, *Techniques in Geomorphology*, Arnold, London, 1966, p. 342.

P. B. MEDAWAR, *Induction and Intuition in Scientific Thought*, Methuen, London, 1969, p. 62.

K. R. POPPER, *Conjectures and Refutations*, 3rd ed., Routledge and Kegan Paul, London, 1969.

J. TRICART, *Precipes et Methods de la Géomorphologie*, Masson, Paris, 1965, p. 496.

Part 2

STRUCTURAL GEOMORPHOLOGY

The influence of structure on geomorphology is profound. Not only are the great features in the morphology of the globe— mountain chains, plateaux, rift valleys, continental margins and the like—predominantly of structural origin and formed by folding, faulting, or warping, but many of the minutiae of landforms, the shape of hills or mountains and the trends and patterns of streams, are controlled by structure through the action of agents of weathering and erosion on complex rock masses.

E.S. Hills, 1963

CHAPTER 2

Major Features of the Planet Earth

A. THE SHAPE OF THE EARTH

The shape of our planet has long been the subject of speculation but not until the space age was it known with any certainty and precision.

Up to the time of the Renaissance, most people thought the earth was either flat or disc-shaped. It is true that as early as the sixth century B.C. the Pythagorean School had asserted that the earth was spherical, but this was more an avowal of faith in the perfection of nature, of which the earth was part, than an inference drawn from observations. Belief in the sphericity of the earth persisted through the ages and formed the basis not only of Columbus' unwise but brave excursion in 1492 but also, to take one example, of Eratosthenes' very close estimation of the size of an arc of the earth based on the results of acute observation, and astute reasoning.

Eratosthenes, who lived early in the third century B.C., noted that the sun was directly overhead at Syene (Aswan), in upper Egypt at midday on midsummer's day. He therefore calculated that Syene was on the tropic and furthermore that by measuring the angle of the sun at midday on midsummer's day at Alexandria, which he thought was located in the same longitude as Syene, he would in effect be measuring the angle subtended by lines joining Syene and Alexandria to the centre of the earth—in other words, the latitudinal difference between the two places (Fig. 2.1). He then estimated the distance between Syene and Alexandria by the time and distance method: the two cities were so many days journey by camel apart and the camels averaged such and such a speed, so that the distance between them was known.

If the distance is X and the latitudinal difference between the two places, as measured on midsummer's day, is $Y°$, then the circumference of the earth (C) can be calculated from $C = \frac{360}{Y} \times X$; or, since Eratosthenes did not use degrees, a certain observed part of a great circle (his measurement indicated that Alexandria and Syene were separated by $\frac{1}{50}$ part of a circle) and the circumference could thus be obtained by multiplying the distance X by 50.

The units of distance used by Eratosthenes were stadia, of which three values are known, though it is commonly accepted that a stadium of 183 m (600 feet) was most widely used. The distance between Alexandria and Syene was calculated as 5000 stadia, and their difference of arc $\frac{1}{50}$ of a great circle (i.e., 7° 12′), so that according to these estimates the circumference of the earth is 250,000 stadia (though other writers, basing their reports on Eratosthenes' calculations, cite a distance of 252,000 stadia: possibly Eratosthenes too preferred a round figure in deference to the perfection of nature) which at 183 m to the stadia gives a circumference of

46,618 km (28,968 miles). The real value is 43,073 km (24,901 miles).

The remarkable proximity of this early calculation is something of a fluke. The measurement of the sun's altitude was 5′ 18″ in error; Alexandria and Syene are not on the same line of longitude, and the estimate of the distance between them was 448 stadia out. Nevertheless, the fact that the calculation was based on an observation and measurement is noteworthy, as is the soundness of the theory on which the calculation was founded. And Eratosthenes' theory assumes, and his calculation is only possible in terms of, a spherical earth.

Notwithstanding this early theory, the sphericity of the earth was not widely accepted until the great voyages of circumnavigation of the Renaissance. And very soon afterwards both observational evidence and theoretical argument suggested that the earth might not be perfectly spherical. [1] French scientists working in South America in 1672 noted that a clock which had kept good time in Paris was behaving

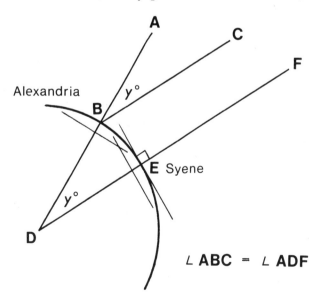

Fig. 2.1 Eratosthenes' calculation of the latitudinal difference between Alexandria (B) and Syene (E) at midday on midsummer's day. The angle Y is measurable and is equal to the angle B–D–E which equals the latitude difference between B and E.

Plate 2.1 The earth from about 157,000 km (98,000 miles) away. North Africa, the eastern Mediterranean and the Arabian Peninsula are clearly visible through the break in the cloud cover in left centre of the photograph. (*NASA Apollo II spacecraft photo, astronauts Armstrong, Collins and Aldrin.*)

erratically and attributed this to varied gravitational pull (which they took to imply variations or eccentricities in the earth's shape) at different points on the earth's surface. Polar flattening of the earth of one part in 230 (compared to the reality of about one in 298) was ingeniously calculated by Newton without leaving his desk. During the eighteenth century the French succeeded in confirming that the earth was an oblate spheroid by measuring the value of a degree of latitude in Lapland and Peru. Despite their measurements, however, their estimates of the amount of polar flattening varied betweeen one in 179 and one in 266. Using essentially the same methods, the degree of polar flattening was gradually refined until in 1957 the generally accepted figure was one part in 297.1. Even then however the precise shape of the earth as a whole was not known.

Then the first artificial satellites and spacecraft were launched and information supplied by their instruments has allowed the definition of the shape of the earth with a precision previously unknown. First, photographs of the earth taken from space confirmed its approximate sphericity (Pl. 2.1). Second, the careful tracking of satellite orbits has permitted a rather precise definition of the *geoid*, that is the earth body considered as sealevel extended in a logical

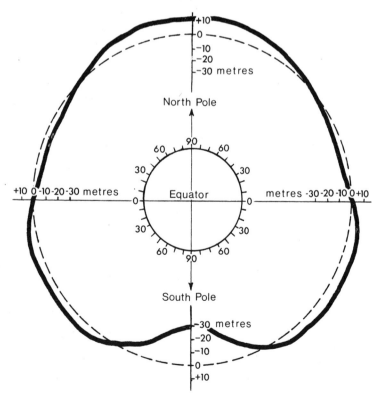

Fig. 2.2a Cross-section through the earth, showing its pear-like shape. (*After D.G. King-Hele, "The Imperfect Sphere". The Listener, 1970, 83, pp. 717–18.*)

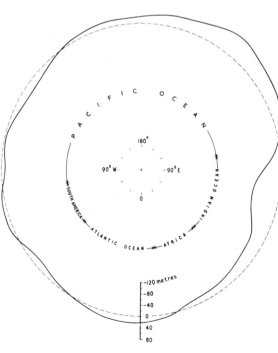

Fig. 2.2b Equatorial section of the geoid. (*After D.G. King-Hele, "Heavenly Harmony and Earthly Harmonics", Q.J.R. Astro. Soc., 1972, 13, pp. 374–5, by permission of D.G. King-Hele.*)

manner beneath the continents and ignoring such comparatively minor features as mountain ranges and ocean deeps.

The pull of gravity varies directly with the product of the two masses involved and inversely with the square root of the distance between them. As the two masses, namely the earth and the artificial satellite, remain essentially of constant mass, any eccentricities in the spherical paths of satellites orbiting around the earth reflect variations in the distance between the orbit and the earth's surface. Thus by measuring such eccentrcities and by making the necessary allowances for other, known, variations in the earth's gravitational field (see Sections 2.C and 2.D), it has been possible to plot the shape of the planet earth more closely than ever before.

The earth is neither a sphere nor an oblate spheroid. It is slightly pear-shaped, the stalk end or stem of the pear being toward the North Pole[2] (Fig. 2.2a). Even this is an over-simplification, however, for there are numerous broad rises and depressions (Figs 2.2a and b, 2.3). The polar flattening of the earth is real enough, and averages one part in 298.2. This means that the equatorial diameter of the earth exceeds the polar diameter by about 49.26 km (26.58 miles). But the polar flattening is not equal, the North Pole being 39.62 m (130 feet) further from the earth's centre than the South Pole. The weight of the Antarctic ice mass is most likely responsible in considerable measure for this anomaly. There is a broad depression some 79.25 m (260 feet) deep south of India and

a hump about 70.1 m (230 feet) high in the region of New Guinea. Britain and Western Europe also occupy a broad rise in the geoid.

The reason for these broad swells and depressions is not known. Though they are most probably related to the internal activity of the earth, there is no obvious correlation between them and major structural features. There is, however, a similarity of pattern between these departures from sphericity and the earth's magnetic field, though the two patterns are separated by about 160° of arc. Whether there is a causal connection between these two patterns is not known, but as the earth's magnetic poles and field vary in time it is interesting to speculate on the possibility of the earth's bumps and hollows migrating in time.

B. THE EARTH'S MAJOR RELIEF

Even a casual inspection of a globe or a map of the earth depicting its relief, or of space photographs of the planet, reveals several major features. The first is the contrast between the continents and the ocean basins. The former occupy 29.2 per cent of the earth's surface and stand, on average, about 840 m above the present sealevel (Fig. 2.4). The oceans and seas cover 70.8 per cent of the surface and have a mean depth of 3800 m below sealevel. The plotting of the frequency of occurrence of various arbitrary topographic

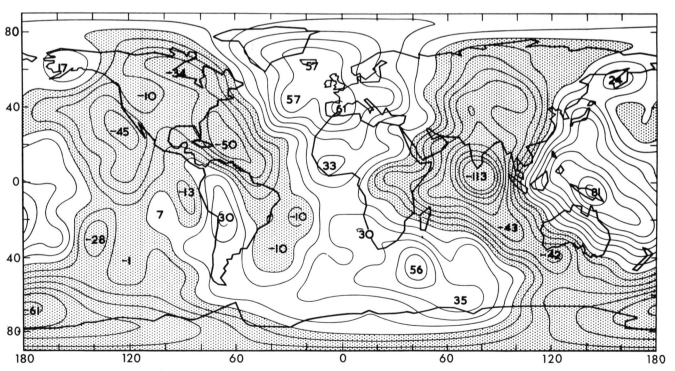

Fig. 2.3 This map of the geoid shows its major broad rises and depressions, in metres. (*After D.G. King-Hele, "Heavenly Harmony and Earthly Harmonics", Q.J.R. Astro. Soc., 1972,* **13**, *pp. 374–5, by permission of D.G. King-Hele.*)

levels reveals maxima at plus 500 m and minus 4700 m with respect to present sealevel (Fig. 2.4).

The second notable feature of the earth's major relief is the Pacific Basin, which occupies about one-third of the entire surface of the globe and which is the largest single feature on earth.

The third concerns the pattern or distribution of continents and oceans (Fig. 2.5). The following are worthy of note:

(a) the predominance of land in the northern hemisphere;

(b) the girdle of land around the north polar basin;

(c) the dominance of ocean in the southern hemisphere;

(d) the three southerly prolongations of land from the northern into the southern hemisphere—South America, Africa and Malaysia-Indonesia-Australasia;

(e) the triangular shape of several of the continents and subcontinents;

(f) the isolated high continent around the South Pole; and

(g) the antipodal arrangement of land and sea.

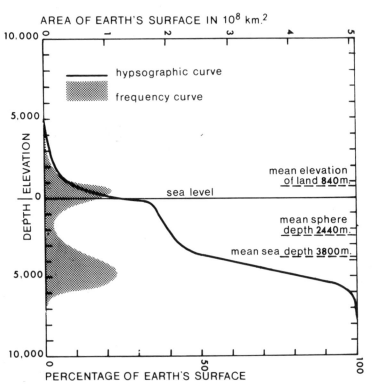

Fig. 2.4   The hypsographic and frequency curves of the earth's surface. (*After P. Lake,* Physical Geography, *2nd ed., Cambridge University Press, 1949, p. 144.*)

The fourth major feature is not so obvious because it is a submarine feature. But rising above the sea-floor is a virtually continuous range of mountains[3] which in many places displays greater relief than the Himalayas.

Many suggestions have been advanced to explain these major relief features. The more important theories are reviewed in Section 2.E. But as many of them involve properties and features observed or deduced from studies of the earth's crust, it is necessary to consider this structure briefly. The first indication of differentiation in the upper layers of the crust arises from the distribution of rock types. Granitic rocks* are widespread both in outcrop and beneath the surface in the continental areas. There are no granite outcrops in the deep ocean basins, where basic rocks (especially volcanic types) are dominant. This suggests that the continents may be compared to rafts of the less dense granitic materials floating in an ocean of denser basic rocks. The layering of the crust is corroborated by a study of the pressure waves generated by earthquakes, which have also allowed inferences as to the earth's deeper structure.

C. STRUCTURE OF THE EARTH

Earthquake shocks are a feature of everyday life in some parts of the world. Japan experiences some 5000 each year. Over the past 2000 years some ten million people have died directly or indirectly from earthquake activities. In Japan in 1556 some 830,500 people died in a single shock.

Earthquakes are caused by fracturing and differential movement (faulting) within the crust. These movements in turn are believed to be the result of a continual disequilibrium between the earth's precise shape (see Section 2.A) and its speed of rotation about the geographical axis. The size of the southern hemisphere bulge of the earth (see Fig. 2.2) varies in some measure with the speed of rotation of the earth. But as the rotational speed is constantly changing, there is a constant tendency for adjustment of shape. However there is a time lag between the speed of rotation and the adjusted surface form: the shape is adjusted to the speed of rotation of about ten years ago. Hence there is disequilibrium and stress in the crust on this account alone.

Pressure gradually builds up in certain zones to about 1.575 kg/mm^2 over an area of about 10,000 km^2 before part of the crust fails. The ground level over a radius of some 50 km begins to tilt down towards the centre of failure—the *focus* of the incipient earthquake—for some months beforehand, and for some hours previous to the shock the tilt is marked. By noting and measuring these changes it is possible to predict the location of earthquakes in the immediate future, at any rate.

Apart from its obvious impact for man (Pls 2.2 and 2.3), the study of earthquakes (*seismology*) provides evidence of the structure of the earth, and particularly of the earth's crust. Most earthquakes are due to faulting in the crust of the earth. The centre or point of origin of an earthquake is called its focus. Most foci are located at shallow depths, the great

---

* For a review of the nature of various rock types, see Appendix 1 which also includes a summary of the more common and important structures found in these rocks and an account of dating methods.

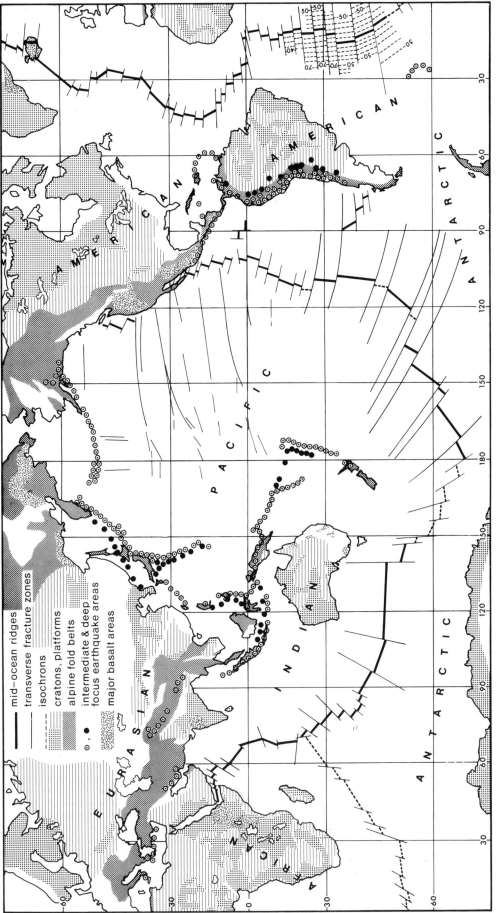

Fig. 2.5 The distribution of the major continental and submarine mountains, major areas of basalt flow, deep and shallow earthquakes, and major plates. Zones of magnetic reversal, with *isochrons* (lines indicating rocks of equal age) are shown for the south Atlantic.

mid-ocean ridges

transverse fracture zones

isochrons

cratons, platforms

alpine fold belts

intermediate & deep
focus earthquake areas

major basalt areas

Plate 2.2   During the Kwanto earthquake of 1 September, 1923, over 127,000 died in the Tokyo and Yokohama areas alone, and many more were injured. There was also great damage to property, much of it caused by fires which were started by the earthquake and which burned out of control because of high winds and the disruption of water supplies. The focus of the earthquake was beneath Sagami Bay, southwest of Tokyo; parts of the sea-floor in the Bay were lifted by as much as 240 m., and some areas near Tokyo were uplifted by as much as 185 m. This picture shows a section of Tokyo devastated by the earthquake and subsequent fire. (*Courtesy Inter. Soc. Educ. Info., Tokyo.*)

Plate 2.3   Railway lines distorted and carriages derailed in northern Honshu during the Toka-chioki earthquake of 16 May, 1968. (*Nat. Res. Centre Disaster Prevention, Tokyo.*)

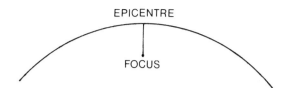

EPICENTRE

FOCUS

Fig. 2.6a   Epicentre and focus of an earthquake.

majority being less than 15 km below the surface, though a few are deep seated; for example, one in Java was estimated to be 600 km deep. Pressure waves spread out from the focus through the body of the earth and are recorded on seismographs.

An elastic body like the earth transmits two types of seismic shock waves:

1. P waves (*primary* or *push* waves) which are caused by particles pushed by the disturbance pushing their neighbours, and so on. The particles move backwards and forwards in the direction in which the wave is progressing, and so form what is called a *longitudinal* wave.

2. S waves (*secondary* or *shake* waves) are caused by the disturbance pulling at particles in a plane normal to direction of wave progression. Such *transverse* or *shear* waves cannot be transmitted by fluids.

Although both P and S waves are propagated together, P waves travel almost twice as fast as S waves, so that a seismograph records the P or primary waves first (hence their name) and the S waves second. They appear as two different phases, and the more distant the recording station from the focus, the greater the phase difference between the two.

The point on the earth's surface immediately above the focus is called the *epicentre* (Fig. 2.6a). The shock waves propagated by any earthquake are recorded on seismographs located in many parts of the world. It has been found consistently that both P and S waves are recorded in a wide arc up to 105° from the epicentre (Fig. 2.6b). Beyond this is a "shadow" zone, in which it was at first thought that neither P nor S waves were received, but within which newer and more sensitive seismographs have now recorded faint and considerably delayed P waves, though S waves remain absent. Between 142° and 180° of arc, P waves are recorded, but again, no S waves.

The most likely explanation of this situation is that the earth has a core which is devoid of strength, and must therefore be either fluid or plastic. Any waves coming into contact with such a core would either be reflected or refracted, thus giving a shadow zone within which no shock waves were felt (Fig. 2.6b). Furthermore, as transverse waves cannot be transmitted by a fluid, this also accounts for the absence of S waves beyond 105° of arc from the epicentre. The reception of faint P waves in the shadow zone suggests that within the fluid or plastic core there is an inner solid core, which causes some P waves to be reflected into the 105°–142° zone (Fig. 2.6b). Thus the picture of the earth which emerges from these earthquake studies is that of a body consisting of a number of shells: a core (with an inner core) and a mantle. The radius of the core is about 3480 km, and the thickness of the surrounding shell 2900 km.

But this outer shell is not uniform, for the time it takes shock waves to travel from the focus to recording station is not directly proportional to the distance between the two. Much depends upon the depth to which the chord linking the two extends, in other words on the depths to which the pressure waves penetrate. Studies of the time travel paths of many earthquake shocks have suggested that there is a marked change in velocity, indicating a marked physical break at about 2900 km (Fig. 2.6c), at the junction of the core and the mantle. Other discontinuities have been located which are less marked than this but nevertheless distinct. P and S waves

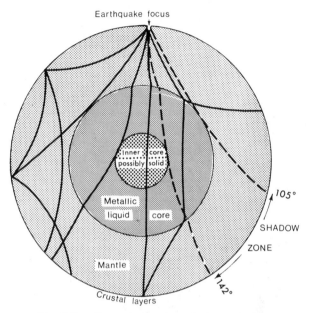

Fig. 2.6b   Cross-section through the earth body showing layered structure and travel paths of pressure waves. (*After J.H. Hodgson, Earthquakes and Earth Structures, © 1964, p. 87. Reprinted by permission of Prentice-Hall, Inc., Englewood Cliffs, N.J., U.S.A.*)

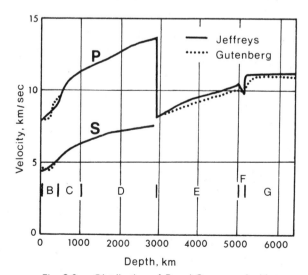

Fig. 2.6c   Distribution of P and S wave velocities throughout the earth according to Jeffreys and Gutenberg. (*After J.H. Hodgson*, Earthquakes and Earth Structures, © 1964, p. 88. Reprinted by permission of Prentice-Hall, Inc., Englewood Cliffs, N.J., U.S.A.)

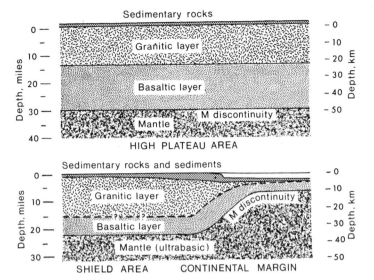

Fig. 2.7 Interpretations of the layering of the earth's crust; the lower diagram is closer to reality. (*After J.H. Hodgson, Earthquakes and Earth Structures, © 1964, p. 85. Reprinted by permission of Prentice-Hall, Inc., Englewood Cliffs, N.J., U.S.A.*)

travelling to seismic stations close (that is, within 20° arc) to the epicentre commonly display three different phases.

As the velocity of a pressure wave varies with the density of the medium through which it is passing it has reasonably been inferred that there are three physically distinct layers close to the earth's surface (Fig. 2.7) by contrasting the behaviour of shock waves travelling through continental masses and those passing only beneath the ocean floors. One layer is of granitic composition and is called the sial, a name derived from its two principal constituents, *si*lica and *al*umina. Below it, though flooring the true oceans like the Pacific and Arctic, is a continuous shell of basaltic composition known as the sima, again a mnemonic word derived from its main constituents *si*lica and *ma*gnesia. Below this again is a shell of ultrabasic rocks the upper part of which, according to some writers,[4] is relatively soft and light, and of critical importance both in reducing seismic velocities and in any consideration of the origin of the earth's major relief (see below). The base of the sima is called the *Mohorovicic* discontinuity, after its discoverer. Often known for the sake of brevity as the M or Moho discontinuity, this is taken as the base of the crust. It varies in depth from one area to another. Beneath the continents the crust averages some 35 km thickness, but beneath mountain ranges like the Andes it reaches 70–80 km, and beneath the ocean floor is only 5–6 km thick. The base of the sial is also named after its discoverer and is called the *Conrad* discontinuity.

Thus the continents appear to be rafts of sial floating on a continuous shell of sima. The sedimentary layers which are such prominent features of many of the continental areas in reality merely form a discontinuous veneer overlying parts of both the sial and the sima on the continents and in the ocean basins.

D. ISOSTASY

The theory of isostasy stems from and rests upon the density layering of the earth's crust. In its geomorphological implications it is one of the most significant concepts concerned with the earth's structure.

The concept of isostasy originated in the geodetic and trigonometric survey of India late in the nineteenth century. The latitude difference between Kaliana and Kalianpur, the former only 111 km (60 miles) from the Himalayas, was calculated both by triangulation and by using astronomical methods: a discrepancy of 5.236″ was found between the two results. However, allowing for an average deviation of the plumb-bob used to determine the vertical in the astronomical method, the difference between the two should have been 15.885″. The explanation offered was that the Himalayas did not exert the gravitational attraction expected of such a large mountain mass because they are underlain by rocks of low density.

This general concept is confirmed by analysis of sialic and simatic rocks, the former having an average density of 2.75–2.90, the latter 2.90–4.75, and by the seismic data (see Section 2.C). It implies that below any unit of the earth's crust there is an elongate pyramid of material which, regardless of its precise length and regardless of whether it underlies an oceanic deep or a high mountain, should be of equal mass in every case. A long (mountain) unit contains enough rocks of below-average density to compensate for the uniformly high density rocks which underlie shorter (ocean floor) pyramids. The continents are higher than the ocean basins but the former are underlain by sialic material of a lower density than the sima which underlies the ocean floors (Fig. 2.7). However, not all, and possibly very few, parts of the crust are in isostatic equilibrium, and on local as well as regional scales isostatic adjustments have occurred and are still in progress.

For example, the Himalayas are still rising, in part at least because of pronounced river dissection and the resulting transfer of load from the mountains to the adjacent Indo-Gangetic plain and neighbouring sea-floors. Continuing orogenesis or compression could, however, be partly responsible. In a statement broadcast in a science review programme in early 1971, two Pakistani scientists claimed that the summit of Mt Everest had risen 55 m in the last twenty years. Howsoever this may be, it is dangerous and unwarranted to extrapolate either forwards or backwards in time from such an estimate, which represents a temporary state of affairs. Consider for example what such an estimate and extrapolation imply in terms of the location in the earth's crust of the sediments which form the crest of the peak say a mere 20,000 years ago: they would have been in a setting so deep as to be impossible from a geological point of view.

However, in the present context the chief importance of the concept of isostasy is that the flotation of continents is theoretically feasible.

E. EXPLANATIONS OF THE EARTH'S MAJOR RELIEF

1. A Cooling and Contracting Earth

At one time it was accepted that the earth originated as a mass of hot gases which gradually cooled, contracted and solidified, the outer shell or crust becoming solid first. The analogy was drawn between the shrinking earth with its mountain ranges and the wrinkled skin of a shrivelled orange.

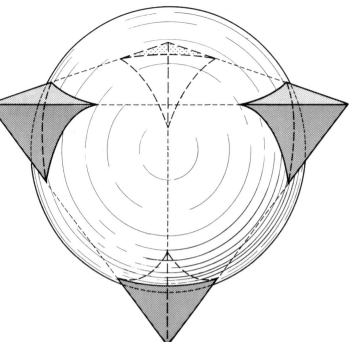

Fig. 2.8 A tetrahedron placed symmetrically in a hydrosphere, showing the corners and edges projecting above the water surface.

It was suggested that the earth had contracted due to the extrusion and cooling of lavas.[5] The contraction theory was the basis of many hypotheses of mountain building. It was also the origin of the Tetrahedral Theory, promulgated by Lowthian Green[6] in 1875 as an explanation for many features of the earth's major relief.

A tetrahedron is a four-sided figure, each face of which is a triangle (Fig. 2.8). A tetrahedron contains the least volume with respect to surface area (a sphere contains the most), so that a cooling earth of decreasing volume might be expected to collapse to such a form. Green conceived that the earth was a roughly tetrahedral mass (the faces tend to bulge) partially immersed in the hydrosphere (Fig. 2.8). Such an hypothesis apparently explains such features as the triangular shape of some land masses, the southerly extensions of three land masses (the protruding edges of the tetrahedron), and the antipodal arrangement of continent and ocean. Unfortunately for this hypothesis, the earth body is not widest in the northern hemisphere (see Fig. 2.2) as suggested by Green, though it has been argued that the location of the bulge changes in time.[7] More important, there are difficulties concerning the postulated cooling and contraction of the earth.

Had the earth in fact originated as a molten mass of gases, it is certain that volatile elements like water and nitrogen would have been boiled off into space. Yet these elements are prominent features of the earth's chemistry. Odd though it may sound, and though the early development of the planet remains quite obscure, some geologists now consider that the earth began its recognisable evolution as a cold solid mass.[8] Through the radioactive decay of various elements included in this mass, the earth has gradually become hotter, causing the melting of rocks and the gravity differentiation of their constituents. The heavier elements have sunk toward the centre of the earth, while the lighter ones have floated to the surface. This is in accord with the findings of seismology and isostasy. The crust as a whole has remained relatively cool, but the heating of the lower crust and possibly the mantle may have permitted not only the flotation of continents but also their lateral movement due to the development of convection currents in the molten layers.

2. The Orocline Concept

Several explanations of the earth's major relief features involve the migration of the continents. It is important to realise that horizontal displacements in the crust have long been demonstrated and accepted: the development of fold mountains involves crustal shortening and hence lateral movements, which in the case of the European Alps amounted to some 15° of latitude or 1850 km (1000 miles). But the lateral migration of tectonic units or blocks during mountain building has also been invoked in the explanation of relief features of regional extent.

Carey[9] for instance, in developing his orocline concept, pointed out that although isostasy presumably limits the depth to which deformation extends below orogenic (fold mountain) belts, there is no such restriction on the amount of lateral deformation. There are many examples of fold mountains which are buckled in plan, the distortion in some cases exceeding 180°. Such markedly-flexed fold mountain belts which are of regional extent, (that is, which are at least

1000 km long) are called *oroclines*. It may be that the *geosynclines*, or major troughs of accumulated sediments, were originally bent in plan; but they may have been essentially linear or at any rate not as buckled as fold mountains are now, and the implications of this second possibility should be examined. Three examples are selected from several cited by Carey.

The Pyrenees are part of the Alpine fold belt which runs from west to east through the length of southern Europe. The rocks of which the Pyrenees are built form a belt of compression characterised by folding and overthrusting. The zone of compression is widest at the eastern or Mediterranean end and narrows to the west, toward the Atlantic (Fig. 2.9a), where the intensity of folding is also less than in the east. If the Pyreneean folds were straightened out and if the Meseta, an ancient massif, remained attached to the sedimentary strata, the greatest unfolding would occur at the eastern end of the mountain range, the Meseta would be rotated in a clockwise direction and the Bay of Biscay would be virtually closed. Moreover, subsidiary structures in western France and the Iberian Peninsula would be brought into alignment (Fig. 2.9b).

In the Pyrenees no major buckling of the orogen in plan is involved, only marked overfolding and compression. However the Western Cordillera of North America displays a 30° bend north of the Gulf of Alaska (Fig. 2.10a). If this flexure were straightened both the Arctic and the North Atlantic ocean basins would be closed. But another suggestion is that these two major features developed during orogenesis as a result of buckling in plan of the geosyncline which contained the Western Cordillera sediments. The then continental nucleus remained attached to the orogens, being horizontally displaced or shattered along fault lines, though in some cases fragments were left behind as a trail of crustal blocks. Such a concept finds some support in the fracture patterns of Arctic North America and in the gross morphology of the North Atlantic region (Fig. 2.10b).

Similar bending and lateral migration of attached continental blocks with regard to the Himalayas and associated Alpine fold mountain belts has been invoked to explain the major outlines of the Arabian Sea and environs (Fig. 2.11a and b).

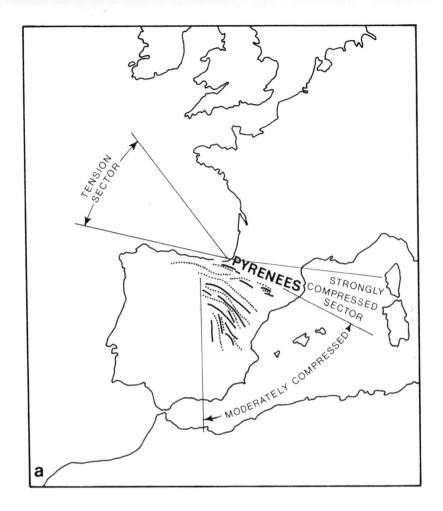

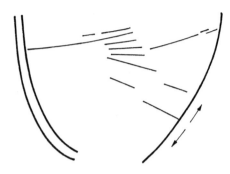

Fig. 2.10a The Arctic and North Atlantic regions as they are now. (*After S.W. Carey, "The Orocline Concept", pp. 255–88.*)

Fig. 2.9a The Pyrenees and Iberian Peninsula as they are now.

Fig. 2.9b The Pyrenees and Iberian Peninsula before compression of the sediments deposited in the geosyncline and opening up of the Biscay Rift. (*After S.W. Carey, "The Orocline Concept in Geotectonics", Pap. Proc. R. Soc. Tas., 1956, **89**, pp. 255–88.*)

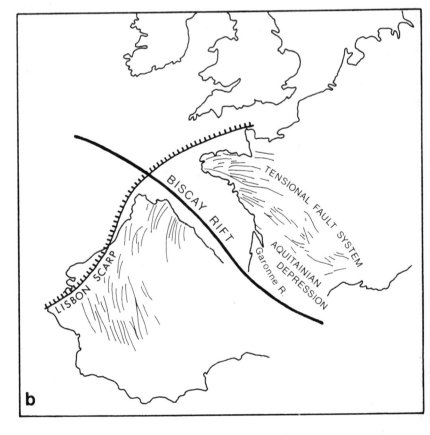

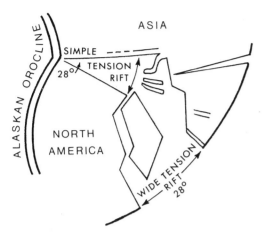

Fig. 2.10b Suggested development of the present situation of the Arctic and North Atlantic regions by flexing of Alaskan orocline and development of a huge tensional rift. (*After S.W. Carey, "The Orocline Concept", pp. 255–88.*)

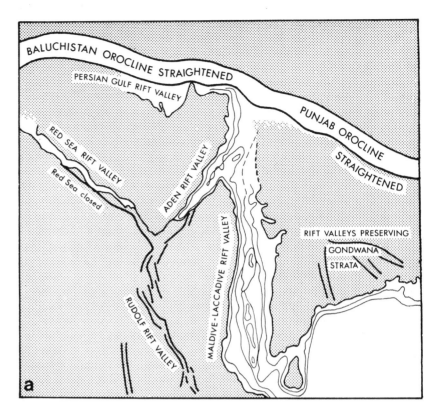

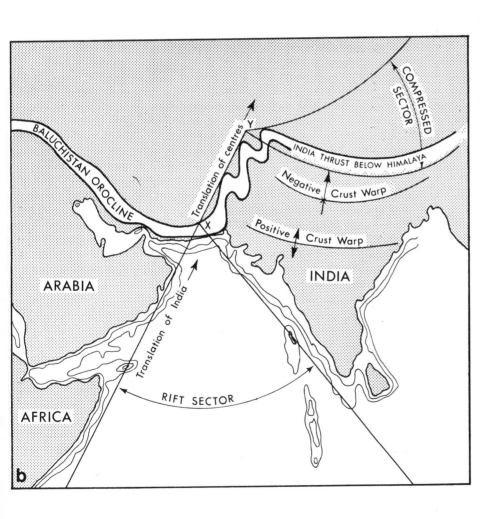

Fig. 2.11a and b Evolution of south and southwest Asia by translocation of India attached to a bent orocline, and the opening up of tensional rifts. (*After S.W. Carey, "The Orocline Concept", pp. 255–88.*)

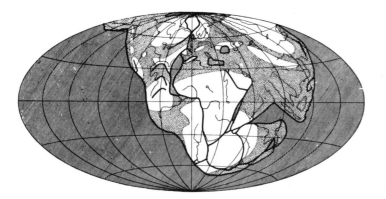

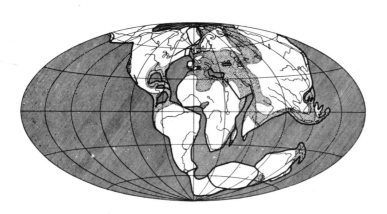

Fig. 2.12 The rifting and drifting of Pangaea into the present continental masses through Mesozoic and Tertiary Stages shown are, from top to bottom, Upper Carboniferous, Eocene and Older Quaternary. (*After A. Wegener,* Origin of Continents and Oceans, *Methuen, 1924, p. 6.*)

3. Continental Drift

Though propounded both in France and in the USA during the nineteenth century, the theory of continental drift is chiefly, and justly, associated with the name of Alfred Wegener,[10] and with that of his eloquent supporter, the South African geologist Alex du Toit.[11] Wegener was initially concerned with explaining past climatic distributions, and in order to achieve a measure of understanding he suggested that the continents had not always occupied their present locations but that they had migrated in the course of geological time, passing through different latitudes and being affected by different climates as they did so. Wegener supposed that during Palaeozoic times the several continental masses as they then existed were aggregated into two major sialic regions. These two ancient continents were called by Wegener *Gondwanaland* (comprising the nuclei of modern Australia, India, Antarctica, Africa and Brazil) and *Laurasia* (northeastern North America, northern Europe and northern Asia). The two were separated by a mediterranean sea called the *Tethys,* and together formed a supercontinent called *Pangaea.* During the late Palaeozoic era the South Pole was located off the southern coast of the ancestral southern Africa and Australia. But during the Mesozoic period the supercontinent was fractured as its several elements drifted apart to become the present continents (Fig. 2.12), the distribution of which is ephemeral since the drift is said to be still going on. It was also suggested that the present high mountain chains could have resulted from the crumpling of the fronts of the advancing continents against the simatic layer through which the sial rafts were drifting.

What sparked Wegener's conception of drifting continents was the closeness of the fit between the west coast of Africa and the east coast of South America, a fit still more remarkable if the 50 fathom (100 m) line is taken as a better representation of the continental margins than the present coasts. Wegener and du Toit marshalled an impressive array of evidence — physiographic, stratigraphic, tectonic, volcanic, palaeoclimatic, palaeontological and geodetic — to demonstrate that there are matching features on what are now opposed coasts and that the continents were formerly much closer together.

Nevertheless, though there was general support for the theory of continental drift from palaeontologists and biologists (for whom the theory apparently reduced problems associated with explaining the dispersal and distribution of plants and animals), most geologists did not accept drift as a working hypothesis until recently. Some of the detailed evidence cited by Wegener and du Toit was either incorrect or equivocal. Awkward questions were asked about the detailed schedule of rifting and drifting proposed by Wegener and necessitated by the evidence, relating for instance to palaeoclimatology. Why did not drift commence until the Mesozoic? What is the origin of those mountain ranges which date from Palaeozoic or Precambrian times? The continents are criss-crossed by major fractures called *lineaments* (Chapter 4) which display preferred and statistically significant trends; these predominantly NW–SE and NE–SW fractures (Fig. 2.13) are evidently part of a global pattern [12] and are of great antiquity. In terms of continental drift they form a tectonic pattern developed in Pangaea. Yet, despite drifting, the present orientation of the fractures implies that the continents have either retained their original alignments or have rotated in all cases through 90°, 180° or 360°.

The source of energy responsible for the massive translocations of the continents, as well as for their initial disruption, was also queried. But the decay of radioactive minerals in the crust and mantle of the earth is a source of heat which may be sufficient to create convection currents powerful enough to rift and move the continents (Fig. 2.14). In any case inability to identify a mechanism does not necessarily rule an idea out of court: when the massive nappe structures and implied translocations of strata in the European Alps were first mooted they were said to be mechanically impossible, yet detailed mapping utterly confirmed the existence of such structures.

Today, however, the displacement of the continents during geological time (though not necessarily in the terms or with the time schedule proposed by Wegener and du Toit) is widely accepted by geologists as a framework into which other geological findings are to be fitted. This is not to say there are no geological difficulties (on the contrary) and, paradoxically, major objections now originate with the biologists; but the general scheme of migrating continents is widely supported. This remarkable turnabout is due to

geophysical, topographical and geological evidence which has come to light during the past two decades. In particular the investigation of palaeomagnetic data and of the ocean floors have provided critical evidence.

4. Palaeomagnetism

One of the principal tools to emerge in the field of palaeogeographic interpretation is palaeomagnetism.[13] When minerals rich in iron and titanium such as magnetite and ilmenite crystallise out from an igneous melt, they acquire a magnetism, called a *remanent magnetism*, which is related to the magnetic field and poles of the earth at the time of their formation. At present, though they are migrating constantly, the magnetic poles are never far from the geographical poles, so that the geomagnetic and rotational axes are closely coincident. One of the basic assumptions underlying the use of palaeomagnetic data for palaeogeographical purposes is that there has always been this close proximity of the two.

Minerals like magnetite which are released from the parent igneous rock by weathering and which are transported and deposited in, say, a river alluvium also become aligned with the magnetic field of the day. Thus by carefully noting the original or field orientation of the rock specimens containing the magnetic crystals and by determining the

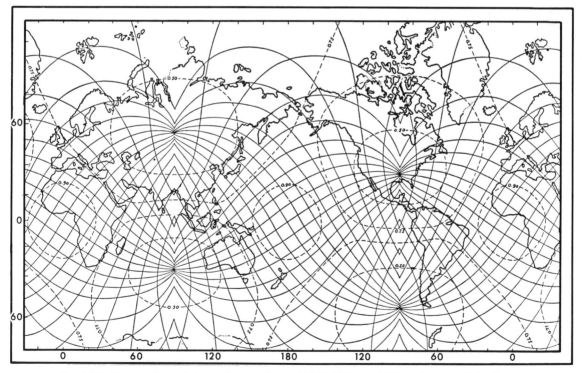

Fig. 2.13 Shear pattern of the earth according to F.A.V. Meinesz. (*"Shear Patterns of the Earth's Crust"*, Trans. Am. geophys. Un., *1947, 28, pp. 1–61.*) The patterns themselves appear to be substantiated by field evidence, but Meinesz' explanation (shift of the poles through 70° latitude along the meridian 90 E) has no sound mechanical foundation.

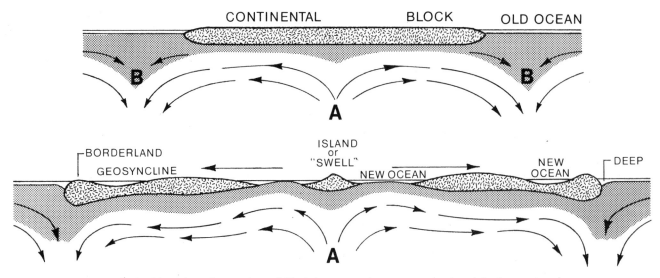

Fig. 2.14   Disruption of a continental block by convection currents developed in the crust and upper mantle: A, updraft; B, downdraft. (*After A. Holmes,* Principles of Physical Geology, *2nd ed., Nelson, 1965, p. 1001.*)

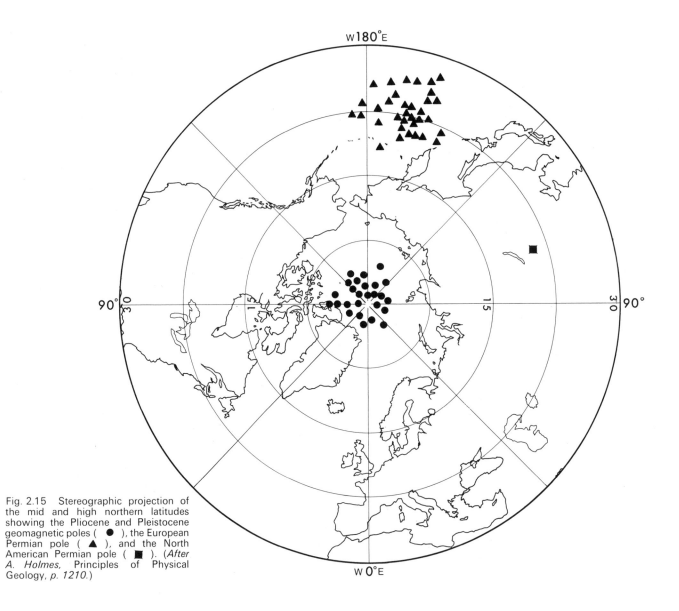

Fig. 2.15   Stereographic projection of the mid and high northern latitudes showing the Pliocene and Pleistocene geomagnetic poles (●), the European Permian pole (▲), and the North American Permian pole (■). (*After A. Holmes,* Principles of Physical Geology, *p. 1210.*)

magnetic field and poles locked into them, it is possible to determine within fairly narrow limits the direction of the magnetic, and hence the geographic, poles at the time of crystallisation or deposition of the minerals, as the case may be. Provided that the geological age of the rocks containing the minerals is definitely known, and if several determinations are made on rocks of the same geological age from different parts of the world, then simple intersection of magnetic directions gives the position of the magnetic poles, and hence the geographic poles, of the day.

This has been done for many rocks of many geological ages and what emerges is that the Quaternary and late Tertiary magnetic poles are close to their present positions. But further back in geological time, there are considerable departures: the Permian poles, for instance, were some 40° of latitude from the present poles (Fig. 2.15). Such a situation may have come about in at least two different ways. The whole crust, and possibly the mantle, may have moved relative to the earth's axis of rotation, and hence to the magnetic axis or dipole. This is called polar wandering and the implication is that the continents should retain their position relative to each other, so that magnetic determinations from rocks of the same geological age and from different continents should all point to the same general areas as representing the locations of the former poles. Thus in Fig. 2.15 the location of the Permian North Pole as indicated from European rocks is shown. But the location of the Permian Pole as determined from North American rocks should of course be the same, if there has been no relative movement of the two continents; and this is found not to be the case (Fig. 2.15).

This indicates that the continents have changed their positions relative to each other. Continental drift of some sort or another has taken place, resulting in a scattering of polar determinations for each geological period. Each continent indicates separate polar positions. It is this evidence that has convinced most geologists that drift is the reality with which other apparent difficulties must be reconciled.

5. Sea-floor Spreading

One of the most remarkable and puzzling facets of palaeomagnetic work has been the discovery that from time to time the earth's magnetic field has suffered sudden and complete reversals.[14] Sequences of rocks containing magnetised particles display similar orientations of magnetic field, but in one layer particles which are north-seeking show the same orientation as particles in succeeding layers which are south-seeking. For instance, such reversals in time are clearly documented in vertical sequences of lava flows. Zones of intrusive volcanic rocks of contrasted polarisation are also known from the ocean floors. In the South Atlantic for instance broad zones of similar polarity occur on either side of the mid-Atlantic Ridge, the sequences to east and west being mirror images of each other (Fig. 2.5). The sea-floor bordering the Carlsberg Ridge in the Indian Ocean displays a similar mirror-image pattern. As reversals of magnetic polarisation occur in time, these broad stripes are in fact time zones. They have been dated by radiometric means: most of the lavas are Cainozoic, though some of Jurassic age have been located in the floor of the Pacific.

As the ridges of the ocean floors are known to be zones of great instability (high terrestrial heatflow, high magnetic anomalies, numerous earthquake shocks) and volcanicity, with many characteristics of faulting and suggestions of rift valleys or graben (see Chapter 4) within them, it has been suggested that the stripes result from repeated, indeed continuous, volcanic intrusions into the gaping tops of the ridges. The migration of lavas from the central ridge area has caused the generation of crust. Thus the Atlantic floor appears to be spreading at a rate of some 2 m every 70 years. Sea-floor spreading may also have caused the pushing apart of continents.

One of the implications of the concept of sea-floor spreading is that if crust is being made, it must either be being destroyed or consumed somewhere, or the circumference of the earth is increasing, or of course both may occur either in balance or with one process outpacing the other. There are indications that parts of the crust may be being pulled down into the mantle in areas like the Tonga Trench and the Kermadec Trench, both in the southwest Pacific. Here, as well as off the west coast of South America, south of Alaska and off eastern Asia, there are gigantic trenches in the sea-floor where crust may be actively subsiding and suffering ingestion.[15] The Tonga Trench is 5000 m deep and 1200 km long, though only 8 km wide. The trench off the South American coast attains depths of almost 8000 m and some trenches are over 3000 km long. They are all areas of pronounced negative gravity anomaly, the crust being very thin beneath them. They are zones of major earthquakes. Parts of the crust seem to be being sucked down into the lower crust (Fig. 2.16). Whether they represent areas of crustal downdraft (see Fig. 2.14) in compensation for its production from the submarine rifts is not clear. One of the problems is that the sediments in the deep trenches off the west coast of South America, for example, are apparently undeformed which they should not be if, as the theory of sea-floor spreading demands, crust is being sucked down into the depth of the crust in the (and other) areas where migrating masses of the crust converge. But if there is no such equilibrium, if destruction of the crust does not at least equal the rate of its production and as the production of crust seems beyond doubt, then the earth must be expanding.

6. An Expanding Earth

The idea of an expanding earth arose from a suggestion by the British physicist Dirac[16] that the value of the gravitational pull of the earth is diminishing in time. One possible explanation for such a decrease would be that the surface of the earth was gradually becoming more remote from the centre of gravity, that is, the centre of the earth. On this basis it has been calculated that in 3.25×10^9 years the earth's circumference may have increased by 4.5 per cent or 2040 km (1100 miles), at a rate of 0.5 mm per annum.[17] The geological implications of the suggestion have also been examined and a good deal of evidence has been marshalled in its support. In particular the system of mountain ranges which occurs on the sea-floor (Fig. 2.5) is cited. The ranges in many areas display graben structures—probably due to tension— high magnetic anomalies, high terrestrial heatflow values, marked seismicity, and much volcanic activity. The mid-Atlantic Ridge is part of the system and as already mentioned this is apparently the source of the volcanic material which is responsible for sea-floor spreading.

Other evidence such as the distribution of flood basalts (see Chapter 6) lends support to the suggestion that the earth planet has expanded progressively through fracturing and drifting (as well as, possibly, continental sliding) since middle Precambrian times[18] (in North American parlance, since the Keweenawan, some 1100 million years ago). In these terms the ocean basins must be seen as having developed throughout geological time and not as being all comparatively young (that is, probably not more than 150 million years old) as suggested by some workers.[19]

7. The Plate Theory

Closely related to the concepts of sea-floor spreading and the expanding earth, and also stemming from defects in the traditional or Wegenerian concepts of continental drift, is the recent suggestion that though there have been continuous and continuing movements of parts of the crust in the horizontal sense, it is not so much, or not only, the continents that have drifted as huge segments of the crust, which are called *plates*, some of which consist partly of sial, partly of sima, though others are composed wholly of sima (Fig. 2.17). The boundaries of the plates are formed by the ranges of submarine mountains or rifts. Where the edges of plates converge, at a slow rate (less than 6 cm per annum), fold mountains are formed. Where two plates converge at speeds greater than 6 cm per annum, one plate slides under the other, forming a submarine trench.[21]

In such *subduction* zones,[22] crust is ingested as sediments are carried downwards along massive thrust zones. The strata become massively sheared and deformed. The *mélange* deposits of the Franciscan Formation near San Francisco—great blocks of rock set in a matrix of finely-sheared strata of the same lithology—may be of such an origin. Moreover, in such zones of plate convergence, former ocean basins may have been gradually eliminated at various times in geological history. Thus some workers consider that the Ural Mountains were the site of such an old ocean basin which disappeared as the opposed continental jaws closed, and the Himalayas may represent a much more recently destroyed ocean basin.

In other areas plates migrate away from each other creating zones of *divergence* where the gaps between plates become filled with volcanic and diapiric rocks upwelling from depth. For instance, Iceland lies athwart the mid-Atlantic rift, a fracture zone where the ocean floor is pulling apart (see Fig. 2.17) with North America moving westward relative to Europe at a rate of about 3 cm per annum. Most of the island is built of volcanic rocks, and Iceland has indeed (see Chapter 6) given its name to a particular type of eruption. The outburst of volcanic activity centred on the island of Heymaey, which was threatened with destruction in January 1973, serves as a reminder that this rifting, instability and volcanism continue.

In yet other regions the plate edges slide by one another in the well-known zones of shearing and transcurrent faulting (see Chapter 4).[23]

The plate theory overcomes or avoids several of the difficulties of the older drift theories. Drift has been in progress throughout geological time. Indeed there is no reason to suppose that these plates and the continents evolved on them have not circled the globe several times during the course of geological history. Not all continents on opposed sides of oceans need have been in very close proximity, though the idea of a single primeval continent remains. The drift is motivated by the internal energy of the earth. It seems

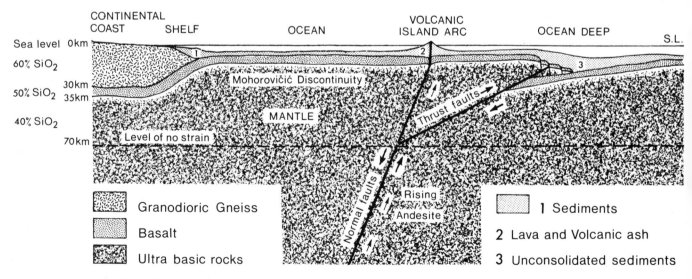

Fig. 2.16   Suggested underthrusting and ingestion of crust in submarine trench, with adjacent island arc. (*After J.T. Wilson, "Geophysics and Continental Growth", Am. Scient., 1959.* **47**, *pp. 1–24.*)

likely, in view of the palaeomagnetic evidence, that the continents may have migrated on their plates, in addition to translocation incurred by movement of the latter: comparable to a passenger walking along the corridor of a moving train.

Combinations of sea-floor spreading and plate tectonics seem capable of explaining the geophysical and geological evidence and the earth's major relief. In these terms, all the other continents have moved relatively away from Africa, and the Pacific is the shrinking remnant of a once enormous ocean. There remain, however, considerable difficulties. Why did all the sial collect in a single continental mass when the balance of the globe would seem to demand its equal distribution? The antipodal arrangement of land and sea is still a problem, though the tendency toward equilibrium of a rotating body may help explain the longitudinal spacing of the continents; the disparity between the extent of land and sea also has to be taken into account in this respect. The problem of lineaments and the biological and palaeontological evidence still has to be accommodated. And the timing of development and movement is still controversial.[24]

Many eminent geologists still express doubts about both drifting continents and the expanding earth, citing geological evidences which are not in accord with these ideas. Gilluly[25] is one who comes to mind, while Meyerhoff concludes that drift, if it occurred, took place very early in earth history—before the early Proterozoic era. Cobbing and Pitcher[27]

have recently discussed evidence from the Peruvian Andes which they consider is inconsistent with the concept of plate tectonics. Even more basic are the objections of Russian geophysicists like Beloussov,[28] who argues that the continents have always been where they are now. Beloussov emphasises vertical movements in the crust in explanation of fold mountain belts, considers that large scale depressed and elevated zones in the crust have never changed position, and believes the continents were formerly larger than they now are. He considers the permanent zones of elevation and depression to be determined by accumulations or blocks of molten basalt at the upper margin of the mantle; these molten rocks migrate upwards, especially up fracture zones and cause volcanicity where they reach the surface and broad swellings of the surface where they do not emerge.

The preceding brief review concerns a problem which, besides being of inherent interest to geomorphologists (though possibly beyond their critical range), also has direct and significant implications for several aspects of the work. For instance, if the earth is expanding and the volume of water on the earth's surface has remained constant or essentially so, then as this volume is being spread over an ever-larger surface area, a long term decline in sealevel is clearly implied. Again erosion, deposition, and the accumulation of volcanic lava or of glacier ice all imply transference of load on the earth's surface, and may be expected to induce isostatic response. Isostatic adjustments are not only significant *per se*

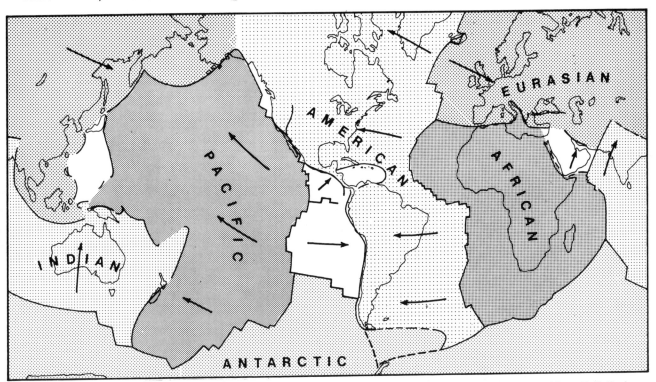

Fig. 2.17 Major plates involved in continental drift. The African plate is shown as being relatively stationary. (*From E. Bullard, "The Origin of the Oceans". Copyright* © *1969 by Scientific American Inc. All rights reserved.*)

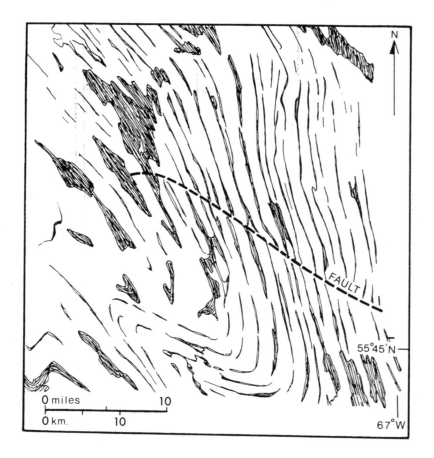

Fig. 2.18 Pattern of folds in inter-bedded sediments and volcanics in the Labrador Trough, Canadian Shield, expressed as ridges and valleys, the latter mainly occupied by lakes (shown in black). The strata are offset along a fault. (*Drawn from an air photograph.*)

Fig. 2.19 Generalised geo-logical map of the Canadian Shield. (*After* Physico-Geographical World Atlas, *Mos-cow* ©, *1964, p. 142.*)

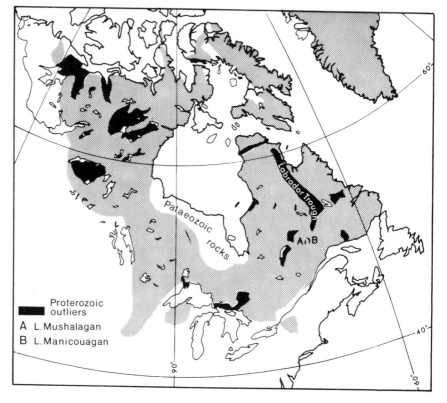

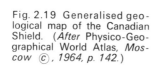

but are also implicit in considerations of development of erosional land surfaces in time. Reference is made to these problems in subsequent chapters. Next, however, the structures and forms of the continents must be examined.

F. TECTONIC REGIONS WITHIN THE CONTINENTS

The continents, however they have achieved their present shapes, sizes and locations, are essentially rafts of sialic material which in places carry a veneer of sedimentary rocks and which in places have been penetrated by volcanic upsurges which have been extruded at the land surface. The continents vary in their complexity, but each is in essence built of three types of structural units.

Each continent and subcontinent has a nucleus of ancient crystalline rocks called a *craton*. The crystalline rocks are exposed over wide areas in what are called *shields* or shield lands, but the cratons are partly buried beneath sequences of sediments deposited in troughs or basins within the crystalline outcrops. Some of these sediments (which frequently include interbedded volcanics) remain flat-lying, but others are more or less intensely folded, and the landforms developed on them vary with the disposition of the strata (see below, Chapter 5). In the Labrador Trough, for instance, which lies within the predominantly granitic Canadian or Laurentian Shield, moderately folded sedimentary and volcanic layers give rise to ridge and valley forms (Fig. 2.18). There are also outliers of Palaeozoic strata in the Manicouagan

Lake–Mushalagan Lake area and in the James Bay region (Fig. 2.19).

The weathering and erosion of the shield lands has, throughout geological time, provided detritus which has been transported to and deposited in the ocean basins. Some accumulated in what came to be major elongate troughs called *geosynclines*, or as thick wedges on continental shelves (Fig. 2.20) which have from time to time been compressed and folded and faulted by intrusions of igneous rocks in batholiths of granite and by upsurges of magma which reach the surface as extrusions of lava. Up to 20,000 m of sediments accumulated in some geosynclines before they became unstable, collapsed and were uplifted to form the modern fold mountain belts or orogens. Removal of large rock masses from these orogens by erosion has induced further uplift in response to isostasy. All the very high mountain ranges of the world (Fig. 2.5) are of Tertiary age, though the worn down but rejuvenated stumps of more ancient orogens form such uplands as the Appalachians, the eastern uplands of Australia, western Scandinavia, northwest Scotland and northwest Ireland. Finally, erosion of either the shields or the orogens causes deposition of sedimentary sequences on the margins of the ancient crystalline massifs or of the eroded orogens. Such areas where flat-lying or only gently warped sediments overlie crystalline or distorted strata are called *platforms*. Many of the world's plains are, from a structural point of view, platforms. Thus the Russian Platform underlies the forests and steppes of central and southern European USSR. The Great Plains of North America are similar, as

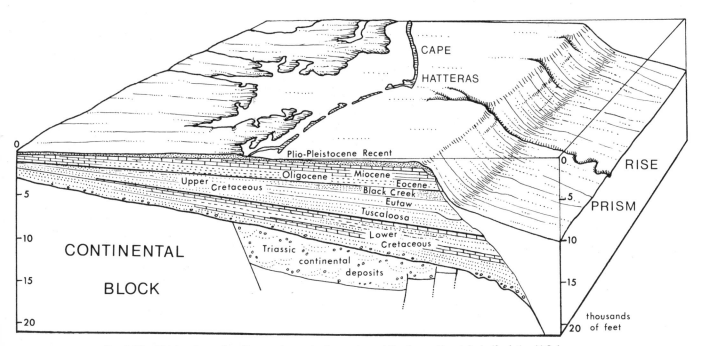

Fig. 2.20   Thick prism of sediments beneath the eastern Atlantic continental shelf of the U.S.A. (*After R.S. Dietz and J.C. Holden, "Miosynclines (miogeosynclines) in Space and Time", J. Geol., 1966, **74**, pp. 566–83.*)

are the central lowlands of Australia underlain by the Great Artesian Basin (*sensu lato*). The deserts of north Africa occupy a platform area between the Atlas Mountains with a Tertiary orogen to the north and a complex cratonic region to the south, though within the desert itself there are numerous and widespread outcrops of crystalline rocks such as those of Precambrian age shown in Fig. 2.21.

G. TECTONIC AND STRUCTURAL LANDFORMS DEFINED

Whatever the tectonic complexity of continents, within each of the basic units some structures occur which are active and which are expressed directly in the form of the land surface, and others which may be regarded as behaving passively and merely forming lines or zones of weakness which can be exploited by external agencies. Forms due wholly and solely to the earth's internal forces are called *tectonic landforms* as mentioned in Chapter 1. Fracturing and uplift or depression of parts of the crust, folding or warping of the earth's surface, and the extrusion of lavas are examples of internal activities in the crust which give rise to tectonic forms. But over wide areas of the earth's surface, rivers and glaciers have discovered and taken advantage of weaknesses in the crust to produce differential weathering and erosion of the surface. "As the zoologist's scalpel dissects the parts of an animal, so do the physiographic agents probe and dissect the earth."[29] Landforms which are due to the exploitation by weathering and erosion of weaknesses in the crust, to the

differential wearing away of rocks which leaves some areas greatly worn down and other areas upstanding, are called *structural landforms*. Joints, faults and folds are lines of weaknesses commonly exploited by external agents, while relatively weak strata are obviously worn away more rapidly than tough rocks. However, the forms developed on such sequences of rocks of contrasted resistance to weathering and erosion (a situation commonly found in thick sedimentary sequences) depend not only on the nature of the strata involved but also on their disposition.

The varied types of tectonic and structural landforms are not restricted in their distribution to any one unit of the continental areas. Nevertheless some patterns can be distinguished. Thus the forms developed on flat-lying strata are most commonly found on platform areas though they also occur on horizontally disposed sedimentary sequences found in basins within the cratons. Forms characteristic of greatly contorted sedimentary rocks are found in intensely folded orogenic belts, those typical of gentle warping and folding in the platform areas or in weakly folded orogens or parts of orogens. Volcanic phenomena occur in relation to young fold mountain belts and faults and faulting are also well represented there. However, not even the cratons are stable and earthquakes and faulting occur there also. Joints are characteristic of all rocks, and are significant everywhere; but they give rise to characteristic suites of forms in sandstone and limestone (both widely represented in platform and orogenic regions) and in areas of granitic outcrop which are common both in the cratonic regions and in the orogens where batholiths have been exposed by erosion.

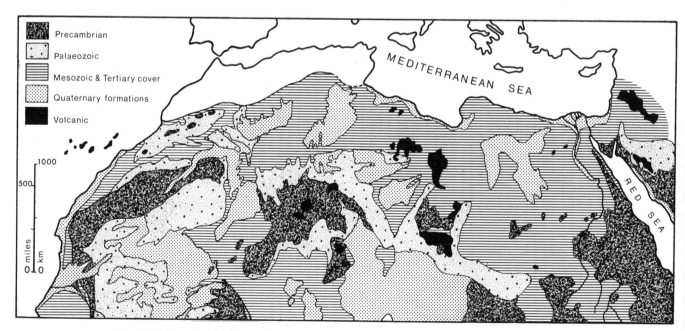

Fig. 2.21 Generalised geological map of the Sahara. (*After N. Menchikoff, "Les Grandes Lignes de la Geologie Saharienne", Rev. Geogr. Phys. Geol. Dyn., 1957, 1, pp. 37–45; reprinted by permission of Masson & Cie, Paris.*)

References Cited

1. D. G. KING-HELE, "The Shape of the Earth", *Sci. Am.*, 1967, **217** (4), pp. 67–76; *idem*, "The Imperfect Sphere", *The Listener*, 1970, **83**, pp. 717–8; *idem*, "Heavenly Harmony and Earthly Harmonics", *Royal Aircraft Establishment Tech. Memo. Space,* 1971, **174**.
2. D. G. KING-HELE, "The Imperfect Sphere", pp. 717–8.
3. J. T. WILSON, "Geophysics and Continental Growth", *Am. Scient.*, 1959, **47**, pp. 1–24.
4. V. V. BELOUSSOV, "Against Continental Drift", *Sci. J.*, 1967, **3**, pp. 56–61.
5. J. T. WILSON, "Geophysics and Continental Growth", pp. 1–24.
6. L. GREEN, *Vestiges of the Molten Globe as exhibited in the Figure of the Earth's Volcanic Action and Physiography*, Part 1, London, 1875.
7. J. W. GREGORY, cited in J. A. Steers, *The Unstable Earth*, Methuen, London, 1932, pp. 4–5.
8. D. P. MACKENZIE, "Drifting Continents", *The Listener*, 1970, **83**, pp. 608–11.
9. S. W. CAREY, "The Orocline Concept in Geotectonics", *Pap. Proc. R. Soc. Tas.*, 1956, **89**, pp. 255–88.
10. A. WEGENER, *The Origin of Oceans and Continents*, Methuen, London, 1924.
11. A. L. DU TOIT, *Our Wandering Continents*, Oliver and Boyd, Edinburgh, 1937.
12. F. A. V. MEINESZ, "Shear Patterns of the Earth's Crust", *Trans. Am. geophys. Un.*, 1947, **28**, pp. 1–61.
13. For review, see A. HOLMES, *Principles of Physical Geology*, 2nd ed., Nelson, London, 1968, pp. 1204–16.
14. See H. H. HESS, "History of Ocean Basins" in *Petrologic Studies: a volume to honor A. F. Buddington*, Geological Society of America, New York, 1962; F. J. VINE and D. H. MATTHEWS, "Magnetic Anomalies over Ocean Ridges", *Nature (Lond.)*, 1963, **199**, p. 947; F. J. VINE, "Spreading of the Ocean Floor: New Evidence", *Science (N.Y.)*, 1966, **154**, pp. 1405–15; A. COX and G. B. DALRYMPLE, "Statistical Analysis of Geomagnetic Reversal Date and the Precision of KA Dating", *J. Geophys. Res.*, 1967, **72**, pp. 2603–14; A. COX, "Geomagnetic Reversals", *Science (N.Y.)*, 1969, **163**, pp. 237–45; E. BULLARD, "Reversals of the Earth's Magnetic Field", *Phil. Trans. R. Soc.*, Series A, 1968, **263**, pp. 481–524; X. LE PICHON, "Sea-floor Spreading and Continental Drift", *J. Geophys. Res.*, 1968, **73**, pp. 3661–97.
15. R. L. FISHER and R. REVELLE, "The Trenches of the Pacific", *Sci. Am.*, 1955, **193**(5), pp. 36–41.
16. P. A. V. DIRAC, "The Cosmological Constants", *Nature (Lond.)*, 1937, **139**, p. 323.
17. J. T. WILSON, "Some Consequences of Expansion of the Earth", *Nature (Lond.)*, 1960, **185**, pp. 880–2.
18. See for instance F. AHMED, "Flood Traps through Space and Time and their Bearing on some Problems of Geotectonics", Proceedings of the International Symposium on "Deccan Trap and other Flood Eruptions", Sagar, M. P., India, 1969, *Bull. Volc.*, 1972, **35** (3), pp. 539–63.
19. See F. J. VINE and D. H. MATTHEWS, "Spreading of the Ocean Floor: New Evidence", *Science (N.Y.).*, 1966, **154** (3755), pp. 1405–15.
20. E. BULLARD, "The Origin of the Oceans", *Sci. Am.*, 1969, **221** (3), pp. 66–75; D. P. MACKENZIE, "Drifting Continents", pp. 608–11.
21. H. W. MENARD, "The Deep-ocean Floor", *Sci. Am.*, 1969, **221** (3), pp. 126–42; E. BULLARD, "Origins of the Oceans", pp. 66–75; D. P. MACKENZIE, "Drifting Continents", pp. 608–11.
22. See R. S. WILLIAMS and J. G. MOORE, "Iceland Chills a Lava Flow", *Geotimes*, 1973, **18**(8), pp. 14–17.
23. See W. R. DICKINSON, "Plate Tectonics in Geologic History", *Science (N.Y.)*, 1971, **174**, (4005), pp. 107–13.
24. See, for example, D. A. ROSS and J. SCHLEE, "Shallow Structure and Geologic Development of the Southern Red Sea", *Bull. geol. Soc. Am.*, 1973, **84**, pp. 3827–48.
25. J. GILLULY, "Continental Drift: A Reconstruction", *Science (N.Y.)*, 1966, **152**, pp. 946–50.
26. A. A. MEYERHOFF, "Continental Drift: Implications of Palaeomagnetic Studies, Meteorology, Physical Oceanography and Climatology", *J. Geol.*, 1970, **78**, pp. 1–51; *idem*, "Continental Drift II: High Latitude Evaporite Deposits and Geologic History of Arctic and North Atlantic Oceans", *J. Geol.*, 1970, **78**, pp. 406–44.
27. E. J. COBBING and W. S. PITCHER, "Plate Tectonics and the Peruvian Andes", *Nature (Physical Science)*, 1972, **240** (99), pp. 51–3.
28. V. V. BELOUSSOV, *Basic Problems in Geotectonics*, McGraw-Hill, London, 1962; *idem*, "Against Continental Drift", pp. 56–61.
29. E. S. HILLS, *Elements of Structural Geology*, Methuen, London, 1963, p. 431.

Additional References

J. H. HODGSON, *Earthquakes and Earth Structure*, Prentice-Hall, Englewood Cliffs, 1964.

D. G. KING-HELE, *Observing Earth Satellites*, St. Martin's Press, London 1967.

H. W. MENARD, "Development of Median Elevations in Ocean Basins", *Bull. geol. Soc. Am.*, 1958, 69, pp. 1179–86.

D. H. TARLING and M. P. TARLING, *Continental Drift*, G. Bell, London, 1971.

J. H. F. UMBGROVE, *The Pulse of the Earth*, Nijhoff, The Hague, 1942.

CHAPTER 3

Fractures in the Crust: Joints

A. GENERAL STATEMENT

Rock strata exposed at or near the earth's surface are fractured to a greater or lesser extent. The fractures are of two types. *Joints* are fractures along which there has been no detectable dislocation, and as used here the term includes bedding and cleavage planes. *Faults*, on the other hand, are fractures along which movement has occured. So far as landform development is concerned, both joints and faults are lines or zones of weakness which are readily exploited by weathering and by erosional agents such as rivers and waves: these forces cause linear clefts or valleys to develop along the fractures in many places.

Thus both types of fracture give rise to structural landforms. However, differential movement along faults gives rise to another group of features which are tectonic, being due in essence to dislocation of part of the crust. Though faults and related landforms are by no means rare, they are not nearly as widespread as joints and the features developed in association with joints. On the other hand, though both types of fracture give rise to relatively minor features, and although some joint patterns are quite coarse (Pl. 3.1) with joint blocks up to 1·5 km diameter developed in granitic rocks in the Everard Ranges and in the Gawler Ranges volcanic porphyry (Fig. 3.1), no joint-controlled forms are of regional or continental scale. Many fault-generated features are. The rift valley complex of central Africa and the Levant, for example, extends through some 5500 km or 52° of latitude (Fig. 3.2).

Although joints and faults have much in common, the landforms evolved on them vary greatly in shape and size. Field evidence suggests that this is partly attributable to variations in the character and pattern of the fractures, but it is also due in considerable measure to the nature of the bedrock in which they occur. Landforms which owe their essential morphology to joints will be discussed first, and their evolution will be illustrated with reference to outcrops of granite, sandstone, limestone, gneiss and schist, and shale.

B. GRANITE LANDFORMS
1. General

As was described earlier (Chapter 2) the present continents are fundamentally granitic, which occurs widely in the shield lands, is commonly exposed in batholiths in deeply eroded orogenic belts, and covers 15 per cent of the earth's land surface. Granite consists of quartz, a feldspar (typically orthoclase), a mica (typically biotite) and minor amounts of other minerals. It is a crystalline rock of low porosity (average 1 per cent) and permeability. That is, there are few voids, and water cannot pass readily through the body of the rock because of its interlocking crystals; but it is characteristically well-jointed and, in some areas, pervious on this account. Granite outcrops give rise to suites or families of characteristic structural landforms.

2. Principal Landforms

Four characteristic and well-known assemblages of landforms are developed on granite. They are shown in Pls 3.2 a-c, 3.3a-d, 3.4a-c, and 3.5.

In many areas, more or less spheroidal boulders occur in clusters or in isolation, standing above the intervening flat or gently-sloping plains. Elsewhere, isolated ridges, ranges or hills of granite rise abruptly from the surrounding flat plains. The first type is known in Australia as a *tor* landscape, though *boulder* is a preferable term[1] (Pl. 3.2a-c). The second is called an *inselberg* landscape, the basic unit of which is the residual hill, either domed (Pls 1.2, 3.3; Fig. 3.3a) or castellated (Pl. 3.4, Fig. 3.3b) in shape (*bornhardt* and *castle koppie* respectively). In addition, some granite areas in the Sinai Peninsula, in the Andes, and in the northern Flinders Ranges are so closely and intensely dissected that the landscapes consist of sharp crusted steep-sided ridges which separate deep valleys of V-shaped cross-section (Fig. 3.3c; Pl. 3.5)—*all-slopes* topography.

A number of problems arise from a consideration of these granite landscapes. Why do the inselbergs and boulders stand above the level of the surrounding plains? Why are some granite outcrops characterised by boulders, others by inselbergs? Why do the shapes and sizes of boulders and inselbergs vary from place to place? Under what conditions do inselbergs develop?

3. Origin of Boulders and Inselbergs

Because of the cover of soil and other superficial deposits, it is difficult in many areas to ascertain the nature of the rock beneath the plains in such boulder and inselberg regions. However, there are enough bores and other excavations to show that, with rare exceptions (see pp. 53–4), the country rock is identical with that exposed in the inselbergs or boulders as far as composition is concerned (as was appreciated by Bain[2] and other early workers). Therefore, the reason the boulders remain upstanding is *not* that they consist of a rock which is mineralogically more resistant.

Road cuttings and quarries in granite and other crystalline outcrops (including basic rocks such as the norite at Black Hill, Murray Plains, South Australia) show that boulders are developed on areas where the joint blocks are reasonably large, and that the intervening flats have formed

Plate 3.1a Prominent joints in granite on Milbank strandflat, Swindle Sound, coast of British Columbia. The most prominent joints run east—west, but a north—south set is also clearly expressed in linear clifts and shallow valleys. Note the drowned coast, numerous lakes and, in the right foreground, Kitasu Hill (262 m), a post—Pleistocene volcanic cone. (*Dept Lands, Forests and Water, B.C. Photo B.C. 501:46, taken 17 September, 1947.*)

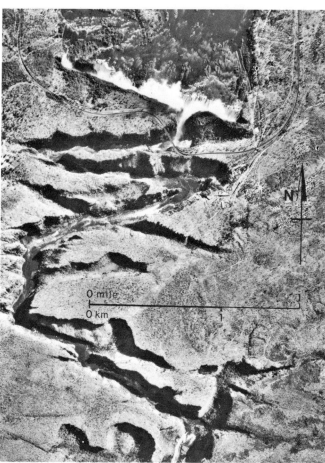

Plate 3.1b Vertical air photograph of the Zambezi River which here at Victoria Falls forms the border between Rhodesia and Zambia. The river below the Falls (about 1·5 km wide and which is a nick point) has eroded a gorge over 120 m deep in flat-lying Mesozoic basalts. The zigzag course of the river is due to its following major ESE—WNW and E—W joint zones in the volcanic rocks. (*Rhod. Dept Lands and Survey, Salisbury.*)

a

Fig. 3.1 Joint pattern in plan: (a) in the
granite Everard Ranges, northern South
Australia; (b) in part of the Gawler Ranges,
S.A., built of a Precambrian volcanic
porphyry, which is mainly dacitic (*Drawn
from an air photograph.*)

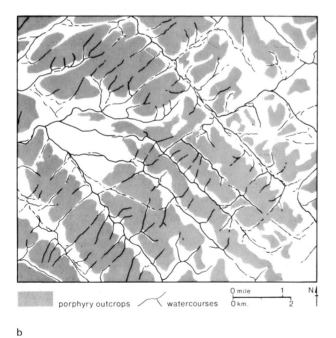

b

Fig. 3.2 The central African-
Levant rift system. (*After E.S.
Hills,* Elements of Structural Geology, *Methuen, London, 1963,
p. 185.*)

Plate 3.2a Little-rounded granite blocks in Rhodes-Matopos National Park, some 12–13km south of Bulawayo, Rhodesia. The uppermost block is called the John F. Kennedy Rock because of its alleged likeness to the late President—an imagination in naming and recognising objects is as necessary in geomorphology as in astronomy. (*Rhod. Nat. Tour. Board.*)

Plate 3.2b Rounded granite boulders (2–4 m diameter) at the Devils Marbles, Northern Territory, Australia. (*C.S.I.R.O.*)

Plate 3.2c Perched granite boulder, Dartmoor, Devon, England. (*C.R. Twidale.*)

Plate 3.3a Smooth granite domes in South West Africa with remnants of sheet structures visible and bouldery disintegration in areas of closer jointing on left. (*J. A. Mabbutt.*)

Plate 3.3b Mono Rock, in the Sierra Nevada of California, standing almost 2200 m above sealevel, showing both thick and thin rock slabs, as well as prominent linear joints, dipping steeply from top left to bottom right. (*U.S. Nat. Park Service.*)

Plate 3.3c   Steep-sided granite dome or Sugar Loaf, Rio de Janeiro, Brazil, developed by differential weathering (chemical alteration under tropical conditions) of compartments of rock of varied joint spacing. Note the vertical fracture on the right, separating a massive slab of rock, which is slightly curved at the top, from the main mass. (*Brazilian Govt*.)

Plate 3.3d Low, smooth granite domes, of large radius and with some bouldery disintegration in evidence, standing above the flat desert plains in the Hoggar Mountains of southern Algeria (central Sahara). In the background are taller and less regular domed forms. (*P. Rognon.*)

Plate 3.4a   Granite inselberg of angular joint blocks and castellated form in the Hoggar Mountains, southern Algeria. Note the prominent bluff and debris slope (an inclined bedrock slope) bounding the residual on the right, and the boulders and gently arched irregular whaleback forms in the foreground. (*P. Rognon.*)

Plate 3.4b   Haytor East, on eastern Dartmoor, Devon, England is a castellated tor or inselberg dominated by orthogonal joints and by cubic or quadrangular joint blocks. The residual rises abruptly from the high moorland plain over which slabs and small blocks of granite are strewn. This compartment is more massive than those surrounding it, and according to some, the differential weathering took the form of contrasted chemical alteration under tropical conditions in the Tertiary, followed by differential erosion by streams and solifluction in the later Cainozoic. But, whatever the precise nature of the processes at work, the resultant form is similar to that developed wholly under tropical conditions, though few British workers would call this feature an inselberg. (*C. R. Twidale.*)

Plate 3.4c   Angular joint blocks form a castle koppie in the Mrewe-Marendellos area, Rhodesia. (*C. R. Twidale.*)

Plate 3.5 This greatly dissected granite outcrop in the north Flinders Ranges, South Australia, at present receives an average of 22 cm of rainfall per annum, but the numerous valleys attest to the effectiveness of rivers in eroding the land surface. It is not clear whether the episodically flowing rivers of a desert régime are responsible or whether the major part of the dissection is to be attributed to more powerful and more regularly flowing streams of a former, wetter period. But apart from a few massive compartments of younger intrusive rock on which domes (X) have evolved, the older granite is so fractured that it is uniformly susceptible to weathering and to erosion: hence the all-slopes topography. (*C. R. Twidale.*)

Plate 3.6a Granodiorite corestones set in weathered granite sand or grus, Snowy Mountains, New South Wales, Australia. Note the influence exerted by joints on the progress of weathering. (*C. R. Twidale.*)

Plate 3.6b Corestone and corestones exposed as boulders in road cutting in granite in D. L. Bliss State Park, Sierra Nevada, California. Note the unstructured nature of the weathered granite or grus surrounding the corestone, and the suggestion of concentric fractures in the partially-weathered granite above the corestone. (*C. R. Twidale.*)

Plate 3.6c Field at Palmer, S. A., strewn with rounded boulders as a result of a stream having washed away the weathered granite which formerly surrounded these residuals. (*C. R. Twidale.*)

Plate 3.6d   Isolated, rounded boulders, the largest 7—8 m in diameter, in the Herbert River valley,
north Queensland. (*C. R. Twidale/C.S.I.R.O.*)

where the joints are close together. On the flat, the granite
is uniformly weathered, for no part of the rock mass is far
from a joint along which water can penetrate. Boulders
occur when the joint blocks are in some areas delineated by
orthogonal joints and are therefore cubic or rhomboidal in
shape. They are a few metres diameter. Their marginal
areas are weathered, but the centre of each block remains
untouched and is occupied by a kernel or *corestone* of fresh
granite shaped like a sphere (Pl. 3.6a-b) or a triaxial ellipsoid
(depending on joint spacing). Such features have been noted
in France,[3] in Portugal,[4] West Africa,[5] South West Africa,[6]
California,[7] Rhodesia and South Africa, [8] and Australia, to
name but a few areas.

The rounded shape derives from the greater susceptibility
to weathering of the corners and edges of cubic and quadrangular blocks where more surfaces are exposed to meteoric
waters (Fig. 3.4a and b).* The sharply defined limit of
weathering between the worn marginal areas of the joint
blocks and the still intact corestone is due to the very low
permeability of granite.

When the land surface is lowered, the weathered rock is
readily removed, but the corestones are too massive to be
transported and so remain strewn over the landscape as
boulders (Pl.3.6c). Some remain as perched blocks (Pl. 3.2c).
Thus it can be suggested, first, that there is a direct relationship between joint spacing and landforms[†] and, second,
that boulders in many areas have developed in two stages:
differential subsurface weathering, followed by differential
erosion (Fig. 3.4a-c).

On a larger scale, also, exposures which permit observations of the structure of the granite beneath the plains strongly
suggest that the spacing and condition of the joints differ from
the joints in inselbergs. For instance, at the margin of
Ucontitchie Hill, northwestern Eyre Peninsula, an exposure
in a reservoir reveals a strong contrast in jointing between
hill and plain (Pl. 3.7). The inselberg is essentially monolithic, for, although many joints can be distinguished, the
joints are so tight that they are virtually impenetrable to
water. Moreover, these joints are at least 5 m apart. Beneath
the plains, on the other hand, the joints are only 70 cm or so
apart, and no part of the rock mass is far from a fracture.
Water has evidently penetrated rather easily into the rock,
for those corestones which do remain are considerably
weathered. Thus it may be suggested that if some *compartments*

---

* In some areas such as Dartmoor hydrothermal fluids and gases
may be responsible for some of the alteration of rock in the marginal
areas of joint blocks (see, for example, J. PALMER and R.A.
NEILSON, "The Origin of Granite Tors on Dartmoor, Devonshire",
Proc. York. geol. Soc., 1962, **33**, pp. 315—40, and E.A. EDMONDS
et al., "Geology of the Country around Okehampton", *Mem.
geol. Surv. Eng. and Wales*, 1968). But the absence of characteristic
hydrothermal minerals and the apparent decrease in extent of
weathering with increasing depth suggest that the general
argument which attributes the peripheral alteration of joint
blocks to meteoric waters is a sound one.

---

† The size of joint block required to survive as a corestone depends
on time—presumably even a large block will be weathered through
if attack continues long enough—and on the resistance of the
particular rock. At Palmer, near Adelaide, corestones of fine,
grey granite are as small as 20 cms in diameter, but on the same
outcrop, joint blocks of a coarser, pink granite have to be about
one metre diameter in order to provide corestones.

Fig. 3.3a Sketch of boulder-strewn domed inselberg in the Hoggar Mountains of southern Algeria. (*Drawn from photograph supplied by P. Rognon.*)

Fig. 3.3b Castle koppie in granite, Mrewe-Marendellos area, Rhodesia.

Fig. 3.3c Conical hills formed by the dissection of a much-shattered granite in the southern Sinai. (*Drawn from a photograph in W.F. Hume,* Geology of Egypt, *Government Printer, Cairo.*)

of the granite are characterised by tight and widely-spaced joints, others by numerous close and open joints, the former will resist weathering, while the latter will not; the former will become inselbergs, the latter will be worn down to plains (Fig. 3.5a and b).

Drilling and excavation in many parts of the world have shown that the present plains are underlain by considerable depths of weathered granite. For example on central Eyre Peninsula there is between 18 and 40 m of weathered granite between the rolling land surface and the fresh rock. In Namaqualand, southwestern Africa, excavations for road fill have revealed at least 5 m of granite, weathered to a white colour by the alteration of the feldspars, underlies the plains between the low granite domes and boulder-strewn hills. Similarly in the Marendellos area east of Salisbury, Rhodesia, quarries have revealed corestones set in weathered granite which extend to a depth of at least 7–8 m and on Goodhope Farm, 24 km east of Salisbury, the weathered granite extends a similar distance beneath the surface.* From this evidence it is inferred that differential weathering

* The writer is grateful to Mr W.D. Purves, Department of Research and Special Services, Salisbury, for conducting him to these Rhodesian sites.

may be going on now, or that, if the weathered rock beneath the present plains is inherited from some earlier, different environment, the rock beneath the plains is more susceptible to erosion than the massive rock which underlies the inselbergs. Thus the inselberg masses are being etched out more and more; and it may be suggested that in the past joint-controlled differential weathering has also caused differential erosion and the development of the present inselberg landscapes. Thus their evolution is considered to be similar to that of boulders, but instead of the joint block being the basic unit of development, large compartments of rock are involved (Fig. 3.5a).

Though this two-stage hypothesis apparently accounts for many structures and is widely accepted as a likely explanation for the development of inselbergs, the field evidence in some areas argues against it. To take one example, at Stone Mountain, Georgia, the plains surrounding the granite dome (Pl. 3.8) are underlain by metamorphic rocks which are mineralogically different from the granite and are more readily weathered.[9] This particular residual can therefore reasonably be attributed to the occurrence of different rock types. Other examples are the domed inselbergs of Mozambique which many years ago were considered to be the exposed irregular tops of batholithic intrusions[10] and in some areas,

a water penetrates down joints

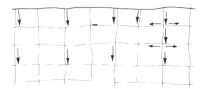

b subsurface weathering guided by joint planes

corestone or kernel — weathered granite

c lowering of land surface; removal of debris; corestones exposed as tor boulders

Fig. 3.4 The two-stage development of boulders: (a and b) differential subsurface weathering controlled by joint pattern; (c) lowering of baselevel and differential erosion, with removal of weathered rock and exposure of corestones as boulders.

Plate 3.7 The hill–plain junction is exposed near the southern extremity of Ucontitchie Hill, northwestern Eyre Peninsula, S. A. In the background is the massive inselberg, with few widely-spaced, tightly-closed joints, and in the foreground, exposed in the reservoir, the deeply-weathered granite with a few corestones which underlies the plains. Note the flared slope (also shown in Pl. 1.6) and the *Rillen* or gutters which score the upper surface of the granite dome. (*C. R. Twidale.*)

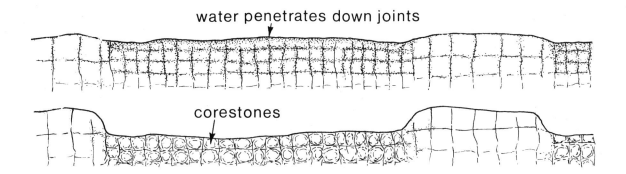

Fig. 3.5a Two-stage development of inselbergs through the joint-controlled differential weathering and erosion of compartments of rock.

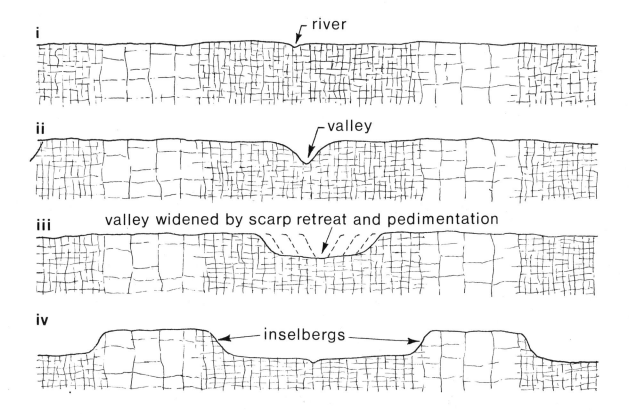

Fig. 3.5b Development of inselbergs by joint-controlled stream dissection, scarp retreat and pedimentation.

Plate 3.8   Stone Mountain, Georgia, a granite dome which is approximately 210 m high and 2·5 km diameter, in plan. Small rivulets of water are coursing down the slopes of this bare rock outcrop, in which few joints are discernible. Similar flows are responsible for the radiating stripes due to the spread and desiccation of mineral salts and organic debris. (*R. Kessel.*)

for instance the small disintegrated domes which occur in the eastern and southern Mt Lofty Ranges near Adelaide, the presence in the small granite outcrops of numerous xenoliths (or contaminant blocks) of the surrounding metamorphic rocks certainly suggest that the roof of the intrusion was not far above the present outcrop.

The great height of some inselbergs is the basis for another objection. According to the simple version of the two-stage theory outlined above, inselbergs should only stand as high as, or less than, the greatest penetration of the *weathering front*, that is, the lower limit of weathering (see Chapter 9). But it has been pointed out that many inselbergs protrude above the adjacent plains to an elevation which exceeds even the greatest known depths of weathering; certainly greater than the extent of weathering reported from the same localities.[11] However, another of the characteristics of inselbergs is that they occur in multicyclic landscapes,[12] that is, landscapes which have passed through more than one cycle or prolonged phase of weathering and erosion (see Part IV). Hence, it is conceivable that the resistant compartments could have been brought into relief during more than one phase of deep weathering and erosion, so that this particular objection is not critical.

More telling is the suggestion that in many areas of southern Africa, for example, there is apparently little or no weathered rock beneath the present plains. It is not true of all areas in southern Africa (see, for instance, earlier comments on deep weathering of the granite near Salisbury and in Namaqualand) but even where the supposition is correct it is, of course, possible that all previously weathered granite has been removed so recently that there has been insufficient time for significant subsurface weathering to have occurred beneath the newly eroded plain. It is equally possible, as King[13] has pointed out, that a significant depth of weathered granite never developed, that erosion equalled or outpaced weathering, and that the inselbergs are the last remnants remaining after joint controlled stream incision and retreat of the valley side slopes (Fig. 3.5b).

Even where a substantial regolith has developed, it is possible to argue that the weathering is a relict from an earlier phase of geomorphological development, and that scarp retreat and pedimentation graded to local baselevels have merely cut across the previously weathered rock; or that the regolith postdates the surface. In other words, it can be asserted that weathering and erosion are separate and genetically unrelated geomorphological events. And though the weathering profiles displayed in the Rhodesian areas mentioned earlier look complete and not truncated, it is not always easy to distinguish a simple from a polygenetic profile. Thus the evidence is equivocal, and can be interpreted in terms of either the two-stage hypothesis, or in terms of scarp retreat and pedimentation. In the latter case, it can be shown that in areas where retreat has been active, it has not proceeded far—only a few score metres in the Valley of the Thousand Hills, Natal—and it must be interpreted in terms of recession toward numerous, closely-spaced foci. On the other hand, in areas like the Everard Ranges (see Fig. 3.1a) there is no evidence of retreat and significant valley development. The closely grouped domes are similar to those standing in isolation, and both are rationally explained in terms of subsurface joint controlled weathering and the subsequent etching out of the weathered rock.

Thus, as is so often the case in geomorphology, it is not possible to differentiate with certainty between these two competing hypotheses. In any event, the two are not mutually exclusive and inselbergs may have formed in one way in one area and in the other elsewhere.

4.   Shapes of Boulders and Inselbergs

Whether the residual remnant of granite is a boulder, a castellated inselberg or a domed inselberg, the shape in detail depends on the geometry of the joint system. Cubic joint blocks are weathered to spherical boulders (Pl. 3.6). Horizontally elongate joint blocks are weathered to give ovoid, log- or barrel-shaped boulders, and vertical elongation results in tall turrets and towers. Rather bizarre forms are also evolved, including some due to the exploitation by weathering of secondary joints within rock masses, producing split boulders (Pl. 3.9a-e).

Similarly, inselbergs in plan consist of square, rectangular, rhomboidal or trapezoidal blocks in various combinations

Plate 3.9a   This granite mass in the Hoggar Mountains, southern Algeria, is subdivided by joints into innumerable tall, thin turrets or spires of rock. Note the bouldery lower slope (debris slope) below the central mass of turrets, and the stream deposits in the foreground. Finer debris has been laid down at higher levels in the banks of the episodically flowing stream, but the bed is occupied by coarse, angular blocks of the traction load. (*P. Rognon.*)

Plate 3.9b   Cathedral Rocks, Sierra Nevada, California: two tall, square columns of monzonite which protrude above the general level of the residual hill. (*C. R. Twidale.*)

Plate 3.9c The "Pile of Logs", a granite dyke subdivided by joints into a series of horizontal columns, Whitsunday Island, Queensland.
(*Dept Nat. Dev., Canberra.*)

Plate 3.9d B o w e r m a n s Nose, an unusually shaped tower of granite blocks on eastern Dartmoor, Devon, England. The feature is about 5 m high. (*C.R. Twidale.*)

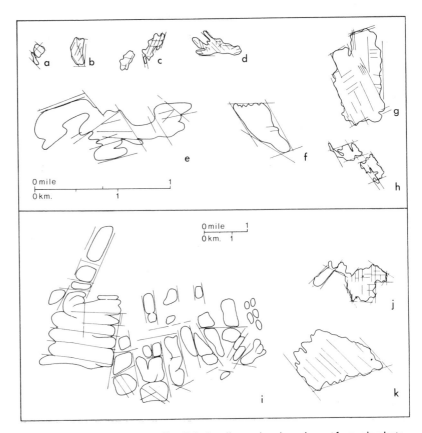

Fig. 3.6 Inselbergs in plan, drawn from air photographs: (a) Chilpuddie Hill, South Australia; (b) Ucontitchie Hill, S.A.; (c) Yarwondutta Rocks, S.A.; (d) Pildappa Hill, S.A.; (e) Hyden Rock, Western Australia; (f) Polda Rock, S.A.; (g) Mt Wudinna, S.A.; (h) Minnipa Hill, S.A.; (i) The Olgas, Northern Territory; (j) Pearson Island, S.A.; (k) Ayers Rock, N.T.

Plate 3.9e Split block of granodiorite, Snowy Mountains, N.S.W. (*C.R. Twidale.*)

Plate 3.10 Vertical air photograph of Ayers Rock, Northern Territory, showing its lozenge shape in plan, prominent near-vertical bedding planes which strike NW—SE and its isolation in the midst of the desert plains. The trend of the suggested delimiting joints is indicated. The "wavy" pattern in the NE corner is due to vegetation. (*Dept Nat. Dev., Canberra.*)

(Fig. 3.6). Ayers Rock, for instance, consists of a single, huge, rhomboidal block of arkosic sandstone, (Fig. 3.6k, Pl. 3.10) and the nearby Olgas of a large number of rectangular or trapezoidal blocks of conglomerate (Fig. 3.6i). The granite inselbergs of Eyre Peninsula, New England (N.S.W.) and of Western Australia consist of either single blocks or a number of blocks which together form a compartment isolated by weathering.

In profile the greatest contrast in the shape of inselbergs arises from the dominance of one of two important joint systems. The castellated type occurs where the widely developed cubic joint blocks, delineated by the *orthogonal joint system,* are massive (Pl. 3.4). In some areas, however, the orthogonal system is subordinate to a set of widely-spaced arched or domed joints which divide the rock into massive slabs or lenses which are called *sheet structure* (Pl. 3.11). Wherever sheet structure is dominant, domed inselbergs or bornhardts (Pl. 3.3) are developed, though the sheets break down into blocks on exposure to the air, so that some domes are boulder-strewn to a greater or lesser degree.

Most workers have long accepted that sheet structure is due to pressure release or erosional offloading. Indeed, many go so far as to call the joints offloading joints.[14] The gist of the pressure release hypothesis, which was promulgated by G.K. Gilbert [15] as a result of his studies in the Sierra Nevada of California, is as follows. Inselbergs and sheet structure are most commonly developed on granitic rocks which crystallised out at great depths under conditions of high pressure. The fact that the granites are now exposed at the surface proves that great masses of rock have been eroded away. The removal of the weight of overlying rock caused the vertical pressures on the granites to be removed and this radial release of pressures is manifested in fractures which developed tangential to the land surface. These are the arched joints. A critical feature of the hypothesis is that it is the form of the land surface which controls the geometry of the joints: the shape of the land surface is the cause, the pattern of joints the effect.

Unquestionably the great appeal of the hypothesis lies in its simple and persuasive logic. The fact that the parallelism between sheet structure and land surface can be interpreted in two diametrically opposed ways is, however, often overlooked and alternative suggestions are not often sought. But before listing some of the weaknesses of the offloading hypothesis and commenting on some of the possible alternatives, two points need to be made. The first is that in a very real sense all joints are manifestations of pressure release,[16] for presumably joints are closed and disappear at depth where hydrostatic pressures are high. The second is that, in addition to sheet structure, a superficial and recent fracturing causes many rocks to be laminated (Pl. 3.12).

Pressure release may well be included among the several possible causes of sheet structure. But the hypothesis of pressure release outlined above, as applied to the massive and deep-seated sheet structure, fails in several respects to explain characteristic field evidence, leaves several problems unexplained, and several expectable consequences of the hypothesis are not found. For example, release of pressure implies expansion, yet sheet structure is typically associated with inselbergs which are evidently preserved not because of the absence of joints, but rather because these joints are tight and closed. There is also local evidence of compression. For

Plate 3.11   Massive sheet structure, some 8–10 m thick, on the south side of Ucontitchie Hill, Eyre Peninsula, S.A. Note the *tafoni* or caverns developed on the underside of the slab, and the minor flares or undulations of the bedrock slope below the sheet structure. (*C.R. Twidale*.)

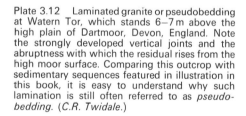

Plate 3.13 Arched slabs of granite with crestal fracture, called A-tents, on Mt Wudinna, northern Eyre Peninsula, S.A. Though their precise origin is controversial, there is no doubt that in general terms they are a manifestation of expansive release in response to compressive stress. (*C.R. Twidale.*)

Plate 3.14a   The Kangaroo Tail, a sheet structure developed on arkosic sandstone on the north-western side of Ayers Rock. Note the abrupt rise of the residual from the surrounding depositional plains, the caverns at the slopefoot, the flared slopes near the Tail, and the ribbing which is related to the steeply dipping strata which comprise the monolithic inselberg mass.
(*C.R. Twidale.*)

example, the raised pairs of slabs (Pl. 3.13) known as A-tents [17] are surely a result of expansion and a manifestation of compressive stress. Wedges of rock developed on the lower exposed edges of sheet structures, and probable imbricate joints within the structures, also point to compression.[18]

On Ayers Rock and the Olgas (both in the desert of central Australia) and in the Colorado Plateau,[19] sheeting is developed on sedimentary rocks which have not suffered metamorphism and which have evidently not been buried at great depths (Pl. 3.14a, b). In the Gawler Ranges (South Australia) sheeting occurs on the dacitic porphyry, which is a Precambrian extrusive rock, and again is unlikely to have suffered deep burial (Pl. 3.14c). If there has been a tendency to expansion, why is it not accommodated along pre-existing weaknesses, for instance along the cleavage of feldspars in the rock or along older orthogonal or columnar joints? In some few localities (for instance in the Tenaya Lake area of the Yosemite National Park, California), sheeting joints dip into hillsides: domed hills are underlain by basin structure (Pl. 3.15, Fig. 3.7), which is difficult to understand in terms of the offloading hypothesis. If the joints are consequent upon the form of the land surface, then the basin structure predates the domed hills, but in many areas such a conclusion is difficult to sustain and in some instances is manifestly not the case: the structure has guided erosion and, hence, the form of the land surface. [20]

Thus there are at least enough anomalies to suggest that explanations other than the pressure release hypothesis should be considered. Some inselbergs (for example, near Rio de Janeiro (Pl. 3.3c) are apparently bounded by faults (Fig. 3.8) and some sheet structure in New England, USA, is definitely associated with faulting (Pl. 3.16). The sheet structure could be developed as secondary shears, the development of which in association with primary faults has been demonstrated in the laboratory. The system of joints which delimits the inselbergs is in some areas geometrically related to major structural features suggesting a causal relationship between the two.[21]

As will be mentioned in connection with diapirs (Chapter 5) many granite masses are lighter than the surrounding country rock and have a tendency to rise; again sheet structure could be interpreted as arched fractures tangential to the uplift. In this case, however, it is difficult to explain the evidence suggestive of compression in inselbergs.

Finally, there is much direct and indirect evidence for strong lateral compression in the crust (cf. Chapter 4). Measurements have shown that lateral compression in the crust is commonly high—greater than would be expected on theoretical grounds. This implies that there is active or residual compression. Thus the arched and domed structures could be interpreted as strain effects resulting from compressive stresses. This would seem to explain not only the sheet structure, but also much detailed field evidence suggestive of compression in the inselbergs, and, indeed, the very preservation of the inselbergs.

However, there is one apparent difficulty. The crests of anticlines and domes are in tension (see Chapter 5) and are therefore susceptible to weathering and erosion. Why, then are compartments of rock—the inselbergs—preserved at such sites? The answer may be that, as has long been recognised and as has been mentioned previously, domed

Plate 3.14b   Sheet structure in massive sandstone, Great Basin, Utah. (*W.C. Bradley.*)

Plate 3.14c   Sheet structure transgressing columnar joints in Precambrian volcanic porphyry, Gawler Ranges, S.A. The arched slope in the foreground is a direct expression of such a fracture. (*C.R. Twidale.*)

Fig. 3.7 Cross-section through domed hill in quartz monzonite, near Tenaya Lake, Yosemite National Park, California.

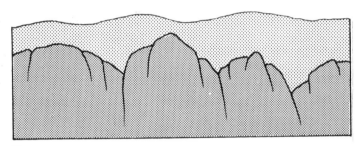

a

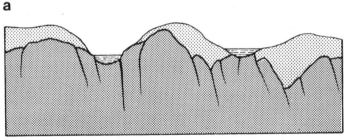

b

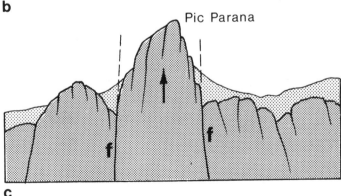

Pic Parana

c

Fig. 3.8 Some of the sugar-loaves of southeastern Brazil have, according to Barbier, developed by a combination of differential subsurface weathering and late Cainozoic faulting: (a) differential subsurface weathering, then (b) differential erosion, and (c) renewal of faulting. (*After R. Barbier,* C.R. Somm. and Bull. Soc. geol. France, *1957.*)

inselbergs occur in multicyclic landscapes which have suffered much erosion.[22] It has been suggested that the inselberg remnants represent not the upper zones of the anticlinal and domed structures, but the lower areas where stress conditions are reversed, and which are indeed in compression and hence less susceptible to weathering and to erosion.

There may be no single interpretation of sheet structure. Pressure release seems the only explanation for the spectacular sheeting found on glaciated surfaces which have suffered marked local loading and unloading (Pl. 3.15). Faulting is certainly associated with sheeting in many areas. But lateral compression (of which faulting may be but one manifestation) offers an explanation for many detailed structures and features, as well as providing a comprehensive hypothesis which explains both sheet structure and the inselbergs with which sheet structure is always associated.

5. Concluding Statement

Inselbergs are an expression of a particular condition in granite and other rocks, namely, few open-joints. This appears to be due to lateral pressures in the crust, which cause certain compartments of rocks to be in compression. These compartments are effectively monolithic and wherever

Plate 3.15 Sheet structure in exposed cirque headwall, Little Shuteye Pass, Sierra Nevada, California. The slabs vary in thickness between 1·5 and 4·5 m. Note the basin-like or synform structure underlying the crest of the domed hill.
(*M King Hubbert, U.S. Geol. Surv.*)

they occur they give rise to upstanding residuals of either castellated or domed form. Such isolated inselbergs are structural features, because they are due to the exploitation of weaknesses in rocks exposed at the earth's surface. They occur in a wide variety of climatic conditions and are not, as earlier writers suggested, restricted to the arid and semiarid tropics.

C. SANDSTONE AND ASSOCIATED LANDFORMS[23]

As will be mentioned in Chapter 4, sandstone forms uplands in many parts of the world, regardless of the prevailing climate. It underlies the principal ridges in such fold mountain belts as the Appalachians, the Cape Fold Belt (South Africa), the Flinders Ranges, the Grampians of Victoria and the James Range of central Australia (Pl. 3.17). It forms plateaux in the Orinoco Basin of South America, in the Sudan, in the Tassili Mountains of north Africa, in Jordan, in the Blue Mountains of New South Wales and the Gilberton Plateau of north Queensland, in the Kimberleys of Western Australia, and many parts of the American West and South-west (Pl. 3.18). In all, sandstone occupies about 15 per cent of the earth's land surface and usually it forms upstanding residual hills. Why is it resistant to weathering and erosion?

Sandstone is a clastic sedimentary rock consisting of small ($\frac{1}{16}$ – 2 mm diameter) quartz fragments bound together by a cement, which may be calcareous (as for example in the

Plate 3.16 Sheet structure increases in thickness in depth and is transversed by faults (f) in the Rock of Ages Quarry, Barre, Vermont. (*C.R. Twidale.*)

Plate 3.17 Sandstone ridges with flatirons (see Chapter 5) and with intervening valleys eroded in shale, James Range, N.T. (*Dept Nat. Dev., Canberra.*)

Miocene calcarenites of the Murray Basin, South Australia), but which in many areas is siliceous, forming a quartzite. Though such cements may impregnate the rock to such an extent that there are few voids, most sandstones, even the quartzites, have a high porosity (average 18 per cent) and permeability. Sandstones are also characteristically well-bedded and jointed, (though the spacing of the bedding varies considerably from place to place) so that they are usually pervious.

The resistant character of sandstone, and especially quartzite, stems from its mineralogical content, permeability and perviousness. Quartz is physically the hardest of the common rock-forming minerals. It rates a hardness of 7 on the 10-unit Mohs' scale. The siliceous cement is similarly durable. No mineral is inert, but quartz is only slightly soluble in water and is, in general, rather inactive chemically. More important, perhaps, is the fact that any water which falls on the surface of a sandstone outcrop readily infiltrates through the rock, either by way of the pore spaces or along the joints. Thus sandstone surfaces in general are not readily attacked by streams, and rivers tend to be widely spaced.

Apart from the ridges and plateaux which vary according to the disposition of the strata, and both of which reflect the essential toughness of the rock, sandstones also form domed inselbergs, such as Ayers Rock, and half domes, as in parts of the Colorado Plateau.[24] Sheet structure is also developed on these forms (Pl. 3.14a and b).

In most sandstone outcrops, however, each fracture is weathered out. Because of the bedding, the outcrops have a layered appearance. The major bedding planes are weathered, leaving the intervening beds outstanding; major vertical joints are also etched out (Pl. 3.19). Within the layers, cross-bedding, which reflects the aeolian or deltaic origin of many of these sediments, is weathered out in intaglio—as are minor bedding planes (Pl. 3.20). In some areas finely bedded and massively structured sandstones are in juxtaposition. In such situations, as in Warrens Gorge in the southern Flinders Ranges where the strata are steeply dipping, the zones of numerous joints are weathered with comparative ease and clefts are developed, leaving the zones of massive rock as upstanding ridges (Pl. 3.21).

In detail, sandstone is weathered and eroded into some peculiar—not to say grotesque—forms. Joint blocks are prominent in the frequently massive bluffs which border plateaux, mesas and buttes, and in the faceted slopes of ridge forms, again particularly on the scarp face. Exploitation of joints leads to the formation of clefts and valleys, which isolate joint blocks, turrets and towers (Pl. 3.22a and b). The joint blocks themselves are weathered into odd shapes (Pl. 3.23a and b, resembling, with the application of a certain amount of imagination, heads, faces, noses, or beehives and other features. These rounded or subrounded blocks are similar to the isolated and, in places, perched boulders developed on granite (cf. Pls 3.2c, 3.23a). Caverns or shelters have

Plate 3.18   Sandstone plateau and bluff, Tassili Mountains, southern Algeria. Note prominent bluff with bedding and vertical joints etched out by weathering, debris slope (in reality a bedrock slope with at most, a veneer of debris) and minor alluvial fans at the base of the scarp. (*P. Rognon.*)

Plate 3.19a Sandstone slope in the Mueller Plateau, northern part of the N.T., with strong joint-control of minor forms not only on the bluff, but also, to a lesser degree, on the plateau surface which is a structural bench. (*Royal Australian Air Force, Crown copyright reserved, Dept Nat. Dev., Canberra.*)

Plate 3.20 Cross bedding in the Navajo Sandstone, Zion National Park, Utah, brought out in intaglio by weathering. (*U.S. Nat. Park Service.*)

Plate 3.19b The Grand Canyon, Grampians, Victoria, with marked joint-control of minor features in the valley sidewall. (*C.R. Twidale.*)

Plate 3.21 Minor ridges and clefts in near-vertical sandstone of varied thickness of bedding, and hence varied susceptibility to weathering and erosion at Warrens Gorge, southern Flinders Ranges, S.A. (*C.R. Twidale.*)

Plate 3.22a   Rectangular joint pattern in sandstone exploited to form the Katherine River Gorge, N.T. (*Dept Int., Canberra.*)

Plate 3.22b   The Three Sisters, minor turrets in sandstone near Katoomba, west of Sydney, N.S.W. Note the plateau bounded by sandstone bluffs in the distance. (*N.S.W. Govt Tour. Bur.*)

Plate 3.23a   Isolated, perched boulders of sandstone near the mouth of the Grand Canyon, Grampians, Victoria. (*C.R. Twidale.*)

Plate 3.23b   A "beehive" in flat-lying sandstone with strong pitting and alveolar weathering, Gilberton Plateau, north Queensland. (*C.S.I.R.O.*)

Plate 3.23c   The "Whales Jaw", a sandstone block in the Grampians, western Victoria, with pronounced cavernous weathering, leaving a shell of rock, the exterior of which displays minute etching of joints. (*J.R.Morrow.*)

Plate 3.23d   The "Jaws of Death", a shelter or cavern in sandstone, Grampians, Victoria. (*Vic. Govt Tour. Bur.*)

developed within the joint blocks and at the base of bluffs, and these too have in some instances been given fanciful names—in the Grampians of Victoria, for example, there are the Whales Jaw and the Jaws of Death (Pl. 3.23c and d). The more compact sandstones, particularly those indurated with a siliceous cement, may display conchoidal fractures and negative exfoliation or flaking on the interiors of caverns. Some well-bedded arkosic sandstones display long, projecting fingers, reminiscent of stalactites, but due to differential weathering (Pl. 3.24).

Sandstone inselbergs like Ayers Rock display sandstone boulders as well as flared slopes identical with those evolved on granite (see Pl. 3.25 and description in Chapter 1). Weathering pits or gnammas are formed on sandstone outcrops on Ayers Rock, on the Arcoona Plateau, in central Tasmania, and in the Flinders Ranges.

Thus, because of its well-developed joint system and its resistant nature, sandstone displays many forms in common with granite. But it is not as vulnerable to chemical weathering as granite and because of its higher permeability, the transition between weathered and unweathered rock is far less smooth and clearly defined. In detail, sandstone also exhibits features such as cross-bedding weathering patterns which basically derive from its sedimentary origin.

D. **LIMESTONE: KARST RELIEF**[25]

1. **General**

Limestone of various types, including dolomite and chalk, is widely distributed in the platform areas and in the fold mountains developed from former geosynclinal accumulations. In all it occupies about 7 per cent of the continental areas. Calcite, which is the principal component of limestone, has a physical hardness of only 3 on the Mohs' scale. Yet in many areas limestone underlies upstanding ridges and high plains and is notable for its lack of surface drainage. Why is this?

Plate 3.24   Slender blades and points due to the delicate weathering of closely-spaced bedding planes and of parts of the intervening beds and laminae of arkosic sandstone give the impression of minor karst forms on Ayers Rock, N.T. (*C. Wahrhaftig.*)

Plate 3.25 Flared slopes in arkosic sandstone near Maggie Springs (Mutitjilda Waterhole), southern side of Ayers Rock, N.T. (*C. Wahrhatig.*)

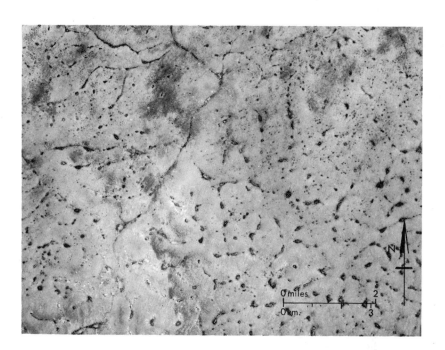

Plate 3.26   Vertical air photograph of southern part of Nullarbor Plain in Western Australia showing numerous dolines and sinkholes, but absence of co-ordinated surface drainage pattern, the only intermittent drainage lines being those linking adjacent dolines. (*Dept Nat. Dev., Canberra.*)

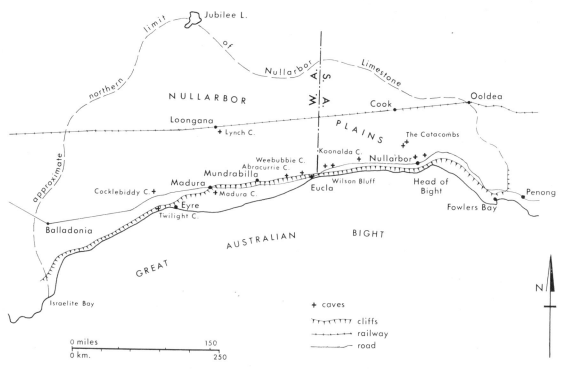

Fig. 3.9   The Nullarbor Plain. (*After J.N. Jennings, "Some Geomorphological Problems", pp. 41—62, by permission of the R. Soc. S. Aust.*)

Plate 3.27a Small solution cup, a shallow cylindrical hollow, in Cambrian limestone, central Flinders Ranges, S.A. (*C.R. Twidale.*)

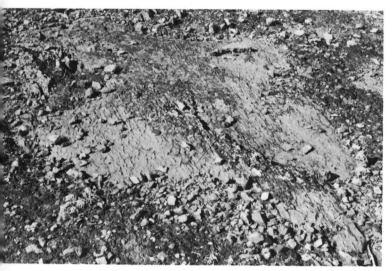

Plate 3.27b Minor solution effects together forming "finger-prints" on a minor limestone pavement, central Flinders Ranges, S.A. (*C.R. Twidale.*)

Plate 3.27c Intricate fretting of lake limestone on borders of Mono Lake, California. The taller columns are algal reefs. (*C.R. Twidale.*)

Limestone is resistant to surface erosion because of its perviousness, in the case of crystalline limestone, and because of its permeability, in the case of chalk. Hence it forms ridges and uplands in many areas. In all types of limestone, however, the chemical nature of calcite causes the development of distinctive landform assemblages.

2. Crystalline Limestone

The most common limestone is wholly crystalline, consisting of interlocking crystals of calcite which give it low porosity and permeability (average 10 per cent). But it is also well-bedded and jointed. It is, therefore, highly pervious and surface waters can easily percolate down through the numerous fractures to the level of the water table. This property is responsible for the single, most outstanding characteristic of limestone terrain—the lack of surface drainage patterns (Fig. 3.9; Pl. 3.26 and also Pl. 5.1). The drainage is said to be *cryptoreic* (i.e., hidden) because most streams and rivers flow underground. Meteoric waters react with the limestone both at the surface and as they penetrate along the joints and bedding planes. Calcite readily reacts with acidulated waters to form the bicarbonate which is soluble. In detail the chemical reaction is complex, and its rate varies with temperature and water flow and turbulence (or the degree of mixing or circulation of the water) and CO_2 concentration, but in very simple form is may be represented:

$$\text{limestone} + \text{carbonic acid} \longrightarrow \text{calcium bicarbonate}$$
$$CaCO_3 + H_2O + CO_2 \longrightarrow Ca(HCO_3)_2$$

At the surface, minor solution forms are developed (Pl. 3.27). Below the surface, fractures are widened and some blocks eventually are so undermined that they collapse. In these two ways, solution and collapse, there develops a suite of forms so characteristic that it has been given a special name: *karst*.

Where the limestone is essentially level-bedded, horizontal or only gently-inclined rock platforms are developed, in most cases corresponding to the bedding planes of the rock. These are the well-known *limestone pavements*, (Pl. 3.28) which consist of straight-sided blocks of limestone called *clints*. One or two metres in diameter and square or rhomboidal in plan, they are separated by clefts up to 5 m deep and up to 60 cms wide at the top, though narrowing in depth. These clefts are called *grikes* (or *karren*, or *lapiés*) and are essentially enlarged joints which delimit joint blocks (Pl. 3.28a-c): the parallelograms are formed on the pavements by the etching out of the grikes (Fig. 3.10), which are deepest at the intersections of joints. [26] Thinly-bedded limestones give rise to flaking in the upper parts of the grikes.

Solution depressions, or pans up to 60 cm diameter and, in some instances, aligned parallel to the local joint trend, are developed on the pavements. Most are shallow and are comparable to the pans developed by weathering on granite outcrops (see Chapter 10), but some are deep and flat-floored and have a cylindrical shape (Pl. 3.27a). In Austria they are called *kamenitzas* and are collecting places for soil and vegetation which produce acids and, hence, increase the rate of enlargement of the hollows. The pavements are also furrowed, the channels in many areas forming co-ordinated dendritic systems similar to those formed by run-off.

Plate 3.28a   Clint, grike and limestone pavements, with vertical joint pattern clearly outlined, Carboniferous limestone, Newbiggin Crags, Westmorland. (*M.M. Sweeting.*)

Plate 3.28b   Clint and grike with pavement greatly dissected by *Rundkarren*, Newbiggin Crags, Westmorland. (*M.M. Sweeting.*)

Plate 3.28c   Clint, grike and limestone pavement, north-western Yorkshire. (*M.M. Sweeting.*)

Plate 3.28d   *Rundkarren* and solution pan (filled with peat, etc.), on limestone pavement, Twistleton Scars, Ingleborough, northwestern Yorkshire. (*M.M. Sweeting.*)

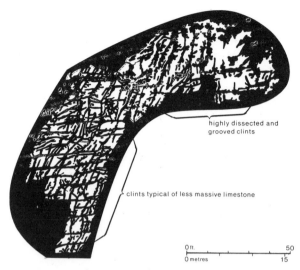

highly dissected and
grooved clints

clints typical of less massive limestone

Fig. 3.10 Plan of joint-controlled clints near Chapel-le-Dales, Derbyshire. (*After M.M. Sweeting, "The Weathering of Limestones", Essays in Geomorphology, p. 186.*)

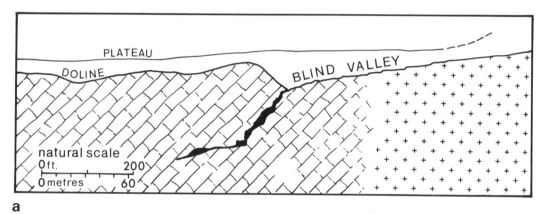

PLATEAU

DOLINE

BLIND VALLEY

natural scale

a

Fig. 3.11a Blind Valley, near Yarangobilly, N.S.W. (*After J.N. Jennings, "Australian Landform Example No. 9. Blind Valley", Aust. Geogr., 1967, 10, pp. 206–07.*)

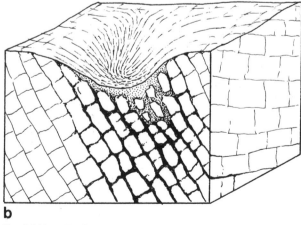

b

Fig. 3.11b (i) Solution doline in section.

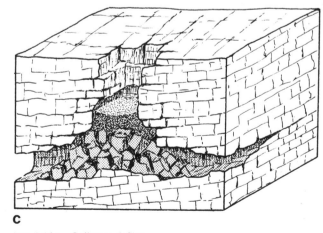

c

Fig. 3.11c Collapse doline.

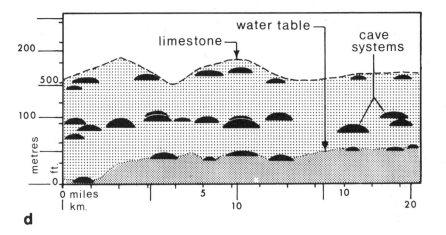

d

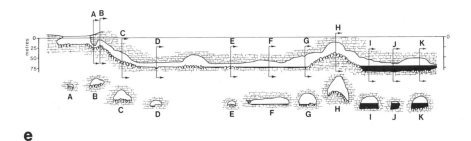

e

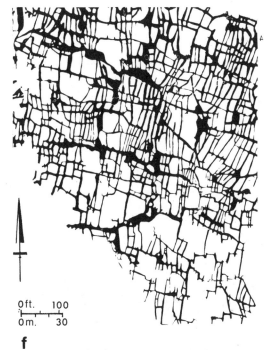

f

Some grooves, however, occur in isolation. They are U-shaped in cross-section and are called *Rundkarren* (Pl. 3.28d). Some are wider at the base presumably due to solution of the longest wetted areas, and some are V-shaped in cross-section. The latter, if small, are called *Rillenkarren* (Pl. 3.29), but the larger runnels, frequently rather rounded in cross-section, though separated by sharp crests, are called *Rinnenkarren*.

Some limestone surfaces, both inclined and horizontal, display numerous closely-packed solution cups, each no more than 2 cm in diameter, which together look like fingerprints formed when the tips of the fingers have been thrust into a thin layer of mud, or other plastic material (Pl. 3.27b). On the bluffs which often delimit these pavements, and on the grikes, vertical grooves are displayed (Pl. 3.29); presumably these are due to the solutional effects of run-off flowing over the surface of the rock. Major vertical joints are enlarged by solution (Pl. 3.30) and any streams of surface waters that may have formed for any reason disappear down these joints. Such streams form valleys, but they are of limited linear extent and terminate abruptly where the waters course down vertical fissures. These valleys lead nowhere and are called *blind-valleys* (Fig. 3.11a). The water thus swallowed by the rock is undersaturated with lime and readily effects the enlargement of major grikes. Large, vertical tunnels, more or less circular or oval in cross-section, are thus formed. They are called *dolines* and, where they are due mainly to solution, *solution dolines* (Fig. 3.11b). But most dolines are *collapse dolines*: due partly to solution, partly to callapse caused by solutional undermining of joint blocks (Fig. 3.11c). The solution dolines are distinguished by the rounded, smooth flutings of the walls and an absence of blocky debris on the

Plate 3.29 *Rillenkarren* formed on Palaeozoic limestone, north Queensland. (*C.R. Twidale/C.S.I.R.O.*)

Plate 3.30 Fluting of Devonian limestone in the Napier Range, Kimberley District, W.A. (*Geol. Surv. W.A.*)

floor, whereas collapse dolines have masses of angular debris strewn over the bottom.

Subsurface waters cause the enlargement of both the horizontal bedding planes and the vertical joints in linear zones where the planes are intersected by major vertical joints, and where, apparently, the water-table formerly stood, or now stands. There are, however, varied opinions as to how these *caves*, which in some areas form complex and extensive systems at various levels within the rock mass (Fig. 3.11d) and which vary in shape in cross section (Fig. 3.11e). were evolved. Some consider that they form by solution beneath the level of the water-table (by *phreatic* or ground-waters). Others are of the opinion that caves result from the solutional actions of diverted surface waters in the zone above the water-table (*vadose waters*). Still others hold the view that corrosion and corrasion by underground rivers (which display meander loops and waterfalls similar to surface streams) play a significant part in cave formation and enlargement.

The presence of large masses of unstratified red clays in many American caves suggests that the deposits were laid down in water and that the caves in which they occur may have formed below the water-table. On the other hand, the distinct elevational zoning of caverns and the relationship of the caverns to erosion surfaces preserved in adjacent areas [27] suggests that they are found in relation to the water-table. This is possibly due to a combination of solution and abrasion at, or more likely just beneath, the surface of the water-table by water that is slow moving, perhaps as slow as 10 m per year. [28] This hypothesis is supported by the observed evidence of smooth solutional fretting in the caves, by the network pattern (Fig. 3.11f) of caves (if descending ground-waters were responsible, cave formation would be limited to a few, major, vertical joints) and by consideration of the chemistry of groundwaters. For it is only at or near the surface of the groundwater zone that the mixing of downward percolating groundwaters results in undersaturated solutions which are still capable of dissolving more lime.

Shifts of the water-table could lead to the development of several levels of caverns such as are found beneath the Nullarbor Plain. [29] Such oscillations of the water-table could also account for the partial burial of pre-existing caves by alluvium carried down into the groundwater zone.

Calcite is precipitated when waters, which have become saturated with carbonate at high partial pressures in zones such as rock crevices and soil interstices through which they have passed, emerge into caves or other openings where pressures are lower. All the reactions involving water, carbon dioxide and calcium carbonate are reversible and it is primarily the diffusion of gas from the migrating waters as they enter caves which induces supersaturation and precipitation; evaporation is rarely significant in this respect.

The precipitation of carbonate takes place from flows and drips of water and since these are repetitive the precipitation gives rise to rocks which consist of layer upon layer of calcite (or some other carbonate). Flowstone, found in the floors of many caves, is a typical example. More importantly from a morphological point of view, carbonate deposition causes the formation of various minor, though frequently spectacular, features collectively known as *speleothems*. The best known are the vertical stalactites (which grow from cave ceilings) and stalagmites (which develop on cave floors) and drip curtains,

Plate 3.31a   The "Judges Wig" stalactites and cave pool, Jewell Cave, Augusta, W.A. (*W.A. Newspapers*.)

Plate 3.31   Stalactites in Treak Cliff Cave, Castleton, Derbyshire. (*Aerofilms Ltd.; Hunting Surveys*.)

Plate 3.32a   Cone karst in the Cockpit country of Jamaica. (*M.M. Sweeting.*)

Plate 3.32b Towerkarst in New Guinea. (*J.N. Jennings.*)

Plate 3.32c   Mogotes or limestone residuals in Puerto Rico. (*W.H. Monroe.*)

though with a little imagination numerous other likenesses can be discerned (Pl. 3.31). One form of particular interest is the helictite, a speleothem which grows eccentrically from stalactites in that it defies gravity by extending upwards at varied and changing angles relative to its host. The carbonate-rich solutions appear to migrate along capillaries in the calcite rock and are thus enabled to move and build out vertically or diagonally.

In some areas, however, the ceilings of caves have collapsed, forming *karst windows*. Eventually, solution and collapse combine to form quite extensive enclosed depressions which are floored by alluvium and called *uvula*. Very large alluviated, flat-floored depressions within the karst areas are called *polje* (pronounced *polyay*). Some of them are partially due to downfaulting.

In the well-vegetated humid tropics, where the rate of limestone solution is high (it may be as much as four times greater than in cooler areas), numerous very deep solution dolines are formed, and, especially in areas of massive, strong limestone, a typical karst landscape called *kegelkarst* is evolved. These deep, open dolines are called "cockpits" in Jamaica— hence "Cockpit Country".[30] These dolines or cockpits merge, leaving scattered, conical residual hills of limestone. This is known as *conekarst* (Pl. 3.32a). In some areas, however, for instance in southern China[31] and in parts of the Antilles,[32] the residuals are steep-sided and are called towers; hence the term *towerkarst* or *Turmkarst* (Fig. 3.12, Pl. 3.32b). In such areas solution depressions extend below the watertable with two results. First, the depressions become alluviated and are extended by lateral river corrasion to form uvulas and poljes. Second, partly as a result of the action of soil moisture at the base, solution causes undermining and steepening of the residual hills (cf., flared slopes on granite), forming towers which stand out in dramatic isolation in some areas which are called variously hums, haystack hills, mogotes and pepino hills (Pl. 3.32c).

In arid climates karst developments are less distinctive. In high latitudes and altitudes, for instance in Spitzbergen, though meltwater from snow can cause some doline development, solution occurs mainly at the surface where extensive karrenfields, lapiés or clints and grikes are formed.[33] Frost action effectively splits the surface exposures of the limestone. In the arid tropics, such as in the Nullarbor Plain, solution is not at present very active. In fact the trend appears to be for collapse to destroy the solutional forms developed during periods of higher rainfall in the recent past.[34]

In New Guinea altitudinal zonation of karst forms has been noted analogous to the latitudinal-climate variations indicated above.[35] In the *tierra caliente*, roughly below 200 m and with 5000–7500 mm annual rainfall, *kegelkarst* with prominent cockpits is developed in several types of Miocene limestone. In this same zone a sort of giant grikeland is displayed—numerous close-set solution corridors breaking up the surface in some areas. At rather higher levels, in the *tierra templada* zone, with 2000–3500 mm rainfall, dolines and towers or pyramids are well developed. Some of the dolines, like the crevices in the lower elevational zones, display joint-control and are elongated along the strike. At higher levels, where rainfall is similar to that experienced in the middle elevations, dolines and pyramids also occur; in addition craggy pinnacles and arêtes, rather like battle-

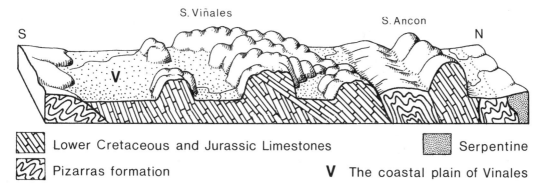

Fig. 3.12a General aspect of towerkarst in Sierra de los Organos, Cuba. (*Reprinted with permission from* Erdkunde, *after H. Lehmann, "Der Tropische Kegelkarst", pp. 130–9.*)

Fig. 3.12b Detail of the towerkarst hills in the Sierra de los Organos, Cuba. (*Reprinted with permission from* Erdkunde, *after H. Lehmann, "Der Tropische Kegelkarst", pp. 130–9.*)

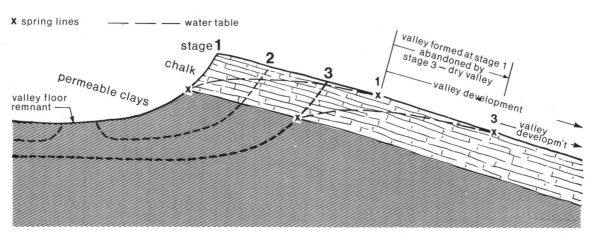

Fig. 3.13 Development of dry valleys in chalk according to C.C. Fagg. *"The Recession of the Chalk Escarpment", pp. 93–112.*)

Plate 3.33 The chalk escarpment of the South Downs in southeastern England, looking westward from near the Devils Dyke. The scarp is capped by the Upper Chalk, but the lower ground is occupied by Lower Chalk, Gault and Upper Greensand. The village in the middle distance is Fulking and the depressions which give the scarp its crenulated appearance are called coombes. The gap which breaches the scarp in the distance is that formed by the River Adur. (*Crown Copyright Reserved, reproduced courtesy H. M. Stationery Office.*)

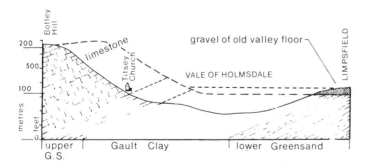

Fig. 3.14 Section through chalk scarp and adjacent vale near Limpsfield, Surrey. (*After C.C. Fagg, "The Recession of the Chalk Escarpment", pp. 93–112.*)

ments surround the doline depressions in this zone, which may be compared to the *tierra fria*.

3. Chalk

Many areas of northern France and southeastern England, though underlain by limestone, do not display typical karst features. The reason is that the limestone is not strong and crystalline, but porous, with poorly-developed joints and bedding planes. This limestone is called *chalk*. Though rather rudimentary swallow holes, solution pits and other karst features are commonly developed, the physical weakness of the chalk causes collapse before any well-developed caves, for example, can be formed.

Because of its propensity for swallowing water, chalk is resistant and forms uplands such as the English Downs, the Chiltern Hills and Salisbury Plain, which are bounded by precipitous scarps (Pl. 3.33). In contrast to the angular, rugged relief developed on crystalline limestone, chalk landscapes are typically rolling due to the widespread development of dry valleys, (that is, typical river valley systems which, however, lack rivers except in their lower reaches). How were these valleys formed in a bedrock which is porous and to some degree pervious? Their origin has given rise to some controversy, for they are clearly not forming now, or at least not developing very actively.

It has been suggested that during the Quaternary ice ages the water contained in the pores of the chalk was frozen, thus rendering the rock impermeable and susceptible to normal stream erosion. With the onset of warm conditions, the ice melted, the rock became porous again, and surface stream action was again of minor importance. However, dry valleys in the chalk occur well beyond the former ice margins, and in any case the valleys seem to predate the Pleistocene.

A higher rainfall could also conceivably have caused a general rise of the water-table, thus allowing the development of valleys. At present, the streams are temporarily active after heavy rains, so that this explanation is possible.

An hypothesis which has, and for good reason, found great favour takes cognisance of the typical structure of the chalk.[36] It is pointed out that the chalk is gently folded and forms broad cuestas, with steep scarp faces and gentler and longer dip-slopes. As the chalk is underlain by impermeable rocks, the level of the water-table depends on the level of the junction of chalk and underlying strata (X in Fig. 3.13). If the scarp were eroded, and receded, the point X would, because of the dip of the chalk, decrease in elevation so that the water-table would be lowered. It may be reasonably asserted that the upper valley sectors would for this reason be abandoned. In this way dry valleys could develop. The recession of the scarp could take place to such an extent that the heads of the dry valleys could be intersected. As there are remnants of former valley floor-levels in the scarp-foot region (Fig. 3.14) this hypothesis has much to recommend it.

4. Other Limestones

Dolomite is not nearly as soluble as is pure limestone, so that it develops rather a poor karst topography. There is, however, a lack of surface drainage. The same is true of the Murray

Plains in South Australia, which are underlain over wide areas by calcarenite, a sandy limestone, with massive bedding and jointing. Dolines, broad, shallow depressions and a few steep-sided flat-floored depressions occur but again, because of weakness of the rock, caves are rare, except in the immediate vicinity of the Murray River, where quite long cave systems have been discovered.

5. Concluding Statement

Karst is best developed on crystalline limestone particularly in areas where such bedrock is pure, flat-lying and thick, but well-bedded and jointed with a deep through drainage, usually developed as a result of deep dissection by major rivers. Typical karst forms are well displayed in such areas as Alabama, Indiana and Kentucky, in New Guinea, Cuba Puerto Rico, Jamaica, Vietnam, China, in the Dinaric region of Yugoslavia, and in numerous smaller areas in the fold mountain belts. The Nullarbor Plain is an extensive outcrop of limestone, but because of the present aridity, the karst is rather degraded, with collapse predominating over solution.

Areas of impure or weak limestone give rise to various degrees of weakly-developed karst. All limestone areas, however, display a lack of surface drainage; the only rivers which cross major limestone outcrops are of an *allogenic* type, that is, they derive their waters from beyond the limestone outcrop and have sufficient volume to survive passage across it.

E. GNEISS AND SCHIST FORMS

Gneiss has a mineral composition similar to that of granite. Instead of an orthogonal system of joints, however, gneiss displays a well-developed cleavage; that is, the rock is divided into slabs by numerous, close, parallel partings. At the surface of outcrops where the cleavage is steeply inclined, tilted slabs of rock separated by clefts or valleys are developed. They arise from the preferential weathering of the cleavage planes and are called monk-stones, *penitent rocks*, (German *Büssersteine*—see Ackermann[37]) or, because they look like the headstones in a neglected graveyard, *tombstones* (Pl. 3.34). Good examples occur in the gneiss and schist areas of the eastern and southern Mt Lofty Ranges in South Australia.

On a larger scale, the foliation and lineation of schists and gneisses are exploited by weathering and streams to form a landscape with a characteristic grain (Fig. 3.15) running parallel to the structures in the country rock. This

Plate 3.34a Minor tombstones or penitent rocks in gneiss near Tungkillo, S.A. (*C.R. Twidale.*)

Plate 3.34b Steeply-inclined metamorphic rocks differentially eroded to form massive slabs in the Bernese Oberland, Switzerland. (*Swiss Nat. Tour. Office.*)

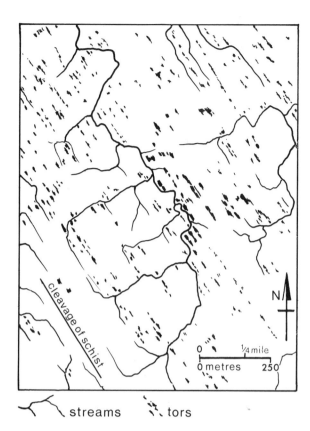

Fig. 3.15 Map of linear relief or *Gefügerelief* in southern Otago, New Zealand. (*Reprinted by permission of the author and the* American Journal of Science, *after F.J. Turner, "Gefügerelief illustrated by 'Schist Tor' Topography",* pp. 802–7.)

streams   tors

Fig. 3.16   Field sketch of schist towers or tors on the flanks of Coronet Peak, near Queenstown, South Island, New Zealand.

Fig. 3.17   Field sketch of residuals of tough "corestones" from the Franciscan *melange* deposits west of San Raphael, California. The residuals vary in size and shape and appear to be randomly distributed.

is called *textured relief*, or *Gefugerelief*, and is well developed in schists in central Otago, New Zealand.[38] These areas are characterised by schist tors or towers, for instance in the Dunstan and Pisa ranges, in the Middlemarch area, and around Queenstown. The "Bayonets", Walter Peak and Cecil Peak, bordering Lake Wakitipu in the South Island of New Zealand, all display such schist towers, as does the summit of nearby Coronet Peak, and the slopes surrounding this eminence (Fig. 3.16).

In the schists and shales of the Franciscan (Jurassic-Cretaceous) rocks of the San Francisco area, the *mélange* deposits give rise to unusual structural landforms. The sequences are greatly disturbed by fault or slip displacements (see Chapter 2) and display large cores of solid rock set in a matrix of rock which, though lithologically similar to that of the cores, is greatly fractured and in consequence very readily weathered and eroded. Hence the cores stand out as residual blocks or hillocks, according to their size. Some of the most spectacular features associated with the *mélange* occur to the north of San Francisco in the Russian River valley, and in the region west of San Raphael, where the solid cores give rise to angular turrets and rounded towers which stand up to 30 m above the hillslopes (Fig. 3.17).

F. LANDFORMS DEVELOPED ON SHALE AND OTHER ARGILLACEOUS ROCKS

Shale and similar rocks occupy over half the land surface. Shales are composed in large measure either of feldspars or of clay minerals, which are chemically very reactive and are thus susceptible to attack by water (see Chapter 9). For this reason alone shale outcrops in general are commonly worn down relatively rapidly compared to sandstone and limestone, and commonly form valleys and plains (see Chapter 5).

Even so, jointing is important to an understanding of these rocks' vulnerability to erosion. Shales have quite a high porosity, but because of the very small size of the pore spaces, surface tension effects prevent ready passage of water through the rock. But such rocks are also commonly fissile, that is, there are numerous bedding planes separating thin layers of rock, and these are further broken by vertical joints. Thus there are numerous avenues along which water can penetrate, and in most cases the outcrops are readily weathered and worn down. However, some locally prominent ridges are underlain by shales which have more massive bedding and jointing.

G. CONCLUSION

Joints exert an important influence on landform development in several widespread rock types. These passive fractures form avenues of weakness which are commonly exploited by weathering and erosion, so that relief comes to reflect the joint pattern. In rate and detail of landform development much depends on other characteristics of the rocks: thus some contrasts may be drawn between the relative significance of joints in permeable rocks like sandstone and impermeable ones such as granite and limestone. The chemical character of limestone gives to karst a distinctive appearance compared to those features of granitic landscapes which are more directly controlled by joints alone. Minor fractures associated with cross-bedding are significant in detail in sandstone forms.

But overall joint patterns impose an impressive unity on landform evolution in many and extensive parts of the world.

References Cited

1. C.R. TWIDALE, *Structural Landforms*, Australian National University Press, Canberra, 1971, pp. 14–17.
2. A.D.N. BAIN, "The Formation of Inselberge", *Geol. Mag.*, 1923, **60**, pp. 97–101.
3. E. DE MARTONNE, *Traité de Géographie Physique*, 9th ed., Colin, Paris, 1951, Vol. 2, pp. 631–2.
4. D.L. LINTON, "The Problem of Tors", *Geogr. J.*, 1955, **121**, pp. 470–87.
5. W. BRUCKNER, "The Mantle Rock ("Laterite") of the Gold Coast and Its Origin", *Geol. Rdsch.*, 1955, **43**, pp. 307–27.
6. J.A. MABBUTT, "A Study of Granite Relief from South West Africa", *Geol. Mag.*, 1952, **89**, pp. 87–96.
7. E.S. LARSEN, "Batholith and Associated Rocks of Corona, Elsinore and San Luis Rey Quadrangles, Southern California", *Mem. geol. Soc. Am.*, 1948, **29**, pp. 113–9.
8. For example in the Valley of the Thousand Hills, Natal and in coastal sections south of Cape Town, and near Marendellos, east of Salisbury.
9. J.G. LESTER, "Geology of the Region around Stone Mountain, Georgia", *Univ. Colo. Stud. geol. Ser.*, 1938, A **26**, pp. 88–91.
10. A. HOLMES and D.A. WRAY, "Outlines of the Geology of Mozambique", *Geol. Mag.*, 1912, **9**, pp. 412–17.
11. L.C. KING, "The Origin of Bornhardts", *Z. Geomorph.*, 1966, **10**NS, pp. 97–8.
12. See E. OBST, "Das Ablusslose Rumpfschollenland in Nord Ostlichen Deutsch-Ostafrica", *Mitt. geogr. Ges. Hamb.*, 1923, **35**; L.C. KING, "A Theory of Bornhardts", *Geogr. J.*, 1949, **112**, pp. 83–7; J. BÜDEL, "Die 'Doppelten Einebnungsflächen' in den feuchten Tropen", *Z. Geomorph.*, 1957, **1**NS, pp. 201–28; C.R. TWIDALE, "Contributions to the General Theory of Domed Inselbergs. Conclusions derived from Observations in South Australia", *Trans. Inst. Br. Geogr.*, 1964, **34**, 91–113.
13. L.C. KING, "Theory of Bornhardts", pp. 83–7.
14. See for instance SOEN OEN ING, "Sheeting and Exfoliation in the Granites of Sermasoq, South Greenland", *Med. Gronland*, 1965, **179**, (6), pp. 1–40.
15. G.K. GILBERT, "Domes and Dome Structures of the High Sierra", *Bull. geol. Soc. Am.*, 1904, **15**, pp. 29–36. For a review of the various hypotheses advanced in explanation of sheet structure see C.R. TWIDALE, "On the Origin of Sheet Jointing", *Rock Mechanics*, 1973, **5**, pp. 163–87.
16. C.A. CHAPMAN, "The Control of Jointing by Topography", *J. Geol.*, 1956, **66**, pp. 552–8.
17. J.N. JENNINGS and C.R. TWIDALE, "Origin and Implications of the A-tent, a Minor Granite Landform", *Aust. Geogr. Stud.*, 1971, **9**, pp. 41–53.
18. C.R. TWIDALE, *Structural Landforms*, p. 247.
19. W.C. BRADLEY, "Large-scale Exfoliation in Massive Sandstones of the Colorado Plateau", *Bull. geol. Soc. Am.*, 1963, **74**, pp. 519–28.
20. G.F. HARRIS, *Granite and Our Granite Industries*, Crosby Lockwood, London, 1888, p. 142; G.P. MERRILL, *Treatise on Rocks, Weathering and Soils*, Macmillan, New York, 1897; C.R. TWIDALE, "Contribution to the General Theory" pp. 91–113; *idem, Structural Landforms*.
21. C.R. TWIDALE, "Contribution to the General Theory", pp. 91–113; *idem, Structural Landforms*.
22. E. OBST, "Das Ablusslose Rumpfschollenland"; L.C. KING, "Theory of Bornhardts", pp. 83–7; C.R. TWIDALE, "Contribution to the General Theory", pp. 91–113; *idem, Structural Landforms*.
23. For a detailed account of landforms developed on sandstones see MONIQUE MAINGUET, *Le Modelé des Grès*, Inst. Geogr. Nat., Paris, 1973.
24. W.C. BRADLEY, "Large-scale Exfoliation", pp. 519–28.
25. For excellent comprehensive accounts see J.N. JENNINGS, *Karst*, Australian National University Press, Canberra, 1971; and M.M. SWEETING, *Karst Landforms*, Macmillan, London, 1972.
26. M.M. SWEETING, "The Weathering of Limestones", in G.H. Dury (Ed.), *Essays in Geomorphology*, Heinemann, London, 1966, pp. 177–210.
27. M.M. SWEETING, "Erosion Cycles and Limestone Caverns in the Ingleborough District", *Geogr. J.*, 1950, **115**, pp. 64–78; M.E. MARKER and G.H. BROOK, "Echo Cave: A Tentative Quaternary Chronology for the Eastern Transvaal", *Dept Geogr. and Envir. Stud. Univ. Witwatersrand Occ. Paper*, 1970, **3**, p. 38.
28. G.W. MOORE, "Limestone Caves" in R.W. Fairbridge (Ed.), *Encyclopaedia of Geomorphology*, Reinhold, New York, 1969.
29. J.N. JENNINGS, "Some Geomorphological Problems of the Nullabor Plain", *Trans. R. Soc. S. Aust.*, 1963, **87**, pp. 41–62.
30. M.M. SWEETING, "The Karstlands of Jamaica", *Geogr. J.*, 1958, **124**, pp. 184–99.
31. H. VON WISSMANN, "Das Karstphänomen in der verschiedenen Klimazonen", *Erdkunde*, 1954, **8**, pp. 112–22.
32. M.M. SWEETING, "Karstlands of Jamaica", pp. 184–99; H. LEHMANN, "Der Tropische Kegelkarst auf der Grossen Antillen", *Erdkunde*, 1954, **8**, pp. 130–9.
33. J. CORBEL, "Les Phenomenes Karstiques en Climat Froid", *Erdkunde*, 1954, **8**, pp. 119–20.
34. J.N. JENNINGS, "Geomorphological Problems", pp. 41–62; *idem*, "Karst in Australia", in J.N. Jennings and J.A. Mabbutt (Eds), *Landform Studies from Australia and New Guinea*, Australian National University Press, Canberra, 1967, pp. 256–92.
35. J.N. JENNINGS and M.J. BIK, "Karst Morphology in Australian New Guinea", *Nature (Lond.)*, 1962, **194**, pp. 1036–8.
36. C.C. FAGG, "The Recession of the Chalk Escarpment and the Development of Chalk Valleys in the Regional Survey Area", *Trans. Croydon Nat. Hist. Scient. Soc.*, 1923, **9**, pp. 93–112.
37. E. ACKERMANN, "Büssersteine—Zeugen Vorzeitlicher Grundwasserschwankungen", *Z. Geomorph.*, 1962, **6** NS, pp. 148–82.
38. F.J. TURNER, "*Gefügerelief* illustrated by 'Schist Tor' Topography in Central Otago, New Zealand", *Am. J. Sci.*, 1952, **250**, pp. 802–7.

CHAPTER 4

Fractures in the Crust: Faults

A. STRUCTURAL FORMS

Linear valleys and escarpments are not uncommon features of the earth's relief. Many are of too great a magnitude to be related to joints (Pl. 4.1) and field evidence shows that they are associated with fractures in the crust called *faults*.

Faults, like joints, are lines or zones of weakness in the earth's crust. Like joints they are readily exploited by weathering and erosion. But other structural forms arise from rocks of contrasted character being displaced along faults.

Linear escarpments may develop as a result of differential erosion of strata brought into juxtaposition by faulting. The scarp is not due directly to earth movements. It is due to the exploitation by erosion of a structural situation brought about by faulting and is called a *fault-line scarp* (Fig. 4.1a). If the scarp faces the downthrown block it is called a *resequent* fault-line scarp, but if, through deep erosion and exposure of strata of contrasted resistance, the scarp faces the upthrown side, it is called an *obsequent* fault-line scarp (Fig. 4.1). In many cases it is very difficult to determine with certainty whether a scarp related to a fault is of fault or fault-line character. A detailed knowledge of the local stratigraphy and tectonic history is required. However, the MacDonald Scarp in the Canadian North West Territories is of fault-line type, and is developed at the junction of the Precambrian sediment of the Et-then Series and the crystalline rocks of the Canadian Shield (Pl. 4.1b).

Dislocation of strata can, after differential weathering and erosion and the development of, say, ridge and valley forms, give rise to discontinuous or offset ridges and valleys related to the same strata (Fig. 4.2 and Pl. 4.2). In similar fashion, faulting of granite massifs, as for instance, near Berridale, NSW determines the distribution of the characteristic granite features (Fig. 4.3). The traces of fault planes and zones are usually straight or only gently curved in plan and in some cases can be traced for hundreds of kilometres. They have been exploited by rivers which, in consequence, are also extraordinarily straight. Examples in Australia include the Georgina, the Darling, the Flinders and Strzelecki Creek.

Major left bank tributaries of the Amazon (Fig. 4.4) appear from their straightness to be fault-controlled, and there is some geological support for this contention.[1] In similar fashion, the regional morphology of the lower Mississippi valley is controlled by a pattern of NW–SE and NE–SW trending faults, which not only delimit the Ozarks on their southeastern side, but also control the location and trend of such rivers as the Ouichita, Red and Arkansas, as well as of the lower Mississippi and its delta.[2]

There are many examples (Pl. 4.3, Fig. 4.4) in which faults influence landform development in a passive way. But many faults are active, and their dislocation has given rise to a number of distinctive tectonic landforms.

B. TECTONIC FORMS

Many linear or only gently arcuate scarps and depressions (Pl. 4.4) are, on field examination, found to be directly related to active faulting in the crust. Some fault dislocations are major and dramatic, involving instantaneous movements of several metres. Thus during the 1906 San Francisco earthquake, the ground was displaced by as much as 5 m in northern California, and there were lateral offsets of 3 m. The available evidence points to displacements of almost 10 m in southern California during the 1857 earthquake.

On the other hand, many fault movements are gradual and of minor magnitude in human terms. In a geological context however, they are cumulatively of great significance. Such almost imperceptible or very slow movements are known as *fault creep*. Thus the Hayward Fault which passes beneath the Cal Stadium at Berkeley, California, is moving at a rate of 0.25 mm per annum. However the dislocations are irregular both spatially and temporally: the average annual fault creep along the Hayward-Calaveras Fault at

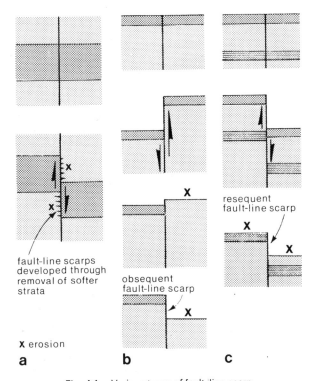

fault-line scarps developed through removal of softer strata

obsequent fault-line scarp

resequent fault-line scarp

X erosion

a   b   c

Fig. 4.1 Various types of fault-line scarp.

Plate 4.1a   The cleft caused by the exploitation by various exogenetic processes of the Hornby Bay Fault, Great Bear Lake region, Northwest Territories, Canada. On the left, Proterozoic sandstone, on the right, Archaean granite. The fault has been traced for some 120 km. (*Dept Energy, Mines and Resources, Ottawa. Photo No. A37834-41.*)

Plate 4.1b   The MacDonald Fault, which has been traced for about 500 km in the Northwest Territories, Canada, trends NE–SW. In this view, looking southwest, the older crystalline rocks of the craton proper underlie the higher block on the left, and sediments of the late Precambrian Et-then Series the downthrow side. The scarp, which attains a maximum height of about 280 m, is an expression of the greater resistance of the crystallines to weathering and erosion, and is not directly due to faulting: it is therefore of fault-line character. Note the numerous lakes, large and small, in this glaciated tundra region. (*Survey and Mapping Branch, Dept Energy, Mines and Resources, Ottawa. Photo A5120-105R, taken 4 August, 1935.*)

Fig. 4.2   Offset ridges on the flanks of the Angepena Syncline, central Flinders Ranges. (*After Geol. Surv. S.A.,* 1-mile Angepena.)

Plate 4.2   This plunging anticline in central Australia has been dislocated by faulting with the result that the resistant strata, and the ridge associated with them, are abruptly terminated. The fault zone is indicated (X), and the amount of throw is represented by A–B. (*Dept Int., Canberra.*)

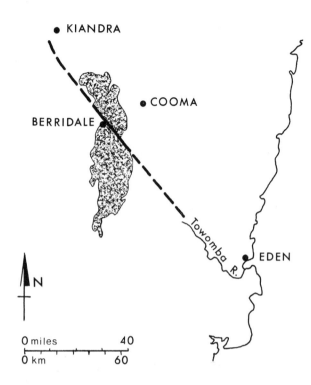

Fig. 4.3 Displacement of the Berridale granite outcrop by about 10 km along the Berridale Fault. Note the Towomba River following along the same line of weakness. (*After I.B. Lambert and A.J.R. White, "The Berridale Wrench-fault: A Major Structure in the Snowy Mountains of New South Wales",* J. Geol. Soc. Aust., *1965,* **12***, p. 27.*)

Plate 4.3a   Minor fault affecting Pound Sandstone in Brachina Gorge, central Flinders Ranges, South Australia. Note the local lowering of the ridge, and the cleft developed along this line of weakness. (*C.R. Twidale.*)

Plate 4.3b   The Great Glen, a long linear valley eroded by rivers and glaciers from an unusually wide fault zone due to marked horizontal displacements in northern Scotland. The lake, or loch, in the foreground is Loch Lochy, the one in the middle distance is Loch Oich. Both are due to glaciers scouring out certain areas of the preglacial river valley eroded along the zone of weakness. (*Crown Copyright Reserved. Reproduced by permission of H.M. Stationery Office.*)

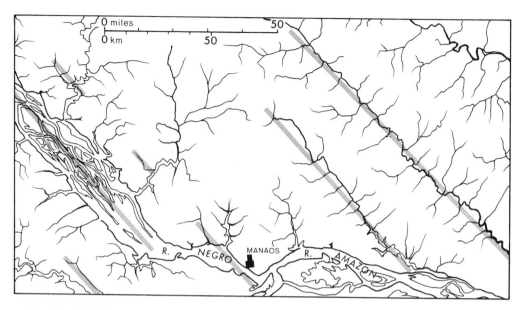

Fig. 4.4   The fault-controlled left-bank tributaries of the Amazon. (*After H. O'R. Sternberg and R.J. Russell, "Fracture Patterns", pp. 380–5, by permission of Int. Geogr. Un. Proc.*)

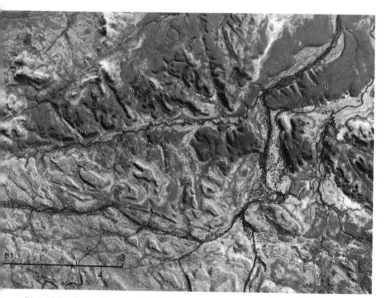

Plate 4.3c This tributary of the West Baines River, near Waterloo H.S. in the Northern Territory and close to the Western Australian border, flows along a fault zone in Cambrian volcanic rocks. Remnants of volcanic plateaux are also discernible. (*Dept Nat. Dev., Canberra.*)

Plate 4.3d These fault zones exploited by rivers and ice and then, in some cases, inundated by the sea, are in coastal British Columbia. Several rectilinear valleys can be seen but the largest (X) is aligned along the Owikeno lineament, a major fault zone. (*Dept Lands, Forests and Water Resources, British Columbia. Photo B.C. 1231:73, taken 2 October, 1950.*)

Plate 4.4 Lake Thingvallavatur and Thingvellir (Plain of the Parliament) in the southwestern fork of the Iceland graben (see Fig. 4.14a). Recent lava flows have been disrupted by normal faults, the principal fracture in the zone being the Almannagja Fault, with a throw of 30 m. Note the numerous secondary fractures and the gaping fissures, and the splinter blocks associated with the main escarpment. (*S, Thorarinsson.*)

Plate 4.5   The Rocky Mountain Trench near the junction of the Fraser and Torpy rivers, with the Rocky Mountains on the left and the Caribou Mountains to the right. The high peak standing above the remarkably even summit level on the left horizon is Mt Robson, 3954 m above sealevel. Note the parallel linear sides of the Trench; the meandering courses of the major rivers; the Fraser valley which is slightly incised below the general level of the floor of the Trench; the rectilinear slopes of the Rocky Mountains; the knife-edged ridges or aretes; and the snowfields and snowline. (*Dept Lands, Forests and Water Resources, British Columbia Photo B.C. 766:37, taken 26 September, 1948.*)

Hollister is 10 mm, but on 17 July, 1971 there was an instantaneous movement of 9 mm—almost the yearly average—along a 6 km section. This sudden relief of the accumulated strain affected not only the Hayward-Calaveras Fault but also the nearby San Andreas Fault, even though there is no surface connection between the two systems. Some fault-generated features are of regional or even continental scale. Many cratons are subdivided into square or rhomboidal blocks by long fracture zones which are composed partly of major joints, but mainly of faults, and are known as *lineaments*.[3] Thus, the Australian and African shield areas display clearly defined lineament-bounded blocks. In the American West, huge fault-delineated blocks, aligned north–south, dominate the landscape. They are separated by downfaulted basins and depressions and, as a result of their hachured representation on early maps, have been likened by C.E. Dutton—an early explorer—to an "army of caterpillars marching to Mexico". The Rocky Mountain Trench, a major structural and topographic feature of the western Cordillera, particularly in its Canadian sector, consists of a number of depressed fault blocks developed in younger rocks, but possibly inherited from movements in the underlying, older basement rocks (Pl. 4.5). Though these fault depressions are discontinuous, they are in alignment and are linked by erosional valleys. Fracture patterns also dominate some continental outlines, as is apparent in the Sinai region (Fig. 4.5, Pl. 4.6), as well as more local forms (Fig. 4.6).

Many of the scarps which delineate blocks are in reality exposed, though rather dissected, fault planes, and the scarps are called *fault scarps*. They are recognisable not only by their linearity (Pl. 4.7), but by preserved slickensides and other geological features, including the displacement of strata (Fig. 4.7). Dissected fault scarps display triangular facets (see Pl. 4.8, Fig. 4.8) though similar forms can also develop in other situations (see Pl. 4.9). They are caused by the removal of part of the fault plane by streams. The V-shaped valleys leave between them inverse Vs, or triangular facets. In some areas, the valley floors are incised well below the lower apex of the V so that the valleys have the form of a wineglass—the open V being the glass, and the gorge below the stem of the glass (Figs 4.9 and 4.10). These are caused in many cases by recurrent movements and rejuvenation of the streams, though lithological control is responsible in some areas where the streams have cut through to a lower sequence which is more resistant to valley widening.

There can be no doubt that fault scarps form directly as a result of earth movements. Earthquakes within historical time have been observed to throw up such scarps (Pl. 4.10) and in many parts of the western USA for instance, linear escarpments are developed in unconsolidated sediments such as fanglomerates and moraines as a result of fault displacements (Fig. 4.11). But whether fault scarps of any magnitude can survive long weathering and erosion, and remain as prominent features of the landscape has been doubted by some investigators. Some have urged that erosion of such features outpaces their rate of development and that, in consequence, the major escarpments related to faults must be of fault-line character.[5] It has also been proposed that some suggested fault scarps are the result of slope recession from fault lines. Thus the broad depression occupied by the Red Sea, for instance, is said to be bordered not by

Plate 4.6 The major relief of the Sinai Peninsula and the adjacent regions is dominated by faults. The Peninsula, some 400 km from north to south, is bordered on the north by the Mediterranean Sea (1), on the west by the Gulf of Suez (2), and on the east by the Gulf of Akaba (3). These last two appear to be bifurcations typical of those found at the termini of many rift valleys. In this case the main graben is occupied by the Red Sea (4). The Akaba arm extends northward as a distinct graben (5) bounded by parallel linear scarps and underlain by thousands of metres of unconsolidated strata. The Dead Sea (6) occupies part of this structure. Note the rugged Red Sea Hills (7), the much dissected granite hills (see Fig. 3.3b) in southern Sinai (8) and the numerous rivers and streams which are spawned in these uplands but die out on reaching the desert plains. (*NASA, Photo 66-H-939/66-HC-1701. Gemini space photograph, astronauts Conrad and Gordon, 12–15 September, 1966.*)

Plate 4.7 The Wellington Fault looking southwest from over Petrone, North Island, New Zealand. The linear scarp delimits the alluviated Hutt Valley in the foreground and Wellington harbour beyond. The city of Wellington is in the distance. (*N.Z. Govt.*)

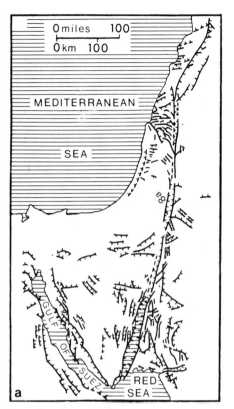

Fig. 4.5 Major faults of the Sinai region. (*After L. Picard,* The World Rift System, *opposite p. 22.*)

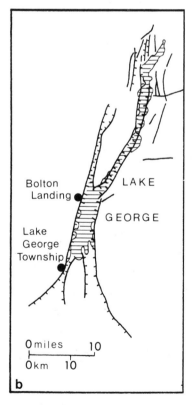

Fig. 4.6 The Lake George Rift, northern New York State.

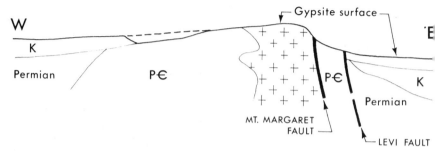

Fig. 4.7 Strata displaced by the Mt Margaret and Levi faults, Peake and Denison Ranges, South Australia. (*After H. Wopfner, "Cretaceous Sediments on the Mt Margaret Plateau and Evidence for Neotectonism",* Q. Notes Geol. Surv. S.A., *1968.*)

Fig. 4.8 Linear fault scarps dissected to form a series of aligned triangular facets: on the west of Ruby Range, Nevada. (*Drawn from photograph in R.P. Sharp, "Basin-Range Structure", 881–919, by permission of the Geological Society of America.*)

Fig. 4.9 Recurrent movement of faults leads to stream rejuvenation and the development of valley-in-valley features in the form of "wineglass valleys".

Plate 4.8   The fault scarp associated with the Awatere Fault in South Island, N.Z., displays prominent triangular facets. The higher ground on the right, on the upthrow side of the reverse fault, is occupied by Jurassic greywackes; the valley and the low hills on the left are underlain by Cretaceous strata (see Fig. 4.13).   (*N.Z. Geol. Surv. Photo NZGS A164C.*)

Plate 4.9   Sandstone ridge of ABC Quartzite in the central Flinders Ranges, South Australia, displays prominent triangular facets, regularly spaced, though there are no strike faults. The facets are eroded by streams which have lowered the shale and siltstone outcrops to the right (west) and have penetrated into the sandstone where the bedding planes have been breached by V-shaped valleys, leaving the triangular facets between.
(*S.A. Dept Lands.*)

Fig. 4.10 Wineglass valley eroded in a fault scarp south of Furnace Creek, Death Valley, California. Note the low scarp in unconsolidated fanglomerate—a low (30–50 cm) fault scarp of recent age.

Fig. 4.11 Recent fault scarps in fanglomerate—alluvial fan sediments—near Badwater, Death Valley, California.

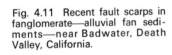

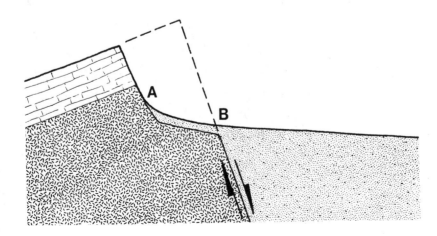

Fig. 4.12 Recession inferentially by a distance of about 1 km (between A and B) of the scarp initiated by the Scandia Fault, New Mexico. (*After Luna B. Leopold, M. Gordon Wolman and John P. Miller,* Fluvial Processes in Geomorphology, *W.H. Freeman and Company, copyright* © *1964, p. 16.*)

fault scarps but by linear escarpments of erosional character. The latter originated in a narrow fault-bordered depression which occupied the structural centre of the area, but which, before the sea entered the region, retreated as a result of subaerial processes—namely, scarp retreat and pedimentation.[6]

While it is true that the observed fault dislocations have been small, it must be borne in mind first that measured or observed displacements do not go back very far in time; and second, that the present rate of activity is not necessarily typical of all geological time. There may have been more active faulting in the recent geological past, just as the Tertiary saw a much higher level of volcanic activity than is taking place now. Seen in a geological context and considering their recurrent nature, even those small displacements which have been observed are adequate to explain many fault scarps. This applies particularly to those features which occur in areas such as the American Southwest where, because of the prevailing aridity, weathering and erosion are comparatively slow and where, in consequence, the preservation of even fragile features is possible (see Fig.4.11). This is not to suggest that the scarps have suffered no erosion, for this is manifestly not so; there has not only been dissection but also considerable recession of some fault scarps (Fig. 4.12). But if a scarp was originally produced by fault dislocation, it is deemed to warrant the name *fault scarp*.[7] Such a definition of fault scarps has in most cases proved satisfactory, though it seems rather extended in the case of escarpments (like that shown in Pl. 4.8) which are associated with reverse faults and which are formed by the marked erosion of the protecting lip of the overthrust block (Fig. 4.13).

Some fault scarps develop in isolation; more commonly patterns of faults and related tectonic scarps subdivide the crust into blocks, of which some are raised and others depressed by movements along the faults. Many of these fault-delineated blocks have a regular shape, being long and narrow and bounded by parallel fault scarps. Raised blocks of this type are called *horsts*, and sunken blocks are called *rift valleys* or *grabens*. The best-known examples of grabens are in the central Africa-Levant region (Fig. 3.2) but others have been described from many parts of the world—Iceland, the Paraiba valley of Brazil, the Lake Baikal depression of central Asia, in the Wasatch Mountains of Utah, the Klamath Lake area of Oregon, Lake George in New York State, the Rhine and Rhone valleys of western Europe (Figs 4.6 and 4.14) and, as an example of a small-scale structure, the Aurès Mountains of Algeria, where grabens are developed in the crests of plunging anticlines (Fig. 4.15). Elongated downfaulted depressions, either graben or fault-angle depressions (see below) which form local basins of interior drainage in the arid American Southwest and in the deserts of Chile, are called *bolsons*.

Examples of horsts include Mt Ruwenzori in central

Plate 4.10 The Meckering earthquake occurred in October, 1968. There was local uplift along fractures in the crust, development of a number of gaping fissures and a discontinuous crenullated scarp which attained heights of 2 m in places and which could be traced in an arc some 47 km long. (*W.A. Newspapers.*)

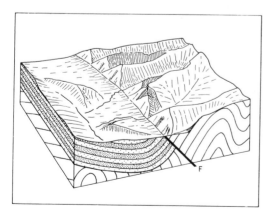

Fig. 4.13 Erosion of the overhanging lip of the overthrust side of the Awatere Fault, South Island, New Zealand, gives a misleading picture of the nature of the dislocation. (*After P. Birot (after C.A. Cotton),* Morphologie Structurale, *Presses Universitaires de France, Paris, 1958, Vol. 1, p. 152.*)

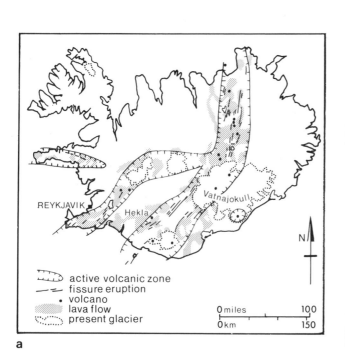

a

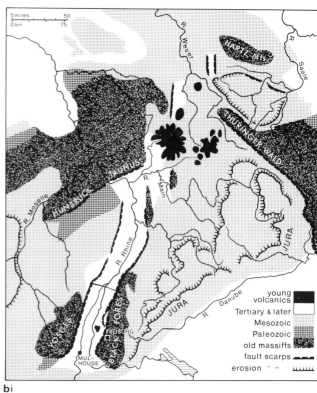

bi

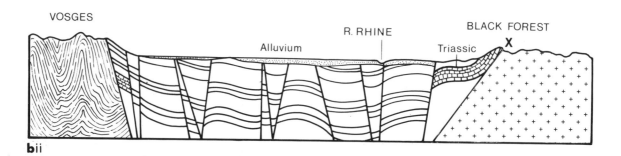

bii

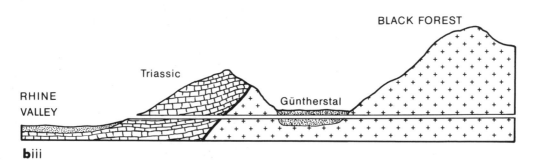

biii

Fig. 4.14 Graben: (a) Iceland; (*after S. Thorarinsson, "The Median Zone of Iceland",* The World Rift System, *pp. 187–211.*) (b) the Rhine at Mulhouse, with detail of scarp near Freiburg: (bi) in plan, (biii) shows detail of eastern fault in location similar to X in (bii); (*After M.R. Shackleton,* Europe, *Longmans, London, 6th ed., p. 276.*)

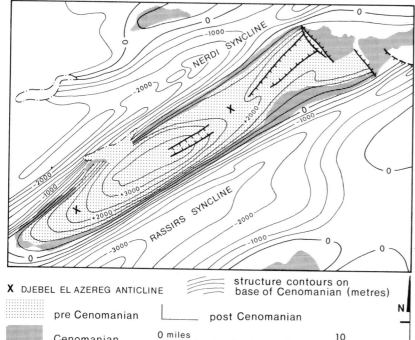

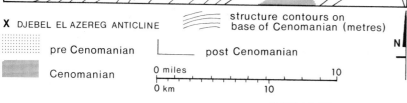

Fig. 4.15 The Aurès Mountains, northern Algeria, showing a graben developed in the crest of an anticlinal structure. (*After R. Lafitte, "Étude géologique de l'Aurès",* Bull. Serv. Carte geol. Alger., *1939 (2e Series Stratig. Descript. Reg.) 15; reprinted by permission of Laboratoire de Géologie du Musée National d'Histoire Naturelle.*)

X DJEBEL EL AZEREG ANTICLINE

pre Cenomanian

Cenomanian

structure contours on base of Cenomanian (metres)

post Cenomanian

N

0 miles 10
0 km 10

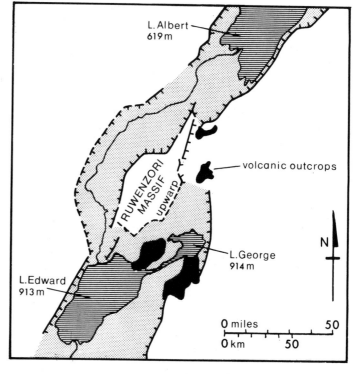

Fig. 4.16 The Ruwenzori Horst, central Africa. (*After A. Holmes,* Principles of Physical Geology, *p. 1070.*)

L. Albert 619 m

volcanic outcrops

RUWENZORI MASSIF upwarp

L. George 914 m

L. Edward 913 m

N

0 miles 50
0 km 50

Africa (Fig. 4.16), the Vosges and Black Forest areas of Europe, parts of the Wasatch Mountains,[8] the Ruby-Humboldt Range in Nevada,[9] and the Mt Lofty Ranges in South Australia[10] (Fig. 4.17).

The origin of grabens (and, by implication, horsts) has given rise to some controversy. Some workers consider that grabens are, in effect, the fallen crests of arches formed by the raising and extension of parts of the crust. There is some observational and experimental support for this theory. For example, the faults bordering both the central African and Rhine rifts appear to be of normal type, and thus to indicate tensional effects (Fig. 4.18a). But other workers have pointed out that most rift valleys are areas of negative gravity anomaly. They are underlain by blocks which are lighter than the bordering areas, and which should not be depressed, unless they are held down by the adjacent blocks which have partly overridden the central block. In this case, the bounding faults must be of reverse type and involve compression (Fig. 4.18b). Again, there is some support not only from the observed gravity anomalies but also from the occurrence within complexes of grabens or horsts like Mt Ruwenzori and Okhor Island (within the Baikal complex). However, volcanoes are commonly found in association with rift valleys and it is difficult to see how magma could make its way to the surface through a zone of compression. Furthermore, in northern Arabia and the Levant it has been found that grabens are underlain by at least 3000 m (and probably 10,000 m) of unconsolidated sediment: evidently as the floor of the structure subsided, detritus was washed and blown in. Similar infillings of unconsolidated sediments have been found in the graben structures of the Wasatch Mountains of Utah (Fig. 4.19). The point is, however, that the presence of such large thicknesses of unconsolidated debris could well account for the negative gravity anomalies of the grabens. Indeed, it may be argued that being zones of tension, denser simatic material has in many places welled up beneath the rift valleys, sometimes reaching the surface as volcanic extrusions, but that the positive gravity anomalies caused by the presence of such rocks just below the surface is more than compensated for by the occurrence of the thick sedimentary sequences in the grabens.

Though either compression or tension may have to be invoked in explanation of a particular graben according to the local evidence, it seems likely that many, probably most, rift valleys are related to tension in the crust.

Other downfaulted areas are less regular in shape. Such irregular fault blocks occur in the South Australian rift or shatter zone, around Port Phillip Bay[11] and in the Kinki depression of central Japan[12] (Fig. 4.20). The Good Friday earthquake of 1964, which affected large areas of southern Alaska, caused a major dislocation of fault blocks that are extremely irregular in shape (Fig. 4.21).[13] Parts of the area were uplifted by 15 m during the earthquake.

Southern Chile and the adjacent submarine floor occupies an extremely unstable area with two parallel meridional zones (Fig. 4.22), each characterised by earthquake activity, recurrent faulting and volcanic activity. During the earthquakes of May and June, 1960, there were marked fault

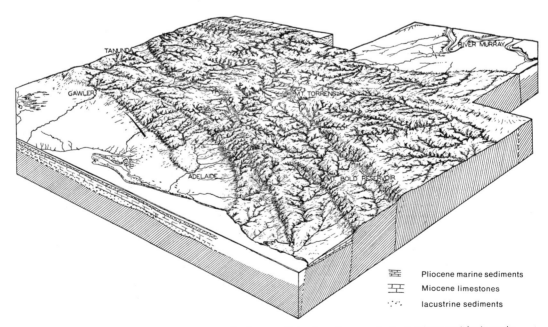

Pliocene marine sediments

Miocene limestones

lacustrine sediments

Fig. 4.17   Block diagram of the Mt Lofty Ranges, S.A., showing arcuate fault scarps and fault angle depressions. (*After R.C. Sprigg, 'Some Aspects'', pp. 277–303, by permission of R. Soc. S. Aust.*)

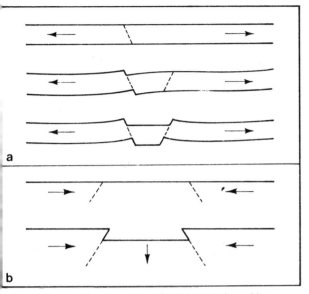

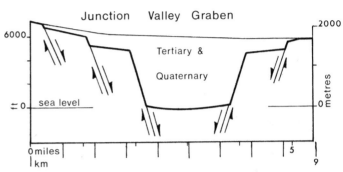

Fig. 4.18 Grabens developed in association with (a) tension in the crest of a broad arch; and (b) compression.

Fig. 4.19 Sections through the Wasatch Mts, Utah, showing normal faults and infilled valleys. (*After K.L. Cook,* The World Rift System, *p. 264.*)

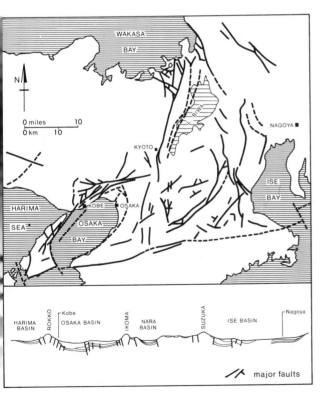

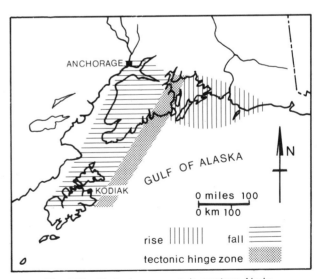

Fig. 4.20 Fault-controlled relief: the Kinki depression, central Japan; (*after N. Ikebe and K. Ishikawa, "Geologic Sketch", pp. 135–48.*)

Fig. 4.21 Vertical movements in southern Alaska caused by the Good Friday earthquake, (*After Grantz et al., "Alaska's Good Friday Earthquake".*)

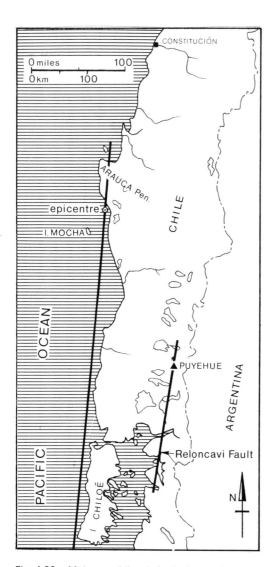

Fig. 4.22 Major meridional faults in southern Chile. (*After J.H. Hodgson,* Earthquakes and Earth Structure, © *1964, p. 38. Reprinted by permission of Prentice-Hall, Inc., Englewood Cliffs, N.J.*)

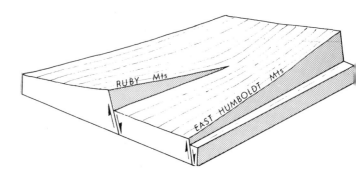

Fig. 4.23 Fault angle depression and scarp associated with a hinge fault, Ruby Range, Nevada. (*After R.P. Sharp, "Basin-Range Structure", pp. 881–919; by permission of the Geological Society of America.*)

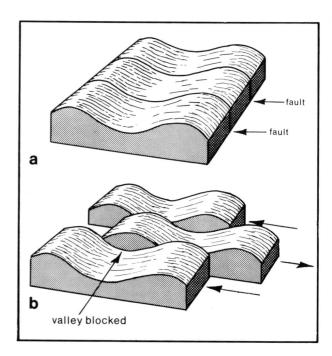

Fig. 4.24 Development of shutterridges by transcurrent faulting.

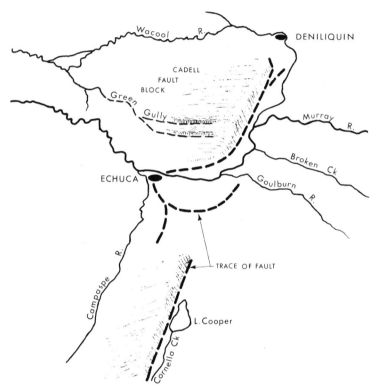

Fig. 4.25 The Cadell Fault Block and diversion of the ancestral Murray near Echuca, Victoria.

displacements: Isla Mocha for example rose by 2·5 m and the western side of Arauco Peninsula by 1·5 m, though to the south the coast subsided between one and two metres. So widespread and pronounced were these changes in land level that one observer commented: "uplift and subsidence along the coastline caused by the earthquake and subsequent tsunamis [tidal waves caused by submarine earthquakes and instantaneous dislocations of the sea-floor] rendered all marine navigational charts of the affected area obsolete." [14]

In some areas, such as the Ruby Mountains of Nevada [15] and the southern Mt Lofty Ranges, [16] the fault blocks have subsided unevenly, so that their surfaces are tilted (see Fig. 4.17). Downthrow on the east side of the Ruby Range averages 1800–1900 m compared to 650 m on the west, so that the block as a whole is tilted to the west. Moreover, a hinge fault has given rise to the fault splinter—the East Humboldt Mountains (Fig. 4.23). Repetition of such uneven movement on adjacent blocks gives rise to *fault-angle valleys* or *depressions*, which are asymmetrical in cross-section, being bounded by the fault scarp of one block and the tilted backslope of the one adjacent.

Dislocation along wrench or transcurrent faults may

cause displacement of ridges, and, if the movement is of an appropriate amount, valleys may be blocked by translocated ridges; these are called *shutterridges* [17] (Fig. 4.24). In some instances drainage within the blocked valleys may be impounded to form lakes. Other lakes, called *sag ponds*, form through unequal total uplift along faults. In some areas drainage is impeded simply by the rise of a fault block across a pre-existing stream line, though in other places streams have been able to maintain their flow across uplifted blocks. Lake Cooper is caused by the rise of the Cadell fault block, near Echuca, in northwestern Victoria, during the late Pleistocene. [18] More important, this uplift caused the River Murray to be diverted, for the most part around the southern end of the main fault block (Fig. 4.25).

Streams may be disturbed also by wrench faulting. Many streams along the Alpine fault zone in New Zealand, in the San Andreas fault zone of California and in western Sumatra, have maintained their courses across fault zones even while lateral movements have taken place along the fault. As a result, they have become offset (Fig. 4.26, Pl. 4.11). Surfaces of low relief and river terraces have also been dislocated in places by faulting (Fig. 4.27).

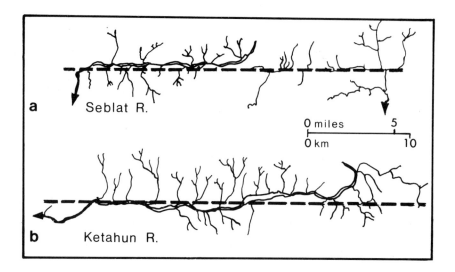

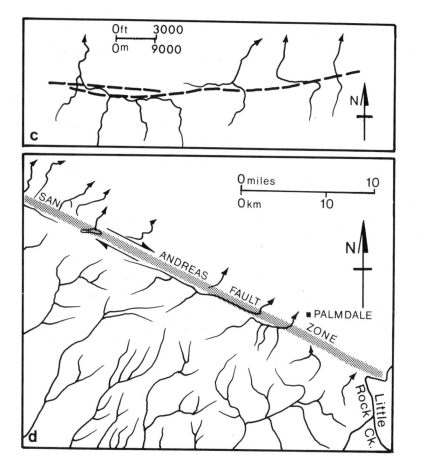

Fig. 4.26 Examples of offset streams: (a) and (b) associated with the Sumatra Fault (see also Fig. 6.1); (*after J.A. Katili and F. Hehuwat, "On the Occurrence of large Transcurrent Faults in Sumatra",* J. Geosci. Osaka Univ., *1967,* **10** *(1), pp. 5–17.*) (c) and (d) associated with the San Andreas system, (c) being related to the Garlock Fault; (*after R.E. Wallace, "Structure of a Portion of the San Andreas Rift in Southern California",* Bull. geol. Soc. Am., *1949,* **60***, pp. 781–806, by permission of the Geological Society of America.*)

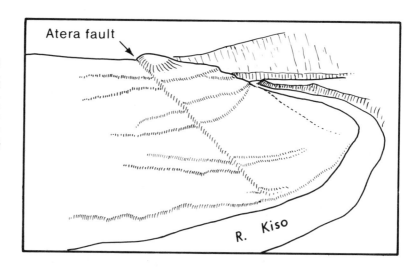

Fig. 4.27 The fault dislocated terraces along the Kiso River, central Japan, due to transcurrent movements along the Atera Fault. (*After A. Sugimura and T. Matsuda, "Atera Fault and its Displacement Sectors",* Bull. geol. Soc. Am., *1965,* **76**, *pp. 509–22, by permission of the Geological Society of America.*)

Plate 4.11 The San Andreas Fault of California is both recurrent and transcurrent; it has moved several times, and continues to do so, mainly in a horizontal sense. Offset streams such as these near Taft, in southern California, are common: as the blocks on opposite sides of the fault have gradually moved laterally, the rivers which cross the fracture zone have been able to maintain their courses, but only by developing the pronounced right-angled turns they now display. The ridges bordering the linear fault zone may be due to local compression and upthrust. (*R.C. Frampton and J.S. Shelton.*)

References Cited

1. H. O'R. STERNBERG and R.J. RUSSELL, "Fracture Patterns in the Amazon and Mississippi Valleys", *Int. Geogr. Union Proc.*, Washington, 1952, pp. 380–5.

2. H.N. FISK, *Geological Investigation of the Alluvial Valley of the Lower Mississippi River*, Corps Eng. U.S. Army, Vicksburg, 1944; H.O'R. STERNBERG and R.J. RUSSELL, "Fracture Patterns", pp. 380–5.

3. E. KRENKEL, *Geologie Afrikas*, Borntraeger, Berlin, 1925, Fig. 4; E.S. HILLS, "Some Aspects of the Tectonics of Australia", *J. & Proc. R. Soc. N.S.W.*, 1945, **79**, pp. 67–91; *idem*, "A Contribution to the Morphotectonics of Australia", *J. geol. Soc. Aust.*, 1955, **3**, pp. 1–15; *idem*, "Die Erdoberflache Australiens", *Die Erde*, 1955, **7**, pp. 195–205; F.A.V. MEINESZ, "Shear Patterns in the Earth's Crust", *Trans. Am. geophys. Un.*, 1947, **28**, pp. 1–61.

4. G.B. LEECH, "The Rocky Mountain Trench" in T.N. Irvine (Ed.), *The World Rift System: Report of Symposium Upper Mantle Project*, Ottawa, Canada, 4–6 September 1965, pp. 307–29. Report published as Geol. Surv. Can. Paper 66–14. (Hereafter referred to simply as *The World Rift System*.)

5. See for example F. DIXEY, "The East African System", *Overseas Geol. and Min. Resour. Bull.*, Suppl. 1, 1956, London.

6. A.J. WHITEMAN, "Formation of the Red Sea Depression", *Geol. Mag.*, 1968, **105**, pp. 231–46.

7. W.M. DAVIS, "Nomenclature of Surface Forms on Faulted Structures", *Bull. geol. Soc. Am.*, 1913, **24**, pp. 187–216; D.W. JOHNSON, "Fault Scarps and Fault Valleys", *Bull. geol. Soc. Am.*, 1939, **2**, pp. 174–7; C.A. COTTON, "Tectonic Scarps and Fault Valleys", *Bull. geol. Soc. Am.*, 1950, **61**, pp. 717–58; *idem*, "Tectonic Relief; with Illustrations from New Zealand", *Geogr. J.*, 1953, **119**, pp. 213–22.

8. K.L. COOK, "Rift System in the Basin and Range Province", in *The World Rift System*, 1966, pp. 246–90.

9. R.P. SHARP, "Basin-Range Structure of the Ruby-East Humbolt Range, North East Nevada", *Bull. geol. Soc. Am.*, 1939, **50**, pp. 881–919.

10. W.N. BENSON, "Note Descriptive of a Stereogram of the Mt Lofty Ranges, South Australia", *Trans. R. Soc. S. Aust.*, 1911, **35**, pp. 108–11; R.C. SPRIGG, "Some Aspects of the Geomorphology of Portion of the Mount Lofty Ranges", *Trans. R. Soc. S. Aust.*, 1945, **69**, 277–303.

11. E.S. HILLS, *Physiography of Victoria*, Whitcombe and Tombs, Sydney, 1940, p. 161.

12. N. IKEBE and K. ISHIKAWA, "Geologic Sketch of the Kinki District, central Japan", *J. Geosci*, 1967, **10**, pp. 135–48.

13. A. GRANTZ, G. PLAFKER and R. KACHADOORIAN, "Alaska's Good Friday Earthquake, March 27, 1964", *U.S. Geol. Surv. Circ.*, **491**, 1964.

14. P. SAINT-AMAND, 1961, p. 3, cited in V. Auer, "The Pleistocene of Fuego—Patagonia", *Ann. Acad. Scient. Fennicae*, 1970, A (III–Geol. & Geogr.) p. 100.

15. R.P. SHARP, "Basin-Range Structure", pp. 881–919.

16. R.C. SPRIGG, "Some Aspects of the Geomorphology", pp. 277–303; M.F. GLAESSNER, "Conditions of Tertiary Sedimentation in southern Australia", *Trans. R. Soc. S. Aust.*, 1953, **76**, pp. 141–6.

17. J.P. BUWALDA, "Shutterridges, Characteristic Physiographic Features of Active Faults", *Proc. geol. Soc. Am.*, 1936, **937**, p. 307.

18. W.J. HARRIS, "Physiography of the Echuca District", *Proc. R. Soc. Vic.*, 1939, **51**, pp. 45–60; J.M. BOWLER and L.B. HARFORD, "Quaternary Tectonics and the Evolution of the Riverine Plain near Echuca, Victoria", *J. geol. Soc. Aust.*, 1966, **13**, pp. 339–54; J.M. BOWLER, "Quaternary Chronology of the Goulburn Valley Sediments and their Correlation in southeastern Australia", *J. geol. Soc. Aust.*, 1967, **14**, pp. 287–92.

General and Additional References

P. BIROT, *Morphologie structurale*, Vol. 1, Presses Universitaires de France, Paris, 1958.

A. HOLMES, *Principles of Physical Geology*, 2nd ed., Nelson, Edinburgh, 1965.

C.R. TWIDALE, *Structural Landforms*, Australian National University Press, Canberra, 1971.

CHAPTER 5

Landforms Developed on Sedimentary Sequences

A. FLAT-LYING STRATA

In many parts of the world the landscape is dominated by flat-topped residual hills. How do these originate? Field observations show that in most, though not all, cases the residuals are underlain by strata which are flat-lying or nearly so. Moreover, the upper, flattish surface of the residuals is underlain directly or at shallow depth by a resistant formation, commonly sandstone. Thus these flat-topped hills are a direct expression of underlying horizontal, or near-horizontal, resistant strata which have suffered limited dissection.

Exposed parts of the former sea-floor are subject to weathering and erosion as soon as they emerge into the wave zone or are exposed to the atmosphere. Emerged sea-floors which are underlain by marine strata and only slightly modified by erosion are called *sediplains* (R.O. Brunnschweiler, cited in Jennings[1]). The present surface of the Nullarbor Plain (Pl. 5.1) is said to be of this type. The thin Miocene marine limestone extends over the length and breadth of the flat plain, 165,000 km[2] in area, having been eroded only a little after uplift.[2] Its preservation is due partly to the prevailing aridity and partly to the propensity of the limestone to swallow water (see Chapter 3). The High Plains of the west-central USA are likewise considered "a structural slope, superficially modified by streams" and with a uniform easterly dip.[3]

More commonly, however, uplifted sedimentary sequences are eroded by streams. If the uppermost strata are unconsolidated or otherwise susceptible to erosion, they are stripped off until a resistant stratum is revealed. This then forms a *structural plain* protected by the resistant formation. But as the rivers cut into the surface, the resistant stratum is left higher and higher above the valley floors; it forms a *caprock* which protects flat-topped hills.

Such hills are bounded by escarpments and are called *plateaux* (Pls 5.2, 5.3a and b). The slopes which delimit the plateaux are of faceted type because of the presence of the resistant stratum, and are maintained at steep inclinations until the caprock is eliminated. In some areas, softer beds persist above the caprock. They are protected from dissection by the resistant rock of the bluff, and give rise to *domed plateaux* (Fig. 5.1a and b). Strata which are only moderately resistant form more gentle slopes, and moreover are more readily dissected by streams. They give rise to upland areas of rolling or undulating relief and are called *high plains*.

The upper surfaces of flat-topped plateaux are not greatly affected by weathering and erosion because of the presence of the caprock, and there is only negligible lowering of the upper surface. The principal means of reducing the areas of the plateaux is by marginal attack. The scarp-slopes are weathered, and worn back through mass movement of debris and rill and stream erosion. Caverns developed at the base of the bluff are instrumental in undermining

Plate 5.1 The Nullarbor Plain is a sediplain of remarkable flatness; so much so that the Transcontinental Railway stretches without deviation for almost 500 km. (*Dept Int., Canberra.*)

Plate 5.2 The dissection of flat-lying Precambrian ironstone formations in the Hamersley Ranges has resulted in the development of plateau, mesa and butte assemblages. Those shown here include typical examples of each of the three forms, plus lower, less well-defined forms of similar type and a pediplained surface. (*Robin Smith Photos.*)

Plate 5.3a   A sandstone capping gives rise to these classic examples of plateaux and mesas bounded by faceted slopes in the Tassili Mountains of southern Algeria. Note the joint control in detail of the form of the plateau and the bluff, and the sandy wadi floor. (*P. Rognon.*)

Plate 5.3b   This massive flat-lying sandstone formation in Jordan has been greatly dissected to give an intricate system of deep valleys and sand plains separating sandstone ridges and towers, some of which are surmounted by conical or pyramidal masses, though others are reasonably planate. (*Hunting Surveys.*)

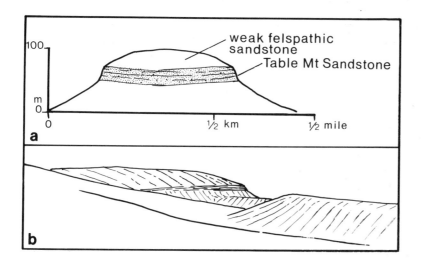

Fig. 5.1a Domed plateaux in southern Africa. (*After L.C. King, "Scarps and Tablelands", Z. Geomorph., 1968, 12NS., pp. 114–5.*)

Fig. 5.1b Domed plateaux in the Arcoona Plateau of South Australia.

Fig. 5.2 Isolated butte of flat-lying sediments in southern Algeria, displaying well-defined bluff and debris slope, with pedimented plain below. The parent plateau is in the background. (*Drawn from a photograph supplied by Hunting Surveys.*)

Plate 5.4 Headward erosion of gullies attacking this granite slope in northwest Queensland has caused the Mesozoic siltsone (which is capped by a ferruginous duricrust of Tertiary age) to be undermined and to collapse. It has been eliminated in some areas and attack from two sides has formed a natural bridge at one point and marked caverns or shelters elsewhere. (*C.R. Twidale/C.S.I.R.O.*)

Plate 5.5a   Sandstone   towers   near
Jabel Ram, Jordan. (*Hunting Surveys.*)

Plate 5.5b   Sandstone   towers   in   the   Tassili,   southern   Algeria.
(*P. Rognon.*)

and causing the collapse of the protective caprock[4] and joints are also exploited (Pls 5.3 and 5.4). Retreat of the bounding slopes causes a reduction in the area of each plateau, and, with the headward extension of the drainage network, the once extensive plateau is subdivided into a number of smaller plateaux. Though the point of transition is subjective, plateaux of only limited extent are called *mesas**　(Pl. 5.2, Fig. 5.2). Erosion of the escarpments continues, and when the maximum diameter of the top of a residual is less than its elevation above the adjacent plain, the mesa conventionally becomes a *butte.*** At this stage, the rounding of the upper part of the bluff, which is of negligible significance on plateaux and mesas, becomes relatively important. The plateau top gives way to a rounded upper slope which may, in some areas, be so well developed that the flat, upper surface is virtually eliminated. Such forms are especially well displayed on thick sequences of lithologically uniform sandstone (Pl. 5.5a and b). Continued erosion of the scarps causes the formation of pillars and towers, many of which exhibit rather unusual forms in detail (Pls 5.6 and 5.7). Eventually, with the elimination of the caprock, the residual takes on a conical shape, and its elevation above the plain is gradually reduced. Until the destruction of the caprock, however, the scarp is maintained at a constant steep inclination.

The morphology of the slopes which delimit the plateaux, mesas and buttes varies with the character of the sedimentary sequence. In areas where the exposed sequence is in broad terms uniformly resistant, as in the Blue Mountains, west of Sydney (Pl. 3.22b), in the Tassili Mountains of the Sahara, or in the Yellowstone National Park of Wyoming, the scarps are of a simple faceted type and consist of an upper, rounded slope; a bluff of steep inclination in which joint-control is prominent in detail; and a so-called debris slope, which in most areas is a bedrock slope with a veneer of detritus, which is of gentler inclination than the bluff and which leads down to the valley floor or adjacent plain (Pl. 3.18). Such simple escarpments reflect structural simplicity. In areas where the sequences are more complex in the sense that strata of contrasted resistance are interbedded, as in some areas of the Kimberleys of Western Australia (Pl. 5.8) and in the Grand Canyon of Arizona (Pl. 5.9), each resistant rock outcrop (e.g. sandstone, dolerite, limestone) gives rise to a bluff and each exposure of weaker rock (e.g. shale, siltstone) to a debris slope, so that the escarpment as a whole presents a stepped appearance and consists of a composite of several bluffs and several debris slopes.

Plateau, mesa and butte assemblages are developed on flat-lying sedimentary sequences in many areas, particularly on the platforms and on undisturbed sedimentary sequences within the cratonic massifs. Examples of the former include the Tassili Mountains of the Sahara, and of the latter, the Mueller plateau of the Northern Territory and the Hamersley Ranges of Western Australia.

But in some areas similar landform assemblages have evolved on folded strata, or on, say, granite bedrock. For

Plate 5.6　Chambers Pillar, a sandstone butte in Australia's Northern Territory. This photograph was taken on the Barclay Expedition of 1911. Note the vertical sides of the butte developed in the more massive sandstone, and the gentler slopes eroded in the finer beds below. (*Dept Int., Canberra.*)

Plate 5.7　The Mukorob (Hottentot for "Finger of God"), an unusual sandstone residual in the Mariental area, South West Africa (*W.S. Barnard.*)

---

* Spanish—table.
** French—knoll, mound or hillock; a residual hill or tower which is, as the term is now used in the English language, flat-topped.

Plate 5.8 Structural plateaux, structural benches and faceted slopes in the Kimberleys, Western Australia. (*Dept Int., Canberra.*)

Plate 5.9 Deep dissection of a thick sedimentary sequence by the Colorado River in Arizona has formed the Grand Canyon, about 1·5 km deep, with complex faceted slopes due to the exposure of rocks of varied resistance, and bordered by plateaux surrounded by conical sandstone peaks. Note the ribbed outcrops in the foreground—a detailed expression of varied rock types and their resistance to weathering and erosion. (*American Airlines.*)

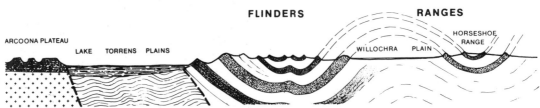

Fig. 5.3 Diagrammatic cross-section through the southern Arcoona Plateau and Flinders Ranges showing the contrasted forms developed on flat-lying and folded sedimentary sequences. Note the relief inversions in the Flinders.

Plate 5.10 These mesa and butte forms in northwest Queensland are formed by the dissection of a laterite or ferruginous duricrust developed on granitic bedrock. (*C.R. Twidale/C.S.I.R.O.*)

instance, in the Victoria River basin of the Northern Territory, in the Western Districts of Victoria and in many other parts of eastern Australia, in the Columbia River catchment of the northwestern USA and in the Drakensberg escarpment of South Africa, basalt flows form the caprock which underlies the plateaux and which protects the faceted slopes. In many parts of Australia, India and other tropical regions, duricrusts (see Chapter 10) of various types form similar hard cappings on folded or crystalline rocks and, when dissected, give rise to these same characteristic landforms (Pl. 5.10, see also Chapter 10).

B. **FOLDED AND WARPED SURFACES AND SEDIMENTARY SEQUENCES**

The Arcoona Plateau of South Australia, particularly in its southern areas, displays domed plateaux, and mesa and butte assemblages characteristic of flat-lying sedimentary sequences. The Precambrian strata include several massive quartzites which form prominent caprocks in various areas. The sedimentary sequence extends and thickens to the east, where it outcrops in the Flinders Ranges (Fig. 5.3). There, however, the strata were laid down in the Adelaide Geosyncline and suffered folding and faulting, whereas in the Arcoona Plateau, underlain by the relatively stable Stuart Shelf (a part of the Westralian Shield), the strata are virtually undisturbed. This contrast in attitude of the beds in the Arcoona Plateau and in the Flinders Ranges is quite apparent in the field (cf. for instance Pl. 5.11a and b). It is also evident in the form of the land surface. In the Flinders Ranges the strata are tilted, ridges and valleys predominate, and there is a direct relationship between the pattern of folding and the pattern of ridge and valley (Fig. 5.4).

The folding of a sedimentary sequence, however, influences far more than dip and strike, and the pattern of outcrop. Folding introduces stresses in addition to those which caused the compression of the strata in the first place. The significance of these stresses extends far beyond the consideration of folded sedimentary sequences (see Chapter 3).

In a sequence of newly-folded strata, the anticlines form ridges and the synclines valleys. The shape of the land surface is a direct reflection of the underlying structures. Though it is difficult to prove the folding of surfaces (as opposed to the folding of strata), it has been possible to do so in some areas. Repeated precise topographical surveys taken over the past half-century in southern California in connection with oil exploration and drilling have revealed quite strong warping of the land surface. The movements are, in some considerable measure, due to the extraction of oil from the underlying strata (see Chapter 26) but some have been related to recorded earthquakes. For instance, there was a moderately strong earthquake in the Long Beach area on 10 March, 1933. It apparently caused the Alamitos plain, east of the city, to be bowed upwards in a gentle arch 18 cm high and about 7 km across (Fig. 5.5a). Between San Bernadino and Victorville, the Mojave desert surface has risen in a broad arch some 20 cm high in about forty years. [5] Similar, seemingly minute vertical movements have been recorded in several parts of Europe, again by means of repeated, precise surveys. Very broad flexing of the land

Plate 5.11a Flat-lying, flaggy sandstones of Proterozoic age in Dutton Bluff, Arcoona Plateau, South Australia. (*C.R. Twidale.*)

Plate 5.11b Folded siltstones and shales of Proterozoic age in the central Flinders Ranges, South Australia. (*C.R. Twidale.*)

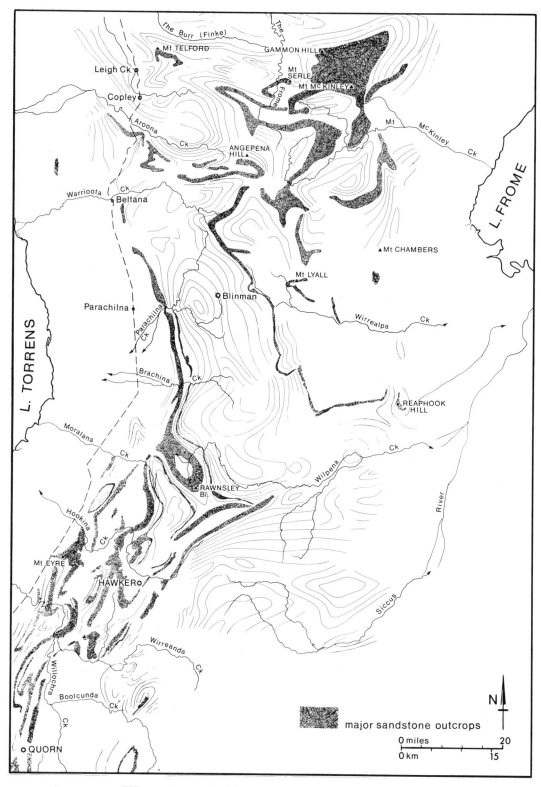

Fig. 5.4   Major fold patterns of the Flinders Ranges, with sandstone formations stippled and regional strike lines shown.

Fig. 5.5a Vertical movements, in feet, near Los Angeles—probably during an earthquake in 1933. The isobases are derived from a comparison of levels taken in 1931 and 1933–34, between which the earthquake occurred. (*After J. Gilluly, "Distribution of Mountain Building" p. 561, by permission of Geological Society of America.*)

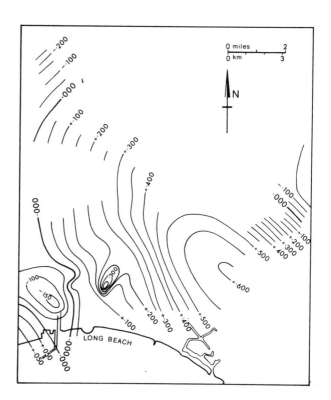

surface has been recorded in France[6] and in the western USSR[7] (see Fig. 5.5b and c) but in Hungary (an area occupying an intermontane basin between active fold mountain belts) the sedimentary plain surface has suffered intricate rises and falls.[8] The flexures, some of which are quite steep, seem to be related to patterns of faulting in the consolidated rocks which underlie the plain (compare for instance the maps of Bendefy[9] and Szentes and Ronai[10]).

Maps demonstrating geologically recent earth movements have been constructed for Japan[11] and for New Zealand[12] using a variety of evidence including levelling, raised and depressed shoreline features, and so on (Figs 5.6, 5.7, 5.8).

No such detailed surveys are available in Australia or in many other parts of the world. However, warping and flexing of the land surface can be inferred from geological and geomorphological evidence. For instance, in the Puna area of the Chilean Andes, the surface of an ignimbrite deposit displays a distinct monoclinal flexure. Ignimbrite is an acid volcanic ash which when deposited blankets the pre-existing landscape to form flat surfaces. Thus the broad

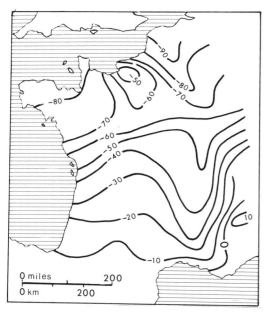

Fig. 5.5b Vertical movements in France, in centimetres, from precise levelling carried out 1857–64 and 1884–93. (*After M. Schmidt, "Neuzeitliche Erdkrustenbewegungen", by permission of* Sber. bayr. Akad. Wiss., math-phys.)

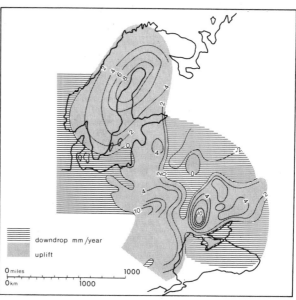

downdrop mm/year

uplift

Fig. 5.5c Present land movements in Scandinavia and European USSR in mm per annum. (*After Y.A. Merscherikov, "Crustal Movements",* Encyclopaedia of Geomorphology, p. 224.)

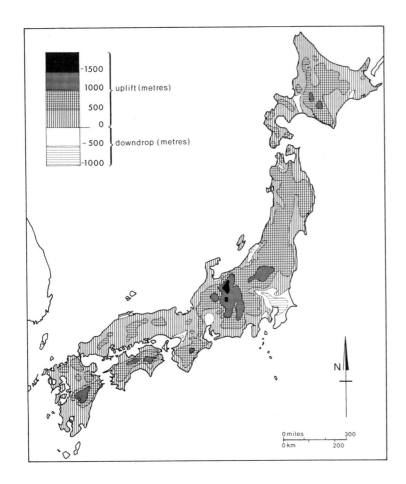

Fig. 5.6 Vertical displacements in Japan during the Quaternary. (*After* Quaternary Tectonic Map of Japan, *Nat. Res. Center Disaster Prevention, Tokyo, 1969.*)

Fig. 5.7 Vertical movements in Japan 1920—28. (*After N. Miyabe, S. Miyamura and M. Misoue, "A Map of Secular Movements", Movements Modernes, pp. 211—2.*)

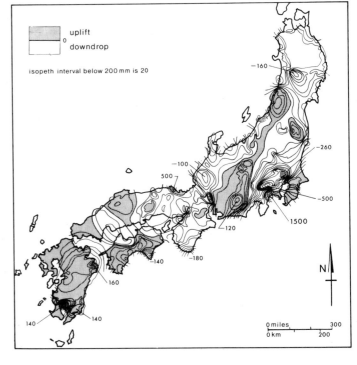

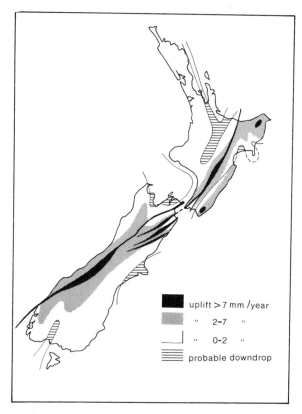

Fig. 5.8 Average rate of uplift and subsidence in New Zealand over the past 10,000 years. (*After H.W. Wellman, "Report on Studies", pp. 34–6, by permission of* Quaternary Res.)

uplift > 7 mm /year

" 2–7 "

" 0–2 "

probable downdrop

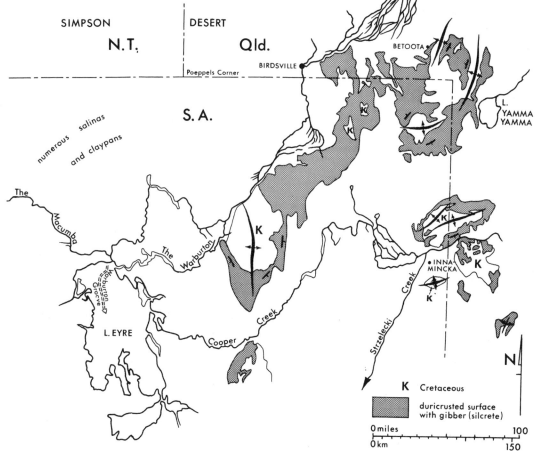

Fig. 5.9 The folded silcrete surface in northeastern South Australia and adjacent parts of Queensland. (*After R.C. Sprigg, "Geology and Petroleum", pp. 35–65, by permission of the R. Soc. S. Aust.*)

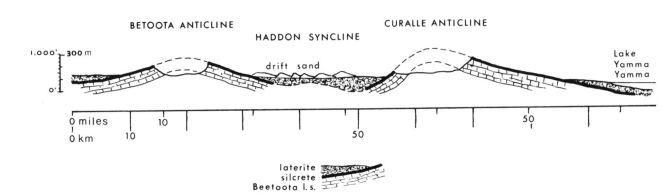

Fig. 5.10   Section through the Betoota and Curalle Anticlines and the Haddon Syncline, showing
the inward-facing escarpment of the breached anticlines and the silcrete surface involved in the
fold structures. (*After H. Wopfner, "On Some Structural Development", pp. 179–94, by permission
of the R. Soc. S. Aust.*)

step now developed in the surface must be due to earth movements, and since field observations[13] show gradual changes of dip in a monoclinal pattern rather than a fracture or series of fractures, warping and not faulting is involved.

The Shaur Hills form a low line of hills standing above the flat flood-plain of the River Karun, in Iran. These hills are caused by the uplift of the Shaur Anticline, which has developed during the past 1700 years.[14] During the first or second century B.C., the then local inhabitants of the area built qanats, or underground water conduits, as well as surface canals to carry irrigation waters. One of the qanats is now distorted in profile along the channel being bowed in an arch 19 m high and 4 km long. A surface channel is now incised below the land surface to a maximum of 4 m. The Shaur Anticline has developed since the construction of the channels—their distortion proves this—and the incision of 4 m by the surface channel is a measure of the amount of uplift, since it was by this amount that the artificial river had to erode its bed in order to maintain its course.

Over wide areas of central and southern Australia a siliceous pedogenic accumulation, known as silcrete (see Chapter 10), formed in the mid Tertiary.[15] Although it occurs in scarpfoot zones, along stream courses and around joint blocks,[16] the really extensive occurrences take the form of massive sheets which apparently formed at or just beneath the land surface. Silcrete forms in valley floors and on other plains of low relief. Though not initially horizontal, it is improbable that the silcrete as originally formed displayed any notable dip. Yet in parts of northeastern South Australia and in the adjacent parts of Queensland and N.S.W. (Figs

5.9 and 5.10), the silcrete surface displays anticlines and synclines, domes and basins, with dips up to 23°.[17] Though considerably dissected, parts of this old land surface, now folded, survive as part of the present landscape. Folded silcrete occurs in the eastern piedmont of the Flinders Ranges where alluvial fans also appear to be warped.[18] Other examples of warped land surfaces are discussed in connection with Pleistocene glaciation and with man's activities (Part IV and V).

Such folded landforms, which owe nothing to weathering and erosion and are due entirely to crustal movements, are therefore tectonic landforms. Inevitably, however, the folded surfaces are attacked by weathering and erosion. It might be thought that because water runs downhill and therefore tends to collect and flow in the initial synclinal depressions, these areas would be eroded most rapidly. In reality it is the crests of the anticlines which are most susceptible to weathering and to erosion. Why should this be so?

The explanation is to be found in the stresses imposed during folding. No sedimentary layer is homogeneous either laterally or vertically, and the weaknesses of any exposed stratum are fully exploited by the various exogenetic forces. In addition, because of folding, the crest of each anticline is placed in tension, and the trough of each syncline in compression. Thus the anticlinal crests are vulnerable because any joints which have developed are open and are easily penetrable by water. In the troughs, on the other hand, joints and other fissures are closed and impenetrable. But at depth these conditions are reversed (Fig. 5.11). Between each zone of tension and compression is a neutral plane

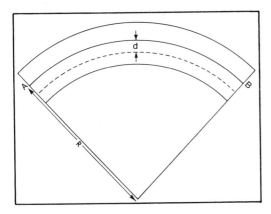

Fig. 5.11 Distribution of strain in an anticlinal fold.

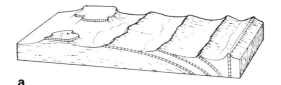

a

Fig. 5.12a Relationship between ridge shape in cross-section and dip of strata. (*After E.S. Hills, Elements of Structural Geology, Methuen, London, 1963, p. 438.*)

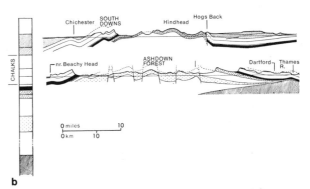

b

Fig. 5.12b Cross-sections through the Wealden Anticline. (*After S.W. Wooldridge and D.L. Linton, Structure, Surface and Drainage in southeastern England, George Philip and Son, London, 1955, pp. 18–19.*)

Plate 5.12 Complex parallel ridges, those in the foreground ribbed, running parallel to each other and straight for some kilometres, and developed in steeply dipping strata, central Australia. (*Aust. Nat. Publ. Assoc.*)

Plate 5.13 Convergent and divergent ribbed ridges and intervening valleys developed on steeply dipping Palaeozoic sediments, James Range, Northern Territory.
(*Dept Nat. Dev., Canberra.*)

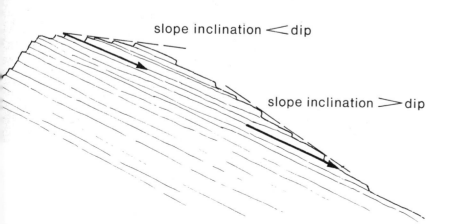

slope inclination $<$ dip

slope inclination $>$ dip

Fig. 5.13 Variations in the relationship between slope inclination and bedrock dip caused by erosion.

Plate 5.14 Simple plunging syncline, James Range, N.T. Note the outward facing escarpments and the cuesta ridges. (*RAAF.*)

Plate 5.15a Vertical air photograph of folded Precambrian sediments, Willouran Ranges, northwestern extremity of Flinders Ranges, South Australia. Note the concentric patterns of ridge and valley, and the creek beds which run across the structural grain (as at X, Y) (see Chapter 21). (*S.A. Dept Lands*)

Plate 5.15b The breached dome of Sheep Mt, near Greybull, Wyoming. Note the steeply dipping strata, especially on the left-hand limb of the structure, the concentric patterns of ridge and valley, the stream cutting across the snout of the structure (see Chapter 21) and the minor basin in left foreground. (*R.S. Shelton.*)

Plate 5.16 The natural amphitheatre of Wilpena Pound, central Flinders Ranges, displays cuestas and homoclinal ridges with outward-facing escarpments. (*S.A. Tourist Bureau*)

along which there is no strain. The degree of strain varies with the degree of curvature of the fold and with the distance of any point above or below the neutral plane:

$$\text{Elastic strain} = \frac{\text{distance from neutral plane}}{\text{radius of fold}}$$

Because of strains induced by folding, the crests of anticlines tend to be weathered and worn away; the structures are breached and the varied strata of the sedimentary sequence are exposed in the core of the anticline. This is the principal significance of folding: outcrops with varied resistance to weathering and erosion are exposed at the surface and are differentially eroded to give ridge and valley topography.

In plan, the pattern faithfully reflects the pattern of folding: straight fold limbs giving rise to linear and parallel to subparallel ridges and valleys (Pls 5.12 and 5.13), arcuate limbs to curved ridges and valleys (Pl. 5.14), plunging folds to convergent and divergent ridges and valleys which narrow and broaden (Pls 5.15a and b, 5.16; see also Fig. 5.4).

In section, the ridges (and hence the intervening valleys) vary largely, though not entirely, according to the local dip of the strata, and are arbitrarily accorded different names.[19] Ridges formed of very gently dipping strata (approximately 3–10°) are distinctly asymmetrical in cross-section and are called *cuestas* (Pls 5.14 and 5.17; Figs 5.10, 5.12). The inclination of the gentle dip-slope closely corresponds to the dip of the strata, and in places consists of a single exposed bedding plane (see Fig. 5.13). Elsewhere, however, erosion introduces complications. For instance, basal scouring by rivers causes slopes to develop which are steeper than the local dip and which truncate several strata, and weathering on upper slopes causes the formation of slopes which are gentler than the local dip (Fig. 5.13). The upper part of the steeper, or scarp, face is formed by a bluff in which joint blocks are prominent. It gives way below to a more gently inclined debris-strewn slope, which leads down to the valley or plain. However, as is also observed on the bounding slopes of plateaux the complexity of this faceted slope varies with local structure. Some scarp slopes for instance have a ribbed appearance due to the exposure of interbedded resistant and weak strata (Pl. 5.18).

Asymmetrical ridges, with steeper dip-slopes (dip of strata 10–30°, though with the erosional complications described in respect of cuestas) are called *homoclinal ridges* (Fig. 1.1f, Pl. 5.17). Ridges underlain by strata so steeply dipping (more than 30°) that their inclination is greater than the average inclination of slopes in the region are symmetrical in cross-section and are called *hogbacks*, (Pl. 5.19, Fig. 5.12b). Steeply dipping dykes and massive veins can also form hogbacks.

Close and regular dissection of any of these types of ridge causes the development of triangular facets on the dip-slope separated by the V-shaped valleys. The facets are called *flatirons* (Pl. 5.20a–c) because those which have curved sides in particular resemble the irons used some 40–50 years ago for pressing clothes.

Despite folding, not all rock strata are breached. In some areas, as, for instance, the Zagros Mountains of Iran (Pl. 5.21), what are clearly arched formations have been exposed and uncovered by erosion of the superincumbent strata. Evidently these are either especially massive and resistant

Plate 5.17 The Stirling Ranges, in the southwest of Western Australia, consist of dissected, gently dipping sandstones. Note the homoclinal ridge in the foreground and cuesta, coincident with gentler dip of beds, in the middle distance. (*Richard Woldendorp, W.A. Dept Int. Dev.*)

Plate 5.18   Ribbed ridge in the central Flinders Ranges. Each "rib" denotes the outcrop of a stratum slightly more resistant than that above and below it. (*S.A. Lands Dept.*)

Plate 5.19a   Dissected hogback ridge developed on strata of moderate dip in northeastern Switzerland. The strata in the hogbacks are dipping to the right of the viewer, those in the ridge in the foreground, to the left, so that the valley between may occupy a syncline. (*Swiss Nat. Tour. Office.*)

Plate 5.19b   Numerous hogback ridges, due to the exposure of steeply dipping resistant strata in the Zagros Mountains, Iran. (*Hunting Surveys.*)

Plate 5.20a Flatirons eroded from massive limestones involved in a plunging anticline, Zagros Mountains, Iran. (*Hunting Surveys.*)

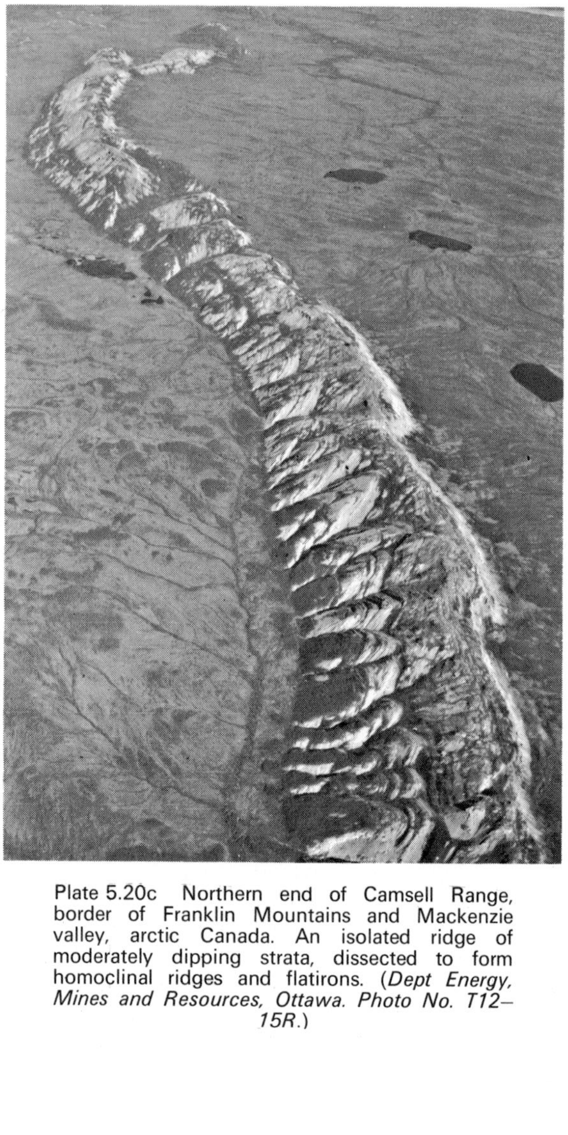

Plate 5.20c Northern end of Camsell Range, border of Franklin Mountains and Mackenzie valley, arctic Canada. An isolated ridge of moderately dipping strata, dissected to form homoclinal ridges and flatirons. (*Dept Energy, Mines and Resources, Ottawa. Photo No. T12—15R.*)

Plate 5.20b Breached anticline with well-developed flatirons on the flanks, Zagros Mountains, Iran. (*Hunting Surveys.*)

Plate 5.21 Limestone dome uncovered by erosion of superincumbent strata, western Zagros Mountains, Iran. (*Hunting Surveys.*)

Plate 5.22 Perched syncline (right) and hogback ridge (left), Zagros Mountains, Iran. (*Hunting Surveys.*)

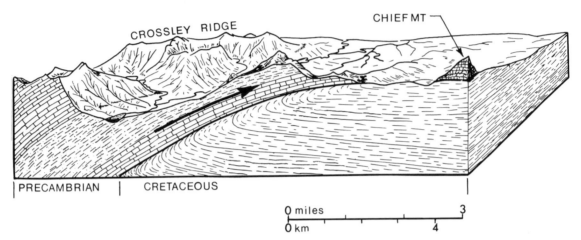

Fig. 5.14   Section through Chief Mountain, Montana. (*Reprinted by permission of Macmillan Publishing Co., Inc., from* Atlas of Landforms, *by Scovel et al.   Copyright 1942 by The Macmillan Company renewed 1970 by Edmund W. King, First National Bank of Ithaca.*)

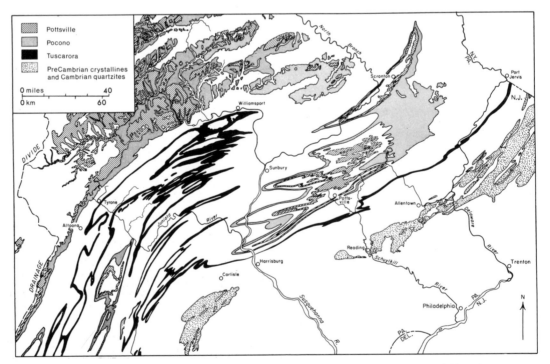

Fig. 15.5   Pattern of ridge-forming sandstones in Pennsylvania and adjacent parts of New Jersey. (*After H.D. Thompson, "Drainage Evolution", pp. 31–62, by permission of the author and N.Y. Acad. Sci.*)

layers, or resistant formations which lie below the neutral plane in anticlinal structures and which are sufficiently tough to survive as exposed humps. In yet other areas, like the Flinders Ranges, anticlinal forms underlie broad valleys and synclinal structures form upstanding ridges (Fig. 5.3; Pl. 5.22). The Berkeley Hills in California are underlain by a sequence of sediments and volcanics of synclinal structure, and the Piketberg, just north of Cape Town, South Africa, is constructed of a syncline in sandstone. Such *relief inversion* clearly implies deep erosion, and in the vicinity of the Willochra basin in the southern Flinders Ranges, for example, the thickness of strata—deduced from the preserved sequences to both east and west of the basin—eroded from above the present topographic basin is of the order of 6000 m.

Ridge and valley assemblages are found in many parts of the world, particularly in the orogenic belts, though in the structurally complex fold mountains like the Western Cordillera of North America, the Andes, the Himalayas and the European Alps, the patterns of folding and, hence, of outcrop distribution are so intricate that the relief pattern is also very difficult to detect. This is particularly the case where there are *nappe* structures and *overfolds*, or where there are thrust faults along which masses of rock or *klippe* have been translocated. Thus Chief Mountain, Montana, is formed by a block of massive limestone which has become isolated by denudation. It is an outcrop of Precambrian rock resting on and surrounded by Cretaceous sediments. It has been carried to this stratigraphically inverse situation along a thrust fault; extensions of the same limestone occur to the west (Fig. 5.14). But Chief Mountain, caused by the isolated occurrence of a resistant mass of limestone, cannot be understood without reference to the structural complexity of this part of the Western Cordillera. Repeated short thrusts along faults led to the repetition of outcrops and of ridges associated with them. In other areas erosion has penetrated through

nappe structures, revealing the underlying rocks in *windows*.

In all the complex fold mountain belts there are similar occurrences of strata, seemingly out of place in that they are not connected to any other nearby outcrops. Strata display pronounced changes of dip along the strike and equally marked changes of strike, all of which are reflected in the landform assemblages. Granitic and metamorphic rocks are exposed by deep erosion and give rise to their characteristic forms. Yet in many areas within these intensely folded regions the sediments are nevertheless readily interpreted in terms of the nature and disposition of the strata.

Undoubtedly, however, the clearest demonstrations of the direct relationship between structure and surface are found in orogens or platform areas, which have been subjected to only mild deformation. Well-known examples include the Ridge and Valley section of the Appalachians[20] (Fig. 5.15), the Paris Basin and southeastern England[21], the Flinders Ranges of South Australia[22] (Fig. 5.4), the Grampians of Victoria[23], the Zagros Mountains of Iran[24] (Pls 5.19, 5.20a and b, 5.21, 5.22), the Atlas Mountains of north Africa and the Cape Fold Belt of South Africa.

These regions all display basically similar assemblages of forms. But they vary in their structural complexity, and hence in the predominance of various ridge and valley forms. For instance, in areas like the northeast of South Australia where the silcrete is only gently folded, there are no hogbacks. These forms are also relatively rare in the Flinders. But in intensely-folded regions like the Alps (Pls 3.34b, 5.19) or the Himalayas, hogbacks outnumber cuestas, though intensity of erosion also contributes to this situation. In general cuestas and homoclinal ridges are the common ridge forms in the platform areas, while hogbacks are more common in the intensely deformed orogens.

But the distribution of the various ridge forms is further complicated by the geological history of the particular region. For instance, in the southern Flinders Ranges (Fig. 5.4)

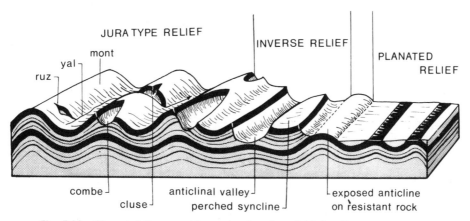

Fig. 5.16 Characteristic assemblages developed on folded sedimentary sequences. (*After M. Derruau, Précis de Géomorphologie, Masson & Cie, Paris, 1958, p. 304.*)

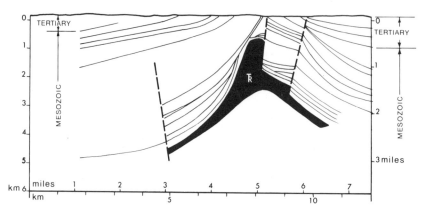

Fig. 5.17   Sections through salt dome at Audignon, France. (*After P. Liechti, "Salt Features in France"*, Geol. Soc. Am. Sp. Paper, **88**, *1968, pp. 83–106.*)

massive sandstone ridges dominate the relief, particularly in the western areas. Such ridges are fewer in the east, however, despite the fact that structurally the latter region is simply a limb of the anticline which straddles the whole upland, with the opposite limb underlying the western parts of the Ranges (Fig. 5.3). The reason is not that the ridges have been worn away in the east, but that they never existed. The source area of the sand was the Westralian cratonic mass to the west (Fig. 5.3). Therefore, unless the old shoreline moved some miles to the east, the sands were deposited close to the shield margin, in what is now the western part of the Ranges, while finer and less resistant shales and siltstones were laid down to the east. The shoreline migrated from time to time, but was more often in a westerly than in an easterly situation. Hence the scarcity of sandstones in the east; and hence the rarity of ridges there. Similar migration of the centre of deposition accounts for the asymmetry of the Paris Basin, where there are at least six outward-facing escarpments of Jurassic limestone on the east of the structure, but only one (of Cretaceous chalk) to the west.

Thus each fold mountain region is, to some extent, individual. Nevertheless specific types can be discerned (Fig. 5.16). Folded regions which have suffered little dissection are compared to the Jura Mountains of the France–Switzerland border region and are said to be of *Jura* type. As dissection increases, inversion of relief develops, and eventually the folds are worn down to subdued relief. Further uplift causes a renewal of erosion and the fold structures are again etched out, but remnants of the old surface of subdued relief remain high in the relief. This is the situation said by many workers to obtain in the Appalachians, and for this reason fold mountain regions with evidence of old erosion cycles high in the relief are said to be of *Appalachian type*.

C. DIAPIRIC STRUCTURES

Some domes are caused by cross-folding, that is, by two sets of opposed horizontal forces (though some geologists, like Beloussov,[25] give more weight to vertical forces). But domes also occur in areas which display no evidence of regional compression, and they seem to be caused by very localised vertical upthrust. Exposed in their cores are rocks of distinctive character, such as rock salts and sulphur, which differ from the strata forming the limbs of the fold. These structures are apparently developed by the vertical injection of material into the crust from below, caused either by magmatic intrusion or by the upwelling (possibly triggered by tectonic movements) of materials with a lower specific gravity than the surrounding rocks. Such gravity-controlled structures are called *diapirs*.

Some granite batholiths are areas of negative gravity anomaly and may rise through the surrounding sediments.[26] This could account for the gneiss domes which occur in parts of the Canadian Shield (Pl. 5.23). The most common example of a diapir, however, is the *salt dome*, typical examples of which have been reported from several parts of the world (Pl. 5.24). They achieve their best topographic expression in arid regions, where the salts are not readily dissolved when they reach the near-surface zone; but salt domes have been described from such humid areas as Louisiana and coastal Texas, as well as from such cold arid areas as the Canadian Arctic e.g., Ellef Ringnes Island. They occur at depth (though they have no topographic expression) in many other areas, for instance France (Fig. 5.17) and north Germany. They are widely distributed over the continental areas.   Where the intruded salts reach the surface, they cause the bending of the strata and give rise to ridge and valley forms similar to those formed as a result of compression and subsequent differential erosion of strata (Pl. 5.25).

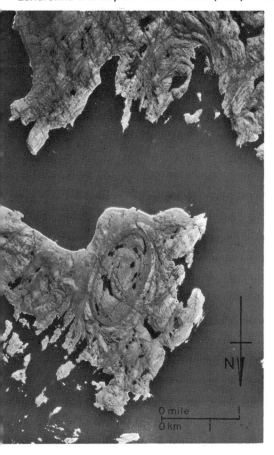

Plate 5.23 Gneiss dome on Baffin Island, arctic
Canada. (*Dept Energy, Mines and Resources,
Ottawa. Photo No. A15459–90.*)

Plate 5.24 Vertical air photograph of salt dome, Qishin Island,
Persian Gulf (*Hunting Surveys.*)

Plate 5.25 Salt dome (Kuh-i-Hamak) southeast of Bushire,
Iran. (*Hunting Surveys.*)

References Cited

1. J.N. JENNINGS, "Some Geomorphological Problems of the Nullarbor Plain", *Trans. R. Soc. S. Aust.*, 1963, **87**, pp. 41–62.
2. Ibid.
3. W.D. JOHNSON, "The High Plains and their Utilisation" in *Rep. U.S. Geol. Surv. 21st Ann.* (1899–1900), Part IV, 1900, pp. 599–768.
4. R.L. JACK, "Geology and Prospects of the Region to the South of the Musgrave Ranges, and the Geology of the Western Portion of the Great Artesian Basin", *Bull. Dep. Mines S. Aust.*, 1951, **5**, p. 72; C.R. TWIDALE, "Some Problems of Slope Development", *J. geol. Soc. Aust.*, 1960, **6**, pp. 131–48.
5. J. GILLULY, "Distribution of Mountain Building in Geologic Time", *Bull. geol. Soc. Am.*, 1949, **60**, pp. 561–90.
6. M. SCHMIDT, "Neuzeitliche Erdkrustenbewegungen in Frankreich", *Sber. bayr. Akad. Wiss., math-phys.*, 1922; E. KAYSER, "Merkwürdige Senkungen des Bodens von Frankreich", *Sber. bayr. Akad. Wiss., math-phys.*, 1922.
7. Y.A. MERSCHERIKOV, "Crustal Movements—Contemporary", in R.W. Fairbridge (Ed.), *Encyclopaedia of Geomorphology*, Reinhold, New York, 1969, pp. 223–7.
8. L. BENDEFY, "Niveauanderungen in Raum von Transdanubian auf Grund Zeitgemasser Feineinwagungen", *Acta tech. Hung.*, 1959, **23**, pp. 167–68.
9. Ibid.
10. F. SZENTES and A. RONAI, *Quaternary Tectonics, Hungary*, (Map), Budapest, 1964.
11. N. MIYABE, S. MIYAMURA and M. MIZOUE, "A Map of Secular Vertical Movements in Japan", in *Mouvements Modernes, Volcanisme et Seismes sur les Continents et les Fonds Oceaniques*, INQUA, Moscow, 1969, pp. 211–2; T. YOSHIKAWA, "On the Relations between Quaternary Tectonic Movement and Seismic Crystal Deformation in Japan", *Bull. Dep. Geogr. Univ. Tokyo*, 1970, **2**, pp. 1–24.
12. H.W. WELLMAN, "Report on Studies related to Quaternary Diastrophism in New Zealand", *Quaternary Res.*, 1966, **6** (1) App. 5, pp. 34–6.
13. S.E. HOLLINGWORTH and R.W.R. RUTLAND, "Studies of Andean Uplift. Part I. Post-Cretaceous Evolution of the San Bartolo Area, North Chile", *Geol. J.*, 1968, **6**, pp. 49–62.
14. G.M. LEES, "Recent Earth Movements in the Middle East", *Geol. Rdsch.*, 1955, **42**, pp. 221–6.
15. R.C. SPRIGG, "Geology and Petroleum Prospects of the Simpson Desert", *Trans. R. Soc. S. Aust.*, 1963, **86**, pp. 35–65; H. WOPFNER and C.R. TWIDALE, "Geomorphological History of the Lake Eyre Basin", in J.N. Jennings and J.A. Mabbutt (Eds), *Landform Studies from Australia and New Guinea*, Australian National University Press, Canberra, 1967, pp. 118–42; C.R. TWIDALE, "Landform Development in the Lake Eyre and Adjacent Depressions", *Geogr. Rev.*, 1972, **62**, pp. 40–70.
16. C.R. TWIDALE, JENNIFER A. SHEPHERD and ROBYN M. THOMSON, "Geomorphology of the Southern Part of the Arcoona Plateau and the Tent Hill Region west and north of Port Augusta", *Trans. R. Soc. S. Aust.*, 1970, **94**, pp. 55—67; J.T. HUTTON, C.R. TWIDALE, A.R. MILNES and H. ROSSER, "Composition and Genesis of Silcrete and Silcrete Skins from the Beda Valley, Southern Arcoona Plateau, South Australia", *J. geol. Soc. Aust.*, 1972, **19**, pp. 31–9.
17. H. WOPFNER, "On Some Structural Development in the Central Part of the Great Artesian Basin", *Trans. R. Soc. S. Aust.*, 1960, **83**, pp. 179–94; R.C. SPRIGG, "Geology and Petroleum Prospects", pp. 35–65.
18. C.R. TWIDALE, *Structural Landforms*, Australian National University Press, 1971, pp. 177–8.
19. See, for example, A. CAILLEUX and J. TRICART, "Le Problème de la Classification des Faits Géomorphologiques", *Ann. Géogr.*, 1956, **65**, pp. 162–85.
20. H.D. THOMPSON, "Drainage Evolution in the Appalachians of Pennsylvania", *Ann. N.Y. Acad. Sci.*, 1949, **52**, pp. 31–62.
21. P. BIROT, *Morphologie Structurale*, Presses Universitaires de France, Paris, 1958, 2 vols; P. PINCHEMEL, *Geographie de la France*, Colin, Paris, 1964, Vol. 1; S.W. WOOLDRIDGE and D.L. LINTON, *Structure, Surface and Drainage in South-East England*, Philip, London, 1955.
22. C.R. TWIDALE, "Chronology of Denudation in the Southern Flinders Ranges, South Australia", *Trans. R. Soc. S. Aust.*, 1966, **90**, pp. 3–28; C.R. TWIDALE, "Geomorphology of the Flinders Ranges", in D.W.P. Corbett (Ed.), *National History of the Flinders Ranges*, Public Library of South Australia, Adelaide, 1969, pp. 57–137.
23. E.S. HILLS, *Physiography of Victoria*, Whitcombe and Tombs, Melbourne, 1940, p. 260; D. SPENCER-JONES, "Geology and Structure of the Grampians Area, Western Victoria", *Mem. geol. Surv. Vic.*, 1965, **25**, p. 92.
24. T. OBERLANDER, "The Zagros Streams", *Syracuse Geogr. Ser.*, 1965, **1**, p. 168.
25. V.V. BELOUSSOV, *Basic Problems in Geotectonics*, McGraw-Hill, London, 1962.
26. M.H.P. BOTT, "Negative Gravity Anomalies over 'Acid Intrusions' and their Relation to the Structure of the Earth's Crust", *Geol. Mag.*, 1953, **90**, pp. 257–67; *idem*, "A Geophysical Study of the Granite Problem", *Q. J. geol. Soc. London*, 1956, **412**, pp. 45–62; E.H. KRANCK, "On Folding Movements in the Zone of the Basement", *Geol. Rdsch.*, 1957, **46**, pp. 261–82.
27. See R.B. MATTOX (Ed.), "Saline Deposits", *Geol. Soc. Am. Spec. Paper*, **88**, 1968.

CHAPTER 6

Volcanoes and Volcanic Landforms

A. GENERAL

Over 24 million km of the continents are covered by volcanic lava (for the sake of comparison, America is about 9·4 million km^2 in area and Australia covers some 7·7 million km^2). There are more than 500 known active volcanoes (Fig. 6.1) and thousands of dormant or extinct volcanoes. Moreover, though many very large volcanoes emerge above sealevel, many more volcanic features of the ocean basins can only be suspected, as yet. Volcanoes may be regarded as the earth's safety valves, for they are essentially holes or fissures through which escape molten rocks, gases and liquids that originated deep in the crust.

Magma from the upper mantle (see Chapter 2) rises into the crust where it forms a reservoir in numerous dikes and sills located 1–2 km below the surface. There, minerals rich in iron and magnesium precipitate out of the molten rock, leaving a residue which is acid. Water vapour also accumulates. Eventually the magma becomes of a sufficiently low density to rise to the surface through zones of weakness in the crust. On its way through the crust the magma is contaminated through stopping or ingesting the surrounding rocks. This in part accounts for the varied, and in some cases quite acid, composition of lavas which have breached the sial, and hence for the varied behaviour and morphology of volcanic extrusions. But the extruded lavas are more acid than the original magma partly because of the fractionation of the latter: for instance the release of basalt at the surface takes certain elements out of the system, which is thereby converted to a composition different from the original.

Only a small proportion of the magma emplaced in the crust actually reaches the surface; indeed some of the basaltic magma which attains the surface has been observed to drain back into the vents whence it came.

Volcanoes are tectonic or constructional forms, due wholly to activities within the earth's crust. They provide numerous, often spectacular and dramatic demonstrations of activities which have played an important part in the evolution of the earth's surface.

B. PRESENT DISTRIBUTION

The distribution of volcanoes and associated rocks and forms is not random, for volcanic features are restricted at present to zones of weakness in the crust; in particular, to the most recent Alpine fold belts or orogens, the great rift or fault areas and the submarine mountain and rift complexes (Fig. 6.1). Thus four important modern zones of volcanic activity can be distinguished:

1. the circum-Pacific zone (including an extension into the Antilles);

2. a zone extending from the Mediterranean through Asia Minor and into the Indonesian archipelago. In the latter area, this zone intersects the circum-Pacific belt and, as a result, is the most active volcanic zone in the world;

3. the African Rift zone; and

4. the submarine mountains and rifts, particularly those of the mid-Atlantic ridge and the Pacific area.

In each of these zones, volcanic activity is merely one expression of a profound instability, for these same areas are also noted for their frequent and commonly disastrous earthquakes. Also, and in local detail, the cones and craters, which are the best known expressions of volcanicity, are aligned along lines of weakness in the local bedrock (Fig. 6.2).

In the past volcanoes have been active in areas which are now relatively stable. For instance, the Precambrian sediments of the Labrador Trough are interbedded with volcanic rocks, the Gawler Ranges of northern Eyre Peninsula (South Australia) are built of Precambrian lavas and ash beds and the Victoria River basin of the Northern Territory is underlain by Cambrian volcanics, though all of these areas are now remote from regions of volcanism. At times in the past, for example in the Tertiary, volcanic activity was also much more pronounced than it is at present.

C. CLASSIFICATION

Each volcanic province has its own characteristics, and both the types of eruptive activity and the shapes of volcanoes vary greatly from place to place. Nevertheless, several general types of volcanic landforms can be distinguished. The classical types of eruption were named after specific localities by Lacroix:[1] *Hawaiian, Strombolian, Vulcanian* and *Peléean* (Fig. 6.3). The Hawaiian type is characterised by freely flowing basaltic lava with comparatively little explosive activity. Only on very rare occasions is lava released from the crater, more commonly pouring out of fissures located in the flanks of the volcanic dome. In the Strombolian type, the lava is less fluid than in the Hawaiian, explosions are more common, and in consequence, more fragmented lava, ash and scoria (*pyroclastic* material) are associated with this type of eruption. The lava associated with Vulcanian eruptions is still more viscous. Vertical explosions are characteristic and common because the lava rapidly solidifies on exposure to the air. There is much fragmentation of debris and considerable distribution of ash. Peléean eruptions are characterised by extremely viscous acid lavas (typically andesite—see Appendix 1); by violent and often devastating explosions many of which are associated with lateral vents which are forced open by the pressure built up in the blocked central wall; and by *nuées ardentes*—glowing clouds of incandescent ash and gas.

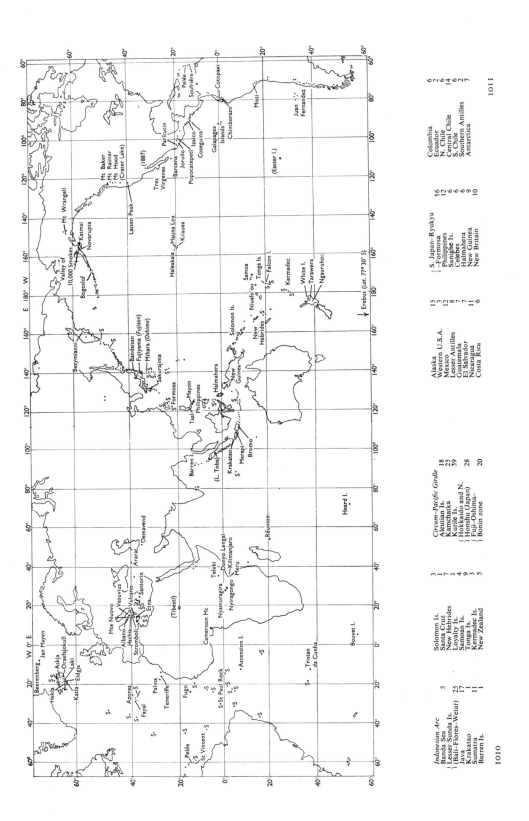

Fig. 6.1   Active volcanoes of the world (S = submarine eruptions). *(After A. Holmes, Principles of Physical Geology, Nelson, Edinburgh, 1968, 2nd ed., pp. 1010–11.)*

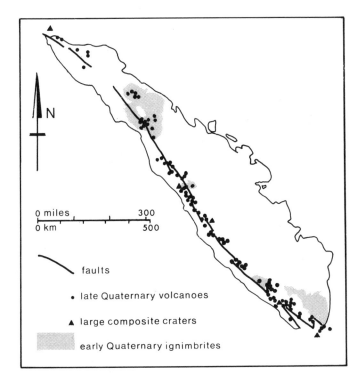

Fig. 6.2 Distribution of volcanoes along the Sumatra Fault. (*After J.A. Katili and F. Hehuwat, "On the Occurrence of Large Transcurrent Faults", pp. 5–17, by permission of J. Geosci., Osaka University.*)

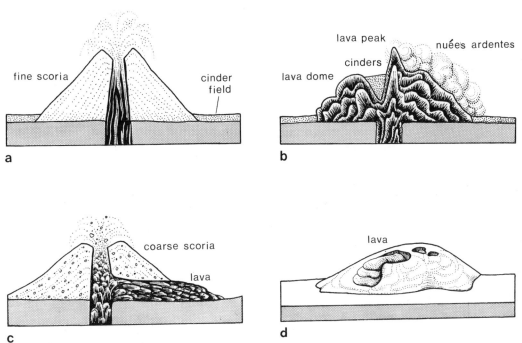

Fig. 6.3 Lacroix's four classical types of volcanic eruption and form: (a) Vulcanian; (b) Peléean; (c) Strombolian; (d) Hawaiian. (*After M. Derruau, Précis de Géomorphologie, Masson & Cie., Paris, 1958, p. 256.*)

Plate 6.1a   1967 eruption of Kilauea, one month old. *(U.S. Dept Int. Geol. Surv.)*

What are the reasons for the contrasted behaviour and morphology of, say, Hawaiian and Peléean volcanoes? The variations in activity and form are related to the chemistry of the lavas: the more acid the lava, the more free silica it contains, the more viscous and explosive the activity, and the steeper the volcanic core or plug associated with the extrusion.

Like all classifications of natural phenomena, these categories are arbitrary. There are many gradations between types: volcanoes form a continuum rather than fall into neat and distinct categories. Moreover, the type of activity and the forms developed in a given province have changed in time, not only at intervals of geological time but also during given phases of eruption, as the composition of the lava available changed. However, the classification places into focus the two extremes of the continuum, the Hawaiian and the Peléean, which are also representative of many actual eruptions.

1. Hawaiian or Basaltic Volcanoes

Lavas of basic composition, such as olivine basalt, are rich in iron, magnesium and calcium, and contain no free silica. Such lavas flow easily, are extruded at comparatively high temperatures (1000–1100 °C) and, though by no means devoid of gases and fluids, contain relatively small amounts compared to acid lavas. Because the lavas are fluid and flow rapidly away from the vents (lavas flowing at speeds of more than 16 km an hour have been noted in Hawaii, and for short periods 1 m per second at Heymaey, Iceland, in 1973), there is little tendency for lava to coagulate around the openings. Hence, though some explosions occur, pressures do not build up to an extent sufficient to cause major blasts.

Plate 6.1b   Lava fountain emptying from a vent in Makaopuhi crater on the east rift zone of Kilauea volcano. Taken March 15, 1967. (*U.S. Dept. Int. Geol. Surv.*)

For this reason, lava flows rather than pyroclastic rocks dominate basic volcanic regions. This is not to suggest that ash and scoria are not present in association with the lava flows, but they are subordinate. Though they give the impression of being dormant for long intervals between periods of activity, continuous observations of some of the Hawaiian volcanoes, and also of Vesuvius, show that appearances are deceptive. There are constant tremors, rise and falls, spurts of steam, other gases and lava (Pl. 6.1a-d); however they are relatively minor compared to those phases when easily discernible changes (marked by explosions, lava outbursts and flows, collapse of crater walls, and so on) take place.[2] (See Pls 6.1 a and b.)

The features typical of Hawaiian volcanic provinces are low angle domes, shields and craters (Pls 6.1c and d, Figs 6.3 and 6.4) which are surrounded by vast spreads of lava plains and plateaux. In many areas, the lava wells up at certain points, apparently through pipes which are marked by craters and more-or-less circular vents at the surface. In many areas, for instance Hawaii, there are clusters of such vents. But in Iceland, though there are cones, some eruptions occur along fissures of considerable length. For example, the fissure involved in the eruption of Laki in 1783 was over 30 km long. For this reason, basalt provinces in which identifiable craters are scarce, and where fissures appear to have been the principal avenues of escape, are sometimes said to be of the *Icelandic* type in contradistinction to the Hawaiian, with numerous craters.

The true lava cone has very gentle slopes which merge gradually with the adjacent lava plains. Mauna Loa, the

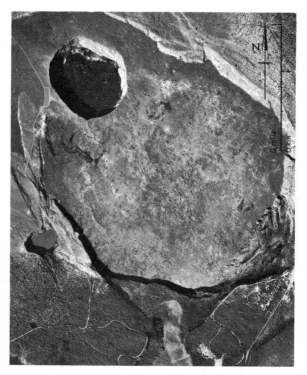

Plate 6.1c Vertical air photograph of Halemaumau, Hawaii (cf. Fig. 6.4b). *(R.M. Towell Corp.)*

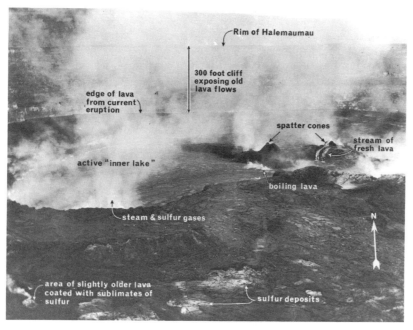

Plate 6.1d Halemaumau "Firepit", after the 1967–68 eruptions. *(U.S. Dept. Int. Geol. Surv.)*

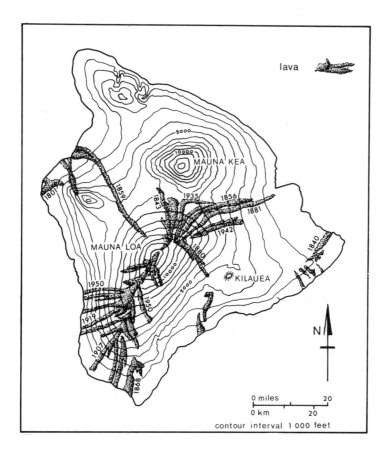

Fig. 6.4a Mauna Loa, Hawaii, showing major craters, associated radial rifts and recent lava flows with dates shown. (*After A. Rittman,* Volcanoes and their Activity, *Wiley, New York, 1962, p. 118.*)

Fig. 6.4b Contour plan of the Kilauea and Halemaumau craters (in metres). *(After J.L. Scovel, E.J. O'Brien, J.C. McCormack and R.B. Chapman,* Atlas of Landforms, *Wiley, New York, 1965, p. 66.)*

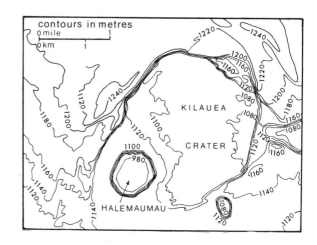

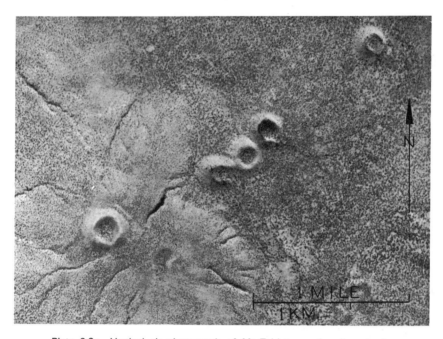

Plate 6.2a   Vertical air photograph of Mt Tabletop, a basalt crater in northern Queensland. Note the secondary or adventitious cones to the northeast, and the radial drainage from the main cone. (*Dept. Nat. Dev., Canberra.*)

Plate 6.2b   Mt Tabletop from the southeast; the adventitious cones are visible on the right skyline. *(C.R. Twidale/C.S.I.R.O.)*

Plate 6.3a Lyttleton Harbour, a breached volcanic crater on Banks Peninsula South Island, New Zealand, with the Canterbury Plains beyond. (*N.Z. Govt Tour. Bur.*)

largest of the Hawaiian islands, is a typical basalt volcano (Fig. 6.4a). It consists of a cone, the slopes of which have a gentle gradient of between 4° and 6 with respect to the horizontal. Although the peak of the cone rises 4170 m above sea-level, what is visible is merely the top of an immense volcano which has a diameter of over 400 km and which stands some 9150 m above the sea-floor. The upward pressure of the magma rising below Mauna Loa has caused the development of radial fissures called rifts (Fig. 6.4a). It is along such weaknesses caused by the volcano itself that secondary or *adventitious* cones have emerged (Pl. 6.2).

Tahiti, Samoa and Mt Etna in Sicily are similar; Banks Peninsula (Pl. 6.3a) in South Island, and White Island in the Bay of Plenty, North Island, New Zealand (Pl. 6.3b, Fig. 6.5, see Black), each consist of twin extinct and overlapping

basalt domes. Lyttleton and Akaroa harbours are formed in the breached and flooded craters of the Banks Peninsula domes.

In oceanic locations, lava flows may reach the sea, causing the formation of great mounds of ash and rock on the beaches, masses of steam and lava rafts. Underwater eruptions have similar effects, but the cone eventually emerges if, as is often the case, the extrusion and build-up are rapid. The ash is just as rapidly eroded. But on land, the lava spreads over very great distances. Extensive basalt plateaux and plains resulting from the extrusion of *flood basalts* occur in Greenland, central Siberia, the Deccan of India, Syria, Arabia, Ethiopia, Iceland (Pl. 6.4), the Columbia Plateau of northwestern USA, the Snake River area of Idaho, Patagonia, and eastern Australia (Fig. 2.5). Less extensive basalt flows occur

Plate 6.3b White Island, an active andesitic volcano, Bay of Plenty, North Island, New Zealand (cf. Fig. 6.5). The newer crater at the southeastern end of the island (near camera) has been deeply breached by erosion. (*N.Z. Govt Tour. Bur.*)

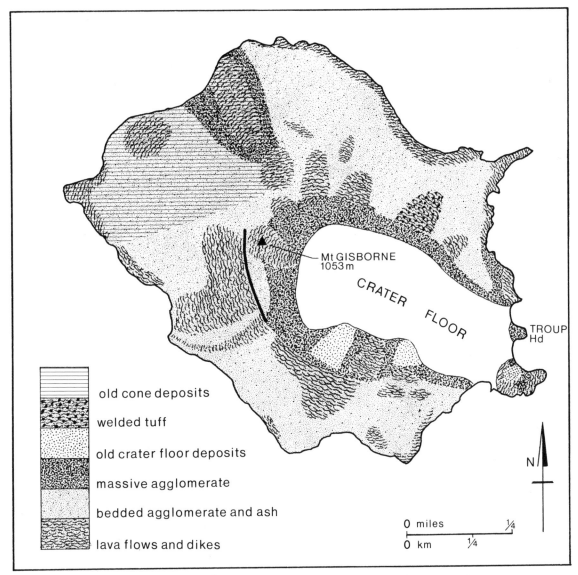

old cone deposits

welded tuff

old crater floor deposits

massive agglomerate

bedded agglomerate and ash

lava flows and dikes

Mt GISBORNE
1053 m

CRATER FLOOR

TROUP
Hd

0 miles   ¼
0 km   ¼

N

Fig. 6.5   Plan of White Island volcanoes, North Island, New Zealand. *(After P.M. Black, "Observations on White Island", p. 160, by permission of Bull. Volc.)*

in many areas, such as Arizona and New Mexico. The Columbia River basalts of the USA cover 400,000 km² and are of late Cainozoic age (Fig. 6.6a). The early Tertiary basalts of the Deccan traps* (Fig. 6.6b) cover at least 650,000 km and, furthermore, extend beneath the sea; it has been estimated that the total area of this flow is of the order of 10 million km². In Australia, basalt plains and plateaux mainly of Tertiary age occur in northern, central and southern Queensland, New England, the Hunter Valley, the

Monaro of N.S.W, and the Western Districts of Victoria.

These lava plains and plateaux are characterised by an absence of surface drainage due to the well-jointed character of the rock; by underground caverns, some of which display lava stalactites formed by still-molten rock dripping from the ceilings of the tunnels (Pl. 6.5); by stony rises (Pl. 6.6) and plains, the rises marking either original lava tongues and flow margins (Fig. 6.7), or subsequently eroded river bluffs; and by soil-filled depressions. If fresh, the lava is blocky and clinkery, with obvious vesicular structure (*aa* lava or structure), or is ropy and twisted (*pahoehoe* lava). It has a high content of ferromagnesian minerals and therefore weathers comparatively quickly. Basalt nevertheless forms plateau-cappings, as for instance in several parts of eastern Australia. In most

* A term resulting from the step-like character of the sides of river valleys cut into basalt regions. In these valley sides are exposed the layers of lava and these give rise to distinct benches or steps. The German word for stairs is *Treppe*, which has been distorted to trap.

Plate 6.4 Aerial view of the Tertiary plateau basalt area of Isafjordur, northwestern Iceland. The plateaux and mesa are underlain by intercalated lava beds, sedimentary strata and soils which are flat-lying in the foreground, but which are gently dipping in the middle distance. The original plateau has been dissected by rivers and glaciers which have exploited various structural lines. (*H. Palsson.*)

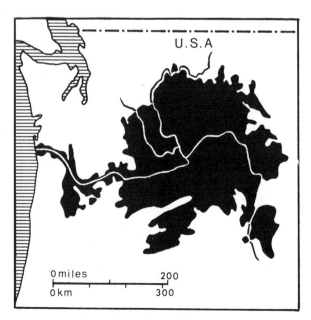

Fig. 6.6a   The Columbia River basalt area.

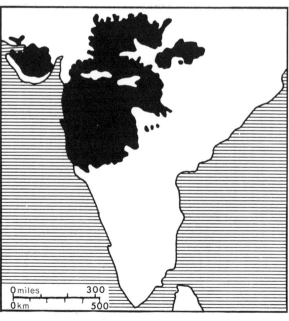

Fig. 6.6b   The Deccan trap.

Plate 6.5 Lava stalactites from a cavern at Suswa volcano, Mountains of the
Moon, Kenya. *(I.S. Loupekine.)*

Plate 6.6 Basalt plain with stony rises (the fronts of lava flows) separating soil-
filled depressions on late Tertiary volcanic plateau, north Queensland. *(C.R. Twidale/
C.S.I.R.O.)*

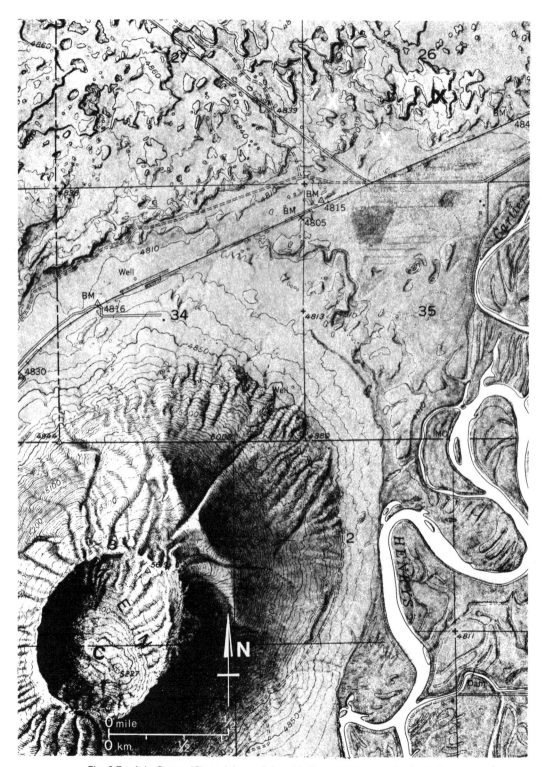

Fig. 6.7   Ash Crater (C) and lava plain with irregular topography including stony rises and lava tongues (X), Menan Buttes, Idaho. Note lack of surface drainage on the lava field, and the meanders and scroll plain of Henrys Fork.
(After U.S. Geol. Surv. Topo. Sheet, Menan Buttes.)

Plate 6.7a   Vertical air photograph of the Tower Hill crater or caldera, western Victoria. *(Dept Nat. Dev., Canberra.)*

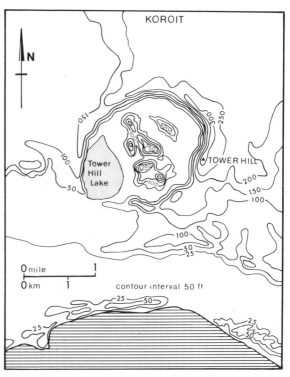

Fig. 6.8   Tower Hill crater, Victoria.

Plate 6.7b   General view of Tower Hill. Note dipping ash layers. *(J.R. Morrow.)*

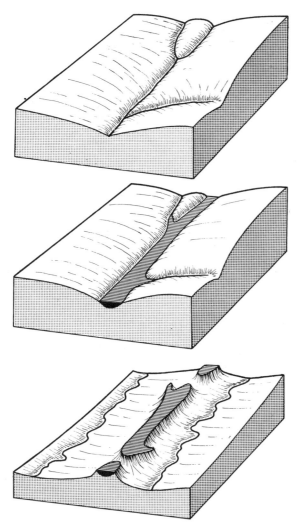

Fig. 6.9 Examples of drainage diversion caused by lava flow. Above—general case. Below—in the Campaspe valley, Victoria, the sequence shown progressing from left to right. *(After C.D. Ollier,* Volcanoes, *Australian National University Press, Canberra, 1969, pp. 124–7.)*

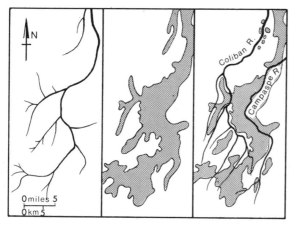

basalt areas, there was some explosive activity, giving rise to piles of pyroclastics in scoriacones and ash cones.

Some volcanic craters, like Tower Hill, Victoria (Fig. 6.8; Pl. 6.7a and b) have been enlarged by explosions and are called *calderas*, though the really large features of this type are found in association with more acid eruptions (see below). Basalt lavas flow freely and mould themselves to the pre-existing terrain, running down valleys and filling depressions.

Basalt flows commonly block pre-existing drainage, so that lakes are rather typical of the lava margins; where a tongue of lava has flowed down a valley, there is often inversion of relief and diversion or disturbance of drainage (Fig. 6.9).[4]

2. Acid (Peleean) Volcanoes

Though spectacular, basalt eruptions, whether of central or of fissure type, are rarely dangerous. The lava flows away and the lava pipes are not blocked. Such is not the case with acid lava, that is, lavas rich in alkalis (particularly Na) and free silica. Such lavas are extruded at temperatures rather lower than the basic flows (about 700–900° C), contain large volumes of gas, and are viscous or sticky. The lava looks as if it has been squeezed out slowly, rather like toothpaste from a tube (Fig. 6.10a). The vents tend to become blocked, pressures build up below the crust until a critical level is attained, and explosions therefore occur. Moreover, because of the plugs of coagulated lava near the surface, these explosions often rupture the flanks of the volcanoes, and dense masses of incandescent gas, volcanic dust and fragmented rock (the infamous *nuées ardentes*—see Pl. 6.8) burst forth and pour down the hillsides, and particularly along valleys and other topographic depressions.[5]

These acid explosive volcanoes are called Peléean, after Mt Pelée in Martinique, French West Indies, where on 8 May, 1902 the volcano suddenly erupted, *nuées ardentes* poured down the hillside and, with a temperature estimated at 1190° C, overwhelmed the nearby town of St Pierre, killing all but two of its 29,000 inhabitants.[6] Mt Pelée at this time also developed a bizarre needle of viscous rock which is again typical of such volcanoes (Fig. 6.10b).

The island of St Vincent, in the Windward Group of the West Indies, experienced an acid-type eruption with the formation of glowing clouds on 7 May, 1902. St Vincent (Fig. 6.11a) consists of an active volcano, Soufrière (Pl. 6.9a), located in the north, and lavas, breccias and other pyroclastic debris related to at least two earlier (late Pleistocene) periods of activity in the south.[7] To the northeast of the present crater, which is occupied by a lake, is part of a *somma* (see Pl. 6.9b), the rim of an older volcano. The present volcano is called Soufrière, and gives its name to areas of hot springs, steam vents, and hot mud.

Andesite lava flows are associated with the volcano, but explosive eruptions scattered ash and lapilla (blocks, bombs and fragments of lava) over the island during a period of several tens of thousands of years during the late Pleistocene. Some 6 m of ash were deposited at Kingstown on St Vincent. The thickness and distribution of these ash deposits (Fig. 6.11b) is inconsistent with the present wind regime—the prevailing east or northeast Trades (Fig. 6.11c)—and it appears that they were spread during a late Pleistocene glacial phase

Fig. 6.10a   A cone of soda-rich lava, Hoggar Mts, southern Algeria. *(Drawn by C.R. Twidale from a photo supplied by Hunting Surveys.)*

Fig. 6.10b   The acid lava needle of Mt Pelée, left—from the east; right—from the south. *(After C.D. Ollier,* Volcanoes, *p. 26.)*

Plate 6.8   *Nuées ardentes*—a glowing cloud of incandescent ash and gas flowing down a volcanic slope during the 1951 eruption of Mt Lamington. *(G.A. Taylor.)*

Plate 6.9a   The crater lake of Soufrière, St Vincent, Windward Isles, West Indies. Note the gullies extending headward toward the rim of the crater. (*Hunting Surveys.*)

Plate 6.9b   Mt Vesuvius, Italy, showing the modern cone rising out of the floor of an older crater (probably formed in an explosion in 79 A.D.). Part of the outer wall—the Monte Somma—remains (left) and has given its name to all such crater wall remnants. (*Hunting Surveys.*)

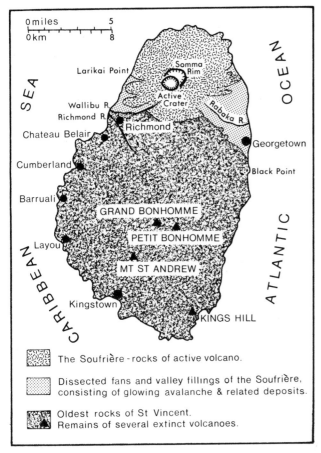

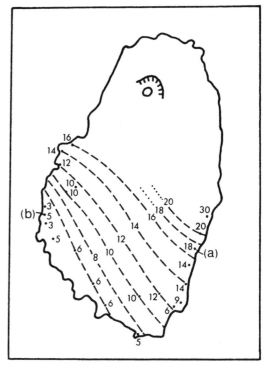

Fig. 6.11a   Map of the island of St Vincent, Windward Isles, showing lava provinces. *(After R.L. Hay, "Origin and Weathering", pp. 65–87, by permission of J. Geol.)*

Fig. 6.11b   Thickness of volcanic ash on St Vincent, in feet. *(After R.L. Hay, "Origin and Weathering", by permission of J. Geol.)*

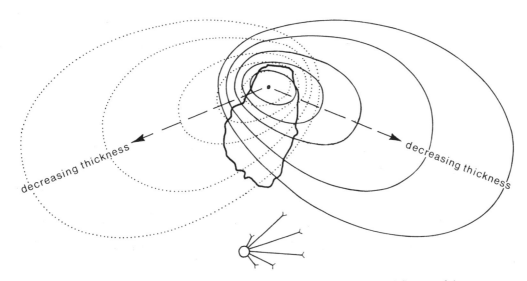

Fig. 6.11c   Distribution of ash on St Vincent as it should be under the influence of the northeast Trades and as it is in reality, apparently under the influence of a west wind regime. *(After R.L. Hay, "Origin and Weathering," by permission of J. Geol.)*

when planetary wind systems were disturbed compared to those of the present day.

During the Recent, there have been several periods of explosive activity when andesitic *nuées ardentes* flowed down the southeastern and southwestern flanks of Soufrière, particularly down pre-existing valleys. The latest eruptions were in 1812 and 1902, and ash was scattered far and wide from the outer margins of the vertical uprush of gas and finely comminuted particles (Fig. 6.12). During the 1902 eruption, ash was spread not only over most of St Vincent but also over the surrounding ocean and neighbouring islands. Almost 1.3 cm fell at Kingstown, St Vincent; on Barbados, some 160 km east of St Vincent, falls varied between 0·5 and 0·9 cm; 2·5 cm fell on a ship 160 km from the eruption and some ash fell on another ship located some 1500 km ESE of St Vincent at the time of the eruption. In the latter case the ash must have travelled through the atmosphere at a rate of some 1130 km an hour.

More recently (August 1968), gas and vapour at a temperature of 788 ° C was blasted from the side of Mt Arenal in Costa Rica, killing seventy-eight people and taking the lives of ten would-be rescuers, who were in a car incinerated by a sheet of flame which suddenly burst from the side of the volcano. Arenal had been dormant for nearly 500 years, but pressures had evidently built up gradually to irresistible proportions and tragedy ensued.

The 1951 eruption of Mt Lamington (Lat 8° 56′S, Long 148° 10′E) in northern New Guinea was closely observed by an able volcanologist and the accounts of the outburst are among the most complete in the literature of volcanology, and are certainly the best of a Peléean eruption.[8]

It is difficult now to believe that until 21 January, 1951 the volcanic nature of Mt Lamington was not appreciated, nor even suspected. The breached crater (Pl. 6.10a), as it was later recognised to be, was seen as merely a large headwater amphitheatre of the Ambogo River. For several weeks before the explosion, landslides were noted on the steeper slopes of the crater, probably as a result of earth tremors. The vegetation near the crater is said to have become brown. On 15 January, a thin column of smoke, like that from a camp-fire, rose from the crater and continued the next day when more landslides were noted. On the 17th, the emission of gases increased and a seismic recording station, fortunately (from a scientific point of view) located only 10 km from Mt Lamington, registered some eight tremors each hour through the day. By the next day, the cloud of smoke was darker, indicating the presence of large quantities of volcanic dust in it, and ash flows were noted in the crater. For the next two days large clouds of dust-laden smoke, accompanied by a continuous roll of earth tremors, poured from the crater. On the Saturday night there was a lull, but this was literally the calm before the storm.

For at 10.40 a.m. on Sunday, 29 January, a huge explosion, which was heard 270 km away, sent a mushroom of dark clouds at least 17,000 m into the air. Only about four minutes later a lethal cloud of glowing gas, incandescent dust and rock fragments poured out of the crater and down the mountainside in all directions, but especially to the north where the old crater was breached and where the *nuées ardentes* poured for 13 km before an indraft of air forced it above ground level. These incandescent clouds killed almost

Plate 6.10a Mt Lamington, looking south into the open north side of the crater up an avalanche valley. Taken January 30, 1951. (*Dept Nat. Dev., Canberra.*)

Plate 6.10b Same view on March 30, 1951, with central dome still developing. (*Dept Nat. Dev., Canberra.*)

Plate 6.10c Close-up of newly developed dome, taken April 8, 1951. (*Dept Nat. Dev., Canberra.*)

3000 people, many of whom died as a result of inhaling hot air and dust.

At 8.45 that evening there was another major explosion involving incandescent lava, followed by a period of quiet which lasted three days. But this phase of inactivity heralded what is surely the most fascinating episode of the entire eruption. On the fourth day after the explosions, minor explosions were renewed and continued for a week. On 3 February a lava dome began to grow out of the floor of the old crater. By 10 March, it had risen 330 m and the extrusion of viscous lava continued (Pl. 6.10b-d). The growth was not continuous, either in space or time. Fingers and needles of lava were squeezed up, in many places only to be destroyed by explosions. They were replaced by protrusions in other places. Also during this period, a flow of lava blocks and ash reached a point 14 km to the north of the crater before it lost its mobility, and there was a small outburst of *nuées ardentes*, which spread 3 km from the crater, also on the northern side (Pl. 6.10e).

For the remainder of 1951 the dome continued to grow. In January 1952 there were many earthquakes, but the summit of the dome reached a height of 63 m above the crater floor. It continued to expand laterally, though not in height, and in fact during the last half of the year when activity in general was on the decrease, the dome lost about 30 m from its summit as a result of various minor explosions. By January 1953, when the eruption of Mt Lamington was virtually over, the new dome stood 30 m above the floor of the old crater, and some 1700 m above sealevel. During this prolonged phase of activity, this predominantly Peléean volcano also displayed characteristics of Vulcanean eruptions. Vertical explosions occurred from time to time, though some erupted unpredictably through fissures or vents in the flanks of the mountain.

Some acid volcanoes obliterate themselves in massive explosions. In addition, the long continued pouring-out of material may cause collapse, and a combination of collapse and explosive activity has given rise to major *calderas*. Lake Toba in Sumatra is considered by many to be the largest caldera in the world (Figs 6.1 and 6.13). During the famous 1883 explosion of Krakatoa, a volcanic island in the Sunda Straits between Java and Sumatra (Fig. 6.1), it is estimated

Plate 6.10d Vertical of dome within old crater of Mt Lamington, taken November 10, 1953. (*Dept Nat. Dev., Canberra.*)

Plate 6.10e Small ash flow on the edge of an area on the north side of Mt Lamington devastated by *nuées ardentes. (Dept Nat. Dev., Canberra.)*

that some 176 km³ of material were poured out and a further 74 km³ engulfed in the subsequent collapse. Only a small part of the original island survived the explosion, which was heard as far away as Perth and Alice Springs. Left behind was a huge cauldron 6·5 km in diameter and 2000 m deep in places.

Crater Lake, Oregon (Pl. 6.11) was formed about 6500 years ago by the collapse of the 4000 m high Mt Mazama.[9] The mountain partly and literally blew its top, and partly collapsed, leaving behind a crater almost 10 km across and 600 m deep (Pl. 6.11, Figs 6.14, 6.15a). During previous volcanic activity and also as a result of the final catastrophe, ash-flows and glowing clouds spread down nearby valleys (Fig. 6.15b).

3. Composite or Stratovolcanoes

It has been mentioned that basaltic volcanoes and plains are built up of layer upon layer of lava. Many other well-known volcanoes are of andesitic composition, are not now as explosive as was Krakatoa, or as Mt Lamington was in the early 1950s, and are also layered. Mt Fuji, many other cones in Japan, Mt Egmont and other volcanoes of North Island, New Zealand, are examples (Pl. 6.12a-c, Fig. 6.16a and b). The cones are built up of layers of lava interbedded with layers of ash and other fragmented or pyroclastic rocks produced by explosive activity in the craters (Pl. 6.13). The cones are called composite or *stratovolcanoes* and attain slopes of up to 30°.

Mt Fuji (Pl. 6.12d) at 3776 m is Japan's highest mountain and is a stratovolcano, the development of which has been elucidated in some detail. Structurally, Mt Fuji consists of three separate volcanoes, each younger structure being superimposed on the older ones (Fig. 6.16a). Since the whole mass of extrusive rocks rests upon folded older rocks involved

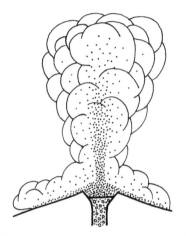

Fig. 6.12 Suggested mode of origin of ash at Soufrière, with the avalanches discharged from the outer part of the cloud which has a dense central ash column. (*After R.L. Hay, "Formation of Crystal Rich flowing Avalanche Deposits of St Vincent, B.W.I." by permission of J. Geol.*)

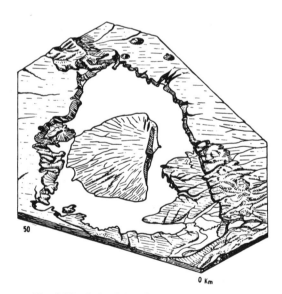

Fig. 6.13   Lake Toba, Sumatra, a depression which is partly tectonic, partly volcanic in origin. The block which forms an island is Samosir and is believed to be part of the former roof of the magma chamber. (*After R.W. Fairbridge (Ed.),* Encyclopaedia of Geomorphology, *Reinhold, New York, 1968, p. 617.*)

Plate 6.11   Crater Lake, Oregon, from the west. Mt Scott, another volcano, is clearly visible beyond the lake. (cf. Fig. 6.14). (*C.R. Twidale.*)

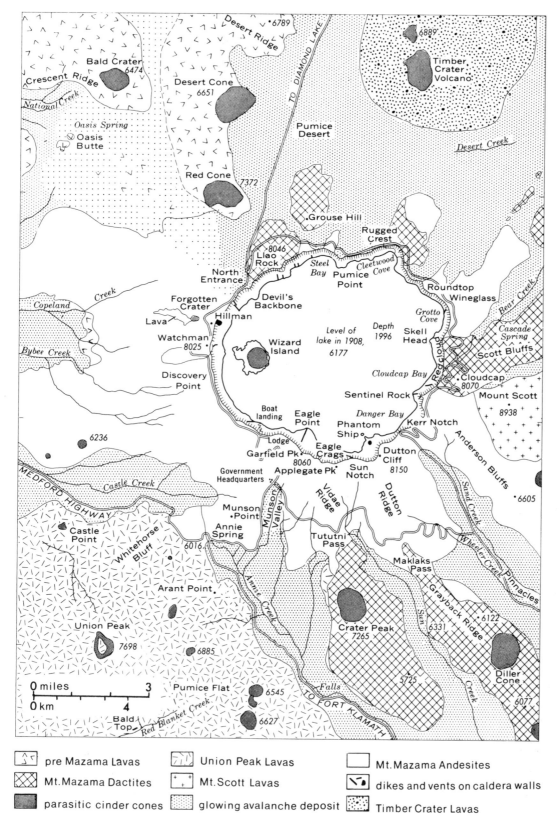

Fig. 6.14 Geological setting of Crater Lake, Oregon. (*After H. Williams, "The Geology of Crater Lake", courtesy of the Carnegie Institution.*)

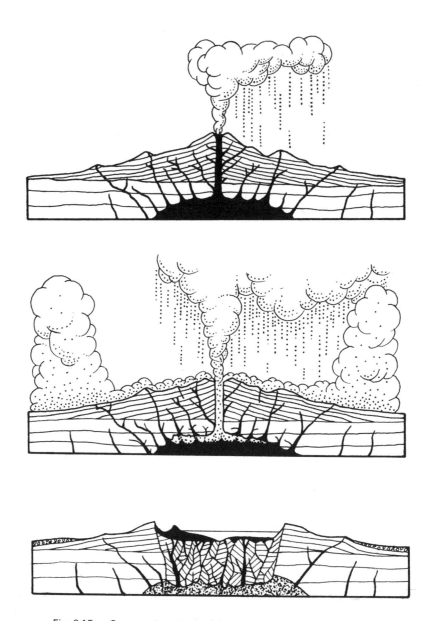

Fig. 6.15a   Suggested mode of origin of Crater Lake. *(After F.E. Matthes, "Crater Lake Map", U.S. Geol. Surv.)*

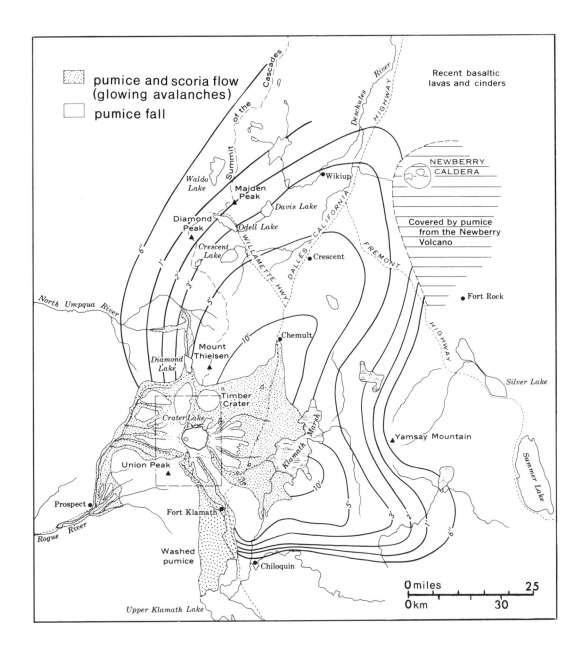

Fig. 6.15b   Distribution of *nuées ardentes* and ash around Crater Lake. (*After H. Williams, "The Geology of Crater Lake", courtesy of the Carnegie Institution.*)

Plate 6.12a   Mt Bandai, composite cone of pyroclastics in central Japan. *(Int. Soc. Educ. Inf. Tokyo.)*

Plate 6.12b   Mt Egmont, a stratovolcano in the North Island, New Zealand, from Kepuni, Taranaka. (*N.Z. Govt Tour. Bur.*)

Plate 6.12c Mt Fuji, a stratovolcano in Honshu, central Japan. (*Int. Soc. Educ. Inf. Tokyo.*)

Plate 6.13   Layers of ash exposed in the inside crater wall at Tower Hill, western Victoria. *(J.R. Morrow.)*

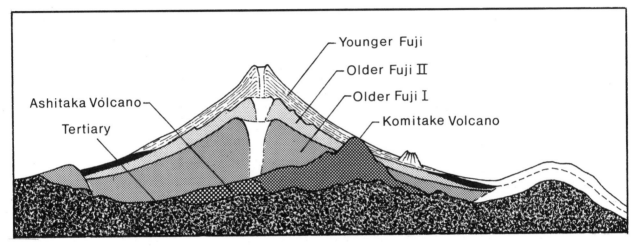

Fig. 6.16a   Section through Mt Fuji, central Japan.

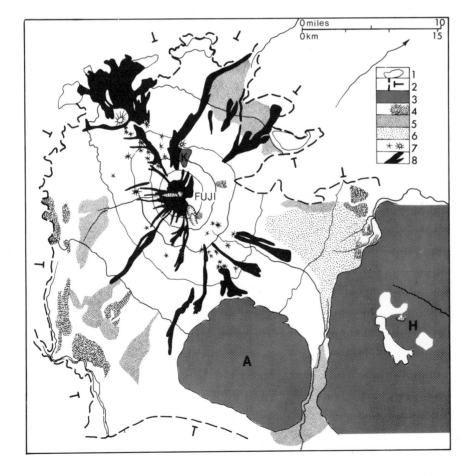

Fig. 6.16b Geological map of Fuji area:
1.   lakes
2.   Tertiary mountains
3.   Komitake, Ashitaka and Hakone volcanoes
4.   mud-flows and other pyroclastics of the Old Fuji volcano
5.   alluvial fans
6.   Gotemba mud-flow deposits
7.   parasitic cones
8.   latite lava flows (*After H. Machida, "The Recent Development", pp. 11–20, by permission of* Geogr. Rep., Tokyo Metro. Univ.)

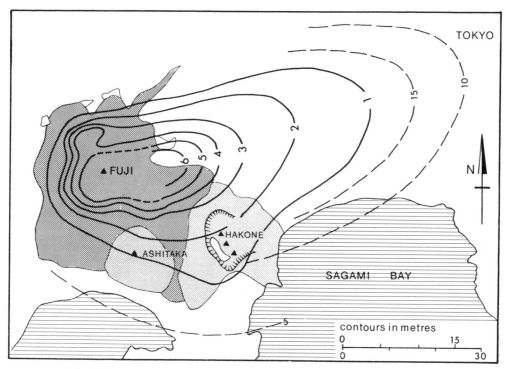

Fig. 6.17   Spread of ash from Mt Fuji, in metres; solid line related to Younger Fuji, dotted line to Older Fuji. *(After H. Machida, "The Recent Development", by permission of Geogr. Rep., Tokyo Metro. Univ.)*

Fig. 6.18   Spread of ash over southern South America following the 1932 eruption of Quisapu. *(After W. Larssen, "Vulkanische Asche", pp. 27–52, by permission Bull. Uppsala Univ. Min. Geol. Inst.)*

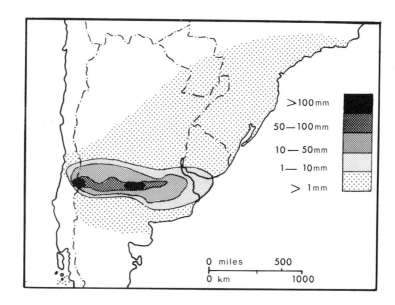

in Tertiary mountain building, even the oldest of the volcanoes, Komitake, together with the other nearby older structures—Ashitaka and Hakone (Fig. 6.16b)—are of Quaternary age. The older volcanoes are Pleistocene, but the two younger volcanoes, Old Fuji and New Fuji, can be dated more precisely because associated .ash layers contain human artefacts and have datable soils developed on them.

The region in which the volcano is located is dominated by westerly winds, so that the ash layers or *tephra* associated with its explosions and other milder phases of activity have been spread mainly to the east (Fig. 6.17). These form stratigraphic layers which enable the phases of volcanicity to be dated. Such chronological studies using tephra are called *tephrochronology*.[11]

The Old Fuji is found to have been built up in two distinct phases separated by some erosion and local faulting. The first phase was explosive and had associated basalt flows; stone implements contained in the ash suggest a date of 25,000 –10,000 years ago. The second Old Fuji phase, which occurred 10,000–5500 years ago, as dated by associated earthenware fragments, was more effusive but still caused the spread of tephra. Then followed a quiescent period which lasted perhaps 4000–5000 years, and which ended with the historical phases of volcanicity. The uppermost two or three layers of the New Fuji phase are known from historical records to have been deposited since 781 A.D. These include Hoei tephra laid down when Mt Fuji was active in 1707.

Ash is associated with many explosive volcanoes. During the explosion of Mt Mazama, ash was spread over a wide area of the surrounding countryside[12] (Fig. 6.15b). Ash from the great Krakatoa explosion of 1883 is said to have travelled round the world in the planetary wind circulation many times and to have stayed in suspension for many weeks. Its presence in the atmosphere interfered with the diffusion of light from the sun and caused a series of particularly brilliant sunrises and sunsets. The suspended dust also made the moon appear blue: hence the derivation of the saying "once in a blue moon", meaning a rare occurrence. The eruption of Quisapu in southern Chile in 1932 spread ash as far afield as Rio de Janeiro (Fig. 6.18) under the influence of the prevailing westerly winds.[13]

Nuées ardentes (Pl. 6.8) are associated with the activity of composite volcanoes, and these, together with *lahars* or volcanic mudflows caused by the downslope movement of water and fine debris originating in the crater, carry detritus from the cones to the surrounding plains. Hence, considerable aprons of volcanic detritus, as well as lava flows, can be deposited around the centres of eruption. Herculaneum was buried by a volcanic mud-flow in 79 A.D.

There are considerable spreads of pyroclastics around Mt Garibaldi, an extinct Pleistocene dacitic volcano, some 60 km north of Vancouver, British Columbia. The associated pyroclastic debris has an interesting distribution for it extends further downslope on the ridges than in the valleys. A suggested explanation[14] is that when the volcano was in eruption, the valleys were occupied by late Pleistocene (Wisconsin) glaciers.

Although extremely acid volcanoes (that is, those of rhyolitic composition) are not common (rhyolite plugs occur in the Sierra Nevada of California for example, but in few other places), spreads of acid ash (ignimbrite) together with

flood basalts are the most common of the continental volcanic rocks. These siliceous lavas appear to be associated with calderas. Some, such as the Bishop Tuff, have been described from California (where they form invaluable time markers, datable by K : Ar ratios,[15] within the early glacial sequences), from Sumatra[16] (Fig. 6.2) and from northern Chile.[17] One spread of ignimbrite occurs around Lake Taupo and in the Rotorua area in the North Island of New Zealand, where as in the other areas mentioned, the acid ash was so compacted on deposition that the ignimbrite is also known as a "welded" tuff. The ash spread at least 60 km from its source. Within the plateau so formed, Lake Taupo occupies a downfaulted depression.

D. OTHER VOLCANIC FEATURES

Some volcanoes, both acid and basic, are so explosive that the lava is fragmented into ash, forming ash cones (Pl. 6.14, Fig. 6.7) such as those in Death Valley, California, some of the Mono Craters and other ash cones (for example at Black Hill) around Mono Lake,[18] also in California, parts of Tower Hill[19] (Pl. 6.7a and b) and Mt Schank in the southeast of South Australia. In some instances, the rock is broken into large chunks or scoria, in which case *scoriacones* (conical accumulations of these fragments) are built up. The pipes of some volcanic vents which lack rims subside and become lakes, or *maars* (Pl. 6.15).

Finally, the long-continued erosion of even large volcanic cones, of whatever composition and origin, results in their total destruction. The last remnants are the central plugs, formerly the volcanic necks, daitremes or gas pipes. Some are finer grained than the surrounding rocks, others filled with rocks of different composition, or are, for other reasons, more resistant than the country rock. The central plugs form dramatic steep-sided turrets, towers or *puys*, many with vertical columnar jointing (Pl. 6.16). The Devils Postpile in California is one which is very well known, and the Devils Tower, Wyoming, and Shiprock, New Mexico are other spectacular examples (Pl. 6.17a-c). In similar fashion, in places the coagulated lava from fissures survives to form

Plate 6.14   Yolnir, a basaltic ash cone off the Icelandic coast, which was built up in 1966. (*T. Einarsson.*)

Plate 6.15   Maar at Keilambete, Victoria. *(C.D. Ollier.)*

Plate 6.16 Wase Rock, a pillar of trachyte of Tertiary age, Nigeria. (*Hunting Surveys.*)

Plate 6.17a Columnar jointing at Devils Postpile, California. (U.S. Dept Int., Nat. Park Service.)

Plate 6.17b The Devils Tower, Wyoming, rises some 180 m above the sedimentary strata on which it rests. Note the magnificent columnar jointing. (U.S. Dept Int., Nat. Park Service.)

Plate 6.17c Shiprock, New Mexico, a basaltic plug from which the original cone has been eroded, standing some 440 m above desert plains. Note the associated radial dykes. (Bureau of Indian Affairs.)

Plate 6.18   Trachyte plug of Tertiary age, Glasshouse Mountains, southeastern Queensland.
(Queensland Railways.)

Plate 6.19a Crater Bluff, Warrumbungles, northern
N.S.W. *(N.S.W. Govt Tour. Bur.)*

Plate 6.19b Radially disposed dykes, the
tallest one known as the Breadknife, Warrumbungles, N.S.W. *(N.S.W. Govt Tour. Bur.)*

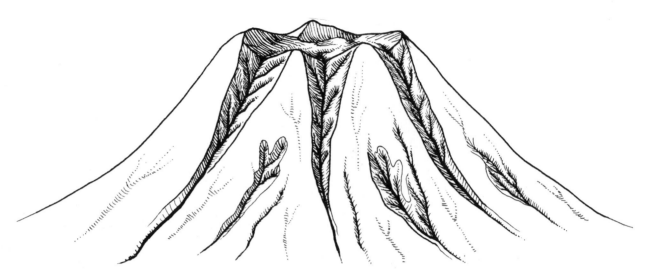

Fig. 6.19   Planeze developed through the dissection and headward extension of streams on a volcanic cone.

long, narrow, but high and vertical-sided residuals. In the Glasshouse Mountains of southeast Queensland (volcanic plugs of mid Tertiary age) and the Warrumbungle Mountains (plugs and fissure fillings of late Tertiary age in northern N.S.W.), such volcanic remnants are prominent landforms (Pls 6.18, 6.19a and b).

The flanks are scored by deep valleys on the sides of which successive layers of lava are exposed. The intervening gently sloping facets are called *planeze* (Fig. 6.19). In many provinces, steam and gas emissions, sulphur springs, boiling mud and geyser activity represent waning volcanic activity (Pls 6.20, 6.21a-c). The vents through which steam and gas are emitted are called *fumaroles*. *Solfatara* (after the town of the same name near Naples, Italy) are also vents through which gases escape, but these are of lower temperature (200° C) than those emerging from *fumaroles* which are generally in the range of 200°–1000° C. *Mofettes* are vents emitting water and carbon dioxide only. These minor, warning phenomena of volcanism are referred to collectively as *solfataric* activity.

E. GENERAL SIGNIFICANCE

The impact of volcanic activity is considerable. It has been estimated that 200,000 people have died through such activities over the past 400 years. Some of this loss of life has resulted directly from volcanic activity. The disaster of St Pierre, Martinique, has already been mentioned; 56,000 people died in a single gigantic explosion at Tamboro in the East Indies in 1815; another well-known tragedy occurred during the eruption in 79 A.D. of Mt Somma, the ancestor of Mt Vesuvius, when the Roman towns of Pompeii and Herculaneum were destroyed (see Pl. 6.9d).

But, more commonly, it is the indirect effects of volcanism which are most disastrous. Only a handful of the 36,000–37,000 who died as a result of Krakatoa's eruption and self-destruction were killed by explosions or ash: most were drowned in the great tidal wave—estimated to have been up to 40 m high—which followed the final collapse of the island. An eruption of the Laki fissure in Iceland in 1783 had little immediate effect, but the ash blown from the fissure destroyed pastures, causing the loss of half the island's cattle, and the death of one-fifth of the human population through famine. The recent explosion of Mt Arenal has caused ash up to 1 m deep to be spread over an area of many square miles: 800 cattle died in the explosion, but 80,000 have had to be slaughtered and sold because of the destruction of pastures, and yet another 100,000 have had to be moved to other areas in order to survive.

On the other hand, volcanic activities bring to the surface new rocks, which are the parent materials for new, rich soils developed as a result of weathering (see Chapter 9).

Plate 6.20   Pohutu Geyser, Rotorua, North Island, New Zealand. *(N.Z. Govt Tour. Bur.)*

Plate 6.21a   Miniature geyser and silica flowers on Warbrick Terrace, Rotorua. *(N.Z. Govt Tour. Bur.)*

Plate 6.21b   The Frog Pool, with bubbling mud, Rotorua. *(N.Z. Govt Tour. Bur.)*

Plate 6.21c   Detail of bubbling mud. *(N.Z. Govt Tour. Bur.)*

References Cited

1. A. LACROIX, *La Montagne Palée et ses Eruptions,* Paris, 1904; *idem,* "La Montagne Pelée après ses Eruptions", *Acad. Sci. Paris,* 1908, pp. 74–93.

2. For detailed accounts of the day-to-day activity of Hawaiian volcanoes, see T.A. JAGGAR, "Origin and Development of Craters", *Geol. Soc. Am. Mem.,* 1947, **21.**

3. P.M. BLACK, "Observations on White Island Volcano, New Zealand", *Bull. Volc.,* 1970, **24** (1), pp. 158–67.

4. Some Australian examples of various scales are described in T.G. TAYLOR, "Physiography of Eastern Australia", *Comm. Met. Bur. Bull,* **8;** C.R. TWIDALE, "Physiographic Reconnaissance of some Volcanic Provinces in north Queensland, Australia", *Bull. Volc. II,* 1956, **18**, pp. 3–23; C.D. OLLIER, *Volcanoes,* Australian National University Press, Canberra, 1970; A. CUNDARI and C.D. OLLIER, "Inverted Relief due to Lava Flows along Valleys", *Aust. Geogr.,* 1970, **11** (3), pp. 291–3.

5. See G.A. TAYLOR, "The 1951 Eruption of Mt Lamington, Papua", *Bull. Bur. Miner. Resour. Geol. Geophys. Aust.,* **38**, pp. 33 *et seq.*

6. For a contemporary account see H.N. DICKSON, "The Eruption in Martinique and St Vincent", *Geogr. J.,* 1902, **20**, pp. 49–68.

7. R.L. HAY, "Origin and Weathering of Late Pleistocene Ash Deposits on St Vincent, B.W.I.", *J. Geol.,* 1959, **67**, pp. 65–87; *idem,* "Formation of Crystal rich flowing Avalanche Deposits of St Vincent, B.W.I.", *J. Geol.,* 1959, **67**, pp. 540–62.

8. G.A. TAYLOR, "1951 Eruption", pp. 33 *et seq.*

9. F.E. MATTHES, "Crater Lake. Commentary on U.S. Geol. Surv. 1:62 500 map, Crater Lake National Park and Vicinity, Oregon", 1956; H. WILLIAMS, "Caderas and their Origin", *Univ. Calif. Pub. Dep. Geol. Surv. Bull.,* 1941, **25**, pp. 239–346.

10. H. TSUYA, "Geological and Petrological Studies of Volcano Fuji. III. Geology of the southwestern Foot of Volcano Fuji", *Bull. Earthquake Res. Inst.,* 1940, **18**, pp. 419–45; H. MACHIDA, "The Recent Development of the Fuji Volcano, Japan", *Geogr. Rep. Tokyo Metro. Univ.,* 1967, **2**, pp. 11–20.

11. S. THORARINSSON, "The Tephra Layers and Tephrochronology" in "On the Geology and Geophysics of Iceland", *Mus. Nat. Hist. Reykjavik Misc. Paper,* 1960, **26**, pp. 55–60.

12. H. WILLIAMS, "The Geology of Crater Lake National Park, Oregon, with a Reconnaissance of the Cascade Range southward to Mount Shasta", *Carnegie Inst. Wash. Pub.,* 1942, **540.**

13. W. LARSSON, "Vulkanische Asche vom Ausbruck des Chilenischen Vulkans Quizapu (1932) in Argentine gesammelt", *Bull. Uppsala Univ. Min. Geol. Inst.,* 1937, **26**, pp. 27–52.

14. W.H. MATHEWS, "Mount Garibaldi, a Supraglacial Pleistocene Volcano in southwestern British Columbia", *Am. J. Sci.,* 1952, **25**, pp. 81–103.

15. G.B. DALRYMPLE. "Potassium-argon Dates of Three Pleistocene Interglacial Basalt Flows from the Sierra Nevada, California", *Bull. geol. Soc. Am.,* 1964, **75**, pp. 753–7; *idem,* "Potassium-argon Dates of some Cainozoic Volcanic Rocks of the Sierra Nevada, California", *Bull. geol. Soc. Am.,* 1963, **74**, pp. 379–90; G.B. DALRYMPLE, A. COX and R.R. DOELL, "Potassium-argon Age and Palaeomagnetism of the Bishop Tuff, California", *Bull. geol. Soc. Am.,* 1965, **76**, pp. 665–74.

16. J.A. KATILI and F. HEHUWAT, "On the Occurrence of Large Transcurrent Faults in Sumatra", *J. Geosci. Osaka Univ.,* 1967, **10**(1), pp. 5–17.

17. S.E. HOLLINGWORTH and R.W.R. RUTLAND, "Studies of Andean Uplift. Part I. Post Cretaceous Evolution of the San Bartolo Area, North Chile", *Geol. J.,* 1968, **6**, pp. 49–62.

18. See for example D.W. SCHOLL, R. VON HUENE, P. ST-AMAND and J.B. RIDLON, "Age and Origin of Topography beneath Mono Lake, a Remnant Pleistocene Lake, California", *Bull. geol. Soc. Am.,* 1967, **78**, pp. 583–600. This gives much useful information and also contains references to the earlier literature concerning the volcanic features of the area.

19. See E.D. GILL, "Evolution of the Warrnambool-Port Fairy Coast and the Tower Hill Eruption, Western Victoria", in J.N. JENNINGS and J.A. MABBUTT, Eds, *Landform Studies from Australia and New Guinea,* Australian National University Press, Canberra, 1967, Ch. 15, pp. 341–64.

General and Additional References

C.A. COTTON, *Volcanoes as Landscape Forms,* Whitcombe and Tombs, Wellington, 1944.

A. RITTMAN, *Volcanoes and their Activity* (trans. E.A. Vincent), Wiley, New York, 1962.

CHAPTER 7

Pseudostructural Landforms

A. GENERAL

Several forms which simulate features associated with the folded sedimentary sequences or with the volcanic eruptions already described are not, however, related to deep-seated activities in the crust. What has caused the development of these features? Less familiar, less commonly recognised, and less well understood, these structures and forms are apparently due not to tectonic activities, but to disturbances in the upper few metres of the earth's crust. These disturbances are caused by a variety of mechanisms, which include the overloading of the crust, local overloading of plastic clays and volumetric changes in superficial layers. Many of these disturbances are referred to in other contexts and chapters. They produce deformations which may easily be mistaken for true tectonic features and which find expression in relatively minor relief forms. These may be, however, more common than we suspect at present. Subsidence, collapse under the influence of gravity, and the impact of extraterrestrial bodies have also been suggested as reasons for the development of forms which could, mistakenly, be interpreted as being of tectonic or of structural origin.

B. CYPTOEXPLOSION FEATURES OR METEORITE CRATERS?

In many parts of the world, spectacular features have been observed in the form of essentially circular or oval craters with raised rims; or shallow, circular or oval depressions with raised central bosses; or broad, shallow depressions without rims. Many of the craters are occupied by lakes. The diameters of the structures vary from a few metres to several kilometres. The New Quebec Crater[1] of Labrador (61° 18'N, 73° 40 W) is 3 km in diameter and 400 m deep (Pl. 7.1). The Ashanti Crater, Ghana (6°30'N, 1° 25'W), is over 8 km in diameter and up to 450 m deep; it is occupied by Lake Bosumtwi[2] which has a depth of 80 m; the crater walls rise 300–400 m above the lake and stand 100–200 m above the level of the surrounding plains. Wolf Creek Crater, Western Australia (19° 10'S, 127° 50'E), is 2.5 km in diameter and has a probable depth of up to 200 m (Pl. 7.2). It is now partially filled in with colluvial and aeolian debris.[3] Gosses Bluff, Northern Territory, is a circular sandstone rim some

Plate 7.1 The New Quebec Crater, 3 km diameter and 400 m deep, is located in the barren tundra of northern Labrador Peninsula. Note the numerous lake basins due to the glaciation of the region. (*Dept Energy, Mines and Resources, Ottawa.*)

Plate 7.2 Wolf Creek Crater, 2 km diameter, in the north of Western Australia. (*W.A.P.E.T.*)

Plate 7.3   Gosses Bluff, in the Northern Territory, some 43 km WSW of Alice Springs. The feature
is some 4 km in diameter. *(Dept Nat. Dev., Canberra.)*

Fig. 7.1 Distribution of tectites (australites)
in South Australia and Australia. Suggested
tracks of showers shown. (*After D.H. McColl
and G.E. Williams, "Australite Distribution",
pp. 154–5, by permission of the authors and
Nature (Lond.).)*

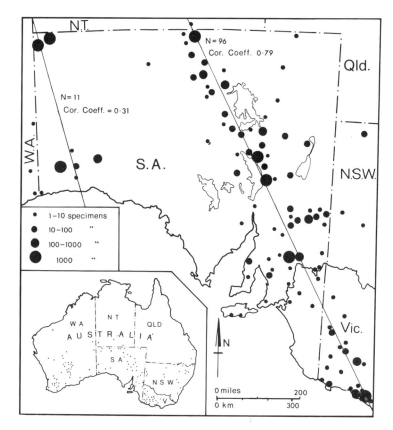

4 km diameter standing some 180 m above the level of the surrounding desert plain (Pl. 7.3).

How did these features originate? A meteorite impact origin has been entertained for many of them.[5] This hypothesis is based on:

1. the similarity of the craters to lunar craters which are interpreted as being due to meteorite impact;
2. comparison with the craters known to have been formed by observed meteorite falls, such as the 30 m diameter craters produced by large iron meteorites at Sikhote Alin, Siberia, in 1947;[6]
3. the absence of recognisable volcanic phenomena, remoteness from volcanic regions, and the evident shallowness of the structures (see, for example, Cook [7] with respect to Gosses Bluff); and
4. the presence of coesite, a dense, high temperature, high pressure form of silica, in some craters; of shatter cones, taken as formed only under conditions of strong impact, in many craters; and of nickel–iron in some craters.

Apart from the possibility that the features listed under the fourth category may form in ways other than by impact or in association with meteorites, perhaps the most telling single argument against the extraterrestrial causation of many of the alleged meteorite craters is that their distribution is not random.[8] On the contrary, it has been alleged that craters occur on known structural trends, for instance the craters southeast of Stuttgart, the Ries and Stenheim basins in southern Germany; the Wells Creek basin structure, Tennessee and Hicks Dome, Kentucky; and the Vredefort Dome of South Africa. In each case it is suggested that the features are aligned along major regional structural axes with which volcanic phenomena are associated. Gosses Bluff is located on a major regional anticlinal trend and a definite distributional pattern has been detected with respect to the Canadian crater forms.[9]

Seemingly contrary to this argument, however, as it applies to the aligned distribution of features, is the observation by McColl and Williams [10] that australites (tectites) display a linear, zoned distribution, presumably because they fell from distinct meteor showers passing over the earth's surface (Fig. 7.1). But if meteor showers are invoked, the alleged impact features in a given zone should all be of the same age, and there is no suggestion that this is the case.

It is certain that small craters such as those referred to in Siberia are due to meteorite impact and almost certain that features such as the Henbury Craters [11] of the southern part of the Northern Territory are of similar origin. Other craters, including forms like the Arizona Meteor (Barrungen) Crater,[12] are susceptible to interpretation either of impact by an extraterrestrial body or of internal (geological) agencies. The weight of evidence in many cases favours an extraterrestrial origin, but caution is necessary, for some of the arguments presented by proponents of the extraterrestrial theory are logically suspect [13] and reasonable alternative arguments have been advanced. For instance Bucher,[14] Goguel, [15] and Currie[16] favour gaseous extrusion (steam produced from the water of magmatic crystallisation such as is associated with modern volcanic activity) or other internal upthrusting as the cause of many of the craters (Fig. 7.2). Such a hypothesis accounts for their essential morphological features, explains their occurrence on major structural axes, their association with volcanic forms, and also the occurrence of circular faults in association with the features.

But, despite many uncertainties and difficulties, and bearing in mind that many of the craters suspected of being of extraterrestrial origin[17] should never have been included in such a list, most of the craters are probably due to meteorite impact.

C. GRAVITY COLLAPSE FEATURES

The Zagros Mountains of Mesopotamia (used here to include parts of both Iraq and Iran) are built of Mesozoic and Tertiary sediments which have been compressed into a series of folds, simple and open in character, but of great amplitude (up to 5000 m). In typical sequences, massive pre-Albian (Cretaceous) limestone forms the cores of the structures, and is overlain and underlain by thick sequences of soft clays and marls. The upland has been, and is, subject to vigorous erosion by rivers which have excavated many deep valleys, principally along the dominant NW–SE strike of the rocks and folds, but also across the structures, forming deep, narrow gorges or *tangs* (see also Chapter 21).

Superimposed on these regional structures however are numerous dislocations and folds alien to the overall pattern. They are not due to tectonism but rather to deep dissection which has induced instability in the flanks of some folds. The

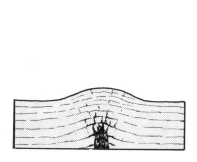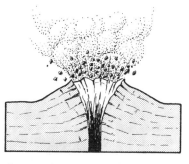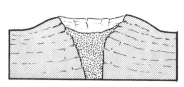

Fig. 7.2 Development of craters by gaseous (steam) upthrusting. (*After K.L. Currie, "Analogs of Lunar Craters", pp. 914–40.*)

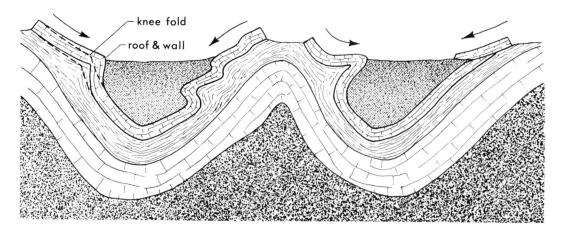

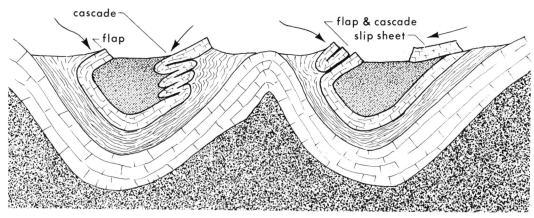

Fig. 7.3 Various types of gravity collapse structure. (*After E.S. Hills,* Elements of Structural Geology, *Methuen, London, 1963, p. 342.*)

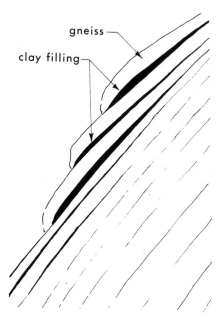

Fig. 7.4 Gravity-induced bulges in the sides of the Torrens Gorge, east of Adelaide, South Australia. (*After D.H. Stapledon, "Geological Investigations", by permission of The Institution of Engineers.*)

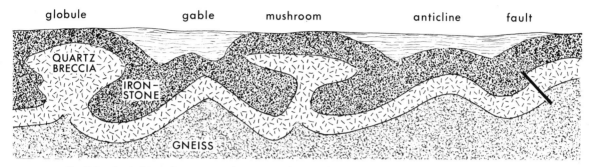

Fig. 7.5 Superficial disturbances in clays in Ghana. *(After W.J. MacCallien et al., "Mantle Rock Tectonics", pp. 257–94. Crown Copyright Overseas Geological Surveys diagram. Reproduced by permission of the Controller, Her Britannic Majesty's Stationery Office.)*

latter now lack the lateral support or buttressing afforded by the overlying sediments. In particular, some of the massive limestone formations fully exposed through the erosion of the overlying argillaceous sediments have become secondarily distorted.[18] The crests of anticlines have subsided, and the flanks have bulged, forming *knee folds*, which develop into *roof and wall structures* (Fig. 7.3). Thus the strata in the fold crests have been disrupted, and those in the flanks of the folds have migrated downslope, in some areas along fault planes forming *slip sheets*, elsewhere as *recumbent folds*, or, where there has been concertina-like compression, *cascades*. In more steeply dipping strata, the erosion of strata in the overturned knee bends and overfolds has left the beds broken and overturned, forming *flaps*. Similar features (although on a minor scale) occur in many deeply eroded areas, for example, the Torrens Gorge[19] near Adelaide (Fig. 7.4). They are also attributed to unbuttressing. Upon excavation of the gorge by the river, vertical pressures became dominant, causing an outward (lateral) arching of the gneisses.

These features have the appearance of tectonic folds, but they are merely superficial structures due to unbuttressing and gravity collapse superimposed on orogenic folding. The presence of soft, plastic clays certainly contributed to the evolution of these various features, which result in odd distributions of resistant strata, and hence of hills and valleys.

D. LOADING AND UNLOADING OF THE CRUST

Reference has already been made (Chapter 2) to the concept of isostasy, and further discussion of the effects of redistribution of load in relation to the waxing and waning of glaciers, and to the distribution of water on the continents, appears in Chapter 23.

E. OVERLOADING OF PLASTIC CLAYS

In many areas underlain by plastic clays, superficial though complex distortion of the strata is in evidence. The structural effects of overloading plastic clays are well displayed near Accra, in Ghana.[20] There, weathered gneisses were shaped by waves into a shore platform; on this, a quartz breccia and other sediments were laid down and a pedrogenic accumulation of pisolitic iron developed on them. The gneisses are unevenly weathered with the lower limit of weathering, or weathering front (see Chapter 9), displaying distinct highs and lows. The strata above the gneiss, though originally flat-lying, are now considerably disturbed, displaying *globules, gables, mushrooms* and *tongues*. In addition, minor faulting has developed (Fig. 7.5). These structures have been attributed to the gravitational pressure exerted by the superior dense ironstone upon the small basins in the weathering front, and forcibly injected into the overlying deposits.

These particular structures, though well displayed in road cuttings, achieve only minor surface expressions, probably because of the inherent strength of the ironstone capping. However, in Northamptonshire in eastern England, genetically similar structures simulate tectonic features but are restricted to the uppermost layers of the crust.[21]

The Northamptonshire ironstone field occurs in a dissected plateau which is located 100–170 m above sealevel and slopes gently down to the east. The valleys, though narrow, are deep; the rivers have cut down 70 m and more below the surface of the plateau which is capped by various resistant members of the Jurassic Oolite groups, prominent among which are limestone, sandstone and ironstone. The Oolite groups are underlain by soft upper Lias clays which have been penetrated by the rivers in many areas. The weight of the overlying limestones and other dense rocks has

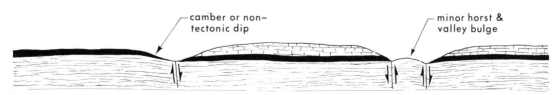

Fig. 7.6 Sections through the Northampton ironstone field showing superficial structures. *(After S.E. Hollingworth et al., "Large Scale Superficial Structures", pp. 7 and 20, by permission of G.A. Kellaway and the Geol. Soc. London.)*

Plate 7.4a   Truncated anticlinal structure in Recent muds, Coorong Lagoon, South Australia. Note the sand dunes in the foreground. *(R.G. Brown.)*

Plate 7.4b   Detail of folds in Coorong muds. *(R.G. Brown.)*

caused the weak plastic clays to flow outwards from beneath the plateau into the valley floors, where they form distinct *bulges* (Fig. 7.6). In some instances, these bulges are bounded by faults which affect the thinner, lower edges of the competent strata above, as well as the clays. Minor horsts have been developed in this fashion.

The flow of clays into the valley has caused subsidence of the limestone at the plateau margins. Water has also percolated through the pervious limestones and permeable sandstones to the upper surface of the impermeable clays. Where it runs toward the valleys it washes out some of the clays, again causing the outer edges of the caprocks to be undermined. For both these reasons, secondary and non-tectonic dips, called *camber*, are well developed in the competent caprocks exposed in valley sideslopes (Fig. 7.6). Cambering, in turn, has led to the development of minor faults running along the contour of the hillslopes: the tension on the cambered rock, resting on an unstable base, exceeds its strength, and fractures have formed along which there has been differential movement.

Gravitational pressure due to differential loading by migrating sand dunes has been cited in explanation of the Parramore Island mounds, 3–5 m high, circular in plan and up to 160 m diameter[22], and a similar mechanism offers the most likely explanation for small anticlinal structures described from the Coorong region in the southeast of South Australia.[23] The anticlines are developed in unconsolidated clays close to the margin of the saltwater lagoon. They are 130–200 m long, 2–3 m wide, and 3–4 m amplitude. The folding is complex in detail (Pl. 7.4a-b). The structures occur on the plain exposed between the lagoon and coastal foredunes, and it seems likely that the build-up of the latter has caused local overloading and vertical compression, which is relieved in the least weighted area: the flat depositional plain (Fig. 7.7). In some places, the upthrust of the clay is accompanied by distinct faulting and slumping of the adjacent dune, suggestive of a transference of load stress.

Plate 7.5 Buckling of edges of salt plates on surfaces of Lake Gairdner, South Australia. *(C.R. Twidale.)*

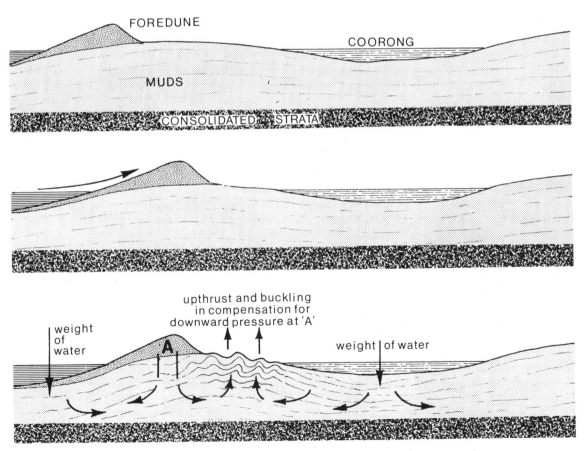

Fig. 7.7 Suggested mode of development of the Coorong folds in Recent muds.

Plate 7.6a Puff and depression of gilgai development in N.S.W.
(*E.G. Hallsworth.*)

Plate 7.6b Microrelief forms due to gilgai development in western Victoria.
(E.G. Hallsworth.)

Plate 7.6c Network of gilgai on cleared Brigalow country, soils developed on basalt,
near Toowoomba, SE Queensland. The rail reserve is about 20 m wide. *(G.D. Hubble and
C.G. Beckman/C.S.I.R.O.)*

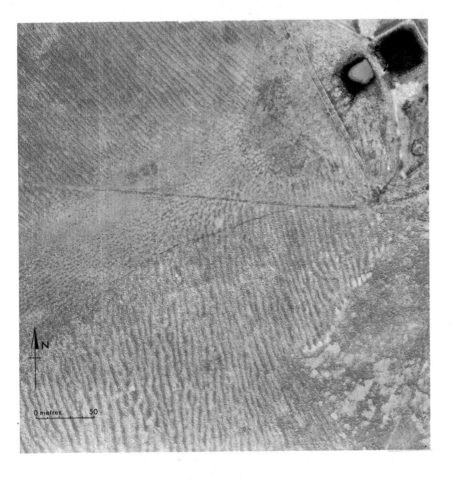

Plate 7.6d Linear gilgai on the Darling Downs, SE Queensland. *(G. D. Hubble and C.G. Beckman/ C.S.I.R.O.)*

F. EXPANSION OF SURFICIAL MATERIALS

Perhaps the most common examples of forms due to volume increase occur on salinas where the crystallisation of salts causes buckling and the development of ridges (Pl. 7.5). In arid and semiarid lands accumulations of salts in soils and in near-surface horizons can apparently cause volume increase and expansion. Lime, for example, can be deposited in a preferred horizon (or horizons) just below the land surface, forming calcrete (see Chapter 10). Continued pedogenic precipitation gives rise to a space problem due to expansion consequent on crystal growth; pressures build up which can only be relieved by upward movement, usually in the form of arching of the calcrete layer. Such structures are not anticlines, because they are not related to deep-seated stresses, but they are commonly referred to as pseudo-anticlines.

They were first described from northeastern Mexico[24] and have also been discovered in the Fitzroy Valley of northern Western Australia.[25] In the latter area they are developed in calcrete. The pseudo-anticlines are seen in the landscape as elongate topographic domes 2–3 m high, 10–50 m across and several hundreds of metres long. The calcrete displays dips of up to 30°. Though the axes of these superficial structures run parallel to a prominent joint direction in the underlying bedrock, they are discordant with the major, deeper structures.

Volume increase consequent upon the conversion of water to ice is responsible for numerous landforms in periglacial regions (Chapter 14), though, in the case of pingos, for instance, other mechanisms have also been suggested.

G. CONTRACTION OR SETTLING

As the more notable landforms due to volume decrease are related to man's activities, they are considered in Chapter 24.

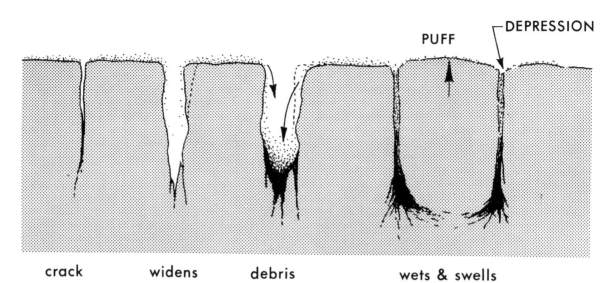

PUFF   ⌐DEPRESSION

crack   widens   debris   wets & swells

Fig. 7.8   Stages in the development of gilgai; the sequence shown progressing from left to right: the initial stage on the left, the fully developed puff and depression on the right. *(After E.G. Hallsworth* et al., *"Studies in Pedogenesis", pp. 1–31, by permission of E.G. Hallsworth and* J. Soil Sci.*)*

H. ALTERNATE EXPANSION AND CONTRACTION

Although contortions of strata caused by alternate expansion and contraction of unconsolidated debris in *gilgai* (Pl. 7.6a-d) and related forms could, at first sight, be mistaken for minor folds, as may the results of frost wedging (Chapter 16), they are identified as pseudotectonic features which are in reality due to mass movements of unconsolidated debris by their superficial distribution, climatic zonation, and in the case of gilgai, their association with clays of particular composition.

The swelling of hydrophilic clays on wetting causes a churning of the soil, with a force sufficient to thrust fence posts from the ground, and gives rise to various patterns of sorted ground. Thus patterned ground (*sensu lato*) is well developed in humid, arid and semiarid regions (as well as cold areas)—in fact in any region with marked wet and dry seasons, regardless of average annual rainfall. In Australia such patterned ground occurs in arid areas of less than 25 cm annual rainfall, as well as in wetter areas like the Darling Downs of southeast Queensland with an annual precipitation which averages 100 cm. The microrelief forms are known variously as melon holes, corduroy soils, crab holes, "Bay of Biscay" soils, or *gilgai*,[26] an Aboriginal word for a small water hole, which is widely used and adopted here.

Several types of gilgai soils have been recognised. The most common gilgai is more or less circular in plan, and consists of a slightly raised puff (after the slightly puffy, almost vesicular character of the surface soils in some parts of the Riverina of New South Wales) and a marginal shelf or depression, generally with several small sink holes (Pl. 7.6b). Some gilgai are arranged in lattices or networks (Pl. 7.6c); some give rise to rectangular depressions like tanks; some display a wavy pattern in plan, with the linear features (Pl. 7.6d) aligned roughly round the contour: they may be elongated along the contour: many, especially in arid and semiarid lands, are stony, with the puff occupied by gibber, and the shelves clear of coarse debris or containing only a few gravel fragments; and some take the form of round or elongated mounds up to 30 cm high developed in turfed soils.

There is a measure of general agreement on the origin of these sorted soils and associated minor relief forms, though it is likely that various mechanisms are involved, either individually or in combination. A correlation has been noted between the clay content of soils (especially montmorillonite), the amount of sodium in the clay lattices, and the development of gilgai. Workers in New South Wales have concluded that the montmorillonite content of the soil, and the exchange capacity (or ability to swell on wetting) of the clay are significant factors in gilgai development. They noted that surface features of the puffs occurred at depth beneath the shelves (for example, stones at the surface, and calcium carbonate nodules in the immediate subsurface of the puff, occur at a depth of several centimetres or, in the shelf area, at a depth of a metre or more) and concluded that a raising of the puff relative to the shelf was involved (Fig. 7.8). Similar evidence of heaving or churning in gilgai soils has been noted in Victoria. A mechanism in explanation of the basic gilgai form (Fig. 7.8) was devised which involved first drying, contraction, cracking and some slumping, followed by wetting, expansion and upthrust.

Working in the western desert of the USA, Springer[27] was concerned with the vesicular character of some gilgai soils, but noted the possibility that although heterogeneous debris would be thrust upwards during wetting and swelling of clay soils, only the finer debris would fall down the cracks developed on drying. In this way, gravel would become concentrated as a surface layer: this is one reason for the stone mantle characteristic of many arid regions (see Chapter 15).

Cold lands display patterned ground similar in many respects to the gilgai developed in extraglacial regions. It takes the form of stone circles, polygons, stripes and steps, some of them several metres in diameter (see Chapter 16). Several explanations have been offered[28] and though there is as yet no general agreement on the point, the basic mechanism certainly involves expansion due to development of ice in the ground. Patterned ground of the cold lands differs from gilgai in that there is no suggestion of structural control whereas specific types of clay are thought to be especially prone to gilgai development.

References Cited

1. K.L. Currie, "Analogues of Lunar Craters on the Canadian Shield", *N.Y. Acad. Sci. Ann.*, 1965, **123**, pp. 914–40.
2. H.P.T. Rohleder, "Lake Bosumtwi, Ashanti", *Geogr. J.*, 1936, **87**, pp. 51–65.
3. G.J.H. McCall, "Possible Meteorite Craters—Wolf Creek, Australia and Analogs", *N.Y. Acad. Sci. Ann.*, 1965, **123**, pp. 970–98.
4. K.A.W. Crook and P.J. Cook, "Gosses Bluff—Diapir, Cryptovolcanic Structure or Astrobleme?", *J. geol. Soc. Aust.*, 1966, **13**, pp. 495–516; D.J. Milton, B.C. Barlow, R. Brett, A.R. Brown, A.Y. Glikson, E.A. Mainwaring, F.J. Moss, E.C.E. Sedmik, J. Van Son and G.A. Young, "Gosses Bluff Impact Structure, Australia", *Science (N.Y.)*, 1972, **175** (4027), pp. 1199–1207.
5. See for instance K.A.W. Crook and P.J. Cook, "Gosses Bluff", pp. 495–516; R.C. Dietz, "Astroblemes", *Sci. Am.*, 1961, **205**, pp. 51–8; *idem*, "Cryptoexplosion Structures: a discussion", *Am. J. Sci.*, 1963, **261**, pp. 650–64; E.M. Shoemaker, "Impact Mechanisms at Meteor Crater, Arizona", in G.P. Kuiper (Ed.), *The Solar System*, Vol. 4, Pt 2, University of Chicago Press, Chicago, 1963; C.S. Beals, "The Identification of Ancient Craters", *N.Y. Acad. Sci. Ann.*, 1965, **123**, pp. 904–14; M.R. Dence, "The Extraterrestrial Origin of Canadian Craters", *N.Y. Acad. Sci. Ann.*, 1965, **123**, pp. 941–69.
6. E.L. Krinov, "Meteorite Craters on the Earth's Surface", in G.P. Kuiper (Ed.), *The Solar System*, Vol. 4, Pt 7, pp. 183–207.
7. P.J. Cook, "The Gosses Bluff Crypto-explosion Structures", *J. Geol.*, 1968, **76**, pp. 123–9.
8. W.H. Bucher, "Cryptoexplosion Structures caused from without or within the Earth? ('Astroblemes' or 'Geoblemes'?)", *Am. J. Sci.*, 1963, **261**, pp. 597–619.
9. K.L. Currie, "Analogs of Lunar Craters", pp. 914–40.
10. D.H. McColl and G.E. Williams, "Australite Distribution Pattern in south central Australia", *Nature (Lond.)*, 1970, **226** (5241), pp. 154–5.
11. A.R. Alderman, "The Meteorite Craters at Henbury, central Australia", *Min. Mag.*, 1933, **23**, pp. 19–32.
12. D. Hager, "Crater Mound (Meteor Crater) Arizona, a Geologic Feature", *Bull. Ass. Am. Petrol. Geol.* 1953, **37**, pp. 821–57.
13. G. Seddon, "Meteor Crater: a Geological Debate", *J. geol. Soc. Aust.*, 1970, **17**, pp. 1–12.
14. W.H. Bucher, "Cryptoexplosion Structures", pp. 597–619.
15. J. Goguel, "A Hypothesis on the Origin of the 'Crypto volcanic Structures' of the central Platform of North America", *Am. J. Sci.*, 1963, **261**, pp. 665–7.
16. K.L. Currie, "Analogs of Lunar Craters", pp. 914–40.
17. Jaquelyn H. Freeberg, "Terrestrial Impact Structures —a Bibliography", *Bull. U.S. Geol. Surv.*, 1960, **1220**.
18. J.V. Harrison and N.L. Falcon, "Collapse Structures", *Geol. Mag.*, 1934, **71**, pp. 529–39; *idem*, "Gravity Collapse Structures and Mountain Ranges, as exemplified in southwestern Persia", *Q.J. geol. Soc. London*, 1936, **92**, pp. 91–102.
19. D.H. Stapledon, "Geological Investigations at the Site for Kangaroo Creek Dam, South Australia", *Inst. Eng. Aust. Site Invest. Symp.*, Paper 2140, Sydney, September 1966.
20. W.J. MacCallien, B.P. Ruxton and B.J. Walton, "Mantle Rock Tectonics. A Study in Tropical Weathering at Accra, Ghana", *Overseas Geol. and Min. Resour. Bull.*, 1964, **9**, pp. 257–94.
21. S.E. Hollingworth, J.H. Taylor and G.A. Kellaway, "Large Scale Superficial Structures in the Northampton Ironstone Field", *Q. J. geol. Soc. London*, 1944, **100**, pp. 1–35.
22. C.I. Cross, "The Parramore Island Mounds of Virginia", *Geogr. Rev.*, 1964, **54**, pp. 502–15.
23. R.G. Brown, "Modern Deformational Structures in Sediments of the Coorong Lagoon, South Australia", *Geol. Soc. Aust. Spec. Pub.*, 1969, **2**, pp. 237–62.
24. W.A. Price, "Caliche and Pseudo-anticlines", *Bull. Ass. Am. Petrol. Geol.*, 1925; **9**, pp. 1009–17.
25. J.N. Jennings and M.M. Sweeting, "Calishe Pseudo-anticlines in the Fitzroy Basin, Western Australia", *Am. J. Sci.*, 1961, **259**, pp. 635–9.
26. E.G. Hallsworth, G.K. Robertson and F.R. Gibbons, "Studies in Pedogenesis in New South Wales. VII. The Gilgai Soils", *J. Soil Sci.*, 1955, **6**, pp. 1–31.
27. M.E. Springer, "Desert Pavement and Vesicular Layer of some Soils of the Lahontan Basin, Nevada", *Proc. Soil Sci. Soc. Am.*, 1958, **22**, pp. 63–6.
28. A.L. Washburn, "Classification of Patterned Ground and Review of Suggested Origins", *Bull. geol. Soc. Am.*, 1956, **67**, pp. 823–66. See also Chapter 14.

Part 3

GEOMORPHOLOGICAL PROCESSES AND CLIMATIC GEOMORPHOLOGY

Davis's great mistake was the assumption that we know the processes involved in the development of landforms. We don't; and until we do we shall be ignorant of the general course of their development.

.John Leighly, 1940

CHAPTER 8

The Concept of Morphogenetic Regions

The morphology of many parts of the earth's surface is largely determined by the character of the crust. Thus in broad view many parts of the American Southwest and of the South Australian Gulfs region are dominated by fault blocks which determine the major relief. Dissected fold belts give rise to distinctive assemblages of forms as do plateau forms related to the occurrence of flat-lying strata. Characteristic features are associated with volcanoes, and jointing and lithology determine landform evolution in many parts of the world.

Although some of these structural landforms (using the term in its broader connotation) owe their character to the nature or behaviour of the crust, all are modified by external agencies to some extent. Structural landforms (*sensu stricto*) are specifically attributable to the exploitation of weaknesses in the crust by such agents as rivers, ice and waves.

Rocks at and near the earth's surface are attacked by atmospheric agencies. They are broken down into fragments by various processes of *weathering*. These particles can be transported by various agencies, causing the land surface to be lowered or *eroded*. The transported debris is eventually deposited, either in the ocean or on the land, to form new strata. Weathering, erosion and deposition are together often referred to as *denudation*, though other terms such as the *morphogenetic system*[1] are also used to embrace this complex of processes which modifies the form of the land surface.

Observations in many parts of the world strongly suggest that the nature of earth sculpture is in large measure determined by climate. For example, vast fields of mobile sand ridges are found only in the tropical arid lands. Coastal foredunes, the ridges of sand found in many places directly inland from beaches, are not as well represented in the humid tropics as in the mid-latitude lands.[2] Pingos, described in a later chapter, are restricted to subarctic regions. Patterned ground is developed in regions subjected to intense frost action; gilgai occur only in areas which experience marked wet and dry seasons. Even karst, an expression of a particular rock type, displays distinctive climatic differentiation (see Chapter 3). Because of this sort of climatic zonation of landforms and landform assemblages, many geomorphologists argue that climate determines the dominant processes at work in a particular area, and that these in turn control the assemblage of landforms evolved there.

The first formal expositions involving a correlation of climate, process and landform (although in a particular context) date from the turn of the century when W.M. Davis[3] described landform assemblages characteristic of "normal" (that is, the moderate or humid temperate), arid (tropical desert) and glacial lands. Distinctive processes and landform assemblages were subsequently claimed for other climatic regions, until, by the 1940s and the early 1950s these suggestions were formally integrated into concepts of *morpho-genetic regions* or *Formkreisen*.[4] Peltier's[5] classification is, perhaps, both the best-known and the most widely used in the English-speaking world. Using annual average precipitation and average temperature range as a basis, he formulated a scheme involving nine theoretical types of morphogenetic regions, in each of which the particular combination of climatic features induced the activity of certain processes which, in turn, produced certain landforms. Of these nine (see Fig. 8.1 and Table 8.1) all but two—the boreal and the maritime—have been generally recognised, though the distinction between the semiarid and savanna types is difficult to define and sustain. Tricart[6] has offered a more detailed scheme which is based on a more refined use of climatic data (Fig. 8.2). It also takes some account of inherited features (in his category 4) and of structural effects (the "elevated" zones of category 13).

Another approach, due to Büdel, considers not so much individual complexes of process but rather the dominant

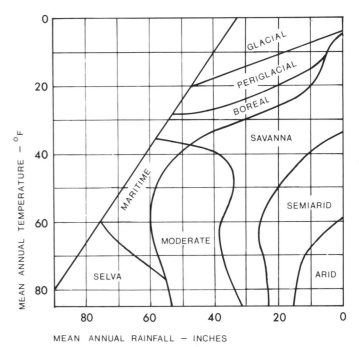

Fig. 8.1   Peltier's morphogenetic regions classified according to temperature and rainfall.

end-products of those processes. Thus, four major zones of landform assemblage can be distinguished (Fig. 8.3) in each hemisphere:

1. an equatorial and savanna zone dominated by planation and hence characterised by sweeping plains;
2. a subtropical zone of pedimentation;
3. a wide mid-latitude zone of pronounced valley development;
4. a high latitude zone dominated by nival and glacial processes.

Such assemblages of forms imply development in time of the evolution of landforms under the influence of, say, river erosion.

One obvious difficulty with such schemes is that many of the landforms described in Part II of this book owe their essential morphology to structure. Outcrops of granite, sandstone, limestone and volcanic rocks and areas occupied by fold mountains or flat-lying strata, together comprise a substantial part of the continental areas, so that they cannot be dismissed as minor elements in the landscape. They occur wherever certain geological conditions exist, and the distribution of geological structures bears no relationship to the present distribution of climatic types. Thus domed inselbergs have been described from such climatically diverse areas as Thailand, Texas, southeastern Brazil, the Sierra Nevada (California), Lapland, Uganda, southern Africa, the Andes of northern South America, coastal Queensland and southern and central Australia.[8] However, though the end product is the same, or similar, the processes responsible may vary from one climatic zone to another. For instance, in modern tropical areas and southwestern England,[9] the weathering responsible for the preparatory differential subsurface weathering of the compartments of rock was accomplished by moisture under warm conditions. The removal of waste was accomplished by streams and rivers, though in southwestern England,[10] frost shattering may have contributed to the weathering, and solifluction (see Chapters 9 and 11) to the evacuation of debris. But whatever the processes involved, the end result — the formation of inselbergs by the exploitation of structural contrasts and weaknesses — is similar. Such comparable features developed by different processes are known as *convergent* forms.

Another problem is that even if climatic conditions control landform development, some such climatically controlled forms are now found in alien territory. Glacial features, for instance, occur in parts of North America and Europe where no glaciers now exist. Forms can be, and in many instances most certainly have been, inherited from periods when the distribution of climatic types was different from that which now prevails (see Part IV). There is ample

Table 8.1
Peltier's morphogenetic regions*

Morphogenetic region	Estimated range of average annual temperature (°F)	Estimated range of average annual rainfall (inches)	Morphological characteristics
Glacial	0–20	0–45	glacial erosion; nivation; wind action
Periglacial	5–30	5–55	strong mass movement; moderate to strong wind action; weak effect of running water
Boreal	15–38	10–60	moderate frost action; moderate to slight wind action; moderate effect of running water
Maritime	35–70	50–75	strong mass movement; moderate to strong action of running water
Selva	60–85	55–90	strong mass movement; slight effect of slope wash; no wind action
Moderate	38–85	35–60	maximum effect of running water; moderate mass movement; frost action slight in colder part of region; no significant wind action except on coasts
Savanna	10–85	25–50	strong to weak action of running water; moderate wind action
Semiarid	35–85	10–25	strong wind action; moderate to strong action of running water
Arid	55–85	0–15	strong wind action; slight action of running water and mass movement

*Reproduced by permission from L. Peltier, "Geographic Cycle in Periglacial Regions", the *Annals* of the Association of American Geographers, Volume 40, 1950.

and incontrovertible evidence that the distribution of the world's climatic zones has changed in time, particularly during the past ten to fifteen million years. In many areas, landforms we see today are *inherited* from previous climates which differed from those which now prevail in these areas. Consequently, as Büdel recognised (Fig. 8.4) the processes moulding the land surface have also varied in time.

The definition of present climates is another difficulty with geomorphological implications. For instance, no desert is rainless and, for reasons which are discussed in Chapter 14, the occasional rains of the arid tropics achieve an amount of work which is disproportionately high compared to their volume. Hence, many deserts display features due to river activity related to the present climatic regime, and not indicative of climatic change: every climate is a complex mixture of diverse elements. Even abnormal storm effects which are experienced perhaps only once a century leave their clear imprint on the landscape.

Peltier's use of the climatic data is much too coarse in another sense also. To state bluntly, as Peltier does (Table 8.1) that running water is of only slight importance in the arid tropics is, in view of eyewitness accounts of desert floods, controversial to say the least. To take another example, Cotton[11] has drawn attention to the contrast between two areas underlain by similar bedrock and experiencing similar

climates in terms of temperature and average annual rainfall. Yet the one, in New Zealand, is much more intricately dissected than the other, located in central Europe. Mortensen has suggested that the contrast is related to the incidence of rainfall in the two areas. In New Zealand, the precipitation is received in winter; in the European locality, in summer. Therefore, rainfall is more effective in the former than in the latter. Hence the greater degree of dissection is due to the closer texture of the drainage network in winter rainfall areas.

Though many processes or agents attain optimal effects in certain climatic conditions, many are significant under a considerable range of environments, and all but a few are active to a minor degree in many of the conventionally defined climatic regions. Glacier ice is restricted in its activities to the high latitudes and to small areas at high elevations. But apart from glaciers, the other major processes are active over quite a range of climatic zones. Thus frost and ice disturb the superficial layers of the crust in subarctic areas and to a lesser extent and degree in mid-latitude areas, particularly those in the interiors of continents. Frosts are experienced in the tropical deserts and freeze–thaw activity may contribute slightly to rock weathering there. Only the equatorial and coastal areas in the tropics are free of frosts.

Again though wind achieves its greatest geomorphological

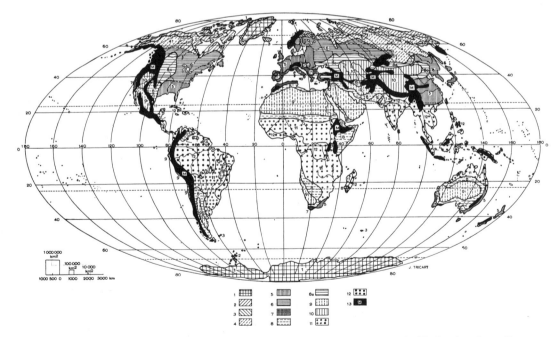

Fig. 8.2 Tricart's subdivision of the world into morphogenetic regions. 1. Glaciated regions. 2. Periglacial (subarctic) regions with permafrost. 3. Periglacial regions without permafrost. 4. Forests or (relict) Pleistocene permafrost. 5. Mid-latitude forests, maritime or with mild winters. 6. As for 5, but with severe winters. 7. As for 5, but with Mediterranean climate. 8. Steppes and semidesert plains. 8a. As for 8, but with severe winters. 9. Deserts and degraded steppes with mild winters. 10. As for 9, but with severe winters. 11. Savanna. 12. Rain forest. 13. Latitudinally anomalous: elevation exerts a controlling influence. (*J. Tricart, "Application du Concept", pp. 422–34, by permission of Tijdschrift van het koninklijk Nederlandsch Aardvikskundig Genootschap.*)

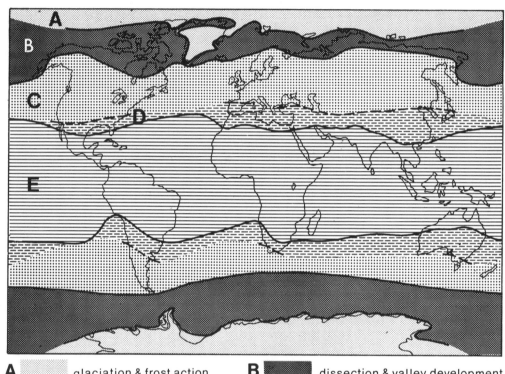

Fig. 8.3 Major latitudinal zones of landform assemblages typical of climatic regions according to Büdel.

A ░░░ glaciation & frost action   **B** ▓▓▓ dissection & valley development

C ⋮⋮⋮ valley development   **D** ≡≡≡ pediments & valleys   **E** ≡≡≡ tropical planation

Fig. 8.4 Latitudinal migration of landform zones during the Cainozoic. Letters refer to key in Fig. 8.3 (After J. Büdel.)

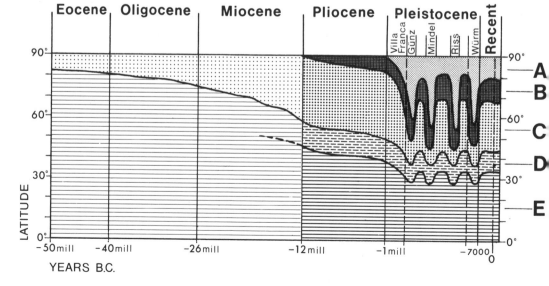

work in the tropical deserts, it is also important in the cold deserts and in some coastal regions. But from time to time, during storms, wind effects transportation of debris even in the humid tropics. Dust has been observed to be carried into the air from ploughed fields by wind squalls even during rain in the Murray mallee, in eastern South Australia.

Water is certainly the most pervasive, widespread and important geomorphological agent. It accomplishes much chemical weathering directly by hydration and hydrolysis; it takes many minerals into solution and acts as a solvent for many salts; subject to temperature oscillations around freezing point, moisture trapped in fractures and in interstitial spaces within rocks changes its form and volume and contributes to rock disintegration. Water is the medium which is responsible for the transportation of many solid particles, for the excavation of valleys, and for the planation of land surfaces. River valleys are eroded in all climatic regions apart from those areas covered by glacier ice, and even there subglacial valleys form under certain conditions. Neither meandering nor braided rivers are restricted to any one climatic region.

In most climatic regions more than one of the major processes has not only been at work in the past but is, from time to time, active at present. In the tropical deserts, for instance, running water as well as wind is responsible for the transportation and deposition of debris, and for the sculpture of the land surface. All landforms are a manifestation of processes which represent the cumulative effect of many small-scale events. They are impossible to follow in detail, though the major forces which have been at work can usually be identified.

Finally, it may be pointed out that the present coasts, though susceptible to analysis in terms of morphogenetic regions (see Chapter 17) are but temporary features, for sealevel has changed greatly in time and has stood at its present level only for the past 6000 years or so.

Thus simple concepts of morphogenetic regions such as those proposed by Peltier and Tricart are at best broad generalisations, and can be said to distract attention—no more—from what forms ought to have developed and were developing under the climatic conditions which prevailed at the time, and thus focus attention on "abnormal" or alien features. They suggest that certain forms which are *assumed* to be expected under the conditions which obtain have *in fact* formed under the assumed process; and this may be unjustified.

Yet with all these reservations distinctive landform assemblages evidently occur in particular climatic regions. Moreover, the complex of processes at work in various areas act at widely different rates. Corbel,[13] basing his findings on an analysis of published stream loads from various parts of the world, suggests that erosion is much more pronounced in mountainous areas than on lowlands (so that the hills are indeed being made low); and that moreover, there are considerable variations from one climatic region to another. The areas subject to most rapid erosion, according to Corbel, are the cold uplands and hot arid mountainous areas, where vegetation cover is sparse. Arid lowlands, however, change but slowly. Of course, factors other than climate have to be taken into account—tectonic uplift, man and so on; but Corbel's analysis is in broad terms corroborated by other findings. For instance, in the USA,[14] the catchment of the Colorado, which is essentially an arid upland, is being lowered over four times as fast as the area drained by the Columbia, which is more heavily wooded and generally better watered (16·57 cm/millenium as against 3·81).

In the following chapters it is proposed to discuss first the work of processes which, if not ubiquitous, are nevertheless widespread in their activities. These processes which operate in a variety of climatic regimes include weathering, mass movements and rivers. There follow chapters concerned with morphogenetic regions where particular complexes of process are at work and which are, as a result, characterised by the presence of distinctive landform assemblages. Tropical lands, both humid and arid, glaciated areas and periglacial regions fall into this category. Finally coastal landforms are considered. These can in many respects be regarded as shaped by similar processes the world over; yet there are sufficient variations to support suggestions that the margins of the continents too can be regarded as displaying climatically induced variations.

References Cited

1. J. Tricart and A. Cailleux, "Conditions anciènnes et actuèlles de la Génèse des Pénéplaines", *Proc. Int. Geogr. Congr. Wash.*, 1955, pp. 396–9; C.A. Cotton, "Alternating Pleistocene Morphogenetic Systems", *Geol. Mag.*, 1958, **95**, pp. 125–36.
2. J.N. Jennings, "The Question of Coastal Dunes in Tropical Humid Climates", *Z. Geomorph.*, 1964, **8** NS, pp. 150–4; *idem*, "Further Discussion of Factors affecting Coastal Dune Formation in the Tropics", *Aust. J.Sci.*, 1965, **28**, pp. 166–7; E.C.F. Bird, "The Formation of Coastal Dunes in the Humid Tropics; Some Evidence from north Queensland", *Aust. J. Sci.*, 1964, **27**, pp. 258–9.
3. W.M. Davis, *Geographical Essays*, Dover, Boston, 1904.
4. J. Büdel, "Die Morphologischen Wirkungen des Eiszeitklimas im Gletscherfreien Gebiet", *Geol. Rdsch.*, 1944, **34**, pp. 482–519; *idem*, "Die Klima Morphologischen Zonen der Polarlande", *Erdkunde*, 1948, **2**, pp. 25–53; L.C. Peltier, "The Geographic Cycle in Periglacial Regions as it is related to Climatic Geomorphology", *Ass. Am. Geogr. Ann.*, 1950, **40**, pp. 214–36.
5. L.C. Peltier, "The Geographic Cycle".
6. J. Tricart, "Application du Concept de Zonalité a la Géomorphologie", *Tijdschr. Kon. Med. Aar. Genootsch.*, 1957, **74**(3) pp. 422–34.
7. J. Büdel, "Morphogenese des Festlandes in Abhangig-keit von den Klimazonen", *Die Naturwissenschaften*, 1959, **48**, pp. 313–18; *idem*, "Klima-genetische Geomorphologie", *Geogr. Rdsch.*, 1963, **7**, pp. 269–86.
8. See H. Wilhelmy, *Klimamorphologie der Massengesteine*, Westermanns, Brunswick, 1958; C.R. Twidale, "A Contribution to the General Theory of Domed Inselbergs. Conclusions derived from Observations in South Australia", *Trans. Inst. Br. Geogr.*, 1964, **34**, pp. 91–133; *idem*, *Structural Landforms*, Australian National University Press, Canberra, 1971, pp. 45–77.
9. D.L. Linton, "The Problem of Tors", *Geogr. J.*, 1955, **121**, pp. 470–87.
10. J. Palmer and R.A. Nielson, "The Origin of Granite Tors on Dartmoor, Devonshire", *Proc. York. geol. Soc.*, 1962, **33**, pp. 315–40.
11. C.A. Cotton, "Fine Textured Erosional Relief in New Zealand", *Z. Geomorph.*, 1958, **2** NS, pp. 187–210.
12. H. Mortensen, "Warum is die rezente Formungsintenstitat in Neuseeland starker als in Europa?", *Z. Geomorph.*, 1959, **3** NS, pp. 98–9.
13. J. Corbel, "Vitesse d'Erosion", *Z. Geomorph.*, 1959, **3** NS, pp. 1–28.
14. According to S. Judson and D.F. Ritter, "Rates of Regional Denudation in the United States", *J. Geophys. Res.*, 1964, **69**, pp. 3395–401.

CHAPTER 9

Weathering Processes

A. GENERAL

Excavations reveal that most rocks near the earth's surface differ either texturally or mineralogically, or in both respects, compared to the parent material from which they are derived. This veneer of weathered rock (which includes soil) is called the *regolith* or *saprolith*. In some areas it is absent; elsewhere it is only a few centimetres thick; but over wide areas of the earth's surface the regolith is a few metres or even several scores of metres thick. In tropical areas several hundreds of metres of weathered rock have been reported. The colour of the regolith varies greatly from place to place and from one level or horizon to another within the profile. Even where the breakdown of the rocks is mainly mechanical, different textures and structures related to weathering processes can be discerned. In some areas, the junction between the fresh and weathered rock is gradational; elsewhere, the two meet abruptly. In either case, the lower limit of effective weathering is called the *weathering front*.[1]

In general terms, weathering may be defined as the disintegration of rocks *in situ* at, or near, the land surface and in the range of temperatures found there. Weathering processes take place at temperatures lower than those involved in metamorphism. Some result in a mechanical or physical breakdown of the rock, without any essential alteration of any of the contained minerals. This is called *physical* or *mechanical* weathering. In other cases, however, the processes achieve the alteration of one or more of the rock-forming minerals: *chemical* weathering. The separation of physical and chemical activities is, however, an arbitrary one, for processes of both types commonly act in concert, the one aiding the activity of the other and together bringing about rock disintegration.

Like most attempts to characterise natural phenomena, this general definition of weathering, though conveying the essence of the range of processes covered by the term, nonetheless leaves something to be desired. For instance, chemical weathering involves alteration of minerals. This alteration in many instances involves the release of some elements or radicles, which are translocated or transported in solution. They may be concentrated at a particular horizon within the regolith or they may be evacuated from the site either to be dispersed over a vast area in the oceans or to be concentrated at other topographic sites far removed from the place of origin. Such evacuation of minerals in solution implies a loss of volume, which in some cases may amount to 50 per cent or more of the rock.[2] Such a volume decrease causes settling of the rock mass and lowering of the land surface, so that it is not possible to differentiate clearly between weathering and erosion in every instance. Similarly, limestone and other rocks are readily attacked by weakly acidulated waters and the products of chemical reaction are removed in solution, causing the development of specific landforms (see Chapter 3).

Just as important, but often overlooked, are the effects of preferred redeposition of the chemicals evacuated in solution. Some are deposited in definite horizons in the regolith, while others are transported in solution to a greater or lesser distance before being redeposited. In either case, silica, alumina, lime, gypsum or iron oxides are accumulated to form hardpans or crusts which render the original rock more, and not less, resistant to erosion. Such induration is an important feature of weathering over wide areas.

B. PHYSICAL WEATHERING

1. General Remarks

Physical weathering involves only the mechanical disintegration of rock masses. Many processes have been cited in explanation of weathering features, both physical and chemical, but few are well founded in observation. Inference from distributional patterns, extrapolations from extreme manifestations of processes, and the application of general physical and chemical principles are the bases of most assertions as to the nature of physical weathering processes. This is not to say that all suggestions are wildly incorrect. Some forms of weathering are apparently limited to particular climatic zones and it is therefore reasonable to endeavour to explain them in terms of processes which are thought to be generated under those climatic conditions. The difficulties with such a procedure are, first, that the range of conditions implied in a given climate is often not clearly known; second, that the occasional unusual climatic event may achieve significant weathering effects; third, that the effects of modern climates are difficult to differentiate from residual effects of past climates; and fourth, that hitherto unknown or neglected processes tend to be overlooked—features are explained in known terms which may be neither correct nor comprehensive.

The following are some of the most commonly cited weathering processes.

2. Insolation

Many weathered rocks consist of loose aggregates of their constituent minerals or they are layered, the layers ranging in thickness between 1 cm and 30 m (Pl. 9.1a and b). In the past, both types of weathering have been attributed to alternations of heating and cooling due to either diurnal or seasonal radiation from the sun.

This suggestion rests on a certain amount of observational evidence and inference, but principally on theoretical arguments. Local and ephemeral, but intense, heating related to bush or scrub fires certainly causes superficial flaking of rocks[3] (see for instance Pl. 9.2a). The ferocious heat generated

Plate 9.1a Spheroidal weathering in gabbro, Giles Complex, northwest of South Australia. *(R.W. Nesbitt.)*

by nuclear explosions has also undoubtedly caused rock disintegration.[4] Man-made fires were used in the quarrying of granite many centuries ago.[5] The soil was cleared from outcrops and large fires built upon them. The intense heat caused the expansion and arching of the rocks in massive slabs, which were then hammered into blocks of a size suitable for transportation.

Loud cracks and reports heard in arid and semiarid tropical regions, where diurnal temperature changes are very great, have often been attributed to the release of pressure consequent upon the heating and cooling, expansion and contraction of rocks. For instance, in his account of the Burke and Wills expedition Alan Moorehead[6] recounts how, according to Wills' diary found after his death, Wills and King, wandering near Coopers Creek on 24 May, 1861, heard the "noise of an explosion, as if of a gun", thus momentarily raising their hopes of rescue. But according to Moorehead, what they heard was "a piece of rock splitting off some distant cliff", implicitly as a result of temperature changes. It is more likely that the noise was made by the sudden failure and collapse of a sand slope on one of the dunes which abound in that region of central Australia, again as a result of heating. As had indeed been known for many centuries[7] such loud cracks, and indeed a variety of noises, can be made by such sand avalanches.

Two separate, but related, theoretical arguments have

Plate 9.1b Onion weathering in granodiorite, near Tooma Dam Site, Snowy Mts, N.S.W. (*C.R. Twidale.*)

been advanced in favour of heating and cooling being an effective weathering process. Many rocks consist of minerals of different colours and of varied coefficients of expansion, which absorb radiation, and so expand at different rates. At night, or in winter, cooling causes contraction. Repeated expansion and contraction causes the creation of stresses between crystals or particles and eventually the rock crumbles. Such a breakdown into individual grains is called *granular disintegration*.

Rocks are also poor conductors of heat: rocks located near the land surface are heated and expand while those some distance beneath the surface remain relatively cool. Stresses are thus set up between the outer and inner zones. The outer gradually breaks away from the parent mass by the development of a fracture or series of fractures more or less parallel to the land surface. Because a roughly spheroidal mass of fresh rock remains in the centre of the joint block, this type of weathering is known as *spheroidal weathering*. In some cases the marginal mass of weathered rock consists of concentric thin flakes. If they are several centimetres thick, the flakes resemble the layers of an onion and the feature is called *onion weathering*. If they are several metres thick, it is known as *sheet structure* (but see also Chapter 3).

These two arguments are initially persuasive, but there are several lines of evidence which suggest that insolation plays only a minor role in rock weathering. First, granular

disintegration, spheroidal weathering, onion weathering and sheet structure have all been observed at considerable depths beneath the land surface. As rocks are notoriously poor conductors of heat, these occurrences are well beyond possible range of diurnal or annual temperature fluctuations, which is reckoned to be no more than a few centimetres at most (hence the shallow burrows dug by desert animals which allow them to escape the intense heat). Second, even in the tropical deserts, observations suggest that other processes are more effective than insolation. In the Egyptian desert near Cairo, granite columns from monuments erected about three thousand years ago have fallen and are partly buried by the sand. At Luxor, the lower parts of some erect columns were buried in mud. According to the insolation hypothesis, those parts of the columns exposed to the fierce heat of the sun's rays should be more weathered than those parts which are buried by the sand and soil, and so protected against the heat. If weathering takes place only very slowly, then at least it would be understandable if neither area exhibited discernible change. But in fact it is the buried and shaded parts of the columns which display pitting and, under the microscope, evidence of alteration of the minerals. In the granite from the exposed faces, no change, either physical or chemical, could be detected.[8] As a result of these and similar observations, it has been argued that even in the desert there is sufficient soil moisture to cause chemical weathering of some

Plate 9.2a   Fire flaking of quartzite resulting from bushfire which swept over flanks of Wilpena Pound, central Flinders Ranges, South Australia, in November 1960. *(C.R. Twidale.)*

of the minerals of which the granite is composed. Insolation is not necessarily ineffective; it is merely that in respect to granite it is slower acting than chemical weathering. Another question is whether the moisture responsible for the weathering is related to the present climatic regime or whether the weathering occurred during an earlier period of rather higher rainfall, of which there is evidence (see Part IV).

Laboratory experiments also suggest that alternate heating and cooling does not of itself cause rapid disintegration of rocks.[9] Rocks have been put through many cycles of heating and cooling equivalent to several thousands of years of natural activity, without discernible change in the rocks when the experiments were carried out under dry conditions. However, in moist air alteration was induced quite rapidly.[10] Again, the suggestion is that moisture is an essential factor in weathering. But laboratory work cannot reproduce the immensity of geological time, and again these results should not be interpreted as indicating that insolation is wholly ineffective.

However, there is reason to doubt that heating and cooling on their own comprise a very effective weathering process. From time to time resort is made to them in order to explain features for which no other explanation is obvious, for example blocky disintegration of quartzite on Mt Conner (Pl. 9.2b) and boulders of a conglomerate on the Olgas, both in central Australia, which are split, in some cases flush with the curved surface of the domed forms.[11] However, it seems that the process is wholly responsible for only a few features, though it may be a contributory factor in the production of several.

Plate 9.2b   Blocky disintegration of quartzite on Mt Conner, N.T. *(C.R. Twidale.)*

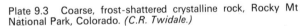

Plate 9.3   Coarse, frost-shattered crystalline rock, Rocky Mt National Park, Colorado. *(C.R. Twidale.)*

3.   Freeze-thaw Activity

Temperature fluctuations about the 0°C mark have a critical effect on water and ice, for, when the temperature falls below freezing point, the liquid changes to a solid in the form of ice. Water contained in either pore spaces or in joints or other crevices in rocks expands by about 10 per cent of its original volume on freezing. Such volume increase exerts considerable pressure on the rocks containing the water and ice. Confined ice generates pressures of some 140 kgm/cm² (2000 lbs/in²), and though the ice in cracks is in some cases incompletely confined, the pressures exerted by freezing water are nevertheless sufficient to shatter many rocks by the process of *frost-splitting* or *-riving*. Rocks which contain water held in pores are broken down to their component particles (cf. granular disintegration), but well-jointed rocks, such as fissile shales, are reduced to a mass of platey fragments or to a field of angular blocks or *felsenmeer* (Pl. 9.3) depending on the thickness of the beds in the country rock. Numerous repetitions of the freeze–thaw process are more effective than a few massive temperature oscillations, because the first freezing opens up a joint and, after thawing, the next developed wedge of ice is broader and thrusts the adjacent blocks of rock aside even more than on the first occasion. Laboratory work[12] confirms that many short range oscillations around freezing point are more effective in rock weathering than are fewer, large amplitude changes.

Thus freezing and thawing is most effective in subarctic regions or in temperate regions at high altitudes, where many oscillations around freezing point are experienced. It is not very effective in arctic regions for two reasons: temperatures are consistently low and rarely rise above freezing, and for this reason water is scarce. But in the right conditions the process is extremely effective: where optimum conditions exist, rocks are shattered so quickly that the process can be demonstrated with ease.

4. Salt Crystallisation

It has been suggested that salts released from the breakdown of minerals, on reaching a site either on the wind or in groundwaters, can exert a pressure sufficient to shatter rocks during crystallisation. Some experimental support for the contention comes from the work of French and British scientists [13] who report that very high pressures are developed during crystallisation. The process is called *Salzsprengung, exsudation* [14] or *salt fretting*.[15] The aggregate of particles held together by salts is called *rock meal*. Crystallisation is induced either by the escape of water from the system, as through evaporation in arid lands, or by changes in temperature. Whether salts can achieve a concentration sufficient to fill pore spaces, and whether they can exert enough pressure to shatter rocks is questionable. The expansive action of salts crystallising out from brine solutions seems capable of rupturing well-jointed, porous, and loosely-cohesive rocks such as sandstone and chalk, and wooden posts (for example, telegraph posts set in salt pans in the American Southwest[17]). It is clearly difficult to differentiate between the effects of crystal growth from solution, and expansion due to hydration (see below). Like frost action, exsudation may be incapable of shattering dense, massive rocks which lack pore space.

However, the process has been invoked in explanation of many *tafoni* (see below); of platforms bordering salt lakes in Western Australia[18] (Fig. 9.1), and other similar features (Pl. 9.4); and of granular disintegration in many areas, but

Plate 9.4 Small platform developed in porphyry block about 0.5 km from shore of Lake Gairdner, South Australia. The white encrustation is sodium chloride, and one argument is that it is the crystallisation of the salt which has caused differential weathering. Others would suggest, however, that it is the water which occasionally covers the salina which is responsible for solutional breakdown of the rock. The scale is given by the penknife. (*C.R. Twidale.*)

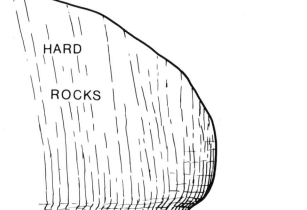

HARD ROCKS

billiard table rock surface of lake

Fig. 9.1 Billiard-table surface or platform and bordering cliff of salt lake in Western Australia, attributed by Jutson to salt weathering or exsudation. (*After J.T. Jutson, "Physiography (geomorphology)", p. 241, by permission of* Bull. Geol. Surv. W.A.)

Plate 9.5a Disaggregated shale overlying disintegrating shale exposed in a road cutting in the central Flinders Ranges, South Australia. The scale is given by the wristwatch. *(C.R. Twidale.)*

especially in Antarctica[19] where other processes are ruled out because of temperatures consistently below freezing. However, in Antarctica, there are special problems for salt crystallisation, because the process depends on a change of state from liquid to solid, and in such low temperatures it is difficult to conceive of liquids existing in the joints and pores of rocks, though former, possibly less severe, conditions may be invoked.

5. Pressure Release

Many rock masses display flakes, layers and thick slabs of rock disposed roughly parallel to the land surface. All have been interpreted by various workers and in various places at various times as due to pressure release. Farmin,[20] for example, considered the flakes and onion skins found around corestones of granite to be due to pressure release within individual joint blocks. The so-called pseudobedding displayed on the granite of Dartmoor is widely attributed to this cause[21] and as mentioned earlier, sheet structure also is so generally seen as a manifestation of pressure release that such joints are called offloading joints by some authors. The arguments against pressure release in respect to sheet structure have been summarised in Chapter 3. Minor superficial fractures observed in many rocks and areas could be attributed to pressure release, but there is little if any diagnostic evidence. The general argument appears sound, but the features can be equally well explained in terms of other processes.

For example, the disaggregation of shales is widely observed in arid regions, such as the Flinders Ranges, and could be related to offloading. Near the land surface, the fissile shales break down into a large number of small, angular fragments (0.5–1 cm diameter) which form a continuous veneer over the outcrop (Pl. 9.5a). The rock remains unaltered. However, it has been noted that dry cores of argillaceous rock brought to the surface in drill stems remain solid and cohesive as long as they are dry. But the clays rapidly disaggregate when wetted. Similarly, the wetting of surface shales, or their wetting and drying, leading to expansion and contraction (see Chapter 7), could lead to disaggregation. It may be that unbuttressing (as with outcrops on slopes), pressure release and possibly superficial insolational effects assist in bringing about the comminution of the shale. Why the fragments are of the size they are is not known.

Again, fractures parallel to the land surface are well displayed in several parts of the Sierra Nevada, California, the Adirondacks, in northeastern USA and on Dartmoor, southwestern England (Pl. 13.12), but frost action exploiting fractures due to other causes, possibly including pressure release, is responsible for their development.

6. Plants and Animals

Burrowing animals such as worms, termites and rabbits contribute to weathering by exposing new surfaces to the air and to circulating waters. Tree roots and the hyphae of

Plate 9.5b Granite block (X) displaced by some 10 cm along the fracture indicated by the growth of this tree trunk, near Palmer, South Australia. *(C.R. Twidale.)*

lichens[22] perform a similar function (Pl. 9.5b). Worms "digest" soils and extract organic material from them. Charles Darwin [23] estimated that worm burrows occupied no less than 15 percent of the top soil and that in southern England they moved 7·623–16·262 tonnes per hectare per annum. More recent estimates essentially confirm this order of activity, though it seems that the activity of worms varies greatly from one area to another. Termites are also extremely active in tropical regions. Nye,[24] working in Nigeria, and Williams[25] in northern Australia, both suggest that termites may move as much as 0.45m³ per hectare annually.

C. CHEMICAL WEATHERING
1. General Remarks

Chemical weathering takes place because minerals which were stable in their original environments—igneous, sedimentary or metamorphic—become unstable when they come within range of the atmosphere and particularly when they come into contact with meteoric waters.[26] New mineral phases are formed which are stable, or essentially so, in the prevailing near-surface conditions.

This briefly and broadly describes a series of reactions which are complex, varied and important and of which only the main outlines are known at present. However, the application of the electron microscope and the neutron probe have provided more detailed information in recent years, though this has not necessarily resolved problems: on the contrary, new complications have been revealed.

Water which has penetrated into and is circulating in surface and near-surface layers is the most important chemical agent affecting rocks, hence the significance of such physical properties of the bedrock as porosity, permeability and perviousness.

2. Solution

"Solution is essential to chemical weathering."[27] Solution is the most common initial phase of many important types of chemical weathering. Solution prepares crystal structures and crystal surfaces for further reactions.

All minerals are soluble to some extent. Some, like rock salts, readily dissolve in water. Others, like limestone which is itself insoluble in water, react with carbonated water to form the bicarbonate, which is soluble. In general terms this may be expressed:

$$CaCO_3 + H_2O + CO_2 = Ca(HCO_3)_2$$

Even quartz is slowly dissolved and, exposed to waters which are slightly alkaline, has a solubility of several parts per million, particularly at high temperatures. The rock platforms developed at the margins of salinas in arid lands (Fig. 9.1) and attributed to salt crystallisation may be due to solutional weathering. The marked weathering that occurs just above the salt surface (Fig. 9.1, Pl. 9.4) on rocks partly buried in the sediments of salinas may be caused by pronounced solution in the high surface temperatures of tropical regions. The increased solubility of silica in alkaline solutions may also explain the solution cups (Pl. 9.6) formed in silcrete (composition 95 per cent plus, SiO_2) near the margin of Lake Torrens, which is encrusted with gypsum and sodium chloride.

Circulating groundwaters dissolve soluble constituents from the minerals with which they come into contact and evacuate them from the system. In each crystal structure or lattice, anions and cations are tightly packed together. Each cation is surrounded by and bonded to ions of opposite

Plate 9.6 Small solution hollow in silcrete, near Beda Hill, southern extremity of Lake Torrens, South Australia. (*C.R. Twidale.*)

electrical charge. Moreover, these ions are of such a size (or ionic radius) that they fit into the lattice. The removal in solution of one constituent of the lattice means that the crystal is in disequilibrium, partly because there is a gap in the structure, partly because the electrical charges are not satisfied. New elements of appropriate charge and radius can be inserted into the structure and new minerals thus formed.

Thus solution is a highly significant weathering process, not only for its direct effects, but also because of the reactions it initiates.

3. Hydration and Hydrolysis

Water reacts with crystal surfaces with significant results. Jenny [28] has pointed out that although the valencies of elements within the crystal lattices are most commonly satisfied, at the crystal boundary there are exposed atoms and ions with unsaturated valencies. When such a surface comes into contact with water, hydration takes place, with water being attracted to the charged surfaces. Dissociation of the water may also occur with OH radicles being bonded to exposed cations *(hydrolysis)* and hydrogen ions to oxygen. Jenny takes the example of a feldspar, a common constituent of many crystalline and sedimentary rocks, in contact with water. The reaction at the crystal surface may be represented as follows.

Table 9.1
Reaction of water with feldspar

Feldspar surface	Water	Altered veneer on crystal

Thus the feldspar acquires a veneer of amorphous colloids, clays or oxides.

However, the natural system is rather more complex than that indicated here, if only because the soil system is not closed, but open. Water is not only constantly entering the system, but water with salts in solution is also leaving it. Hence the reactions are not strictly reversible, for the precise nature of the weathering product depends on the extent or degree to which soluble salts are evacuated. Thus the hydrolysis of a potash feldspar may produce either illite or kaolinite clay, according to whether the potash produced in the reaction is taken completely out of the system or not. If the potash is evacuated:

$$\begin{aligned}
\text{potash feldspar} + \text{water} &\rightarrow \text{kaolinite} \\
2KAlSi_3O_8 + H_2O &\rightarrow Al_2Si_2O_5(OH)_4 \\
&\quad + \text{silica} + \text{potash} \\
&\quad + 4SiO_2 + KOH.
\end{aligned}$$

But if some potash remains in the soil suffering alteration:

$$\begin{aligned}
\text{potash feldspar} + \text{water} &\rightarrow \text{illite} \\
3KAlSi_3O_8 + 2H_2O &\rightarrow KAl_2(Al, Si_3)O_{10}(OH)_2 \\
&\quad + \text{silica} + \text{potash} \\
&\quad + 6SiO_2 + 2KOH.
\end{aligned}$$

Many common rock-forming minerals react with water. Biotite for instance expands and takes in water between its layers. Al is substituted for Si in the lattice, iron is oxidised, and without losing its essential structure, the biotite is gradually converted to hydrobiotite, vermiculite and chlorite. Feldspars may also suffer hydration in their transformation to clays of various types; however, there is some suggestion that what is involved is not so much hydration and cation exchange within the original lattice, but rather solution of the feldspar and recrystallisation of clays.

Feldspars are very common constituents of several common rocks. They react with water to form clays:

$$\text{feldspar} + \text{water} \rightarrow \text{clay} + \text{solution.}$$

Apart from their economic significance, clays are easily eroded by streams and easily translocated both within and from soil profiles, and thus can lead to volume decrease in some situations. Some have hydrophylic properties which cause volume change and the development of gilgai features. They are therefore of considerable geomorphological significance as evidenced by the erosion of valleys and basins in argillaceous strata in many fold mountain belts (see Chapter 5).

Just which clay mineral is produced in a given reaction depends in large measure on the environment in which it takes place. In theory the reaction sequence, starting with the potash feldspar orthoclase, is:

orthoclase → illite → kaolinite → gibbsite (as in bauxite).

But the progress of this chain reaction is dependent on continuous leaching and desilication, that is, substitution of Al for Si. Erosion may interrupt the reaction when it has reached, say, the kaolinite stage, and it would be that mineral which is transported. If weathering and erosion were in equilibrium, the process would only ever reach that stage. Again, the reaction may be stopped at a particular stage, perhaps because of dessication caused by climatic or seasonal change. Climate is also an important factor influencing the course of weathering. For instance according to Tardy[29] in the humid tropics biotite is altered to chlorite, vermiculite and kaolinite, while plagioclase feldspar, though ephemerally changed to vermiculite and montmorillonite, and occasionally attaining gibbsite, most commonly and persistently gives rise to kaolinite. In the arid tropics it is montmorillonite which is the stable end-product of biotite and plagioclase weathering. And in humid temperate areas these minerals pass to vermiculite and montmorillonite phases with the feldspar eventually being transformed to kaolinite.

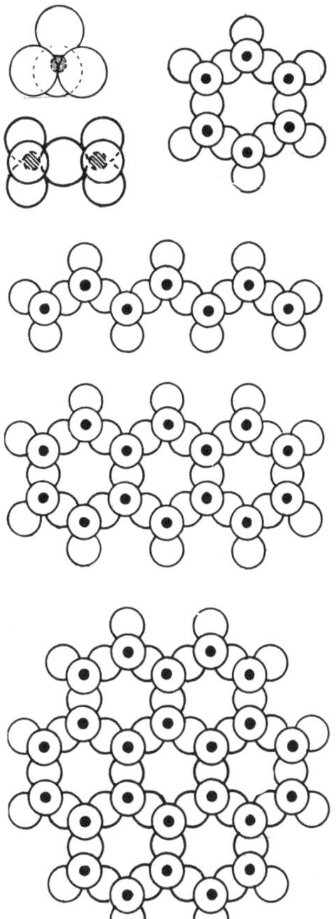

4. Substitution

The silicates, overwhelmingly the most common constituents of rocks, possess tetrahedral, linked chain and sheet structures (Fig. 9.2). They are susceptible to the removal of elements in solution and the substitution of other atoms. They are said to have a high exchange capacity. Perhaps the most important and widespread substitution is Al for Si. The most readily exchangeable cations of the silicates are those such as Na, K and Ca which are of appropriate size and balance the excess charge brought about by the substitution of Al for Si in the silicate lattice.

The dissociation of water leads not only to the hydrolysis of minerals, but to the release of hydrogen ions. Soils and clays with high concentrations of such hydrogen ions are acid (low pH). Hydrogen ions, because of their very small ionic radius and high energy, are extremely active in substitution, in entering and breaking down crystal structure and thus in causing alteration. It has been suggested that in the humid tropics, water in the capillary pores of soils becomes dissociated and acts as an acid. The development of acid clays is also aided by the roots of plants. The presence of organic ions is said to enhance the transformation of kaolinite to gibbsite, and the roots of plants are also thought to be significant in chemical weathering. Keller and Frederickson[30] have suggested that the roots of plants are particularly rich in soil minerals (Fig. 9.3). The roots of more primitive plants, that is, plants low on the evolutionary scale, have the highest concentrations of hydrogen ions. Thus lichens are particularly effective in rock weathering; as mentioned earlier the hyphae of these plants can penetrate into cleavages and along crystal

Fig. 9.2 Tetrahedral and linked chain structures of the silicate minerals. The open circles represent oxygen atoms; silicon is indicated by dots or shaded circles.

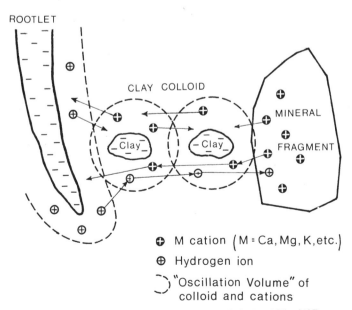

Fig. 9.3 Ion exchange in the root zone of plants. *(After W.D Keller and A.F. Frederickson, "Role of Plants", pp. 594–608, by permission of W.D. Keller and Am. J. Sci.)*

interfaces causing both physical shattering and chemical change. It has been suggested that blue algae are capable of mobilising iron and other salts, even in desert conditions. [31] Plants are certainly capable of extracting minerals from the soil and concentrating them in their tissues. [32] It has been shown that some ears of wheat consist of 15 per cent SiO_2 by weight. [33] After the death of the plant, the concentrated minerals are returned to the soil surface.

5. Carbonation, Oxidation and Reduction

Finally, besides the other effects achieved directly by contact with rock-forming minerals, water also carries gaseous carbon dioxide which reacts with certain minerals, such as calcite, to form carbonates or bicarbonates (see p. 186). In well-drained and aerated soils, gaseous oxygen combines with a wide range of minerals to form various oxides. For example, ferric iron oxides, though present in only very small amounts in most cases, give the reddish colours so typical of arid and semiarid lands. Only a small percentage of iron oxide in a soil gives it a bright red colour, though the ratio of FeO to organic contents is also important.[34] Electrons are added to minerals in oxidation and, by definition, reduction involves the loss of electrons. Reduction occurs in poorly drained and anaerobic sites and gives soils their grey greenish colours.

6. Examples of the Course of Chemical Weathering

It might be thought that the course of weathering would be apparent merely by tracing the changing mineralogy of a weathering profile from the surface to the bedrock contact. In some instances the general sequence of change can be traced by such a procedure, but complications are introduced by many variables, the most important of which are related to the fact that soils are open systems.

Mohr and Van Baren [35] have suggested that in such humid tropical areas as Indonesia, the soluble salts of Ca, K and Na, plus some silica in fresh volcanic ash, are the first elements to be leached. FeO is oxidised to Fe_2O_3 and MnO to MnO_2. Where drainage is good, kaolin is formed, but where it is poor, montmorillonite results. Leaching of bases continues and the depth of weathering increases. The removal of soluble salts results in the concentration of Fe_2O_3 and Al_2O_3 in the near-surface horizons, and silica is precipitated close to the weathering front, just above the parent material.

Studies of the weathering of basalt in New South Wales [36] show that olivine was the least stable mineral in the parent rock, and that pyroxenes were altered to montmorillonite, though in near-surface areas where there is intense leaching the montmorillonite was, in turn, changed to other clays—namely halloysite, kaolinite and poorly crystalline montmorillonite. The oxidised part of the profile was rich in goethite and haematite (Fig. 9.4) forming a laterite.

In the intertropical regions, kaolin formed by the alteration of feldspars is desilicified to form gibbsite (bauxite). Thus, at Weipa in north Queensland, the country rock of kaolinitic sandstone (90 per cent quartz, 10 per cent kaolin) has been extensively leached under the prevailing hot monsoonal conditions (annual rainfall 150 cm). The silica

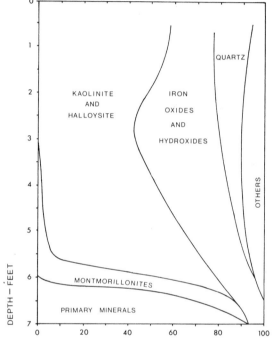

Fig. 9.4 Variation of mineralogy of a weathered basalt in depth at Bathurst, N.S.W. *(After D.C. Craig and F.C. Loughnan, "Chemical and Mineralogical Transformations", pp. 218–34, by permission of* Aust. J. Soil Res.*)*

has been removed (Fig. 9.5) giving a thick deposit of lateritic bauxite. [37]

Acid crystalline rocks subjected to hydration and hydrolysis are also weathered to clays. Plagioclase is the least stable component, biotite the next to be altered, then orthoclase and finally, much later, quartz. Albite is altered to halloysite plus a soluble salt, sanidine to illite, and illite in turn to kaolinite. Biotite is hydrated to form vermiculite, chlorite and related products. The quartz suffers slow solution. But, with the removal of solubles, the overall concentration of silica increases [38] (see Fig. 9.6).

In general then, it can be said that water, by removing some constituents, leaves behind a residue which becomes more and more enriched in the less-soluble materials plus oxides and hydroxides. In the tropics, laterite and bauxite are typical residues, whereas outside the tropical lands, sand and clay are characteristic end-products of weathering. The materials transported in solution are deposited either in the profile at the source site, or elsewhere more-or-less distant from the source rock. There they form pans and crusts (see below).

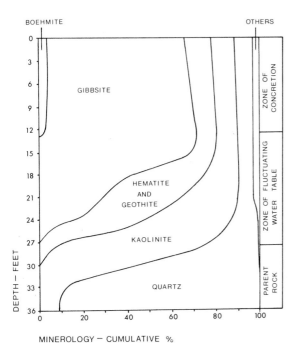

Fig. 9.5 Variation of mineralogy of a bauxite at Weipa, Queensland. *(After F.C. Loughnan and P. Bayliss, "The Mineralogy of Bauxite Deposits", pp. 209–17, by permission of the Mineralogical Society of America.)*

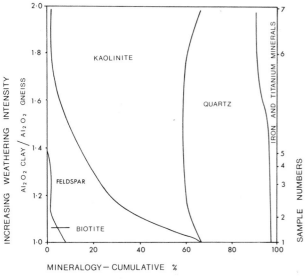

Fig. 9.6 Variation of mineralogy in the weathered Morton Gneiss, Minnesota. *(After F.C. Loughnan, Chemical Weathering, p. 101.)*

D. FACTORS CONTROLLING THE TYPE AND RATE OF WEATHERING

Many simple observations indicate quite clearly that the type and rate of weathering vary greatly from place to place on the earth's surface. Many tectonically undisturbed rock outcrops stand higher than do those adjacent to them (see Chapter 5). This surely argues that the latter are more vulnerable to weathering and hence to erosion than are the former. Situations similar to that depicted in Fig. 9.7 and Pl. 9.7 are commonplace. A sequence of, say, shales is intruded by a vein of quartz. When these two are exposed to weathering, the shale is weathered far more rapidly than the quartz, with the result that, though the quartz reef superficially crumbles to some extent into angular fragments, it nevertheless remains upstanding to form more-or-less prominent linear outcrops, or perhaps, if it is sufficiently massive, a ridge. Such observations suggest that the type of rock, or its *composition*, can influence its susceptibility to weathering.

Of the common rock-forming minerals, the order of susceptibility to chemical attack is generally the same as the sequence of crystallisation from an igneous melt: olivine is the least stable, followed in order by augite, lime soda, plagioclase, biotite, potash feldspar, muscovite and quartz. Thus a shale, with a significant percentage composition of feldspars, is more readily weathered than the quartz vein. As indicated earlier in the discussion of sandstone terrain, rocks which are composed largely of quartz crystals or fragments are both physically durable and chemically tough. On the other hand, rocks like basalt, which contain a high proportion of minerals of basic composition and are from the unstable end of the vulnerability spectrum, rapidly decay on exposure to the atmosphere, and particularly on contact with meteoric waters.

However, the factors which govern the progress of weathering are complex and there are many examples of basic rocks rich in the supposedly susceptible minerals which for one reason or another are relatively resistant and upstanding. For instance near Fresno, California, granites in the foothills of the Sierra Nevada have been worn down to form what are locally called granite "flats",* whereas basic outcrops in the same zone form prominent hills, and in the nearby ranges proper, biotite schists form rugged uplands.

As discussed below, water is the most potent of all weathering agents. Hence the means it has of entering a rock mass is important. The significance of such properties as faults, various types of joints, porosity and permeability in differential weathering has already been discussed (Chapters

---

* An interesting example of varied perception, for the flats in reality display an undulating or rolling relief. But in contrast with the adjacent Sierra Nevada, the plains are indeed of low relief. The reverse side of the coin is provided by Mt Brown and Mt Fort Bowen in the Carpentaria plains of northwestern Queensland, which though only a few metres high nevertheless stand abruptly above the monotonous flatness of the surrounding plains and no doubt to the local inhabitants warrant the term "mountain".

Fig. 9.7   Sketch of differential weathering of shale and vein quartz.

3 and 4). Weathering of bedding planes gives rise to such spectacular features as those shown in Pl. 9.8a and b. Fine grained rocks are particularly susceptible to chemical attack because, compared to coarse grained rocks, they have a large area of crystal surfaces and, as these surfaces have high free energy, they are prone to reaction with other suitable elements which are in circulation in the system. Coarse textured rocks, on the other hand, are relatively resistant because of their smaller exposed areas of crystal face. Thus both the *structure* and *texture* of rocks is important in controlling the rate of weathering.

Rocks exposed in cold climates are frequently shattered and broken up into angular masses, but apart from a certain surface discolouration, possibly due to colonisation by lichens and mosses, they are not greatly altered. In the humid tropics, however, rocks are so rotted that any original structures and textures are lost. The rocks are broken down to particles of small size, except where there has been accumulation of iron oxides and alumina in the weathering profile to form massive carapaces on the land surface. Thus *climatic conditions* significantly control both the type and the rate of weathering.

Chemical weathering is particularly effective in the humid tropics where temperatures are consistently high, where moisture is constantly available and where there is abundant vegetation. Solution, hydration and hydrolysis are important, and though the nature of these processes does not vary from one climatic region to another, the reactions take place most rapidly here. Chemical weathering in the humid tropics may be as much as three to four times as rapid as in temperate regions (depending, to some extent, on the parent material under attack). However, the decrease in viscosity of capillary water held in the pore spaces of the weathering profile, and its more rapid circulation due to the constant accession and flushing of water in the surface layers of the soil, to some extent balances this tendency. But overall, weathering is both rapid and intense in the humid tropics. In consequence, even where there is strong erosion, weathering outpaces the removal of debris on slopes and a thick regolith is commonplace. Even on steeper slopes (60°) there are up to 40 cm of soil in places.

Materials rich in alumina, iron oxides and water are the common end-products of weathering here, in contrast to the sand and clay which constitute the regolith in temperate regions. Weathering also penetrates deeper in the humid tropics than anywhere else. This is indicated diagrammatically in Fig. 9.8 and is amply borne out by observations in the present tropical world: 50 m in Nigeria,[39] 200 m in south-eastern Brazil,[40] and though these are exceeded outside the tropics (300 m in N.S.W.[41] and 1500 m in the USSR[42]), the latter may well be inherited from times when humid tropical conditions obtained in these areas.

Physical disintegration caused by temperature oscillations, on the other hand, are, or theoretically should be, most pronounced in the arid lands. In the cool arid regions of the subarctic (and equivalent high altitude areas) the freeze–thaw mechanism certainly appears to be effective, to judge by the vast amount of angular scree and blockfields

Plate 9.7 Broken linear outcrops of limestone extending along strike near Blinman, central Flinders Ranges, South Australia. (*C.R. Twidale.*)

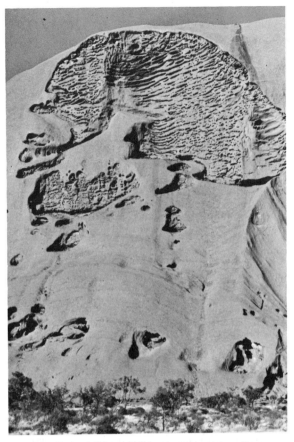

Plate 9.8a "The Brain", an odd weathering effect due to the weathering out of near-vertical bedding plains on the flanks of Ayers Rock, Northern Territory. (C.R. Twidale.)

Plate 9.8b A similar effect is achieved by the weathering both of bedding planes and the pitting of the laminae of micaceous sandstone on the Arcoona Plateau, South Australia. (C.R. Twidale.)

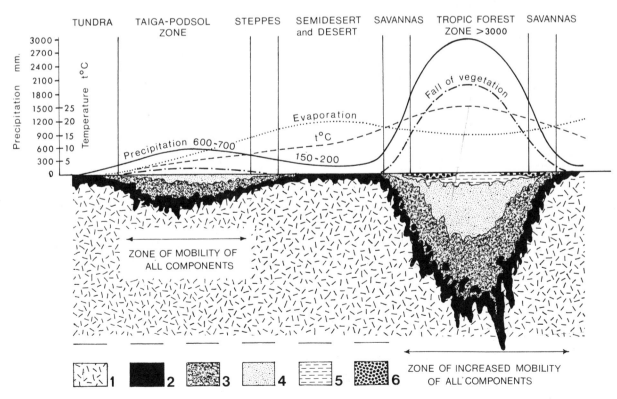

Fig. 9.8 Diagrammatic representation of variation of weathering in latitudinal zones, assuming tectonic stability. (After N.M. Strakhov (trans. J. Fitzsimons), Principles of Lithogenesis, Oliver and Boyd, Edinburgh, Vol. 1, 1967.)

strewn over the landscape. The effects of temperature variations in the tropical deserts are, as already indicated, open to question. Salt crystallisation is effective, if anywhere, in the arid regions again, but such structural factors as porosity are also significant in controlling the effectiveness of the mechanism.

Variations in the rate at which *erosion and deposition* occur in different parts of the earth's surface determine the time available for the various processes at work in the regolith to run their course. For instance, in some of the sequential changes caused by the hydration and desilication of clays, the reaction may have attained, say, the illite stage; but if erosion occurs the chain is broken. The relative rates of weathering and erosion may be such that the reaction never proceeds beyond that point at that particular locality. Erosion causes the exposure of new areas of bedrock to weathering agencies: there is said to be a *renewal* of weathering. Rapid deposition, on the other hand, can cause disequilibrium between weathering and surface accretion and result in the burial of soils (Fig. 9.9).

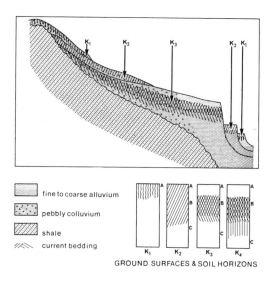

▦	fine to coarse alluvium
⣿	pebbly colluvium
▨	shale
⩘	current bedding

GROUND SURFACES & SOIL HORIZONS

Fig. 9.9 Buried soils in the Nowra district, N.S.W. (*After B.E. Butler, "Soil Periodicity in Relation to Landform Development in southeastern Australia", in J.N. Jennings and J.A. Mabbutt (Eds),* Landform Studies from Australia and New Guinea, *Australian National University Press, Canberra, 1967, p. 235.*)

The work of such creatures as earthworms and termites has already been mentioned. Man also influences weathering in a number of ways. For instance, by clearing the land man has caused accelerated soil erosion and has thus disturbed the equilibrium between erosion and weathering. Removal of vegetation in semiarid lands permits increased evaporation of soil moisture, so that salts are drawn to the surface, thus changing the type of soil and weathering there. Man causes bushfires and thus flaking. His atomic bombs cause superficial weathering. [43] His factories and cars emit chemicals into the atmosphere which affect the type and rate of weathering in particular localities. For instance, the southwestern side of Malham Tarn House in northwestern England is more weathered than are the other sides of the building. [44] This is attributed to its facing the prevailing winds which blow from industrial areas and which are heavily charged with sulpur dioxide. The stones of which Cologne Cathedral are built are suffering such severe corrosion that the whole fabric of the building is endangered.

Slope and topography influence the movement of water and debris and again the nature of a soil, its drainage characteristics and hence its mineral composition. Thus on a slope a *catenary* [45] sequence of soils is developed. Soils on the upper slopes are well drained, whereas those at the slope foot may be waterlogged. Soil and rock particles are washed down the hillslope so that the lower sectors may have a different mineralogy and texture from those on the hillcrest.

So many factors influence weathering processes that it is extremely difficult to ascertain the average rate or rates at which they progress. Limestone gravestones in northwestern England have taken between 250 and 500 years to weather to a depth of 2·5 cm. On the other hand, it is reported that some of the minerals of granite used for public buildings in Rio de Janeiro were altered in about thirty years and in Madagascar, biotite displayed alteration after only a decade of exposure to the elements. Ash falls associated with the 1883 eruptions of Krakatoa (see Chapter 6) already exhibit strong leaching of alkalies and some removal of silica. At Soufriere at St Vincent, in the West Indies, "normal" soils were developed on volcanic ash only thirty years after its deposition in 1902. Weye, [46] working in El Salvador in central America, estimates that it had taken 5000 years to weather dacitic ash to a depth of 1 m.

But interesting as such figures undoubtedly are, it is very difficult to extrapolate figures of geological significance from them. In a deeply weathered profile of laterite developed on Cretaceous bedrock, did the duricrust evolve in a relatively brief space of time when, say, climatic conditions were suitable, or has there been slow, continuous development since the bedrock was exposed at the land surface?

References Cited

1. J.A. MABBUTT, " 'Basal Surface' or 'Weathering Front' ", *Proc. geol. Soc. London*, 1961, **72**, pp. 357–8.
2. B.P. RUXTON, "Weathering and Subsurface Erosion in Granite at the Piedmont Angle, Balos, Sudan", *Geol. Mag.*, 1958, **45**, pp. 353–77; A.F. TRENDALL, "The Formation of 'Apparent Peneplains' by a Process of Combined Lateritisation and Surface Wash", *Z. Geomorph.*, 1962, **6** NS, pp. 51–69.
3. E. BLACKWELDER, "Fire as an Agent in Rock Weathering", *J. Geol.*, 1927, **35**, pp. 134–40.
4. T. WATANABE, M. YAMASAKI, G. KOJIMA, S. NAGAOKA and K. HIRAYAMA, "Geological Study of Damage caused by Atomic Bombs in Hiroshima and Nagasaki", *Jap. J. Geol. & Geogr.*, 1954, **24**, pp. 161–70.
5. H. WARTH, "The Quarrying of Granite in India", *Nature (Lond.)*, 1895, **51**, p. 272.
6. A. MOOREHEAD, *Coopers Creek*, Reprint Society, London, 1963, p. 128.
7. W. THESIGER, *Arabian Sands*, Longmans Green, London, 1959, p. 166; C.R. TWIDALE, "Singing Sands" in R.W. Fairbridge (Ed.), *Encyclopaedia of Geomorphology*, Reinhold, New York, 1969, pp. 994–6.
8. D.C. BARTON, "Notes on the Disintegration of Granite in Egypt", *J. Geol.*, 1916, **24**, pp. 382–93; E. BLACKWELDER, "Exfoliation as a Phase of Rock Weathering", *J. Geol.*, 1925, **33**, pp. 793–806; *idem*, "The Insolation Hypothesis of Rock Weathering", *Am. J. Sci.*, 1933, **26**, pp. 97–113.
9. E. BLACKWELDER, "Exfoliation as a Phase" and "The Insolation Hypothesis"; D.T. GRIGGS, "The Factor of Fatigue in Rock Exfoliation", *J. Geol.*, 1936, **44**, pp. 783–96.
10. D.T. GRIGGS, "The Factor of Fatigue", pp. 783–96.
11. C.D. OLLIER, "Insolation Weathering: Examples from Central Australia", *Am. J. Sci.*, 1963, **261**, pp. 376–81.
12. A.S. POTTS, "Frost Action: Some Experimental Data", *Trans. Inst. Br. Geogr.*, 1970, **49**, pp. 109–24.
13. See P. BIROT, *Le Cycle d'erosion sous les Différent Climats*, University of Brazil, Rio de Janeiro, pp. 99–100; A. GOUDIE, R.U. COOKE, and I. EVANS, "Experimental Investigation of Rock Weathering by Salts", *Area*, 1970, **4**, pp. 42—8.
14. J.T. JUTSON, "Physiography (geomorphology) of Western Australia", *Bull. Geol. Surv. W.A.*, 1934, **95**, pp. 254–6.
15. H.W. WELLMAN and A.T. WILSON, "Salt Weathering, a Neglected Geological Agent in Coastal and Arid Environments", *Nature (Lond.)*, 1965, **205**, pp. 1097–8.
16. A. GOUDIE *et al.*, "Experimental Investigation".
17. See for example E. BLACKWELDER, "Geomorphic Processes in the Desert" in "Part V, Geomorphology", in "Geology of Southern California", *Calif. Div. Mines Bull.*, 1950, **170**, pp. 11–9.
18. J.T. JUTSON, "Physiography of Western Australia", p. 241.
19. H.W. WELLMAN and A.T. WILSON, "Salt Weathering", pp. 1097–8.
20. R. FARMIN, "Hypogene Exfoliation of Rock Masses", *J. Geol.*, 1937, **45**, pp. 625–35.

21. R.S. WATERS, "Pseudo-bedding in the Dartmoor Granite", *Trans. geol. Soc. Cornwall*, 1954, **18**, pp. 456–62.
22. See E. JENNIE FRY, "The Mechanical Action of Corticolous Lichens", *Ann. Bot.*, 1926, **40**, pp. 397–417; *idem*, "The Mechanical Action of Crustaceous Lichens on Substrata of Shales, Schist, Gneiss, Limestone and Obsidian", *Ann. Bot.*, 1927, **41**, pp. 437–60.
23. C. DARWIN, *The Formation of Vegetable Mound through the Action of Worms, with Observations on their Habits*, Appleton, London, 1881.
24. P.H. NYE, "Some Soil Forming Processes in the Humid Tropics. I. A Field Study of a Catena in the West African Forest", *J. Soil Sci.*, 1954, **5** (1), pp. 7–21; *idem*, "Some Soil Forming Processes in the Humid Tropics. IV. The Action of the Soil Fauna", *J. Soil Sci.*, 1955, **6** (1), pp. 73–83.
25. M.A.J. WILLIAMS, "Termites and Soil Development near Brocks Creek, Northern Territory", *Aust. J. Sci.*, 1968, **31**, pp. 153–4.
26. F.C. LOUGHNAN, *Chemical Weathering of the Silicate Minerals*, Elsevier, London, 1969. Much of this review of chemical weathering closely follows Loughnan's work.
27. Ibid., p. 61.
28. H. JENNY, "Origins of Soils", in P.D. Trask (Ed.), *Applied Sedimentation*, Wiley, New York, 1950, pp. 41–61.
29. Y. TARDY, personal communication.
30. W.D. KELLER, and A.F. FREDERICKSON, "Role of Plants and Colloidal Acids in the Mechanism of Weathering", *Am. J. Sci.*, 1952, **250**, pp. 594–608. See also W.D. KELLER, "Mineral and Chemical Alluviation in a Unique Pedologic Example", *J. Sed. Petrol.*, 1961, **31**, pp. 80–6.
31. F. SCHEFFER, B. MEYER and E. KALK, "Biologische Ursachen der Wüstenlachebildung", *Z. Geomorph.*, 1963, **7** NS, pp. 112–19.
32. For review, see T.S. LOVERING, "Geological Significance of Accumulator Plants in Rock Weathering", *Bull. geol. Soc. Am.*, 1959, **70**, pp. 781–800.
33. J.T. HUTTON and K. NORRISH, personal communication.
34. See for example H.C.T. STACE, "Chemical Characteristics of Terra Rossas and Rendzinas of South Australia", *J. Soil Sci.*, 1956, **7**, pp. 280–93.
35. E.C.J. MOHR and H. VAN BAREN, *Tropical Soils*, Van Hoewe, Amsterdam, 1959.
36. D.C. CRAIG and F.C. LOUGHNAN, "Chemical and Mineralogical Transformations accompanying the Weathering of Basic Volcanic Rocks from New South Wales", *Aust. J. Soil Res.*, 1964, **2**, pp. 218–34.
37. F.C. LOUGHNAN and P. BAYLISS, "The Mineralogy of Bauxite Deposits near Weipa, Queensland", *Am. Miner.*, 1961, **46**, pp. 209–17.
38. F.C. LOUGHNAN, *Chemical Weathering*, pp. 93–101.
39. M.F. THOMAS, "Some Geomorphological Implications of Deep Weathering Patterns in Crystalline Rocks in Nigeria", *Trans. Inst. Br. Geogr.*, 1966, **40**, pp. 173–93.

40. R. Barbier, "Aménagements Hydroelectriques dans le Sud de Brésil", *C.R. Somm. et Bull. Soc. geol. France*, 1957, **6**, pp. 877–92.
41. C.D. Ollier, "Some Features of Granite Weathering in Australia", *Z. Geomorph.*, 1965, **9** NS, pp. 285–304.
42. V.N. Razumova and N.P. Keraskov, 1963, cited in C.D. Ollier, *Weathering*, Oliver and Boyd, Edinburgh, 1969, p. 121.
43. T. Watanabe *et al.*, "Geological Study of Damage", pp. 161–70.
44. A. Raistrick and O.L. Gilbert, "Malham Tarn House: Its Building Materials, their Weathering and Colonisation by Plants", *Field Stud.*, 1963, **1**, pp. 89–115.
45. G. Milne, "A Provisional Soil Map of East Africa", *Amani Mem.*, 1936, **28**; G.C.T. Morison, A.C. Hoyle and J.F. Hope-Simpson, "Tropical Soil-vegetation Catenas and Mosaics. A Study in the Southwestern Part of the Anglo-Egyptian Sudan", *J. Ecol.*, 1948, **36**, pp. 1–84; P.H. Nye, "Some Soil Forming Processes", pp. 7–21.
46. R. Weye, "Beitrage zur Geologie El Salvadors", *Neues Jahrb. Geol. Palaeont. Min.*, 1954, pp. 49–70.

General and Additional References

Birkeland P.W., *Pedology, Weathering and Geomorphological Research*, Oxford University Press, New York, 1974.
Keller, W.D., *Principles of Chemical Weathering*, Lucas Bros., Columbia, Miss., 2nd ed., 1957.
Reiche, P., "Survey of Weathering Processes and Products", *Univ. N. Nex. Pub. Geol.*, 1950, **3**.

CHAPTER 10

Geomorphological Significance of Weathering

A. GENERAL

Weathering is of vital significance to humanity, if only because it is an essential part of soil formation. Furthermore, weathering accumulations of certain minerals such as bauxite, iron, nickel and kaolin, are of international significance; while the very widespread pedogenic accumulations of limestone (as well as of laterite and ferricrete) due to weathering processes are locally significant in many parts of arid and semiarid lands, for example in Australia, because this rock is the only available source of roadmetal. From a geological point of view, weathering is an essential preliminary to transportation and deposition, and is thus a precursor to the development of new sedimentary strata. The state of the particles transported obviously affects the nature of these new layers. Thus the erosion and transportation of detritus from a lateritic profile results in a sediment in which kaolinitic clays and iron oxides are common.

From a geomorphological point of view, however, weathering is important in three major ways: in the breaking down of rocks, in the formation of duricrusts, and in the development of particular landforms.

B. PREPARATION OF ROCKS FOR TRANSPORTATION

Weathering is responsible for the reduction or breakdown of rocks to a size in which they are amenable to transport. The wind is incapable of moving particles greater than sand size (though claims to the contrary have been made with respect to playa stone tracks on some of the salinas of southern California[1]) and though streams in flood are known to have shifted very large blocks, even major rivers are usually very restricted in the calibre of debris they can transport. Glaciers can move large blocks and the local transporting power of storm waves is legendary, but these dramatic examples should not be allowed to obscure the fact that the vast bulk of debris carried by these agencies, ice waves, running water and wind, is of relatively small size. That it is comparatively fine is due to the work of various weathering agencies.

The importance of size and cohesiveness of debris so far as stream transport and erosion are concerned is well illustrated in gullies due to man-induced accelerated erosion (see Chapter 24). In many cases intermittent streams have cut readily and rapidly through unconsolidated clays, but have had little effect on even slightly weathered shales below. Weathering is thus a vital preliminary to transport and erosion, and without the activity of the various weathering processes, much less denudation would take place.

C. DURICRUSTS

An important family of landforms owes its origin to the dissection of indurated horizons, crusts and layers formed during weathering. Such *duricrusts* are also important as morphostratigraphic horizons (or time markers) in some areas, and are locally significant from a hydrological standpoint; they therefore warrant brief consideration here.

The salts released by chemical attack and transported in solution may be translocated vertically (either up or down the profile, according to the precipitation:evaporation ratio and drainage conditions) and redeposited either at a particular level or in an horizon at essentially the same site. Alternatively, they may be redeposited some distance downslope, or, if carried to the lower reaches of the drainage system, in an area far removed from the source and even in the sea. The redeposition of such salts in terrestrial environments results in their concentration and the induration of the host rock. Yet other salts may be concentrated *in situ* by the removal of more soluble constituents of the rock. But in each case a hard layer is formed. Such indurated layers were called duricrusts by Woolnough,[2] though Lamplugh[3] had earlier recognised and named several specific types. Duricrusts rich in baryte, gypsum and titanium oxide have been recognised, but the most widespread and important ones are undoubtedly laterite, bauxite, silcrete and calcrete.

Laterite is forming today in the humid intertropical and adjacent regions. However, in relict form, it is widespread in such areas as central and southern Australia, including Tasmania, and it has also been described in relict form from eastern Africa and from western areas of the same continent.[4] Laterite was first described from India but since that time the term has been applied to rocks of several types, so that several definitions of laterite are extant. There has been a tendency for geologists to label as laterite any rock with an iron-rich horizon, whereas pedologists have usually required in addition a kaolinised mottled and pallid zone below the ferruginous horizon. Moreover various origins have been attributed to laterite and the term has been extended well beyond the original descriptions.[6] However, the definition of Alexander and Cady[7] is possibly the most apt. They consider that laterite is a highly weathered material, rich in secondary oxides of iron or alumina, or both. It is virtually devoid of bases and primary silicates, though it may contain large amounts of quartz and kaolinite. It is either hard, or hardens on exposure to wetting and drying.

In Australia the typical laterite profile[8] consists of a sandy soil, the A-horizon, which rests, or originally rested, on a sheet rich in iron oxides and alumina perhaps 5 m thick

Plate 10.1   Laterite profile showing tough ferruginous zone and kaolinised zone beneath; Hoddy Well, Western Australia. *(C.G. Stephens.)*

Plate 10.2   Nodular bauxite underlain by kaolinised zone at Weipa, north Queensland. *(E.A. Rudd.)*

(Pl. 10.1). This massive sheet displays pisolitic and vesicular structure, and is in many places distinguished by a faint bedding—a number of discontinuous fractures, in many instances gently curved, running parallel or subparallel to the land surface. The material is characteristically brown, yellow and white in colour, though the exposed surface typically displays a veneer of rather shiny, brown iron oxides. Below this iron-rich sheet is a kaolinised zone up to 30 m thick, and characteristically white, though with splashes of red and yellow, and in some areas, small, irregularly shaped masses of chalcedony.

Some workers have subdivided this kaolinised zone into an upper, mottled zone and a lower, pallid zone according to their field appearance, but the two are not everywhere present and in some places they occur in reverse sequence. Howsoever this may be, the parent rock occurs beneath the kaolinised zone. The sandy A-horizon is readily erodible and therefore is commonly absent; and there are many local variants of this "typical" profile.

Bauxite (Pl. 10.2) is a duricrust which has suffered intense desilication and which is rich in alumina (see pp. 186–187). It forms in the same climatic conditions and is of similar origin to laterite.

Many hypotheses have been advanced to explain laterite. Some early workers in India, impressed by the vesicular structure and its widespread occurrence on the Deccan, thought it was a volcanic rock. Others thought it due to deposition. But though there are controversial points and probably more than one way in which morphologically and chemically similar rocks evolve,[9] it is now generally agreed that laterite is a weathering profile due to strong leaching of soluble salts in conditions highly suitable for oxidation. Desilication of clays is very advanced. Heavy rainfall, high temperatures, and abundant vegetation all aid these chemical processes, so that laterites are developed in the humid tropics.

The reason or reasons for the concentration of iron oxides and alumina in the B-horizon is controversial, but many workers consider it to be related to a fluctuating water-table; this may be the reason for the exceptional development of laterite in monsoon lands. The hardening which occurs on drying is spectacular. Wet, the rock can be dug with a spade and trimmed with a trowel; on drying, however, it hardens to a brick-like consistency. Hence its use as a building stone in southern India, and its name, derived from the Latin *later*, a brick. This induration is not related to any change of composition, but to the development of crystalline continuity between the iron oxides.[10] Such hardening can be disastrous for agriculturalists. Lateritic soils kept in cultivation retain good structure because they are kept moist, but during World War II in West Africa some of the normal agricultural areas were temporarily abandoned, the soils dried and hardened and are now useless.

Ferricrete* is another duricrust rich in iron oxides. But it exists in isolation, without the sandy A-horizon and the

---

* Ferricrete was the name proposed by Lamplugh [11] for iron duricrusts in general; in the use of the suffix—*crete*—this is correct, but the term laterite was then, and remains, so widespread that it cannot easily be superseded. However, the word ferricrete is useful in the context suggested here.

kaolinitic C-horizon. The structures of ferricretes show quite clearly that many of them are of depositional origin. They are commonly coarsely clastic, were deposited in valley floors (though, because of their toughness, they now commonly cap ridges—an example of relief inversion) and in some areas they are derived from the dissection of laterites.

Silcrete is the silica-rich duricrust—also known as billy, wethers, greywethers and sarsenstone. Silcrete is widespread in central and some areas of southern Australia. It has been described from South Africa,[12] central Africa,[13] Western Australia[14] and in southern England and northern France in relict form. It has not been possible to demonstrate conclusively that silcrete is forming anywhere today, though there have been suggestions from southern Queensland not only of several phases of post Cretaceous silcrete formation, but of contemporary development. What has been called a silcrete has recently been reported from Wisconsin[15] but from the description given this relict duricrust of tentative Miocene age is closer to a laterite rather than a silica-rich weathering development.

Silcrete has not been satisfactorily defined. As most commonly used at present, the term embraces a complete weathering profile which includes a silica-rich horizon (Pl. 10.3a). But occurrences of silica-rich rock in isolation are also referred to as silcrete. In central and southern Australia the characteristic silcrete profile consists of a silica-rich horizon (95 per cent plus, SiO_2, with quartz grains set in a matrix of fine quartz, or in a few cases, opaline silica) overlying a kaolinised

zone which, in turn, merges with the parent material. In some areas there is a ferruginous-rich horizon just above the weathering front. The silica-rich horizon is compact and typically displays columnar structure. The vertical sides of the columns are commonly grooved. Vesicular and whorled occurrences are also common. The silica-rich rock, which may be up to 3 m thick and which is in some places overlain and underlain by nodular fragments of silcrete, varies in colour: grey, white, brown, yellow, red and combinations of these. The rock has a vitreous appearance and, where dense (that is, not vesicular), a good conchoidal fracture. It has a typical ring, and a pungent smell when hit with a hammer.

In central Australia (Fig. 10.1) the extensive sheets of silcrete developed during the early mid Tertiary,[16] the most recent geological evidence pointing to a range between the Eocene and early Miocene inclusive. However, silcrete with an opaline matrix accumulated just west of Lake Eyre during the Pleistocene[18]. Around Port Augusta and south towards Adelaide, the silcrete has accumulated in valleys and in scarpfoot zones. The same is true of the southern Flinders Ranges. The silcrete from scarpfoot zones, particularly near the southern extremity of Lake Torrens, where it forms more-or-less thin skins around joint blocks (Pl. 10.3b), has a peculiar mineralogy for it contains a high percentage of TiO_2.[19]

The origin of silcrete is obscure and controversial. In some areas and particularly in scarpfoot situations,

Plate 10.3a   Horizontally disposed silcrete dissected to plateau and mesa forms, south of Alice Springs, Northern Territory. Note bluff coincident with silicified zone and the gullied debris slopes eroded in the kaolinised zone. *(C.S.I.R.O.)*

Plate 10.3b Skin of silcrete (left) up to 30 cm thick formed on a quartzite block (right) at the foot of Beda Hill, southern Arcoona Plateau, South Australia. (*C.R. Twidale.*)

the possible source of silica is not difficult to identify: in some areas the run-off from the quartzite hills could have, over the years, carried enough silica in solution to account for the silcrete. But what causes it to be redeposited in particular localities is debatable. Many silcrete occurrences are found in or adjacent to former lacustrine settings, as for instance in the southern Flinders Ranges at Kanyaka; some silcretes like those of the lower Beda Valley, southern Arcoona Plateau, contain well-rounded pebbles and boulders; and water-table effects may be part of the reason for the redeposition of the silica. Alternatively, certain plants may have acted as accumulators in the past, just as some do now (for instance some cereals[20]). The plants then, as now, were probably more profuse in the better-watered scarpfoot zones of arid or semiarid lands.

Those silcretes of widespread extent which occur in topographic lows (valley floors, basins) appear to be of accumulational origin, though the chemical processes involved are controversial,[21] and may vary from place to place. Some have formed as silica-rich gels in former lakes.[22] Since its formation, the silcrete of central Australia has been folded and dissected (see Fig. 5.10 and Pl. 10.3c); it is not known whether it ever formed a continuous extensive sheet, or whether it developed only in the many valley floors and caused relief inversion. But it has been suggested[23] that silica derived from the leaching of soils and formation of laterite in the areas peripheral to the Lake Eyre Basin (Fig. 10.1), was carried toward the centre of the drainage basin and there precipitated as the river waters evaporated in the arid climate.

However, it seems possible that the thin skins of silcrete which occur in the Beda area in the southern Arcoona Plateau of South Australia[24] are of residual origin—the residue of relatively inert silica left behind after all else had been leached out in solution from clays, shales and other rocks, all of which contain silica in some form or another. This hypothesis is based on the relatively high concentrations of minerals even more insoluble than silica—oxides of

Plate 10.3c Homoclinal ridges eroded in folded and dipping silcrete, northeast S.A. Note the tough silicified zone (e.g. foreground) and the gentler slopes on the kaolinised material. (*H. Wopfner, Geol. Surv. S. Aust.*)

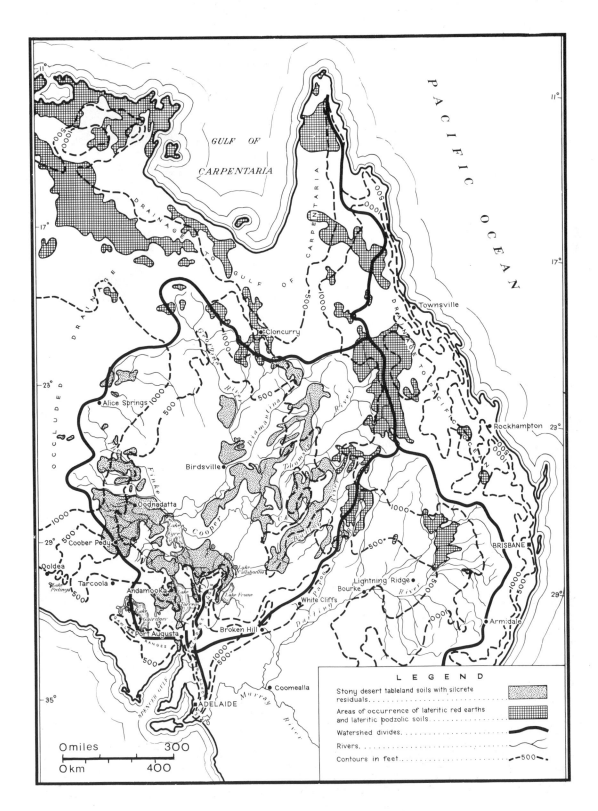

Fig. 10.1   Map of eastern Australia, showing distribution of laterite and silcrete.
(Based on work by C.G. Stephens.)

Plate 10.4 Nodular calcrete near River Murray, between Waikerie and Overland Corner, S.A. (*Miss R.M. Thomson.*)

Plate 10.5 Massive sheet of calcrete, near Virginia, Adelaide Plains, S.A. (*C.R. Twidale.*)

Plate 10.6 Limestone developed as travertine in the bed of an intermittently flowing creek, forming a resistant layer and an overhanging lip in the waterfall. Gorge Creek Valley, eastern Mt Lofty Ranges, S.A. (*C.R. Twidale.*)

titanium, phosphate and cerium—in the matrices of these skins and the evidence of etching of the silica grains within these thin carapaces.

Whatever the chemical details, the association of silcrete with plains and areas which have enjoyed diastrophic quiescence suggests that silica accumulation has been a slow process which has occupied lengthy periods of geological time.

Calcrete has a widespread distribution in arid and semiarid lands. It is known by various names in different parts of the world: kunkar, kankar, nari and caliche, are the best known. But calcrete, the term proposed by Lamplugh,[25] seems apt and consistent with other usage. It is a pedogenic accumulation rich in lime. Indeed on a world scale calcretes average almost 80% calcium carbonate, with quartz, dolomite and minor amounts of clay (frequently palygorskite) making up the rest.[26] Initially, small nodules of lime form in the soil. These increase in size (Pl. 10.4) as layers of lime are formed around the original nodule, so that many calcrete lumps display a concentric structure when split open. The nodules grow and coalesce to form honeycomb calcrete, more-or-less massive sheets, and eventually hardpans,[27] which extend over wide areas, attain several metres thickness, and display a semblance of layered structure in places (Pl. 10.5). In some areas of Texas and northwestern Australia, the accumulation of lime has created a space problem and caused buckling and the development of pseudo-anticlines (see Chapter 7).

Lime accumulation continues in the arid and semiarid lands, as is evidenced by the replacement of decayed plant roots by lime (for example, in dune sands in the Simpson Desert and in the Adelaide district). Lime accumulation also occurred in the past. The Devonian cornstones of Scotland are probably ancient calcretes. Many calcrete sheets formed in southern Australia during the Pleistocene, possibly rendering dubious the various attempts to emulate Arellano's work in Mexico[28] using calcrete as a stratigraphic marker.[29]

Calcretes probably form in several ways. Some appear to be due to sheet floods of water charged with carbonates drying up and precipitation of their lime at or near the plain surfaces over which they have flowed. Others may have been precipitated in lakes, possibly aided by organisms, and some are due to carbonate precipitation *in situ* under the influence of water table oscillation.[30] Capillary rise is widely held to be important in the precipitation of some calcrete, but carbonate deposition because of variations in pore pressure is also significant.

Travertine is lime precipitated by springs and in the channels of streams through the drying up of the river waters (Pl. 10.6). Such deposits have a very different distribution—linear as opposed to extensive sheets—from true calcrete.

Stalactites and related forms are other examples of features due to lime precipitation (see Chapter 3). *Aeolianite* is, as its name suggests, a wind-borne sediment. It has a coastal distribution, is widespread in the coastal areas of Western and South Australia which are transverse and face the dominant westerly winds, and consists of dune sands impregnated by lime. It weathers to give a weak karst, with solution cups, etc., and with well-developed calcrete. The lime is thought to have derived from the seabed during glacial phases of the Pleistocene when the upper margins of the continental shelf were exposed. The shells of marine animals were broken up by waves and the fragments were blown onshore where the lime was washed into the dune sands of the coastal areas and into other porous rocks of the coastal zone.

The dissection of any of these duricrusts gives rise to plateau, mesa and butte forms (Pls 5.10 and 10.3a) where the duricrust caprock is flat-lying, or to cuestas, homoclinal ridges and hogbacks where it is tilted or folded (Pl. 10.3c). In other words, the assemblages are similar to those eroded from horizontally disposed or folded sedimentary sequences (see Chapter 5). There are differences in detail according to the nature of the parent material subjected to duricrusting. For example, whereas the forms developed on varied flat-lying sequences frequently exhibit structural benches on the bounding scarps (Chapter 5), dissection of duricrust surfaces down to the parent rock reveals structures and forms which are characteristic of the latter—joint clefts in schists, boulders of granite, and so forth, as the case may be.

But in broad view, dissected duricrusts have the same protective effect as resistant sedimentary sequences, and the landform assemblages vary according to the disposition of the caprock. There are, however, unusual effects such as that caused by a travertine capping which has been described from near Dahran, in Arabia.[31] The travertine was deposited in a winding river valley, and, after stream rejuvenation, proved to be more resistant to the prevailing weathering and erosional agents than the surrounding divides. This, the former valley floor, became upstanding and now stands above the surrounding plains as a winding ridge—a spectacular example of relief inversion (Fig. 10.2).

Duricrusts also retard the infiltration of water into the subsoil and on this account are an important factor influencing the rate of surface run-off in arid lands; hence the relative effectiveness of rivers and streams in these areas. Their importance in this regard is not, however, restricted to the tropical deserts. The disastrous floods which caused loss of life and extensive damage to property at Lynmouth in north Devon in 1952 were caused basically by an intense

Fig. 10.2 Relief inversion caused by travertine duricrusting of a river bed (A) in Arabia and subsequent erosion of adjacent areas (B). *(After R.P. Miller, "Drainage Lines", pp. 432–8, by permission of J. Geol.)*

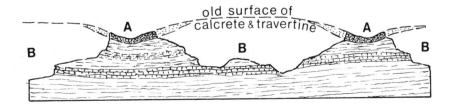

Plate 10.7 Differential weathering in granite, Haytor West, Dartmoor, England. The more massive and resistant upper granite is the "Giant" granite, the finer, less resistant rock below is the "Blue" granite. (*C.R. Twidale.*)

Plate 10.8a   A weather pit or gnamma in granite, Pildappa Hill, northwestern Eyre Peninsula, S.A. (*C.R. Twidale.*)

Plate 10.8b Flat-floored pan in granite on Haytor West, Dartmoor, England. (*C.R. Twidale.*)

rainstorm following wet weather. But particularly near the margins of the Exmoor uplands drained by the River Lyn, the presence of iron hardpans at a depth of, perhaps, only 30 cms in the soils of Exmoor caused run-off to be very rapid, and contributed significantly to the magnitude and rapid rise of the flood.[32]

D. DEVELOPMENT OF SPECIFIC LANDFORMS

Many erosional landforms manifestly owe their origin to differential erosion consequent upon differential weathering (Pls 9.7, 9.8, 10.7). Indeed a strong case can be made for suggesting that most erosional landforms are basically due to the exploitation by erosional agents (of which rivers are the most important) of zones which are rendered vulnerable by weathering and are fundamentally zones of structural weakness. Many examples of such structural landforms have already been described in Part II of this book, and more are cited in the following chapters, but a few features clearly due to differential weathering may be mentioned here, in order to exemplify the general case as well as certain problems of interpretation. Both minor and major features, some of restricted occurrence and some of wide distribution, are discussed.

Several minor forms are well developed by the weathering of granitic rocks, though they are not restricted to such petrological environments. *Weather pits* or *gnammas*, for example, are widely reported from granite inselbergs and whalebacks, but also from sandstones, for instance in the southern Flinders Ranges and Tasmania. They occur in many climates and take several forms (Pl. 10.8). Basically, however, the features originate through the differential weathering of the local bedrock, though it is not yet clear whether such weathering takes place wholly when the country rock is exposed to the atmosphere or whether it occurs in part in the groundwater zone while the surface is still buried. But there is no doubt that water standing in joint clefts and in pools which are initiated by differential weathering[33] enlarges the original limited depressions. The granite sand so formed (in the case of granitic bedrock) may be flushed out by periodic run-off, or it may accumulate in the bedrock hollow, harbouring moisture, and if vegetated, humic acids, which further aid the weathering of the depression. The precise form of the gnamma depends partly on the local structure of the granite and partly on the inclination of the slope on which the weathering occurs. Thus where the granite is essentially uniform and unstructured, the fact that water persists longest in the floor of the depression is of greatest significance, and hemispherical pits or even flask-shaped hollows are formed (Fig. 10.3). However, where the upper zones of the granite are laminated (or where other bedrock is horizontally bedded) water tends to penetrate along the fractures more rapidly than it weathers through the rock layers, and relatively shallow and wide flat-floored pans develop. Both of these are formed on the comparatively flat upper surfaces of inselbergs and other outcrops: on the steeper marginal slopes, water weathering from original depressions produces armchair-shaped hollows.

Though some gnammas occur in isolation, on many outcrops they are linked by gutters, many of which run along joints. The gnammas and gutters together form a rudimentary drainage pattern on the upper slopes of the inselbergs and where the run-off from these bare rock surfaces pours over the steep, even overhanging, bounding slopes, the water causes disintegration of the rock and also washes away the loosened particles to form distinct, regularly spaced grooves, gutters or *Rillen (Granitrillen, Silikatrillen)* (Pl. 10.9a). There is some evidence (Pl. 10.9b) that such linear grooves are initiated by subsurface weathering.

The evolution of *flared slopes* by subsurface weathering of the scarpfoot zone and subsequent erosion of the weathered debris has been discussed at some length (see Chapter 1). Linear depressions or moats which occur at the base of cliffs and which have been reported from several areas of the intertropical zone[34] are of a similar origin, as are the narrow troughs found at the base of bouldery granite inselbergs in the Mojave Desert of southern California, in the Granite Mountains between Baker and Amboy and on the base of granite residuals on Eyre Peninsula and of the arkosic Ayers Rock.[35] However, sub-surface transport of debris (either in solution or by flushing) may be significant. Southeast of Alice Springs, the writer has noted duricrusted mesa remnants surrounded by salinas (Pl. 10.9c) which are probably of similar type.

On granite inselbergs in the Wheat Belt of Western Australia and at Ayers Rock (Fig. 10.4a and b), flared slopes

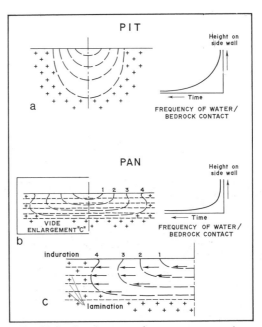

Fig. 10.3   Development of gnammas or weather pits: (a) pit; (b and c) pans. *(After C.R. Twidale and Elizabeth M. Corbin, "Gnammas", pp. 1–20, by permission Rev. Geomorph. Dyn.)*

Plate 10.9a *Granitrillen* on the flared bounding slopes of Pildappa Hill, northwestern Eyre Peninsula S.A. (*C.R. Twidale.*)

Fig. 10.4a Sketch of Kokerbin Rock, near Merredin, Western Australia, showing merging of flared slope on granite and cavernous forms.

Plate 10.9b This groove on Pildappa continues beneath the natural soil surface (dashed line). (*C.R. Twidale.*)

Fig. 10.4b Sketch of similar juxtaposition of flared slope and cavern near Maggie Springs on the south side of Ayers Rock. (*Drawn from photograph supplied by Miss R.M. Thomson.*)

Plate 10.9c Salinas surrounding duricrusted residuals some 30 km ESE of Alice Springs, N.T. (*C.R. Twidale.*)

give way to caverns or *tafoni* (Pl. 10.10a). Tafoni are developed on boulders and inselbergs in many parts of the world but particularly in arid, semiarid and coastal regions. They are best developed on the undersides of boulders, and on the undersides of the exposed edges of sheet structures (Pl. 10.10b), but also occur on vertical faces. There is no doubt that tafoni are initiated by weathering in areas of comparative abundance of moisture such as shady areas and the scarpfoot zone. Tafoni, at present located above the plain and soil level, may occur in horizontal or aligned rows, suggesting that they represent old groundwater zones. For instance, on Ayers Rock which stands some 350 m above the adjacent plain, there is a distinct row of caverns on the south face some 15–20 m above the present plain level, but others occurring in isolation represent bedrock zones which are particularly susceptible to weathering for mineralogical or textural reasons.

But once initiated, the hollows have clearly been further enlarged. The weathering processes responsible are still active in some cases, as indicated by the presence of flakes and grains of rock adhering only loosely to the ceiling and walls or strewn over the floor of the shelters.

The processes are, however, difficult to identify. Temperature oscillations are ruled out by the comparatively small range experienced within the sheltered caverns. Salt weathering has been suggested,[36] particularly in Antarctica and cooler regions where tafoni are well developed (Pl. 10.11). Chloride may be released by the weathering of accessory granite minerals such as phlogopite. It combines with Na and crystallises disrupting the rock. Many of the flakes of granite loosely attached to the ceiling of tafoni are apparently unaltered mineralogically.

But the development of the tafoni involves two distinct tendencies: the hollowing out of the fresh rock, and the development of the hard skin or desert varnish protecting the outside of the block and forming a visor (Pl. 10.12). The desert varnish, which allows the preservation of the prominent visors and tough outer skins of blocks, is very thin (less than 100 μ, yet it consists of two layers, the inner minor one rich in SiO_2 and Al_2O_3 (though with traces of Fe and Mn), and the main or outer skin, rich in FeO and MnO.[37]

The origin of the skin is not clear, but it could be due to the capillary attraction of salts released by waters penetrating to the surface during dry spells. In this case there should be not only an outer zone of accumulation, but also an inner zone of depletion. Such a depleted zone is in evidence in some localities but not everywhere. There is also some evidence that the concentration of minerals takes place beneath the land surface in certain zones surrounding corestones.[38] The accumulation of salts at the surface could also, in part, be due to the activities of lichens; again there is some observational evidence that this is so.

The origin of boulders was briefly considered in Chapter 3, and in Chapter 9 reference was made to spheroidal and onion weathering (Pl. 9.1). Such features were at one time attributed to heating and cooling until it was realised that they occur at depths well beyond the range of diurnal and annual temperature variations, and they were then attributed instead to chemical weathering. It was supposed that hydration of the rock-forming minerals in granite caused volume increase, which induced the rupture of the rock in

Plate 10.10a Complex alveolar weathering in cavern developed in granite, Remarkable Rocks, Kangaroo Island, S.A. (*S.A. Govt Tour. Bur.*)

Plate 10.10b Cavernous weathering on the underside of a massive granite sheet, Ucontitchie Hill, Eyre Peninsula, S.A. (*C.R. Twidale.*)

Plate 10.11   Alveolar weathering has formed minor *tafoni* in these granite blocks in Taylor Dry Valley, Antarctica. *(T.L. Péwé.)*

Plate 10.12   Cavern with pronounced visor, Ayers Rock, N.T. Note exploitation of bedding plane left of centre. *(C.R. Twidale.)*

more-or-less thin flakes or layers. As the water penetrated further and further into the rock, the process was repeated and so developed a series of layers in concentric arrangement around a corestone. However, it has also been suggested that some chemical changes in rocks can involve little or no volume increase because minor textural features remain undisturbed.[39] Another problem is that the rock of the weathered zone around the corestones in some areas is not altered, merely broken into flakes.

Some onion weathering (Pl. 9.1b) is caused by chemical weathering of layers of minerals which are an original feature of the cooling magma.[40] Other corestone development may result from tectonic movements.[41] But the origin of layered arrangements in weathered rocks in many instances remains obscure.

E. THE PIEDMONT ANGLE

Two examples of landforms of rather greater magnitude which are caused in considerable measure by weathering processes are discussed here and in the section which follows. The first is the piedmont angle, which is the sharp break of slope that characteristically separates hill and plain in semiarid lands.

The piedmont angle is cut in bedrock, and is clearly distinguished from the angular junction made with bedrock hillslopes by the surface of alluvial deposits which bury the lower slopes of residual uplands. For instance in the Everard Ranges (Fig. 3.1), ridges consist of groups of granitic domes standing 300 m and more above the adjacent alluviated plains. Ayers Rock, an isolated sandstone inselberg in the desert of the southern part of the Northern Territory, stands abruptly above a plain built up of alluvial and windblown deposits which are at least 100 m thick only 0.5 km from the base of the Rock. Similarly, the inselbergs shown in Pl. 10.13 look like true "island-mountains" because they rise through and above a "sea" of alluvium deposited by the meandering rivers.

Plate 10.13 "Island-mountains" (of limestone) standing above an alluvial plain in Burma. (*Hunting Surveys.*)

Plate 10.14a   Piedmont angle between a bouldery granite nubbin or low hill and pediment surface (termite mounds in middle distance—sand too shallow near the residual hill), north of Cloncurry, northwestern Queensland. *(C.R. Twidale/C.S.I.R.O.)*

Plate 10.14b   The piedmont angle takes the form of a narrow zone of curvature at the base of this sandstone ridge (Ragless Range) in the southern Flinders Ranges, S.A. The conical hill in the distance is the Devils Peak, near Quorn. *(C.R. Twidale.)*

But the sharp break of slope in many areas is formed in bedrock. On well-jointed rocks like limestone and granite, the piedmont angle lives up to its name and it is possible to indicate precisely where hill and plain meet (Pl. 10.14a). More commonly, however, the feature takes the form of a narrow zone of curvature, concave upwards, separating hill and plain (Pl. 10.14b).

The origin of the piedmont angle has given rise to some controversy. Early workers in the American Southwest[42] attributed the abrupt break of slope to faulting and subsequent scarp recession (Fig. 10.5—see also Fig. 4.12). Fault delineated bolsons are common in this area, but even if faulting is allowed as initiating the angle, no explanation was offered for its maintenance during scarp retreat. In other parts of the world the angle is well developed in areas where no faulting is in evidence — bordering domed inselbergs in northwestern Eyre Peninsula (Pl. 1.2), nubbins in northwest Queensland (Pl. 10.14a) and in uninterrupted sedimentary sequences bordering the Flinders Ranges.

An American geomorphologist, D.W. Johnson,[43] endeavoured to explain the piedmont angle in terms of stream erosion. He argued that where streams debouch from arid uplands, they deposit much of their load and divert themselves along the mountain front, adopting a meandering habit as they do so. Such streams erode the base of the mountain front and thus sharpen the break of slope between hill and plain. There are several obvious objections to this hypothesis. The first relates to the logic of the argument, which, in order to induce stream deposition and thus the diversion of the river along the mountain front, invokes a break of slope between hill and plain, that is, the feature to be accounted for. Then, rivers debouching from uplands in arid and semiarid regions rarely run along the scarpfoot, but continue out towards the centre of the drainage basin. Many adopt a distributary habit for a few kilometres on leaving the upland, but nevertheless do not markedly deviate from the original course. Third, if the streams meander or formerly meandered along the scarpfoot, the mountain front should be scalloped by meander bluffs; such forms have not been found in these situations. Fourth, the piedmont angle is as well developed where no streams leave the upland. Fifth, the piedmont angle borders small residuals such as those shown in Pl. 10.14a, which are too small or too pervious to generate streams.

The South African geologist and geomorphologist, L.C. King,[44] attributed the pronounced change of slope at the scarpfoot to a change of process. On the steeper scarpfoot run-off is turbulent, but on the smooth pediment fronting the upland it is laminar, and the piedmont angle is the result of the change of flow. As with Johnson's argument, there is the possibility of a chicken-and-egg type of confusion. The change of process is attributed to the very feature the change is alleged to explain. In any case, it is very doubtful if laminar flow persists or is widespread in nature if only because of the roughness in detail of even the seemingly smooth surfaces of pediments.

Evidence from Australia points to the piedmont angle being caused initially by structural contrasts, either lithological or structural, and to these weaknesses being exploited and emphasised by differential scarpfoot weathering and erosion.[45] With granite residuals the structural contrast is between compartments of massive, well-jointed granite (Fig. 3.5a): the development of flared slopes is merely a special case of scarpfoot weathering and erosion (Chapter 1). In areas like the Flinders Ranges, the break of slope is primarily of lithological origin. Sandstone and limestone form ridges, and shale and siltstone form valleys and plains (Chapter 5). But field observation here and elsewhere shows that intense weathering—and erosion—of the bedrock is

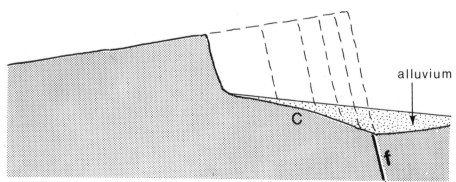

Fig. 10.5 The piedmont angle as due to fault dislocation and scarp recession. Note the convex subdebris slope (C).

Plate 10.15a   Scarp foot depression in the western piedmont of the central Flinders Ranges, north of Brachina, S.A. The scarp face is in Cambrian limestone dipping left. The (higher) range beyond, visible through the valley on right, is of sandstone. The plains are underlain by siltstone and shale. Note the pediment remnant (P) with the reverse scarp facing the main range, the depression between the remnant and the range (D) and the curved piedmont angle beyond. *(C.R. Twidale.)*

Plate 10.15b   Scarp foot valley developed in zone of pronounced chemical weathering caused by wash from the slope above and its subsurface infiltration, southern Arcoona Plateau, S.A. The plateau on the left is capped by quartzite, the hill to the right by silcrete. Note the kaolin (white) exposed in the mesa in the middle distance. *(C.R. Twidale.)*

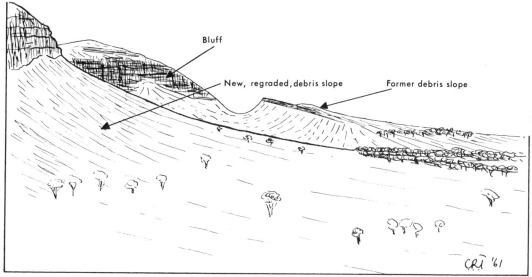

Fig. 10.6   Field sketch of the Lizard, on the south side of Mt Conner, Northern Territory, showing the erosion of the former gentle slope and the development of a col on the intensely weathered rock.

limited to the scarpfoot zone and in the Flinders it has mainly affected the argillaceous strata (Pl. 10.15a). Here, run-off from the nearby ridge slopes, as well as subsurface moisture from the ridge aquifers, contributes to the scarpfoot rotting.

But in areas of essentially flat-lying strata (Pl. 10.15b) where there is no possibility of pronounced lateral variations in lithology, similar marked scarpfoot weathering and erosion are in evidence. Mt Conner, N.T. (Fig. 10.6), has already been noted, and there are similar examples at the base of the Great Escarpment near Van Rhynesdorp, Cape Province, South Africa. In this last area remnants of older, more gentle slopes with preserved weathered rock can be observed at some sites but nearby there is only evidence of the pronounced scarpfoot erosion, all the weathered debris presumably having been removed.

The presence in the piedmont zone of numerous remnants of older pediment levels (Fig. 10.7), which increase in elevation and in the inclination of their upper surfaces away from the present scarpfoot, suggests that there has been both slope lowering and retreat in the zone and that the present piedmont angle has gradually been etched out and sharpened from the contact between the resistant and the weaker strata.

Of course, the occurrence of a fault zone aids subsurface weathering and at the mouths of major rivers there is limited lateral corrasion. Moreover, in areas of impermeable rock where stream density is high and where braided streams are common, lateral corrasion is sufficient to undercut slopes. This occurs for instance in parts of the James Range in the

Northern Territory (Pl. 10.16) and bordering Death Valley, California, for instance in the clays around Zabriskie Point (see also Chapter 19). But the hypothesis of differential weathering and erosion of structural junctions appears to account for the development of the piedmont angle in most areas.

F. STEPPED TOPOGRAPHY

Irregularly stepped topography is clearly developed over considerable areas of the Sierra Nevada northeast of Fresno, California (Fig. 10.8). It has evolved on granitic bedrock and the explanation offered[46] essentially involves peculiarities of granite weathering. As indicated in the discussion of flared slopes,[47] granite is peculiar in that it is vulnerable to attack by moisture but is relatively immune to other weathering agencies; and the least that can be said is that chemical weathering of granite by moisture is far more rapid than physical weathering of granite exposed to the atmosphere. This was appreciated by Bain who, writing half a century ago on the origin of inselbergs in Nigeria, noted that in areas with marked wet and dry seasons chemical weathering is pronounced, especially along joints. He went on,

> The result is that hollows due to slight surface irregular-ities and joint cracks soon become deepened by this means [i.e. by chemical attack] while the rocks between weather by exfoliation only.[48]

This clearly implies that the latter is much less rapid in its action than the former.

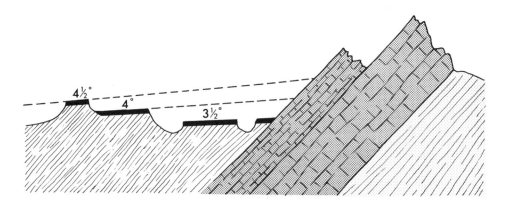

Fig. 10.7 Pediment remnants in the scarpfoot zone at Brachina adjacent to the Flinders Ranges, South Australia.

Plate 10.16 Intricately dissected shales in the James Ranges, N.T., with braided channels in the valleys. The latter are widened through undercutting when these rivers are in flood. (*C. Wahrhaftig.*)

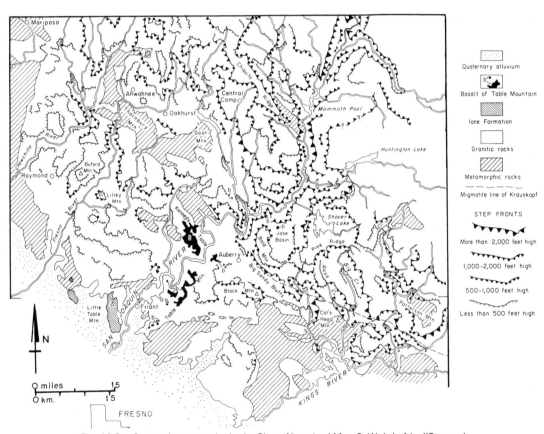

Fig. 10.8   Stepped topography in the Sierra Nevada. *(After C. Wahrhaftig, "Stepped Topography", pp. 1165—90, by permission of C. Wahrhaftig and the Geol. Soc. Am.)*

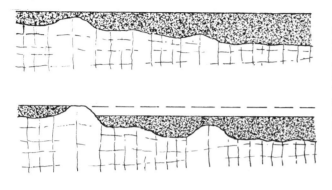

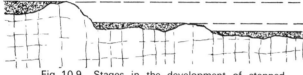

Fig. 10.9 Stages in the development of stepped topography on granite. The weathered granite is shown as stippled. *(After C. Wahrhaftig, "Stepped Topography", pp. 1165–90, by permission of C. Wahrhaftig and the Geol. Soc. Am.)*

In the Sierra Nevada irregular, though broadly linear, zones of massive granite blocks have withstood weathering and hence erosion better than have the intervening areas. These resistant zones form ridges which are irregular and discontinuous in plan. They surely reflect contrasts in jointing in the bedrock, either that now exposed or that which once was located above the present ridge areas, and which has since been eroded. But the ridges act as local baselevels for the streams which flow at right angles across them and so the stepped relief is developed (Fig. 10.9).

G. CONCLUDING STATEMENT

Weathering is a vital component of the geomorphological complex. It prepares material for transportation. It is responsible for the formation of duricrusts. It imposes a certain unity of landform development on rocks which are mineralogically contrasted, so that hollows and gutters, for instance, are similarly developed on granite, sandstone, limestone, and so on.

References Cited

1. See for example J.F. McAllister and A.F. Agnew, "Playa Scrapers and Furrows on Racetrack Playa, Inyo County, California", *Bull geol. Soc. Am.*, 1948, 59, p. 1377; T. Clements, "Wind-blown Rocks and Trails on Little Bonnie Claire Playa, Nye County, Nevada", *J. Sed. Petrol.*, 1952, 22, pp. 182–6; L. Kirk, "Trails and Rocks Observed on a Playa in Death Valley National Monument, California", *J. Sed. Petrol.*, 1952, 22, p. 173–81; J.S. Shelton, "Can Wind Move Rocks on Racetrack Playa?", *Science*, 1953, 117, pp. 438–9; G.M. Stanley, "Origin of Playa Stone Tracks, Racetrack Playa, Inyo County, California", *Bull. geol. Soc. Am.*, 1955, 66, pp. 1329–50.
   For experimental work in this field see A.T. Grove and B.W. Sparks, "Le Déplacement des Galets par le Vent sur La Glace", *Rev. Géomorph. Dyn.*, 1952, 3, pp. 37–9. But see also S.A. Schumm, "The Movement of Rocks by Wind", *J. Sed. Petrol.*, 1956, 26, pp. 284–6.

2. W.G. Woolnough, "The Influence of Climate and Topography in the Formation and Distribution of the Products of Weathering". *Geol. Mag.*, 1930, 67, pp. 123–32. See also A. Goudie, *Duricrusts in Tropical and Subtropical Landscapes*, Clarendon Press, Oxford, 1973, for a general account of duricrusts.

3. G.W. Lamplugh, "Calcrete", *Geol. Mag.*, 1902, 9, p. 575.

4. J.W. Pallister, "Slope Development in Buganda", *Geogr. J.*, 1956, 112, pp. 80–7; W. Bruckner, "The Mantle Rock ("Laterite") of the Gold Coast and its Origins", *Geol. Rdsch.*, 1955, 43, pp. 307–27.

5. F. Buchanan, *A Journey from Malabar through the Countries of Mysore, Canara and Malabar*, Vol. 2, East India Co., London, 1807, pp. 436–60.

6. See J.A. Prescott and R.L. Pendleton, "Laterite and Lateritic Soils", *Comm. Bur. Soil Sci. Tech. Comm.*, 1952, 47; R. Maignien, "Review of Research on Laterites", *Nat. Resour. Res.*, 1966, IV, UNESCO; for a review of characteristics see this paper and also S. Sivarajasingham, L.T. Alexander, J.G. Cady and M.G. Cline, "Laterite", *Adv. Agron.*, 1962, 14.

7. L.T. Alexander and J.G. Cady, "Genesis and Hardening of Laterite in Soils", *U.S. Dep. Agric. Tech. Bull.*, 1962, 1282.

8. See for example R. Prider, "The Lateritized Surface of Western Australia", *Aust. J. Sci.*, 1966, 28, pp. 443–51.

9. See R. Maignien, "Review of Research on Laterites".

10. L.T. Alexander and J.G. Cady, "Genesis and Hardening of Laterite".

11. G.W. Lamplugh, "Calcrete", p. 575.

12. J.J. Frankel and L.E. Kent, "Grahamstown Surface Quartzites (Silcretes)", *Trans. geol. Soc. S. Afr.*, 1937, 40, pp. 1–42.

13. H. Bassett, "Silicification of Rocks by Surface Water", *Am. J. Sci.*, 1954, 252, pp. 733–5.

14. R. Prider, "The Lateritized Surface of Western Australia", pp. 443–51.

15. G.H. Dury and J.C. Knox, "Duricrusts and Deep-weathering Profiles in southwestern Wisconsin", *Science (N.Y.)*, 1971, 174 (4006), pp. 291–2.

16. See H. Wopfner, "On Some Structural Development in the Great Australian Artesian Basin", *Trans. R. Soc. S. Aust.*, 1960, 83, pp. 179–94; N.F. Exon, T. Langford-Smith and Ian McDougall, "The Age and Geomorphic Correlations of Deep Weathering Profiles, Silcrete and Basalt in the Roma-Amby District, Queensland", *J. geol. Soc. Aust.*, 1970, 17, pp. 21–30; this last paper includes references to a protracted debate on the age of silcrete in the *Australian Journal of Science*; H. Wopfner and C.R. Twidale, "Geomorphological History of the Lake Eyre Basin", in J.N. Jennings and J.A. Mabbutt (Eds), *Landform Studies from Australia and New Guinea*, Australian National University Press, Canberra, 1967, Ch. 7, pp. 118–42.

17. H. Wopfner, R. Callen and W.K. Harris, "The Lower Tertiary Eyre Formation of the Southwestern Great Artesian Basin", *J. geol. Soc. Aust.*, 1974, 21, pp. 17–51; but see also C.G. Stephens, "Laterite and Silcrete in Australia", *Geoderma*, 1970, 5, pp. 5–52.

18. H. Wopfner and C.R. Twidale, "Geomorphological History" in *Landform Studies*, pp. 118–42.

19. J.T. Hutton, C.R. Twidale, A.R. Milnes and H. Rosser, "Composition and Genesis of Silcretes and Silcrete Skins from the Beda Valley, southern Arcoona Plateau, South Australia", *J. geol. Soc. Aust.*, 1972, 19, pp. 31–9.

20. See T.S. Lovering, "Geological Significance of Accumulator Plants in Rock Weathering", *Bull. geol. Soc. Am.*, 1959, 70, pp. 781–800; L.H.P. Jones and A.A. Milne, "Studies of Silica in the Oat Plant. I. Chemical and Physical Properties of the Silica", *Plant and Soil*, 1963, 18, pp. 207–20; L.H.P. Jones, A.A. Milne and S.M. Wadham, "Studies of Silica in the Oat Plant. II. Distribution of the Silica in the Plant", *Plant and Soil*, 1963, 18, pp. 358–71; L.H.P. Jones and K.A. Handreck, "Studies of Silica in the Oat Plant. III. Uptake of Silica from Soils by the Plant", *Plant and Soil*, 1968, 23, pp. 79–96; J.T. Hutton and K. Norrish, personal communication.

21. See for example J.J. Frankel and L.E. Kent, "Grahamstown Surface Quartzites", pp. 1–42; H. Wopfner and C.R. Twidale, "Geomorphological History" in *Landform Studies*, pp. 118–42.

22. A.A. Öpik, "Mesozoic Plant-bearing Sandstones of the Camooweal Region and the Origin of Freshwater Quartzite", *Bur. Min. Resour. Tech. Rep.*, 1954.

23. C.G. Stephens, "Silicretes of central Australia", *Nature (Lond.)*, 1964, 203, p. 1407.

24. J.T. Hutton, C.R. Twidale, A.R. Milnes and H. Rosser, "Composition and Genesis", pp. 31–9.

25. G.W. Lamplugh, "Calcrete", p. 575.

26. See A. Goudie, "The Chemistry of World Calcrete Deposits", *J. Geol.*, 1972, 80, pp. 449–63; and A. Goudie, *Duricrusts in Tropical and Subtropical Landscapes*.

27. See F. Netterberg, "Calcrete in Road Construction", *Nat. Inst. Road Res. (Pretoria)*, 1971, Bulletin 10.

28. A.R.V. Arellano, "Barrilaco Pedocal, a Stratigraphic Marker ca 5000 B.C. and Its Climatic Significance", *Rep. 19th Int. geol. Congr. Algiers*, 1952, 1953, Pt. VII,

pp. 53–76.

29. J.B. FIRMAN, "Quaternary Period", in L.W. Parkin (Ed.), *Handbook of South Australia Geology*, Geological Survey of South Australia, Adelaide, 1969, Ch. 6., pp. 204–33.

30. See A. GOUDIE, *Duricrusts in Tropical and Subtropical Landscapes*, pp. 121 *et seq.*

31. R.P. MILLER, "Drainage Lines in Bas-relief", *J. Geol.*, 1937, **45**, pp. 432–8.

32. L.F. CURTIS, personal communication.

33. C.R. TWIDALE and ELIZABETH M. CORBIN, "Gnammas", *Rev. Géomorph. Dyn.*, 1963, **14**, pp. 1–20.

34. See R.W. CLAYTON, "Linear Depressions *(Bergfussnie- derungen)* in Savannah Landscapes", *Geogr. Stud.*, 1956, **3**, pp. 102–26; B. DUMONOWSKI, "Comments on the Origin of Depressions surrounding Granite Massifs in the eastern Desert in Egypt", *Bull. Acad. Pol. Sci.*, 1960, **8**, pp. 305–12; J.A. MABBUTT, "The Weathered Land Surface in central Australia", *Z. Geomorph.*, 1965, **9** NS, pp. 82–114.

35. T. OBERLANDER, personal communication.

36. H.W. WELLMAN and A.T. WILSON, "Salt Weathering, a Neglected Geological Agent in Coastal and Arid Environments", *Nature (Lond.)*, 1965, **205**, pp. 1097–8.

37. R. LE B. HOOKE, HOUNG-YI CHANG and P.W. WEIBLEN, "Desert Varnish: an Electron Probe Study", *J. Geol.*, 1969, **72**, pp. 275–88.

38. C.R. TWIDALE and ELIZABETH M. CORBIN, "Gnammas", pp. 1–20; R.G. WOLFF, "Weathering of Woodstock Granite near Baltimore, Maryland", *Am. J. Sci.*, 1967, **265**, pp. 106–17.

39. See C.D. OLLIER, "Spheroidal Weathering, Exfoliation and Constant Volume Alteration", *Z. Geomorph.*, 1967, **11** NS, pp. 103–8; C.R. TWIDALE, *Structural*

Landforms, Australian National University Press, Canberra, 1971, pp. 36–9.

40. See for example J. BARBEAU and B. GEZE, "Les Coupoles Granitiques et Rhyolitiques de la Region de Fort-Lamy (Tschad)", *C.R. Somm. et Bull. Soc. geol. France*, 1957 (Ser. 6), **7**, pp. 341–51; H. WILHELMY, *Klimamorphologie der Massengesteine*, Westerman, Brunswick, 1958, p. 109; C.R. TWIDALE, *Structural Landforms*, pp. 31–6.

41. C.R. TWIDALE, *Structural Landforms*, pp. 37–8.

42. A.C. LAWSON, "The Epigene Profile of the Desert", *Univ. Calif. Pub. Geol.*, 1915, **9**, pp. 23–48.

43. D.W. JOHNSON, "Rock Planes of Arid Regions", *Geogr. Rev.*, 1932, **22**, pp. 656–65.

44. L.C. KING, "The Pediment Landform: Some Current Landforms", *Geol. Mag.*, 1949, **86**, pp. 245–50; *idem*, "Canons of Landscape Evolution", *Bull. geol. Soc. Am.*, 1953, **64**, pp. 721–54.

45. C.R. TWIDALE, "Origin of the Piedmont Angle, as evidenced in South Australia", *J. Geol.*, 1967, **75**, pp. 393–411.

46. C. WAHRHAFTIG, "Stepped Topography of the southern Sierra Nevada, California", *Bull. geol. Soc. Am.*, 1965, **76**, pp. 1165–90.

47. See C.R. TWIDALE, "Steepened Margins of Inselbergs from northwestern Eyre Peninsula, South Australia", *Z. Geomorph.*, 1962, **6** NS, pp. 51–69.

48. A.D.N. BAIN, "The Formation of Inselberge", *Geol. Mag.*, 1923, **60**, pp. 97–101.

General and Additional Reference

P.W. BIRKELAND, *Pedology, Weathering and Geomorphological Research*, Oxford University Press, New York, 1974.

CHAPTER 11

Mass Movement of Debris

A. GENERAL

A few kilometres south of Mt Babbage, in the northern Flinders Ranges, the local schistose bedrock is intruded by a vein of milky quartz which forms a prominent linear outcrop running at a slight angle across the face of the schist slope (Pl. 11.1). Below the outcrop, the slope is strewn with small, angular fragments of the quartz. There are no streams or stream channels indicative of occasional linear flows of water down the incline, and in any case the white quartz is spread over the slope too widely and evenly to have been distributed by rills or streams. Yet the fragments have clearly been transported downslope by some agency or other. How and why have these particles migrated?

In other areas, for instance around Yankalilla south of Adelaide, large masses of rock and soil have travelled downslope *en masse*, in some instances for a distance of a few metres, in others for several scores of metres. Not only do the masses of displaced material stand above the general surface level, but they can be traced to the depressions or scars higher up the slopes whence they came (Pl. 11.2). What has caused these masses of rock and soil to move downslope?

These are but two of many indications that masses of weathered debris and rock travel downslope, primarily under the influence of gravity but possibly aided by other agencies. Such *mass movements* of debris are in aggregate responsible for the transportation of large volumes of material and for a number of specific landforms.

All these are manifestations of instability or failure within the debris or rock in the near-surface zone of slopes. Some factors operating on slopes induce stresses and instability; others produce resistance and stability. Failure occurs when the strength of the material on slopes, whether it be weathered debris or cohesive rock, is exceeded by the force of gravity. For mass movements of material, which may be differentiated from the mass transport of debris by such media as running water and ice, are primarily due to gravity: other factors influence the stability of slopes but in the last analysis it is gravity acting upon materials which have lost adequate cohesion that causes the mass downslope movement of debris.

Although in theory it is possible and useful to classify various types of mass movement according to morphology and genesis, in practice "their rigorous classification is hardly possible".[1] Mass movements are commonly complex and one which is initiated as one sort of migration may be transformed downslope into another.

Nevertheless it is useful to consider a number of broad types. One is *creep*, by which is implied the slow permanent deformation of slope debris, and of which several subtypes have been recognised either in theory or in practice. Then there are *frozen ground features* which are mainly considered in the context of periglacial environments in Chapter 16. *Slides and slips* are relatively rapid downslope and outward movements of debris involving failure of part of the surficial rock or debris and separated from the stable part of the slope by a definite plane or planes of separation. *Flows* result from flood conditions: there is differential movement of debris within the mobile mass and the proportion of solids to liquid is of the order of 1:1. These flood induced flows are known as *muren* in the European Alps.[2]

All of these involve some lateral translation of debris. In addition there are vertical movements involving *collapse* and *subsidence*.

The difficulties of hard and fast classification may be illustrated with reference to the Saint Jean-Vianney landslide of southern Quebec in which thirty-one people were killed. The slippage took place in the Champlain Clay on May 4, 1971. Clays of volume $6.9 \times 10^8 \mathrm{m}^3$ ($9 \times 10^6 \mathrm{yd}^3$) slid from an area of $268,000$ m^2 ($350,000$ yd^2) and liquefied as an earth flow in the Riviere aux Vases. It travelled downstream a distance of 2880 m as far as the Saguenay River, with sufficient momentum to wash away a bridge constructed over the tributary near its junction with the Saguenay.[3]

Plate 11.1   Quartz reef with spread of angular fragments on the slope below, north Flinders Ranges. *(C.R. Twidale.)*

Plate 11.2a Landslip south of Normanville, southern Mt Lofty Ranges, South Australia; note the rucked front and gap left at the head of the slip. In addition to a decline toward the camera there is also a slope down from right to left; hence the resultant movement down to bottom left as indicated by the numerous upthrust layers in that quarter of the slip. *(C.R. Twidale.)*

Plate 11.2b   Earthflows south of Yankalilla, southern Mt Lofty Ranges, South Australia. (*C.R. Twidale.*)

B. CONDITIONS CONDUCIVE TO MASS MOVEMENTS

On every slope there is a tendency for shearing stresses to develop: there is a tendency everywhere for either particles or masses of debris to slide over the underlying material. These stresses increase proportionately with slope inclination and the density of the slope debris. Hence, mass movements of debris are particularly active and important where certain conditions obtain.

1. Tectonic Instability

In areas subject to earthquake shocks or to regional tilting, for example, the horizontal acceleration of the shock and the earth waves which traverse the affected region can cause momentary instability adequate to trigger off downslope movement of debris masses which are already close to the critical shearing stress. Thus earth tremors in Peru in June 1970 caused massive landslides[4] which caused havoc to property and great loss of life, partly by their direct action, partly by blocking rivers and impounding ephemeral lakes that later burst through the barriers of unconsolidated debris. The earthquake which shattered the San Fernando, southern California, region on 9 February, 1971, triggered more than 1000 landslides in a single square kilometre of the hills and mountains north of the township. Most occurred in areas of ground rupture but failure was common also in surficial deposits and on sedimentary sequences.[5]

2. Water

A sudden accession of water due to heavy rains or to rapid melting of snow and ice can fluidise debris sufficiently for it to move downhill. Many argue that the accession of water increases pore pressure in rocks and decreases friction between particles, so increasing the likelihood of failure and the shearing dislocation. On the other hand it has also been suggested that water in small amounts increases friction[6] and decreases lubrication and thus inhibits shearing. It may also have a buoyant effect on the overlying soil.[7] But there can be no doubt first that a sudden accession of water increases the mass of the debris, and second that it increases fluidity. Such effects are pronounced in humid regions, and especially tropical areas subject to cyclonic downpours.[8] About 23 cm of rain was received in twenty-four hours, on Exmoor, north Devon, during the night of 15 to 16 August, 1952, most of it in the space of five hours. The water was not able to penetrate into the slaty bedrock and the gently sloping moor tops were soon awash with water. Not only were there disastrous floods in the valley of the Lyn, but many earth-flows and highly fluidised debris avalanches formed as a result of these catastrophic rains.[9]

Two spectacular landslides occurred after heavy rains during the spring of 1971 (when the soil was saturated with meltwater from snow and ice) in eastern Canada. One at Saint Jean-Vianney has already been mentioned. The other is the South Nation River landslide which developed in eastern Ontario on a slope some 24 m high in the Leda Clay, which is notorious for the development of slides and slips. The landslide left a scar 640 m × 490 m which covered an area of 28 ha (70 acres). It filled a valley over a sector 2450 m long to such a depth that the river would have had to form

a lake 11 m deep before the barrier formed by the landslide debris would have been topped.[10]

Heavy rains also induce mass movements in another way, for the impact of raindrops on unprotected land surfaces causes considerable detachment of soil particles and their downslope transport. On the other hand, intense rains may also cause compaction of the surface soil and induce rapid run-off. Thus the condition of the vegetation cover influences the erosion of the land surface. On ploughed land over 410 tonnes of soil per hectare may be in suspension at any moment during an intense rainstorm as a result of raindrop impact.[11]

Spectacular mass movements of debris also occur in arid regions after heavy rains. The weathered rocks, lacking the protection of a vegetation cover, are extremely vulnerable to rains and run-off, and the quickly formed debris avalanches and mud-flows are all the more notable because of the relatively slow rate of change of the land surface characteristic of many areas.[12]

3. Ground Ice

The development of ice in the shallow subsurface (see Chapter 16) has similar effects. Water in the soil freezes and expands. The direction of least resistance is normal to the slope, so that debris particles are thrust out at right angles to the local land surface. When the ice melts or the slope becomes locally so steep as to become unstable (Fig. 11.1), the fragments fall vertically under the influence of gravity and, in this way, and especially where there are numerous cycles of freezing and thawing, large masses of debris are made to migrate downslope. The occurrence of permanently frozen ground (permafrost) also has important effects. In summer the upper few centimetres or metres of soil thaw, forming a mush of rock and water resting on the upper surface of the permafrost. The latter provides a gliding or slipping

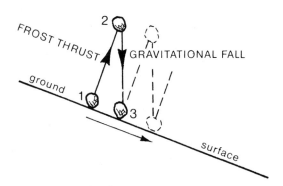

Fig. 11.1 Downslope movement of particles is induced by frost heaving which acts normal (that is, at right angles) to the slope surface and causes instability.

surface over which debris slush very readily moves under the pull of gravity. Hence in cold lands, and especially subarctic regions with considerable spells of above-freezing temperature, ice-induced mass movements are so important that they have acquired a special name: *solifluction* (initially used in a general sense to imply any soil flow[13] but in practice now restricted to flow of soil and debris in cold lands).

4. Structure

Certain geological structures are particularly conducive to mass movements. For instance, a sequence of flat-lying sediments may include some strata which are more permeable or pervious than others, so that water tends to accumulate above the impermeable members. Water running above the interface may cause slight erosion (see section on camber in Chapter 7), or slumping because the clays are thixotropic (that is, they have the property of changing from solid to fluid very rapidly under ephemeral stress). The pseudotectonic features caused by the occurrence of plastic clays in overloaded situations has been described in Chapter 7. Again, strata which dip toward an unbuttressed slope are more prone to slippage than are rocks which dip into a slope.

As described in Chapter 9, weathering has created thick regolithic veneers on the land surface in many areas, particularly in humid regions. Clays are a very common weathering product. Some clays are thixotropic, and again, even comparatively small rainfalls and consequent increase in weight can produce instability of slopes. For example, allophane, a clay produced by the weathering of volcanic rocks of intermediate composition (for example, in several parts of Japan), is thixotropic on wetting. Other clays are hydrophilic, that is, they include in their lattices cations of Na or K, both of which have the ability to take up and discard water molecules according to the availability of

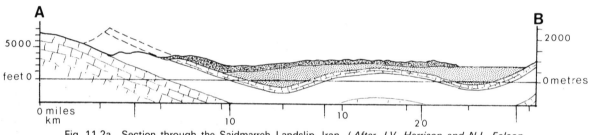

Fig. 11.2a Section through the Saidmarreh Landslip, Iran. (*After J.V. Harrison and N.L. Falcon, "The Saidmarreh Landslip", pp. 42–7 and "An Ancient Landslip", pp. 269–309, by permission J. Geol.*)

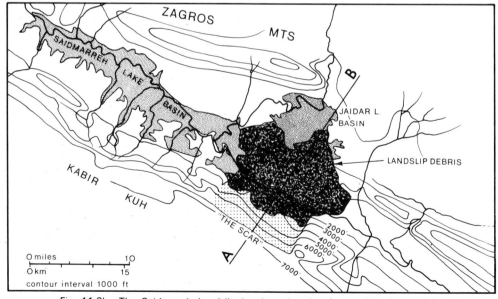

Fig. 11.2b The Saidmarreh Landslip in plan, showing former lake basins, lake sediments deposited in lakes impounded by the slip. (*After J.V. Harrison and N.L. Falcon, by permission J. Geol.*)

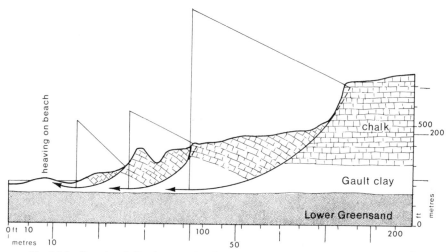

Fig. 11.3 Rotational slips in coastal cliffs near Folkestone, Kent. (*After W.H. Ward, "The Stability of Natural Slopes", Geogr. J., 1945, 105, pp. 170–91.*)

moisture in the environment. Consequently, in areas of marked wet and dry seasons these clays take up and shed water, and as a result swell and contract, causing the development of various forms of gilgai (see Chapter 7) and producing a transitory but, in some cases, critical steepening of slope sufficient to induce downslope movements.

The weathered clayey Franciscan mélange deposits of the San Francisco area are especially prone to landslides and landslips, which are a major factor in slope development in the area and which are, or should be, an important consideration in construction work on such outcrops. However, the dangers of such slope movements are commonly neglected. At Willetts, some 210 km north of San Francisco, for instance, roads constructed in relation to a housing development have been cut across the tops of slopes, reactivating old slides and initiating new ones so that whole slopes are now in motion. One such landslide has ruptured the concrete retaining wall of a small dam constructed to ensure a water supply for the anticipated new inhabitants of the area.

Excavations in clay in humid climates have to be undertaken with great care, for landslips develop if the slopes are too steep. For instance in the excavation of the Gaillard Cut, part of the Panama Canal, over 38 million m³ had to be excavated after the initial channel had been made in order to stabilise slopes. Similar landslips were experienced during the excavation of the Kiel Canal.

5. Undermining

Local erosion causes slopes to be undermined, thus encouraging slumping or collapse. The development of caverns or tafoni is instrumental in many areas in causing cliff collapse while undercutting by rivers and streams and by wind-driven waves at the margins of the ocean or of lakes is also important in this respect. For instance, the Saidmarreh Landslip (Fig. 11.2a) Iran, involving a mass of rock 14 km × 300 m is attributed to undercutting by the Saidmarreh River[14] which was blocked, its waters forming an ephemeral lake (Fig. 11.2b). Undercutting of the banks of the Ouse, in the English Fenland, has from time to time caused instability and failure, with the result that large masses of rock have slipped with a rotational movement.

Plate 11.3 This landslide, which occurred on November 1, 1967 at Oodomari, Mie Prefecture, Japan, was caused by heavy rains; it swept aside a considerable forest. (*Nat. Res. Center Disaster Prevention, Tokyo.*)

The end result is a slope less steep than the critical inclination. In similar fashion, undermining of the chalk cliffs near Folkstone causes rotational slips and incidentally the repetition of outcrops (Fig. 11.3), though faulting (see Chapter 4) is not involved.[15]

In some areas several factors conducive to the development of mass movement occur in combination, and these regions are particularly affected by this means of debris transportation. For instance, in the young fold mountain belts of the world, erosion has produced steep slopes and moreover, earth tremors are frequent and the rocks are shattered and pervious. Around the margins of the Pacific Basin, in areas like New Guinea, the South American Andes and southern Japan, such unstable areas receive heavy rainfall and the rocks are deeply and intensely weathered under the prevailing conditions of abundant moisture, high temperatures and luxuriant vegetation. Hence, landslides and landslips are frequent and dangerous in such areas, and landslides and mud-flows are important factors in mountain sculpture[16] in Japan (Pl. 11.3), which is unstable, deeply dissected and subject to heavy rains. These land movements feed large masses of debris into the valleys which have been temporarily choked, though the rivers rapidly erode through the masses of unconsolidated debris.

Again, in central Otago and the Banks Peninsula, New Zealand, many hillslopes are scarred and hummocky due to the extensive development of slides and flows. Conditions for mass movements of debris are very good. In central Otago, for example, there is a bedrock (a mica schist) which is widespread and which weathers to a clayey debris; in the west, at any rate, precipitation is heavy; there are summer meltwaters from the snowfields, tectonic instability, deep erosion and steep slopes, as well as the destruction of vegetation by man. All these are conducive to the formation of landslides and earth-flows.

C. TYPES OF MASS MOVEMENT

1. Creep and Slopewash

Creep is a term used to denote shallow seasonal downslope migrations of particles on or within the slope mantle. Its most obvious manifestation is the bending of steeply inclined strata in a downslope direction (Pl. 11.4). Creep occurs in the regolith where seasonal variations of temperature and moisture are experienced, and where there are as a result volume changes sufficient to cause local and temporary instability of the particles in the regolith. Propensity to creep increases with slope angle and also with the incidence of colloidal clays but particles move downslope on even very gentle slopes. Rates of movement vary but up to 10cm in a year has been reported.[17] Talus creep is similar in principle to soil creep but involves larger fragments on slopes up to 35°. It is especially pronounced in subarctic and mountainous regions where freeze-thaw activity is frequent.

On very gentle slopes it is very difficult to comprehend fully how frictional forces between particles can be overcome and some of the movement attributed to creep may well be due to slopewash,[18] that is, wash by diffuse flows or discontinuous and ephemeral sheets of water. Particles may also be put into motion by raindrop impact. But whatever the processes involved particles do demonstrably move downhill (Pl. 11.1).

2. Frozen Ground Features

A combination of permafrost, ice-heaving and surface thaw produces several types of mass movement. In addition, freezing of soil water causes organisation of debris resulting in *patterned ground* (see Chapter 16).

3. Landslides, Earthflows and Mudflows

Landslides are comparatively rapid movements of material

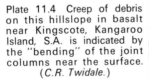

Plate 11.4 Creep of debris on this hillslope in basalt near Kingscote, Kangaroo Island, S.A. is indicated by the "bending" of the joint columns near the surface. (*C.R. Twidale.*)

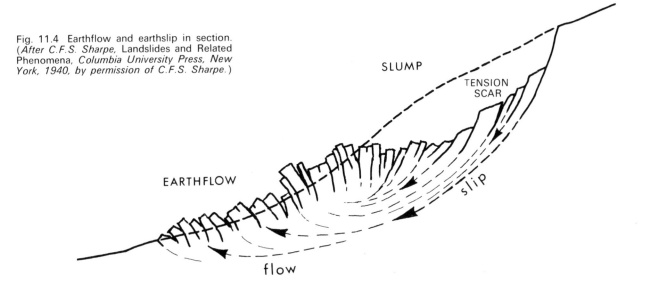

Fig. 11.4 Earthflow and earthslip in section. (*After C.F.S. Sharpe*, Landslides and Related Phenomena, *Columbia University Press, New York, 1940, by permission of C.F.S. Sharpe.*)

SLUMP

TENSION SCAR

EARTHFLOW

slip

flow

over distinct planes along which shear failure has occurred. The planes are located at shallow depths and run roughly parallel to the slope surface. The mass moves downward and outwards, leaving a distinct gap or tension scar at the head of the slide (Fig. 11.4; Pl. 11.2a). There are slickensides visible on the shear plane. And where the mobile debris overrides the still stable slope there is commonly a distinct raised lobe of material (Pl. 11.2a). Because of the presence of the tension scar it is frequently found that water drains into the depression and saturates the translated material downslope from it, inducing further movement. Thus continued instability is not unknown once a slide has been initiated.

Rock, slab and debris slides are particular types of landslide. Some of those involving coarse debris call for further comment. It has been found that some of the largest known are apparently fluidised or lubricated, not by water as is the case with earthflows and mudflows, but by air. [19] In these the debris is unsorted and in some areas it travels not only downslope but for some distance up the opposite sides of the valleys. Eyewitness accounts suggests that the debris travelled at well over 160 km/hr and at times approached 480 km/hr. Blasts of air associated with these large debris flows ripped the leaves and twigs from nearby trees, suggesting that, as the debris advanced on its cushion of air, the weight of debris squeezed out air at high speeds. The air, however, was replaced by the advancing tumbling wave of rock (cf. the hovercraft principle). In 1903 at Turtle Mountain, near Crowsnest Pass on the border between Alberta and British Columbia, a mass of limestone 0·8 km^2 and 130–180 m thick slipped from the slopes and rushed down hill, partially burying the township of Frank (Fig. 11.5). The mass, travelling on a cushion of air, extended about 4 km across the valley and about 120 m up the opposite, rising slope. It has been estimated that some 34–35 million m of debris were involved. The Madison Canyon slip (Montana), involving a huge volume of debris, was caused by the

Hegben Lake earthquake of 1959 and was, in broad outline, similar to the Frank slip; the Elon, Switzerland, slide of 1881 was of similar character. The Blackhawk landslide [20] of prehistoric date occurred on Blackhawk Mt, part of the San Bernardino Range of southern California. The summit of the mountain is marble which slipped downslope, forming a huge tongue of debris 10–30 m thick, 3 km wide and 8 km long (Pl. 11.5).

In some slides the sliding planes are arcuate and the masses of mobile material form rotational slips which may be simple or complex (Figs 11.3, 11.4). These develop mainly in clay or shale. Their failure is rapid.

Mudflows and earthflows consist of poorly sorted coarser debris mixed with clays, and involve differential movement as is indicated by distinct flow lines visible at the surface of the forms. Such flows include climatic earthflows induced by heavy rains or sudden accession of water (Pl. 11.6), lahars or volcanic mudflows, and bog bursts.

Bog bursts are due to the sudden accession of water, particularly in raised bogs, that is, bogs which are located high in the relief and which are swollen due to bacterial action in the decaying organic matter accumulated in the bog. Swelling increases as water, say from heavy rains, enters the bog and eventually the pressure builds up inside the mass until the semi-solid outer crust is ruptured, releasing the semifluid peat inside. In a major bog burst in County Kerry, Ireland, in 1896, nearly 50 million m^3 of peat were discharged.

Mud- and earthflows are most significant, however, in the humid tropics. The blanket of vegetation certainly stabilises the surface, but because of heavy rains, deep weathering and the common production of colloidal clays, great rafts of soil and vegetation commonly slide and flow down hillsides; the trees, for instance, remain *in situ* and are apparently unaffected by their change of location. Such flows and slides are, of course, especially frequent on steep slopes and are responsible for quite enormous lowerings of

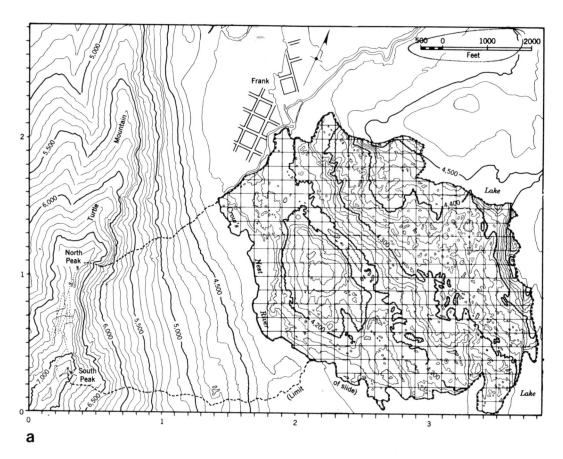

a

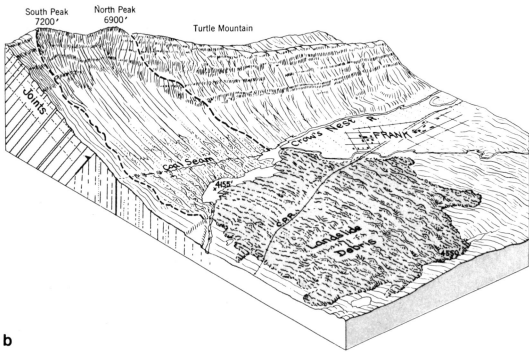

b

Fig. 11.4   The landslide near Frank, Alberta, (a) in plan, and (b) shown in block
diagram. (*After A,N, Strahler,* Exercises in Physical Geography, *Wiley, New York,
1969, pp. 205 and 207.*)

Plate 11.5 The Blackhawk landslide, southern California, from the north. The lobe is about 3 km wide and the front of the slide is some 16–20 m high. (*R.S. Frampton and J.S. Shelton.*)

the earth surface. In Hawaii, for instance, some 200 land-slides and earth-flows developed over a period of eight years in an area of steep slopes (inclination about 40°) some 40 km[2] in extent, effected a transport of debris equivalent to the removal of 1 cm of soil from the whole area each year.[21] This is ten times the rate estimated for slopes near Madison, Wisconsin.[22]

Even in temperate regions these types of mass movement are common. In the Yankalilla area south of Adelaide, for instance, the Permian glacigene (i.e., glacial and fluvio-glacial) sediments are unconsolidated, the loose sands being interbedded with clayey horizons which retard water pene-tration and thus allow fluidisation to develop during heavy rains (Pl. 11.2b).

Mudflows are especially well represented in areas which lie downslope from uplands that carry snow in winter. Sudden melting of snow in spring and early summer produces large volumes of water which initiate damaging mud-flows. Part of the small resort town of Wrightwood, 70 km northeast of Los Angeles in southern California, was buried by a mud-flow in 1941.[23] Conditions were particularly conducive to the development of such a river of mud, for, in addition to meltwaters from the high snow-fields, shattered schists from the San Andreas fault zone are markedly wea-thered and provided debris for the flow, and steep slopes are common in the uplands above the town. The mud-flow (Fig. 11.6, Pl. 11.7), consisting of 25–30 per cent water by weight and carrying large boulders with it, ran some 24 km over gradients which, in the upper source area, were as steep as 32°, but which decreased to as little as 0.6° on the adjacent plains (Fig. 11.6b). The flow ran an average of 3 m per second, with a maximum of the order of 5 m per second.

Even in arid areas, mudflows are initiated by occasional heavy rains and leave a lasting impression on the landscape.

Plate 11.6 Mudflow induced by summer melt, Tintina Trench, Glenyon Range, Yukon Territory. The river in the fore-ground is the Pelly. (*Dept Energy, Mines and Resources, Ottawa, Photo T10–134R.*)

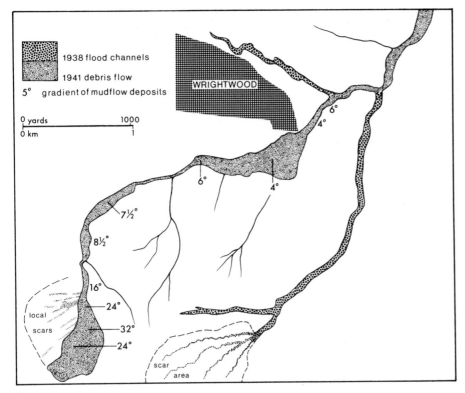

a

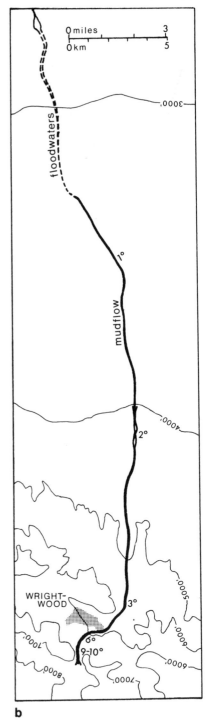

b

For example, Blackwelder[24] has described mudflows which consist of unsorted debris, with fronts up to one metre high, from several parts of the American West and Southwest, including Utah and southern California. In the Owens Valley of southern California, for instance, it is claimed that granite boulders up to 15 m diameter have been transported by mud-flows some 6–10 km from the base of the mountains over the surface of the gently sloping plains.

4. Collapse and Rock Falls

The unbuttressing and undermining of slopes by weathering and erosion causes the collapse of rock masses (Pl. 11.8). The collapsed masses in some instances spread over the slope below, forming screes, talus aprons or blockstreams. Some of the larger blocks tumbling down the slopes gouge the rock and soil surface, accomplishing considerable erosion in the long term.[25] Major collapses, however, give rise to considerable talus cones and scree slopes. Such major failures cause avalanches of debris, analgous to snow avalanches in cold lands, (see Chapter 16) to flow down the slopes a greater or lesser distance.

D. CONCLUDING STATEMENT

Mass movements are important in all climatic regions, but more important in some areas than in others. They achieve their greatest effects in the periglacial or subarctic regions and the humid lands. Nevertheless, similar features develop wherever similar environmental (geological, climatic, alluvial) conditions obtain. The mudflows of the selvas are similar to those of arid regions and temperate regions. Patterned ground is evolved in hot, as well as in cold, desert lands. Avalanches are triggered off by earthquakes regardless of the climatic context.

Though many forms of mass movement are spectacular and rather dramatic (if only for the human suffering they entail), the most important developments from a geomorphological point of view are probably the virtually imperceptible but consistent effects of slopewash and creep.

Fig. 11.6 Mudflow near Wright-wood, southern California, (a) showing local detail and (b) showing the path followed by the flow from the mountains on to the adjacent plain. The figures indicate the local slope in degrees. *(After R.P. Sharp and L.H. Nobles, "Mudflow of 1941", pp. 547–60, by permission Geological Society of America.)*

Plate 11.7 This house was largely inundated by the Wrightwood (southern California) mudflows of 1941. Only the roof gables stand above the deposit. (*R.P. Sharp.*)

Plate 11.8 Pronounced lateral migration of this small creek during heavy rains and flooding in early winter 1968, caused the outside of the bank to be undercut and eventually to collapse, taking with it part of the Lincoln Highway, north of Port Lincoln, South Australia. (*J.J. Noyce.*)

References Cited

1. J.N. HUTCHINSON, "Mass Movement", in *Encyclopaedia of Geomorphology*, (ed. R.W. Fairbridge), Reinhold, New York, 1968, pp. 688–96.

2. See Q. ZARUBA and V. MENCL, *Landslides and their Control*, Elsevier, New York, 1969.

3. F. TAVERNAS, J-Y. CHAGNON and P. LA ROCHELLE, "The Saint Jean-Vianney Landslide: Observation and Eyewitness Accounts", *Can. Geotech. J.*, 1971, **8**, pp. 463–78.

4. G.E. ERICKSEN and G. PLAFKER, "Preliminary Report on the Geologic Events associated with the May 31, 1970, Peru Earthquake", *U.S. Geol. Surv. Circ.*, 1970, **639.**

5. D.M. MORTON, "Seismically Triggered Landslides in the Area above the San Fernando Valley" in "The San Fernando, California, Earthquake of February 9, 1971", *U.S. Geol. Surv. Pub. Paper*, **733**, pp. 99–104. (Several other descriptions of seismically induced surficial earth movements are included in this report.)

6. See D. BRUNSDEN, "Ever moving hillsides", *Geogr. Mag.*, 1971, **43**, pp. 759–764.

7. See L.W. GARBER, "Relationships of Soils to Earth Flows", *J. Soil & Water Conserv.*, 1965, **20**, pp. 21–3.

8. See, for example, C.L. So, "Mass Movements associated with the Rainstorm of June 1966 in Hong Kong", *Trans. Inst. Br. Geogr.*, 1971, **53**, pp. 55–65.

9. JOYCE GIFFORD, "Landslides on Exmoor, caused by the Storm of 18th August, 1954", *Geography*, 1953, **38**, pp. 9–17.

10. W.J. EDEN, E.B. FLETCHER and R.J. MITCHELL, "South Nation River Landslide, 16 May 1971", *Can. Geotech. J.*, 1971, **8**, pp. 446–51.

11. W.D. ELLISON, "Erosion by Raindrop", *Sci. Am.*, 1948, 179, pp. 40–5.

12. See for instance J. CORBEL, "Vitesse d'Erosion," *Z. Geomorph.*, 1959, **3** NS, pp. 1–28.

13. J.G. ANDERSSON, "Solifluction, a Component of Subaerial Denudation", *J. Geol.*, 1902, **14**, pp. 91–112.

14. See T. OBERLANDER, "The Zagros Streams", *Syracuse Geogr. Ser.*, 1965, **1**; also J.V. HARRISON and N.L. FALCON, "The Saidmarreh Landslip, south-west Iran", *Geogr. J.*, 1937, **89**, pp. 42–7; *idem*, "An Ancient Landslip at Saidmarreh in southwestern Iran", *J. Geol.*, 1938, **46**, pp. 296–309; see also R.A.

WATSON and H.E. WRIGHT JR., "The Saidmarreh Landslide", *Geol. Soc. Am. Spec. Paper*, 1970, **123**, pp. 115–39.

15. See W.H. WARD, "The Stability of Natural Slopes", *Geogr. J.*, 1945, **105**, pp. 170–91.

16. H. MACHIDA, "Rapid Erosional Development of Mountain Slopes and Valleys caused by Large Landslides in Japan", *Geogr. Rept. Tokyo Metropo. Univ.*, 1966, **1**, pp. 55–78.

17. A. RAPP, "Karkvegge: Some Recordings of Mass Movements in the northern Scandinavian Mountains", *Buil. Prriglacj.*, 1962, **11**, pp. 287–309; S.A. SCHUMM, "Seasonal Variations of Erosion Rates and Processes on Hillslopes in western Colorado", *Z. Geomorph. Supp.*, 1964, **5**, pp. 215–38; M.A.J. WILLIAMS, "Prediction of Rainsplash Erosion in the Seasonally Wet Tropics", *Nature (Lond.)*, 1969, **222**, pp. 763–5.

18. M.A.J. WILLIAMS, "Prediction of Rainsplash Erosion", pp. 763–5.

19. P.E. KENT, "The Transport Mechanism in Catastrophic Falls", *J. Geol.*, 1966, **74**, pp. 79–83; R.L. SCHREVE, "The Blackhawk Landslide", *Geol. Soc. Am. Spec. Paper*, 1968, **108**.

20. R.L. SCHREVE, "The Blackhawk Landslide".

21. C.K. WENTWORTH, "Soil Avalanches on Oahu, Hawaii", *Bull. geol. Soc. Am.*, 1943, **54**, pp. 53–64.

22. R.F. BLACK, "Slopes in southern Wisconsin (U.S.A.), Periglacial or Temperate?", *Biul. Periglacj.*, 1969, **18**, pp. 69–82.

23. R.P. SHARP and L.H. NOBLES, "Mudflow of 1941 at Wrightwood, southern California", *Bull. geol. Soc. Am.*, 1953, **64**, pp. 547–60.

24. E. BLACKWELDER, "Mudflow as a Geologic Agent in Semiarid Mountains", *Bull. geol. Soc. Am.*, 1928, **39**, pp. 465–80.

25. E. BLACKWELDER, "The Process of Mountain Sculpture by Rolling Debris", *J. Geomorph.*, 1942, **5**, pp. 325–8.

General and Additional References

C.F.S. SHARPE, *Landslides and Related Phenomena*, Columbia University Press, New York, 1940.

R.F. LEGGET, *Cities and Geology*, McGraw-Hill, New York, 1973.

CHAPTER 12

Running Water
and Some Related Forms

A. GENERAL

Many parts of the earth's surface are scored by valleys in which streams or rivers flow. The size of streams, their spacing and pattern in plan all vary widely from one part of the earth's surface to another. For instance, some rivers are sinuous in plan (Pl. 12.1) whereas others are relatively straight but consist of a multitude of channels (Pl. 12.2). Why do rivers flow in valleys and why are there such variations in pattern?

There have always been perceptive observers who noted the deposits laid down by rivers and streams—the well-known phrase "Egypt is the gift of the Nile" refers to the vital renewal of soil fertility brought about by the annual floods and accretions of alluvium in the valley of Africa's greatest river—and who have deduced that rivers not only carry a load, but that they are capable of acquiring these loads themselves, that they erode the land surface and that they excavate the valleys in which they flow. For example, in 1802 Playfair wrote,

Every river appears to consist of a main trunk, fed from a variety of branches, each running in a valley proportioned to its size, and all of them together forming a system of vallies, communicating with one another, and having such a nice adjustment of their declivities, that none of them join the principal valley, either on too high or too low a level; a circumstance which would be infinitely improbable, if each of these vallies were not the work of the stream that flows in it. [1]

Though observations invalidate those parts of this argument concerning the relative sizes of rivers and their valleys and the accordance of tributary junctions, the winding patterns of many valleys and of the rivers that flow in them fully substantiate the conclusion embodied in the final sentence of Playfair's statement, namely that rivers are responsible for the excavation of the valleys in which they presently flow.

Nevertheless, this view was not always widely held. Many scientists took a cataclysmic view of valley formation, believing that rivers merely flowed in pre-existing valleys caused by downfaulting of the crust; only a century

Plate 12.1 Mackenzie Valley, North West Territory, with a meandering tributary, the Carcajou, in the foreground. In the background the Mountain River emerges from the Mackenzie Mountains. Note the meandering River Carcajou, with point bars (P), river bluffs (B) and terraces (T); oxbows (O), scroll patterns (X) and chutes (C) clearly seen in the foreground. (*Dept Energy, Mines and Resources, Ottawa Photo No. T 4–157R.*)

Plate 12.2   Broad, braided and comparatively straight channels of the Neales River west of Lake Eyre, South Australia. *(C.R. Twidale.)*

ago Colonel George Greenwood engaged in acrimonious public correspondence in order to defend his views and those of other fluvialists, like Hutton and Playfair, against "Lyell and all comers".

Nowadays, however, it is widely accepted that, except for those few streams which do run in graben, rivers have eroded their own valleys. How do rivers achieve this work? For what precisely are they responsible? And what factors govern their variations? Basically, the energy of rivers derives from the sun and from the pull of gravity, which is in turn due to the earth's internal energy sources. For the sun is the engine which drives the hydrological cycle and which, among other things, is responsible for rainfall and hence run-off; and it is the kinetic energy resulting from the flow of water down slopes under gravity which endows the rivers with energy.

B. THE HYDROLOGIC CYCLE

Of the precipitation received at the earth's surface, some is evaporated back into the atmosphere either directly or indirectly as evapotranspiration. Some infiltrates into the subsurface, whence some returns to the surface by way of either springs and seepages or by capillary attraction. Some groundwater, however, is stored in the interstices of rocks as underground pools or lakes for many thousands or even millions of years, for instance in parts of Texas, in the downfaulted valleys of southern California, beneath the north African deserts and possibly in the Great Artesian Basin of Australia.[2] Some precipitation is also stored at the surface as snow or ice for a few months or for several hundreds of thousands of years. But precipitation which is not stored, which does not percolate underground, and which is not evaporated, flows over the land surface as run-off: ephemeral sheets of water or, more characteristically, as rills, rivulets, streams and rivers.

Thus the hydrological cycle may in global terms and in the context of geological time be represented:

RF	=	RO	+	ET	+	G	+	US	+	GS.
Precipitation	=	Run-off	+	Evapotranspiration	+	Circulating groundwaters	+	Underground storage	+	Glaciers and snowfields.

In this chapter that part of the precipitation which runs over the land surface—the run-off—is considered. Moreover, we are mainly concerned here with concentrated flows in defined channels, that is, with rivers and streams.

C. STREAM ENERGY AND FLOW[3]

Water flows downhill under the influence of gravity. Each river or stream therefore has a certain amount of potential energy which depends on the weight and head of water.

This potential energy is converted to kinetic energy by downhill travel. In simple terms the kinetic energy of a stream is proportional to half the mass of water times the square of the velocity:

$$Ek = \frac{M}{2}V^2 \tag{12.1}$$

where Ek is the kinetic energy, M the mass and V the velocity. Most kinetic energy is dissipated by the heat of friction. According to Rubey,[4] between 96 and 97 per cent of stream energy is used in this way. But energy not lost through frictional drag is used for erosion and transportation.

Thus the kinetic energy of a stream depends on its mass and upon its velocity, which varies according to stream gradient, channel characteristics, and the volume and viscosity of the water. Because of the velocity component in the kinetic energy formula, it is clear that in considerable measure velocity determines stream energy.

The volume or discharge (Q) of a stream is related to its cross-section area and its mean velocity:

$$Q = AV \qquad (12.2)$$

where A is the cross-section area (Fig. 12.1). The rate of water passing through successive cross-sections is constant provided that the system is closed, that is, that there is no inflow or outflow between the two sections:

$$A_1 V_1 = A_2 V_2 = Q \qquad (12.3)$$

This is due to the fact that water is not compressible, and implies that if, say for structural reasons, the cross-section area of a stream channel increases, then stream velocity decreases proportionately. If, on the contrary, the cross-section area decreases, velocity increases: hence the rush of water in rivers passing through narrow gorge sections. This variation of velocity with cross-section area is known as the Venturi or Bernouilli effect.

The velocity of a stream can be expressed in general terms in a number of ways, the most important of which are the Chezy and Manning formulae, both of which are empirically derived, and both of which show velocity as varying with slope, hydraulic radius of channel and a number of factors such as cross-section form of channel, roughness of bed of channel and straightness of channel, which influence frictional losses. The Manning formula is a refinement of the Chezy expression.

The zone of maximum velocity is centrally located in many river channels (Fig. 12.2) though there are variations according to the width; depth ratio of the channel.

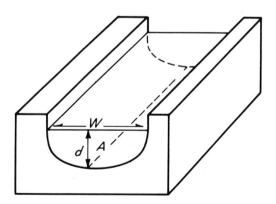

Fig. 12.1 The cross-section area of a stream. W = width; the wetted perimeter is the length of channel where the river is in contact with bed and banks; A is the cross-section area of the stream; and d the maximum depth. The hydraulic radius is A/wetted perimeter and is roughly equal to mean depth. (*From Streams, their Dynamics and Morphology by M. Morisawa. Copyright 1968. Used with permission of McGraw-Hill Book Company.*)

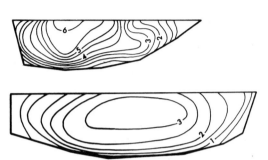

Fig. 12.2 Distribution of velocity in the cross-section of a stream.

The velocity of a stream of water determines not only the energy level, but also the type of flow, which has important implications for transport and erosion of debris. Given absolutely smooth conditions and low velocities in the laboratory, flow is *laminar*, that is, thin sheets of water slide over one another rather like a deck of cards being pushed unequally from one end. But when and where stream velocity exceeds critical value, flow becomes *turbulent*, which means that disordered eddies or swirls are superimposed on the overall flow. These eddies are important for moving particles and particularly for lifting them from the low velocity zones close to the stream bed (Fig. 12.2) and into areas of more rapid flow. Temperature, viscosity, density, depth of water and channel characteristics all influence velocity and, hence, the type of flow which develops. The expression most commonly used to distinguish between laminar and turbulent flow is a Reynolds number, N_r:

$$N_r = p\frac{VR}{\mu} \qquad (12.4)$$

where p is density, V is mean velocity, R is hydraulic radius and μ is viscosity. Flow is laminar for small values of the Reynolds number, and turbulent for high values. In natural streams the number is usually around 500 (varying between 300 and 600) so that flow in streams is almost always turbulent.

Stream velocity varies not only from area to area within the cross-section (Fig. 12.2) but also along the stream profile or thalweg. The latter is commonly concave upward in shape, so that the gradient decreases from source to sea. For this reason it was assumed that stream velocity decreased downstream. Many riverine features were described in these terms and explained on the assumption that there was a downstream decrease in velocity. Thus the alluvial plains which are well developed in the plains sectors were attributed to a decline in energy level consequent on decreased gradient and velocity. Rivers in their plains sections were described as sluggish. The scarcity of coarse debris in the lower reaches of some large rivers was attributed to the inability of the flow to transport boulders, gravel and so on.

But gradient is only one factor determining stream energy and velocity, and many of these apparent indications of downstream decrease in stream velocity are either invalid or can be explained in other terms. Observations of surface flow, as indicated in Fig. 12.2 do not directly indicate the true maximum velocity of a river. The deposition of alluvia in the flood plains of rivers is largely due to lateral migration of the streams. The absence of coarse debris in the lower reaches of major rivers like the Congo is due to its being unavailable—weathering is so efficient in the humid tropics that no gravel survives to be supplied to the river. Elsewhere, attrition accounts for the absence of coarse debris in the plains sector of major rivers.

Measurements of stream velocity at several points along the channels of several rivers in the USA, wherever possible taking readings in the zone of higher velocity perhaps one-third of the way between surface and bed, suggest that although there are local variations along each river course (for structural reasons or junctions with tributaries) stream velocity in general increases downstream.[6] This is of course just the opposite of what had previously been assumed (Fig. 12.3).

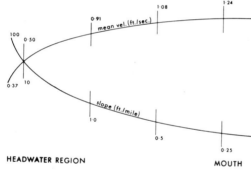

Fig. 12.3 Distribution of velocity along a stream profile. *(Adapted from G.H. Dury,* The Face of the Earth. *(Harmondsworth, 1959) p. 76. Reprinted by permission of Penguin Books Ltd.)*

The explanation of this seemingly paradoxical situation is probably related to variations in some of the other factors which determine stream velocity, and particularly to the roughness factor or channel characteristics, which can account for the bulk of the kinetic energy of a stream through their influence on dissipation of energy in frictional losses. Because the river tends to flow in its own alluvium in its lower course, but over rough outcrops in the upper reaches, frictional losses are greater in the latter than in the former, and this alone more than compensates for the un-doubted loss of energy due to loss of head, slope, or gradient. In addition, plains streams carry a high proportion of their load in suspension; this reduces turbulence, producing an increase in kinetic energy, and hence an increase in stream velocity.

Stream volume and velocity both vary in time as well. Even rivers which run perennially have stages of flood, phases of high and low water, phases of rapid and compara-tively sluggish flow and there are all gradations between such *perennially* flowing rivers and those which are intermittent

and ephemeral in habit. *Intermittent* streams are seasonal; at some periods they are running, at others dry; but there is a measure of regularity about them, possibly because they are in some degree dependent on groundwater sources for their discharge. *Ephemeral* streams, however, run at irregular in-tervals following rains, but are dry for long periods; they are quite unreliable and unpredictable. To some extent stream regimes are dependent on rainfall regimes. Thus ephemeral streams are typical of arid lands, the only exceptions being *exotic* or *allogenic* rivers like the Indus or Nile which rise in heavy rainfall areas and enter the desert regions with a volume adequate to ensure survival, despite the net loss of water incurred during passage across the arid regions. Streams like the Murray are allogenic but the snowfields which supply the river with water are not voluminous and the river experiences considerable fluctuations, not only from season to season, but also from year to year.

Such variations in discharge cause considerable changes in the form of the river channel. Work in the USA[7] has shown that both the width and the depth of a river channel varies with discharge:

$$W \propto Q^a \qquad (12.10)$$
$$D \propto Q^b \qquad (12.11)$$

where W is channel width, D channel depth and Q discharge. It was found that the powers a and b have mean values of about $\frac{1}{2}$ and $\frac{1}{3}$ respectively, values confirmed by the work of Blench.[8] Large rivers, that is, rivers with large discharges, have large width to depth ratios and are wide and shallow (see also McKinlay[9] at Fig. 10). But the type of sediment load carried also affects channel form. Many lightly laden streams are deeper and narrower than those with large loads, which tend to display flat beds. Leopold and Maddock[10] have suggested further that increasing width of channel is associated with an increase in bed load and a decrease in the volume of suspended load (see below).

The importance of stream discharge, velocity, volume and calibre of load in relation to channel morphology emphasises the importance of the occasional floods which occur on all rivers (though they are more extreme (or depart more from the mean flow) in some climates than others) in shaping river channels and river valleys. The work of Hjülstrom[11] shows that velocity is the most important though by no means the only factor determining whether the stream erodes its bed, transports its load, or deposits its load (Fig. 12.4). During floods large volumes of water and debris pour down the channels at high velocity and are to a greater or lesser degree in disequilibrium with the surrounding land surface. They cause drastic, often dramatic, changes in the valleys; much geomorphological work is accomplished in a short space of time. The more average levels of activity in the valley take many years to obliterate the effects of floods. Yet when evaluating flood effects (and this applies equally to other flood or storm effects in different media—coastal, wind or rain storms, as well as river floods) both the frequency and magnitude of forces have to be taken into account. Wolman and Miller have suggested that the biggest flood of a millen-ium is of less significance in geomorphological terms than the largest flood of a few decades, for the latter, although less devastating, nevertheless occurs a few times each century:

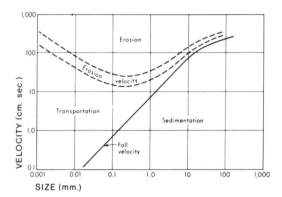

Fig. 12.4 Relationship between erosion, transport and deposition (sedimentation) in a stream and grain size and velocity. *(After F. Hjulstrom, "Studies of the Morphological Activity", pp. 221–57, by permission of Uppsala Geol. Inst. Bull.)*

Analyses of the transport of sediment by various media indicate that a large portion of the "work" is performed by events of moderate magnitude which recur relatively frequently, rather than by rare events of unusual magnitude.[12]

D. EROSIONAL PROCESSES

Some of the available stream energy is used for eroding the bed and banks of the channel in which it flows. The mere flow of water, the *hydraulic force*, is responsible for some erosion. The water can prise off slabs and particles of rock loosened by some other agency, particularly those that are well bedded with the dip downstream (Fig. 12.5). A river also carries sand, gravel and boulders, which it has either acquired through its own activities or which are brought to the river by mass movements. These are used as the tools of erosion; turbulent eddies swirl pebbles around and eventually gouge out potholes (Pl. 12.3). Horizontal vortices armed with sand and other debris in the overall flow can certainly cause the development of grooves in soft materials[13] (Pl. 12.4a) and possibly also in resistant rocks like granite. For instance in the channel of the Umgeni River, in the Valley of the Thousand Hills, Natal, the writer has noted granite bedrock which is grooved, scalloped and smoothed (Pl. 12.4b). There is, in addition, a general scouring and scarifying of the stream bed and banks by the water flow armed with debris: *abrasion* or *corrasion*. River waters accomplish some *solution* or *corrosion*, and it has been further argued that water can erode potholes and grooves, without the aid of stones and pebbles, by swirling about—a process known as *evorsion*. The transported load (boulders, gravel) suffers impact and fracturing or *attrition*. Pebbles and boulders tend to develop a characteristic shape in cross-section[14] (Fig. 12.6a) which, together with the imbricate structure they display (Fig. 12.6b), distinguishes them from beach shingle.

Rivers also erode by a process known as *cavitation*, which requires further explanation.

At a constriction in the stream channel velocity must increase, for water is not compressible, and this raises the level of the kinetic energy. As the specific weight of water, the density, the head of water and the total energy of the stream remain constant, the pressure of water must decrease. And if pressure decreases to the vapour pressure of water, bubbles form. When the channel widens, however, pressure increases as the velocity and kinetic energy component decrease. The bubbles collapse and shock waves of surprising force (around ships' propellers such collapse can cause the metal to be shattered) spread to the bed and banks. The very strong stresses so generated cause erosion by this process of bubble development and collapse related to varied pressure conditions in the fluid. Obviously cavitation is most active where there are, perhaps for structural reasons, variations in channel width, and in consequence sectors of high velocity.

Plate 12.3 Pot-hole with rounded boulders and sand still in the hollow, cut in granite in the bed of the Umgeni River, Valley of the Thousand Hills, Natal. *(C.R. Twidale.)*

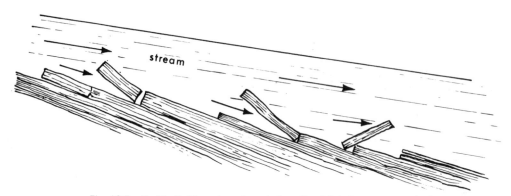

Fig. 12.5   Hydraulic lift and erosion of plate-like debris from a stream bed.

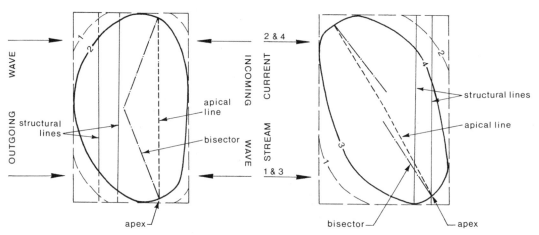

Fig. 12.6a   Variation in shape of stream- and wave-transported pebbles.
(According to P. Lenk-Chevitch, "Beach and Stream Pebbles", pp. 103–8, by permission of J. Geol. *and University of Chicago Press.)*

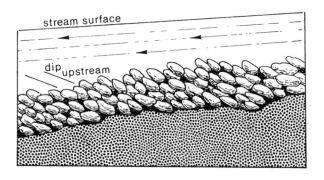

Fig. 12.6b   Imbricate structure in river-deposited pebble bed.

Plate 12.4a A series of shallow parallel grooves in southern Eyre Peninsula, South Australia, scoured in unconsolidated soil by a short-lived flow of water. The scale is given by the coin. (*C.R. Twidale.*)

Plate 12.4b Grooves and scallops cut by the Umgeni River, Natal, in granite bedrock. The river flows from left to right. (*C.R. Twidale.*)

E. TRANSPORTATION OF DEBRIS

Debris eroded by the river or carried to the river by other agents may be deposited and remain stationary until such time as it is reduced to a size amenable to transportation, or it is carried along by the flow of water. Several types of transport have been recognised (Fig. 12.7). Some salts derived from chemical attack on rocks either by river waters or by groundwaters are carried in *solution*. This dissolved load has been calculated to amount to some 3900 million tonnes, representing a significant part of the total denudation of the continents.[15]

Some debris has such a mass that it can only be rolled along the channel bed—the *bed* or *traction* load. Some material can be picked up and carried forward in *suspension*, while other particles progress in a series of leaps and bounds in the process known as *saltation* whereby large particles are lifted, travel a short distance, and settle. A given particle may be transported in suspension or saltation, or as part of the traction load, at various times depending on the velocity of flow in the stream. Thus a sand grain may be part of the bed load at low water, but may be carried in suspension at times of flood.

The force necessary to put debris in motion is called the *critical tractive force*, and the velocity at which this force operates is the *erosion velocity*. The critical tractive force depends on grain diameter, but with larger grains particularly, it is stream velocity that is critical regardless of depth and gradient.

The velocity of a stream determines the size of the largest grain a stream can carry as part of the bed load. This size is known as the *competence*. The *capacity* of the stream, on the other hand, is the maximum amount of debris a stream can carry and is determined by stream volume.

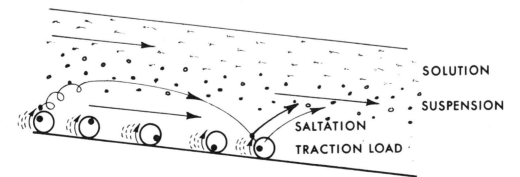

Fig. 12.7 Methods of transport in a stream.

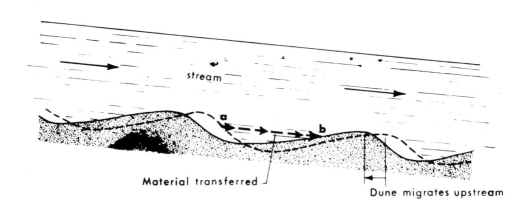

Fig. 12.8   Formation of antidune—the upstream movement of dune form through downstream transport of sand particles.

The bed or traction load of a stream varies both in size and volume. In some sections the bed is clean and the local bedrock is exposed to potholing or scouring, as the case may be. Elsewhere, however, considerable volumes of debris, mostly transported when the river was in flood and dumped with the ebb of water, cover the channel floor. The river flow effectively sorts the debris, for the larger the fragment, the more difficult it is for the stream to lift and transport it. The volume and, hence, the mass of a boulder or pebble varies with a cube function while the cross-section area of the boulder upon which the flow of water exerts pressure is related to a square of the radius. Thus larger boulders are moved downstream less rapidly than are smaller fragments, and, as a result, there is bypassing and sorting (the perfection of which is, however, marred by attrition and the trapping of finer particles among the coarser).

The bed sediments are moulded into various forms some of which are very familiar—some less so. Ripples, aligned transverse or normal to the direction of flow (which may be highly irregular because of the development of turbulent eddies), develop at low stream velocities, and then with increasing velocity (and changes in other factors—see below) dunes, plane surfaces, standing waves and antidunes form in turn. The dunes may be a few centimetres high and several metres apart, but in the Mississippi dunes have been recorded which are 12–13 m high, and a hundred or so metres apart. The antidunes are particularly interesting because, although the debris moves downstream, the dune forms migrate upstream because debris from the steeper downstream slope of one dune is carried forward to be plastered on the gentle backslope of the next dune downstream (Fig. 12.8). Bogardi[16] considers that the distribution of these bed features in space and in time varies according to a number of factors which include particle size, the shear velocity, the mean depth of flow, wetted perimeter and hydraulic radius (see Fig. 12.9).

Saltation, according to many workers, is the dominant mode of stream transport,[17] though others, for instance Kalinske,[18] have attributed greater importance to the rolling and sliding of particles along the stream bed. Turbulent updrafts are an essential feature of saltation, both in streams of water and in streams of air, because in order to be carried forward, a particle first has to be lifted out of the zone of low velocity near the stream bed and into the higher zone of more rapid flow. Any particle exposed in the bed sediment-layer is subjected to turbulent velocity fluctuations, and flow around the particle creates a wake or zone of low pressure behind it. Velocity fluctuations and pressure variations together cause agitation of the particle and eventually it is lifted. It is carried forward for a greater or lesser distance, until gravitational pull becomes dominant over other forces acting on the particle, and it falls to the stream bed. Thence, it may ricochet off into another bound, or it may explode into another mass of particles and help lift them into motion.

In the middle of one of its forward leaps, the particle is indistinguishable from one which is in true suspension. A grain in suspension is influenced by various forces. First there is the resistance of fluid to a sphere falling through it. This varies with the surface area of the grain, the viscosity of the fluid and the velocity of fall. On the other hand, there is a buoyant force which tends to hold the sphere in the fluid, and this varies with the volume of the sphere, the density of the fluid and the force of gravity. This buoyant force is countered by a force which pulls the grain down, and this varies with the volume of the sphere and gravity and with the density of the grain. These upward and downward tending forces are equal with a particle in suspension. In general, the rate of settling of a spherical particle depends primarily upon the density and size of the particle. However, most grains are not spheres, nor are they of regular shape, and this factor alone introduces complications. Particle shape is all important. Small plate-like clay particles, for instance, remain in suspension for long periods. Furthermore, water flow is not equal everywhere, for there are turbulent updrafts and downdrafts superimposed on the overall flow. For this reason larger particles especially are lifted, carried a short distance, and then allowed to settle (saltation).

But in general for small grains, the settling velocity varies with the square of the grain diameter. For large grains, settling velocity varies with the square root of the grain diameter.

Thus in these several ways (solution, suspension, saltation and traction) rivers carry their loads. The Mississippi, for instance, which drains about 41 per cent of the total area of the USA and is one of the world's greatest rivers, carries about 400 million tonnes of silt and gravel to the sea each year. Ninety per cent of this is carried in suspension.[19]

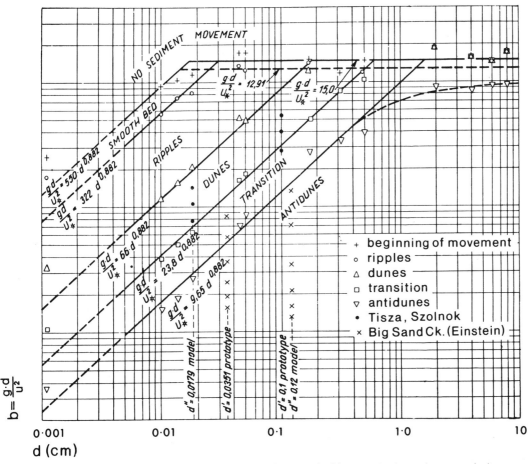

Fig. 12.9 Bogardi's diagram showing relationship between bed form, grain size and stream velocity. On the ordinate scale the parameter b is expressed by $= \frac{g.d}{u^2}$, where g is the gravitational constant, d is the particle size and u the boundary shear velocity. (*After J. Bogardi, "Hydraulic Similarity", pp. 417–45, by permission of Acta. Tech. Hung., 1959.*)

F. RIVER DEPOSITION

1. General

The load carried by rivers is eventually deposited. Deposition is caused by a number of factors: decreased slope or discharge, the latter caused by changes in climate; vegetation; man's activities; increased calibre of the load, which causes the competence of the stream to be decreased; a change in bedrock and hence in channel characteristics; or an increase in the length of the valley, either through headward erosion, seaward extension, or, most effectively, through the development of meanders. In all cases a decreased gradient is the result.

2. Deltaic Deposition

Many rivers deposit debris at their mouths and extend out to sea or into lakes (Pl. 12.5) by building deltas, with topset, foreset, and bottomset beds (Fig. 12.10). The delta of the Tigris–Euphrates system, for instance, has advanced over 300 km during the last 5000 years (Fig. 12.11a). But the Mississippi delta appears to be reasonably stable so far as total area is concerned, though its shape has altered even during historical and recent times (Fig 12.11b and c). Geophysical work has shown that the delta is an area of subsidence, and as fast as debris is deposited at the river mouth, the crust sinks so that, despite a vast annual accretion, there is no overall advance of the land margin. There is an equilibrium between deposition, subsidence and the area of the delta. A similar balance has been achieved at the seaward margin of the Nile delta, where marine erosion is balanced by deposition of debris from the Nile. However, this equilibrium has recently been disturbed by the building of a vast silt trap in the form of the Aswan Dam, so that the same volume of debris is not now deposited at the coast and there is serious erosion of the delta front.

In plan various forms of delta are developed (Figs 12.11b and 12.12), ranging from the true fan shape to the bird's foot. In each case the river deposits are laid down rapidly as the stream velocity decreases. Channels become choked and the flow is diverted, so that distinct lobes can be traced, each due to deposition during a particular phase of flow. In the Mississippi delta, for instance, no fewer than seven lobes have been constructed during the past 5000 years (Fig. 12.11b). Deltaic deposits laid down at the mouths of rivers accumulate in considerable volume, reflecting in part local

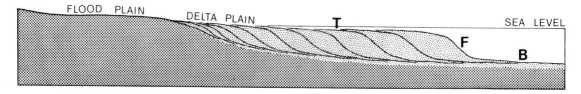

Fig. 12.10   Deltaic bedding. T, topset; F, foreset; B, bottomset.

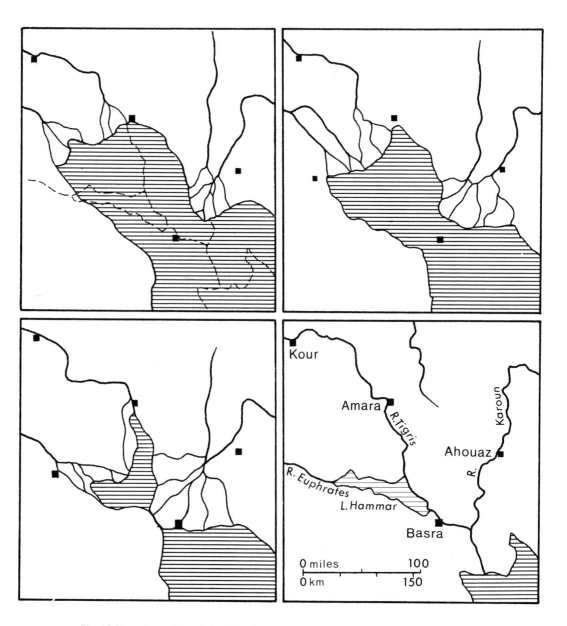

Fig. 12.11a   Extension of the Tigris-Euphrates delta during historical times. *(After R.O. Whyte, "Evolution of Land Use in south-western Asia", in L.D. Stamp (Ed.),* Arid Zone Research XVII—A History of Land Use in Arid Regions, *UNESCO, Paris, 1961, p. 95. Reproduced by permission of UNESCO.)*

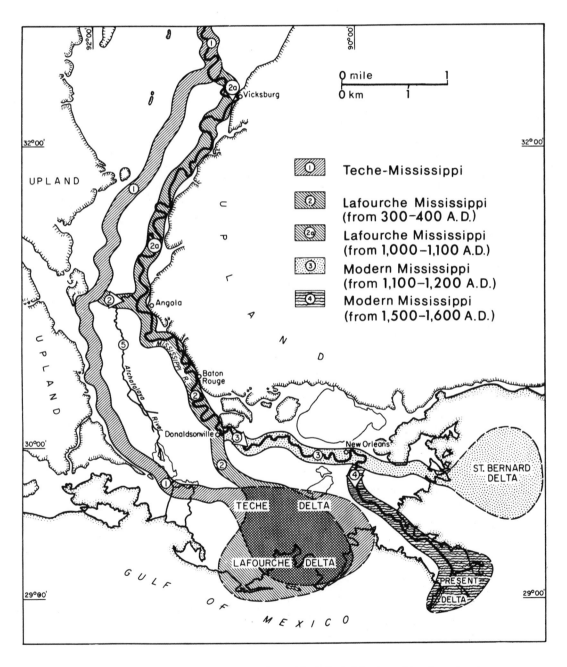

Fig. 12.11b   Variation in the shape of the Mississippi delta in prehistoric and early historic times. *(After H.N. Fisk,* Regional Geomorphology of the United States, *Wiley, New York, 1965, p. 61.)*

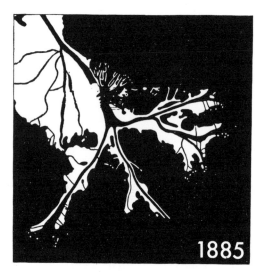

Fig. 12.11c   Variation in the shape of the Mississippi delta between 1885 and 1935. (*After J.M. Coleman and S.M. Gagliano,* cited in *J.H. Vann,* A Geography of Landforms, *Brown, Dubuque, 1971, p. 103.*)

Plate 12.5   A small delta deposited by the River Jordan where it debouches into the Dead Sea. *(Hunting Surveys.)*

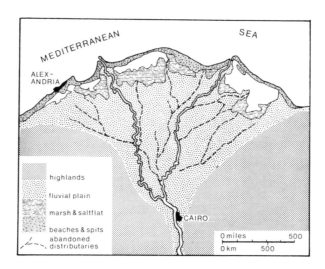

Fig. 12.12a   Nile Delta.

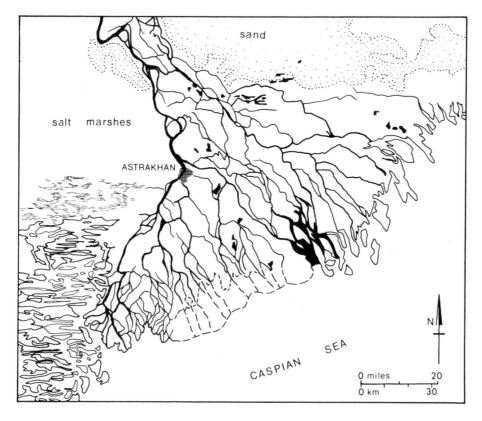

Fig. 12.12b   Volga Delta.

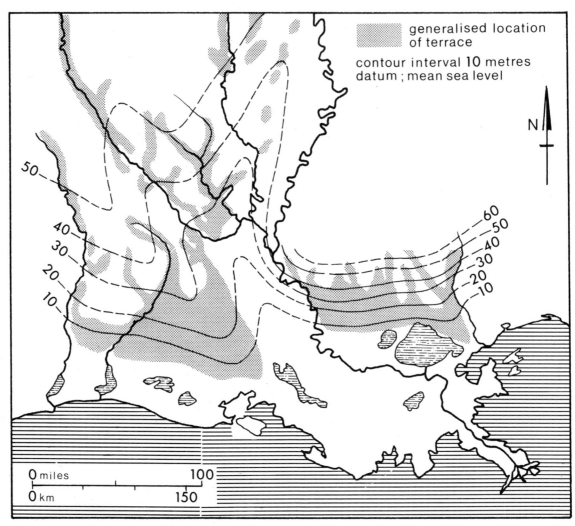

Fig. 12.13a   The warping of the Prairie Terrace in the lower Mississippi valley due to isostatic subsidence in the delta region. *(After H.N. Fisk,* J. Geomorph., *1942.)*

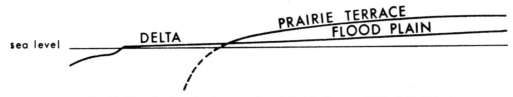

Fig. 12.13b   Section showing warping of Prairie Terrace. *(After H.N. Fisk,* J. Geomorph., *1942.)*

subsidence in response to depositional loading, and in part glacioeustatic variations of sealevel, that is, changes due to the waxing and waning of ice sheets and glaciers. Beneath the Mississippi delta, there are some 300 m of late Quaternary and Recent sediments alone, and the delta is the surface expression of a huge protuberance some 80 km wide, covering some 1800 km², which extends out from the coast near the edge of the continental shelf. The annual increment of sediments amounts to 5 million tonnes (45 million tons) or 2500 million m³ (0.06 miles³). The weight of sediments causes isostatic depression which is not restricted to the immediate vicinity of the area of deposition. Terraces of the lower Mississippi valley, for example, have been warped as a result of the depression of the delta area to the south (Fig. 12.13a and b).

3. Alluvial Fans

Some alluvial deposits are laid down, particularly in the tropical arid and semiarid areas, in the piedmont zone, where the deposits form alluvial fans.[20] Such accumulations occur in cold climates[21] and in temperate regions such as New Zealand as well. Typical examples are shown in Pls 12.6a and b.

Alluvial fans are built of deposits the character of which varies with the source area. But most display pronounced lensing due to the frequent lateral shifting of the stream channels, and there are in consequence marked changes in size of deposit both vertically and laterally. The fan area varies according to the size of the drainage basin to which it is related, and the gradient of the fan with the discharge of the contributing stream, and the volume and calibre of the deposits laid down. Some fans are very wide measured in a direction normal to the upland front. Fans up to 160 km across have been described in the literature. The gradient of the surface slope of such large fans is very gentle, and commonly as low as 1°. But other fans are narrow and steep, forming cones with gradients of as much as 15°; and there are all gradations between these two extremes.

Why is there such pronounced deposition in many areas at the hill-plain junction? Alluvial fans and aprons form where rivers leave the constricted single channel they occupy in upland regions and debouch on to the plains (see Fig. 12.14a). It is most commonly asserted that the alluvial fan deposits, are laid down because at the hill-plain junction there is an abrupt decrease in gradient, and hence a loss of energy. Moreover it is argued that waters percolate subsurface so that the river loses volume. This last is undoubtedly an important factor in some areas but that the first suggestion concerned with change of slope is not the essential explanation is suggested by two observations. First, gradients similar to those found in the piedmont zone of river sectors extend some distance into the upland on some streams. Second, in some areas, for example on the western side of the Flinders Ranges, alluvial fans occur not at the hill-plain junction, but below the point where the rivers leave the piedmont zone (into which they are slightly incised) and where they spread on to the alluvial plain (Fig. 12.14b). Alluvial fans are deposited where the river can, on leaving the confines of an incised channel, develop distributaries so that waters spread on to the plain. The area of wetted perimeter increases, there is a larger frictional drag and hence a loss of

Plate 12.6a Oblique aerial photograph of two of the fans, *X* and *Y*, of late Pleistocene age fronting the Flinders Ranges, north of Port Augusta, South Australia. (*G. Williams.*)

Plate 12.6b Alluvial fans developed on the margin of a breached anticline in the Zagros Mountains of Iran. Note the snow-capped ranges in the distance. (*Hunting Surveys.*)

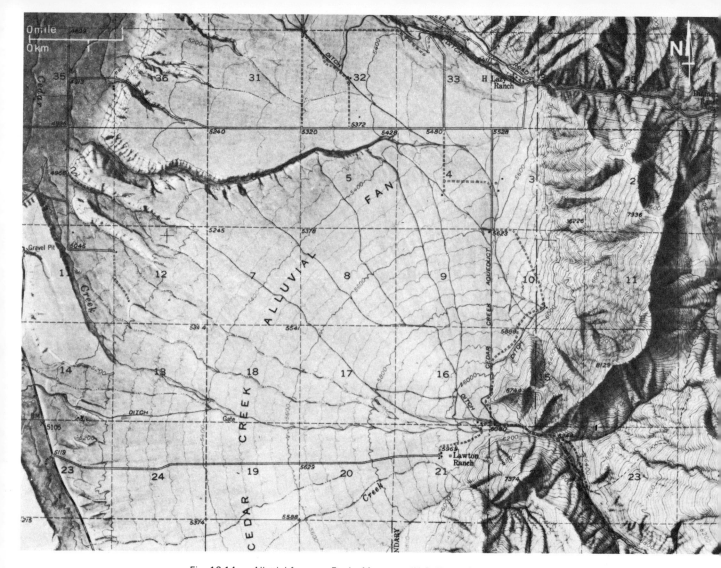

Fig. 12.14a   Alluvial fan near Ennis, Montana. (U.S. Topo. Series, Ennis Lake.)

velocity. For this reason, part of the stream load is deposited with the coarser debris near the apex of the fan and the finer near the toe. This, in turn, further increases the wetted perimeter. Channels are choked and distributary channels are formed, again involving energy loss. Further deposition causes these channels to be choked and the flow to be diverted almost continuously. Hence, the calibre of sediments changes rapidly in a lateral direction in alluvial fans, though there is a general decrease of calibre of debris from the apex to the toe. The surface of the fan is built up by river deposition, but for a variety of possible reasons the river may cut down through its own deposits, giving rise to dissected alluvial fans of various forms. [22]

4. Meandering Streams and Point Bar Deposits

The most important alluvial deposits are those laid down in river valleys. In some valleys about six years' supply of transported debris is "stored" in the alluvial deposits which floor many river valleys. Alluvial debris is gradually spread over many, but not all, river valleys. For instance, in the valley of the lower Mississippi, in view of the large load one might anticipate that the river would build up its bed and valley floor, yet observations over many years show that there is no such aggradation. Part of the explanation is that, despite its large volume of debris, the discharge of the river (up to 56,620 m³ —or 2 million feet³ —of water in flood) is so

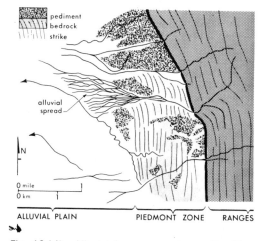

Fig. 12.14b Alluvial fan or spread near Brachina, located not at the hill-plain junction but where the river leaves its constricted incised channel to spread on to the plain.

gigantic that the Mississippi is really a clear stream.[23] As is shown in the Table 12.1, the weight ratio of sediment load to water is low at all times, and the river is able to carry the detritus to the sea.

Table 12.1
Weight ratio of sediment load to water

River	Parts per million
Mississippi average	550–600
Mississippi in flood	2600
Mississippi at low water	50
Missouri in flood	20,000
Lower Rio Grande	40,000
Colorado	40,000

In the Hwang Ho (northern China), the weight of sediment at times equals the weight of water. These figures illustrate again the overwhelming importance of floods in shaping the land surface: when in flood, the Mississippi, for instance, transports about four and a half times the suspended load it carries under average conditions of flow and over fifty times the amount it carries at low water. Floods, and indeed storms in general, leave their imprint on the land surface for years after the event.

Because some rivers have built up their beds and levees to such an extent that they stand above the level of the adjacent flood plains (the lower Hwang Ho is an example), until relatively recently it was generally accepted that most flood plain deposits were of the *overbank* type: they were laid down when the river burst its banks and flooded out over the flood plain. Work in the USA[24] has shown, however, that many flood plain deposits are due to the lateral migration of the stream, and though some modern workers consider that there is a tendency to overestimate such processes, there can be little doubt that these deposits are predominant in most valleys. Thus in order to understand why these valley floor deposits are laid down, it is necessary to discuss why rivers migrate laterally and why they develop curves or meanders.

All natural rivers have irregular courses and it seems to be an inherent property of running water to develop sinuous courses. River curves, usually perfectly regular and sinuous (though in places oddly angular) in plan (Pl. 12.7a) are well developed in the flood plains of rivers and in coastal mud flats (Pl. 12.7b).

The river banks are eroded on the outside and downstream sides of each curve, forming an undercut bluff, though during high velocity flows the river tends to follow a straighter course and to erode on the inside of curves.[25] Under normal conditions there is concurrent deposition on the inside of each curve at the point of each spur. These shoals are called *point bars* and the deposits are called *point bar deposits* (Pl. 12.8a—c). The latter consist of coarse debris, in many cases gravel, pebbles and boulders, which displays an imbricate structure (Fig. 12.6b). The point bars are commonly separated from the adjacent slip-off slope by flood channels called *chutes* or *shutes* (Pl. 12.1). The

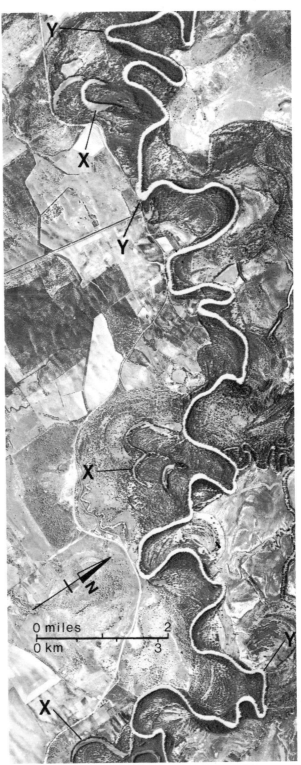

Plate 12.7a The meandering course of the Murrumbidgee River at Balranald, N.S.W. Note the oxbow lakes (X) and the angular bends in the river (for example, Y).
(Dept Nat. Mapping, Canberra.)

Plate 12.7b Sinuous course developed by Arm Creek as it flows through the coastal mud flats to the Gulf of Carpentaria. Note the subsidiary dendritic drainage patterns (*R.A.A.F.*)

Plate 12.8a   Point bar and under-cut bluff on Collins Creek, near Queenstown, South Island, New Zealand. Note slumping of turf on undercut bank. *(C.R. Twidale.)*

Plate 12.8b   The River Add, Argyll, Scotland, showing winding course, incipient cut-offs, point bar deposits (white) on the insides of the river curves, and abandoned channels. *(Hunting Surveys.)*

Plate 12.8c   The Hurunui River, South Island, New Zealand, showing a broad flood plain floored by alluvium deposited by the river as its channels migrate from side to side. *(N.Z. Govt. Tour. Office.)*

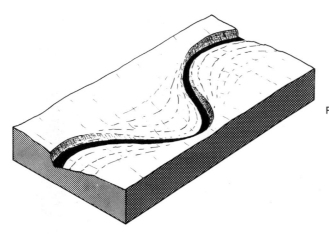

Fig. 12.15a Winding river with undercut bluffs and step-off slope

river gradually migrates laterally until its curve reaches a maximum radius, which is usually 14—17 times the width of the river at any given point; when the intervening necks of land are cut through and the former river loops abandoned, the well-known *oxbow lakes* develop (Pls 12.1, 12.7a and b, 12.8b). The river changes course and also migrates along its valley, sweeping downstream in a series of curves (Fig. 12.15a and b), and paring away the upstream sides of the interlocking spurs. In many places quite fantastic meander patterns have evolved in the course of time. (Fig. 12.16).

Eventually a flat-floored *plain of lateral corrasion* underlain by alluvial deposits (a complex of point bar deposits laid down during successive sweeps of the river), is developed.*

When the river has excavated a plain of lateral corrasion of such a width that the river curves can develop freely without necessarily eroding the bluffs in order to achieve their maximum radius, the curves are called river meanders, and the zone in which they flow is called the *meander belt*. The widening of the river valley does not cease with the attainment of meanders, for the meander belt also migrates from side to side undermining the bounding bluffs and causing the lateral extension of the flat-floored plain. Especially extensive flat plains of fluviatile aggradation occur where (because of a relative subsidence of the land) rivers have been able to build considerable thicknesses of alluvium and submerge any inequalities of pre-existing relief beneath a blanket of alluvium (Pl. 12.9a) or where many rivers debouch from a highland region (Pl. 12.9b). Basically, however, these alluvial spreads are due to the rivers' winding habits.

Some of the most spectacularly sinuous valleys occur not in alluvial valleys but in upland regions. Deep gorges describe smooth curved courses and are called *intrenched meanders* (Pl. 12.10; Fig. 12.15b) with a slip-off slope on the inside of the curve and an undercut on the outside. Such meanders display close adaptation to structure. As in the meanders of alluvial valleys, streams pare away the outside of valley walls, and form valleys with interlocking spurs in which intrenched river curves evolve. These continue to develop until adjacent parts of the same curve cut through the intervening neck of land and truncate the loop, forming a *meander core* and an *abandoned meander loop* (Fig. 12.17). Such intrenched meanders and abandoned meander loops are a normal autogenic development and in no sense imply anomalous drainage patterns as has been suggested by some workers (see Chapter 21).

---

* It may be mentioned that biological agencies, notably the beaver, may be responsible for the damming of many streams and hence for the deposition of alluvium along extensive valley tracts in northern Europe and North America—see Chapter 25.

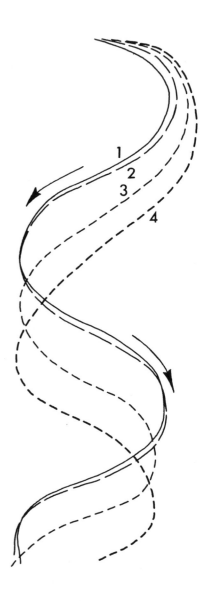

Fig. 12.15b Downstream sweep of meanders.

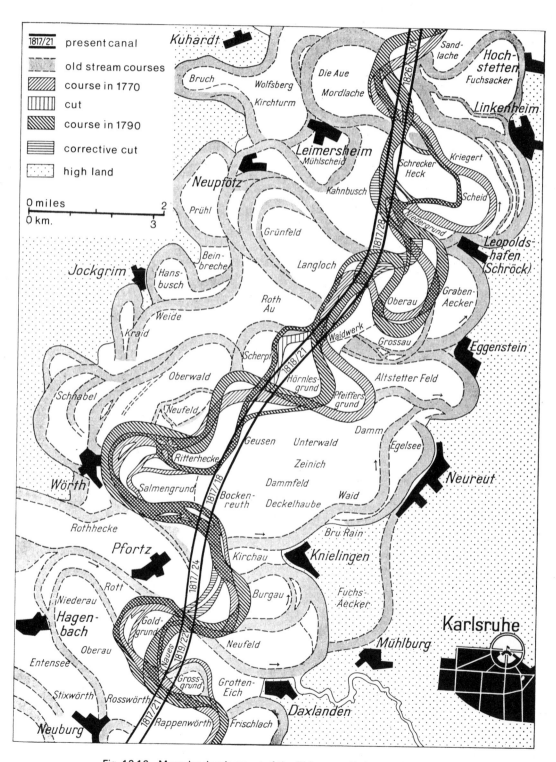

Fig. 12.16   Meander development of the Rhine near Karlsruhe, Germany.

Plate 12.19a The alluvial plain of the Georgina near Bedourie, Queensland. (*C.R. Twidale.*)

Plate 12.9b The Canterbury Plains, South Island, New Zealand, with the Rangitata River in the left foreground. Note the quite remarkable flatness of the plains and the snow-capped Southern Alps in the distance. These uplands are the source of the numerous heavily-laden rivers which have deposited the boulders, gravels and other deposits which underlie the Plains. (*N.Z. Govt Tour. Office.*)

Plate 12.10 Intrenched meanders and interlocking spurs developed by the Karun River as it flows through an anticlinal structure in the Zagros Mountains, Iran. Note the undercut bluff (left) and the gentler slip-off slope. (*Hunting Surveys.*)

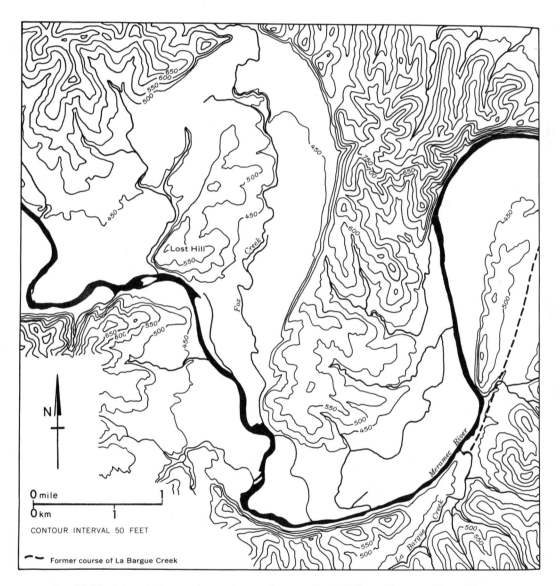

Fig. 12.17 Intrenched meanders and meander core (Lost Hill) at Meramec River, Minnesota.
(*After G.H. Dury, "Principles", p. 43.*)

Whether the river curves and loops develop in areas of high or low relief, the fundamental mechanism as manifested in the mode of development seems to be the same. The only contrast is that in upland regions the banks and bluffs are higher, so that in order to achieve the same amount of lateral migration, the river has to transport more debris in such a context than it does on an alluvial plain.

5. Causes of Meandering

Laboratory experiments[27] show beyond doubt that the size of the river loops increases as stream discharge is increased, as stream gradient is increased (at least within certain limits) and with increased angle of attack or degree of non-parallelism between current and banks. Such work also shows that as sediment is released into the flow, loops or meanders develop more rapidly; but sediment is not the cause of meandering, for streams or glaciers composed of meltwaters from the ice display such sinuosities, and flows of distilled water released on plate glass also form meanders.[28]

The winding and meandering habit of rivers is obviously an important factor in landscape development. As yet, however, there is no comprehensive and reasoned explanation of meander development. Many facts are known about meandering streams, but the basic cause of meandering is still not understood. It is known[29] that there is a fairly constant relationship between stream discharge (Q) at bankfull stage, the width of the river, amplitude or radius of meanders and the wave length of the meanders. All are proportional to the square root of the discharge ($Q^{\frac{1}{2}}$). The channel is deep in the curves and comparatively shallow in the intervening straight sectors, so that the channel consists of alternations of shoals, bars or riffles, and pools. All else being equal, the more sinuous the stream, the deeper it is. It has been suggested that the sinuous course represents the form of least work for the river. The curves tend to sine-generated form, so that the curvature is the least abrupt possible at all points and frictional losses are minimal.[30]

Chance obstructions in a channel, the earth's rotation and excess of energy and load have all been cited as basic causes of stream meandering, but all have been found wanting. Obstructions, such as trees and rocks, certainly divert flows unequally, but this is a subordinate effect, for streams of pure water display a meandering tendency. Other obstructions, such as rock bars, and even clay layers, actually restrict meandering by inhibiting lateral movement. If the Coriolis force has any effect on rivers, as has been suggested for more than a century, rivers in the northern hemisphere should have a tendency to erode their right banks more than their left; and *vice versa* in the southern hemisphere. There is no evidence that this is so. The load of a stream has little or nothing to do with meandering: its transfer appears to be an effect rather than a cause.

The development of pools and shoals (or riffles), and of highs and lows on the bed relief, which develop in any natural flow of water, appears to be related to meander evolution. Work in the USA has shown that even in relatively straight stream sectors, pools develop at regular intervals on the bed, and that the spacing is similar to that of the wave length of meanders appropriate to the discharge of the stream in question. It may be suggested, though it is mere speculation, that the deposition of regularly spaced shoals appears to be

analogous with wave development, which occurs wherever one medium passes over another: air over water, air over sand, or water over sediment. In such boundary conditions, waves of one sort or another seem to form. It may be that wave motion in streams creates areas of high and low velocity, and massive standing waves, in which erosion (or non-deposition) and deposition occur respectively. Deposition of shoals could give rise to channel widening, to an increase in wetted perimeter and to loss of efficiency. Hence, bank erosion could be concentrated in the areas of high bed velocity and hydraulic efficiency: this is where river curves are initiated and evolve into meander loops. On the other hand, of course, it may be argued that the deposition of shoals is the result (not the cause) of the initiation of meanders: that deposition is likely to take place in the straighter, wider sections between loops, where stream velocities are lower.

6. Braided Streams

Many streams do not have a meandering pattern in plan, but are *braided*, that is, they consist of not one but many interconnected channels separated by debris islands, which are often vegetated (Pls 12.2 and 12.11). Such braided streams are generally straighter than nonbraided channels, and observations show (Fig. 12.17) that braided streams are characterised by steeper gradients and/or higher discharges than meandering streams.[31] Because of variations in these and other factors, the pattern of the channel may vary from one section of the river to another (Pl. 12.12). However many channels there are, the two lateral channels are generally the largest. Many streams of the arid plains display a distributary habit (Pl. 12.13) and some rivers are both braided and meandering and are called *anastomose* streams.

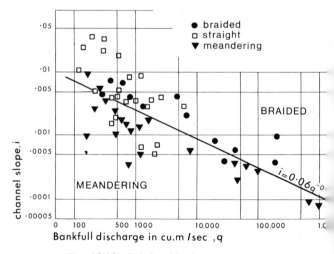

Fig. 12.18 Relationship between stream discharge, gradient, and the occurrence of braiding and meandering streams. (*After L.B. Leopold and M.G. Wolman, "River Meanders", p. 59.*)

Plate 12.11   Broad bed of the Einasleigh River, north Queensland, showing vegetated bars which divide the channel into several subchannels. (*C.R. Twidale/C.S.I.R.O.*)

Braided streams are common in arid, semiarid, and humid tropical regions, and are also characteristic of those rivers fed by meltwaters from glaciers, as for instance in Greenland and the South Island of New Zealand (Pl. 12.14). In the latter, high summer discharge is the crucial factor determining the pattern in plan, though the draining of periglacial lakes (see Chapter 23) causes similar floods (which are called *jokhulhaup* in Iceland) and broad braided channels and depositional plains.[32]   In the humid tropics, in such rivers as the Amazon, stream discharges are high, but an important contributory factor is the rapidity with which shoals, charac-

teristic of all rivers, are vegetated and stabilised by vegetation. The braided streams of the more arid regions reflect the high seasonal or episodic flows, but again the colonisation of the debris islands by trees and shrubs during the periods of nonflow is an important factor in perpetuating the pattern.

Many streams display both meandering and braided sectors and presumably these charges reflect local discharge and gradient.   Many river curves in arid regions display a series of chutes across the point bars, indicating the tendency of braiding and straightening during periods of high discharge.

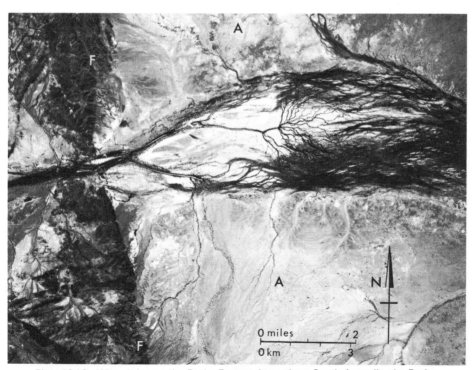

Plate 12.12   When it leaves the Peake Ranges, in northern South Australia, the Peake River consists essentially of a single channel, but on reaching the Lake Eyre plains, it bifurcates and then divides and divides again into a large number of small intermeshed channels. The broad, flat-floored bed is cut below a gibber capped alluvial apron (A) which fronts the fault scarp (F) delimiting the upland on its eastern side. (*Vertical air photograph courtesy S.A. Dept Lands.*)

Plate 12.13 Vertical air photograph of distributary channels and lagoons in the northeast of South Australia. (*S.A. Dept Lands.*)

Plate 12.14 Braided and anastomo
river, western Otago, New Zealar
(*C.R. Twidale.*)

References Cited

1. J. PLAYFAIR, *Illustrations of the Huttonian Theory of the Earth*, Cadell and Davies, London, 1802, p. 102.
2. See for instance G.W. MURRAY, "The Waters beneath the Egyptian Desert", *Geogr. J.*, 1952, **118**, pp. 443–52; J.N. JENNINGS, "Water Policy for the Great Artesian Basin", *Geogr. Stud.*, 1956, **3**, pp. 127–32.
3. Much of the material in this section is derived from MARIE MORISAWA, *Streams, Their Dynamics and Morphology*, McGraw-Hill, New York, 1968, and from L.B. LEOPOLD, M.G. WOLMAN and J.P. MILLER, *Fluvial Processes in Geomorphology*, Freeman, San Francisco, 1964. These books also contain references to the detailed journal literature.
4. W.W. RUBEY, "Geology and Mineral Resources of the Hardin and Brussels Quadrangles, Illinois", *U.S. Geol. Surv. Prof. Paper*, 1952, **218**.
5. After MARIE MORISAWA, *Streams*, p. 38.
6. L.B. LEOPOLD, "Downstream Change in Velocity in Rivers", *Am. J. Sci.*, 1953, **251**, pp. 606–24.
7. See for example L.B. LEOPOLD and T. MADDOCK, "The Hydraulic Geometry of Stream Channels and some Physiographic Implications", *U.S. Geol. Surv. Prof. Paper*, 1955, **252**.
8. T. BLENCH, *Regime Behaviour of Canals and Rivers*, Butterworth, London, 1957.
9. D.G. MCKINLAY, "Physics of Fluvial Transport of Deposition", in E.G. Hallsworth and D.V. Crawford (Eds), *Experimental Pedology*, Butterworth, London, 1965, pp. 317–39.
10. L.B. LEOPOLD and T. MADDOCK, "Hydraulic Geometry".
11. F. HJULSTROM, "Studies of the Morphological Activity of Rivers as illustrated by the River Fyries", *Uppsala Geol. Inst. Bull.*, 1935, **25**, pp. 221–527.
12. M.G. WOLMAN and J.P. MILLER, "Magnitude and Frequency of Forces in Geomorphic Processes", *J. Geol.*, 1960, **68**, pp. 54–72.
13. See M. LUGEON, "Sur un Nouveau Mode d'Erosion Fluviale", *C.R. Acad. Sci. Paris*, 1913, **156**, pp. 582–4; E. BLACKWELDER, "Grooving of Rock Surfaces by Sand Laden Currents", (Abs.), *Bull. geol. Soc. Am.*, 1933, **44**, p. 167.
14. P. LENK-CHEVITCH, "Beach and Stream Pebbles", *J. Geol.*, 1959, **67**, pp. 103–8.
15. D.A. LIVINGSTONE, "Chemical Composition of Rivers and Lakes", *U.S. Geol. Surv. Prof. Paper*, 1964, **4409**.
16. J. BOGARDI, "Hydraulic Similarity of River Models with Movable Bed", *Acta. tech. Hung.*, 1959, **23**, pp. 417–45.

17. G.K. GILBERT, "The Transportation of Debris by Running Water", *U.S. Geol. Surv. Prof. Paper*, 1914, **86**; P. DANEL, R. DURAND and E. CONDOLIAS, "Introduction à l'Etude de la Saltation", *Huille Blanche*, 1953, **6**.
18. A.A. KALINSKE, "Criteria for Determining Sand Transport by Surface-creep and Saltation", *Trans. Am. geophys. Un.*, 1942, pp. 639–43.
19. G.H. MATTHES, "Paradoxes of the Mississippi", *Sci Am.*, 1951, **184** (4), pp. 18–23.
20. See E. BLISSENBACH, "Geology of Alluvial Fans in Semi-arid Regions", *Bull. geol. Soc. Am.*, 1954, **65**, pp. 175–90.
21. See, for example, J.M. RYDER, "The Stratigraphy and Morphology of Periglacial Alluvial Fans in south-central British Columbia", *Can. J. Earth Sci.*, 1971, **8**, pp. 279–98.
22. See E. BLISSENBACH, "Geology of Alluvial Fans", pp. 175–90.
23. G.M. MATTHES, "Paradoxes of the Mississippi", pp. 18–23.
24. J.H. MACKIN, "Erosional History of the Big Horn Basin, Wyoming", *Bull. geol. Soc. Am.*, 1937, **48**, pp. 813–94; M.G. WOLMAN and L.B. LEOPOLD, "River Flood Plains: Some Observations on their Formation", *U.S. Geol. Surv. Prof. Paper*, 1957, **282C**.
25. See M.G. WOLMAN and L.B. LEOPOLD, "River Flood Plains".
26. S.A. SCHUMM, personal communication.
27. J.F. FRIEDKIN, *A Laboratory Study of the Meandering of Alluvial Rivers*, U.S. Waterways Experimental Station, Vicksburg, Mississippi, 1945.
28. L.B. LEOPOLD and M.G. WOLMAN, "River Meanders", *Bull. geol. Soc. Am.*, 1960, **71**, pp. 769–94; W.F. TANNER, "Helicoidal Flow, a Possible Cause of Meandering", *J. Geophys. Res.*, 1960, **65**, pp. 993–5.
29. See for example G.H. DURY, "Principles of Underfit Streams", *U.S. Geol. Surv. Prof. Paper.*, 1964, **452A**. R.A. BAGNOLD, "Some aspects of the Shape of River Meanders", *U.S. Geol. Surv. Prof. Paper*, 1960, **282E**.
30. L.B. LEOPOLD and W.B. LANGBEIN, "River Meanders", *Sci. Am.*, 1966, **214** (6), pp. 60–70.
31. L.B. LEOPOLD and M.G. WOLMAN, "River Patterns: Braided, Meandering and Straight", *U.S. Geol. Surv. Prof. Paper*, 1957, **282B**.
32. S. THORARINSSON, "The Ice-dammed Lakes of Iceland with particular reference to their Values as Indicators of Glacier Oscillation", *Geogr. Ann.*, 1939, **21**, pp. 216–42.

CHAPTER 13

Valleys and Valley Side Slopes

A. GENERAL

Fluvial processes dominate the shaping of most of the earth's land surface. Many land masses and all uplands are scoured to a greater or lesser degree by running water, and extensive plains have been formed as a result of river deposition. This is as true of the arid deserts as of humid regions (Pl. 13.1). Even on icecaps and other glaciers, summer meltwaters from time to time cut channels in the ice surface, and in the sandy deserts, though most rainwater infiltrates into the subsurface, some seeps into the interdune corridors where it forms short and ephemeral streams which convey waters to playa depressions.

In many areas systems of valleys have been eroded by streams and rivers, but elsewhere running water is evidently responsible for the development of gently sloping planate surfaces called pediments. Moreover on a comparative basis running water is not as important in shaping the land surface is some really humid areas as it is in arid lands. What are the reasons for these contrasts?

B. VALLEY DEEPENING, GRADE AND BASELEVEL

The various erosional processes described in the previous chapter cause the beds of rivers and streams to be worn down. Such lowering of channel floors is not a continuous process, for each stream adjusts its flow to varying conditions of discharge and load; each stream is self-regulating and tends to a condition of *grade* which implies that the river adjusts its gradient and channel conditions (width, roughness, length) to changing circumstances in such a way that the river energy is adequate to carry the available load. For instance, the Aswan High Dam traps much of the 136·2 million tonnes (134 million tons) of sediment which previously spread annually over the floor of the lower Nile valley. The river below the dam is relatively free of silt and the river has scoured its bed, eroded its banks and undermined bridges. These activities have had the effect of increasing channel roughness, thus decreasing both stream velocity and the transportation potential of the river. If the average calibre of available debris decreases through attrition of the load, the stream will require a lower velocity to transport that debris: hence the river may again erode its bed and banks or develop a narrower and deeper channel. Increased load leads to the development of a steeper gradient; gentler gradients are formed in response to decreased energy demands. In the long term, however, each stream erodes its bed.

Such lowering cannot proceed indefinitely, because a sufficient gradient must be maintained to permit flow. This lower limit of stream incision is called the *baselevel*, which may be visualised as an imaginary line extending at a gentle

Plate 13.1 The slopes of this peak on the island of Abu Musa, Persian Gulf, are scored by watercourses which carry debris to the flat alluvial floors of valleys and plains. *(Hunting Surveys.)*

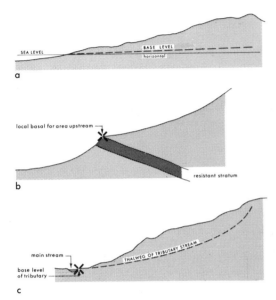

Fig. 13.1 (a) Regional baselevel; (b) an example of local baselevel provided by a resistant stratum; (c) local baselevel determined by the junction of tributary and mainstream.

inclination from sealevel to the margin of the land mass inland. It is not a horizontal plane (Fig. 13.1a–c) for a gradient, however slight, must be maintained in order for a river to flow. Sealevel may be regarded as the ultimate baselevel of erosion to which all land masses would be graded, given long-continued stable conditions. But it, like regional and local baselevels, is temporary, because sealevel (see Chapter 23) has oscillated in the past and as glaciers melt or expand will again do so in the future. A resistant stratum (Fig. 13.1b) forms the local baselevel for the stream above it;

the stream sector upstream from the barrier cannot in essence erode below its level; though erosion may locally take the bed to a lower level, the surface of the stream must slope down toward the level of the rock bar. But even resistant strata are slowly worn down, the local baselevel is lowered, and the stream sector above adjusts to its changing elevation. Similarly, the local baselevel of a tributary stream is the elevation of its junction with the main stream (Fig. 13.1c), and as the latter is lowered, so the tributary profile is adjusted.

Lake Eyre is the regional baselevel for some 1·3 km² of central Australia. Its bed lies some 16–17 m below sealevel and if its rivers continue to extend headward, they should eventually erode back to the coast, allowing the sea to penetrate into the interior of the continent, which would become flooded. The deposition of large volumes of sediment would build up the level, though the combined weight of sediment and seawater would cause isostatic depression and further sinking. Such subsidence is in any case going on at present because of faulting,[1] so that this baselevel is ephemeral.

Rivers cut down to their baselevel, constantly adjusting the detailed morphology of their beds as they do so in order to maintain equilibrium between the available load and energy. If only lowering of stream beds were involved, rivers would flow in more-or-less deep ravines as wide as the river channel. Manifestly this is not the case. Most rivers flow in valleys that are very much wider than they are deep, though in some cases, due to the human tendency to exaggerate vertical distances, the opposite appears to be true.[2] But even the valleys which are so deep that they are called gorges or canyons (or whatever the local term is) on measurement are found to be wider than they are deep. For example, the Barron Gorge, above Cairns in north Queensland, is over 300 m deep but 2·5 km wide in the same sector. The Werribee Gorge, in Victoria, is 200 m deep but some 500 m wide in the same sector. The Grand Canyon (Pl. 5.9), between the Kaibab and Coconino plateaux, Arizona (Fig. 13.2), is some 1380 m deep but some 12 km wide. Even in the inner (granite) gorge, the valley is clearly wider than it is deep. What agents at work on the valley sideslopes so effectively bring about their erosion?

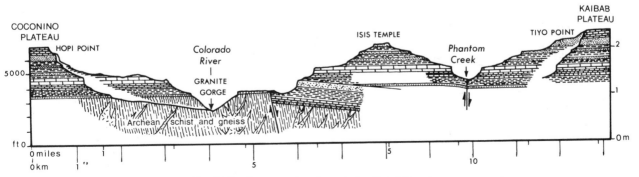

Fig. 13.2 Cross-section through the Grand Canyon.

Plate 13.2 Unpaired terraces (X1, X2 and X3 at different levels) in the valley of the Hope River, South Island, New Zealand. Note the terraces on the other side of the river. *(C.R. Twidale.)*

C. VALLEY WIDENING

Tributaries certainly contribute to the erosion of the valley side slopes, as do gullies formed on the slopes. Weathering and mass movements of debris (earth-flows, landslides, avalanches) together aid the process, particularly in certain areas whose conditions are favourable for mass movements. In central Otago, New Zealand, for instance, slides and flows are very important factors in the shaping of valley side slopes.

However, although each of these processes in places acts independently of the main stream which is responsible for the primary excavation of the valley, observations suggest that the behaviour of the trunk stream greatly influences their effectiveness. As mentioned earlier, meandering seems to be an inherent and characteristic feature of rivers. In many river valleys it can be seen that river curves are developed with the channel banks undercut on the outside and downstream sides of the curves, and with compensating deposition in the form of point bars on the insides of the curves (Pl. 12.8a). Gradually the winding river pares off

the previously interlocking spurs (Pl. 12.10) to form a flat-floored valley bounded by river cliffs (Fig. 13.3). Even after the river curves have attained their maximum amplitude commensurate with the prevailing discharge and even after the process of self-beheading and oxbow development has commenced, the valley floor continues to be widened—apparently because the *meander belt* or zone within which the meanders develop continues to sweep from side to side. The river valley is widened because the river impinges on the base of the bounding bluffs, causing them to be undercut, to collapse and to recede. As the meander curves migrate downstream, this paring process sweeps down the valley side; though not necessarily completely, for during floods sudden and rapid changes in the river course take place, leaving some spurs projecting.

It is notable that many valley floors have a stepped appearance, with distinct flats bounded by bluffs standing above the present flood plains. Some, though not all, retain fluvial deposits suggesting that they are of riverine origin. These flats are called *terraces* and they form because, at the

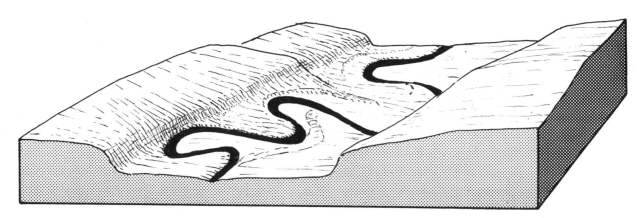

Fig. 13.3 Flat-floored valley of lateral corrasion bordered by river bluffs.

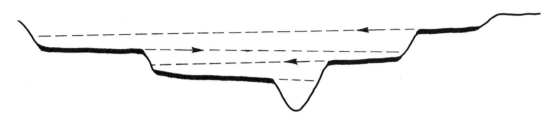

Fig. 13.4   Development of unpaired river terraces by lateral stream migration during slow incision.

same time as the river migrates from side to side within its valley floor, it continues to erode its bed toward baselevel. Thus, as a river moves laterally from one side of a valley, it leaves behind a gentle slope essentially cut in bedrock, but bearing a veneer of point bar deposits. When the river migrates back to this side of the valley it has cut down a few metres and is at a lower level (Pl. 13.2, Fig. 13.4). The undercut bluff advances and in places eliminates the old valley floor but remnants survive in many places and form flats stranded above the flood plain. Such old valley floor remnants are called river terraces if they stand above flood level. Those underlain essentially by bedrock and formed by river incision during lateral stream migration are called unpaired or unmatched terraces because they occur at alternate heights on opposite sides of the valley (Fig. 13.4).

River bluffs undermined by river erosion eventually collapse, partly because of direct lack of support and partly because, as mentioned previously (Chapter 11), mass movement of debris on slopes is facilitated by steep slopes. Tributaries and gullies are also affected by the migrations of the main stream, for their courses are effectively shortened as the main stream swings toward them, and extended as it moves away. For instance, in the Torrens Gorge it can be observed that those tributaries which join on the outside of the curves of the Torrens are greatly shortened and are noticeably incising their beds. Those that join on the inside of the curves, where the Torrens is moving away from them, on the other hand, are extending their courses by the deposition of debris (Fig. 13.5).

D. SLOPE MORPHOLOGY AND BEHAVIOUR

1. Faceted and Graded Slopes

In some valley systems the bounding slopes all appear to be of similar morphology and inclination. Elsewhere slopes vary both in shape and overall inclination, and in particular appear to steepen up-valley. Why do these varied slope morphologies and hence valley shapes in cross-section exist?

Morphologically, slopes are of two types, the *faceted* and *nonfaceted* or *graded* (Fig. 13.6). The faceted slope comprises several distinct units which meet in angular discordance: the upper slope, the bluff or free face, the debris slope or constant slope, and the lower slope (Pls 13.3a and b). The typical graded slope consists of an upper convex slope or crest and a lower concavity or footslope, which meet either at the point of inflexion or in a rectilinear slope called the side slope (Pl. 13.4a). The vital difference between the two slopes is the presence of the bluff in the faceted type, but in broad view, the contrast in assemblages is between the rolling, rounded terrain characterised by graded slopes, and the angular, rugged country of areas where faceted slopes predominate (Pls 5.3, 5.9, 13.3, 13.4a-b). In some areas virtually all slopes are faceted, with distinct and prominent bluffs, but elsewhere slopes are smooth or graded. In some regions underlain by what appears to be similar bedrock, slopes vary in their inclination, whereas in others all the slopes are of virtually identical form and inclination. What are the reasons for these contrasts?

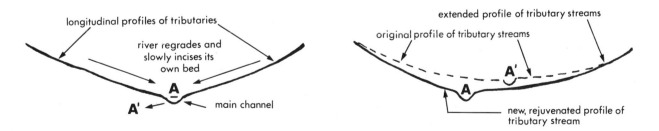

Fig. 13.5   The effect of lateral stream migration on the gradients and behaviour of tributary streams.

Plate 13.3a   Faceted slopes in granite, south of Cloncurry, northwest Queensland, with well-developed bluff, block-strewn debris slope, and pediment plain. *(C.R. Twidale/C.S.I.R.O.)*

Plate 13.3b   The Bluff, a faceted slope some 15 km north of Quorn in the southern Flinders Ranges, South Australia. The bluff is coincident with an outcrop of dipping sandstone and the sandstone blocks are scattered over the debris slope below. *(C.R. Twidale.)*

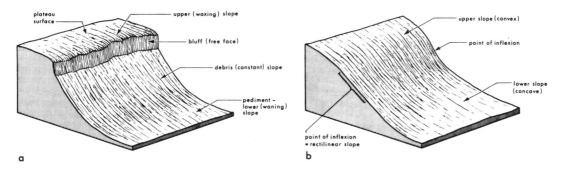

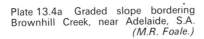

Fig. 13.6 (a) Faceted slopes; (b) graded slopes.

Plate 13.4a Graded slope bordering Brownhill Creek, near Adelaide, S.A. *(M.R. Foale.)*

Plate 13.4b Assemblage of graded slopes eroded in shale and siltstone in the Pertnjara Hills, Northern Territory. *(Dept Nat. Dev., Canberra.)*

2. Slope Variations within River Valleys

Some slope variations are clearly attributable to structural factors. For instance, near Adelaide the Torrens flows through a series of shallow gorges (with faceted slopes) and open basins (bounded by graded slopes), the locations of which are coincident with the outcrop of quartzite formations and argillaceous bedrock respectively (Fig. 13.7a). In the Grand Canyon (Pl. 5.9) the slopes are faceted and characteristically stepped, with prominent structural benches formed on outcrops of resistant sandstone (Figs 13.2, 13.7b). In many fold mountain belts, the ridges (cuestas) and the intervening valleys (see Chapter 5) are asymmetrical, the variations in opposed slopes in large measure reflecting the local disposition of strata (Fig. 13.7c).

Some indication of other factors which control the morphology and evolution of slopes can be derived from a consideration of areas of uniform bedrock in which the two basic morphological types occur side by side. This situation is found in many river valleys. For example, for much of its course in South Australia the River Murray flows in a

narrow, moderately deep trench eroded through calcareous sandstones (calcarenite) of Miocene age (Pls 13.5a and b, Fig. 13.8). The bedrock is flat-lying and massively jointed. The climate is hot and arid to semiarid. The river, which has aggraded its bed so that the bedrock floor is concealed beneath a few metres of alluvium, pursues a winding course. Intrenched meanders are well developed, and there are good examples of undercut bluffs and slip-off slopes.

The valley sidewalls display both faceted and graded slopes, as well as all gradations between the two extreme types. Moreover, the distribution of the morphological types is not random, and there is some evidence that the distribution has changed in time.

At regular intervals in its course, the Murray impinges against, and is actively undercutting, the bordering river cliffs. At these sites the slopes are faceted and are dominated by bluffs (Pl. 13.5b). Indeed, at the very point of undercutting, the slopes consist almost wholly of a bluff; but a little distance upstream and downstream debris slopes are present, and the cliffs are restricted to the upper parts of the profile and are not as extensive. Still further from the areas of impingement the bluffs diminish proportionally as the debris slope increases by extending higher and higher up the slope (Fig. 13.8). Midway between any two points where the river impinges the cliffs, the debris slope is seen to extend far up the slope; the rounding of the upper angle of the bluff, which causes the formation of the upper slope, has progressed to such an extent that the bluff is either represented by a minor break of slope, or has been eliminated. The faceted slope has been replaced by one which is graded.

Such gradations, from boldly faceted slopes dominated by the bluff to rounded and comparatively subdued graded slopes, occur systematically and repeatedly along both sides of the Murray valley. Why? The explanation for these variations is not to be found in lithology or other structural factors, for the Miocene bedrock is, to all intents and purposes, homogeneous. The answer must be sought in the processes at work on the slopes. At this stage, even without identifying such processes in detail, it is clear that the steepest slopes and the highest bluffs occur where the river is actively and vigorously eroding the base of the slope (Fig. 13.8, Pl. 13.5b) and that, conversely, the smoothest and most gentle slopes are found where the river is furthest removed from the slope, where there is apparently no major disproportion in the distribution of erosive forces on the slope and where the slope is more-or-less equally and evenly attacked. Thus the location and intensity of processes on a slope is evidently an important factor.

A second important conclusion derived from observations on the Murray and in many other river valleys, large and small, is that the location of intense attack must change in time as the river curves migrate downstream, and so it appears possible that the type of slope evolved at a given site may alter according to the proximity of the greatest local source of energy, the river. This suggestion is borne out by the discovery, in archaeological excavations[3] (at Y in Pl. 13.5a) located a short distance from the present points of impingement of the river against the valley sidewalls, of what are apparently truncated bluffs a few metres in advance of the present bluff which are now buried beneath the debris slope (Fig. 13.9a). Thus it may be suggested that the two

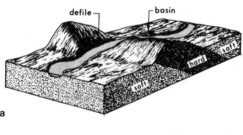

a

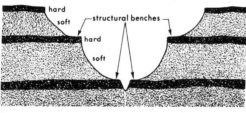

b

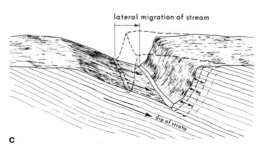

c

Fig. 13.7 (a) Basin and defile forms; (b) structural benches; (c) asymmetrical valley due to uniclinal shifting.

Plate 13.5a Vertical air photograph of the River Murray at Walkers Flat, South Australia. X and Y refer to sites of Plates 13.5b and 13.6.

Plate 13.5b   High bluff of Miocene calcarenite on the outside of a river curve at X in Pl. 13.5a.
(C.R. Twidale.)

morphological types of slope not only merge in space, but also can be transformed one to the other in time as a result of changes in energy distribution.

The systematic development of the various observed forms can readily be visualised (Fig. 13.9b). First imagine the simple bluff found where the river is impinging directly on the base of the slope. The rocks with which the river comes into contact are worn away, the cliff is undermined and eventually collapses and recedes. Such cliff recession apparently outpaces the rounding of the top of the bluff caused by attack on two exposed faces; as fast as such rounding and development of the upper slope takes place, the cliff is undermined and the results of the weathering are lost. The undermined and collapsed blocks of sandstone accumulate to some extent at the base of the bluff and afford it some protection, but the flow of the river, especially in flood, is eventually sufficient (after due attrition of the blocks) to evacuate this debris. But there is a constant downstream migration of the river curves which causes the eventual abandonment of the cliff by the river. Debris from the bluff is no longer evacuated, or at least not so readily. A certain amount of fine debris may be flushed from the base of the slope by local washes, by subsurface waters running to the river, and by the few tributary streams of the Murray in this South Australian sector, but by and large the debris remains in place.

Such a cycle of processes and events related to the downstream migration of the river apparently accounts for the systematic and repeated variations in valley side slopes along the course of the lower Murray. But what of those valleys in which the inclination of the valley side slopes increases toward the valley head (Fig. 13.10)? Because of the lateral migrations of rivers, valley floors tend to become wider and flatter downstream. Hence, the further downstream a particular

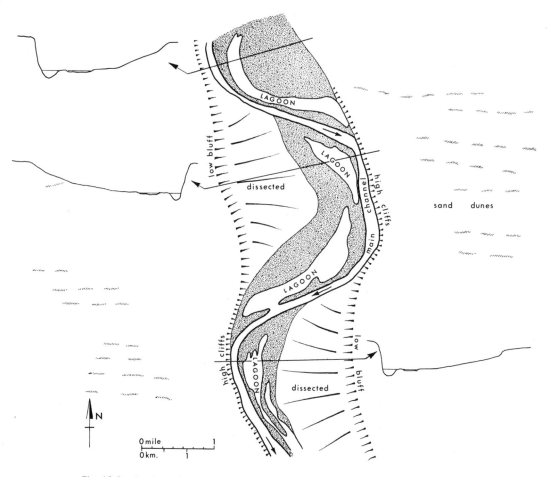

Fig. 13.8   Section of the Murray valley near Walkers Flat, S.A. (see also Pl. 13.5), showing winding in plan, systematic variation in location of river with respect to the bluffs which delineate the trench, and typical cross-sections at sites indicated.

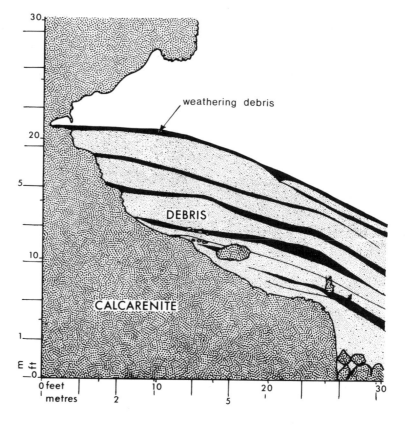

Fig. 13.9a Section through archaeological excavation near Walkers Flat, S.A. The cliff has receded both back and up the slope. (*After C.R. Twidale, "Effect of Variations", pp. 177–91, by permission of Z. Geomorph.*)

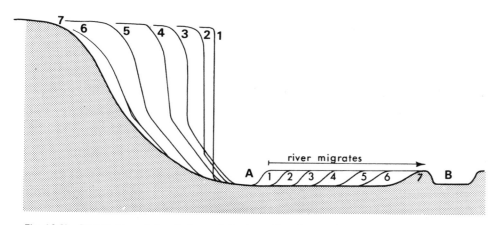

Fig. 13.9b Systematic variation in form of the bounding slopes of the Murray valley according to intensity of basal slope attack.

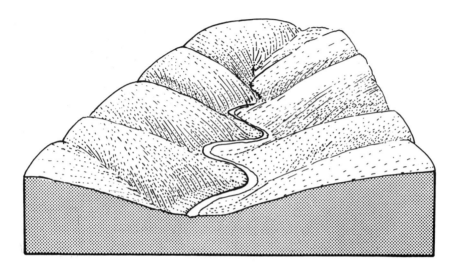

Fig. 13.10 Block diagram showing up-valley steepening of
graded side slopes.

valley side slope is located, the less frequent are the intervals
when the river or stream impinges on the base of the slope.
Thus in such situations there are long periods when the
various processes active on the slope itself are at work
transporting debris downslope. Much of this debris is not
evacuated (hence the alluvial fans and cones so commonly
found at the base of such valley side slopes, especially where
the debris is coarse—see for example Pl. 12.8c); nor is there
direct stream erosion of the base of the slope, except at infrequent intervals. The tendency of bounding slopes to become
more gentle downstream is a function of the frequency of
stream attack or of the slope budget (see below).

3. Effects of Debris Accumulations

Observations at sites such as that shown in Fig. 13.9a suggest
that debris accumulations have important effects on slope
development. First, the lower parts of the bluff buried by
the debris are protected from erosion, though not necessarily
from weathering. Second, as the wedge of debris increases
in size, the area of cliff contributing material to the accumulation decreases, and as this lesser amount of debris is
spread over an increased area of debris slope, the rate of
encroachment of debris on the cliff decreases. Thus, as has
been deduced by several workers,[4] there is a tendency for
the subdebris bedrock slope to develop a convex form (Fig.
13.11). This tendency is clearly manifested in excavations
which expose the subdebris slope. Third, a zone of intense
weathering is frequently in evidence at the upper limit of
the debris slope where air and moisture retained in the
unconsolidated debris meet the bedrock. Here are commonly
found numerous small caverns which merge to form major
caves or shelters (Pl. 13.6). Their extension, like erosion by
the river described earlier, causes undermining of the cliff.

As the debris slope is gradually built up, this zone of
intense weathering migrates higher and higher up the cliff
so that the bluff is gradually decreased in size by this weathering from below. Concurrently, weathering also attacks the

exposed upper angular edge of the bluff, which becomes
rounded; thus the bluff is also reduced at its upper edge,
and eventually the bluff is eliminated. Fourth, the accumulation of debris introduces a new factor into slope development. The inclination of the debris slope, where it consists
of aggregates of unconsolidated material, is controlled not
only by the intensity and type of attack acting on it, but also
by the size and shape of the constituent fragments, and the
more angular they are, the steeper is their angle of rest and
the resultant slope.

Gradually, then, the debris slope, spearheaded by the
development of caverns at the base of the bluff, extends
upwards. The upper slope develops as the bluff gradually
disappears and the faceted slope is replaced by a smooth,
graded slope.

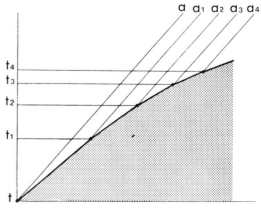

Fig. 13.11 Development of convex subdebris
slope through retreat of bluff and non-evacuation
of debris.

Plate 13.6   Small caverns at the rear of a large shelter at the bluff/debris slope
junction at Walkers Flat (Y in Pl. 13.5a). The bedrock is Miocene calcarenite and
the scale rule propped against the wall is marked in inches. *(D. Casey.)*

4. Areas of Mixed Slopes

As indicated above, an important factor in slope development
is the debris which is transported to the foot of the slope,
whether the latter is graded or faceted. Thus, in the Murray
valley there is, as described, a systematic variation of slope
form along each side of the valley. Graded slopes are
commonplace on the slip-off slopes found on the insides of
river curves. But as the next river curve migrates from
upstream, the energy distribution changes. The main
stream comes closer, tributary streams are effectively
shortened and their gradients are increased. Their erosional
potential is also increased and they cut down into the slope
which is made steeper (convex). This tendency to basal
steepening becomes even more marked when the main
stream impinges on the foot of the slope (Pl. 13.7). Its erosion
causes the paring off of the lowest part of the slope and the
development of a bluff there. This is enlarged as the slope
is more and more undermined, and in time, the slope consists
entirely of a single, simple towering cliff.

Thus at any given site the slope form varies according
to the intensity of forces at work on the slope, and in the
Murray valley (and any other valley occupied by a meandering and migrating river) this varies in time, so that slopes are
converted from graded to faceted and *vice versa* according
to their location with respect to the river.

Just as slope morphology varies along the length of a
river valley, so contrasted slope types occur within the same
region. Thus, in northwest Queensland, the Carpentaria
plains are underlain by Cretaceous sediments which are
predominantly argillaceous and which are only very gently
flexured. Interbedded thin limestones give rise to subdued
cuestas in places, but as a whole the Mesozoic sequence is
weak and susceptible to weathering and erosion. During

the mid Tertiary, however, the peneplain eroded in the
sediments was subject to lateritisation and remnants of this
low relief, dissected weathering surface survive on major
interfluves.[5] They are greatly dissected, but in several areas
the duricrust-capped remnants dominate the landscape, for
instance in the Kynuna plateau on the old divide between
the Gulf and interior drainage basins, and in the west and
southwest of Normanton. In such situations the resistant and
weaker rocks occur side by side: the resistant duricrust on
the plateau remnants, and the essentially unaltered sediments
on the present plains below.

Where the laterite persists, it gives rise to simple, faceted
slopes (Pl. 13.8) consisting of a bold bluff in which the
laterite is exposed and which is being undermined through
the development of caverns; and of a more gently inclined
debris slope strewn with blocks of laterite from the bluff
above which is subject in many places to very marked erosion
in the form of gullying. Many of these shallow valleys head
back to the base of the bluff and assist in the formation of
caverns and the recession of the bluff through their removal
of debris.

A.A. Opik[6] has described evidence from the Camooweal
area of northwest Queensland, adjacent to the Northern
Territory border, which indicates that the size and location
of the bluff varies in time according to intensity of erosional
attack, though the evidence is fragmentary. A faceted
limestone slope bounding a plateau remnant displays a low
bluff which is demonstrably being undermined by the
development of caverns, the interiors of which display a
coating of iron oxides presumably derived from a lateritic
encrustation which is preserved on the plateau above. On
the debris slope, however, there are two small, imcomplete,
iron-encrusted hemispheres which are comparable to those
at the base of the bluffs. Do they represent the bases of

Plate 13.7 The channel of the Upper Para, near Hermitage, Mt Lofty Ranges, S.A., has impinged
on the base of this valley side slope, causing it to be steepened. Slumping of the undercut bank is
obvious, but the entire lower slope has been regraded and steepened, with the result that a low bluff
has formed in mid-slope. (*C.R. Twidale.*)

former bluffs? It is arguable that as the plateau slopes
retreated, they became more and more distant from the
main energy lines — the streams — and that as the area of
plateau was reduced, the amount of run-off and moisture
affecting the slope decreased; in this way the attack of the
lower slope diminished and the bluff decreased in elevation.
Thus, as is demonstrated on the Murray valley, the size of
the various slope elements changes in time as well as spatially.

The location of intense erosion on the debris slope varies
in time due to the slumping of coarse debris from the bluff.
On the Kynuna plateau, and in many other areas of caprock
protection, a particular mode of slope development called
gully gravure[7] is in evidence. When a part of the bluff collapses
due to undermining, the coarse debris falls and flows a
greater or lesser distance down the debris slope, depending
on the inclination of the latter and the size and shape of the
detritus. The debris naturally gravitates to the lowest points
of the debris slope, namely the gullies and minor valleys
(Pl. 13.9). But as the coarse debris is difficult to evacuate, it
protects the floors of the valleys and the streams now develop
on, and head back in to, the interfluvial areas between the

gullies. Such areas are eroded most rapidly. Thus there is
local inversion of relief. But when the bluff next collapses
it is the newer depressions which receive the protective debris.
The divides or former valleys are again vulnerable to gradual
weathering and evacuation of the coarse detritus and are
eventually eroded.

Intermittent erosion of the debris slope causes the caprock
to be gradually undermined and reduced in area. Eventually
it is eliminated. Henceforth the slopes must be lowered,
though once they have adjusted to no longer receiving the
supply of protective caprock debris, the graded slopes
maintain the same inclination.

Finally, as an example of the self-regulating mechanisms
at work on slopes, reference may be made to the detailed
minor changes which occur on debris slopes. In the Flinders
Ranges, for instance, on a prominent faceted slope called the
Bluff, some 16 km north of Quorn, many small terracettes
and scarps are displayed on the debris slope. In the field
many are seen to be related to blocky debris which has fallen
to the slope from the sandstone bluff above. Each block
forms an obstacle impeding the downslope creep and wash of

Plate 13.8   The Kynuna plateau, northwest Queensland. The plateau, mesa and butte assemblage is caused by the dissection of a laterite duricrust which caps the residuals and forms the bluffs. The rivers have, however, cut down through the regolith to the former weathering front (so that the plain below the plateau is of *etch* character—see Chapter 19) into the unweathered but unresistant Cretaceous shales and siltstones in which they have carved a plain of subdued, rolling relief. (*C.R. Twidale/C.S.I.R.O.*)

Plate 13.9 Gullies cut in a siltstone bluff near the southern extremity of Lake Torrens, S.A. Blocks and boulders from the alluvium above have poured down the gullies forming a protective veneer to the channel floors. (*C.R. Twidale.*)

debris, which accumulates upslope of the block forming a small terrace. But the area just below the block is deprived of debris and is therefore degraded by run-off. Eventually the block is undermined and tumbles downslope; the unconsolidated debris is returned to the flow of material downslope; and the slope is restored to its former condition.

What is implied here is that each slope tends toward an equilibrum from and inclination under the prevailing lithological and climatic conditions. Departures from the "normal" conditions induce changes, but these processes are self-regulating and the slope eventually returns to its quasi-stable form. There are minor morphological oscillations around a norm, as it were, but in the long term, slope form and inclination are constant for long periods.

Such concepts of equilibrium and of loss and gain are also useful in the analysis of graded slopes.

5. Development of Graded Slopes

The wash and downslope creep of debris causes the recession and decline of the slope to such an extent that the wedge of debris at the base of the slope is gradually eliminated as the slope recedes and the debris slope becomes a gently inclined bedrock slope mantled with a veneer of debris. The sigmoidal form of the slope is developed and maintained by weathering, wash and the mass movement of debris, with the steepest slope sector located approximately in mid-slope.

The nature of the processes at work on such graded slopes has given rise to considerable discussion and controversy. On the upper rounded slopes, rills and streams are not in evidence, and various forms of mass movement, including creep and wash, have been suggested as the processes responsible for producing the convex crests of hills typical of many regions.[9] This is generally agreed, though it has been argued that weathering reduces projections and thus produces rounded surfaces.[10] However, as Fenneman[11] pointed out, if wash and creep are principally responsible for shaping the bedrock surface, the slope should become increasingly steep downhill, because a greater and greater mass of debris passes over the slope at lower and lower levels (Fig. 13.12). In fact, steepening proceeds only to the point of inflexion, after which gradients decrease. Birot[12] has endeavoured to explain this contrast in terms of increased impermeability and concurrent increase in mobility of debris on the lower slopes, but Baulig[13] suggests that the sigmoidal shape of many graded slopes is due to the dominance of different processes on different parts of the slope: mass movements on the upper sector, and rills and streams on the lower.

Rills, streams and gullies cutting into the base of the slope induce steepening in their headwater zones. These ephemeral watercourses appear as lines of concentrated drainage on the slope, but do not develop steeper and steeper gradients as they progress downslope, as might be expected in view of their increasing volume and erosional potential. They cut down relatively markedly for a time, but they cannot cut down below the level of the base of the slope—the local baselevel of erosion. Hence they develop a concave profile which is reflected in the form of the entire lower slope.

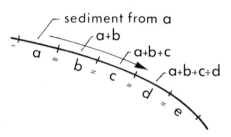

Fig. 13.12 Variation in the volume of debris passing over slope units.

6. Slope Equilibrium and Disequilibrium

A rather different deductive approach to graded slopes has been suggested by Ahnert,[14] who agrees with other workers that weathering and mass movements of debris go far to explain the morphology, but who draws attention to the results of variations in the rate of waste production, downslope transport and evacuation from the toe of the slope.

Ahnert argues that rock waste is produced by weathering, is transported, and is redeposited at a lower level on the same slope or evacuated from the site, usually by rivers. Ahnert's thesis is that the form of a slope depends on the relative rates at which this production, transportation and deposition of waste takes place. He assumes that the accumulation of a mantle of waste can inhibit further weathering, and points out that activities on one part of a slope affect all other slope sectors.

Thus if A_s equals the rate at which waste arrives at the foot of the slope, W_s the rate of waste production on the slope by weathering, and Rpf the potential rate of waste removal at the slope foot, the following cases can be envisaged.

Where A_s equals both W_s and Rpf at all points, then the slope remains the same in all respects. It is in equilibrium (Fig. 13.13a). Where Rpf equals A_s, but is less than W_s, then there is a build up of debris at the foot of the slope, and the slope inclination decreases. For example the graded slopes on the inside of the river curves of the Murray, cited earlier in this chapter, are of this type. In this way A_s decreases and equilibrium is restored (Fig. 13.13b). At the point of transition between zones 1 and 2 the potential rate of waste removal in zone 1 ($Rp1$) is less than the rate at which waste arrives at that point from upslope (A_2).

Where on the other hand Rpf is equal to A_s, but greater than W_s, (Fig. 13.13c) material is evacuated faster than it is supplied, slope inclination is increased, removal of waste facilitated, the mantle becomes thinner so that weathering increases until equilibrium is again restored.

Slope behaviour can also be interpreted in a qualitative way. Whether downwasting or backwearing prevail, whether slope decline or retreat is dominant, depends on the slope budget.[15] It has been suggested that where, as in temperate regions with an even distribution of soils, vegetation, weathering and erosion, baselevel control is crucial, slopes soon decline. This is suggested by the varied slope inclinations in these areas. Where, as in the arid tropics for instance, soils and vegetation are scarce on upper slopes, or where there are durable caprocks, scarp foot weathering and erosion become significant. Here slopes are maintained at the maximum inclination commensurate with stability and parallel retreat takes place. Here the piedmont angle is developed and maintained,[16] scarp foot valleys are eroded, and hillslopes tend to similarity of form and inclination.

This contrast in behaviour between downwasting and backwearing is obviously of significance in overall landscape developments, as will be apparent when the origin of pediments is discussed in the next chapter, and when the peneplain and pediplain models are evaluated in Chapter 19.

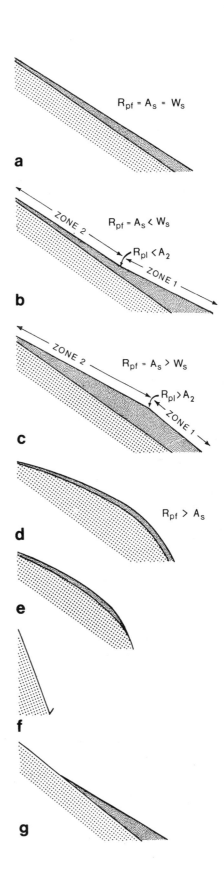

Fig. 13.13 Slope forms resulting from variations in weathering, transport and evacuation of debris on slopes. Equilibrium conditions indicated in text in cases a–c; in d, $R_{pf} > A_s$, and the slope remains mantled with waste during downcutting; in e, bedrock exposed at slope base during intensive downcutting; in f, long-continued incision of stream results in constant and complete removal of waste over entire slope; in g, increased rate of weathering of bare crest compensates for increased removal of debris from slope foot. *(After F. Ahnert, "The Role of the Equilibrium Concept", pp. 23–41, by permission of Frank Ahnert.)*

References Cited

1. H. WOPFNER and C.R. TWIDALE, "Geomorphological History of the Lake Eyre Basin", in J.N. Jennings and J.A. Mabbutt (Eds), *Landform Studies from Australia and New Guinea*, Australian National University Press, Canberra, 1967, Ch. 7, pp. 118–43; C.R. TWIDALE, "Landform Development in the Lake Eyre Region, Australia", *Geogr. Rev.*, 1972, **62**, pp. 40–70.

2. D.W. JOHNSON, "Streams and their Significance", *J. Geol.*, 1932, **40**, pp. 481–97.

3. C.R. TWIDALE, "Effect of Variations in the Rate of Sediment Accumulation on a Bedrock Slope at Fromm's Landing, South Australia", *Z. Geomorph. Supp.*, 1964, **5** NS, pp. 177–91; D.J. MULVANEY, G.H. LAWTON and C.R. TWIDALE, "Archaeological Excavation at Rock-shelter No. 6, Fromm's Landing, South Australia", *Proc. R. Soc. Vic.*, 1964, **77**, pp. 487–516.

4. O. FISHER, "On the Disintegration of a Chalk Cliff", *Geol. Mag.*, 1866, **3**, pp. 354–6; A.C. LAWSON, "The Epigene Profile of the Desert", *Univ. Calif. Pub. Dep. Geol.*, 1915, **9**, pp. 23–48; O. LEHMANN, "Morphologische Theorie der Verwitterung von Steinschlagwanden", *Vierteljahrsschr. Naturforsch. gesellsch. Zurich*, 1933, **87**, pp. 83–126; C.R. TWIDALE, "Evolution des Versants dans la Partie Centrale du Labrador-Nouveau Québec", *Ann. Geogr.*, 1958, **68**, pp. 54–70.

5. C.R. TWIDALE, "Chronology of Denudation in northwest Queensland", *Bull. geol. Soc. Am.*, 1956, **67**, pp. 867–82; *idem*, 'Geomorphology of the Leich-hardt-Gilbert Area, northwest Queensland", *C.S.I.R.O. Land Res. Ser.*, 1966, **16**.

6. A.A. ÖPIK, personal communication.

7. K. BRYAN, "Gully Gravure—a Method of Slope Retreat", *J. Geomorph.*, 1938, **3**, pp. 89–107.

8. For review see C.A. COTTON, "The Erosional Grading of Convex and Concave Slopes", *Geogr. J.*, 1952, **118**, pp. 197–204.

9. G.K. GILBERT, "The Convexity of Hilltops", *J. Geol.*, 1909, **17**, pp. 344–51.

10. W. PENCK, *Morphological Analysis of Landforms* (trans. H. Czeck and K.C. Boswell), Macmillan, London, 1953, pp. 141–3.

11. N.F. FENNEMAN, "Some Features of Erosion by Unconcentrated Wash", *J. Geol.*, 1908, **16**, pp. 746–54.

12. P. BIROT, *Essais sur quelques Problemes de Morphologie Generale*, Inst. Alta Cultura, Centro Est. Geogr., Lisbon, 1949, pp. 17–33.

13. H. BAULIG, "Le Profil d'équilibre des Versants", *Ann. Geogr.*, 1940, **49**, pp. 81–97; *idem, Essais de Geomorphologie*, Publ. Fac. Lettres, Univ. Strasbourg 114, Soc. d'Edition, Les Belles Lettres, Paris, 1950, pp. 125–47.

14. F. AHNERT, "The Role of the Equilibrium Concept in the Interpretation of Landforms of Fluvial Erosion and Deposition", *Proc. Symp. l'évolution des Versants (Liège)*, 1967, **40**, pp. 23–41.

15. J. TRICART, "L'évolution des Versants", *L'inform. geogr.*, 1957, pp. 108–15; C.R. TWIDALE, "Some Problems of Slope Development", *J. geol. Soc. Aust.*, 1960, **6**, pp. 131–48.

16. C.R. TWIDALE, "Origin of the Piedmont Angle as Evidenced in South Australia", *J. Geol.*, 1967, **75**, pp. 393, 411. See also Chapter 10.

CHAPTER 14

Rivers and Running Water in the Tropics

A. GENERAL

What has already been stated of rivers and valleys is of general application, but in the tropics and subtropical regions conditions of extreme humidity and aridity are experienced. The effects of running water in these areas are not, however, what might be expected.

B. THE HUMID TROPICS

In the *humid tropical regions* (selvas) temperatures are consistently high and there is abundant moisture and luxuriant vegetation, save where the forest has been cleared by man. Chemical weathering is rapid and intense. Few fragments of a size greater than sand survive weathering. The rivers lack pebbles and boulders and thus lack the tools used to abrade the bed and banks. Pot-holing is not common. It is worth emphasising that the common absence of coarse debris in the lower reaches of rivers like the Congo and Amazon does not imply a low capacity, but merely reflects the unavailability of coarse material.

Other factors also mitigate against strong stream action in these very high rainfall areas. The vegetated slopes and banks resist erosion. The flow of the rivers is remarkably even, for much rainfall is intercepted by vegetation and soil and only slowly infiltrates into the rivers. Extremes of flood and drought, typical of the arid regions for instance, are lacking; there is little disequilibrium between the river and its inter-fluves. Hence, though the rivers of the humid tropics transport much fine debris and large volumes of salts in solution (though the amount of silica carried in solution in rivers and in groundwater is surprisingly constant the world over[1]), they are not as dominating as might be anticipated.[2] Because of the activities of landslides and earth-flows, erosion of the valley sides commonly extends to the very crests of the inter-fluves. The latter therefore tend to be knife-edged and, in areas of high relief, all-slopes topography is well represented. Rectilinear slopes (in New Guinea of 35–40° inclination) are common. Wasting and erosion appear to be equal at all parts on such slopes, implying that they maintain a constant inclination and that they either recede without change of shape or gradient, or maintain constant forms, inclination and location.[3]

Thus though precipitation is high and there is commonly a dense network of permanent rivers and streams which have incised deep valleys in the selvas regions, running water is not the dominating component of the morphogenetic system in the selvas regions. Indeed degradation of the tropical humid lowlands is apparently among the slowest known.[4] The reason for this seeming anomaly is that the slopes are well clothed and protected by vegetation: the land surface is in equilibrium with the climate. But in the tropical and subtropical desert regions there is enormous variation between drought and flood, and there is only sparse vegetation. Hence, climate and land surface are from time to time in marked disequilibrium, and for this reason, paradoxical though it seems, running water is relatively more important here than in the humid lands.

Though the desert lands extend beyond the confines of the tropics they are, for the sake of brevity, usually referred to below as the tropical deserts.

C. THE ARID TROPICS

1. Water in the Tropical Deserts

Though the peculiar landform assemblages of tropical deserts are of aeolian origin (see Chapter 15) there is no doubt that run-off is of consider-time (Pl. 13.1) despite the low average rainfall in the deserts.

There are many eyewitness accounts of desert floods. For example, Wells describes the rise of a flood in a watercourse near a mine in upper northern Egypt in the autumn of 1902. There had been no rain within a hundred kilometres of the mine and at 6.30 a.m. there was a mere trickle of water down the valley; but "an hour later the whole watercourse was a mighty stream, over 300 m wide and from one to two and a half metres deep rushing past the mine".[5] The Battle of Tit (in which the French subjugated the Tuaregs of the north African deserts) was interrupted and the pursuit of the defeated tribes prevented by a storm which flooded the wadis in March 1902. According to some writers, more people are drowned than die of thirst in the Sahara. The deserts of northwest Peru are swept by violent floods only two or three times a century, but landforms and human activities are profoundly influenced by these rare, but catastrophic, floods. Bosworth reports that:

> The flood of 1891 was brought about by daily torrential rains in February and March. The water poured down the deep mountain valleys and immediately overflowed the quebradas [deep, trench-like canyons cut across a plateau below the mountains] uniting as a great sheet, which rolled on to the sea. Many drainage channels were changed, and an immense load of stones was swept from the mountains to the plain below. For two years afterwards the country was moist and green, so that herds of cattle and goats were introduced, and also cotton planted. In the next two years, the country dried up completely, the vegetation died and the cotton gin became a curio.[6]

Similarly, much erosion and deposition was accomplished by the floods which affected central Australia in May and June, 1967,[7] and in 1974–5.

The effectiveness of the occasional rains of the present climatic regime in inducing run-off and erosion is not basically due to a greater incidence of intense rains. Storms in which large volumes of rain fall in a short period of time undoubtedly occur in arid and semiarid regions but the available evidence suggests that such falls are no more frequent and intense·than those experienced in temperate regions.[8] It is the nature of parts of the desert surface which bestows particular effectiveness to the rains.

Run-off following rains is not heavy, for evaporation losses are high and in the sandy desert there is rapid sub-surface percolation. But run-off is rapid in some parts of the desert terrain. The common development of pedogenic horizons, mantles and crusts (the *croûtes désertiques*) retards or prevents subsurface infiltration of water. The sparse vegetation cover provides little in the way of an umbrella effect to protect the soil surface against raindrop impact, though this is compensated in some regions by the occurrence of a gibber mantle. There is little vegetation either to retard surface flow or to bind or protect the upper soil layers. River flow tends to be of short duration: only° rarely are rainfall and run-off sufficiently prolonged to allow rivers to reach the sea or the centre of the local drainage basin. More commonly these rivers extend a short distance out on to the plains from the uplands (where rainfall is higher) before the waters seep underground to form a subsurface flow. But the river load, save that in solution and some very fine sediment, cannot be carried underground and the watercourses of arid lands are commonly strewn with mineral and organic debris. Hence there is usually ample debris at the surface and the rivers of arid regions do not lack the means of erosion. Uprooted logs are particularly potent tools.

During such brief periods of activity, rivers undoubtedly scour and enlarge and extend existing drainage systems. New channels are initiated. The soil moisture which results from modern rains contributes to weathering processes and to the development of gilgai in some areas. But the question remains as to how many of the valleys or wadis of the deserts, how much of the chemical alteration evidenced there, and how much of the gilgai development observed there is due to essentially modern processes, and how much to similar processes induced by the heavier rains of former times?

There is no doubt that some fluvial features of the desert originated some time ago. For instance, prominent alluvial fans which front the desert uplands of Arabia are attributed to a humid phase of some 7000 years ago[9] when ancient civilisations flourished there and in other parts of south-western Asia. The fans which front the southern Flinders on their western side are of late Pleistocene age.[10]

Others take the view that modern riverine processes are effectively eroding the land surface near major river valleys, but on the interfluves between—and in arid lands these are frequently very extensive—run-off is of negligible significance. This may be an underestimate of the effectiveness of running water in moulding the desert surface at the present time, but it is highly relevant to the interpretation of the fluvial forms which undoubtedly exist in the tropical deserts.

No part of the arid lands is entirely devoid of water: from time to time even the driest of deserts experiences the effects of running water. Of course in some areas modern run-off may merely retouch forms which have essentially developed under the influence of more profuse rains and run-off in the recent past, for as is described in Chapters 22 and 23, at times in the Pleistocene the modern tropical deserts experienced rather higher rainfall. At present it is not possible to evaluate the relative effects of past and present rainfall in such areas. Suffice it to say that the rate of development of erosional forms in such arid lands is so slow that to some extent it must all be inherited in the strict sense, though it is more likely a question of the rate at which features are evolving today compared to the past, not a matter of a definite and distinct change of process.

2. Pediments

Many desert uplands are bordered by smooth, gently sloping erosional surfaces called fringing pediments, and similar remarkably smooth and very gently sloping surfaces constitute a significant element of the great plains of the north African and other deserts. What are these pediments and how are they formed?

Views differ as to what a pediment is. The most widely accepted definition which is soundly based both in the field and in historical precedent, is as follows.[11] Pediments are smooth, undissected surfaces of erosion inclined at an angle of 3–4 to the horizontal (Pl. 14.1), though they may be as steep as 10° and as little as $\frac{1}{2}$° or less. Where there are residual remnants of uplands the pediments characteristically meet these in a sharp break of slope called the piedmont angle (see also Chapter 10 and Pl. 14.2).

Pediments display little dissection and present a smooth profile parallel to the mountain front (Fig. 14.1). They commonly carry a veneer of debris, though this generally thickens downslope to merge with the alluvial fill of the adjacent valley or basin.

Most of these characteristics are accepted by most workers. One which is in dispute is the piedmont angle, which some regard as critical,[12] but which is seen by others as a local variation. Thus Dury attaches little significance to the feature and defines a pediment as "A degradational slope, cut across rock in place, abutting on a constant slope at its upper end, and decreasing in gradient in an orderly fashion in the down-slope direction …".[13] This is open to a number of objections. For instance planate surfaces which would be accepted as pediments by most workers may be convex or rectilinear in profile normal to the upland front.[14] The lack of relief in profile parallel to the upland is not mentioned. And, as is discussed elsewhere (Chapter 19) it is the piedmont angle which seems to be the key to understanding the accepted distribution of pediments.

Pediments form extensive plains of remarkably low inclination in the Sahara and in the area to the west of Lake Eyre in central Australia. Residuals which stand above the pediments are traditionally called inselbergs but, as argued in Chapter 3, inselbergs are really structural forms which

Plate 14.1   Dissected pediments north of Brachina Gorge, on the western flank of the central Flinders Ranges, S.A. The pediment, capped by a gibber mantle, dips at 3—4° from the base of the ranges. Beyond is a lower ridge in limestone and beyond that again a higher sawtooth ridge in sandstone. Higher pediment remnants are visible in the middle distance. (*C.R. Twidale.*)

Plate 14.2   Pediments around the Pinnacles, Broken Hill, New South Wales. (*C.R. Twidale.*)

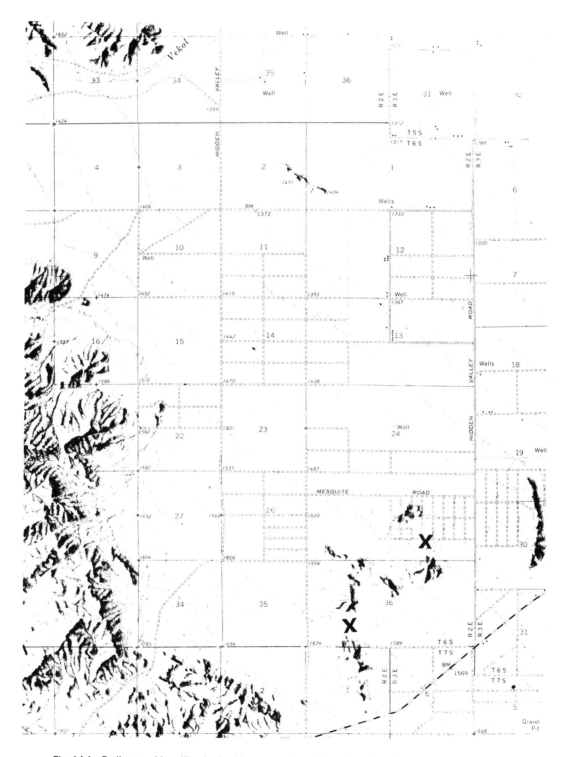

Fig. 14.1   Pediment with gullies, backed by mountains and merging downslope with alluvial spreads
or bajadas. X = pediment pass. (*U.S. Topo. Map, Antelope Peak Quadrangle, New Mexico.*)

occur in a wide variety of climatic settings. Pediments are found in the arid tropics but are perhaps best developed in the semiarid tropical zones.

However it would be incorrect to conclude that all planate surfaces in the arid tropics are pediments, or that pediments are restricted to the tropics. On the one hand, rolling topography which descriptively is evocative of a Davisian peneplain (though there is no evidence as to whether it results from downwasting or from backwearing) is well developed on comparatively weak sedimentary sequences in central Australia and near Amman, in Jordan (see Pls 13.9, 19.1). Moreover pediments occur in central Alaska and they are found in the valleys bounded by high, seasonally snow-capped peaks and are themselves blanketed by snow in winter in such parts of the American West as Utah, fronting the Rockies around Boulder, Colorado, and in western Nevada (for example the so-called Lewis Terrace near Aldrich Station near Yerington, Nevada).[15] It is asserted that they exist in southern England,[16] though the validity of this claim rests on the definition of what a pediment is.

3. Lithologic Control of Pediment Development

In the Carpentaria plains of northwest Queensland, rolling topography of the peneplain type is well developed on unresistant Cretaceous sediments.[17] Within this semiarid region, the plains extend both to the east and west from the Cretaceous outcrops on to the crystalline rocks (particularly granitic rocks) of the Isa and Einasleigh uplands, where the typical bedrock plain is a smooth, gently sloping, undissected pediment with a veneer of sand debris (Pl. 19.6a).

Lithological considerations thus introduce complexities,[18] for as indicated with respect to northern Queensland, pediments are especially well formed on granite outcrops

and in most regions a disproportionately high percentage of pediments occur on such bedrock. Yet in the southwest of Western Australia (for example, around Hyden and Corrigin) and on Eyre Peninsula in South Australia, the weathered granite has been dissected to form rolling lowlands, though pediments fringe the numerous granite inselbergs of these regions.

However, pediments are by no means restricted to granitic terrain. In northwest Queensland and on the north side of the Harts Range of central Australia, pediments are formed on basic crystalline outcrops (amphibolites) and bordering and within the Flinders Ranges they exist on argillaceous sediments (Pl. 14.1). In Death Valley, California, there are smooth, sloping surfaces which are much dissected and are interpreted by some as remnants of once contiguous pediment surfaces (Fig. 14.2) which cut across folded but only poorly lithified Pliocene fanglomerates of the Furnace Creek Formation.[19]

4. Origin of Pediments

Pediments were at one time attributed to wind erosion [20] but it is very doubtful (see Chapter 15) whether sand blasting is as widely effective as is implied in the extensive development of pediments. The optimal development of these forms in the semiarid, rather than arid, regions also argues against wind erosion and points to their formation under the influence of running water. However, it may appear paradoxical to attribute these arid zone landforms to the work of rivers and streams, and quite apart from the question of the effectiveness or otherwise of running water in shaping the land surface in these regions, there is the problem of the age of the features. Are the pediments forming today or are they inherited from past periods of higher rainfall? Modern deserts display pronounced effects due to palaeo-climates,

Fig. 14.2 Sketch of "pediment" remnants which are cut across folded strata and which carry a mantle of gravel, just south of Furnace Creek Ranch, Death Valley, California.

which are however difficult to evaluate precisely, but it seems reasonable to suggest that the pediments, though initiated under climatic conditions different from those which obtain at present, nevertheless continue to evolve.

How, then, are pediments formed? Many workers, notable among whom is L.C. King,[21] have considered that pediments are the planate surfaces left behind by the recession of escarpments. But in view of the climatic distribution of pediments, what is it about arid and semiarid regions which preferentially induces scarp retreat as opposed to the lowering of slopes? The problem may be approached first through the little diagnostic evidence that is available, and second by way of general argument.

Although in the Valley of the Thousand Hills, near Durban, Natal, there is clear evidence of scarp retreat in granite from the valleys of such rivers as the Umgeni, the amount is only a matter of a few metres or at most a few score of metres, not the hundreds of metres or even kilometres envisaged in King's hypothesis. Similar evidence of only very minor recession of granite scarps has been noted on the inselbergs of the northwestern Eyre Peninsula. As outlined in Chapter 3, bornhardts are intrinsically structural forms and their borders are determined by major joints or by the junctions between well-jointed, massive compartments of rock; their dimensions have not altered greatly in time. There is some field evidence to support this view (see for instance Pl. 3.7).

This argument is also relevant to another piece of evidence adduced in support of the scarp retreat hypothesis. In many areas pediments merge with each other in cols between inselbergs (see Fig. 14.1). These cols are called *pediment passes*[22] and they have been interpreted as being due to the headward regression of scarps and the upslope advance of pediments, with the residual eventually being eliminated and the headward advancing pediments from opposite slopes coalescing. However, the features can be explained equally well in terms of compartment weathering and the exploitation of joints by weathering and erosion. There is no necessity to suggest scarp retreat: weathered granite could well be eroded by rills and streams which lower the land surface leaving the fresh rock compartments in relief as nubbins or inselbergs.

Indeed, there is some evidence that in a few areas, at any rate, slopes in the piedmont zone have not retreated in recent geological times but have declined. In the Brachina region in the western piedmont of the Flinders Ranges, higher dissected pediment remnants are slightly more steeply inclined with respect to the horizontal than are their counterparts at a lower level, suggesting a gradual decline of slopes in the piedmont region (Fig. 10.7). But in most areas there is no evidence as to whether slopes have wasted back or downwards. The only guidance on the point is to be derived from general considerations of slope development.

In theory the critical difference between the peneplain and pediplain concepts is the contrasted behaviour of slopes. In the peneplain the end-product allegedly results from a dominance of slope decline; in the pediplain, from scarp retreat. What are the possible reasons for this contrasted behaviour over time? Some were indicated in the account given earlier (Chapter 13) of the development of slopes in the

Murray valley and other locations, where faceted and graded slopes occur side by side. Many writers have either assumed or inferred a marked climatic control of slope behaviour, with decline typical of temperate areas and retreat (of both faceted and graded slopes) typical of the arid and semiarid tropics.

Examination of the landscapes depicted in Pls 19.1 and 19.5 shows that graded slopes occur in arid and semiarid regions at the present time, though whether they formed in aridity is perhaps another question. Faceted slopes are well represented in the tropical desert and semidesert regions, though they are commonplace (given the right structural setting) in all parts of the world. Nevertheless it is undeniable that faceted slopes and evidence of slope retreat can be observed in abundance in the hot desert regions of the world. Why should this be so?

As indicated in Chapter 13, steep slopes are eroded and maintained where there is strong attack on the lower slope elements. Differential erosion, such as between the upper and lower slopes, can take place for either or both of two reasons. First, structural factors, such as the occurrence of a resistant capping to a slope which occurs in many areas including temperate lands gives rise to comparatively marked erosion of lower slopes. It is not suggested that there is higher proportion of primary caprocks in arid and semiarid lands than elsewhere, but for climatic reasons they are less susceptible to disintegration than in the humid tropics. Moreover, gibber and dried clays, which do not form in humid conditions, form effective protective mantles. And, partly because of their location adjacent to the humid tropics and the reality of climate change, partly because of existing climatic conditions, duricrusts which form tough carapaces to the land surface are widely developed and preserved in the arid and semiarid tropics.

Second, as was explained in Chapter 10, in the arid and semiarid tropics, wash from the slopes does not commonly extend far beyond the piedmont zone. Instead it infiltrates into the subsurface and there weathers the bedrock, contributing significantly to the development of the piedmont angle and particularly to the regrading, steepening and maintenance of the backing scarps.[23] Thus for both structural and climatic reasons the dominance of scarp retreat in the arid tropics can readily be explained and justified.

Such deductive analyses supported by field observations indicate the significance of the relative effectiveness of weathering and erosion on the upper and lower slopes respectively in determining slope behaviour, or in other words the slope budget.[24]

But if slopes tend to be self-regulating, what of the suggestion that slopes may decline? There is no doubt that soil and vegetation cover are very important in retaining moisture, hence in influencing the distribution of weathering over the land surface, and hence in producing rounded hill crests.[25] It may be argued that temperate humid conditions favour a distribution of soils and vegetation that is more even than that found in arid regions, that faceted slopes are therefore frequently formed in arid lands, and that graded slopes become a more common feature of the land surface in the later Cainozoic, when the continents were clothed in grasses and other ground cover.

The graded slopes which occur in arid lands carry only a sparse cover of vegetation (Pls 19.1, 19.5) but are all apparently of similar inclination. Hence, if there is only slow evacuation of debris from the slope toe upper slope wasting exceeds that of the lower slope so that slopes are lowered and relief amplitude decreases. But if there is a tendency to equilibrium, the only reason for a systematic variation in the inclination of valley sides slopes along the length of the valley in a uniform lithological situation is as suggested in Chapter 13, a systematic variation in the distribution of energy along the valley: the forces responsible for the weathering and erosion of divides remain constant (provided structural factors are unchanged along the divide) but the effectiveness of evacuation from the footslope may vary. Although the river increases in volume downstream so does the available load. Because of baselevel control, however, it is a matter of observation that in the absence of structural restriction, rivers widen the valley floors and become more sinuous downstream. Thus the effectiveness of debris evacuation by the stream from the toe of the slope decreases downstream because, whereas the stream is at the floor of the slope virtually everywhere in the narrow upper valley, lower down it seldom impinges on the scarp. Hence there is time for gentler, graded slopes to evolve downstream.

Thus there is some rationale in support of the suggestion that climate exerts a broad control on slope behaviour and to some, as yet uncertain extent, on slope morphology.

But there are many variations of slope morphology in any region or locality. Some local factors have been discussed in earlier chapters (see for instance Chapters 5 and 13) and as described earlier in this chapter, structural factors impose considerable differences in slope morphology within the same climatic region.

5. Processes Involved

The processes invoked in explanation of pediment formation, involve either running water or subsurface weathering.

Sheetwash, or the flow of thin laminae of water over planate surfaces, has been invoked in explanation of pediments in southern Africa by L.C. King,[26] who considers that such flows are capable of paring off thin layers of rock. Wavy soil and vegetation patterns in parts of eastern Africa have been explained in terms of sheetwash[27] and the results of such surface flows can be seen in the shape of small debris ridges which are aligned approximately along the contour and are arranged concentrically, with respect to topographic highs, on pediments in the Flinders Ranges. Sheetwash of local character has also been invoked in explanation of low debris ridges on gentle silcrete slopes in western Queensland.[28] But though they help to maintain and to shape the pediment surface in detail, such flows, plus the work of rills, seem scarcely capable of initiating the gross form. Pediments carry a mantle of debris and such sheetwash and shallow flows (though they modify the finer surficial debris in detail) do not touch the bedrock below the mantle.

Planation in detail or on a small scale by means of weathering processes is suggested in evidence from the Flinders Ranges and from central Australia. In the southern Flinders, sandstone residuals and boulders display miniature flared slopes and small platforms. They have been compared to the forms developed on granite and attributed to soil-controlled moisture weathering.[29] Likewise, Mabbutt[30] has suggested that stone and soil mantles in central Australia are capable of inducing the development of relatively smooth weathering fronts: the subsurface bedrock plains covered by debris mantles which are called pediments. Evidence from Eyre Peninsula supports this contention. There platforms cut in fresh granite occur both in isolation and in bordering inselbergs. In places they are 750 m across, though most are narrower. These platforms are rock pediments (*glacis d'érosion, glacis de dénudation*) which have formed beneath the regolith as a result of solution, hydration and hydrolysis. They are exposed weathering fronts that demonstrably extend beneath the contemporary regolith and have been revealed through the stripping of grus and the lowering of the plains.

Because of the difficulty introduced by the mantle which in many areas has been transported and is not merely a regolith *in situ*, it has been suggested with respect to the pediments which border the valleys and plains of the Flinders Ranges that they were essentially moulded during scarp retreat at a stage of development when the bedrock surface was at the scarpfoot, and where the wash from the hillslope attained maximum effectiveness.[31] Subsequently the bedrock surface was modified by subsurface flushing and weathering beneath the debris mantle. Twidale[32] emphasises that the remarkable smoothness of the pediment surface in detail is due to the deposition of debris on the pediment bedrock surface, for it is this mantle which fills in depressions and blankets rises in the admittedly low bedrock surface that is due to wash and the work of rills.

In the Flinders and other uplands, rivers and streams are so widely spaced with respect to the extensively developed pediments that lateral stream corrasion cannot be held responsible for their formation. Elsewhere, however, the bedrock is so impermeable and the stream pattern so closely spaced that such a process warrants serious consideration. In parts of the James and Krichauff ranges, Northern Territory, for instance, the numerous streams are braided from inception and when in flood undoubtedly corrade laterally, truncating spurs and forming plains of lateral corrasion which coalesce to form plains of pediment type which are covered with a spread of alluvial gravel (Pl. 10.16).

This process is probably also responsible for the formation of the narrow plains, now dissected and above present stream beds, which are found in parts of Death Valley, California (Fig. 14.2). The gently sloping surfaces are cut across steeply dipping Pliocene fanglomerates of the Furnace Creek Formation, and carry a mantle of cobbles and gravel identical with that found flooring the beds of the present narrow valleys. These narrow bevels can be related in the mind's eye and considered as remnants of a once continuous planate surface. But what is obvious and plausible is often fallacious and such a conclusion may be unwarranted. The present streams are closely spaced and are separated by the bevelled ridges which carry the remnants on their summits (see Fig. 14.2). It is equally reasonable to interpret the bevels as former stream beds which, because of their mantle of gravel, were protected against dissection consequent upon the tectonic uplift which frequently occurs in this region (see Chapter 4, Figs 4.11b and c). The streams cut into the unprotected, unconsolidated bedrock exposed at the channel margin, eventually eroding the interfluves and leaving the stream beds high and dry. In these terms there never was a planate pediment surface of any considerable lateral extent, but merely a series of flat-floored stream beds separated by high but readily erodible interfluves. The present situation has arisen through uplift in the form of fault displacements (for which the area in question is well known), and what may be described equally well as either gully gravure (see Chapter 13) on a grand scale, or a relief inversion (Fig. 14.3a-c).

Significant as these several processes and mechanisms are, they apply to specific localities and regions. Any satisfactory general explanation of pediment development should take account of their climatic distribution and their preferred development on granitic rocks in many areas. A number of factors are probably involved.

For instance, Birot[33] has drawn attention to the distinct bimodal calibre of weathered granite—blocks on the one hand, and granite sand or grus (see Chapter 3) on the other. He goes on to suggest that in dry hot climates the low volume streams incise their beds only slowly, but slopewash proceeds apace so that wide, shallow, flared valleys result. The divides separating the shallow, flat-floored valleys (or pediments) consist of bouldery nubbins and the sharp break between hill and plain, according to Birot, reflects the location at which run-off is of sufficient volume and velocity to carry the available sand, and thus lower the land surface. It would perhaps be mistaken to consider that rivers of the arid and semiarid tropics are always of low volume; on the contrary they display great variation in discharge over time, and are especially effective geomorphologically in brief periods of flood. Moreover, the nubbins are especially joint-controlled inselbergs and the piedmont angle is probably due to exploitation of structural contacts by weathering and erosion. Nevertheless the nature of valleys developed in the arid tropics is worth further consideration.

In arid and semiarid climates not all rains are sufficient to produce run-off with enough volume and duration to reach the defined channels and cause them to run. Many flows are limited to wash and rill work on the divides. On this account, the divides could be lowered relative to the infrequently flowing stream beds, particularly if the latter ran to basins of internal drainage in which the local baselevel tended to rise due to infilling of the playa depression. This would be true particularly of weathered (that is, weak) bedrock, such as weathered granite and other readily altered rocks like shale and siltstone.

Weathering of the bedrock near the stream channels could induce differential erosion of the broad divides and the development of scarp retreat into the interfluves from the stream lines. Erosion by rills and wash to the local baselevel (the stream bed) produces a gently sloping bedrock surface —the pediment—extending away from the stream line. Because of the crystalline nature and low porosity of granite, there is a sharp junction between weathered and fresh rock and for this reason differential erosion at the weathering front is particularly pronounced on this bedrock.

Some subsurface flushing also occurs on granite, and marked evacuation of the products of granite alteration has been noted.[34] As already mentioned, some of the smoothness of pediments is undoubtedly due to the deposition of a mantle of debris over the bedrock surface and some to the action of rills and wash on already weathered bedrock:[35] but long-continued slow weathering and occasional wash probably has to be invoked in explanation of the extensive flat pedi-plains of the Sahara.

6. Concluding Statement

The question of pediment development is complex and no single explanation applies to all pediments: several different processes can apparently produce similar landforms. In some regions lateral stream planation is most probably responsible for narrow planate remnants regarded as former pediments, though because of the ubiquitous presence of a mantle, it seems likely that erosion by wash and rills in the scarpfoot area, plus mantle-controlled weathering and subsurface flushing together are responsible for many pediments. In the case of the extensive and unbelievably level pediments of the north African deserts, it is necessary to suppose that they have developed under the influence of such processes over a very long period of time.

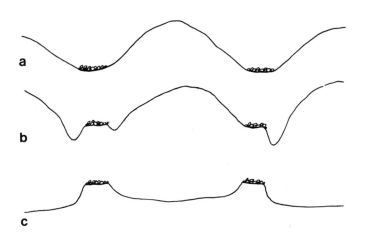

Fig. 14.3a–c Possible development of pediment remnants in Death Valley, California. In (a) river valleys are cut into the unprotected rock, leaving the bed load gravels untouched until in (c) the former river channels stand high in relief.

References Cited

1. S.N. Davis, "Silica in Streams and Groundwater", *Am. J. Sci.*, 1964, **262**, pp. 870–91.

2. I. Douglas, "The Efficiency of Humid Tropical Denudation Systems", *Trans. Inst. Br. Geogr.*, 1969, **46**, pp. 1–16.

3. F.W. Freise, "Erscheinungen des Erdfliessens in Tropenwaldes, Beobachtungen aus Brasilianischen Küstenwalden", *Z. Geomorph.*, 1935, **9**, pp. 88–98; idem, "Inselbergs und Inselberg-landschaften in Granit und Gneissgebiete Brasiliens", *Z. Geomorph.*, 1938, **10**, pp. 137–68; B.P. Ruxton, "Slopewash under Mature Primary Rainforest in northern Papua", Ch. 5 in J.N. Jennings and J.A. Mabbutt (Eds), *Landform Studies from Australia and New Guinea*, Australian National University Press, Canberra, 1967, pp. 85–94.

4. J. Corbel, "Vitesse d'Erosion", *Z. Geomorph.*, 1959, **3**, pp. 1–28.

5. See J.F. Wells, cited in W.F. Hume, *Geology of Egypt*, Government Printer, Cairo, 1925, p. 85.

6. T.O. Bosworth, *Geology of the Tertiary and Quaternary Periods in the north-west Part of Peru*, Macmillan, London, 1922.

7. G.E. Williams, "Flow Conditions and Estimated Velocities of some central Australian Stream Floods", *Aust. J. Sci.*, 1969, **31**, pp. 367–9.

8. R.J. Russell, "The Desert Rainfall Factor in Denudation", *Rep. 15th Int. Geol. Congr.*, 1936, Pt. 2, pp. 337–72.

9. G.F. Brown, "Geomorphology of western and central Saudi Arabia", *Proc. Int. Geol. Congr., Copenhagen*, 1960, Pt. 2, pp. 150–9.

10. G.E. Williams, "Glacial Age of Piedmont Alluvial Deposits in the Adelaide Area, South Australia", *Aust. J. Sci.*, 1969, **32**, p. 257; and personal communication.

11. For review of literature on and summary of morphological characteristics of pediments see B.A. Tator, "Pediment Characteristics and Terminology", *Ass. Am. Geogr. Ann.*, 1952–3, **42**, pp. 295–317; **43**, pp. 47–53.

12. See W.J. McGee, "Sheetflood Erosion", *Bull. geol. Soc. Am.*, 1897, **8**, pp. 87–112; C.R. Twidale, "Origin of the Piedmont Angle as evidenced in South Australia", *J. Geol.*, 1967, **75**, pp. 393–411.

13. G.H. Dury, "A Partial Definition of the Term *Pediment* with Field Tests in Humid-Climate areas of Southern England", *Trans. Inst. Br. Geogr.*, 1972, **57**, pp. 139–52.

14. See for instance J. Gilluly, "Physiography of the Ajo Region, Arizona", *Bull. geol. Soc. Am.*, 1937, **48**, pp. 323–48. Also C.R. Twidale, "Geomorphology of the Leichhardt-Gilbert Area, northwest Queensland", *Land Res. Ser.*, 1966, **16**, reference to the Georgetown uplands, where these are pediments cut in granite which steepen downslope. The Brachina pediment (see C.R. Twidale, "Hillslopes and Pediments in the Flinders Ranges, South Australia", in J.N. Jennings and J.A. Mabbutt (Eds), *Landform Studies from Australia and New Guinea*, Australian National University Press, Canberra, 1967, pp. 95–117) also steepens slightly at its toe.

15. See for instance C. Wahrhaftig, "Geologic Quadrangle Maps. Healy, Fairbanks, Alaska", *U.S. Geol. Surv.*, 1970. For examples of Colorado pediments see "Golden", "Louisville", "Ralston Butte", "Eldorado Springs", *U.S. Topographic map sheets*; for sections of the Lewis Terrace see D.I. Axelrod, "Mio-Pliocene Floras from west-central Nevada", *Univ. Calif. Pub. Geol.*, 1956, **33**, Fig. 2; see also C.M. Gilbert and M.W. Reynolds, "Episodic Late Cainozoic Deformation in the east Walker River area, western Nevada", *Abs. geol. Soc. Am. 67th Ann. Mtg*, (Cordilleran Section), 1971, **3**, p. 124.

16. G.H. Dury, "A Partial Definition".

17. See C.R. Twidale, "Geomorphology of the Leichhardt-Gilbert Area".

18. See H. Baulig, "Pénéplaines et Pédiplaines", *Bull. Soc. Belge d'Études Géogr.*, 1956, **25**, pp. 25–58; (see also translation by C.A. Cotton in *Bull. geol. Soc. Am.*, 1957, **68**, pp. 913–30.

19. J.F. McAllister, "Geology of the Furnace Creek Borate Area, Death Valley, Inyo County, California", *Calif. Div. Mines and Geology*, Map Sheet and Commentary 14, 1970.

20. See for instance J.T. Jutson, "The Physiography (Geomorphology) of Western Australia", *Bull. Geol. Surv. W.A.*, 1934, **95**; C.R. Keyes, "Deflative Systems of the Geographical Cycle in an Arid Climate", *Bull. geol. Soc. Am.*, 1912, **23**, pp. 537–62. J. Walther also attributed the erosion of plains to the wind but within a few years changed his mind in favour of water erosion.

21. L.C. King, "The Pediment Problem: Some Current Problems", *Geol. Mag.*, 1949, **86**, pp. 245–50.

22. A.D. Howard, "Pediment Passes and the Pediment Problem", *J. Geomorph.*, 1942, **5**, pp. 2–31, 95–136.

23. See C.R. Twidale, "Origin of the Piedmont Angle"; idem, "Hillslopes and Pediments".

24. J. Tricart, "L'Évolution des Versants", *L'Inform. Geogr.*, 1957, pp. 108–115; C.R. Twidale, "Some Problems of Slope Development", *J. geol. Soc. Aust.*, 1960, **6**, pp. 131–47.

25. R.J. Russell, "Geological Geomorphology", *Bull. geol. Soc. Am.*, 1958, **6**, p. 5.

26. L.C. King, "The Pediment Problem", pp. 245–50; see also, J.L. Rich, "Origin and Evolution of Rock Fans and Pediments", *Bull. geol. Soc. Am.*, 1935, pp. 999–1024.

27. W.A. MacFadyen, "Vegetation Patterns in the Semi-desert Plains of British Somaliland", *Geogr. J.*, 1951, **116**, pp. 199–211; J.E.G.W. Greenwood, "The Development of Vegetation Patterns in Somaliland Protectorate", *Geogr. J.*, 1957, **123**, pp. 465–73.

28. C.R. Twidale, "Landform Development in the Lake Eyre Region, Australia", *Geogr. Rev.*, 1972, **62**, pp. 40–70.

29. C.R. Twidale, "Origin of the Piedmont Angle", pp. 393–411.

30. J.A. Mabbutt, "Mantle-controlled Plantation of Pediments", *Am. J. Sci.*, 1966, **264**, pp. 78–91.

31. C.R. Twidale, "Hillslopes and Pediments" in *Landform Studies*, pp. 95–117.

32. C.R. Twidale, "Origin of the Piedmont Angle", pp. 393–411; idem, "Hillslopes and Pediments".

33. P. Birot, *Essai sur quelques Problèmes de Morphologie Générale*, Inst. Alta Cultura Centro Estudos Geogr., Lisbon, 1950, pp. 88–91.

34. B.P. Ruxton, "Weathering and Subsurface Erosion in Granite at the Piedmont Angle, Balos, Sudan", *Geol. Mag.*, 1958, **95**, pp. 353–77.

35. See also G.E. Williams, "Characteristics and Origin of a Precambrian Pediment", *J. Geol.*, 1969, **77**, pp. 183–207.

CHAPTER 15

Wind Action on Land Surfaces

A. GENERAL

Air is a fluid which behaves in much the same way as running water. However, because air is of a much lower density and viscosity than water, a flow of air (wind) is more limited in its activities than a flow of running water. In some regions, this contrast between the physical properties of the two media is compensated by the increased range of wind compared to rivers, for flows of air affect all parts of the land surface whereas flows of water are most effective in restricted linear zones: streams and rivers. The very high average and maximum velocities achieved by wind, compared to stream flows, also increase its potential as an erosive agent.

But, having pointed out the factors which make wind capable of erosion and transportation in some areas, the significance of the physical contrasts between air and water must be emphasised, for it is crucial. Quartz, a common constituent of the earth's crust and of the regolith, is only 2·65 times as heavy as an equal volume of water, but is roughly 2000 times as heavy as air. Water also has a greater viscosity than air.

Clearly then, rivers are capable of lifting and carrying much coarser debris than the wind. Thus, though rivers can roll very large blocks and boulders along their beds during floods, all but the finest gravel remains immobile in the face of even the strongest winds. Both vegetation and coarse debris disturb the airflow so that pronounced turbulence develops in the lowermost layer of air, and much of the forward motion of the air is dissipated. For these reasons, wind achieves direct geomorphological significance in relatively few environments, namely in tropical and polar deserts and in some coastal regions. In high latitudes, the prevailing cold, in addition to aridity, inhibits vegetation growth. In the hot deserts slow rates of weathering assist in the preservation of protective mantles of gravel. In coastal areas, strong winds, salt spray and shifting sands all limit plant growth.

Wind influences the land surface to a minor extent in all areas, even those that are well covered by vegetation (for example during storms or during ploughing) but long-lived features due to aeolian action are observed only in the arid tropics and subtropics and in coastal areas. And only in the arid low latitude areas is the direct work of wind significant and extensive enough to be considered a major geomorphological agent imposing its characteristic imprint on the land surface.

B. THE WORK OF WIND IN THE ARID TROPICS

1. General Remarks

The nature of the land surface in the tropical and subtropical deserts varies greatly from place to place. The uplands are readily explained in structural terms, for there are fault-block mountains (as in the southwestern USA where the enclosed elongated downfaulted valleys are known as *bolsons*), dissected fold mountains (MacDonnell Ranges of central Australia), plateaux (Tibesti highlands of northern Africa and Hamersley Ranges of Western Australia), and volcanic forms (as in the Tassili of the central Sahara), similar to their humid land counterparts. But the desert plains are not of uniform character. In some areas, as in the riverine plains of Iraq and Iran and in many parts of the Lake Eyre and Lake Torrens basins of Australia, there are vast spreads of clays. Elsewhere, for instance in Sturts Stony Desert, Australia, and the Tanezrouft of the Sahara, a mantle of stones forms a carpet over the plain surface. Very large areas of the tropical arid lands are occupied by sands, but the forms moulded from them vary greatly: fields of linear sand ridges here, crescentic dunes there, rather irregular and amorphous accumulations in some areas, and yet elsewhere quite flat sand surfaces. Why does the nature of the desert plains vary so markedly from place to place? Why is sand deposited in some areas and not in others? Why do sand accumulations vary in shape from one place to another?

The nature of the desert plains, the distribution of the clay, stony and sandy deserts are not entirely related to present processes. These are determined by gross structure and drainage, both past and present. For instance, in central Australia the clay plains are located in the lower reaches of the Lake Eyre drainage basin, which is in large measure a tectonic feature with the downfaulted Lake Eyre salina (Pl. 15.1) as its focus, and with its margins in many areas determined by upwarped ridges and upfaulted blocks.[1] The rivers flowing to or toward Lake Eyre drain 1·3 million km[2] of central and north-eastern Australia and they bring to the basin sand winnowed, transported northward and moulded into dunes by the wind (Fig. 15.1). Again, an ancient lake, Lake Cahuilla, the precursor of the Salton Sea (Fig. 15.2), was the receptacle in which were deposited the sediments from which the Algodones Dunes of southern California were possibly derived.[2] Gibber plains are well developed marginal to many desert uplands, but particularly adjacent to those in which sandstone and limestone are prominent. The nature of any outcrop obviously influences the nature of the regolith weathered from them and hence the nature of the desert surface. Thus flaggy sandstone tends to break down to plate-like fragments and to produce a desert pavement (Pl. 15.2a); orthogonally jointed sandstone or quartzite produce blocky fragments (Pl. 15.2a) as does silcrete (Pl. 15.2b), but many rocks are reduced to sand which can be shaped into dunes.

The prime reason for variations in the nature of desert plains, then, is the source and distribution of debris. But the wind itself, as is shown both by field observations and by laboratory experiments, is capable of sorting the debris exposed at the surface.

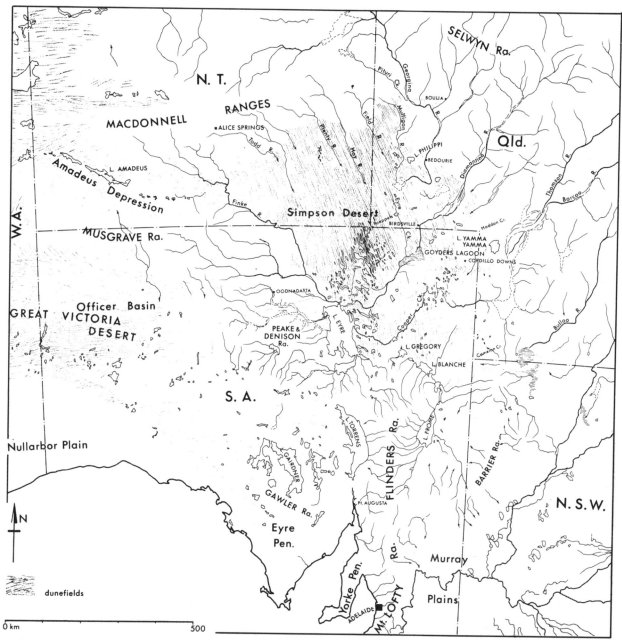

Fig. 15.1   The Lake Eyre catchment, showing principal salinas, rivers and dune fields.

Plate 15.1 The northwest corner of Lake Eyre showing fault or fault-line features. *(RAAF.)*

Plate 15.2a Platey gibber or pavement derived from flaggy sandstone, southern Arcoona Plateau, South Australia. (*C.R. Twidale.*)

Plate 15.2b Silcrete gibber near Bedourie, western Queensland: clay accumulation in valley floors —hence vegetation cover. (*C.R. Twidale.*)

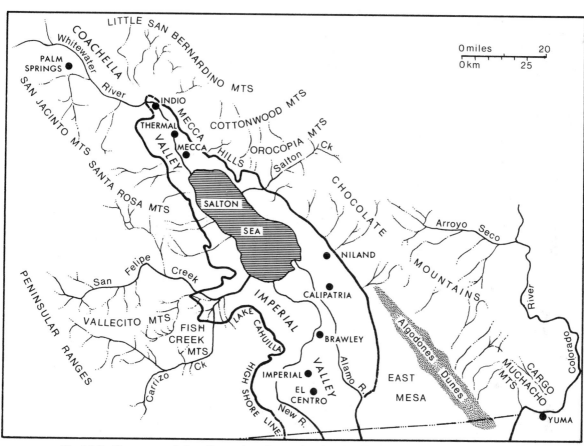

Fig. 15.2   The present Salton Sea and environs, showing the Algodones dune field and the high shore line of the Pleistocene Lake Cahuilla. *(After R.M. Norris and K.S. Norris, "Algodones Dunes", pp. 605–20, by permission of the Geological Society of America.)*

2. Particles in Motion[3] : the Sorting Action of the Wind

Though clay particles are small and of low mass, they are not readily lifted by the wind both because they are plate-shaped and thus tend to pack down well to form a smooth surface, and because many clays are bonded by ionic charges. But in some desert areas, salt crystallisation disturbs the packing of the clay particles; elsewhere, as in the Gobi Desert, frost action achieves the same effect. Initially many clay particles are lifted from the surface and into the air by turbulent eddies, which develop as a result of the passage of the air over roughened surfaces; some are lifted by distinct vortices (Pl. 15.3) or minor eddies called dust-devils. Sand is moved in the same way and because it is more susceptible to direct observation at the scale of individual particles, it has been studied in detail in the laboratory.[4]

Each sand grain exposed at the land surface lies in a zone of low velocity caused by the frictional drag due to the sand

grains themselves, pebbles, vegetation and other obstructions to the air-flow near the ground. However there is a rapid increase in velocity a short distance above the surface,[5] after which it remains essentially steady (Fig. 15.3). The precise nature of the velocity gradient above the ground varies with the character of the surface (Fig. 15.3). Turbulent eddies, which involve a dissipation of energy, are formed partly because of the roughness of the surface which induces pressure contrasts between the upwind and lee side of any obstacle, and partly because of the large temperature differences commonly developed between the air and the sand surface in hot desert regions (temperatures of up to 84°C have been measured on sand surfaces in Africa). It might be thought that an increase in wind velocity would diminish or even eliminate this zone of near-surface turbulence and low velocity, but in reality the converse is true: a general increase in wind velocity causes an increased velocity gradient just above the ground, because turbulence is increased as clay or sand grains become agitated, and the wind velocity near the

Plate 15.3 Dust devil in Nebraska. *(R.L. Heathcote.)*

surface (that is, closer than about 0·3 cm) actually decreases.

Yet this near-surface turbulence which poses the main problem to transport is also the means whereby clay and sand particles are put in motion. Sand grains are lifted out of this low near-ground velocity zone and follow short parabolic paths downwind. The initial lift is provided by numerous turbulent updrafts of air, which provide an upward suction. The height to which the grains are lifted varies according to their mass and to the force of the updraft. The larger vortices are capable of lifting not only dust, but also leaves, twigs and other debris high into the air. They are probably responsible for lifting the small animals which are occasionally reported as falling from the sky in the tropical deserts.

The wind carries clay particles high into the air (Fig. 15.4). Winds of even moderate velocity cause a dust haze to develop because fine debris (less than 0·08 mm diameter) is carried in suspension in the air. The so-called sandstorms of the desert region are, in reality, dust storms. The removal or evacuation of solid particles by the wind is called *deflation*, which is, in quantitative terms, a very important aspect of desert erosion. It has been estimated that in the western USA alone almost one thousand tonnes (850 tons) of dust are carried on average over 2400 km each year. In a great storm in 1895, 15–40 kgm of dust per hectare (4–10 tons per acre) were deposited in Indiana alone. Dust from the Sahara was deposited on England in July, 1968.[6] Dust from the Australian deserts is occasionally carried across the Tasman Sea to New Zealand.

Fine silt and dust from the Gobi Desert of central Asia are carried southeastwards to northwestern China, where they form a widespread deposit, 140–300 m thick. This deposit blankets the pre-existing irregular land surface and forms a high plain, which, however, has now been dissected by the Hwang Ho and its tributaries to form the Ordos Plateau and related features. Such deposits of wind-blown dust are called *loess*. The reason for the great volume of loess derived from the Gobi Desert is that the region, though very hot in summer, suffers low temperatures in winter so that frost heaving and similar activities disturb the packing of the silts and clays and thus render the desert surface vulnerable to wind erosion.

Some wind-blown dust is deposited within the deserts themselves and, if concentrated by rivers, gives rise to fertile soils such as those exploited in the Negev by the Nabateans some 2000 years ago. Such loess derived from the tropical deserts is sometimes called *hot loess* to distinguish it from that derived from glacial outwash (known as *cold loess*) such as that deposited over a pre-existing landscape in the Mississippi valley.[7]

Sand grains 0·15–0·3 mm diameter are too large to be carried in suspension like dust, but they are picked up when the wind attains a critical velocity and turbulence and, once above the near-surface layer of low velocity, they are carried forward in the general air-flow. The critical or drag velocity at which sand grains are just put in motion is expressed by:

$$Vcr = A\sqrt{\frac{\sigma - Q}{Q}gd} \tag{16.1}$$

where Vcr is the critical drag velocity, A a constant approximately equal to 0·1, σ the density of the sand, Q the density of the air, g the acceleration due to gravity and d the diameter of the grains involved. This equation holds for particles

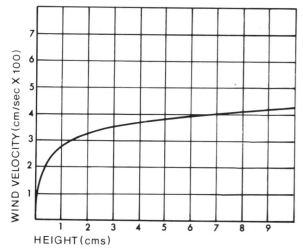

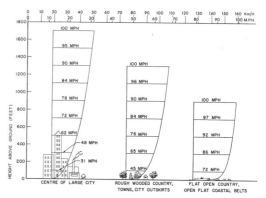

Fig. 15.3 Velocity gradients above the ground (a) general gradient determined in the laboratory (*after R.A. Bagnold, "Transport of Sand", pp. 409–38.*) and (b) over terrains with different roughness characteristics. (*After A.G. Davenport, "Wind Loads as Structures"*, Nat. Res. Counc. Canada, Dev. Build. Res. Tech. Paper, *1960, 58.*)

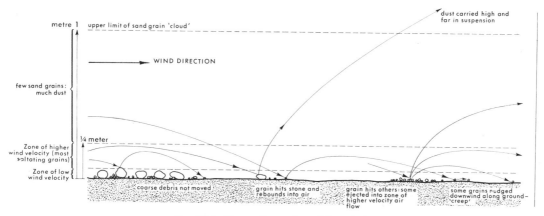

Fig. 15.4 The sorting action of the wind.

0·2 mm diameter and greater.[19] It applies also to particles of smaller size, but as the diameter decreases, the value of the constant has to be increased.

Sand grains are rarely lifted more than 1 m above the ground and once raised by turbulence are immediately subject to a number of forces: the forward momentum of the air-flow, the resistance of the air to the descent of the particle, the pull of gravity, and local eddies and currents. But gravity inevitably and rather quickly triumphs, and the particle descends to the surface. Thus each particle, once uplifted, begins a descent to the surface. Its forward velocity is close to that of the wind, Vw, and its downward speed to that of the settling velocity, Vs. Hence the angle at which it strikes the ground is given by:

$$\tan = \frac{Vs}{Vw} \qquad (16.2)$$

and in practice is usually a little less than $15°$ to the horizontal (Fig. 15.5).

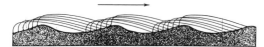

Fig. 15.5 Travel paths or parabolas of saltating sand grains.

When a particle hits the surface, it may ricochet off and proceed on another leap if it hits a rock outcrop. Quartz is an elastic medium and readily bounces. On the other hand, a particle may descend to a surface composed of similar grains. Some grains are ejected into the air, possibly out of the zone of low velocity and into the general forward air-flow. Thus sand grains progress downwind spasmodically in leaps which take the form of asymmetrical parabolas (Fig. 15.5) and which may be up to 8 m long. This process is known as *saltation*.*

Some sand grains, however, are not lifted but are merely nudged forward by the forward motion of the descending grain. Such *surface creep*, or reptation, of sand grains accounts for perhaps 25 per cent of the forward flow of sand. But the major part—75 per cent or more—of the forward movement of sand grains is accomplished by means of saltation. Because of the explosive effect of descending sand grains, saltation is self-perpetuating and self-augmenting unless and until wind speed drops below the critical drag velocity.

The velocity of wind required to move stationary sand, the *static threshold*, varies according to the size of the sand grains exposed at the surface, the turbulence of the wind near the ground, and with the length of the exposed surface (a factor comparable to the fetch of waves—see Chapter 18 below). It has been noted that the initial sand movement always occurs at the downwind end of an exposed sand surface and that the grains are put in motion further and further upwind as the wind velocity increases. In general, however, it may be stated that a wind of 16–20 km/hr is needed to put sand grains into motion.

* Latin: *saltare*—to jump

Table 15.1

Effect of stone cover and distance from source of eroding material on soil accumulation.*

Distance from source in inches	Stoniness (stones/ft²)				
	g soil/ft²				
	0	5	10	15	20
30	0·013	0·94	6·41	45·27	35·38
85	0·013	0·73	5·91	5·78	4·82
130	0·032	2·42	7·90	3·38	2·72
180	0·020	1·24	4·14	1·99	1·79
Mean	0·020	1·33	6·09	14·11	11·18
% of soil blown	0·011	7·10	32·40	75·00	59·50

Difference for significance between means: 5 % = 2·57; 1 % = 3·48; 0·1 % + = 4·52.

After Pandastico and Ashaye,[9] p. 386.

Thus, by means of saltation and creep, sand is sifted from regolith covers or from spreads of alluvial deposits and is moved downwind.

However, larger particles cannot be shifted by the wind.* They remain *in situ* and become more and more concentrated as the finer debris is winnowed out. The coarse debris—gravel, larger blocks and boulders—forms a protective carapace to the land surface (Pl. 15.2). Such surface layers of stones are known by various names in different parts of the world: *croûtes désertiques*, desert armour, desert pavement (where the stones are flat), gibber, *reg, hamada* and *serir*,[8] though with respect to the last two, there is a tendency to differentiate between finer serir and coarser hamada.

Running water also helps in the formation of gibber by washing the finer particles from the surface debris, and it is able to evacuate particles of greater calibre than the wind can shift. In debris of mixed calibre, but which contains large volumes of clays (especially hydrophilic clays), wetting and drying also assist materially in the concentration of stones at the surface. In road-cuttings and gully sections it can be seen that the stones are a surface occurrence; coarse fragments are rare in the underlying clay mass. Yet the superficial cover of stones does not have the appearance of an alluvial deposit laid down over the clays. It has been suggested, and there is observational evidence of various stages in these developments, that soil churning in effect sorts the soil particles, not only in arid lands but anywhere where there are marked wet and dry seasons (see Chapter 7). In a mixture of gravel and clays the latter expands when wetted. The

* Note however the possibility, under special conditions, of the wind shifting the very large boulders on the surface of some playas (for example, the Racetrack Playa in southern California). See Chapter 10.

whole mass including the stones is forced upwards. When the dry season ensues, the clays shrink and crack and fines are thus able to fall back to zones below the surface. However, the gravel is too coarse: it becomes wedged in the crack, and is thus not only prevented from returning to depth, but blocks the way for other stones. With several cycles of wetting and drying, the stony fraction of a mass of debris, whether alluvial or weathered, becomes concentrated at the surface. So the gibber mantle which is such a widespread type of desert surface can originate in several ways.

As the concentration of stones at the surface increases, the work accomplished by the wind may change, according to laboratory work carried out by Pandastico and Ashaye. [9] With no stones protecting the surface and disturbing the air-flow, virtually none of the soil carried by the wind was deposited, at least within 460 cm (180 in) of the source area; but with 15 stones (average diameter 3·8 cm and not affected by the wind) to the square foot (1 to 62 cm²), no less than 75 per cent of the blown particles were deposited. Rather surprisingly, with a concentration of 20 stones to the square foot (1 stone to 47 cm²) the percentage of mobile debris deposited fell to just under 60 per cent (Table 15.1), possibly because of the difficult and comparatively slow rate of erosion beneath the cover. Another interesting aspect of the experiment was that as the stoniness of the surface increased, the clay content of the surface layer decreased (Fig. 15.6). This may have

been because increased disturbance of air-flow and decreased saltation caused larger amounts of sand to be trapped, though there was some suggestion that, even under these conditions, clay particles could be sucked into the air-flow.

However, the effect of the gibber mantle is clearly not the same everywhere and at all times: depending on its concentration at a given time and place, it may be protective or induce the deposition of debris, or it may permit the removal of fines and thus itself become more concentrated.

3. Erosion by Wind

Though wind is most important in the tropical deserts for the transportation and deposition it achieves, some erosion also occurs.[10] The most important erosional effects pass virtually unnoticed, for the lowering of the land surface due to the aeolian transport of fine particles is a gradual process. Yet deflation is frequent, and in the long term must effect a considerable wearing away of the land surface.

The most spectacular effects of erosion by the wind are carried out by sand blasting. As already described, the wind is able to carry sand grains in saltation close to the ground surface. The passage of wind laden with sand grains achieves minor polishing and abrasion. In particular, the sides of stones are polished and faceted by sand-laden wind being diverted around such minor obstacles, and eventually stones of triangular cross-section, called *dreikanter* (German for "three sides") are formed. The general name for such a wind-faceted stone or pebble is *ventifact*. Of slightly greater magnitude are the mushroom rocks or *zeugen* formed by the sand blasting of rocks, especially weaker strata, which are exposed near the ground level where sand blast action is pronounced. Great whalebacked and streamlined ridges elongated parallel to the dominant wind direction occur in a few areas, notably the American Southwest,[11] central Asia and Iran. These are called *yardangs*[12] and although they vary in size (some in Iran are up to 200 m high),[13] they all occur in relatively soft outcrops, particularly in exposed unconsolidated sediments. They are attributed to wind scouring in parallel zones and, once initiated, the upstanding ridges tend to be undercut (Fig. 15.7). Similar wind-eroded ridges (developed in Pleistocene lake clays, and called *wind-rift* dunes (Fig. 15.8)) as they are veneered by sand, occur near the southeastern extremity of Lake Eyre.[14]

Most wind-eroded forms are found in weak strata but both Hume and Russell[15] have described features due to sand blasting in resistant rocks. Hume has described fluted limestone pavements and vertical cliffs from Egypt, and bearing in mind the nature of the country rock, the possibility of solutional effects cannot be overlooked. But the grooves and furrows on the vertical walls run parallel to the ground surface and not vertically as would be the case if solution were responsible. The forms illustrated (see Hume's Pls XXVII and XXIX and pp. 65–73) are identical to those described by Russell from near Palm Springs, in the southern Mojave Desert, California, where there is incontrovertible evidence of aeolian fluting in fresh, unweathered gneisses and schists. There, the well-named Windy Point is a col in an east-west ridge which flanks a graben of similar trend. Fixed sand dunes and the protective metal shields placed on wooden telegraph posts indicate that strong westerly winds blow in the narrow

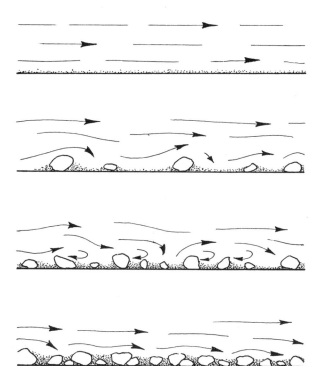

Fig. 15.6 Model showing increased small particle accumulation and clay content with increasing number of stone obstacles. *(After E.B. Pandastico and T.I. Ashaye, Experimental Pedology.)*

Fig. 15.7 Sketch of wind-fluted and basally-undercut yardangs in unresistant lacustrine deposits, at Rogers Dry Lake in the Mojave Desert, California. (*Drawn from photograph by Eliot Blackwelder reproduced in C.A. Cotton, Climatic* Accidents, *Whitcomb and Tombs, Christchurch, 1947.*)

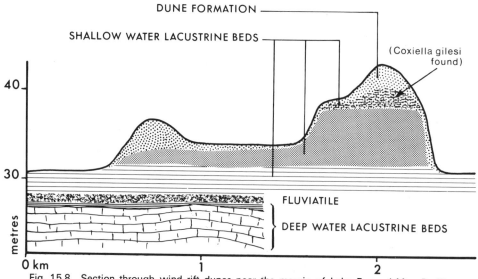

Fig. 15.8 Section through wind-rift dunes near the margin of Lake Eyre. (*After D. King, "Quaternary Stratigraphic Record at Lake Eyre", pp. 93–103,* by permission of the Royal Society of South Australia.)

Plate 15.4a Fluted gneiss blocks at Windy Point, Southern California. Note protective metal collars on wooden posts. *(C.R. Twidale.)*

elongate valley and the moulding of sand accumulations on the lower slopes of the ridge below the col indicate that these winds are further funnelled in a southeasterly direction through the col. A variety of evidence points to strong sand blasting in and just below the col. The surfaces of the gneiss and schist outcrops are polished, fluted and smoothed (Pl. 15.4a and b), to give an effect rather similar to that displayed on the beds of some rivers (Pl. 12.4a); but this site is well above any stream channels. Bushes and shrubs in the col are shaped and truncated in patterns consistent with sand blasting and erect telegraph posts as well as some which are lying on the slope display pitting on those surfaces exposed to the north-westerly winds.

Some alleged examples of wind erosion do not stand up to close scrutiny or are at least capable of alternative explanation. For example in Death Valley, California, a few

kilometres south of Furnace Creek, there is a basalt block shaped like an irregular hour-glass (Fig. 15.9) and advertised as due to wind erosion. The site is open, with no strong funnelling of winds, and despite the abundance of sand it is doubtful whether it is moved about very much. Moreover, where the bases of this and other nearby basalt blocks are exposed, it can be seen that the naturally subsurface part of the residual displays fretting and weathering which is characteristic of attack by moisture, and it is likely that the shaping of the hour-glass form is attributable to subsurface moisture weathering at a time when the land surface was at or close to the top of the residual.

In any case these are minor features. The only widespread and major landform due to wind erosion, however, is the deflation hollow. There are many claypans and salinas, large and small, in the tropical deserts. There is no doubt that most

Plate 15.4b Detail of fluted gneiss at Windy Point. *(C.R. Twidale.)*

Fig. 15.9 Field sketch of hour-glass-shaped basalt residual in Death Valley, California. Allegedly shaped by sand blasting, it may have been caused by subsurface moisture weathering.

of the larger depressions are, in whole or in part, due to tectonism (for instance, Lake Eyre, Lake Torrens, Lake Frome, in Fig. 15.10). Elsewhere, they can be attributed to solution of the underlying calcarous bedrock. But some, including the great Kharga Depression of the Libyan Desert, and the thousands of small depressions in the interdune corridors of the Simpson (Fig. 15.1) and other sand ridge deserts, can be due only to wind scouring of unprotected alluvial surfaces. After rains and floods, large volumes of sediment are deposited in the desert basins. Water spreading between the dunes deposits alluvium in small local basins of interior drainage. When dry, these surfaces which are unprotected by vegetation are vulnerable to wind scouring and small deflation hollows are formed which become the foci for later drainage and deposition. These later deposits are again deflated and so the hollow extends laterally. However, the depth to which the wind can erode is effectively controlled by the level of the water-table, for the wind cannot pick up moist sediments, and so erosion ceases as the water-table is approached.

Localised deflation is also responsible for the differential lowering of sand sheets, resulting in the formation of numerous hollows or *fulje* enclosed by residual sand ridges, and together forming what has been called *alveolar* relief. Such features are fairly widespread in parts of the Soviet desert where they are, in part at any rate, attributed to powerful vertical updrafts related to strong surface heating;[16] and in northern Eyre Peninsula, in South Australia, again apparently in rather better watered areas. Erosional sand features due to the action of the wind may be more common than is generally thought, for the residual ridges can at first glance be confused with sand dunes which are due to aeolian deposition.

C. AEOLIAN DEPOSITION

1. General

In addition to the long distance transport of fine silt and clay mentioned in connection with deflation, the wind is capable of more local movement, mostly of sand although clay, salt and gypsum are also involved in some areas. Sand put in motion by the turbulent wind is, after a brief period of mobility, deposited before again being picked up by wind. Wind transported sand grains are characteristically rounded and pitted. They are well size-sorted, and fragments of mica, which are easily broken up by attrition, are absent from aeolian deposits.

Sand accumulations are widespread in the tropical deserts and have also been described from the semidesert regions. In some areas these sand deposits form featureless *sheets*. For example, the Selima sand-sheet of the Sudan Egypt border region consists of a veneer of sand only 50 cm thick overlying the bedrock. The plain which it mantles is remarkably smooth and has a gradient of only 50–60 cm per km. Its origin is unknown.

More commonly the sands are moulded to give some relief, either through preferential deposition of sediment, or through differential erosion of an existing sand-sheet. In many areas the surface form of the sand-sheet is irregular, as for example over much of the Kalahari Desert of southwestern Africa. There are rather vague swells and intervening depressions, but no apparent regularly repeated pattern such as characterises the *dune field* which is the landform assemblage unique to the tropical deserts. Bagnold has written of these fields of sand dunes:

> ...instead of finding chaos and disorder, the observer never fails to be amazed at the simplicity of form, an exactitude of repetition, and a geometric order unknown in nature on a scale larger than that of crystalline structure.[17]

But not all parts of the tropical deserts display dunes, and some actively forming dunes have been observed outside the tropics and the really arid lands. Moreover, the patterns of the dunes vary greatly from one region to another, and some dunes are demonstrably mobile, while others are just as clearly fixed. What are the reasons for these variations?

2. Mobile Dunes

Fields of sand dunes are the characteristic feature of the tropical deserts, though they do not extend over all the desert surface and other forms are also well developed.

Such dune fields are composed of sand ridges which are of varied form but all of which move under the influence of the prevailing wind or winds (Pl. 15.5a-d). Sand ridges up to 150 m high cover large areas of the deserts and crescentic dunes are also well known. Reticulate patterns are also widespread and in the Libyan Desert transverse and longitudinal dune fields are superimposed and intertwined, giving box-like patterns (Pl. 15.5). In parts of Arabia and the Sahara great dune massifs (again up to 150 m high and of varied morphology), are prominent. And there are many gradations between, and variations of, these basic types.

Mobile dunes occur in areas where vegetation is scarce or absent—in other words, in areas where there is minimal

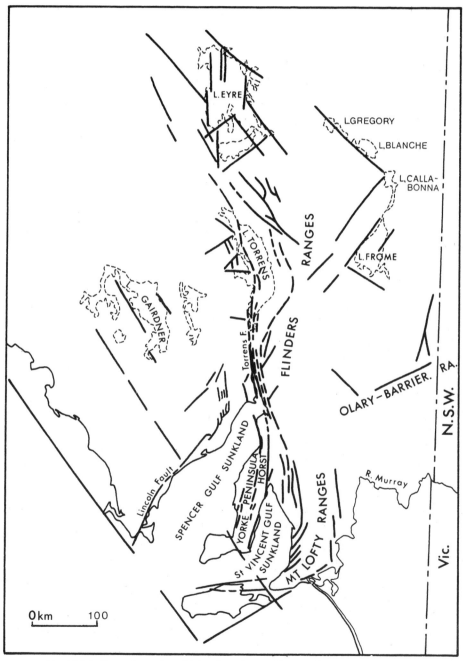

Fig. 15.10   Tectonic setting of lakes Eyre, Frome, Torrens, Blanche and Gregory.

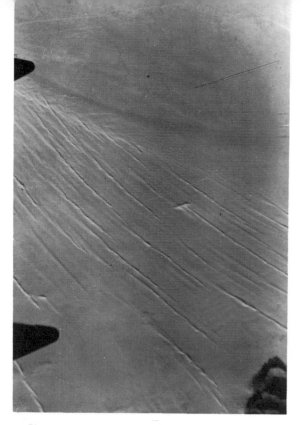

Plate 15.5a Parallel sand ridges or longitudinal dunes in the Libyan Desert. The dominant wind direction is from top left to bottom right. *(C.R. Twidale.)*

Plate 15.5b Complex sand ridges in the Libyan Desert, with large transverse ripples and arcuate transverse ridges in the intervening corridors. *(C.R. Twidale.)*

Plate 15.5c Part of the Libyan Desert showing complex longitudinal ridges at bottom left, dome dunes in the middle distance, and an alluvial plain at top right. *(C.R. Twidale.)*

Plate 15.5d Box-like pattern of transverse and longitudinal dunes in the Libyan Desert. The sand ridges running from top left to bottom right are of longitudinal character, those at right angles to them are of transverse type. *(C.R. Twidale.)*

interference with the air-flow near the ground. There also has to be a source of sand, and this varies from place to place. In some areas the weathering of local bedrock provides the necessary raw material from which dune ridges are moulded. For example, the Stormberg Sandstone is said to be the source of the Kalahari sands, though there is some suggestion that the material may have been through more than one cycle of transportation and deposition before being moulded by the wind. In many areas it is certainly apparent that the dune sands have had a complex history. In many of the American deserts, debris was carried to the local bolsons by streams, both past and present; the wind has winnowed out the sand from the rather ill-sorted fluvial deposits and concentrated it in the form of dunes at the northern (lee) extremities of such meridional bolsons as the Panamint Valley, California. In some areas, former lake basins are the source from which sand is derived for the construction of dune ridges. As mentioned previously, in southern California the former Lake Cahuilla is the source of the Algodones dunes. In central Australia, debris was and is brought to the arid centre by a huge river system and was deposited in broad alluviated valleys and numerous shallow lakes. During marked aridity, as at present and in the recent past, the wind sorted the sand from these riverine and lacustrine spreads and formed the dunes which are so characteristic of the area and which, in many areas, can be demonstrated to overlie the alluvia (Fig. 15.11). Modern rivers continue to supply sand and other debris, so that dune formation can still be observed and the cycling of debris continues; but the bulk of the sand of which the dunes are built was transported to the Lake Eyre depression some thousands of years ago, during late Cainozoic pluvials. [18] The dominant southerly winds have spread the debris northward from the major area of debris accumulation around Lake Eyre into the Simpson Desert. When strong winds are blowing, the sand of which the dunes are built can be seen to be in motion. The aeolian origin of the dunes is confirmed not only by their internal structures, notably cross-bedding, but also by the absence of any other feasible agency. They cannot be, for example, beach ridges or giant ripples due to wave action, as Sturt [19] surmised, for the sea has been absent from central Australia since the Cretaceous whereas the

dunes of that region have demonstrably formed during the Recent, and indeed continue to form at the present time.

As aeolian features, dunes should be aligned in a pattern which bears some geometrical relationship with the local wind regime. Comparison between dune patterns and wind roses suggests three major types of dune:

1. *transverse dunes:* oriented normal to the dominant wind direction, that is, to the most common wind which, in the environmental conditions that obtain, is capable of moving the particles exposed at the land surface;

2. *longitudinal dunes:* aligned parallel or subparallel to the dominant winds. In reality, this is commonly a vector or resultant which reflects both the frequency and strength of two or more non-opposed winds. Discrepancies between such mean winds and dune trend are characteristic [20] though this may reflect the paucity of wind data from the desert proper, rather than any physical process.

   Strong winds which blow obliquely to the dominant winds result in the development of lateral bars attached to the main ridge, and opposed strong winds result in a decrease in the rate of sand advance downwind under the influence of the main force;

3. *nonaligned dunes:* nonlinear and consist merely of mobile mounds activated by several strong winds from opposed directions.

Transverse dunes consist of chains of dunes with wavy crests, a gentle upwind slope and a steeper lee slope which is, in fact, an avalanche slope. Some of the transverse chains are continuous over long distances. In some areas they form a series of distinct, but still linked, crescentic forms (*mréyé*—see Pl. 15.5). Elsewhere, and apparently in areas of sand shortage, the ridge disintegrates into a series of isolated crescentic dunes—the well-known *barchans* (Pl. 15.6a–c)—the horns of which extend downwind. Barchans also develop and exist in isolation.

The gentler slopes of both elongate transverse dunes and barchans display sand ripples which are caused by the passage of air over the sand surface, and which are, in other words, wave forms developed at the interface between the two media. The ripples move downwind, coarse sand being concentrated

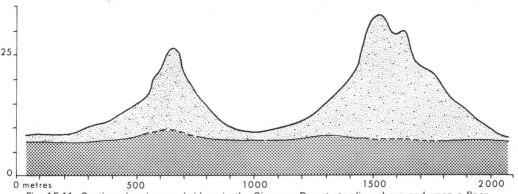

Fig. 15.11 Section showing sand ridges in the Simpson Desert standing above and upon a floor of alluvium. (*After J. A. Mabbutt and Margaret E. Sullivan, "The Formation of Longitudinal Sand Dunes: Evidence from the Simpson Desert", pp. 483–7, by permission of J. A. Mabbutt and Geogr. Soc. N.S.W.*)

Plate 15.6a   Isolated barchan in the coastal Namib Desert, South West Africa.   (*W.S. Barnard.*)

Plate 15.6b   Barchans west of the Salton Sea, California.   (*J.S. Shelton.*)

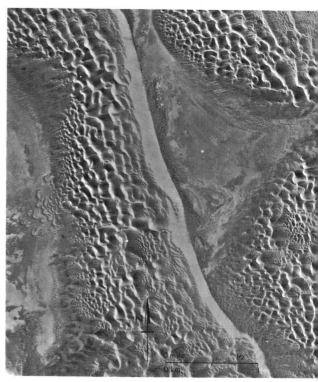

Plate 15.6c   Isolated barchans and transverse dunes, some with complex minor ridges superimposed on major dunes, or draas, are displayed in this vertical air photograph of part of the Arabian Desert. (*Hunting Surveys.*)

on their crests and finer material in the intervening swales or depressions. Very large ripples, up to 60 cm high and with a wave length of 20 m or more, have been observed in the Libyan Desert[21] and it has long been known from laboratory work that the spacing of sand ripples in considerable measure reflects wind speed and the calibre of the sand available,[22] though saltation distance governs the spacing of smaller ripples. Whether such factors also influence the development of transverse dunes is not clear, but it could explain why sand is preferentially deposited in some areas and not in others nearby.

In addition to ripples and dunes, very large transverse mounds called *draa* (Pl. 15.6c) have also been observed in the sandy deserts. Wilson   has indeed suggested that these three, ripples, dunes and draas, represent three prominent wave lengths of desert sand waves representing stages in the growth of such waves. He attributes their development to wind regime: long continued strong winds produce sets of large ridges, and if the strong winds are succeeded by gentler ones then minor waves are superimposed upon the larger, though if they are prolonged the larger ridges eventually disappear and are succeeded by a field of smaller ridges the wave length of which primarily reflects wind velocity.

Transverse dunes, which extend over large areas of the Soviet and of the North African deserts, move downwind as sand is driven up the gentler windward slope of the crest, then tumbling down the slip or avalanche face to come to rest. The angle of inclination of the entire avalanche slope reflects the shape and size of the sand grains, but is never steeper than 33 .

Where there is an unusually heavy accession of sand, definite flows develop on the avalanche face. It is commonly asserted [24] that a reverse eddy develops in the lee of barchans and other transverse dunes (Fig. 15.12) and this is confirmed by work in the wind tunnel. But it is not supported by all observational evidence; on the contrary complex and variable air movements are reported in these situations.[25]

Because of these two contrasted movements on the windward and the lee slopes of transverse dunes, quite distinct depositional layers are formed within the sand accumulation, though the lee slope (avalanche slope) structures are eliminated because of the overall downwind movement of the sand mass. Hence the bedding of transverse dunes, including barchans, is dominated by gentle upwind dips.   In barchan deposits there is a considerable local variation in the strike (Pl. 15.7) of the bedding due to the curvilinear shape of the strata deposited as sand is blown around the main sand mass and into the horns, which point downwind.

Barchans march across the desert plains at considerable rates (Fig. 15.13a and b). Such movements have been plotted, for instance in the Peruvian Desert and in southern California.[27] According to Beadnell [28] downwind movement of barchans averages 17 m per annum near the Kharga Oasis in Egypt. During a seven-year period of observation (1956–63) the movement of forty-seven barchans on the western side of the Salton Sea in southern California averaged 26–27 m per annum, ranging between 15 and 43 m.[29]

There has been some suggestion that the dunes grow during movement until they attain a maximum size, after which they remain stationary until reshaped into new forms.[30] However, in some areas at least, it appears that the rate of movement varies in part according to the supply of sand. For example, in the fifteen years up to 1956 the dunes near the Salton Sea moved only 16 m downwind per annum, compared to an average of 26–27 m for the next eight years when there was an increased supply of sand.[31]

But work in both the American Southwest and Peru [32] confirms Beadnell's conclusions that the rate of downwind movement depends considerably on the height of the dune, and attempts have been made to relate height of slip face and rate of movement in a mathematical expression. Finkel[33] plotted the reciprocal of D (distance travelled in unit time) against the height of the slip face H. But a considerable number of plotted points lay well outside the resulting curves, possibly because of factors such as sand supply, vegetation and the influence of local topography and winds on sand movement.

Barchans are characteristic of both the North and South American desert regions, and are also commonplace in the USSR, North Africa and Arabia. Giant barchans, with many small barchans formed on the long upwind slope, have been described from both Arabia and the USSR. Though barchans occur in the coastal desert or Namib of Southwest Africa (Pl. 15.6) they are absent from the Kalahari interior.

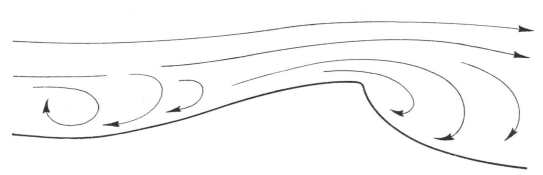

Fig. 15.12 Section showing wind circulation over a transverse or barchan dune.

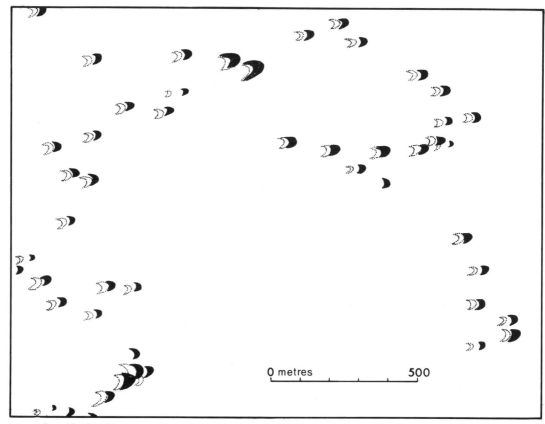

Fig. 15.13a Movement of barchans in Peru in a three-year period. (*After H.J. Finkel, "Barchans of southern Peru", pp. 614–47, by permission of J. Geol. and the University of Chicago Press.*)

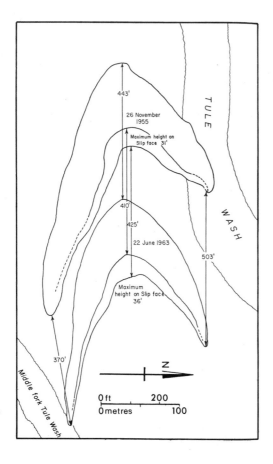

Fig. 15.13b Movement of barchan over eight-year period in Mojave Desert. (*After R.M. Norris, "Barchan Dunes of Imperial Valley", pp. 292–306, by permission of J. Geol. and the University of Chicago Press.*)

Plate 15.7 Cross-bedding exposed in eroded barchan, Kelso Dunes, Mojave Desert, California. (*R.P. Sharp.*)

Fig. 15.14 Crescentic coastal dunes near Port Hedland, W.A. (*Drawn from photograph provided by M. Pitt.*)

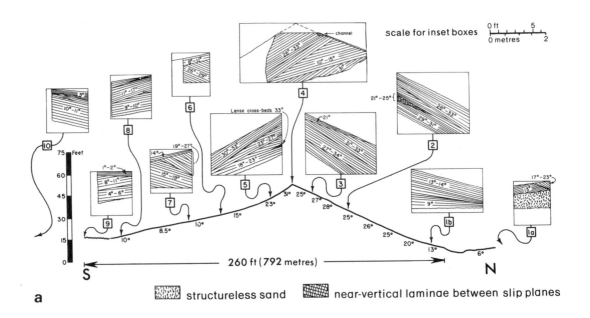

a

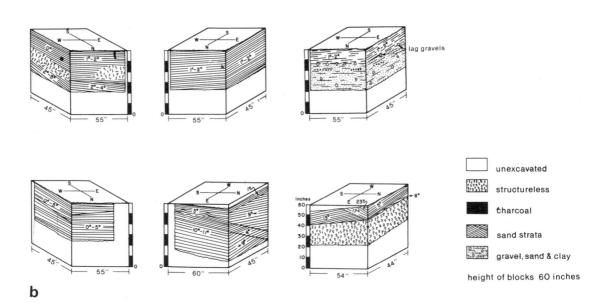

Fig. 15.15 (a) Aeolian cross-bedding in a longitudinal or seif dune in the Libyan desert, and (b) aeolian bedding in interdune corridor in the same area. (*After E.D. McKee and G.C. Tibbitts, "Primary Structures of a Seif Dune", pp. 5–17, by permission of J. Sed. Petrol.*)

b

In Australia they are absent or very scarce; they occur in coastal areas of Western Australia (Fig. 15.14), and near Eucla on the Nullarbor Plain but how durable and mobile they are is not clear. Crescentic forms have been observed by the writer in southwest Queensland, but they are probably stationary.

In general, this distribution appears to reflect wind regime, for barchans and transverse dunes occur in areas dominated by a single strong wind direction—unidirectional wind regime. In Peru opposed winds blow, but the only effect of the subordinate wind is to retard the rate of dune advance. In the southwestern USA, the local relief characterised by north-south trending ridges causes distinct funnelling of the wind.

Longitudinal dunes are elongated sand ridges (Pls 15.5a, 15.8a-c) which in some instances extend unbroken for scores, or even a few hundred kilometres. It has been suggested [34] that longitudinal dunes indicate sparse sand supply, whereas transverse ridges evolve where sand is abundant. However, examination of wind roses constructed from data taken at nearby recording stations suggests that such dunes form under the influence of a bidirectional wind regime, that is, two strong winds blowing from roughly the same quarter, or at least the same hemisphere. Though the relationship of the trend of these sand ridges to wind regime is very complex, [35] the essential, if general, validity of this suggestion is confirmed first by detailed observations of the same dunes at short intervals, revealing that the slip slope changes orientation according to the direction of the prevailing wind (Pl. 15.9a-b). Second, observations suggest an overall sand movement in a direction, which is the vector of the two major winds. [36] Furthermore, the internal structures of such longitudinal or seif dunes [37] consist of two sets of steeply dipping coarse strata disposed in nearly opposite directions (Fig. 15.15). Sand is driven up one flank and tumbles over the crest to form the slip slope. When the wind direction swings through, say, 60°–90°, sand is driven up the erstwhile avalanche slope and spills over the earlier upwind slope which is now a slip slope in respect of the new wind direction. Thus two sets of thin layers are evolved. This is in contrast both to the interdune areas, where the layers are flat-lying, and to transverse dunes, where only one set of gently dipping beds is generally preserved.

Some longitudinal dunes extend unbroken for hundreds of kilometres, diverging and converging in places to form Y-junctions (Pl. 15.10), but essentially maintaining a parallelism with each other. However, some are short with blunt upwind terminations and pointed downwind extensions; hence the application of the term *seif* or *sword-shaped* dunes. Some complex chain-like and reticulate linear dunes have also been observed (Pl. 15.11).

The crests of longitudinal dunes are of varied morphology. Some are sharp; others are broad and undulating with vague swells and depressions. Still others display barchans aligned diagonally with respect to the trend of the parent sand ridge. Some particularly broad, low chains of dunes in Arabia display undulating topography on their crests and give an overall impression of the rolling Downlands of southern England, and hence are called "downs". [38]

Both Bagnold[39] and Féderovitch[40] considered that longitudinal dunes may develop from barchans under the influence of cross winds (Fig. 15.16).

When considering the genesis of the sand ridges of the Simpson Desert, Madigan[41] argued that the winds responsible for moulding the dunes were organised presumably in horizontal spirals, helicoidal flows or vortices, which differentially eroded sand from a thick cover of alluvial material and deposited it in the "dead" areas between the air-flows. Bagnold[42] and Glennie[43] made similar suggestions in a general context and Féderovitch[44] suggested this origin for the longitudinal sand dunes of the Soviet deserts.

In the Simpson Desert, central Australia, the ridges evolve from lunette-like mounds deposited on the lee sides of playas and alluvial plains, [45] whence the debris is carried by rivers, lake waves and then by the wind, much as lunettes are formed (see below). In some cases small transverse ridges form on these mounds which are aligned normal to the dominant winds (Fig. 15.17) but linear ridges, moulded and aligned by the two prominent winds, extend downwind from the mounds (Pl. 15.12a-c). Initially the sands are a drab grey, like those of the mounds, but within a few kilometres they have acquired their characteristic red colour, probably due to the release of iron oxides during the weathering of clays deposited with the sands. Although there are still many playas within the Simpson Desert, problems of sand supply

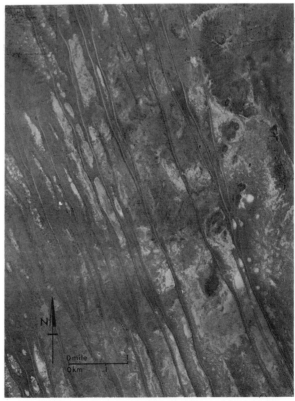

Plate 15.8a Widely spaced sand ridges with claypans in interdune corridors, western Queensland. *(Dept Nat. Dev., Canberra.)*

Plate 15.8b   Longitudinal dune in the southern Simpson Desert, Northern Territory.
(*C.R. Twidale/C.S.I.R.O.*)

Plate 15.8c   The same dune, from within the interdune
corridor. *(C.R. Twidale/C.S.I.R.O.)*

a

Plate 15.9   The same longitudinal dune in the Simpson Desert seen from the south (a) on 28 August, 1963, the wind having blown from the southeast; and (b) on 19 September, 1963, the wind having meanwhile blown from the southwest.   (*H. Wopfner.*)

b

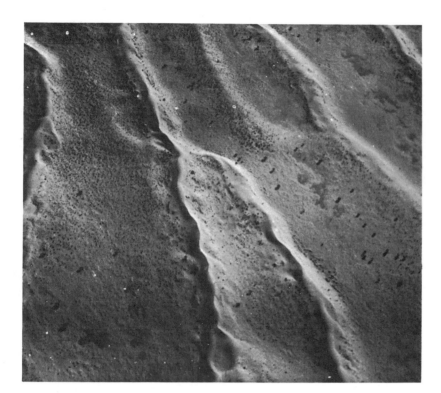

Plate 15.10 Y-junction of Simpson Desert dunes. *(Photographed bt C.T. Madigan and supplied by R.C. Sprigg.)*

Fig. 15.16 Development of longitudinal or seif dune from barchan subjected to a crosswind, according to Bagnold.

Plate 15.11 Lines of chain dunes, northeast of Wiluna, central W.A. *(Dept Nat. Dev., Canberra.)*

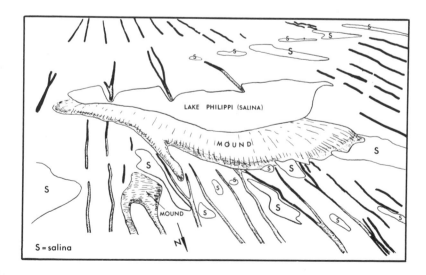

Fig. 15.17a Map showing mound developed on the lee or northern shore of Lake Phillipi, western Queensland.

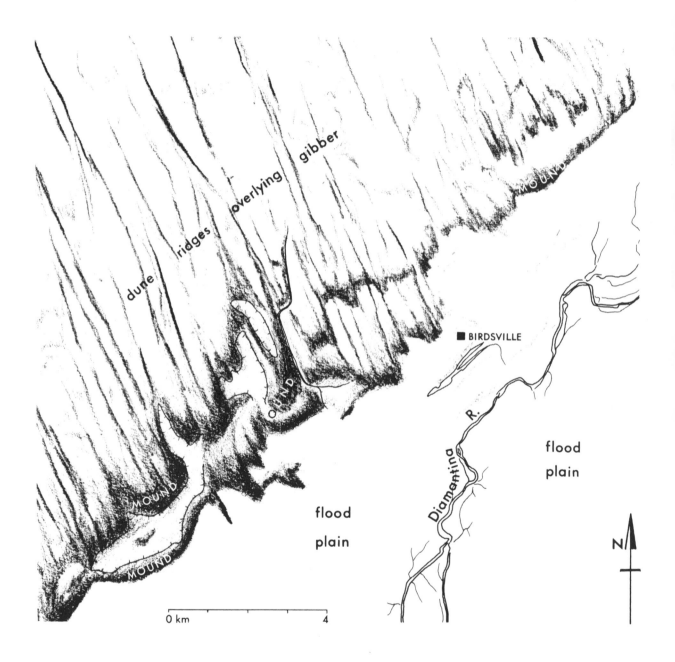

Fig. 15.17b Debris mounds on the northern or lee side of the
Diamantina River valley near Birdsville, southwest Queensland.
(See also Pl. 15.12.)

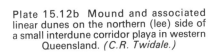

Plate 15.12a Vertical air photograph of the Diamantina (X) flood plain in northeastern S.A., showing mound on the northern (lee) side (M) with sand ridges extending north from it; Goyders Lagoon (L) with numerous interlaced channels and white sand ridges (R). (Cf Pl. 16.15)
(S.A. Lands Dept.)

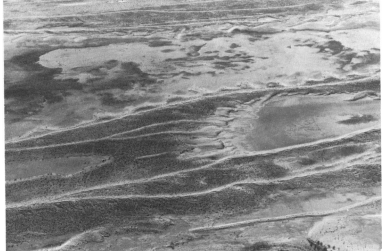

Plate 15.12b Mound and associated linear dunes on the northern (lee) side of a small interdune corridor playa in western Queensland. *(C.R. Twidale.)*

Plate 15.12c Debris mound and longitudinal sand ridges extending northward from it, on the northern side of the Diamantina flood plain near Birdsville, southwest Queensland. *(C.R. Twidale.)*

Plate 15.13 Longitudinal sand ridges advancing northward over the Diamantina flood plain near Birdsville, southwest Queensland. (*C.R. Twidale.*)

Plate 15.14 Dune near Motpena, Lake Torrens plains, S.A. The crest migrates back and forth within a range of perhaps 100 m under the influence of the wind of the time. (*C.R. Twidale.*)

and the short time available for the downwind extension of dunes suggest that, as mentioned previously, the ridges may have derived (at least in part, and especially in the southern Simpson) from more extensive and numerous spreads of riverine and lacustrine deposits of late Pleistocene age.

There is little doubt that in the southern and eastern parts of the Desert especially the sand ridges are migrating in a NNW direction over the underlying alluvial and lacustrine plains (Pl. 15.13). In the north and northwest of the Simpson there is some evidence to suggest that sand transport has been only local and that the ridges are due to the reworking and moulding by the wind of local sedimentary and regolithic veneers; but even here the areas studied are limited and there is in any case also evidence of dunes overlying alluvial materials.[46]

Basically then logitudinal sand ridges are initiated because of turbulence in the airflow near the ground induced by irregularities and obstacles there. Their initiation can be simulated in the wind tunnel (Fig. 15.18).

Longitudinal dunes dominate the Australian deserts, though in some of the better vegetated parts, for instance around Ayers Rock, reticulate and other patterns are prominent. Reticulate patterns and chain dunes develop in areas of variable wind and abundant sand (for instance, in river valleys) and in slightly better vegetated regions. Seif and longitudinal dunes also cover extensive areas of the north African deserts, Arabia, and the Soviet arid zones. They occur in the southwest of the Kalahari, but are unknown in the Americas.

Nonaligned dunes are not common, but have been described from Libya and Arabia and also occur on the Lake Torrens plains of Australia (Pl. 15.14). They consist of mounds of sand up to 130 m high which take several forms (domes, pyramids, star shapes) and which oscillate in location, migrating with the wind of the day. In Arabia, where strong winds blow from several directions, *dome* dunes, for instance, migrate under the influence of whichever wind is blowing, though the overall or long-term movement is to the east.[47]

3. Fixed or Immobile Dunes

Many masses of sand and clay have accumulated around obstacles such as shrubs (Pl. 15.15), outcrops and ridges, which disturb the air-flow by diverting the wind over and around them. Secondary eddies and currents are set up, and zones of low velocity, or dead areas, are produced. It is commonly found that if the height of the obstacle is designated h, the area of accumulation extends $1\cdot5\ h$ on the windward side and $6\ h$ on the lee side (Fig. 15.19). Debris such as sand, silt or salt accumulates in these shadow zones in front of, behind, or, in the case of shrubs, actually within the obstacle. Silt, however, which is being carried in suspension, is not greatly susceptible to such localised deposition. These accumulations are known by various names: sand shadow, rebdoux, tamarisk mound, *sablé mammelonné*. According to some writers, some of the giant pyramidal dunes of the true desert are relatively stable and are produced by the deflection of wind on a large scale from major obstacles such as ridges and ranges.

Some fixed dunes are restricted to one location because their source of debris is limited. Thus, in many parts of the

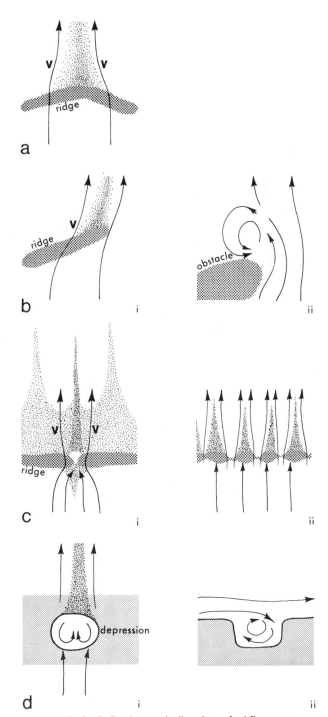

Fig. 15.18 Deflection and diversion of airflow over obstacles and deposition of sand in their lee as shown in wind tunnel experiments. V, vortex; (a) and (b) show deflection over crescentic and diagonal obstacles; (c) distortion of angular through gaps or cols; (d) the effect of a depression.

Plate 15.15   Fixed dune—sand accumulated among the roots and lower branches of
a shrub on the Lake Torrens plains, S.A. (*C.R. Twidale.*)

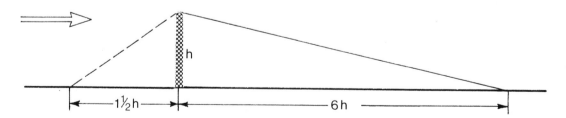

Fig. 15.19   Deposition of debris around an obstacle.

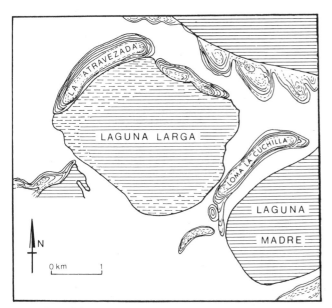

Fig. 15.20 Clay dunes in Texas. (*After W. A. Price, "Clay Dune", in R. W. Fairbridge (Ed.)*, Encyclopaedia of Geomorphology, *Reinhold, New York, 1969.*)

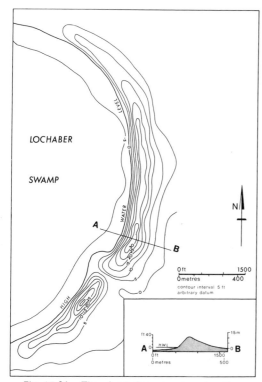

Fig. 15.21   The Lochaber swamp lunette. (*After Elizabeth M. Campbell, "Lunettes in southern Australia", pp. 85–109, by permission of the Royal Society of South Australia.*)

semiarid world, dunes have been formed by wind-blown sand from the desiccated beds of rivers. Outside the bare channel bed, there is sufficient vegetation to prevent wind transport of debris, so that the dunes remain essentially stable, until such time as the wind and rain are able to spread the sand accumulation and eliminate the dune forms.

Some fixed dunes occur on the lee side of playa depressions, which are the sources of the debris from which the dunes are constructed. They are typical of semiarid regions (though, as is described below, they also develop in arid areas) and again there is sufficient vegetation in the surrounding areas to inhibit or prevent any downwind migration of the feature, once it has formed near the lake or playa margin. These dunes bordering playas are known as *clay dunes* in Texas (Fig. 15.20) but because of their characteristic shape, are known as lunettes in Australia;[48] they are referred to as *bourrelets* in the French literature.[49]

Lunettes are crescentic in plan (Pl. 15.16; Fig. 15.21) and are asymmetrical in cross-section (Pl. 15.17), with the lake slope steeper than the lee slope and with the crest of the ridge close to the shore. They are up to 21 m high, over 1 km wide and 6·5 km long. Constructed of silt, sand, gypsum or clay, lunettes occur consistently on the lee shore of the playas to which they are related. Thus they occur on the eastern side of playas in southeastern Australia. Judging from the buried soils displayed in some of the lunettes, they have formed spasmodically in the recent past. But in some areas they are still developing. In places the sandy lunettes display cross-bedding, and the mineralogy of the dunes is always closely, though not always exactly, comparable with that of the sediments exposed in the adjacent depression.

Lunettes were first noted in 1840, in northern Victoria, by the explorer Major Mitchell, but it was not until a century later that they were formally named, described and explained. Hills,[50] working on lunettes in northern and north-western Victoria, where they happen to be built almost wholly of silt, considered that they were formed in relation to pre-existing lake depressions. He argued that dust in the atmosphere coagulates due to wetting as it passes over the shallow lakes, and that the wetted sediments—too heavy to be carried in suspension—fall to earth on the lee side of the lake to form a mound which becomes fixed by vegetation. Since an essential feature of the mechanism is that there should be water in the lake depression, this is called the "wet" hypothesis and it has been taken to imply that lunettes denote humid conditions. However, several features of lunettes are difficult to understand in terms of the wet theory. It cannot apply to those lunettes which consist of sand, for the latter cannot be carried in suspension. It does not explain the cross-bedding observed in some lunettes; it does not explain the definite shape in cross-section, and it does not account for the close sedimentological similarity between the lunette and the lake depression. Furthermore, in this theory, the lunette and the nearby playa are seen as chronologically distinct and separate.

With these difficulties in mind, Stephens and Crocker[51] suggested the alternative "dry" theory. They argued that the playa depression and the lunette are genetically related, and that wind abrasion or deflation caused erosion of areas devoid of vegetation, the eroded material being trapped a short distance downwind to form distinct mounds or lunettes.

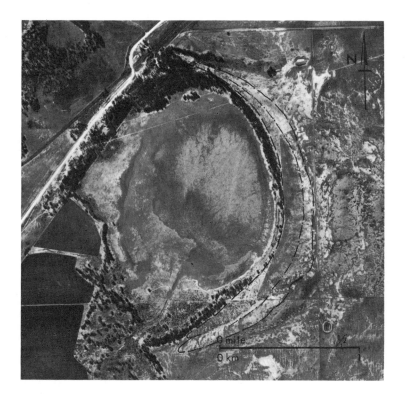

Plate 15.16 Vertical air photograph of Huttons Lagoon, mid-north of S.A. with a lunette (crest and limits marked) on its western side. *(S.A. Lands Dept.)*

Plate 15.17 Lunette at Bool Lagoon, South East of South Australia, showing steeper slope facing lagoon (right) and gentler or lee slope (left). *(Elizabeth M. Campbell.)*

In this view, lunettes develop under dry conditions in contra-distinction to Hills' theory; in effect, lunettes are taken to imply aridity. It has subsequently been suggested that salt crystallisation in the sediments of the lake bed induces roughness of clay and silt surfaces and that this materially assists the development of turbulence, and hence, erosion. [52] It may be suggested further that much drainage is endoreic (that is, interior) in areas of seasonal rainfall and river flow, particularly those areas which have recently emerged from arid phases; and that, in consequence, numerous small, shallow lakes form after rainfall and run-off. The alluvial spreads associated with such basins of interior drainage lack vegetation and are vulnerable to wind action.

The deflation hypothesis accounts for much of the observed evidence. The cross-bedding of some lunettes is due to aeolian deposition. The sediments of the lunettes are derived from the nearby lake depression and are therefore of similar composition. The crest of the lunette is developed close to the lake margin which is the source of the deposits. Nevertheless problems remain. Some lunettes occur on the lee side of

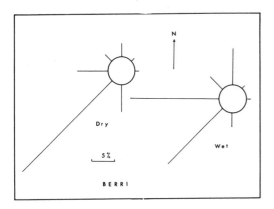

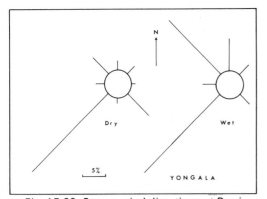

Fig. 15.22 Strong wind directions at Berri and Yongala, S.A., showing discrepancy between dry season wind roses and lunette location on east and southeast sides of playa depressions.

salinas, and it is difficult to understand how sediment from beneath the salt can be eroded and built up into lunettes. Again, if the lunettes are due simply to deflation, then they should be oriented in relation to the dry season strong winds. Campbell [53] has shown that in southern South Australia the dunes are more closely related to wet season wind directions (Fig. 15.22).

Twidale and Campbell [54] were both impressed by the longshore transport achieved by waves in even shallow lakes, and suggested that during winter (the wet season in southern South Australia), wind-driven waves drifted sediment and other debris (see also Woods [55]) to the lee shore, and that the sediment was then blown from the beaches both by winter and summer winds to the areas beyond the shore where it was trapped by vegetation and accumulated to form lunettes. Bowler [56] noted gravel in the lower parts of lunettes in Victoria which he attributed to wave transport and deposition. Thus the alignment of the mounds is seen as more closely related to winter, or wet season winds. Sediment is transported to the lee shore when the lake carries water, and dry season encrustation by salt is unimportant with regard to lunette development. Another conclusion stemming from this modification of the deflation hypothesis is that lunettes have no long-term climatic significance: they develop in climates with a markedly seasonal distribution of rainfall, rather than in either arid or humid climates. However, this is contrary to the findings of Bowler, who believes that lunettes have a limited time distribution and that they are significant as indicators of past climates. [57]

Lunettes are to be compared to coastal foredunes (see below) in the sense that they may develop under a wide range of climatic conditions, provided that there are both phases when debris supply can be renewed and others when deflation of that debris can take place. Thus the mounds from the arid areas of central Australia described earlier are comparable to lunettes but are not so regularly shaped possibly because of infrequent periods of construction and long intervals of degradation.

Several aspects of lunette distribution remain which at present are not open to rational explanation. For instance, not all playas in a given region have lunettes associated with them: some have lunettes at their margins, some do not, and there seems to be no consistent reason for the difference. Furthermore, there is no consistent relationship between the size of playa and the size of the associated lunette. Some large playas have large lunettes, but other large playas display no lunette form; on the other hand, some small playas have large lunettes at their margins. The answer does not seem to be related to the nature of the deflated material, nor to wind regime. However, in broad view and in many detailed cases, a combination of wave transport and deflation accounts for the salient features of this dune form which is, as mentioned, of more than passing interest when the evolution of true desert dunes is considered.

In semiarid areas where there is considerable vegetation cover and particularly in areas which were formerly dune fields, large, fixed ridges of sand have developed. Because many of the dunes are curved or U-shaped, with the open end of the U pointing upwind, they are called *U-dunes*, though *parabolic dune* is probably to be preferred in view of the variation in form which they display (Pl. 15.18). Such

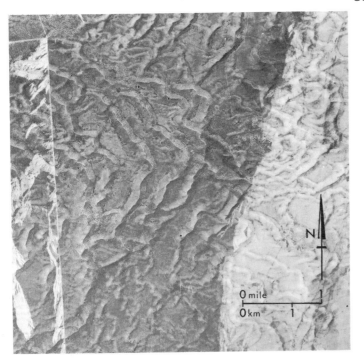

Plate 15.18 Vertical air photograph of parabolic dunes in the Murray mallee, S.A. Dominant wind westerly. The contrasted line reflects a fire burn. *(S.A. Lands Dept.)*

fixed dunes cover immense areas of the Murray plains in South Australia and the western areas of Victoria. They are also abundant in depressions on northern Eyre Peninsula, which have received accretions of alluvium. Though open U-dunes probably predominate, many are linked to form rake dunes *(dunes en rateau)*; many are interlaced to give a chain effect; and some, especially on northern Eyre Peninsula, take the form of irregular, concentric rings of sand ridges (Pl. 15.19). More regularly shaped parabolic dunes have been described from the Navajo region of Arizona [58] where they occur amidst mobile dunes and where they appear to be associated with areas of greater vegetation cover.

In their formation and morphology these parabolic dunes are similar to coastal foredunes which are developed in relation to coastal beaches and which are discussed in this context in Chapter 17. For the moment suffice it to say that they appear to be due as much to wind erosion as to deposition.

Plate 15.19 Air photograph of linear and concentric patterns of fixed dunes on northern Eyre Peninsula. *(S.A. Lands Dept.)*

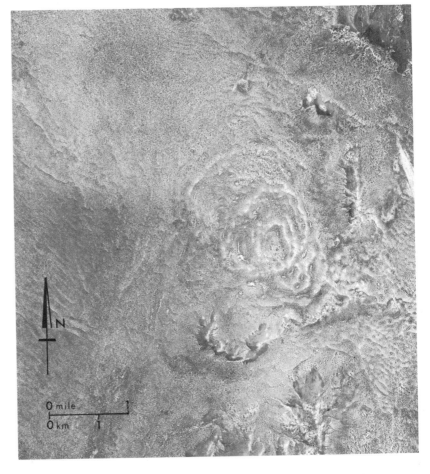

References Cited

1. H. Wopfner and C.R. Twidale, "Geomorphological History of the Lake Eyre Basin", Ch. 7 in J.N. Jennings and J.A. Mabbutt (Eds), *Landform Studies from Australia and New Guinea*, Australian National University Press, Canberra, 1967, pp. 118–43.

2. R.M. Norris and K.S. Norris, "Algodones Dunes of south eastern California", *Bull. geol. Soc. Am.*, 1961, **72**, pp. 605–20.

3. The basic text for this area of study is R.A. Bagnold, *Physics of Blown Sand and Desert Dunes*, Methuen, London, 1941.

4. R.A. Bagnold, *Physics of Blown Sand*; *idem*, "The Transport of Sand by Wind", *Geogr. J.*, 1937, **89**, pp. 409–38.

5. R.A. Bagnold, "The Transport of Sand", pp. 409–38.

6. A.F. Pitty, "Particle Size of the Sahara Dust which fell in Britain in July, 1968", *Nature (Lond.)*, **220**, pp. 364–5.

7. See for example E.L. Krinitzsky and W.J. Turnbull, "Loess Deposits of Mississippi", *Geol. Soc. Am. Spec. Paper*, 1970, **94**.

8. See R.U. Cooke, "Stone Pavements in Deserts", *Ass. Am. Geogr. Ann.*, 1970, **60**, pp. 560–577.

9. E.B. Pandastico and T.I. Ashaye, "Demonstration of the Effect of Stone Layer on Soil Transport and Accretion", in E.G. Hallsworth and D.V. Crawford (Eds), *Experimental Pedology*, Butterworth, London, 1956, pp. 384–90.

10. Note that earlier workers such as C.R. Keyes and J.T. Jutson (see Chapter 20) believed the wind to be capable of planation on a continental scale, as did S. Passarge (*Die Kalahari*, Reiner, Berlin, 1904) in his earlier writings.

11. E. Blackwelder, "Yardangs", *Bull. geol. Soc. Am.*, 1934, **45**, pp. 159–66.

12. S. Hedin, *Central Asia and Tibet*, Vol. I, Hurst, London, 1903.

13. H. Bobek, "Features and Formation of the Great Kawir and Masileh", *Univ. Tehran, Arid Zone Res. Centre Pub.*, 1959, **2**; *idem*, "Nature and Implications of Quaternary Climatic Changes in Iran", in *Changes in Climate*, UNESCO, Paris, 1963, pp. 403–13; *idem*, "Zur Kenntnis der Sudlichen Lut", *Mitt. Osterreich Geog. Gesell.*, 1969, **III**, pp. 155–92.

14. D. King, "The Quaternary Stratigraphic Record at Lake Eyre North and the Evolution of Existing Topographic Forms", *Trans. R. Soc. S. Aust.*, 1955, **79**, pp. 93–103.

15. W.F. Hume, *Geology of Egypt*, Government Printer, Cairo, 1925, pp. 65–73; R.J. Russell, "Landforms of San Gorgiono Pass, southern California", *Univ. Calif. (Berkeley) Pub. Geogr.*, 1932, **6**, pp. 37–44.

16. B.A. Féderovitch, "L'Origine du Relief de Deserts de Sable actuel", in *Essais de Géographie*, Acad. Sci. USSR, Moscow, 1956, pp. 117–29.

17. R.A. Bagnold, *Physics of Blown Sand*, p. xvii.

18. See C.R. Twidale, "Landform Development in the Lake Eyre Region, Australia", *Geogr. Rev.*, 1972, **62**, pp. 40–70 and *idem*, "Evolution of Sand Dunes in the Simpson Desert, central Australia", *Trans. Inst. Br. Geogr.*, 1972, **56**, pp. 77–109.

19. C. Sturt, *Expedition into central Australia*, Vol. 1, Boone, London, 1849, pp. 380–1.

20. Muriel Brookfield, "Dune Trends and Wind Regime in central Australia", *Z. Geomorph. Supp.*, 1970, **10**, pp. 121–53.

21. R.A. Bagnold, *Physics of Blown Sand*.

22. V. Cornish, *Waves of Sand and Snow, and the Eddies which make Them*, Unwin, London, 1914.

23. I. Wilson, "Sand Waves", *New Scientist*, 1972, **53** (788), pp. 634–37.

24. See for instance, V. Cornish, *Waves of Sand and Snow* and H. Landsberg, "The Structure of the Wind over a Sand Dune", *Trans. Am. Geophys. Un.*, 1942, pp. 237–9.

25. H.J.L. Beadnell, "The Sand Dunes of the Libyan Desert", *Geogr. J.*, 1910, **35**, pp. 379–92; R.P. Sharp, "Kelso Dunes, Mojave Desert, California", *Bull. geol. Soc. Am.*, 1966, **77**, pp. 1045–74.

26. E.D. McKee and G.C. Tibbitts, "Primary Structures of a Seif Dune and Associated Deposits in Libya", *J. Sed. Petrol.*, 1964, **34**, pp. 5–17.

27. See H.J. Finkel, "The Barchans of southern Peru", *J. Geol.*, 1959, **67**, pp. 614–47; R.M. Norris, "Barchan Dunes of Imperial Valley, California", *J. Geol.*, 1966, **74**, pp. 292–306; J.S. Shelton, *Geology Illustrated*, Freeman, San Francisco, 1966, p. 198.

28. H.J.L. Beadnell, "The Sand Dunes of the Libyan Desert", pp. 379–92.

29. J.T. Long and R.P. Sharp, "Barchan Dune Movement in Imperial Valley, California", *Bull. geol. Soc. Am.*, 1964, **75**, pp. 149–56.

30. G.W. Murray, "The Egyptian Climate: an Historical Outline", *Geogr. J.*, 1951, **117**, pp. 422–34.

31. See J.T. Long and R.P. Sharp, "Barchan Dune Movement", pp. 149–56.

32. J.T. Long and R.P. Sharp, "Barchan Dune Movement", pp. 149–56; H.J. Finkel, "The Barchans of southern Peru", pp. 614–47.

33. H.J. Finkel, "The Barchans of southern Peru", pp. 614–47.

34. J.T. Hack, "Dunes of the western Navajo Country, Arizona", *Geogr. Rev.*, 1941, **31**, pp. 240–63.

35. Muriel Brookfield, "Dune Trends and Wind Regime", pp. 121–53.

36. F.N. Ratcliffe, "Soil Drift in the Arid Pastoral Areas of South Australia", *C.S.I.R. Pamphlet*, 1936, **64**; *idem*, "Further Observations on Soil Erosion and Sand Drift, with Special Reference to southwestern Queensland", *C.S.I.R. Pamphlet*, 1937, **70**; H. Wopfner and C.R. Twidale, "Geomorphological History of the Lake Eyre Region", pp. 118–43.

37. E.D. McKee and G.G. Tibbitts, "Primary Structures of a Seif Dune", pp. 5–17.

38. R.A. Bagnold, "Sand Formations in southern Arabia", *Geogr. J.*, 1951, **117**, pp. 78–86.

39. R.A. Bagnold, *Physics of Blown Sand*.

40. B.A. Féderovitch, "L'Origine du Relief de Déserts", pp. 117–29.

41. C.T. Madigan, "The Australian Sand Ridge Deserts", *Geogr. Rev.*, 1936, **26**, pp. 205–27; *idem*, "The

Simpson Desert Expedition, 1939, Scientific Reports 6: Geology—the Sand Formations", *Trans. R. Soc. S. Aust.*, 1946, **70**, pp. 45–63.

42. R.A. Bagnold, "The Surface Movement of Blown Sand in Relation to Meteorology" in *Desert Research*, Res. Counc. Israel Spec. Publ., 1953, **2**, pp. 89–93.

43. K.W. Glennie, *Desert Sedimentary Environments*, Elsevier, Amsterdam, 1970.

44. B.A. Féderovitch, "L'Origine du Relief de Déserts", pp. 117–29.

45. D. King, "The Sand Ridge Deserts of South Australia and Related Aeolian Landforms of the Quaternary Arid Cycles", *Trans. R. Soc. S. Aust.*, 1960, **83**, pp. 93–108; H. Wopfner and C.R. Twidale, "Geomorphological History of the Lake Eyre Region", pp. 118–43; C.R. Twidale, "Evolution of Sand Dunes".

46. Dorothy Carroll, "The Simpson Desert Expedition, 1939, Scientific Reports. 2. Geology—Desert Sands", *Trans. R. Soc. S. Aust.*, 1944, **68**, pp. 49–59; R.L. Folk, "Longitudinal Dunes of the Northwestern Edge of the Simpson Desert, Northern Territory, Australia. 1. Geomorphology and Grain Size Analysis", *Sedimentology*, 1971, **16**, pp. 5–54. But see also J.A. Mabbutt and Margaret E. Sullivan, "The Formation of Longitudinal Sand Dunes: Evidence from the Simpson Desert", *Aust. Geogr.*, 1968, **10**, pp. 483–7.

47. R.A. Bagnold, "Sand Formations in southern Arabia", pp. 78–86; D.A. Holm "Dome shaped Dunes of central Nejd, Saudi Arabia", *C.R. 19th Int. Geol. Congr.*, 7: *Deserts actuels et modernes*, 1952.

48. G.N. Coffey, "Clay Dunes", *J. Geol.*, 1909, **17**, pp. 754–5; E.S. Hills, "The Lunette, a new Landform of Aeolian Origin", *Aust. Geogr.*, 1940, **3**, pp. 15–21.

49. J. Boulaine, "La Sebkha de Ben Ziane et sa 'Lunette' ou Bourrelet", *Rev. Géomorph. Dyn.*, 1954, **5**, pp. 102–23.

50. E.S. Hills, "The Lunette", pp. 15–21.

51. C.G. Stephens and R.L. Crocker, "Composition and Genesis of Lunettes", *Trans. R. Soc. S. Aust.*, 1946, **70**, pp. 302–12.

52. J. Boulaine, "La Sebkha", pp. 102–23; J. Tricart, "Une forme de Relief Climatiques; les Sebkhas", *Rev. Géomorph. Dyn.*, 1954, **5**, pp. 97–101; J. Tricart, "Influence des Sols Salés sur la Deflation Éolienne en Basse Mauritanie et dans le Delta du Senegal", *Rev. Geomorph. Dyn.*, 1954, **5**, pp. 124–132; P.G. Macumber, "Lunette Initiation in the Kerang District", *Mining and Geol. J.*, 1970, **6**, pp. 17–18.

53. Elizabeth M. Campbell, "Lunettes in southern Australia", *Trans. R. Soc. S. Aust.*, 1968, **92**, pp. 85–109.

54. C.R. Twidale, *Geomorphology, with special reference to Australia*, Nelson, Melbourne, 1968, pp. 233–4; Elizabeth M. Campbell, "Lunettes in southern Australia", pp. 85–109.

55. J.E.T. Woods, *Geological Observations in South Australia principally in the District southeast of Adelaide*, London, 1862, pp. 28–9.

56. J.M. Bowler, "Australian Landform Example No. 11. Lunette", *Aust. Geogr.*, 1968, **10**, pp. 402–4.

57. J.M. Bowler and L.B. Harford, "Quaternary Tectonics and the Evolution of the Riverine Plain near Echuca, Victoria", *J. geol. Soc. Aust.*, 1966, **13**, pp. 339–54; J.M. Bowler, "Quaternary Chronology of Goulburn Valley Sediments and their Correlation in southeastern Australia", *J. geol. Soc. Aust.*, 1967, **14**, pp. 287–92.

58. J.T. Hack, "Dunes of the western Navajo Country", pp. 240–63.

Additional and General Reference

R.U. Cooke and A. Warren, *Geomorphology in Deserts*, Batsford, London, 1973.

CHAPTER 16

Snow and Ice

A. GENERAL

At high latitudes and altitudes some precipitation is received in the form of snow. In these areas the snow persists for all or part of the year, and in wide areas of the polar regions it lasts for such long periods that it may be regarded as permanent. As snow banks up, it is transformed into ice under the pressure of superincumbent snow layers. This compaction process involves the transformation of powdery snow to glacier ice, and may extend over several centuries. Water molecules migrate within the crystal lattices of the ice, and there is an increase in density from $0.01-0.25$ gm/cm^3 for powdery snow to $0.8-0.9$ gm/cm^3 for crystalline ice.[1]

The ice masses so formed may remain stationary, or, if they are sufficiently dense or are perched on steep slopes, they move downhill under the influence of gravity. The merging of many such individual glaciers, such as those which originate in cirques, snowcaps and icecaps, gives rise to streams of ice, or glaciers. These valley glaciers too may coalesce to form an ice sheet such as that which occupies most of the Antarctic continent.

In the subarctic regions and at high elevations frost riving is an important means of rock weathering. Chemical alteration appears to be minimal. The importance of river erosion is in the absence of any significant long term measurements difficult to estimate. What is clear is that rivers in periglacial regions are, because of the winter freeze and summer thaw, strongly seasonal in their activity. In consequence of the high summer discharges subarctic streams are commonly braided.

The assemblage of landforms found in areas subject to pronounced freezing, but not covered by glacier ice can be considered in terms of wind action, the effects of snow, of ground ice, and of solifiual stream features.

B. AEOLIAN EFFECTS IN COLD REGIONS

Many subarctic areas are in reality cold deserts. The prevailing arid and cold conditions are not conducive to the growth and survival of vegetation: these are the tundra areas. The land surface is not protected against the effects of a virtually unimpeded airflow, and the wind therefore achieves significant, though, because of the character of much of the surface, ephemeral results.

In the arctic and antarctic deserts, dry snow is moulded by the wind into various forms which are similar in morphology and origin to those developed on clean sand. Ripples, transverse waves and *crescentic drifts*, which appear analogous to barchans and which display stratification, are formed as a result of the frictional drag of air passing over the snow surface.[2] Snow dunes accumulate around obstacles, and drifts with strongly developed overhangs or cornices on the lee side are also formed (Pl. 16.1). Bizarre accumulations on branches and on the canopies of trees ("snow mushrooms") have been described.[3] Very strong winds cause differential erosion and deposition of snow surfaces, forming ice caverns and bridges though preferential melting gives similar results. The most spectacular forms developed by wind erosion, however, are the elongated parallel ridges, called *sastrugi* or *zastrugi*, which have been compared to the yardangs excavated by wind erosion in tropical areas (see Chapter 15). The longitudinal sand dunes of the Simpson Desert in central Australia were also originally compared to zastrugi by Madigan,[4] who had experienced the blizzards and snow forms of the Antarctic.

These ephemeral snow features have been observed wherever snow accumulates. For example, they form during blizzards in areas which have a snow cover for only part of the year. They are best developed, however, in areas of permanent snow and ice, notably in Antarctica where there is no vegetation to impede the air-flow, and where wind velocities are commonly very high: truly what Mawson has called *"The Home of the Blizzard"*.

True dunes are also shaped by the wind on unprotected sedimentary surfaces in periglacial conditions. For instance,

Plate 16.1 Snow cornices on the Piz Bernina (4052 m), Switzerland. *(Swiss Nat. Tour. Office.)*

parabolic dunes, as well as deflation hollows, have been described from the Kaamasjoki-Kiellajoki river basin in Finnish Lapland.[3] They are moulded from sediments laid down in a proglacial lake, and the dunes are not related to the modern wind regime; that is, they are fossil or relict forms.

Wind armed with ice crystals is also apparently capable of eroding rocks, for both wind-polished and wind-faceted stones or *ventifacts* comparable to those found in tropical deserts have been described from cold regions. They have been described in northern Greenland by Nichols[6] and in Iceland by Derruau.[7] King[8] has also reported facets "several feet long" cut by the wind in a basaltic outcrop on the south coast of Iceland. Boye[9] has described elongated depressions which he calls microyardangs, allegedly eroded by wind, in vegetated unconsolidated deposits in northeast Greenland, and Nichols[10] has reported wind-eroded earth hummocks and frost mounds from northern Greenland (Fig. 16.1). Boye[11] is one of several workers to report dust storms in Greenland in late summer, when the exposed deposits, unprotected by vegetation, are dry and vulnerable to wind action.

uneroded mounds   eroded mounds

Fig. 16.1   Eroded frost-heaved mounds or hummocks in Greenland. *(After R.L. Nichols, "Geomorphology of Inglefield Land", by permission of Robert L. Nichols.)*

C. **SNOW**

Linear or semicircular depressions (Fig. 16.2) up to 100 m across are typical of many cold lands and since they occur in areas which have not suffered glaciation, they cannot be attributed to the work of glaciers. They are thought to be caused by weathering and erosion related to snow patches, and are called *nivation hollows*.[12] Existing hollows and valleys (due for instance to differential weathering or to stream erosion) are enlarged, and structural weaknesses are exploited to form new depressions. Surface snows exposed to the sun may be melted, and the snow which is in contact with bedrock (which heats up more rapidly than the snow) is also melted. The meltwater runs down the back of the patch and the rock in contact with the basal snow is subject to the expansive force of freezing water. It thus becomes shattered. The meltwaters probably evacuate some of the debris as they flow out of the hollow, though some transportation is accomplished by solifluction either beneath the snow patch, or at times when the snow has melted.

Some debris may be moved downslope with sliding snow masses. The snow patches themselves apparently move downslope, for blocks and other fragments contained in the patch scratch the underlying surface as they are dragged over it. Blocks are also found some distance away from their place

of origin and have been attributed to snow patch sliding— the movement of snow and ice over the bedrock—in both the USA[13] and Australia.[14] Debris derived from the cliffs or slopes above snow patches falls on to the latter and slides over the surface to the toe of the drift. There it accumulates and because of the melting of the snow is exposed as low mounds or ridges called *protalus ramparts*.

The most widespread, obvious and effective means of carrying debris by snow, however, is by way of avalanches, which are typical of many mountainous regions because they form most readily on steep slopes. On slopes exceeding 50°, snow rarely accumulates sufficiently to generate major avalanches though small flows and slides of snow—called *sluffs*— are common. On slopes below about 30°, and certainly below 25°, avalanches are rare, but on slopes of 30°–50° thick blankets of snow accumulate and can give rise to avalanches. These may be triggered by a further heavy snowfall, or by the collapse of a cliff (possibly due to frost shattering), or by an earth tremor. The snow slides and pours downslope, usually along well-used and well-known paths or channels, causing great devastation, and in some tragic cases, loss of life. Such avalanches would create enough damage if they consisted only of snow, but armed with blocks of rock, uprooted tree trunks and other debris they are very destructive. After the snow has melted the debris remains in long trails or ribbons, called *debris trails* or *boulder tongues* according to their shape and composition.

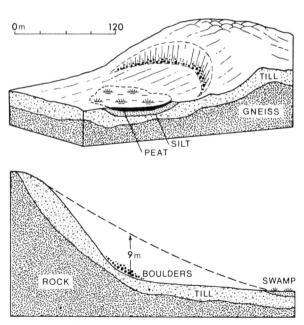

Fig. 16.2 Perspective view and section through nivation hollow. *(After E.P. Henderson, "Large Nivation Hollows", pp. 607–16, by permission of J. Geol. and University of Chicago Press.)*

D. GROUND ICE

Many subarctic and arctic areas display hummocky micro-relief, and regular patterns of coarse and fine debris. In addition, there is often evidence of debris flow in the form of lobes and tongues of debris. These and many other minor landforms are due to the presence of ice in the ground, which, according to Davies,[15] is the most consistent feature of periglacial or nival* (that is, cold, nonglacial) areas. Several of the features associated with the presence of ice in the regolith have been discussed previously.

When water freezes to form ice, it undergoes a volume increase of about 9 per cent, which is sufficient to shatter most porous and fissile rocks which contain a good deal of moisture. Also, migrating water molecules adhere to the surfaces of the ice crystals; areas of low moisture pressure develop and this induces further diffusion of water molecules toward the initial ice concentrations. Thus ground ice becomes both segregated[17] and unevenly distributed in the surface layers of rock and soil.

Ground ice which forms diurnally is called *pipkrake* or needle ice and it is, of course, confined to the near-surface horizons. Because groundwater is unevenly distributed, so is pipkrake, and so are the various minor features owing their origin to this ground ice. Soil heaving due to ice growth is not the same everywhere: some areas rise relative to others. In some places this steepens slopes enough to induce instability of the surface and downslope movement of individual particles or of masses of debris. Particles of soil and rock form long ribbons of debris running downslope. The downhill passage of debris is interrupted by outcrops and barriers of vegetation and small terraces are built behind these. The build-up continues until debris pours through gaps in the outcrops or until it overrides and flattens part of the vegetation wall.

Elsewhere, joints and crevices are the sites of quite massive ice development, and considerable *ice wedges* are formed (Pl. 16.2). Typically 1–5 m deep and 30–50 cm wide at the surface, examples up to 10 m deep and 10 m wide are known. They grow at 0·5–1·5 mm per annum and exert pressure on the adjacent unconsolidated soil sediments which may become folded or convoluted, though similar distortion may be caused by the passage of masses of debris or ice over the unconsolidated materials.

Ground ice is also capable of sorting soil and rock particles according to their size to form patterned ground of various types. Evidently some kind of churning action in the near-surface layers is involved.[18] One result is either a series of nets, stripes, polygons and circles (Pl. 16.3a—e) which are quite regular in plan and which tend to become elongated downslope on steeper inclines, or a general patchy appearance due to the imperfect or irregular development of such forms.

In explanation of this patterned ground it may be argued that the growth of ice needles beneath stones (which provide insulation against both heating and drying) causes the latter to be thrust upwards and become unstable. They migrate laterally and become concentrated in distinct rings. Once this tendency toward sorting begins, the centres of the polygons, which contain much moisture in the interstices of the fine materials, become arched on freezing, thus assisting the continued lateral migration and sorting of the coarse debris.

Plate 16.2 Inactive ice wedge exposed by placer mining at Wilbur Creek, near Livengood, Alaska. Note distortion of sediments caused by expansion of wedge. *(T.L. Pewe.)*

---

* For discussion of the use of this term and suitable alternatives, see Linton.[16]

Plate 16.3a Sorted polygons in dry lake bed, involving quite coarse blocks (scale 16 cm), central Labrador. *(C.R. Twidale.)*

Plate 16.3b Small sorted polygons (scale 16 cm), east side of Korsbjerg massif, 1 km south of Ansgar, Greenland. *(A.L. Washburn.)*

Plate 16.3c   Small nonsorted stripes on northwest side Danevirke Hills, Greenland (16 cm rule in centre).
(*A.L. Washburn.*)

Plate 16.3d   Large sorted stripes on north side Hesteskoen Ridge, Greenland. The dark stripes are about
50 cm wide. (*A.L. Washburn.*)

Plate 16.3e Frost polygons on the lake-studded Hudson Bay Lowland, near Churchill, Manitoba. The Hudson Bay Railway can be seen in the middle distance. (*Dept Energy, Mines and Resources, Ottawa. Photo No. A1426—66.*)

Fig. 16.3 Distribution of permafrost in the northern hemisphere. (*After R.F. Black, "Permafrost—a Review", pp. 839—55, by permission of the Geological Society of America.*)

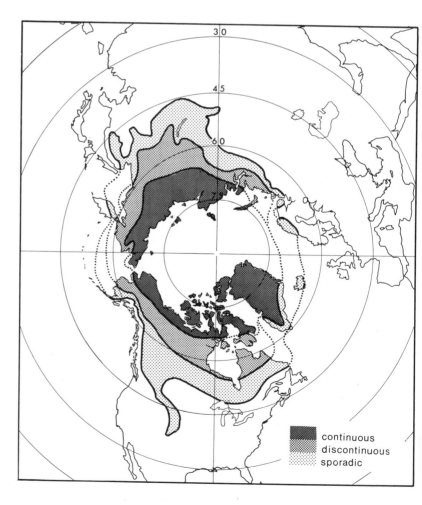

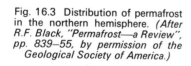 continuous
discontinuous
sporadic

On the other hand, during the summer the thaw of surface ice may bring water temperature to 4°C (39°F), at which water is at its densest. It thus sinks, and lighter water rises to replace it, initiating a circulation. Laboratory experiments have shown that polygons form on the rising currents; but it is very doubtful indeed whether such currents can move materials of the calibre involved in some of the subarctic patterned ground. Following along much the same lines, as the surface layers are first to dry in summer, these heavier layers tend to subside, while the lower, moisture-laden layers tend to rise, with similar results.

Important features develop in arctic and subarctic regions as a result of the presence of perennially frozen ground, *permafrost* or *tjale*.[19] Permafrost (either continuous, discontinuous or sporadic) occupies some 26 per cent of the continents. It is widespread in the northern hemisphere (Fig. 16.3), and is present also in Antarctica, as well as in the southern Andes. It attains considerable thicknesses in some areas, persisting to 610 m at Prudhoe Bay, on the Arctic coastal plains of Alaska. It has been reported to depths of 1400–1450 m in the Upper Markha River Valley of Siberia. These great thicknesses are developed in the zones of continuous permafrost; in the sporadic zone thicknesses greater than 30 m are rare. Permafrost is absent or its upper surface is depressed beneath major water bodies, like large rivers and lakes, where the water forms an insulating layer. Such a gap or window in the permafrost horizon is known by the Russian term, *talik*.

The most important geomorphological effects of permafrost stem from the formation of an impermeable horizon just beneath the land surface. Summer meltwaters cannot infiltrate into the subsurface, and instead run off into rivers or, more commonly, gather in local depressions to form the swamps, marshes and lakes so typical of subarctic regions. The well-named Lake Plateau of Labrador, for instance, appears to consist more of water surface than of land.

Permafrost develops from the surface downwards, and pockets of still liquid water may persist in the process, for the freezing point is lowered under the high pressures developed as a result of confinement by permafrost. Eventually, however, as the front of the permafrost advances and the volume of the water pocket decreases, *cryostatic pressures* develop to such an extent that the water is forced up to the surface. There, exposed to the low air temperatures and no longer under high pressure, the water freezes, forming an ice mass or *naled* (a Russian term). This sort of mechanism has been used in explanation of one of the most spectacular features of the cold lands, namely the ice-cored hill or *pingo*.

E. PINGOS

Pingos are characteristically of conical shape. Relict, inactive pingos have been recorded from central Alaska, the coastal plain of Alaska, and from the floor of the Beaufort Sea in the Canadian Arctic;[20] active, modern pingos are found in central Alaska and coastal Greenland, and in northern Siberia, especially the Indigirka and other delta, estuary and alluvial areas. But well over 1400 pingos, a very large proportion of the total so far reported, occur in the Mackenzie delta in Canada's Northwest Territory. In this region they are restricted to alluvial areas and are especially abundant and well developed in shallow lake settings. A few have been

reported from valley floors. Though fragile in construction and youthful in geological terms (carbon 14 dates of material from the sediments of which the features are built indicate that they are only a few thousand years old[21]), pingos evidently endure some hundreds, or even thousands of years.

Typical pingos are shown in Fig. 16.4a and Pl. 16.4, and excellent accounts of these features are due to Müller, Mackay, and Mackay and Stager.[22] They range between 30 and 600 m in diameter, and it has been noted that extension of the diameter above this range is accompanied by a decrease in the altitude of the particular pingo. They are smooth in outline, and the majority are oval or elliptical in plan. Some are elongate, with straight sides and rounded ends, and resemble ancient burial mounds. A few pingos which occur on river flats are sinuous or irregular in plan. Pingos are characteristically asymmetrical in profile, the difference in gradient between opposite sides being of the order of 20°. The steepest slope observed on the flank of a pingo is 45°, and the slopes are straight or convex. More than 75 per cent of all pingos described to date have smooth, rounded summits, though the larger specimens display breached crests which Mackay describes as "star-shaped craters surrounded by cuestas which seemingly open out as rupturing progresses, like the petals of a budding flower"[23] (Pl. 16.5). In such breached summits, ice cores, or several discrete wedges of ice, are commonly exposed. But when this core ice melts, the central part of the pingo collapses. Ice lenses forming the cores of pingos have also been exposed by wave erosion of the mound sediments (Fig. 16.4b).

Some pingos are associated with radial and concentric fracture patterns. The overburden of sediments is one half to one third as thick as the height of the pingo, the rest being taken up by ice cores which, though they extend well below the level of the surrounding plain, do not extend to great depths. Freshwater shells and peats contained in the mound sediments indicate that the latter are of lacustrine origin. The sediments are domed, with dips up to 53°, and in a few places display minor faulting.[24]

Three major hypotheses have been advanced in explanation of pingos. These are the open hypothesis, the closed hypothesis, and what might be termed the gravitational hypothesis. It will be noted that the various hypotheses involve mechanisms which have been cited earlier in the explanation of other superficial pseudotectonic forms.

The pingos of Alaska, Greenland and northern Siberia have been interpreted as due to hydrolaccoliths bowing up the overlying strata. Water flowing from the adjacent uplands under strong hydraulic pressure is said to migrate beneath the permafrost cover. Where there is a weakness in this cover, for example a window of thin or absent permafrost beneath a lake or river, the water thrusts upwards, doming the overlying sediments and when it penetrates the cold, permanently frozen layer, itself freezes and expands, causing further arching. Since this hypothesis involves water originating some distance away from the pingo, and is thus an open system, it is called the *open* hypothesis[25] and may well be valid for the areas cited. The preferential development of pingos in lakes or in lake beds is possibly due to the development of windows in the permafrost beneath the water bodies. However, the open hypothesis does not explain the concentration of pingos in the Mackenzie delta.

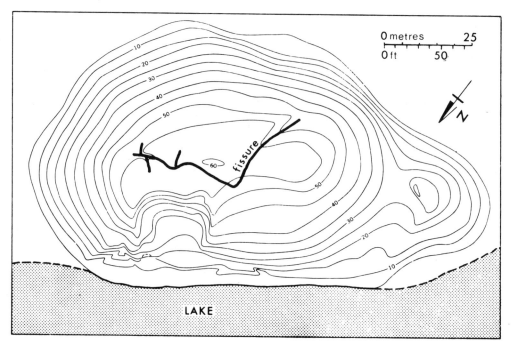

Fig. 16.4a   Contour plan of pingo in the Mackenzie delta, North West Territory. Contours in feet. (*After J.R. Mackay and J.K. Stager, "The Structure of some Pingos", pp. 360–8, by permission of Geogr. Bull.*)

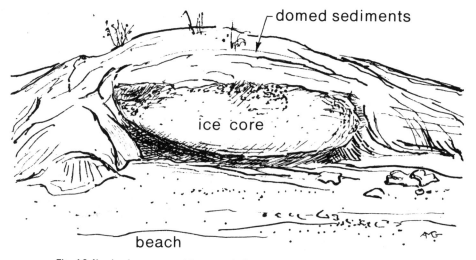

Fig. 16.4b   Ice lens exposed in core of pingo at McKinley Bay in Mackenzie delta. (*Drawn from photograph in J.R. Mackay, "Pingos of the Pleistocene", pp. 21–63.*)

Plate 16.4   Pingo 24m high, near Hesteskoen, Greenland. (*A.L. Washburn.*)

Plate 16.5   Ruptured pingo 30m high. Cape Franklin area, central East Greenland. (*F. Muller.*)

The Mackenzie delta pingos have been attributed to an interpermafrost mechanism developed beneath lakes. It is pointed out that lake waters provide insulation, and thus that the soils and unconsolidated alluvia beneath large lakes (at least 300 m diameter and 30 m deep) are unfrozen, in contrast to the surrounding permafrost (Fig. 16.5a). All lakes are ephemeral, and as these subarctic lakes are filled with sediment and organic material, they lose their protective or insulating function, and permafrost forms beneath the lake bed. In this way, a deep unfrozen lens of soil and moisture becomes surrounded by permafrost, or is trapped within the permafrost (Fig. 16.5b). As the permafrost extends, volume expansion consequent on freezing imposes cryostatic pressure on the water contained in the interstices of the unfrozen soil. Eventually, this water is forced upwards (in the direction of least resistance) to the surface of the shallow lake or the old lake bed. The sediments there are domed, and a lens of ice forms by the freezing of the expelled water (Fig. 16.5c). As the mechanism involves a closed cell of unfrozen soil and moisture, this explanation is widely known as the *closed* hypothesis.[26] However, though it offers an explanation for the association of lake and pingo, like the open hypothesis it does not explain why these landforms are so abundantly developed in one area—the Mackenzie delta. Similar

geological and geomorphological conditions obtain in, for example, the deltas of the great Siberian rivers. Though bulging ridges and ice mounds have been described from the Lena delta, they have been attributed to permafrost and fossil ice. Nowhere, apparently, do the conical, spectacular and unmistakable shapes of the pingos occur in such great profusion in the USSR or in southern Alaska, as they do in the Mackenzie delta.

The closed hypothesis is widely accepted but a rather different interpretation of the Mackenzie pingos, in which their association with lakes has been—as it were—turned upside down, has recently been put forward. Seismic studies have shown that the Mackenzie delta is a region of subsidence (Fig. 16.6). It has been suggested[27] that as it sinks, the permafrost formed at or near the surface passes through the 1°C isothermal surface, and into higher temperature zones. The permafrost melts, and the rock or soil particles, formerly held in a rigid matrix of ice, lose strength on their conversion to a mush. Compaction of these water-logged sediments takes place under the weight of newly-deposited overlying sediments. This causes the expulsion of the interstitial water, which migrates laterally until a constriction (such as a tectonic or bedrock high) is reached. Thus gravity causes water to be squeezed out of minor basins of subsidence, and to be forced upwards where there are barriers to its lateral movement, or where two opposed migratory flows meet. The water is forced upwards, through the permafrost, notably above the tectonic highs where the sinking permafrost may be bent and weakened, to the surface where pingos are formed.

Why then the association of pingos and lakes? According to the tectonic hypothesis, it is not the lakes which indirectly initiate pingo development, but rather the converse. The upwelling of relatively warm waters causes what has been called a "permafrost holiday": the permafrost is melted and the soil volume decreases; the updoming of the pingo sediments causes subsidence in adjacent areas, and for both these reasons lakes are formed.

The gravitational hypothesis at this stage fails to explain why pingos are not profusely developed in the great deltas of northern Siberia and southern Alaska. It may be, however, that these regions are not of negative tectonic character or such complex subsurface structure as the Mackenzie delta. The occurrence of lacustrine sediments in the pingos shows, however, that the lakes predate the conical hills, and thus argues against this gravitational hypothesis. Moreover, published maps which show the distribution of pingos in the Mackenzie delta[28] disclose no obvious linear patterns.

Thus, on present evidence, the closed hypothesis appears most satisfactory to explain the field evidence.

F. THERMOKARST

The degeneration of ground ice, and especially permafrost under the influence of warmer climates, gives rise to pitted and hummocky topography (Pls 16.6 and 16.7) which is called *thermokarst* because it in some respects resembles the collapse features associated with limestone terrain (Chapter 3) and because it is basically due to temperature changes in the near-surface layers of rock and soil. The melting of ground ice, which is unevenly distributed, causes differential

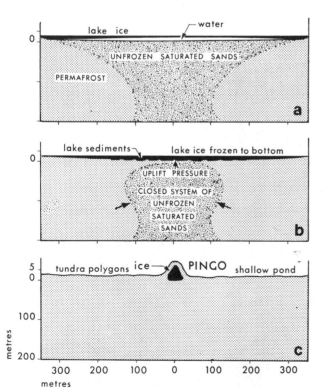

Fig. 16.5 Development of pingos according to the "closed" hypothesis. *(After J.R. Mackay, "Pingos of the Pleistocene", pp. 21–63, by permission of Geogr. Bull.)*

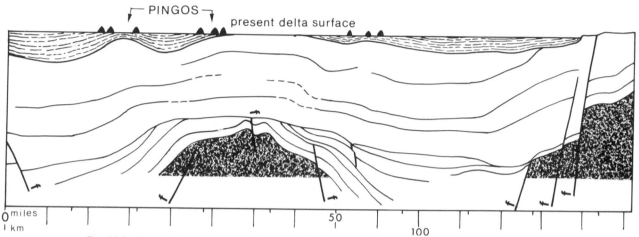

Fig. 16.6 Diagrammatic cross-section through the Mackenzie delta, showing fault blocks, perma-frost and (according to Bostrom) location of pingos above major fault zones. (*After R.C. Bostrom, "Water Expulson", pp. 586–72, reproduced from the* Journal of Glaciology *by permission of the International Glaciological Society and R.C. Bostrom.*)

Plate 16.6 Thermokarst developed alongside gravel highway where forest has been cleared 20km southwest of Lena River, eastern Siberia, USSR. (*T.L. Pewe.*)

Plate 16.7   Recent thermokarst pit, Maya, eastern Siberia, USSR.   (*T.L. Pewe.*)

volume decrease and the settling and subsidence of some areas. Pits, dry gullies and small valleys, hummocks and closed depressions, many of them with lakes, are the result. Very large areas of the northern USSR and Alaska are occupied by such landform assemblages, and relict thermokarst has been reported from Poland. Most thermokarst features are manifestations of climatic change, but on a local scale forest clearance and even the erection of buildings can alter the thermal characteristics of the near-surface layers, and cause thermokarst features to develop.

G. SOLIFLUAL STREAM FEATURES

Frozen ground forms a gliding surface over which unstable debris can readily slide or slip. The regolith is made unstable partly by frost heaving, partly by the release of water during the summer melt which causes the surface soil and weathered rock to lose cohesion. The result is mass movement on a grand scale. Slopes in periglacial regions are alive and active compared to slopes in temperate areas. Although the rates of downslope movement vary greatly from place to place and time to time, movements due wholly or partly to creep induced and facilitated by ground ice include one of 10 cm in 24 hours[29] and one, of surface pebbles, of 70 cm in a year on a 21° slope.[30]

The term applied to such flowage under gravity of masses of waste saturated with water is *solifluction*, which literally means "soil flow" and which therefore could be applied, as its originator realised,[31] to mass movements in a wide range of climatic conditions. However, though there is considerable latitude in its usage, the term is usually applied to movements involving either the flowage of water-saturated masses or the creep of solid waste under the influence of frost heaving.

Solifluction deposits which are characteristically involuted and which are known in southern England as *head*, form lobes and terraces on hillslopes (Pl. 16.8). Masses of debris slide and flow downhill, and stones and vegetation which tend to be sorted into distinct zones as is riverborne debris, form barriers in front of the flows: hence there are stone-banked and turf-banked terraces. So mobile is some of the soliflual debris that streams of stones, blocks and mud, rivers of coarse material in a matrix of water and mud, flow down the floors of valleys, and down hillslopes. Thus there are distinct streams of basalt blocks derived from cappings of that rock covering granite slopes in the Monaro district of southern N.S.W. (Fig. 16.7). The fines have been washed from some of the streams of originally mixed debris leaving behind a concentration of coarse debris in *stone streams*. Talus and scree cones (Pl. 16.9) are commonplace in upland periglacial regions and are due to the accumulation of coarse frost-riven debris on lower hillslopes or at the outlets of local and ephemeral streams. Frost-action is also thought to be responsible for the great upland fields of residual, coarse and usually angular blocks known as *Felsenmeer*.

Some blockstreams are found to have cores of ice. Each such *rock glacier* (Pl. 16.10) consists of a tongue or ribbon of angular blocks with a core of ice located within the mass. They are well-named for they are due to the flow of a frozen mass of rubble which has moved downhill in a fashion closely akin to that displayed by real glaciers. Their movement has however been limited by the loss of water and ice.

Solifluction flows are also evidently responsible for some erosional features. For instance, *shaved surfaces* (Fig. 16.8) considered to be due to the erosional work of soliflual debris over weathered rock[33] have been described from the Wellington area of New Zealand. Rock benches or platforms

Plate 16.8a The whole of this headwater amphitheatre of East tributary of Nome Creek, Tolovana, Yukon, is occupied by solifluction lobes. *(T.L. Pewe.)*

Plate 16.8b Small solifluction lobe terrace in central Labrador. *(C.R. Twidale.)*

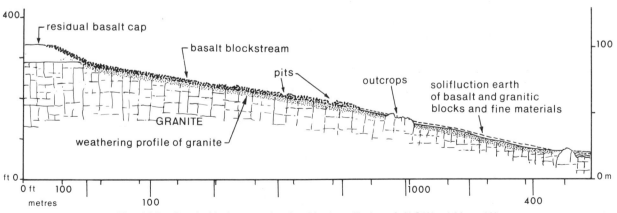

Fig. 16.7 Basalt blockstream in the Monaro district of N.S.W. *(After J.N. Jennings, "Australian Landform Example No. 13. Periglacial Blockstream", Aust. Geogr., 1961,* **11**, *pp. 85—6, by permission of J.N. Jennings and Aust. Geogr.)*

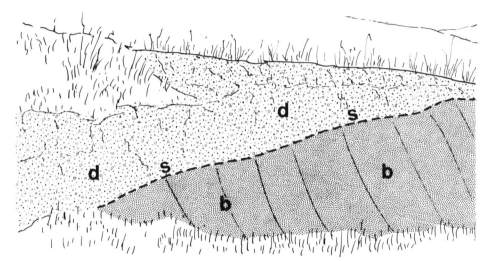

Fig. 16.8   Shaved surface (s) in section, d = solifluction debris, b = bedrock.

Plate 16.9   Screes at Wast Water, Lake District, northwest England. (*Hunting Surveys.*)

thought to be caused by solifluctional abrasion have been described from subarctic regions such as Alaska and Labrador. Some of the flats which surround the tors and other residuals of Dartmoor and Exmoor are alleged to be of this character.[34] Gently sloping and rough in detail, though planate in broad view, in some cases these features (which are the *goletz terraces* of the Russian literature) merge downslope with solifluction terraces (that is, terraces of depositional or constructional origin). The erosional platforms are called *altiplanation benches* or *terraces*[35] and are said to be formed through frost shattering of the bedrock and subsequent removal downslope of the debris in solifluction flows (Pl. 16.11). However, why they should acquire their planate form is difficult to understand unless there is some sort of groundwater or ground ice control of the processes involved in their formation. For example, could the benches mark the lower limit of summer melting of ground ice, and thus in effect mark the lower boundary of rocks subjected to alternations of freezing and thawing?

CONCLUSIONS

In general, then, areas subject to frost and snow are subjected to relatively rapid changes of the land surface. They are active or alive, with superficial changes taking place almost perceptibly. Frost shattering can easily be demonstrated. In the right lithological setting, frost shattering is so rapid that it can be shown to take effect in one winter, or in a shorter time span. Downslope movements due to solifluction are also more rapid than in temperate lands.

Plate 16.10 Rock glacier, Tangle Lakes area, south side Alaska Range, central Alaska. *(T.L. Pewe.)*

Plate 16.11   Altiplanation terraces developed on andesite on the south side of Indian Mt, Hughes, Alaska. *(T.L. Pewe.)*

References Cited

1. P.V. Hobbs, "Snow: Metamorphism of Deposited Snow", in R.W. Fairbridge (Ed.) *Encyclopaedia of Geomorphology*, Reinhold, New York, 1969, pp. 1025–8.
2. V. Cornish, "On Snow Waves and Snow Drifts in Canada", *Geogr. J.*, 1902, **20**, pp. 137–75; *idem*, *Waves of Sand and Snow and the Eddies which Make Them*, Unwin, London, 1914.
3. M. Seppälä, "Evolution of Eolian Relief of the Kaamasjoki-Kiellajoki River Basin in Finnish Lapland", *Fennia*, 1971, **104**, pp. 5–88.
4. C.T. Madigan, "The Australian Sand Ridge Deserts", *Geogr. Rev.*, 1936, **26**, pp. 205–27.
5. D. Mawson, *The Home of the Blizzard*, Heinemann, London, 1915.
6. R.L. Nichols, "Geomorphology of Inglefield Land, North Greenland", *Med. Grønland*, 1969, **188**.
7. M. Derruau, "Les Formes Périglaciaires du Labrador-Ungava centrale comparees a celles de l'Islande centrale", *Rev. Geomorph. Dyn.*, 1956, pp. 11–16.
8. C.A.M. King, "The Coast of south-east Iceland near Ingolfshofdi", *Geogr. J.*, 1956, **122**, pp. 241–6.
9. M. Boyé, *Glaciaire et Périglaciaire de l'Ata Sund, Nord Oriental Groenland*, Hermann, Paris, 1950.
10. R.L. Nichols, "Geomorphology of Ingelfield Land".
11. M. Boyé, *Glaciaire et Périglaciaire*.
12. See for example E.P. Henderson, "Large Nivation Hollows near Knob Lake, Quebec", *J. Geol.*, 1956, **64**, pp. 607–16.
13. J.L. Dyson, "Snowslide Striations", *J. Geol.*, 1937, **45**, pp. 549–57.
14. A.B Costin, J.N. Jennings, H.P. Black and B.G. Thom, "Snow Action on Mt Twynham, Snowy Mountains, Australia", *J. Glaciol.*, 1966, **5**, pp. 219–28.
15. J.L. Davies, *Landforms of Cold Climates*, Australian National University Press, Canberra, 1968, p. 18.
16. D.L. Linton, "The Abandonment of the Term 'Periglacial' ", in *Palaeoecology of Africa and of the Surrounding Islands and Antarctica*, Balkema, Cape Town, 1969, Vol. 5.
17. S. Taber, "Frost Heaving", *J. Geol.*, 1929, **37**, pp. 428–61; *idem*, "The Mechanisms of Frost Heaving", *J. Geol.*, 1930, **38**, pp. 303–17.
18. See A.L. Washburn, "Classification of Patterned Ground and Review of Suggested Origins", *Bull. geol. Soc. Am.*, 1956, **67**, pp. 823–66. *See also* Chapter 7.
19. R.F. Black, "Permafrost—a Review", *Bull. geol. Soc. Am.*, 1954, **65**, pp. 839–55.
20. J.M. Shearer, R.F. McNab, B.R. Pelletier and T.B. Smith, "Submarine Pingos in the Beaufort Sea", *Science (N.Y.)*, 1971, **174**, pp. 816–18.
21. F. Müller, "Analysis of some Stratigraphic Observations and Radiocarbon Dates from Two Pingos in the Mackenzie Delta Area, N.W.T.", *Arctic*, 1962,

15, pp. 278–88.
22. F. Müller, "Beobachtungen über Pingos", *Med. Grønland*, 1959, **153**; J.R. Mackay, "Pingos of the Pleistocene Mackenzie Delta Area", *Geogr. Bull.*, 1962, **4** (18), pp. 21–63; J.R. Mackay and J.K. Stager, "The Structure of Some Pingos in the Mackenzie Delta Area, N.W.T.", *Geogr. Bull.*, 1966, **8**, pp. 360–8.
23. J.R. Mackay, "Pingos of the Pleistocene", p. 27.
24. J.R. Mackay and J.K. Stager, "Structure of Some Pingos", pp. 360–8.
25. E. de K. Leffingwell, "The Canning River Region, northern Alaska", *U.S. Geol. Surv. Prof. Paper*, 1919, **109**.
26. A.E. Porsild, "Earth Mounds in Unglaciated Arctic northwestern America", *Geogr. Rev.*, 1938, **28**, pp. 46–58; J.R. Mackay, "Pingos of the Pleistocene", pp. 210–63.
27. R.C. Bostrom, "Water Expulsion and Pingo Formation in a Region affected by Subsidence", *J. Glaciol.*, 1967, **6**, pp. 568–72.
28. J.R. Mackay, "Pingos of the Pleistocene", pp. 21–63, at Figure 1.
29. P.J. Williams, "An Investigation into Processes Occurring in Solifluction", *Am. J. Sci.*, 1959, **257**, pp. 481–90.
30. J. Smith, "Cryoturbation Data from South Georgia", *Biul. Periglacj.*, 1960, **8**, pp. 73—9.
31. J.G. Andersson, "Solifluction, a Component of Sub-Aerial Denudation", *J. Geol.*, 1906, **14**, pp. 91–112.
32. C. Wahrhaftig and A. Cox, "Rock Glaciers in the Alaska Range", *Bull. geol. Soc. Am.*, 1959, **70**, pp. 383–436; Victoria Guiter, "Une Forme Montegnarde: le Rock-glacier", *Rev. Géogr. Alpine*, 1972, **60**, pp. 467–87.
33. C.A. Cotton and M. Te Punga, "Fossil Gullies in the Wellington Landscape", *N.Z. Geogr.*, 1955, **11**, pp. 72–5.
34. A. Guilcher, "Nivation, Cryoplanation et Solifluction Quaternaires dans les Collines de Bretagne occidentale et du Nord de Devonshire", *Rev. Géom. Dyn.*, 1950, **1**, pp. 53–78; M. Te Punga, "Altiplanation Terraces in southern England", *Biul. Periglacj.*, 1956, **4**, pp. 331–8.
35. H.M. Eakin, "The Yukon-Koyukuk Region, Alaska", *Bull. U.S. Geol. Surv.*, 1916, **631**.

Additional and General References

C. Embleton and C.A.M. King, *Glacial and Periglacial Geomorphology*, Arnold, London, 1968.
A.L. Washburn, *Periglacial Processes and Environments*, St Martins, London, 1973.

CHAPTER 17

Glaciers and Their Work

A. GENERAL STATEMENT

In cold lands which receive adequate precipitation accumulated snow is compacted to form ice which moves downslope under gravity as streams of ice or *glaciers*. How do glaciers affect the landscape?

Modern ice masses include icecaps, cirque, valley and piedmont glaciers. Snow accumulates and suffers compression under the weight of the overlying snow, first to *névé* (also called *firn*) which is a mush of mixed snow and ice, and then to ice—a wholly crystalline material. Much snow and ice accumulates on high plateaux to form *icecaps*, whence it flows outwards by way of gaps between the mountains (Pl. 17.1) to lower levels. Much accumulates also in depressions on mountain sides; this ice flows outwards too, rounding out the depressions as it does so to form armchair-shaped hollows, or cirques (Pl. 17.2). Some *cirque glaciers* exist in isolation but many extend downslope and coalesce as they flow downstream to form *valley glaciers* (Pls 15.2, 17.3), rivers of ice flowing down constricted and confined courses in pre-existing valleys. Those which coalesce in their lower reaches below the level of the intervening peaks give rise to *dendritic* glacier systems (Pl. 17.4). Some, as in Antarctica, merge near the source area and inundate many of the mountains, forming what are called *transection glaciers*

(Pl. 17.5). Where glaciers emerge from uplands onto, say, a coastal plain, they coalesce to form a *piedmont glacier*. Many examples of icecaps and cirque and valley glaciers are known, but piedmont glaciers are rare. The Frederikshaab glacier in Greenland and the Bering glacier of Alaska are examples, but the best known is the Malaspina glacier of southern Alaska[1] (Fig. 17.1).

In Antarctica numerous icecaps, cirque and valley glaciers merge to form the largest ice-sheet in the world, and though it occupies a different topographic setting, the smaller Greenland ice-sheet is similar. Together they contain almost 2·24 per cent of the world's total supply of water (compared with 97·1 per cent in the ocean basins, less than 0·02 per cent in rivers and lakes, and 0·001 per cent in the atmosphere). They are important, not only because of their great volume and area, but also because similar ice-sheets formerly extended over much of America and Europe; thus, investigations of the behaviour of these two modern ice-sheets provides important information for the interpretation and understanding of the effects of the former sheets.

With an area of some 9 million km^2, the Antarctic ice mass is the largest in the world (Fig. 17.2a and b). It is seven times as large as Greenland, exceeds 2000 m in thickness over 75 per cent of its area and, existing as it does in an

Plate 17.1 The Grantland Mountains, northern Ellesmere Island, showing ice caps with tongues of ice flowing into the fjord-like Piper Pass valley. The glaciers have blocked the drainage in the valley, forming ice-dammed lakes. *(Dept Energy, Mines and Resources, Ottawa. Photo No. T401L—86.)*

Plate 17.2 The Coast Mountains in the Pacific Ranges, British Columbia. Mt Waddington (3994 m) is in the centre of the picture. The Radiant Glacier (left) rises in a cirque on the flanks of Mt Waddington and merges with the Scimitar Glacier which displays lateral and medial dirt bands—future moraines. The small terminal moraine (X) is caused by the advance of a small tributary glacier (right—Y). *(Dept Lands, Forests and Water Resources, B.C. Photo No. BC 551076.)*

Plate 17.3 Bernina Range, in the Grisons, Swiss Alps, with the Morteratsch Glacier. Note glacier junction crevasses, and icefall near junction indicating a steep subglacial floor. *(Swiss Nat. Tour. Office.)*

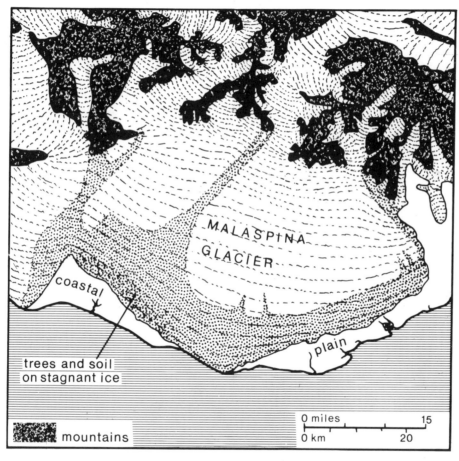

MALASPINA GLACIER

coastal

plain

trees and soil
on stagnant ice

mountains

0 miles 15

0 km 20

Fig. 17.1 The Malaspina Glacier is fed by several valley glaciers, the most important
of which is the Seward. On reaching the coastal plain, these unite to form a piedmont
glacier 110 km long with an area of almost 4000 km , which rises some 500 m above
its margin. The surface is undulating, though rough in detail, being broken by crevasses
and morainic ridges. The ice is incredibly contorted. Numerous lakes are impounded
at the glacier margin by ice and morainic ridges, many of which are forest covered.
Surface streams generated by summer meltwaters disappear into the ice and flow as
englacial and subglacial streams before emerging at the ice margins where they have
deposited a considerable outwash plain.

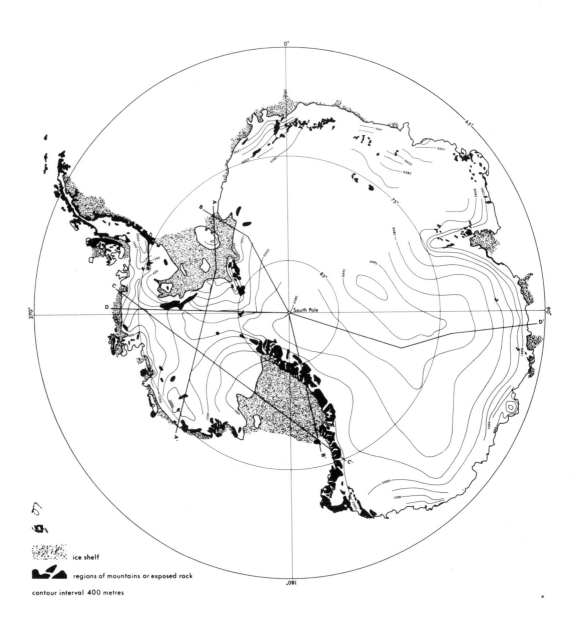

Fig. 17.2a   The Antarctic ice-sheet in plan.

SECTION: WEDDELL SEA TO HOBBS COAST

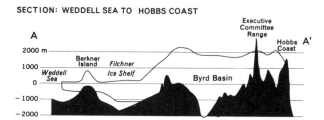

SECTION: WEDDELL SEA TO ROSS ISLAND

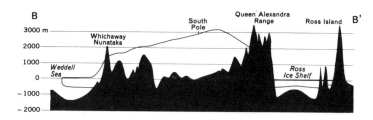

SECTION: BELLINGSHAUSEN SEA TO PRINCE ALBERT MOUNTAINS

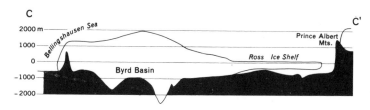

SECTION: BELLINGSHAUSEN SEA TO MIRNY

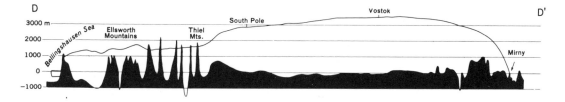

Fig. 17.2b   The Antarctic ice-sheet in section.

Plate 17.4 Disraeli Fjord and Glacier, Ellesmere Island, with sharp-ridged mountains and nunataks. *(Dept Energy, Mines and Resources, Ottawa. Photo No. T400R—80.)*

Plate 17.5 Sharp-crested nunataks standing above the general level of the thick ice cap in Antarctica. *(Dept Foreign Affairs, Canberra.)*

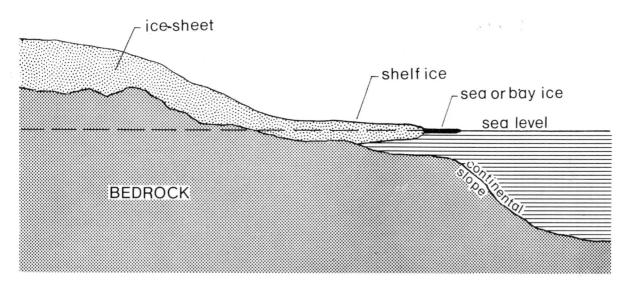

Fig. 17.3 Section through ice shelf off Antarctica.

essentially ice-free world, exerts a significant influence on sealevel, climate and other aspects of the environment. Not all of Antarctica is ice, as mountain ranges protrude above the level of the glaciers *(nunataks)*.

High as some of the nunataks are, they are only the uppermost parts of huge mountains largely buried by the ice-sheet. The ice is very thick, rising to well over 9000 m above sealevel. The subglacial relief is irregular, particularly in West Antarctica, though it is more subdued in the East (Fig. 17.2a and b).

But these rock outcrops cover a relatively small proportion of the whole area, for ice streams flow out toward the ocean in all directions and even extend scores of kilometres out over the ocean as the massive sheets of ice known as *ice shelves*[2] (Figs 17.2a and b, and 17.3) which, on breaking up, form tabular icebergs (Pl. 17.6a). Both here and in the Arctic, sea-ice also forms (Pl. 17.6b).

The Greenland ice mass is an elongate lens of ice (Fig. 17.4) extending over an area of 2.8 million km². As is the case in Antarctica, the bedrock floor in places extends below present sealevel; the topography of the floor is subdued, and similar to that in East Antarctica. Nunataks form a discontinuous coastal ring around the ice dome which rises to over 3000 m above sealevel, the crest lying closer to the east than to the west coast. The ice-sheet flows outwards through gaps in the encircling rim of mountains to reach the sea, where the glaciers float out a short distance before breaking up to form icebergs.

At least some of the ice is relict in nature, as tests have disclosed that ice from a depth of some 1500 m below the surface is nearly 10,000 years old and the basal ice may be up to 100,000 years old; similar to the age of the lowest Antarctic ice. There are small ice-free areas both on the east and west coast. A more extensive area, Pearyland, in the far north of Greenland, is ice-free because of its very low precipitation.

B. GLACIER MOTION

Glacier ice moves outwards from the centre of accumulation and downslope at rates varying from a few centimetres to several metres a day. A small Greenland glacier has been observed to travel at 6 m per day. In Alaska, the Yanert Glacier travels at 8–10 m per day. Measurements and estimates made on an annual basis reveal similar variations: valley glaciers in the European Alps move at speeds of between 30 and 150 m per annum, in the Himalayas between 200 and 1300 m each year and in Greenland between 1000 and 3000 m per annum. This flow is motivated by gravity, but the weight of ice pushing behind (the hydraulic head) can cause uphill movement over short sectors. Glaciers exhibit differential movement, both in space and time.

As in rivers, movement along the length of a valley glacier is governed in some measure by the nature of the channel: flow is faster through narrow sectors than through broad basins. Apart from these local variations, however, velocity tends to increase downstream from the head of the glacier to just below the néve limit, from which point the velocity decreases to the snout (Fig. 17.5a and b). In the upper reaches, it has long been observed that the ice tends to move downwards from the surface into the body of the glacier, and that conversely there is an upward movement in the lower areas (Fig. 17.5b).

Terminal movement is at a maximum in summer when meltwaters are most abundant. Movement at the head is greatest in winter when the greatest amount of snow accumulation takes place. In addition to these normal variations, waves, surges, or sudden unaccountable phases of movement several times more rapid than normal are also known.

A consideration of boundary layer conditions suggests that the centre of a valley glacier should move more rapidly than the margins and base which are in contact with the bedrock (Fig. 17.5c). Simple observations show that this is

Plate 17.6a Edge of tabular iceberg in Antarctica, showing annual growth layers of ice. (*Dept Foreign Affairs, Canberra.*)

Plate 17.6b Yukon coastal plain showing sea ice and floes in the Beaufort Sea. The plain has been glaciated and is characteristically hummocky, with numerous lakes. Note the meanders and oxbow lakes developed in the stream which runs across the field of view parallel to the coast from middle left to bottom right. (*Dept Energy, Mines and Resources, Ottawa. Photo No. T29R025.*)

generally true but the increase in velocity from margin to centre is not gradual, as might be anticipated. Instead, the velocity gradient is restricted to a quite narrow marginal zone, with the bulk of the ice moving at a uniform rate as a stream within the glacier (Fig. 17.5d). This is known as *blockschollen* flow. Because of the marked velocity gradient in the marginal zone, strong shearing results in the development of numerous *crevasses* (Fig. 17.5e).

Due to frictional drag, the basal area of ice is retarded. Thus, the upper area most commonly moves more rapidly than the lower (Fig. 17.5f): a pipe placed vertically in a glacier leans forward in time. Such a pattern of flow is called *gravity flow* and observations on modern glaciers suggest that this is by far the most common type of behaviour. It was alleged that glaciers could flow at greater velocities at depth than at the surface,[3] partly because of eddy-like patterns formed by glacial striae, indicating plasticity of the basal ice, and partly because of alleged discrepancies between the intake of glaciers from snow and ice on the one hand, and their outflow as measured by surface velocities near their outlets (income exceeded the volume of outgoing ice) on the other. But physical considerations are against the hypothesis which, apart from some measurements taken on the small glaciers in Greenland, has never been demonstrated in the field.

However, although basal ice apparently achieves a plastic state (see below), the concept of *extrusion* flow is mechanically difficult, and now the most commonly accepted theory of glacier motion is due to Nye.[4] Nye's theory is based on the assumption that the ice is plastic and that the glacier flow is such that at all times the glacier maintains a critical thickness for the prevailing conditions. The glacier becomes deformed, largely through the development of slip lines and fault planes, to accomodate the irregularities of the subglacial relief.

In general, where there is a concave bottom or loss of ice, flow is *compressive* (or passive) as the glacier and the slip planes run from the base to the surface of the glacier. Where, on the other hand, the subglacial floor is convex or there is an addition of ice, the glacier maintains a critical thickness by *extending* (active flow), again due to the development of shear planes which run from the surface down to the base of the ice mass. This is in accord with the observation reported above that ice moves downwards near the head of the glacier and upwards near the snout. An interesting corollary is that the main work of erosion occurs in relation

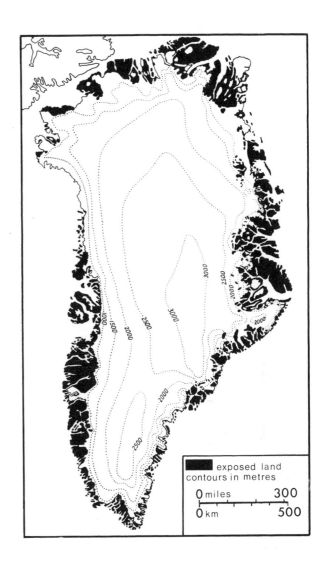

Fig. 17.4 Greenland icecap in plan (above) and in section (below). *(The latter after R.H. Hamblin et al., "British North Greenland Expedition 1952–4. Scientific Results", Geogr. J., 1956, **122**, pp. 203–37.)*

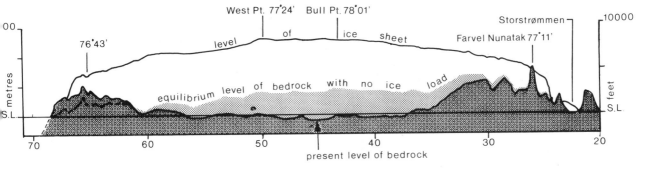

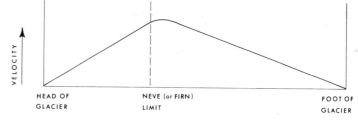

Fig. 17.5a Velocity distribution in a valley glacier in longitudinal section.

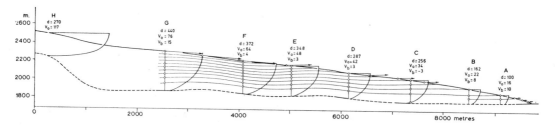

Fig. 17.5b Movement of a valley glacier (the Saskatchewan glacier) in longitudinal section showing flow lines and velocity of flow along profile. (*After M.F. Meier, "Mode of Flow of Saskatchewan Glacier, Alberta, Canada", US Geol. Surv. Prof. Paper, 1960,* **351**.)

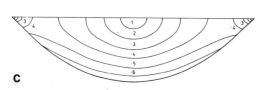

c

Fig. 17.5c Theoretical velocity distribution in a channel four times as wide as it is deep. (*Adapted from J.F. Nye, "The Flow of a Glacier in a Channel of Rectangular, Elliptic or Parabolic Cross-section", J. Glaciol., 1965 (5), pp. 661–90, by permission of J. Glaciol and J.F. Nye.*)

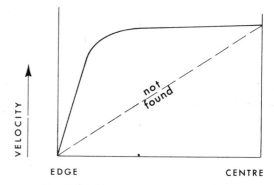

Fig. 17.5d Movement of a valley glacier in transverse section—blockschollen flow.

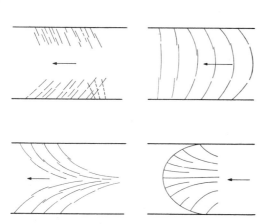

Fig. 17.5e Common crevasse patterns. Top left: radial; top right: splaying; bottom left: transverse; bottom right: radial, with older rotated crevasses indicated by dashed lines. (*After R.P. Sharp,* Glaciers, *University of Oregon, 1960.*)

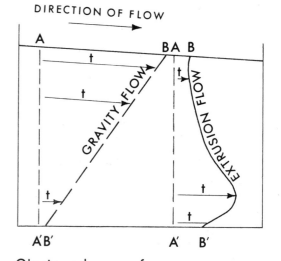

Glacier advances from

A-A′ – B-B′ in time t

Fig. 17.5f Movement in valley glacier in vertical section.

to compressive flow where debris is carried up to the surface from the base of the glacier.

In detail, and although several mechanisms have been suggested,[5] intergranular yielding seems to be the most important mode of achieving differential movement within the glacier. This involves slippage along cleavage planes in the ice crystals in certain shear zones. Evidence of such dislocations has been observed in small cirque glaciers where *rotational* flow occurs (see below).

Most of the actual downslope movement of glaciers as a whole seems to be accomplished by basal or longitudinal slipping—the sliding of the whole ice mass over the underlying floor. Such movements are probably helped by the presence of meltwaters (hence maximum glacier movement near the snout in summer), but the mechanics of the process are obscure. Several theories have been suggested involving pressure melting and stress distribution in the basal ice. But none of the elaborate mathematical models seem adequate to explain the basal slippage observed in glaciers. Lliboutry[7] has proposed that glaciers might detach themselves from the bed, leaving cavities between ice and bedrock, with the glacier sliding over the high points in the latter. This seems at odds with field observations which suggest that, possibly because of bottom melting, the ice and bed load maintain virtually constant contact with the subglacial floor.

C. GLACIAL EROSION

1. Means of Erosion

That glaciers erode cannot be denied. If they achieve nothing else, they undoubtedly emphasise the preglacial relief. The numerous descriptions of surfaces scratched or striated by rocks carried along in the ice, of surfaces plucked of loose fragments by moving ice, of grooves and deep basins many of which are now occupied by lakes, and of debris of various grades carried far from its source, testify to this. The least that can be said is that masses of moving ice act as huge bulldozers which transport weathered and unconsolidated debris.

Nevertheless it is fair to add that in many areas, such as the Canadian Shield, the land surface has been modified, rather than radically altered, by the very large masses of ice that occupied the area. Preglacial soils have survived the passage of glacial ice in some areas, possibly through being located in depressions where ice stagnated, or through being frozen to form permafrost before the glacier arrived. Ice stagnation on a large scale has been suggested as a result of studies at the margin of the Greenland ice-sheet,[8] where, in view of the vast area covered by the ice and the known geological complexity, it might be expected that large volumes of far-travelled (or exotic) rocks might be brought by the ice from the central to the marginal areas, and dumped there to form moraines. In fact, in some areas only very small moraines, built of angular local rocks, are found. To account for this it has been suggested that because of its occurrence in a broad basin, a large part of the Greenland ice is stagnant, only the upper part being mobile, causing erosion near the coast (Fig. 17.6).

The enormous weight of glacier ice must cause significant crushing of the rock in the glacier bed, and must also facilitate the incorporation of debris into the ice. Shearing of the ice may also drag debris into the ice mass.

Apart from this general bulldozing effect of glacier ice, boulders carried in the basal ice abrade the underlying bedrock and cause quarrying. Ice also *plucks* small fragments of rock by freezing them into the main body of the glacier, though it may be more a matter of loosened blocks being carried forward by the general downslope movement rather

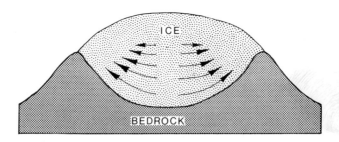

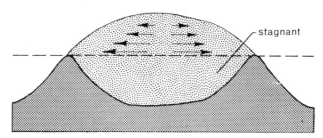

Fig. 17.6 Boyé's notion of stagnant basal ice and mobile upper ice in the Greenland ice cap.

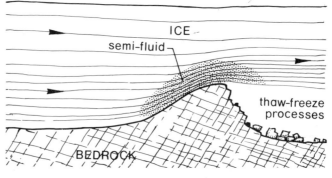

Fig. 17.7 Development of *roche moutonée* according to H. Carol. ("The Formation of Roches Moutonées", pp. 57–9, reproduced from the Journal of Glaciology by permission of the International Glaciological Society.)

Plate 17.7   McGregor plateau of the Canadian Rockies, looking east across the Angusmac Creek to a spectacular fluted or "drumlinised" surface eroded by ice in bedrock, and with marked grooving of depositional surface in right foreground.   Both bedrock and constructional forms trend 40° E, parallel to the known direction of ice movement in the area.   Note the steepening of the streamlined forms.   (*Dept Lands, Forests and Water Resources, B.C. Photo No. BC 761–70.*)

than plucking as such, as this takes place mainly on the downstream side of protruding outcrops, and in well-jointed rocks.

Glaciers certainly achieve contrasted effects on different sides of outcrops.   There is a general streamlining and polishing and rounding of the upstream surface, a process known as *mamillation*, while on the downstream faces there is quarrying and plucking.   This produces *roche moutonnées*, stoss and lee effects, and spectacular parallel grooving (Fig. 17.7, Pl. 17.7), depending on the size of the feature.   Carol[9] has explained this contrasted development in terms of the pressure condition of the ice in different areas.   He made direct observations of the consistency and speed of the ice both upstream and downstream from obstacles such as outcrops.   Where the lower layers of ice passed over rock barriers, they were not only transformed from a brittle state to a semiplastic condition (according to Carol of a consistency like cheese), but the rate of flow doubled.   Carol suggested that, because of the pressure exerted on the ice by the rock, the ice changes condition and is no longer capable of erosion

on this uphill or upstream slope—the ice is partially melted and any rocks and sand grains are not firmly pressed against the underlying rock surfaces as they would be by brittle, crystalline ice.   But on the slope, there is more space, the ice is no longer under such great pressure, and the water produced by the squeezing of the obstacle refreezes.   Here, then, is an area of freeze–thaw activity and of plucking of joint blocks (Fig. 17.8).

2.   Resultant Forms

The principal effects of glacial erosion as it affects pre-existing river valleys were succinctly described over seventy years ago by Harker[10] who, working in the Cuillen Hills in Skye which are underlain by a uniform mass of gabbro, described an assemblage of forms which we now recognise as being typical of glaciated valleys.   The valleys are U-shaped in cross-section, and their upper reaches open out into cirques, many of which contain lakes or tarns.   The valleys deepen downslope (as in Loch Coruisk), and in many cases they descend in a series of steps.   Tributary valleys are

Plate 17.8 Cirque with tarn (left) separated from U-shaped valley, Hawes Water, Cumberland, with Riggindale Beck in the U-shaped valley. Monuments separate the two valleys and lakes occupy the glacially-scoured valley floor in the foreground. *(Hunting Surveys.)*

discordant or *hanging* with respect to the main valley, which they join by way of waterfalls. The one important feature formed by valley glaciers and not mentioned by Harker is the truncated spur.

The *cirque*, also known as *cwm* or *corrie*, is one of the most common and striking of all glacial forms in upland regions (Pls 17.2, 17.8). Rather characteristically, this landform has not yet been satisfactorily explained although several hypotheses have been advanced. A cirque is an armchair-shaped hollow consisting of steep head- and sidewalls, a basin, and a threshold or raised rim which separates the basin from the downhill slope of the valley (Fig. 17.8a). The headwall may be up to 1000 m high and, even where the cirque glacier is no longer present, there is little scree at its base. Where a glacier occupies the hollow, there is commonly either a deep crevasse (called a *bergschrund*) in the firn near the backwall or a shallow break (*Randkluft*) between the glacier and the backwall.

For many years it was held that the bergschrund played a critical role in the development of cirques. Early in this century Willard Johnson descended a bergschrund and was impressed by the amount of shattered rock there. He argued[11] that basal sapping or plucking at the base of the bergschrund was responsible for cirque development: the rock fragments were carried in the base of the glacier and were instrumental in gouging out a basin in the underlying bedrock. Although this theory stood virtually unquestioned for many years, serious objections can be levelled against it. Why, for instance, does the threshold survive erosion by the heavily laden cirque ice? Temperature observations taken in bergschrunds[12] have shown that there is very little variation there, indicating that the freeze–thaw action is not likely to be marked. Furthermore, the bergschrund is not deep

(about 30 m), making it difficult to understand how backwalls up to 1000 m high can be formed, unless it is argued that the wall was developed as the glacier waxed and waned, and the bergschrund migrated up and down the wall. Even so, one is faced with the problem of the base of the wall and the basin, indicative of strong erosion, being formed by a very small glacier.

Lewis[13] suggested that summer meltwaters soaking into the backwall and freezing there could effect some shattering, but the lack of temperature fluctuation argues against this hypothesis. Subsequently, he advocated that the fracture of rocks which have been subjected to pressure release was important in this context.[14]

It has also been suggested that the basin form of cirques is related to the type of flow prevalent in cirque glaciers. Study of flow lines, dirt layers and the distribution of velocity in cirque glaciers (Fig. 17.8b), all indicate that rotational flow prevails[15] in these ice masses. It may be argued that the rotation of the ice about an axis located some distance above the cirque glacier in effect scoops out the hollow and is responsible for the characteristic hollow shape. The rampart or raised rim which intervenes between the cirque and the valley below, and which in many cases impounds the small lakes or tarns, is comprehensible in these terms.

But there is a danger of confused argument here. Is it the rotational flow of the glacier which causes the hollow, or the hollow which induces rotational flow? Clearly, if the former argument has any validity it should be possible to suggest reasons for the preferred accumulation of snow (and ice) in these upland regions in the first place. Fortunately it is possible to do so. Some of the initial areas of accumulation were probably old valley heads which received snow and ice from the adjacent slopes and where small glaciers were

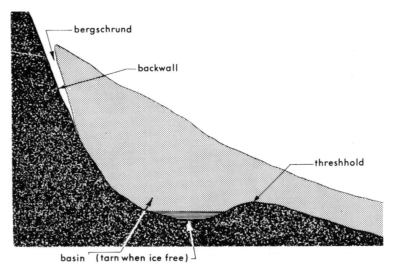

Fig. 17.8a Longitudinal section through a cirque glacier, and,
superimposed, a cirque free of ice, with lake or tarn.

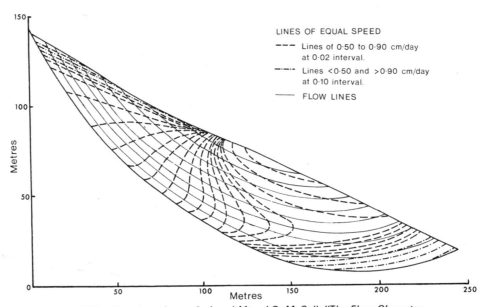

LINES OF EQUAL SPEED

– – – Lines of 0·50 to 0·90 cm/day
        at 0·02 interval.

–·–·– Lines <0·50 and >0·90 cm/day
        at 0·10 interval.

——— FLOW LINES

Fig. 17.8b Flow in a cirque glacier. (*After J.G. McCall, "The Flow Characteristics", pp. 39–62, by permission of the Royal Geographical Society.*)

formed. Others could have been nivation hollows (Chapter 16). Once they became sites of snow and ice accumulation, they were enlarged, and especially deepened,[16] by glacial scouring to their present armchair-shaped form.

Where a number of cirques have developed, the mountain mass is gradually reduced with the development of *monuments* between adjacent cirques (Pls 17.2 and 17.8) and where frost shattering has helped or is aiding glacial erosion, knife-edged ridges called *arêtes* (Pl. 17.9a) are formed. Three or more cirques eroding back toward the same area cause the formation of *pyramidal peaks* (Pl. 17.9b), of which the Matterhorn is a well-known example.

Valley glaciers are fed by cirque glaciers and flow as great streams of ice in valleys. Glaciated troughs differ from valleys excavated by rivers in several ways: they are straighter and lack interlocking spurs, that is, they have truncated spurs. They have steeper sides and flatter floors, or display a catenary or U-shaped profile in cross-section, in contrast with the V-shaped river valleys of mountain regions (Pls 17.2, 17.8 and 17.10a-c). The profiles are set into pre-existing V-profiles which are represented by distinct shoulders above the U-shaped valley (Fig. 17.10, Pl. 17.10a). Why do glaciated valleys have these breaks of slope in their sidewalls? And how has ice, which assumes plasticity under pressure, eroded these deep troughs?

Many glaciated valleys are flat-floored and trough-shaped in cross-section (Pl. 17.11a and b), largely because of postglacial deposition of alluvium from heavily laden braided glacial meltwater streams (Pl. 12.14). The Yosemite valley of California is a typical example (Pl. 17.12, Fig. 17.9) with over 600 m of outwash and lake sediments overlying an ice-scoured valley floor. But whether U-shaped or trough-like, a more-or-less catenary bedrock profile is preserved, the contrasted morphology being due to varied depth of erosion, narrowness of valley and degrees of subsequent infilling.

Projection of profiles from the preserved valley shoulders (Pl. 17.10a, Fig. 17.9) shows that considerable deepening of the valleys has been accomplished by the glaciers. This is achieved as much by the bulldozing and evacuation of weathered debris from the preglacial valley floor and lower slopes as by the quarrying of crushed unweathered rock.

Frost shattering in the period immediately preceding the arrival of the glacier would prepare the valley floor for bulldozing by the ice (Fig. 17.10). Hence, the glaciated valley is in many essential respects of etch type (see Chapter 19), with the erosional process removing weathered debris and exposing the unweathered rock profiles.

However, such bulldozing cannot entirely account for truncated spurs. Certainly the interlocking spurs of river valleys (see Pl. 12.10) could be markedly eroded at their toes because it is there that weathering is most pronounced. But the occurrence of such features as Half Dome, bordering the glaciated Yosemite Valley in California (Pl. 17.12), surely argues considerable lateral erosion by valley glaciers. This may seem surprising in view of the low velocity of the marginal areas of such glaciers, but these zones are very heavily charged with debris — the glacier's tools of abrasion — derived from the nearby screes and frost-shattered valley side-slopes.

Deepening of the main valleys to a greater extent than the tributaries, which contained only smaller glaciers (or no glaciers), caused tributary valleys and streams to be discordant with, and perched well above, the main valleys. They are known as *hanging valleys* (Pl. 17.11a).

The bedrock floors of glaciated valleys are also irregular, with many basins and steps (or *riegel*), which as Bakker[17] pointed out, probably represent irregularities in the preglacial or interglacial weathering front. Some of the basins are occupied by lakes (Pls 17.8, 17.10b) which tend to be elongated and strung out like beads on a string (*paternoster* lakes) though some are continuous. The Finger Lakes of northern New York State are good examples (Fig. 17.11). This group consists of eleven major lakes arranged in radiating orientations; they are undoubtedly due to glacial scouring of preglacial valleys, though some glacial initiation of minor valleys has been claimed. The area in which they occur was overridden by more than 1000 m of ice at least twice during the Pleistocene (in Illinoian and Wisconsin times—see Chapter 21) and there is little wonder, therefore,

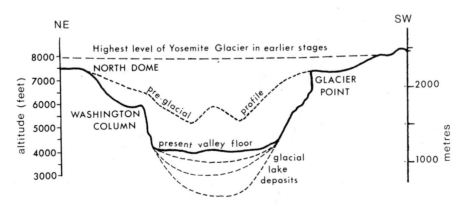

Fig. 17.9   Cross-section through glaciated U-shaped valley, Yosemite Valley, California.

Plate 17.9a Nunataks with arêtes, near the mouth of the Beadsmore Glacier, Antarctica. *(R. Oliver.)*

Plate 17.9b Mitre Peak and Milford Sound, Otago, New Zealand. *(N.Z. Govt Tour. Office.)*

Plate 17.10a Ferpecle Glacier in the Valais of Switzerland, showing the Matterhorn, a pyramidal peak, in the background. The glaciated valley shows well-defined valley shoulders. Note the crevasses, and the dam in the fore-ground. *(Swiss Nat. Tour. Office.)*

Plate 17.10b   U-shaped valley, eroded in granite hills, Glen Tarbert, Argyllshire.   Looking west from Loch Linnhe to Loch Sunart.   *(Hunting Surveys.)*

Plate 17.10c U-shaped valley of Tinae Creek, which leads down to Stave Lake, Coast Mountains, Pacific Range, B.C., with Mt Baker (3288 m), a volcanic cone in the Cascades of Washington State, USA, in the right distance. *(Dept Lands, Forests and Water Resources, B.C. Photo No. BC 499–82.)*

that weathered rock formed beneath the old valley floors has been extensively bulldozed away, so that in two cases the lake beds now lie beneath sealevel (Seneca, 160 m and Cayuga, 17 m).

In detail, the floors of some glaciated valleys are pitted with deep cylindrical hollows (analogous to the pot-holes of river channels) which are called *moulins*. They have indeed been attributed to meltwater streams plunging through the glacier to the bedrock floor beneath and there scouring out a minor plunge pool. But if the glacier is moving it is difficult to visualise even such marked erosion acting for sufficient time to produce the observed results, and it may be that such moulins form in relation to stagnant, and not active, glaciers, or that they have been eroded as pot-holes in beds of sub-glacial streams.

Near the coast, the U-shaped valleys of the glaciated uplands merge with *fjords*, which are long, narrow glacial troughs that have been drowned by the sea (Pls 17.9b, 17.13, Fig. 17.12a-e). They are developed from preglacial river valleys, the pattern of which in some areas was strongly guided by major joints and faults (Fig. 17.12b). Each fjord possesses a lip or threshold at its mouth. Many fjords are very deep, due to glacial scouring. Sogne Fjord in western Norway, for instance, is 1308 m deep in places (Fig. 17.12c), but has a lip only 160 m below sealevel. Chatham Strait, Alaska, is 878 m deep and Finlayson Channel, British Columbia, 780 m; Messia Channel, in south Chile, is 1288 m deep and Skelton Inlet in Antarctica achieves a depth of 1933 m. Many fjords are in west coast situations and are backed by precipitous ranges: the ice masses, fed by abundant precipitation, must have flowed rapidly and caused much erosion (Fig. 17.12d). But where they entered the sea, the glaciers must have floated. Thus they ceased to erode the bedrock, and this is said to be the reason for the lip found at the outer margin of the fjords.

However, Crary[18] has suggested with respect to Skelton Inlet, where there are two significant deeps separated by highs or lips along the longitudinal profile of the fjord (Fig. 17.12e), that the irregularities in bottom topography are due

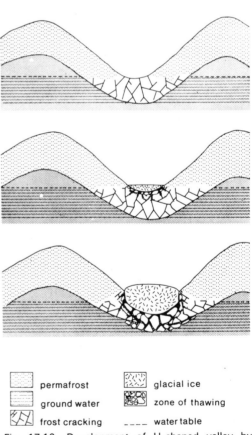

	permafrost		glacial ice
	ground water		zone of thawing
	frost cracking		water table

Fig. 17.10 Development of U-shaped valley by frost shattering and subsequent glacial bulldozing. *(After J. Tricart and A. Cailleux, Traite de Geomorphologie, 1962.)*

Plate 17.11a Lauterbrunnen valley in the Swiss Bernese Oberland, with percipitous sidewalls and alps above, flat-floored valley and hanging valley. *(Swiss Nat. Tour. Office.)*

Plate 17.11b Lyttle Flat, Upper Holyford valley, Otago, N.Z., showing steep valley sides, flat-floored talus cones and pyramidal peak in background. *(N.Z. Govt Tour. Office.)*

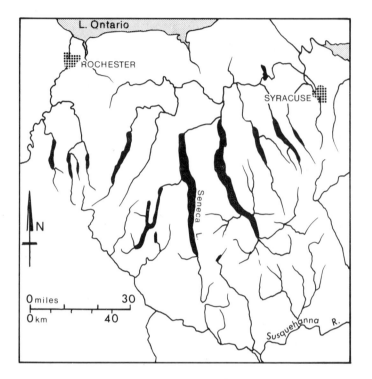

Fig. 17.11 The Finger Lakes of northern New York State.

Plate 17.12   The Yosemite Valley, California, a glaciated and alluviated valley bordered by dramatic granite domes, including Half Dome (X) and El Capitan (Y). Cathedral Rocks are opposite the latter, on the south side of the valley. *(U.S. Geol. Surv.)*

Plate 17.13 Cascade Inlet, a fjord tributary to Dean Channel in the Kitimat Ranges, B.C. The channel is straight for 25km in this view alone, and is bordered by steep slopes eroded in granite. *(Dept Lands Forests and Water Resources, B.C. Photo No. B.C. 513: 94.)*

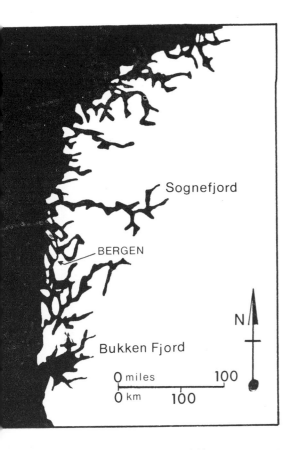

Fig. 17.12a Fjords in plan—coast of Norway.

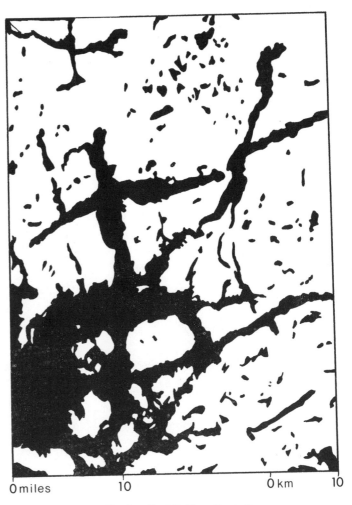

Fig. 17.12b Detail of Bukken Fjord, Norway.

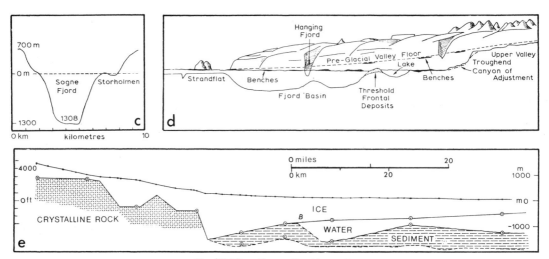

Fig. 17.12c   Sogne Fjord in section.

Fig. 17.12d Some common features of fjords.

Fig. 17.12e   Skelton Fjord, Antarctica, and the origins of highs or lips according to Crary.
("Mechanism for Fjord Formation", pp. 911–29, by permission of the Geological Society of America
and A.P. Crary.)

Fig. 17.13 Schematic representation of glacial and fluvioglacial
features at the terminus of a
glacier. A, outwash apron with
kettle ponds; R, recessional moraine; F, new outwash fan; f, old
outwash fan; S, braided stream;
D, debris-laden ice; I, clean ice;
M, medial moraine; L, lake; V,
valley sides; Sn, snow patches;
m, lateral moraine.
(From Principles of Geology,
*Third Edition, by J. Gilluly, A.C.
Waters and A.O. Woodford, W.H.
Freeman and Company. Copyright
C 1968, p. 264.)*

to deposition, not erosion. He points out that the ice is advancing and is thickest at the true coast and that the amount of sedimentation beneath the ice is determined largely by the thickness of the ice. Thus if there were phases of ice advance, the location of maximum deposition should vary along the fjord, resulting in highs and lows.

3. Relative Effectiveness of Glaciers and Rivers

The assemblage of forms in glaciated valleys has been interpreted in several ways. Some consider the forms are due to marked erosion by glaciers. Others believe they are due to the protective action of ice which has preserved them from erosion by rivers and streams. It is now generally agreed that ice erodes: the question is how rapidly, relative to running water.

The hanging valley, for instance, can be explained in two ways. The erosionists argue that larger glaciers would occupy the main or trunk valley which would therefore be deepened relative to the tributary valleys. Protectionists on the other hand suggest that glacier ice would persist longer in the hanging valley, thus protecting it from river work which effectively lowers the main floor, thus causing the discordant junction to develop.

Similarly, valley steps are attributed by erosionists either to the ice picking out weaknesses in areas of bedrock, or to differential erosion, as, for instance, at the junction of glaciers. The protectionists, on the other hand, believe that the areas above the steps have been protected by ice, while those below have been exposed to river action and are therefore lower: each valley step marks a halt in the glacier retreat.

Although the protectionists have been able to marshal some evidence in support of their thesis, in general, it does not provide a satisfactory explanation of the facts.

D. GLACIAL AND FLUVIOGLACIAL DEPOSITION

Erosion is typical of glaciated uplands and deposition of glaciated lowlands, though neither process is wholly localised in its activities. Thus depositional features such as lateral and end moraines are found in glaciated valleys.

1. Moraines

As well as the rock scoured from the valley sides, debris falls to the glacier surface from the slopes above, so that the margins of valley glaciers carry a more-or-less thick veneer of debris which becomes incorporated in the glacier and which later becomes the *lateral* moraine (Pls 17.2a and 17.14). Those lateral moraines which have been formed only very recently may take the form of knife-edged ridges, with the rectilinear slopes reflecting the angle of rest of the mixed debris of which the feature is built (Pl. 17.14c). Where two glaciers coalesce, these lateral moraines merge to form a linear zone of very dirty ice which later becomes the *media* moraine. Pauses in the retreat of the glacier are marked by *end* or *terminal* moraines. The moraines comprise weathered debris and isolated blocks (*erratics*) gathered by the moving ice.

Landforms due to glacial deposition fall into two broad categories, according to the nature of the deposits. Some forms are developed wholly as a result of deposition by the glacier, while others are due to the deposition of debris by meltwaters derived from the glacier. The former are *glacial* deposits, the latter *fluvioglacial* (Fig. 17.13).

Possibly the least spectacular, but certainly the most widespread, type of moraine is the *ground* moraine, variously known as till, drift or boulder clay (hence, *till plain, drift sheet*). The ground moraine forms a discontinuous veneer over the preglacial land surface, though it appears to have formed in various ways. It characteristically displays an irregular form (Pls 17.15a and b) though in a few areas streamlined hills, which are oval in plan and are called *drumlins*, are well developed. Ground moraine covers wide areas of glaciated lowlands in the northern Great Plains of the USA and the Canadian Prairies, in the north European plains and eastern England. The true ground moraine is hummocky and pitted by *kettle holes*. These are depressions left behind after the melting of ice blocks and rafts contained within either the *ablation* or the ground moraine (see below), or formed as a result of the removal of ablation moraine by

Plate 17.14a Fairweather Ranges and St Elias Range, B.C., with the Ferns Glacier displaying medical moraines. Note also icefalls, crevassed ice, and numerous pyramidal peaks, including, just to right of centre, Mt Fairweather (4663 m). (*Dept Lands, Forests and Water Resources, B.C. Photo No. B.C. 688:16.*)

Plate 17.14b Glaciated valley with granite domes on the left, and with a large lateral moraine on the right. The moraine has impeded drainage so that a number of shallow lakes have been formed. Rocky Mountains, Colorado. *(C.R. Twidale.)*

Plate 17.14c A lateral moraine of the Tasman Glacier, South Island, New Zealand. The moraine has been modified by mass movement of debris causing downslope migration of debris, which is now at its angle of repose or rest, and the development of a sharp knife-edged crest. The glacier beyond the moraine displays a well-developed ablation moraine, and also a medial moraine. *(C.R. Twidale.)*

Plate 17.15a Typical hummocky till plains in the great central plains of North America. *(U.S. Geol. Surv.)*

Plate 17.15b View north over Lower Matanuska valley, Cook Inlet, Susitna Lowland, Alaska. In the foreground is a complex of eskers and crevasse systems. On the left is an outwash plain with many lakes. The mountains in the background are fronted by a fault-line scarp. *(US Geol. Surv.)*

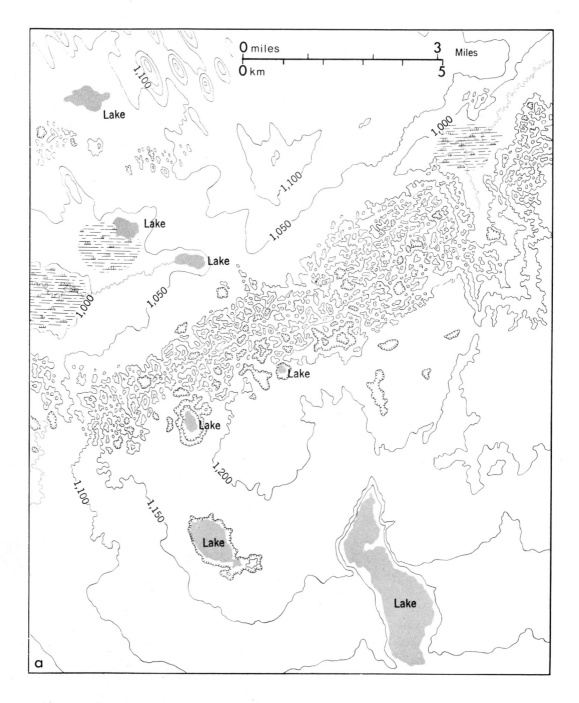

Fig. 17.14 Pitted end moraine: generalised model with contours, kettle holes and irregular relief.

meltwater streams plunging to the glacier base or developed as the result of the partial collapse of the roof of subglacial or englacial stream channels. They are commonly occupied by small lakes called *kettle lakes* or ponds (Fig. 17.14). *Recessional* moraines marking pauses in the retreat of the ice are also deposited as are other ridges of unsorted debris which mark the sites of major crevasses into which debris has been washed (Pl. 17.16, Figs 17.15a-b).

The ground moraine consists mainly of debris plastered against subglacial obstacles by the advancing ice, lodged in depressions, or simply dumped by the glacier. But a considerable volume of debris may have been formed by the surface melting (*ablation*) of the ice and the concentration of included debris (the *ablation* moraine, a good example of which occurs on the lower part of the Tasman Glacier, New Zealand (Pl. 17.14b) or by dumping of the contained debris[19] during the stagnation and melting of glaciers *in situ*. Thus from western Canada, Gravenor and Kupsch[20] have described assemblages of knobs, hummocks, ridges and depressions formed by the dumping of moraine from an inactive ice mass. The direction of ice motion exerted no control over the grain or trend of the deposited material.

2. Origin of Drift

The interpretations of glacial drift have occasioned much discussion,[21] but it seems unavoidable that some layered glacial deposits should be interbedded with true unsorted and unstratified tills as a result of undermelting of the ice. This has led to the extreme position adopted by R.G. Carruthers[22] of attributing all the glacial deposits of the British Isles to a single glaciation.

Carruthers believed that undisturbed stratified drifts, which most workers consider to be of interglacial origin, are in reality of subglacial or englacial type, and the whole question revolves around the interpretation of definitely layered but somewhat distorted drifts deposited by glaciers. Part of the difficulty is that ice-sheets have deposited debris in different ways. Some is laid down by moving ice, while a great deal is associated with ice which has rapidly disintegrated *in situ*. Some glaciers have decayed in small intermontane basins and valleys, like the Vale of Eden in north-western England,[23] where conditions differ greatly from those which obtained on the Great Plains of North America.

But there are many local variations. Observations of modern glaciers show that the distribution of morainic material in and on them is very varied. Some, as in Iceland, and the lower Tasman Glacier, New Zealand, are covered with debris; others are quite clean on the surface. Small glaciers may display large end moraines, while others are lacking any such development. In all likelihood, surface melting is responsible for many, perhaps most, of the drifts, but some are due to subglacial melting.

3. Drumlins

Drumlins (Pl. 17.17a and b) stand up to 60 m above the level of the surrounding plains, are elongated in the direction of local ice movement, and are built mainly of moraine (though some (rock drumlins) have cores of bedrock, and

Plate 17.16 Ice lobe with end moraine (or push ridge?) in Antarctica. Granite hills with scree slope beyond. (*R.L. Oliver.*)

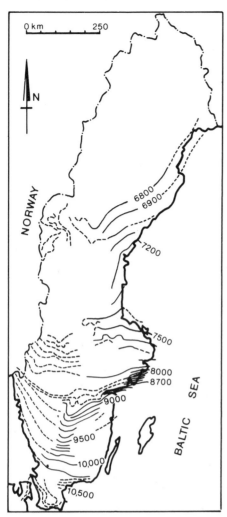

Fig. 17.15b Recessional moraines in Sweden. *(After K. Rankama (Ed.), The Quaternary (The Geologic Systems), Wiley, New York, 1965, Fig. 14.)*

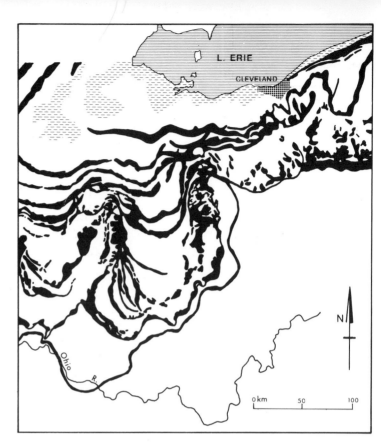

Fig. 17.15a End moraines in Ohio. *(After R.P. Goldthwait et al., Fig. 5, p. 87. Reprinted by permission of Princeton University Press.)*

Fig. 17.16b Distribution of drumlins in Northern Ireland.

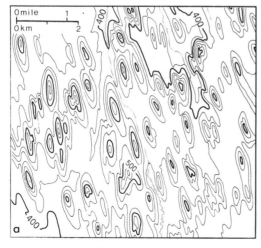

Fig. 17.16a Streamlined form of drumlins.

Plate 17.17a   Drowned drumlin connected to the shore by a tombolo,
at Chester, Nova Scotia. *(Dept Energy, Mines and Resources, Ottawa.
Photo No. A3618:63.)*

Plate 17.17b   Drumlin topography, Northern Ireland. *(Geol. Surv. N. Ire.)*

many display various types of organised structure—they are not merely random deposits of boulder clay). Drumlins have a length to breadth ratio of between 2·5 and 4·1 (Fig. 17.16a). Their form may be likened to that of an inverted spoon with the rounder, blunter and steeper end (the *stoss* slope) facing the direction from which the glacier advanced, and the gentler, streamlined slope (or *tail*) facing downstream. Drumlins occur in groups (Fig. 17.16b) or fields, giving "basket of eggs" topography in areas which, while they are close to the glacial limit, are nevertheless clearly removed from it.

The streamlined appearance of these features[24] shows that they were formed by moving ice, but they are not, as has been suggested, evidence of multiglaciation, the drift of one glaciation being moulded by the next advance of ice. If this were so, there would be no drumlins associated with ground moraine of the latest

(Wisconsin—Weichsel) glaciation, yet drumlins of Wisconsin age occur in Wisconsin and Minnesota, in northern New York State (where there are thousands of drumlins), in the northern Great Plains, in Northern Ireland (Pl. 17.17b) and in the Vale of Eden. Drumlins develop where the ice is still moving strongly but is beginning to spread as it extends towards its limits. Zones of low pressure apparently develop within the spreading glacier and local deposition occurs at these zones with the deposits being moulded and streamlined by the still-moving ice.

4. Fluvioglacial Forms

Much glacially transported debris is carried by glacial meltwaters and streams just before final deposition (Pl. 17.18). Such debris is sorted and stratified, in contrast to the purely glacial material, and is moulded to rather different forms. Plains of such fluvioglacial material form *outwash plains* or *sandurs* close to the ice front (Figs 17.1 and 17.13; Pls 17.14 and 17.19). Observations in Iceland[26] suggest that some sandurs may have been deposited in considerable measure by the floods associated with the draining of proglacial lakes (see Chapter 22) which are and were impounded between the ice mass and the land. Such floods may also be responsible for the so-called Channeled Scablands, a much-dissected plain in Washington, northwestern

Plate 17.18 Aerial view of snout of Franz Josef glacier, South Island, New Zealand. Immediately below the ice is a valley train deposited by a braided meltwater stream. Several dissected lateral and terminal moraines are visible (see also Figs. 17.15 and 17.16). *(N.Z. Govt Tour. Bur.)*

Fig. 17.17 Explanatory diagram for Plate 17.18: A—glacier, B—braided river channel, C—terminal moraines, D—lateral moraines, E—moraine, F—meander bluff.

USA, though there is debate as to whether a single gigantic flood is responsible, or whether long-term dissection by meltwaters[28] should be invoked in explanation of this area of interlocking channels cut in basalt and loess.

Outwash plains are commonly found in association with terminal moraines. They display kettle lakes of similar origin to those of the ground moraine. Many extend long distances down-valley from the glacier snout, forming *valley trains* such as those developed downstream from many glacier snouts in South Island, New Zealand[29] (see Pl. 17.18). Where the glacier flows close to bedrock walls in constricted valleys, outwash deposits may be dumped in a narrow wedge, between ice and rock, to form a *kame*, which includes much glacial debris as well as water-sorted material. The flattish upper surface of such deposits are called *kame terraces*. Many good examples border the present Lake Wakatipu, South Island, New Zealand. *Eskers* are sinuous ridges of sorted sand and gravel which generally display rudimentary stratification. They are best developed on glaciated lowlands, and are common in Finland, Labrador, the Canadian Northwest Territories and central Ireland (Pl. 17.20).

Eskers appear to develop in several ways. They are considered by some workers[30] to represent the outwash of streams debouching into proglacial lakes. But some in Scandinavia display water-deposited sediments overlain by

Plate 17.19 Glacial outwash plain (Skeidararsandur) in southern Iceland, with braided river channels. *(S. Thorarinsson.)*

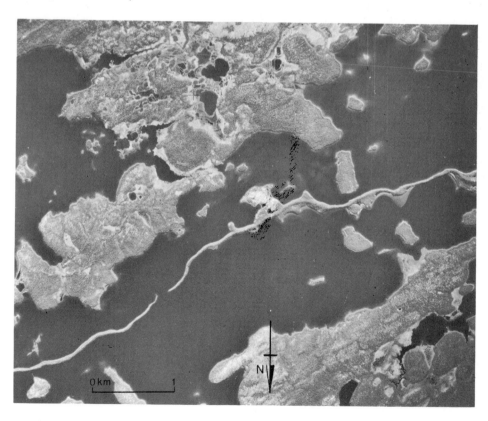

Plate 17.20 Vertical air photo of esker in Ashuanipi Lake, central Labrador. (*Dept Energy. Mines and Resources, Ottawa. Photo No. A12060—69.*)

0 km   1

N

boulder clay, indicating a subglacial origin.[31] Lewis[32] has described an esker forming in a subglacial tunnel in Norway and being exposed as the glacier front recedes (Fig. 17.18), and it seems certain that some, at least, form as valley trains in stream channels located upon, within, or beneath the ice and so bear little or no relation to the underlying geology and topography.

The outwash plains of glaciers are especially susceptible to deflation and contribute material to sheets of loess which are widespread in northern Europe and the USA (*cold* loess, as compared to the *hot* loess derived from the tropical deserts).

Finally, lakes are common in glaciated lowlands. Some areas, like Finland and the lake plateau of central Labrador,

seem to consist of more water than land, despite the fact that many of the former lakes have been infilled with debris and now form marshes. The formation of marshes and of shallow lakes is greatly aided by the presence of permafrost which hinders or prevents subsurface percolation of water. Lakes arise in such circumstances either through erosion or deposition. Areas of weathered rock (or other areas of weakness) are scoured out to form deep depressions. Many of the large lakes of Otago, New Zealand, like Wakatipu, Pukaki, Te Anau and Manapouri, are of the order of 400 450 m deep. On the other hand, glacial deposits block drainage lines. In both cases, hollows or basins filled with water are created.

Fig. 17.18 Esker built out from snout of glacier at Boderbreen, Norway. *(Drawn from photograph in W.V. Lewis, "An Esker in Process", pp. 314–9.)*

References Cited

1. See I.C. Russell, "The Malaspina Glacier", *J. Geol.*, 1893, **1**, pp. 219–45.
2. See for example C. Swithinbank, "Ice Shelves", *Geogr. J.*, 1955, **121**, pp. 64–76.
3. For critical review, references and discussion, see G. Seligman, "Extrusion Flow in Glaciers", *J. Glaciol.*, 1951 (1) pp. 12–21.
4. J.F. Nye, "The Mechanics of Glacier Flow", *J. Glaciol.*, 1952 (2), pp. 82–93.
5. See C.A. Cotton, *Climatic Accidents*, Whitcombe and Tombs, Wellington, 1967, pp. 126–36.
6. H. Weertman, "On the Sliding Glaciers", *J. Glaciol.*, 1957, (3), pp. 33–8; *idem*, "The Theory of Glacier Sliding", *J. Glaciol.*, 1964 (5), pp. 287–303; L. Lliboutry. "Une Théorie du Frottement du Glacier sur son Lit", *Ann. Geophys.*, 1959, **15**, pp. 250–65; *idem*, "Subglacial Supercavitation as a Cause of Rapid Advances of Glaciers", *Nature (Lond.)*, 1964, **202**, p. 77.
7. L. Lliboutry, "Une Théorie du Frottement", pp. 250–65.
8. M. Boyé, *Glaciaire et Périglaciaire de l'Ata Sund, Nord Oriental Groenland*, Hermann, Paris, 1950.
9. H. Carol, "The Formation of Roches Moutonnées", *J. Glaciol.*, 1947 (1), pp. 57–9.
10. A. Harker, "Ice Erosion in the Cuillen Hills, Skye", *Trans. R. Soc. Edin.*, 1901, **40**, pp. 221–52.
11. W.D. Johnson, "The Profile of Maturity in Alpine Glacial Erosion", *J. Geol.*, 1904, **12**, pp. 569–78.
12. W.R.B. Battle, "Temperature Observations in Bergschrunds and their Relationship to Frost Shattering", in W.V. Lewis (Ed.), "Norwegian Cirque Glaciers", *Royal Geogr. Soc. Res.*, 1960, Ser. 4, pp. 83–95; W.R.B. Battle and W.V. Lewis, "Temperature Observations in Bergschrunds and their Relationship to Cirque Erosion", *J. Geol.*, 1951, **59**, pp. 537–45.
13. W.V. Lewis, "A Meltwater Hypothesis of Cirque Formation", *Geol. Mag.*, 1938, **75**, pp. 249–65.
14. W.V. Lewis, "Pressure Release and Glacial Erosion", *J. Glaciol.*, 1954 (2), pp. 417–22.
15. See J.G. McCall, "The Flow Characteristics of a Cirque Glacier and their Effect on Glacial Structure and Cirque Formation", in W.V. Lewis (Ed.), "Norwegian Cirque Glaciers", *Royal Geogr. Soc. Res.*, 1960, Ser. 4, pp. 39–62.
16. W.A. White, "Erosion of Cirques", *J. Geol.*, 1970, **78**, pp. 123–6.
17. J.P. Bakker, "A Forgotton Factor in the Interpretation of Glacial Stairways", *Z. Geomorph.*, 1965, **9** NS, pp. 18–34.
18. A.P. Crary, "Mechanism for Fjord Formation indicated by Studies of an Ice-covered Inlet", *Bull. geol. Soc. Am.*, 1966, **77**, pp. 911–29.
19. J.G. Goodchild, "The Glacial Phenomena of the Eden Valley and the western Part of the Yorkshire Dale District", *Q. J. geol. Soc. London*, 1875, **31**, pp. 55–99.
20. C.P. Gravenor and W.O. Kupsch, "Ice Disintegration Features in western Canada", *J. Geol.*, 1959, **67**, pp. 48–64.
21. See for example S.E. Hollingworth, "Origin of Glacial Drift", *J. Glaciol.*, 1951 (1), pp. 430–7; W.V. Lewis, "Origin of Glacial Drift", *J. Glaciol.*, 1951 (1), pp. 433–5.
22. R.G. Carruthers, "On Northern Glacial Drifts", *Q. J. geol. Soc. London*, 1939, **95**, pp. 304–5; *idem*, "The Secret of the Glacial Drifts", *Proc. York. geol. Soc.*, 1947–8, **27**, pp. 43–57 and 129–72; *idem*, "Origin of Glacial Drift", *J. Glaciol.*, 1951 (1), pp. 431–3.
23. J.G. Goodchild, "Glacial Phenomena of the Eden Valley", pp. 55–99; S.E. Hollingworth, "The Glaciation of western Edenside and adjoining Areas of the Drumlins of Edenside and the Solway Basin", *Q. J. geol. Soc. London*, 1931, **87**, pp. 281–357.
24. R.J. Chorley, "The Shape of Drumlins", *J. Glaciol.*, 1959 (3), pp. 339–44.
25. J.G. Speight, "Late Pleistocene Historical Geomorphology of the Lake Pukaki Area, New Zealand", *N.Z. J. Geol. Geophys.*, 1963, **6**, pp. 160–88.
26. S. Thorarinsson, "The Ice-dammed Lakes of Iceland, with Particular Reference to their Values as Indicators of Glacier Oscillations", *Geogr. Ann.*, 1939, **21**, pp. 216–42.
27. J.H. Bretz, "Glacial Drainage on the Columbia Plateau", *Bull. geol. Soc. Am.*, 1923, **34**, pp. 573–608; *idem*, "The Channeled Scablands of Columbia River", *J. Geol.*, 1923, **31**, pp. 617–49; J.H. Bretz, H.T.U. Smith and G.E. Neff, "Channeled Scablands of Washington: New Data and Interpretations", *Bull. geol. Soc. Am.*, 1956, **67**, pp. 957–1050.
28. I.S. Alison, "New Version of the Spokane Flood", *Bull. geol. Soc. Am.*, 1933, **44**, pp. 657–722.
29. See J.G. Speight, "Late Pleistocene Historial Geomorphology", pp. 160–88, for excellent illustrations.
30. N. Shaler, "On the Origin of Kames", *Proc. Boston Soc. Nat. Hist.*, 1894, **23**, pp. 36–44; W.B. Wright, *The Quaternary Ice Age*, Macmillan, London, 1937, pp. 35–41.
31. C.M. Mannerfelt, "Nagra Glacial Morfologiska Formelment", *Geogr. Ann.*, 1945, **27**, pp. 1–239.
32. W.V. Lewis, "An Esker in Process of Formation, Boverbreen, Jotunheimen, 1947", *J. Glaciol.*, 1949 (1), pp. 314–19.

General and Additional References

C. Embleton and C.A.M. King, *Glacial and Periglacial Geomorphology*, Arnold, London, 1968.

L. Lliboutry, *Traité de Glaciologie*, Masson, Paris, 1965, 2 vols.

J. Tricart and A. Cailleux, *Traité de Géomorphologie. 3. Le Modele Glaciaire et Nival*, C.D.U.S., Paris, 1962.

J.K. Charlesworth, *The Quaternary Era*, Arnold, London, 1957.

R.F. Flint, *Glacial and Pleistocene Geology*, Wiley, New York, 1957.

CHAPTER 18

Coastal Processes and Landforms

A. GENERAL

Wind-driven waves are ceaselessly at work shaping the coast, and the coastal zone changes remarkably quickly both in detail and in gross.[1] Yet the variety of forms observed on coasts suggests either that the action of waves is far more complex than is generally thought to be the case, or that factors other than waves contribute to the development of coastal forms. What, for example, is the reason for the morphological contrast between the rocky cliffed coasts of such areas as southwestern England, eastern Australia, the Great Australian Bight or coastal British Columbia, and the flat, comparatively regular coasts of the eastern USA, eastern England and northern USSR?

Moreover, there are contrasts within these broad categories. The cliffed coast of the Great Australian Bight (Pl. 18.1) is an unbroken arc of cliffs up to 100 m high which continues for some 2700 km, though as shown in Fig. 3.10 it is fronted by a coastal plain in one sector. The cliffed coast of British Columbia, on the other hand, is greatly and deeply indented (Pl. 3.1a). On this coast and off the coast of Norway there are rock platforms several kilometres wide, but elsewhere platforms at the cliff foot are much narrower —commonly a few scores of metres at most. Platforms also occur at not one, but several distinct levels. Again, on flat coasts, sand dunes are common behind beaches, but they do not occur, or are at least not well developed, in the humid tropics. In some areas there are sandy beaches, in others mud flats, in some areas lagoons, in others salt marshes, and still elsewhere mangrove swamps.

Thus both in broad view and in detail, coasts present great morphological variety. However, a basic distinction can be made between coasts dominated by erosional forms and those where depositional features are most common.

Plate 18.1 The unbroken cliffs of the great Australian Bight with the darker Nullarbor Limestone above and the Wilson Bluff Limestone below. Note the slumped mass of limestone blocks due to undercutting by waves and the extraordinary flat Nullarbor Plain which, however, displays a gentle rise to the north. (*S. Cheshire.*)

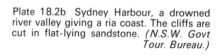

Plate 18.2a Vertical air photograph of a drowned coast at Yampi Sound, northern Western Australia. The general trend of the shore runs parallel to the structural grain, that is, NW–SE, as for instance between D–D. Here the coast is of parallel, Dalmatian or Pacific type. But within embayments the coast runs at a high angle to the structural grain, for instance around P, and here it is of Atlantic or transverse type. *(Dept Nat. Dev., Canberra.)*

Plate 18.2b Sydney Harbour, a drowned river valley giving a ria coast. The cliffs are cut in flat-lying sandstone. *(N.S.W. Govt Tour. Bureau.)*

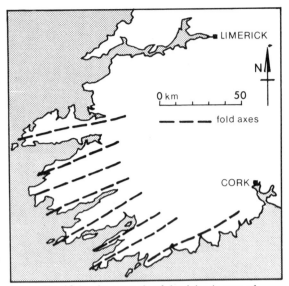

Fig. 18.1a An example of the Atlantic type of coastline, with long indentations—the coastline of southwest Ireland, where the structural grain has been brought strongly into relief by the drowning of valleys. This is also an example of a ria coast.

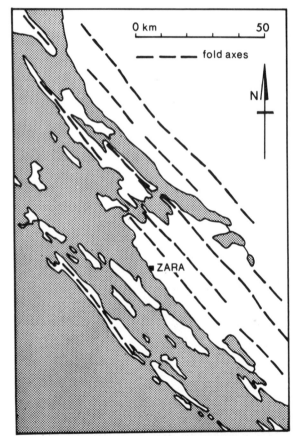

Fig. 18.1b The Pacific type of coast is characterised by a structural grain which parallels the coast and is exemplified by the coast of Dalmatia, in Yugoslavia.

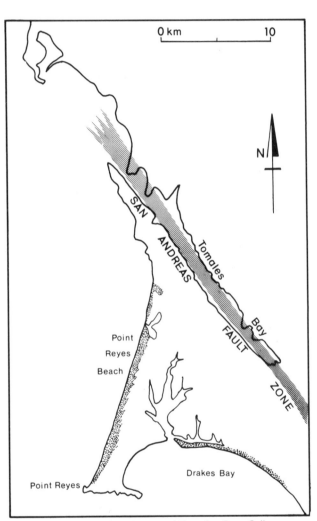

Fig. 18.2 Sketch map of Tomales Bay, California, showing the long narrow inlet which is located on a fault zone. Note also the straight beach on the western side of the Point Reyes Peninsula and the curved beach on the southern side.

B. THE FACTORS AND FORCES AT WORK
1. Structure

Both in broad view and in detail, and as has been emphasised in other contexts, the most important single factor influencing coastal morphology may well be *structure* (including the nature and disposition of the rocks and the nature of tectonic movements affecting the coastal zone). The varied trend of strata *vis-a-vis* the coast is responsible for the contrast between the so-called *Pacific* type, with the strike of the bedrock and the alignment of the coast approximately parallel, and the *Atlantic* type of coastline where the strike is at a high angle to the coast (Pl. 18.2a, Fig. 18.1a and b). Fault movements in some areas directly affect the disposition and shape of the coast (see for example Fig. 18.2). Relative movements of land and sea cause the development of drowned (fjord and ria coasts—Pls 17.4, 17.13, 18.2a and b) or flat coasts (Pl. 18.3). In some areas glacial deposition has profoundly influenced the gross morphology of coasts, though there has been postglacial modification by waves (Pl. 17.17a).

Joints, faults, and weaker outcrops are etched out to form clefts, which are called *geos* (Pl. 18.4) or embayments, between ridges and promontories. The commonly

Plate 18.4 Geo eroded along a major joint complex in granite near Stenhouse Bay, southern Yorke Peninsula, S.A. *(C.R. Twidale.)*

Plate 18.3 A flat, even, aggradational coast near Kingston, in the South East district of South Australia. The whole area was raised relative to sealevel during the later Pleistocene and the old littoral zone, with coastal foredunes, has been left high and dry and up to 100 km inland. But there is also recent and continuing aggradation. Note the active foredunes (A) behind the present beach, the coastal lagoons, and the older beaches and dunes located further inland. *(S.A. Lands Dept.)*

found alternation of headland and bay, of cliffed promontory and embayment with bayhead beach, is basically a reflection of structure. Areas of considerable relief which have suffered subsidence form indented irregular coasts due to the drowning of previously formed valleys (Fig. 18.1a). Uplift, on the other hand, causes the old sea-floor to be exposed, and many such coasts are smooth and flat (Pl. 18.3).

Thus, structure in the broadsense is a fundamental factor in coastal landform development. The rocks exposed at the coast are, however, subject to weathering and erosion. Though several agents are involved, wind-driven waves are undoubtedly the most important at work on the coast.

2. Waves

In the open ocean, *waves* are merely shapes. Wave forms advance, but the forward motion of the water particles amounts to only about 1 per cent of the wave velocity. Waves originate with the frictional drag of moving air (wind) on a calm sea; ripples are created, the wind presses against the windward faces of such irregularities and wavelets are formed, These in turn, coalesce to form waves.

The *wave length* is the distance from crest to crest (or from trough to trough); the *wave height* is the vertical distance from trough and crest; and the *wave period* is the time it takes for two successive crests to pass a given point (Fig. 18.3)

The ability of the wind to initiate waves depends upon its force or velocity, the length of time it blows and the extent of open water over which it blows unimpeded (a measure known as the *fetch* of waves). Waves up to 30 m high have been observed at sea in storms, but there is a limit beyond which the wind destroys, rather than increases the size of waves. In storms, the pressure applied by the wind to waves is very great, and when the wave crest becomes a wedge of water extending through less than 120° of arc, and the height of the wave is about one-seventh of its length, the wave front is steepened, the thickness of the wedge of water is decreased still further and eventually it is blown off, forming a breaking wave in open water: the phenomenon known as "white horses".

Waves generated and maintained directly by the wind are known as *forced waves*. Waves formed by the impact of a stone on the surface of a pond are of forced type, but just as these waves radiate out from the point of impact, decreasing in amplitude and increasing in wave length as they do so, forced or wind-driven waves at sea spread as *free waves* or *swell* beyond the area of origin, and beyond the area in which the wind responsible for the waves is blowing.

As waves approach the shore, they undergo critical changes in nature and morphology, both in section and in plan, as they enter shallow water. Although the wave period does not change, the wave length becomes shorter and the wave velocity decreases. The wave height decreases slightly when the wave first enters shallow water. Thereafter, however, wave height increases rapidly until the wave breaks. Because of this concurrent increase in wave height and decrease in wave length, the steepness of the wave increases although while the wave height: wave length ratio remains less than 1:7, the form remains stable. Gradually, however, the wave form changes. A sharp, narrow crest is developed. The intervening troughs are wide and flat. The orbital motion of particles in the waves changes from circular to elliptical, the long axis of the ellipse being parallel to the seabed.

The waves break when they enter water the depth of which is one third of the wave height. However, low waves break in even shallower water, which is only as deep as the wave height.

The waves lose some energy due to friction with the sea-floor in shallow water. But the breaking of waves is due primarily to changes in the movement of particles within the wave. As noted earlier, the orbital paths followed by the water particles increase in length and become elliptical as the waves enter shallow water. Because the wave period is constant, the particles follow longer paths and the velocity therefore increases. However, the wave velocity (as opposed to the particle velocity) decreases and, when the orbital velocity exceeds the wave velocity, the water mass overtakes the wave form and the wave breaks.

When a wave breaks there is a lateral translation of water up to the beach—the *swash*. Some of this water (not all, because of percolation into the beach) which is thrown up on to the beach returns to the sea as the *backwash* (Fig. 18.4).

3. Wave Refraction

Waves approaching the shore obliquely (Fig. 18.5a) suffer changes in plan, when they enter shallow water. One section of the wave enters shallow water before the other. This part therefore breaks while the other continues on. The orientation of the wave is thus changed, and the wave suffers *refraction*, the overall effect of which is to bring the waves more and more nearly parallel to the shore. A wave approaching an irregular or indented shoreline is similarly refracted (Fig. 18.5b), with the waves being moulded to the submarine offshore contours and to the shape of the coast. There is, however, an important additional effect: wave energy is concentrated on headlands, and dispersed in bays (Fig. 18.5b), with the result that shorelines tend to be made smooth in plan. In addition to refraction, waves also undergo reflection and diffraction. *Reflection* occurs when a wave

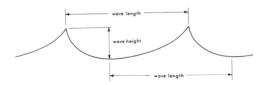

Fig. 18.3 Dimensions of a wave.

Fig. 18.4 Swash and backwash
on a beach.

hits against a seawall or any obstruction—natural or man-made—rising out of deep water. The wave "bounces" off, suffering a reversal of movement, with the angle of reflection equal to the angle of incidence. *Diffraction* takes place in the lee of islands, where waves change direction and disperse.

4. Destructive Work of Waves

Anyone who has watched huge waves of water hurled against the shore by the wind during a storm will readily acknowledge their potential as erosive agents. Certainly, waves pound against the shore with great force. There are many well authenticated examples of the power of and the damage wrought by wind-driven waves, especially storm waves. One of the best known, which bears repetition, concerns the harbour breakwater at Wick in eastern Scotland. This structure was capped by an 85 tonne block of concrete secured to the foundation by iron rods 8 cm in diameter. The total weight is estimated to have been of the order of 133 tonnes; yet, during a great storm in 1872, the whole mass was lifted bodily and dumped in the harbour. The breakwater was rebuilt and a larger capping, this time weighing 266 tonnes, was put in its place, but this, too, was removed *en bloc* during a later storm.

The spray thrown up by breaking storm waves has been measured as travelling at speeds of up to 110 km/hr, while small jets attain velocities of up to 270 km/hr. The damage effected by such high velocity spray can well be imagined. A considerable amount of coastal erosion is accomplished by such hydraulic effects, with the sheer weight and velocity of the water fracturing the rocks exposed at the shore.

The abrasive effect of boulders, pebbles and sand hurled against the shore by waves is also great, and the alternate expansion and contraction of air trapped in rock crevices also causes fracturing (as well as ghostly wailings). Once they are fractured and fragmented, rocks are plucked away and accumulate at the foot of the cliff to undergo further attrition and, eventually, to be evacuated along the coast.

The effects of coastal erosion are obvious in times of storm; quite staggering amounts of erosion have been accomplished by waves, particularly of course in areas where unconsolidated rocks are exposed at the coast. Toward the end of the last century, the average annual loss from the coast of Holderness, in East Yorkshire, an area of unconsolidated, glacial ground moraine, was nearly 2 m. At Pakefield, near Lowestoft (also in eastern England) one point on the coast recorded a loss of 200 m between 1883 and 1947, an average of nearly 3 m per annum. Almost 1·5 km of volcanic ash was eroded from the Indonesian island of Krakatoa between 1883 and 1928, a rate of almost 40 m per year. A distinct, though narrow, shore platform (Pl. 18.5) was eroded in the space of one year in the ash of Surtsey, a volcanic island born off the coast of Iceland in 1963.

Admittedly these are exceptionally high rates accomplished in special circumstances. Coastal erosion is generally much slower. Nevertheless it would be unwise to underestimate its effects. At Hallett Cove, south of Adelaide, for instance, the serrated platform exposed in the intertidal zone (Pl. 18.6) is in places almost 70 m wide, and taking the present sealevel as having been essentially stable for the

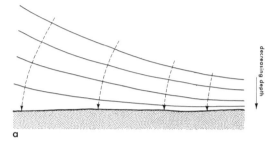

Fig. 18.5 Refraction of waves, in (a) approaching the shoreline obliquely, and in (b) approaching an indented coast.

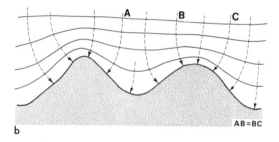

Plate 18.5 Wave-cut platform on the south-west flank of the lava shield of Surtsey, which was built up out of the sea in 1963. The platform was eroded in one year and the boulders of the shingle beach were presumably rounded during the same interval. In the background is the volcano Yolnir, built up in 1966 and after the eruption ceased, broken down by waves in a mere four months. *(G. Kjartansson, 6 May, 1966.)*

Plate 18.6   The curvilinear cliff here at Hallett Cove, south of Adelaide, S.A., reflects the bedding planes in the folded Precambrian strata. A serrated platform is well developed in the intertidal zone below the cliff, and there was a shingle beach in the foreground at the time the photograph was taken. *(C.R. Twidale.)*

Plate 18.7 Gully erosion in a cliff of Triassic marl near Sidmouth, Devon. Note the slumped debris at the base of the cliffs. (*Hunting Surveys.*)

Fig. 18.6 Varvau, in the Tonga Islands, is built of limestone and displays a pronounced sealevel notch. (*Drawn from photograph in R.W. Fairbridge,* Encyclopaedia of Geomorphology, *p. 654.*)

past 6-7000 years (and it may well have been less—see Chapter 22) this represents a rate of erosion in these consolidated mudstones of almost 1 m per century.

5. Other Agents at Work on the Coast

So impressive are the obvious effects of wave attack that other forces were overlooked in the past; they may still be underrated. For the first quarter of this century the erosional forms observed on the coast were, with few exceptions, attributed to wave attack: the feature known now by the neutral or noncommittal term *shore platform* was called a wave-cut platform. Yet on many coasts there are features which attest to the activity of other agents. Gullying and landslips significantly contribute to cliff recession in many coastal sectors (Pls 18.1, 18.7). In southern Victoria[2] and on southern Yorke Peninsula, platforms of the intertidal zone are extensively preserved in bayhead situations (Pl. 18.8a) where wave energy is dispersed by refraction. Though these bayhead platforms are usually covered with sand or seaweed and algal slime, they are undoubtedly extensive and common. Moreover, in some areas (see below) platforms occur at many levels within the tidal zone. Clearly, either wave activity is complex, enabling platforms to be developed simultaneously at different levels, or there 'are structural

effects or tectonic disturbances or processes other than waves which are active in the development of shore platforms. Or, of course, several of these possibilities may apply in some areas.

Again, the cliff-foot notch (Pl. 18.8b) is widely attributed directly to wave action and it is clearly forming near the high water-mark in many areas.[3] But in the interpretation of notches which are beautifully developed on limestone islands (such as those illustrated in Fig. 18.6) the role of solution tends to be underplayed because of the chemical difficulties presented by seawater which is normally saturated with lime. It is relevant that similar forms are also developed in limestone (Fig. 18.7) at the edges of pools just within the range of waves but in sheltered situations, which strongly suggests that solution weathering plays a considerable part in their development.

In arid and semiarid climates, limestone exposed at the coast displays the same minor karst features (solution cups, fingerprints, lapies, minor dolines) which have been noted from humid regions and which attest to the effectiveness of solution (Pl. 18.9). Some of this is undoubtedly due to groundwater percolating to the coast, but much occurs in the intertidal and spray zones and is thus due in part at least to the chemical action of seawater. The chemical

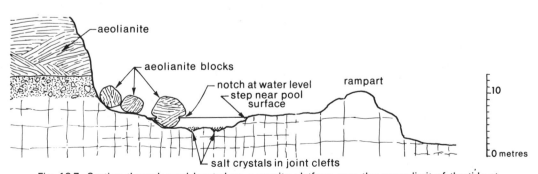

Fig. 18.7 Section through pool located on a granite platform near the upper limit of the tide at Hillock Point, southern Yorke Peninsula, South Australia. A limestone block which has tumbled into the pool from the cliff above shows solutional effects at pool level.

Plate 18.8a Shore platform in aeolianite with cliffs of similar material. Note the extremely planate platform with its cover of marine plants. Near Portsea, Victoria. *(E.S. Hills.)*

Plate 18.8b Notch cut at the base of a aeolianite cliff near Cape Northumberland in the southeast of S.A. Note the numerous platforms with marginal rims, each located at slightly different levels just offshore (X). *(C.R. Twidale.)*

Plate 18.8c Spray Point, Koonya, Sorrento, Victoria, with cliffs in aeolianite, and in the foreground a planate shore platform in the same rock. The outer edge of the platform has broken away to form a near-vertical low tide cliff. *(E.S. Hills.)*

and physical action of water left behind at low tide or thrown up as spray, though not thoroughly understood, is also clearly responsible for the development of hollows morphologically similar to weather pits, or gnammas (see Chapter 10) which extend and coalesce to form distinct benches. Water thrown up as spray, for instance, collects in clefts and depressions in the exposed bedrock. There the water weathers the rock with which it comes into contact, partly by solution and partly by hydration and similar processes. Some hollows develop a hemispherical shape (cf. pits, see Chapter 10), but more commonly (and possibly because of the general soaking and partial weathering of the upper layers of the bedrock) broad, shallow, flat-floored depressions (cf. pans) evolve (Pl. 18.10a). Such depressions extend and join, and when the intervening walls are wholly consumed, a stepped but flattish bench or platform remains (Pl. 18.10b).

Seawater is normally saturated with lime and it is therefore difficult to understand how it can effect further solution of lime. Careful examination of the pH of coastal pools has shown that there are systematic variations of the CO_2 concentration of the water throughout the day, partly as a result of the photosynthetic activities of algae and seaweeds which live in the pools. The pH is raised during the day and there is emission of CO_2 at night when the water is undersaturated with $CaCO_3$. Diurnal temperature variations also lead to nocturnal undersaturation in CO_2.

The process whereby pool waters weather the rock with which they are in contact is complex, and probably varies from one environment to another. In humid tropics, for instance, where rain falls almost daily (or in areas where there are many wet days) rainwater is probably important in the solution of limestone. Because it is not saturated with lime, it is capable of taking the bicarbonate into solution. In addition it tends to remain as a distinct layer near the surface of the pool because of its lower density. Most weathering therefore occurs at a layer near the pool surface: hence what is called here *pool weathering*.[4]

Rocks other than limestone are also affected by seawater and display pitting (Pl. 18.10b-d). There is some experimental evidence to suggest that many rock-forming[5] minerals are more soluble in alkaline than in fresh water though some of the evidence is contradictory and temperature conditions appear to be crucial. But many rocks contain minerals which are vulnerable to solutional attack by seawater. Aeolianite, for instance, which is widely exposed on the coasts of southern Australia, consists of lime plus quartz; as seawater is not saturated in silica, solution is possible, causing the rock to crumble. On some coasts, for example the east coast of Otago (Moeraki and Palmerston areas), at Cape Paterson on the Victorian coast south of Wonthaggi, and the Californian coast near Davenport (some 70 km south of San Francisco) there are pits, basins and hollows which are raised above the level of the surrounding platforms or other rock surfaces. The floors and sides of these elevated pits which are developed in arkosic sandstone and mudstone are coated, and presumably protected, by a thin skin of iron oxide (in Victoria it is limonite). The pits constitute a minor example of relief inversion—the floors of the pits, originally set below the general local ground level, are now above the general level (Pl. 18.11). It may be argued that the pools of water protected the hollows by keeping them wet, whereas the adjacent rock surfaces, subject to alternate wetting and drying, were more rapidly weathered and worn down. On the other hand the production of distinct iron-rich zones within the rock argues some mobilisation and preferred precipitation of these substances. In the Davenport area the greater precipitation of iron oxides in the floors, as opposed to the sides, of the hollows may be taken to suggest precipitation from pools of water which gradually dry up.

Associated with such spherical precipitation zones in the the Moeraki and Palmerston areas of New Zealand are boulders up to 40 cms diameter (Pl. 18.11b), the origin of which is related to chemical weathering, but the details of which remain obscure.

Salt crystallisation, which is evident in most dry coastal pools, may assist in loosening rock fragments, especially in the tropics.[6] But this does not occur to a great extent everywhere, because along the South Australian coast, for instance, salt crystals present in the pools are not evenly distributed over the pool floor but are concentrated in the lower pool levels.

Plate 18.9 Coastal lapies in aeolianite, Gym Beach, Yorke Peninsula, S.A. *(C.R. Twidale.)*

Plate 18.10a Flat-floored pans eroded in aeoline, Gym Beach, Yorke Peninsula, S.A. (*C.R. Twidale.*)

Plate 18.10b Granite platform and shallow pool with seaward rampart of more coarsely jointed granite, Hillock Point, southern Yorke Peninsula, S.A. (*C.R. Twidale.*)

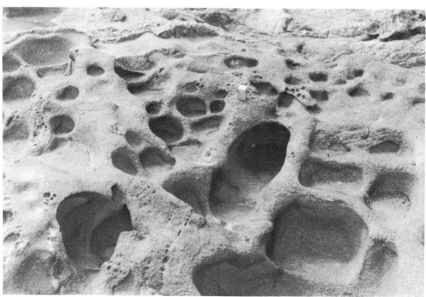

Plate 18.10c Pitted gneiss at the edge of gypseous Lake Greenly, southern Eyre Peninsula, S.A. (*C.R. Twidale.*)

Plate 18.11a Limonite-encrusted basin standing above the level of the shore platform at Shag Point, near Palmerston, South Island, New Zealand. (*C.R. Twidale.*)

Plate 18.10d Superficially pitted or honeycombed steeply inclined surface in arkose near Bean Gulch (Arroyo de los Frijoles), 65km south of San Francisco, California. (*C.R. Twidale.*)

Plate 18.11b Spheroidal ball, one of many formed on the shore platforms at Shag Point and near Moeraki, South Island, New Zealand. (*C.R. Twidale.*)

Organisms may contribute significantly to the disintegration of rocks on the coast. Observations on the coast of southern Yorke Peninsula, South Australia, show that an algal slime covers the surface of the granite bedrock in pools of seawater. When the slime is scraped off, particles of rock are detached with it, whereas on exposed surfaces nearby, the granite is cohesive and it is not possible to remove particles of rock.

Quite apart from the photosynthetic effects of plants, other effects are achieved by organisms on the coast. Boring animals accomplish rapid erosion and facilitate weathering. The seagrape *Hormosira* which is found in abundance on the coast of southern Australia forms a distinct matting on many platforms. On the one hand this organism protects them from the full force of wave attack (the rampart or rim commonly found on such platforms may in part be due to such protection, though the fact that the edges of platforms are constantly wet is perhaps more significant). On the other hand it ensures that the platforms are never dry, even at low tide and on many shores such as the east coast of Otago, it effectively traps sand and other debris and contributes to the formation of minor beach forms. Vegetation effectively fixes the sand blown from the beaches to form coastal foredunes and stabilises muds in marshes and swamps (see below).

C. EROSIONAL FORMS ON THE COAST

Typical assemblages of forms apparent on many rocky coasts are shown in Figs 18.8 and 18.9.

The *cliff* is very much an expression of the local structure (Pls 18.1, 18.6 and 18.12a-d). Blocky or bouldery cliffs are characteristic of granite exposures, though sheet structure is also well developed in places. In areas of sedimentary outcrop, the nature and disposition of strata is directly reflected in cliff morphology. Resistant rocks form precipitous cliffs and weaker strata form more subdued forms, though factors other than structure, notably the intensity of the erosional forces at work on various parts of the cliff, are significant in this respect.

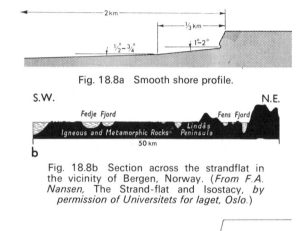

Fig. 18.8a Smooth shore profile.

Fig. 18.8b Section across the strandflat in the vicinity of Bergen, Norway. (*From F.A. Nansen,* The Strand-flat and Isostacy, *by permission of Universitets for laget, Oslo.*)

Fig. 18.8c Stepped shore profile.

The *notch* is not always present at the base of the cliff, but in many areas is well developed, (Pls 18.8, 18.12d; Fig. 18.9).

Stacks (Pls 18.12d, 18.13a-d, Fig. 18.9) and offshore islands reflect local structure in considerable measure, though in many cases these residuals are markedly asymmetrical in cross-section, the seaward facing slope being more rocky and precipitous than that which faces the shore (Pl. 18.13a, see also Pl. 18.14).

Cliffs, notches and stacks are widespread features of many rocky coasts, but the nature and complexity of the more-or-less planate components of the profile varies greatly from area to area. In many temperate regions like the coast of northern California and western Europe there is a gently inclined submarine slope, 1-2° gradient within a few hundreds of metres of the shore but decreasing to ½ - in deeper water (Fig. 18.8a). But elsewhere this smooth

Plate 18.12a Blocky granite cliff north of Bowen, Queensland. (*Dept Nat. Dev., Canberra.*)

Plate 18.12b Sheet structure exposed in cliffs on Pearson Island, Great Australian Bight. (*C.R. Twidale.*)

Plate 18.12c Structural benches eroded along gently dipping bedding planes and backed by vertical cliffs in Mesozoic sediments near Davenport, south of San Francisco, California. (*C.R. Twidale.*)

Plate 18.13a The Twelve Apostles, prominent stacks eroded from flat-lying sediments off the Victorian coast near Port Campbell. Note the steep seaward facing cliffs (left) and the cliff-foot notches. *(Tour. Dev. Authority, Victoria.)*

Plate 18.13b Stack some 140 m high eroded from Old Red Sandstone off the Orkney Islands, northern Scot-land. *(H.M. Geol. Sur., Crown copy-right reserved.)*

Plate 18.13c Stacks off the arcuate Whakapohai Beach, Haast Road, South Island, New Zealand, as seen from Knights Point. Note the microrelief—berms and runnels—on the beach indicated by the distinct drainage gutters. (*NZ Govt. Tour. Office.*)

Plate 18.13d The Needles, stacks in steeply dipping and resistant chalk at the western end of the Isle of Wight, southern England, with chalk cliff and shingle beach. (*H.M. Geol. Surv., Crown copyright reserved.*)

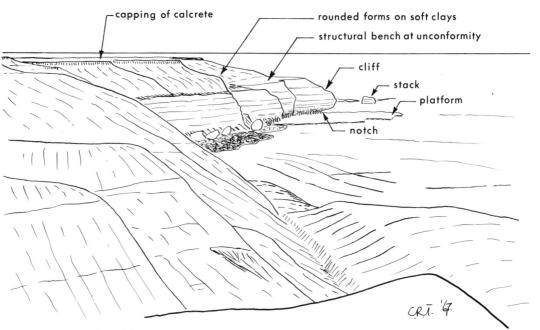

Fig. 18.9 Sketch of Blanche Point, south of Adelaide, S.A., showing characteristic forms of a coastline suffering erosion: cliff (with convex slopes on the upper clay sediments and a vertical slope on the limestone below), a cliff-foot notch, platform and stack. (See also Pl. 18.12d.)

transition from the coast to the continental shelf is interrupted by the presence near the shore of essentially flat platforms. In periglacial regions there is commonly a very wide shelf called the strandflat (Fig. 18.8b). In many tropical regions, and particularly in the Pacific and Australasian regions a multiplicity of relatively narrow platforms is displayed within and relatively close to the tidal zone (Fig. 18.8c). Such stepped profiles are known from temperate regions but are relatively rare there, whereas they are common in the arid tropics.[8]   Why are these contrasted profiles developed?

In general terms the inshore sector of the gently sloping smooth profile is comprehensible in terms of simple wave attack: the greatest amount of wave agitation and hence of kinetic energy occurs near the upper limit of wave activity, where the waves are in the deepest water, so that the greatest abrasion should occur near the upper limit of the tide. And the amount of abrasion should decrease progressively within the tidal range so that a seaward slope can reasonably be expected. But how the deeper and more gentle profile has developed is far from clear. It lies below the generally accepted limit of effective wave activity[9] and the only possibility seems to be that it formed in relation to a lower sealevel.

Turning to coastal platforms mention may first be made of the *strandflat*, the extraordinarily broad platform bordering the coasts of Norway, Iceland, Spitzbergen, Greenland and parts of British Columbia. It extends for over 60 km off

the coast of Norway (Fig. 18.8b), and is breached by troughs which are extensions of major fjords. Numerous reefs—the *skaargard*—stand above the platform. Several distinct levels have been distinguished, notably at 30–40 m and 15–18 m above sealevel and at 10 m below. Tongues of the strandflat extend into the mouths of major fjords.

Though the Norwegian stradflat has been regarded as a submerged peneplain[10]—that is a surface of low relief shaped by rivers (see Chapter 19)—it is certainly of coastal origin, and its distribution suggests that its great width is related to frost and ice. Its relationship to fjords suggests that the platform and the glacial trough may have formed during the same period. Nansen[11] considered that ice accumulated on the cliff just above the high tide mark and was derived from fresh water and not from seawater. This he took to indicate the activity of freezing and thawing near the cliff-foot, which was thereby undermined and worn back, with the debris evacuated by waves.

However, the origin of the strandflat remains controversial.

The *higher platforms*, that is, those close to or beyond the highest tide level but within the spray zone, are clearly not caused by the mechanical abrasion of waves, yet they are actively extending. They have been attributed to pool weathering, that is, to chemical and physical breakdown (through alternate wetting and drying) of the rock marginal to spray pools as well as to biochemical effects. The nature of the processes is not clearly understood but as water trapped in crevices causes the rock with which it is in contact to disintegrate, the crevice is enlarged laterally and eventually becomes a pool or depression. Several such depressions, each flat-floored (Pl. 18.10a), but at various levels, merge to form a platform or a series of platforms above high water mark and each in reality consists of a multiplicity of levels. The absence of wave activity poses problems, for the rock which has been weathered to form the depressions and platforms has obviously been transported away from the sites. Possibly some is removed by the wind during storms, or by run-off and seepage during heavy rains.

The *intertidal platform* was widely considered to be simply due to wave action, but observations show that not one, but several closely related but distinct, platforms occur in a zone where several processes are demonstrably active. The lower platforms are washed by waves almost constantly, but others near the upper limit of the tidal range are only rarely touched, save by spray. On the lower platforms, pot-holes with pebbles still in them suggest that the waves accomplish some abrasion.[12] But platforms in gneiss are developed at the margins of shallow lakes where the waves are not powerful (Pl. 18.15), suggesting that other agencies can shape platforms. Moreover, in some areas, for instance southern Victoria and southern Yorke Peninsula, intertidal platforms are well developed in bayhead situations where wave action is minimal. On Yorke Peninsula, platforms in granite near the upper limit of the tidal range are being extended by the pools weathering. (Fig. 18.7, Pl. 18.10b).

In some areas the shore platform is gently sloping close to the cliff-foot, forming a *ramp*, but in many localities, the planate rock benches extend to the base of the cliff (Pl. 18.8, 18.12d). In detail the surface of the platforms is not smooth. Even in areas where horizontally bedded strata are exposed, minor irregularities are developed. Where the platform cuts across steeply inclined strata, weaker members are etched out and the platform is *serrated* (Pls 18.16, 18.11b); where the platform is cut in well-jointed rocks such as granite, joint planes and compartments of finely and coarsely jointed rock are differentially eroded; sheet structure is exploited and expressed in some areas and the shore is strewn with blocks and boulders. Some platforms, especially those formed in limestone, display a distinct rim at their margins; in some cases, ridges also traverse the otherwise planate bench.

Another factor complicating the interpretation of platforms is structure. Many years ago Bartrum,[13] suggested that the waves in the intertidal zone are not acting on a uniform structure. This is not to suggest that platforms are everywhere coincident with resistant strata—though they are in the Sydney area [14] and elsewhere along the N.S.W. coast[15] (see also Pl. 18.12c)—but that the waves can more readily remove the strata weakened by chemical weathering above the zone of groundwater saturation. This zone is roughly coincident, according to Bartrum[16] and to Cotton,[17] with the high tide level. However, it is arguable that near the coastal fringe where the water-table oscillates in sympathy with tidal variations, weathering due to wetting and drying results in a zone vulnerable to weathering by wind etc., and hence to wave attack. Be this as it may, a platform known as the "Old Hat" type (Fig. 18.10) is developed near the upper limit of the tidal range by weathering. In addition, Bartrum claimed that storm waves are responsible for the erosion of another platform again at about high tide level: the storm wave platform. But it is difficult, in practice, to distinguish between those platforms due primarily to wave action and those near the upper limit of the tidal range which are formed basically by pool weathering and which are swept occasionally by storm waves performing the important function of evacuating debris. Furthermore, on several parts of the South Australian coast, platforms are

Plate 18.14. The Investigator Group of islands in the Great Australian Bight off the South Australian coast. True "island mountains" or inselbergs of granite, these islands display sheet structure to perfection (Pl. 18.12b) and also a pronounced asymmetry with the strongly cliffed west-facing side which faces the long westerly fetch from the Bight (right in the photograph). *(Keith P. Phillips.)*

Plate 18.15 Platform cut in gneiss at the edge of Lake Greenly, southern Eyre Peninsula, S.A. Though the lake is very shallow (about 2 m at most), sand and pebbles have nevertheless been moved and rounded by wave motion. *(C.R. Twidale.)*

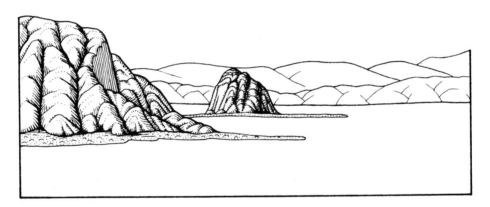

Fig. 18.10 The "Old Hat" type of shore platform. (*After C.A. Cotton, "Levels of Planation", pp 98–110, by permission of* Z. Geomorph.)

Plate 18.16 Detail of the serrated platform at Hallett Cove, S.A. (cf Pl. 18.6). (*C.R. Twidale.*)

developed above or near the upper limit of the sea, but below the structurally determined spring line suggesting that the water-table is not vital to the formation of high platforms.

That solution and biochemical effects generally are active is evident in many places. The activity of such processes rather than wave attack may help explain why intertidal platforms are frequently well developed and preserved in bay-head situations, though the fact that similar features (though more likely due to wave abrasion) on headlands are narrow in comparison, due to rapid wave undermining, should also be borne in mind.

Platforms of the Australasian type have long been noted in the scientific literature and have given rise to much argument. Some platforms occur above even the highest tide level but are within the spray zone. The number of platforms within the tidal zone varies but whether there be one or six all are remarkably level. Two basically different interpretations have been placed on such flights of platforms. Some workers, particularly the so-called Western Australian School of whom Fairbridge is the leading representative, consider that the platforms have all formed in essentially the same way, that they are all intertidal platforms, and that they have formed at different times in relation to different stands of the sea. The occurrence of platforms at different levels on the same coast in these terms argues changes in the relative levels of land and sea (see also Chapter 23). Even quite minor elevational differences between platforms are interpreted as evidence of changes in the level of the sea.[18] Others, however, notably Victorian workers of whom Hills[19] is best known, see the platforms as developing simultaneously at different levels, but under different processes and in relation to the present level of the sea.

Which of these two theories is correct? There is no doubt that the relative levels of the land and sea have changed greatly during geological time and that sealevel fluctuated within fairly wide limits during the Pleistocene. The only questions are, by how much and when? It is equally apparent that platforms are actively developing at present at several levels on some coasts.

But whatever their detailed morphology and location these intertidal features all exhibit a curious characteristic: it is that though they are all actively developing at the present time, all are also being dissected by waters thrown up by waves returning to the ocean, as is shown by the development of numerous gulches, and all are being undermined at their seaward margins, as witness the presence of the low tide cliff which leads down to either a submarine slope or to a lower, submerged platform which is termed the ultimate platform by some workers.

Possibly the most reasonable and comprehensive explanation of these varied platforms and profiles is due to Hills[20] who, though long an advocate of the effectiveness of pool weathering and the simultaneous and modern evolution of several platforms at different levels on the coast, also offers an explanation linking the stepped and the graded erosional shore profiles. Hills points out that all intertidal platforms of the Australasian type are suffering destruction and suggests in effect that their degree of preservation depends on the efficacy at any particular site of wave attack on the one hand and pool weathering on the other. Thus in hot arid areas where there is wetting and drying, weathering

ensures the development of planate benches in the tidal zone, but in temperate and humid regions this is not so, and waves are the dominant agency shaping the coast.

D. DEPOSITIONAL FORMS ON THE COAST

Not only do waves and weathering produce a great deal of debris on the coast, but in addition rivers bring a large volume of material worn from the land to the margin of the sea. Some debris is deposited at river mouths in the form of deltas (Chapter 12), but observations of marked and traceable particles show that much is transported along the coast, sometimes over considerable distances. This is accomplished by the process known as *longshore drifting*.

Waves are refracted and, for this reason, tend to approach more and more nearly parallel with the coast, and though local parallelism between waves and coast is frequently achieved waves breaking obliquely to the shore are also commonplace. Thus, the sand and other debris picked up by breaking waves are carried forward in the swash obliquely up the beach. Some of the water soaks into the beach but some washes directly down the surface following the steepest gradient, and taking debris with it. These particles are picked up by the next breaking wave and are thus taken further along the beach in a zig-zag motion known as *beach drifting*, the overall direction of which is determined by the strength and duration of winds: debris travels along the shore in the direction of the prevailing winds (Fig. 18.11).

A distinction can be made between longshore drift and beach drift. The former refers to the transport of debris along the coast under the influence of large waves and takes place some distance offshore. Away from the minor irregularities of coastal configuration, it constitutes a continuous flow of sand and other debris along the coast. Beach drift, on the other hand, is more local. Waves breaking on the beach cause movement of debris only within local embayments. Such near-shore debris cannot pass around headlands until considerable volumes of debris accumulate at the downwind end of the beach or until the beach is affected by strong destructional waves capable of combing down the beaches and transporting the sand, in the undertow, into deep water to join the deep water drift. Compared to waves, currents play a minor role in the transport of debris

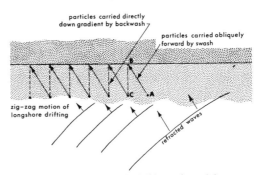

Fig. 18.11 Beach drifting of particles under the influence of oblique waves.

Plate 18.17 Chesil Bank or Beach is, as the name suggests (*cesil* is an Anglo-Saxon word meaning pebble), a shingle beach on the coast of southern England. The spit, which encloses a shallow lagoon known as the Fleet is almost 30 km long and the size of the pebbles increases eastwards from about 1–2 cm diameter at the northwestern end to 5–8 cm at the eastern extremity. The beach faces gales from an open stretch of water between Ushant and Start Point. The beach orientation is determined and shingle has been driven up the beach by SW gales from the Atlantic and easterlies from the Channel are sufficiently powerful to transport the smaller pebbles westwards though they cannot move the larger ones. The highest ridge on the beach was thrown up during a great storm in 1852, and still stands. (*Hunting Surveys.*)

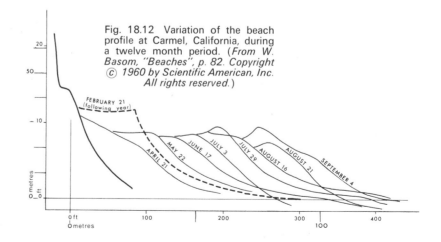

Fig. 18.12 Variation of the beach profile at Carmel, California, during a twelve month period. (*From W. Basom, "Beaches", p. 82. Copyright © 1960 by Scientific American, Inc. All rights reserved.*)

but in restricted passages where locally high velocities are attained they are responsible for moving silt and clay. However sand and coarser material can be moved only by longshore drifting under the influence of wind-driven waves and swell.[21] The evidence upon which this conclusion is based includes the following points.

1. Beaches and other coastal features built by deposition occur above high water mark, and thus beyond the reach of the currents.
2. Only in very special circumstances can sand, for example, be moved by tidal currents and, even then, it needs to be placed in suspension by waves so that currents can move it.
3. All depositional forms found on sea coasts are found also on the shores of lakes, where there is no possibility of strong currents.
4. In the laboratory, all known coastal depositional forms can be simulated by using wave action only.

Much of the debris carried by beach drifting is deposited by waves to form beaches, which are particularly well developed in embayments. Detritus derived from the erosion of headlands is carried into bays and accumulates there, the composition of the beach material bearing a close relationship to the nature of the bedrock in the headlands. Thus the beaches of volcanic islands are composed of fine basalt sand, those of coral reefs of coral (calcium carbonate) sand and those fronting aeolianite cliffs are also of lime. Some beaches are of shingle (Pls 18.5, 18.13d, 18.17), some are almost straight (Pl. 18.3), others are arcuate (Pl. 18.13c), and others though arcuate are asymmetrical in shape. They probably result from wave refraction around headlands.

The beaches themselves change greatly in detail from day to day, and from hour to hour [22] (Fig. 18.12). They are built up during calm weather when constructional waves are active and are combed down and even removed during storms when short period destructional waves are active. Individual ridges, or *berms*, which have been built up on the exposed beach and each of which represent a period of construction, may be wholly or partly destroyed. Beach ridges built of sand and organic debris mark former high tide levels, and wave attack can result in the formation of a low cliff in the beach sand (Pl. 18.18). On some coasts the beach ridges are vegetated (for instance, the cheniers of the Gulf coast in the USA).

Within the tidal zone there are many minor sand features. A broad, sandy rise with a depression beyond and breached at intervals by shallow gutters is commonly developed near the upper tidal limit. The rises are called *balls* and the depressions *lows*, or alternatively *ridges* and *runnels* respectively. The runnels, up to 1 m deep, may contain water, but more commonly waves wash over the ridge, and the water runs along the low until it reaches a gutter by which it returns to the sea.

Cusps and ripple marks are common beach features. *Cusps* are crescent-shaped depressions ranging in length from a metre to several scores of metres, and in depth from a few centimetres to a metre or more. The "horns" of the crescents point seawards. Although they are clearly related to wave action, the origin of cusps is little understood. *Ripple marks*, too, are formed by wave action: oscillatory wave motion creates horizontal vortices which cause differential deposition and erosion, and the development of ripple marks, analogous to those formed by the wind passing over sand.

On every beach there are two simultaneous, but opposed, movements; water and debris is carried up the beach toward the land by the swash, while both are also carried seaward by the backwash. Where the onshore wind is particularly strong, a hydraulic head may be created by the water being piled up, so to speak, by successive waves. In order to escape seawards, the water tends to concentrate into a series of distinct currents, called rip currents, which flow away from the beach along the sea-floor. These powerful local currents excavate *rip channels*. In the zone where waves break for a particular period of time, distinct ridges or bars which are oriented parallel to the waves form on the sea-floor. The onrushing waves are carrying debris toward the shore; on the other hand, debris is being carried seaward by backwash and, on occasions, by rip currents. The encounter between retreating and advancing waves not only makes the latter break, but in addition causes deposition of debris, forming a *breakpoint bar*. There are often two or more such bars, one corresponding to high tide, one to low tide level. They are ephemeral features, but survive on protected coasts.

On coasts with gentle offshore gradients, larger features called *barrier bars* are formed by waves which break some distance from the coast. Those off the east coast of the U.S.A. and in the southern Baltic (Fig. 18.13) extend for long distances. Bars are called *spits* when they become tied or attached to the mainland or to an island. Once in shallow water they are supplied with debris by longshore drift, the

Plate 18.18 Cliff 2 m high cut by waves in sand of the former beach, southern Yorke Peninsula, S.A. (*C.R. Twidale.*)

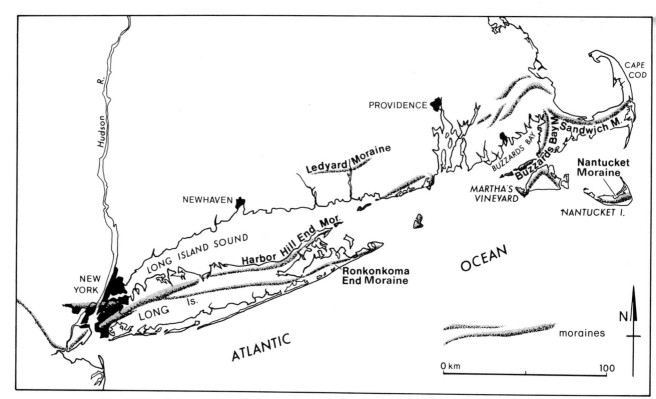

Fig. 18.13a Spits on the south shore of Long Island, New York State. Note also the major end moraines which form the backbone of the island. *(In part after C.A. Kaye, "Illinoian and early Wisconsin Moraines of Martha's Vineyard, Massachusetts"*, U.S. Geol. Surv. Prof. Paper, *1964,* **501-C,** *pp.134–9.)*

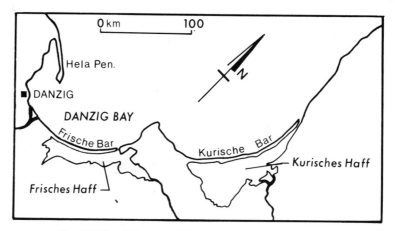

Fig. 18.13b   Spits east of Danzig: curved baymouth bars enclosing two large embayments on the south Baltic Coast.

material immediately in front of the barriers being thrown up by wave action. Thus, as waves are generated by the wind, many spits and bars are oriented normal to an important wind direction though such factors as fetch and the configuration of the coast as a whole also influence their alignment in detail. The refraction of waves around offshore islands for instance can lead to deposition and the formation of spits called *tombolos* which link the island to the mainland (Pl. 18.19).

Not all winds influence the trend of spits and bars: onshore winds of a strength greater than Beaufort Force 4 (28–29 km/hr) are those which effectively mould the coast. When fetch is equal in all directions, or when maximum

fetch and the resultant of strong wind directions are co-incident, the coast is aligned at right angles to the resultant of the winds. However, when the fetch and resultant are not coincident, the coast is oriented at an angle normal to the line between these two (Fig. 18.14a and b).

On some coasts spits and bars are built up at a high angle to each other. This is because on some coasts two or more winds significantly effect the shaping of the coast, and if these blow from different quarters, bars aligned at contrasted angles are built up. They enclose lagoons which develop into swamps, marshes and eventually low-lying plains (see below) and form *cuspate forelands* [23] which, like spits, enclose lagoons, marshes and lowlands (Figs 18.15, 18.16).

Spits, like bars, commonly display secondary spits or *laterals* which have a different orientation from the parent spit and which are built up waves from other, less significant directions caused by secondary winds and waves, or by refracted waves. Several specific types of spit are developed, but all are related in their orientation to the dominant wind, after due allowance has been made for refraction. Spits change location and morphology quite rapidly and the development of several in historical times has been traced in some detail (see for instance Fig. 18.16a; see also Pl. 18.20).

Spits and bars create relatively sheltered water. In such areas, and on other sheltered parts of the coast (particularly those near the outlets of major rivers which bring large

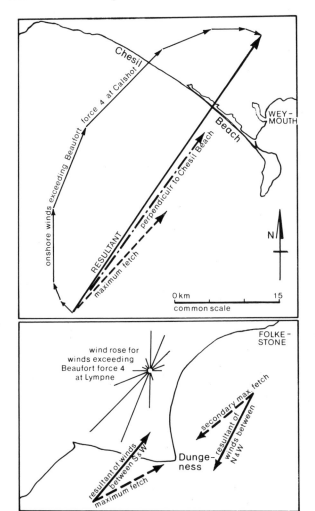

Fig. 18.14a The orientation of Chesil Beach and Dungeness with respect to fetch and the origin of strong winds. (*After A. Guilcher,* Coastal and Submarine Morphology, *Methuen, London, 1958, p. 183.*)

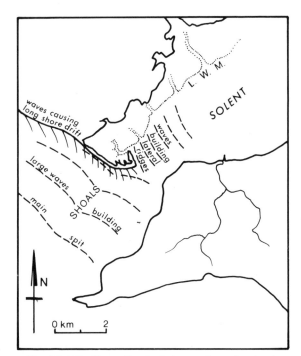

Fig. 18.14b Hurst Castle Spit, southern England, showing the lateral spits and their relationship and that of the main spit with wave directions. (*After P. Lake, Physical Geography,* Cambridge University Press, *1954, p. 281.*)

Plate 18.19 Greyhound Rock is an island eroded in sedimentary rocks on the Californian coast some 80 km south of San Francisco. It is linked to the mainland by a tombolo deposited where waves refracted round the island meet. (*C.R. Twidale.*)

Plate 18.20 Spurn Head, on the east coast of England, and on the north side of the Humber Estuary, is built of debris which has been driven by longshore drifting southward along the coast of Holderness, east Yorkshire. The debris is built southward and curved to the west by refracted waves and by easterly and southeasterly winds. Several incipient secondary spits (which are ephemeral, as is the location and morphology of the main spit—see G. de Boer, "Spurn Head", pp. 71–89) are visible. There is historical evidence to suggest that the rate of southerly growth is of the order of 12 m per annum though this does not continue indefinitely, for the neck is periodically breached, and the spit partially destroyed, after which rebuilding and extension begins over again. (*Hunting Surveys.*)

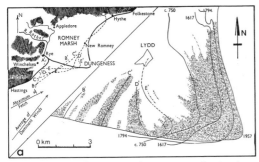

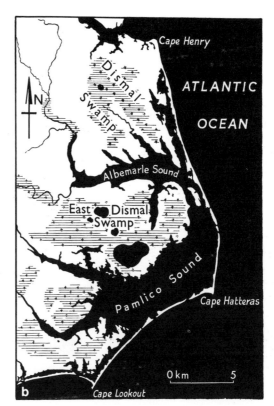

Fig. 18.15a Dungeness is a cuspate foreland on the coast of southern England. Various beach ridges developed during its development are discernible. Suggested stages in the evolution of the feature are shown. *(After W.A. Lewis, "Formation of Dungeness', pp. 309–24 and W.G.V. Balchin "Past Sea-levels at Dungeness", pp. 258–77.)*

Fig. 18.15b Cape Hatteras, a large cuspate foreland on the east coast of the USA. *(From A. Holmes, Principles of Physical Geology, Nelson, Edinburgh, 1965, p. 829.)*

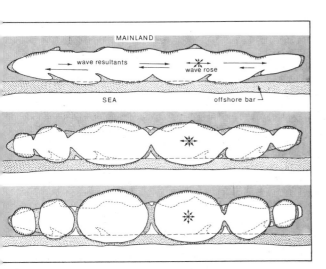

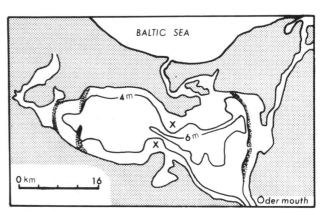

Fig. 18.16 Development of spits and cuspate spits in lagoons under the influence of wave resultants: (a) the general case; (b) in the Stettin Gulf, East Germany, where there are two opposed submarine cuspate forelands (X) in the middle of the lagoon and submarine bars at each end of the Gulf. *(After V.P. Zenkovitch, "On the Genesis of Cuspate Spits", pp. 269–77, by permission of* J. Geol. *and the University of Chicago Press.)*

Plate 18.21 Bolinas Lagoon—an arm of the sea now separated from the sea by a curved spit just north of San Francisco, California. (*C.R. Twidale.*)

Plate 18.22a Small isolated coastal plain surrounded by the Exmoor hills and cliffs at Porlock, north Devon. The former embayment was cut off by a shingle bar and subsequently filled in by debris carried from the surrounding hills. *(C.R. Twidale.)*

Plate 18.22b Westward Ho! is an accumulation of sand and mud behind a shingle ridge which projects NNE across the Taw-Torridge estuary in north Devon (Fig. 17.18). A small lake, Sandy Mere, is trapped behind the ridge which was driven forward by westerly wind-generated waves during the latter half of the last century. However it is now virtually stationary. There are recurved spits at the distal (near) end of the ridge. Refraction of waves round the end of the spit can clearly be distinguished as can the towns of Westward Ho! (right) and Appledore (left). (*Hunting Surveys.*)

volumes of debris to the sea), mudflats, marshes, swamps and lagoons are formed (Pl. 18.21). These are eventually filled in to form a flat coastal plain (Pl. 18.22). In temperate regions the sand and mud are stabilised and accumulated by plants, particularly the samphire, until the floor of the tidal lagoon is built up above low water mark, and then above high tide level. Debris from the land, as well as that brought by the sea, contributes to this build-up. In the tropics, bars and tidal flats are colonised by mangroves, the aerial roots of which form a splendid trap and stabiliser of the silty surface. Great mudflats are accumulated behind the mangrove region (Pl. 18.23), and besides bringing sand and silt to the coast, major rivers also develop fantastic drainage patterns when breaching the swamps. In arid and semiarid regions the formation of baymouth bars and spits causes former areas of the sea to become isolated. Natural salt pans develop: water is which evaporated, leaving behind dry salt lakes. Wave action in such coastal lakes causes the development of beaches and spits and the eventual subdivision of the lagoon into a chain of smaller lakes[24] (Fig. 18.16).

E. COASTAL FOREDUNES

Coastal foredunes, which are a type of fixed dune, are closely related to wave deposition in the form of beaches and bars. They are not unchanging, for tongues of sand can and do migrate downwind and inland from the coastal areas under certain circumstances (Pls 18.24, 18.25), but they do not extend far before they become colonised by vegetation and thus stabilised. Like lunettes they do not migrate far from the source of the debris of which they are built because their general setting is not one of aridity. Indeed, their humid and vegetated environment leads to variations in structure which, though minor, are nevertheless significant in distinguishing such coastal dune deposits from their desert analogues where the sediments are preserved in stratigraphic sequences. The aeolian, cross- or false-bedding characteristic of coastal dunes display disturbances due to root penetration; and the presence of moisture leads to a loss of cohesion between sand grains sufficient to induce frequent minor slips and slumps which interrupt the normal pattern of lamination.[25]

Coastal foredunes have been closely studied in some areas, and their origin is reasonably well understood.[26] They can be observed in process of formation at the present time and stages in their evolution can be reconstructed and plotted. Wind-driven waves carry debris, including sand, to the shore where it forms beaches. Onshore winds can lift this sand in the same way as the material is lifted in desert regions, though wetness can inhibit the wind's activity. The sand is carried by saltation a short distance beyond the beach

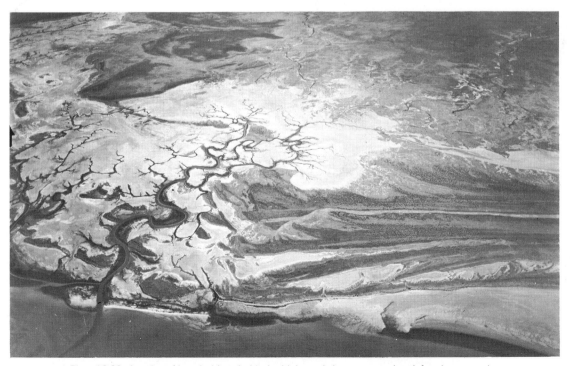

Plate 18.23 A series of beach ridges behind which muds have accumulated, forming extensive saline and tidal flats, on the southern shore of the Gulf of Carpentaria, just west of Parker and Bayley points. Note the peneplain beyond the littoral zone (see Chapter 20 for further reference to peneplain) and the river meanders within the tidal zone. (*RAAF.*)

Plate 18.24 Vertical air photograph of coastal parabolic dunes just southeast of the Murray mouths, S.A., showing their inland extensions under the influence of the prevailing and onshore south-westerlies. (*S.A. Lands Dept.*)

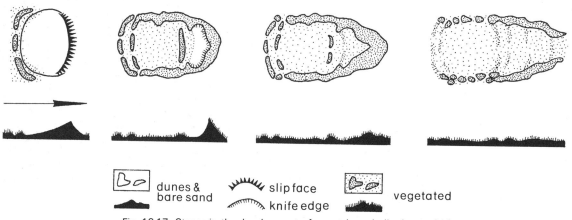

Fig. 18.17 Stages in the development of coastal parabolic dunes. (*After S.Y. Landsberg, "The Orientation of Dunes", pp. 176–89.*)

where the air-flow is disturbed by vegetation and the sand is deposited. The sand accumulations are, in turn, colonised by vegetation such as Marram grass and are gradually stabilised as hummocky ridges running parallel to the beach. Several such linear ridges are developed in some areas.

The beach and foredunes are naturally in equilibrium and human disturbance of the system (as by the carting of dune sand) is a significant cause of beach erosion. Sand supplied to the beach by drifting and by waves is carried from the beach to dunes. Waves replenish the beach supply during periods of constructional activity. But during storms, breakers comb the beach and also undermine the toes of the foredunes, which collapse and return debris to the beach.

Unusually strong winds scour the areas between the clumps of vegetation and also transport very large volumes of sand from the beach, thus overwhelming the existing vegetation cover which may be observed to be wholly or partially buried. The sand then becomes free moving, and it is moulded into parabolic or U-shaped dunes, which are not usually aligned with their long axes parallel to the shore. In some areas lakes are impounded in the depressions between the dunes.[27]

The evolution of parabolic dunes is described below, and reflects varied conditions and sand mobility. Because of proximity to the water-table and because of the consequent relative abundance of vegetation, the base of a mound of sand thrown up by onshore winds is comparatively stable. But the upper parts are vulnerable to wind action and sand is driven downwind, leaving behind two short areas pointing upwind (Fig. 18.17) in contrast to the downwind pointing horns of a barchan. In some areas continuation of this process leads to the formation of trailing arms several kilometres long. Pauses in downwind progress due to seasonal wetness or to lack of strong winds are marked by small cross-bars, colonised and stabilised by vegetation[28] (for instance in parts of Denmark, at Fowie in Scotland, and at the Kurische Nehrung on the south shore of the Baltic). As sand is left behind in the trailing arms, the volume of the main mass decreases unless, of course, there is replenishment from the beach. More and more of the sand mass comes within reach of the vegetation supported by moisture from the water-table, so that movement decreases as the dune becomes more and more stabilised. During storms, however, the associated strong winds cause erosion of the sand between shrubs, and such blowouts extend down to the water-table. As the wind channels through the cols formed by the blowouts, the whole of the central mass may be eroded leaving only the long linear ridges of the parallel trailing arms oriented parallel to the dominant wind.

Landsberg[29] suggests that the axes of the parallel dunes and the trailing arms are aligned with respect to the resultant of the winds which is capable of shifting sand in the particular locality, taking account of local topography, the condition of the vegetation and so on. An important factor to be borne in mind, however, when the geometry of these coastal dunes is under consideration is that the beach is the source of the sand.[30] Onshore winds are therefore of greater significance than those which blow offshore. Thus, as with desert dunes, many factors influence the shape and trend of coastal dunes.

Coastal foredunes commonly achieve heights of 30–50 m but on the Natal coast (South Africa) and on the east coast of Queensland they attain heights of 250–300 m.[31] In many parts of southern and southwestern Australia the dunes are impregnated and indurated with lime derived from the shells of marine organisms broken in the surf, probably during periods of low sealevel which occurred during the glacial phases of the Pleistocene (Chapter 23). But their essential morphology and origins remain clear (Pl. 18.26). The shell fragments were concentrated on the beaches and then blown inland in the normal way. Some such lime-rich dunes

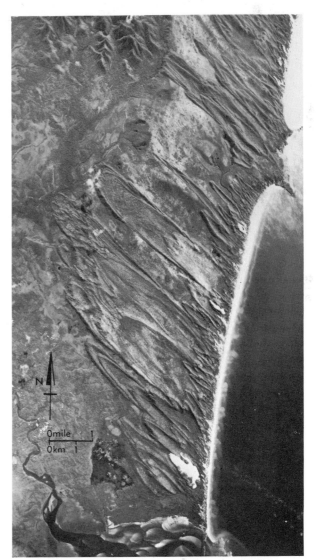

Plate 18.25 Vertical air photograph of parabolic dunes formed under the influence of onshore southeasterlies near Port Clinton, some 65 km north of Rockhampton, Queensland. (*Dept Nat. Dev., Canberra.*)

Plate 18.26 The backslopes of these old coastal foredunes in aeolianite are preserved near Pondalowie Bay, Yorke Peninsula; the old frontal areas have been eroded by the sea. (*C.R. Twidale.*)

extend up to 100 km inland in areas which are not tectonically disturbed and much further in areas that are. The lime-indurated dunes are called aeolianite dunes. A magnificent sequence of such features is preserved in the South East of South Australia, and, though several attempts have been made to establish their chronology, the dunes remain an enigma, from this point of view.[32]

In their general development, coastal foredunes are analogous both to lunettes and to the debris mounds developed

Plate 18.27 Climbing dune at Royston Head, Yorke Peninsula. At the time of observation it fell just short of the cliff top in old dune sand which overlies gneisses exposed by the sea in the foreground. (*C.R. Twidale.*)

on the lee sides of playas in desert regions. However, they display a greater range of form as a result of more extensive source areas, stronger wind activity, and hence greater variability of development. One notable feature of coastal foredunes is that they are only poorly developed in or are absent from the humid tropical regions,[33] even in contexts which would elsewhere seem to favour their development. It has been suggested that the frequent rains experienced in such regions and the consistently high moisture content of beach sands there critically decreases their mobility.[34] But this is not the whole story, for Jennings[35] has shown that high velocity winds are generally less frequent in the humid tropics than in the subtropics and temperate regions, and he suggests that the absence of coastal foredunes in these regions is mainly due to this factor. This again illustrates the validity of Wolman and Miller's[36] assertion that it is the fairly frequent, moderately intense forces which determine the landforms of a particular region: one might point to the intensely strong winds associated with tropical cyclones and suggest on that account that strong winds are experienced; but they are of insufficient frequency to accomplish much in the way of moulding even such fragile materials as beach sands.

Most coastal dunes occur close to sealevel, their toes being just above high tide level, but in some parts of southern Australia, there are *climbing dunes* in which sand migrates from the beach to the cliff top (Pl. 18.27), and dunes occur on top of coastal cliffs which, in King Island, Bass Strait, are up to 55 m high and which extend as high as 70 m above sealevel bordering the Great Australian Bight.[37] Such cliff-top dunes occur above cliffs up to 100 m high at the southern extremity of Yorke Peninsula, South Australia.

Cliff-top dunes may be explained in terms of their migration from the shore of a lake or lagoon or arm of the sea behind the coast proper (Fig. 18.18a-d) implying the effectiveness of offshore, and not onshore, winds. Many lines of evidence argue against such a hypothesis, the most common being that cliff-top dunes occur where no such inland source of sand exists. Furthermore, the structures and general orientation of the dunes suggest an origin on the real coast. The dunes stranded on cliff tops could also have developed through the landward advance and subsequent erosion of the lower dune areas by waves and other marine agencies, or through normal aeolian activity in relation to a higher stand of the sea. But on the whole, the most

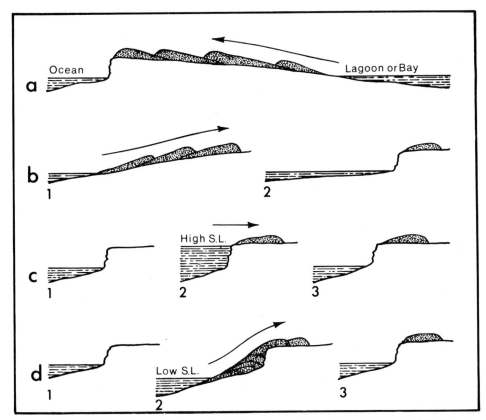

Fig. 18.18 Possible explanations of cliff-top dunes: (a) in relation to a source inland from the dune; (b) in relation to a receding coast; (c) in relation to higher sealevel; (d) in relation to lower stands of the sea. (*After J.N. Jennings, "Cliff-top Dunes", pp. 40—9, by permission of* Aust. Geogr. Studies *and J.N. Jennings.*)

satisfactory explanation in terms of the field evidence is that the dunes developed on a narrow coastal plain in relation to a lower sealevel; that because of the large volumes of sand (and shell) exposed to wind action, the dunes grew so high that they overtopped the cliff; and that subsequently, with a rise in sealevel, the lower parts or toes of the dunes were eroded, leaving the upper areas isolated on top of the cliffs (Fig. 18.18c). This not only accounts for the cliff-top dunes, but is also consistent with their being indurated with lime in some areas and with the extension of aeolianite below sealevel in the same areas as the cliff-top dunes occur.

F. CONCLUDING STATEMENT

The coastal zone appears at first sight to constitute such a characteristic assemblage of forms moulded by such distinctive processes that it transgresses climatic boundaries and provides an example of azonal geomorphological features. But even a casual inspection either regionally or in detail reveals how important is structure in controlling coastal morphology. Climatic control has been emphasised in the discussion of platforms, but coral reefs (see Chapter 25) and coastal foredunes are two other coastal features which display a climatic distribution. Thus the notion implicit in the idea of morphogenetic regions applies also on the coast though it is true to say that the idea has not yet fully been explored in this context. A pioneering effort in this direction is due to Davies[38] who has attempted to classify coasts on the basis of their wind and tide regimes. He has identified storm wave, swell and low energy coasts. The storm wave environments are characterised by strong cliffing and marked mechanical abrasion of platforms, and by coarse depositional features such as shingle beaches and spits. The swell environments, according to Davies, display development of platforms due to chemical activity, as well as large constructional features. The low energy areas exhibit barrier beaches, spits and other constructional features. There are obvious defects in this scheme but it is nevertheless an original first attempt.

But the coastal zone typifies the whole of the land surface, inasmuch as not only structure and process are reflected in the shape of the surface: the changes in time in these two, the history of the zone, is also manifest. Historical geomorphology is discussed in the next sequence of chapters.

References Cited

1. See for instance C. Thompson, "Submarine Erosion off the Holderness Coast", *Geol. Mag.*, 1923, **60**, pp. 313–17; A.H.W. Robinson, "The Harbour Entrances of Poole, Christchurch and Pagham", *Geogr. J.*, 1955, **121**, pp. 33–50; W.W. Williams, "An East Coast Survey: Some Recent Changes in the Coast of East Anglia", *Geogr. J.*, 1956, **122**, pp. 317–45; *idem*, *Coastal Changes*, Routledge & Kegan Paul, London, 1960; G. de Boer, "Spurn Head: Its History and Evolution", *Trans. Inst. Br. Geogr.*, 1964, **34**, pp. 71–89.

2. E.S. Hills, "Shore Platforms", *Geol. Mag.*, 1949, **86**, pp. 137–52; *idem*, "A Study of Cliffy Coastal Profiles based on Examples in Victoria, Australia", *Z. Geomorph.*, 1971, **15** NS, pp. 137–80. This last paper includes splendid descriptions and illustrations of a wide range of erosional coastal forms.

3. See E.P. Hodgkin, "Rate of Erosion of Intertidal Limestone", *Z. Geomorph.*, 1964, **8** NS, pp. 385–92; E.S. Hills, "Shore Platforms and Wave Ramps", *Geol. Mag.*, 1972, **109**, pp. 81-8.

4. See C.K. Wentworth, "Marine Bench-forming Processes. I. Water-level Weathering", *J. Geomorph.*, 1938, **1**, pp. 5–32; *idem*, "Marine Bench-forming Processes. II. Solution Benching", *J. Geomorph.*, 1939, **2**, pp. 3–25; E.S. Hills, "Shore Platforms", pp. 137–52; see also C.A. Kaye, "Shoreline Features and Quaternary Shoreline Changes, Puerto Rico", *U.S. Geol. Surv. Prof. Paper*, 1959, **317B**; *idem*, "The Effect of Solvent Action on Limestone Solution", *J. Geol.*, 1957, **65**, pp. 35–46.

5. J. Joly, "Experiences sur la Dénudation par Dissolution dans L'eau Douce et dans L'eau de Mer, VIII", *Inst. Geol. Cong. (Paris)*, 1901, **2**, pp. 774–84; R. Revelle and K.O. Emery, "Chemical Erosion of Beach Rock and Exposed Reef Rock", *U.S. Geol. Surv. Prof. Paper*, 1957, **260**.

6. J. Tricart, "Problèmes géomorphologiques du Littoral oriental du Brésil", *Cah. océanogr. C.O.E.C.*, 1959, **11**, pp. 276–308; *idem*, "Experiences de Désagregation de Roches granitiques par la Crystalisation du Sel Marin", *Z. Geomorph. Suppl.*, 1960, **1**, pp. 239–40; J.M. Coleman, S. M. Gagliano and W.G. Smith, "Chemical and Physical Weathering on Saline High Tidal Flats, northern Queensland, Australia", *Bull. geol. Soc. Am.*, 1966, **77**, pp. 205–6.

7. W.C. Bradley, "Submarine Abrasion and Wave-Cut Platforms", *Bull. geol. Soc. Am.*, 1958, **69**, pp. 967-74.

8. See E.S. Hills, "Shore Platforms" for references to earlier works. For an example of a shore platform in a temperate setting see T. Eastwood, *Northern England*, British Regional Geography Series, H.M.S.O., London, 1955, Pl. IIIA.

9. R.S. Dietz, "Wave Base, Marine Profile of Equilibrium and Wave Built Terraces—a Critical Appraisal", *Bull. geol. Soc. Am.*, 1963, **74**, pp. 971–90.

10. H.W. Ahlmann, "Geomorphological Studies in Norway", *Geogr. Ann.*, 1919, **1**, pp. 3–148, 193–252.

11. F. Nansen, *The Strand-flat and Isostasy*, Skrifter utgit av Videnskapsselskapeti Kristiania I Math., Naturvidensk, Klasse, 2 vols., Kristiania, 1922.

12. C.K. Wentworth, "Potholes, Pits and Pans: Subaerial and Marine", *J. Geol.*, 1944, **52**, pp. 117–30.

13. J.A. Bartrum, "'Abnormal' Shore Platforms", *J. Geol.*, 1926, **34**, pp. 798–806; *idem*, "Shore Platforms", *Rep. ANZAAS*, 1935, **22**; *idem*, "Shore Platforms—Discussion", *J. Geomorph.*, 1938, **1**, pp. 266–72.

14. J.T. Jutson, "Shore Platforms near Sydney, New South Wales", *J. Geomorph.*, 1939, **2**, pp. 237–50.

15. E.C.F. Bird and O.F. Dent, "Shore Platforms on the south Coast of New South Wales", *Aust. Geogr.*, 1966, **9**, pp. 207–17.

16. J.A. Bartrum, "'Abnormal' Shore Platforms", pp. 798–806.

17. C.A. Cotton, "Levels of Planation of Marine Benches", *Z. Geomorph.*, 1963, **7** NS, pp. 98–110.

18. See for example R.W. Fairbridge, "Geology and Geomorphology of Point Peron, Western Australia", *Proc. R. Soc. W.A.*, 1950, **34** (for 1947/8), pp. 35–72; *idem*, "Limestone Coastal Weathering", in R.W. Fairbridge (Ed.), *Encyclopaedia of Geomorphology*, Reinhold, New York, 1969, pp. 653–7; W.T. Ward and R.W. Jessup, "Changes in Sealevel in southern Australia", *Nature (Lond.)*, 1965, **205**, pp. 791–2.

19. E.S. Hills, "Shore Platforms", pp. 137–52.

20. E.S. Hills, "Shore Platforms".

21. W.V. Lewis, "The Effect of Wave Incidence on the Configuration of a Shingle Beach", *Geogr. J.*, 1931, **78**, pp. 129–37; *idem*, "The Evolution of Shoreline Curves", *Proc. geol. Soc. London*, 1938, **49**, pp. 107–27; J.N. Jennings,, "The Influence of Wave Action on Coastal Outline in Plan", *Aust. Geogr.*, 1955, **6**, pp. 36–44.

22. W. Bascom, "Beaches", *Sci. Am.*, 1960, **203** (1), pp. 80–94.

23. See for instance W.V. Lewis, "The Formation of Dungeness", *Geogr. J.*, 1932, **80**, pp. 309–24; W.V. Lewis and W.G.V. Balchin, "Past Sea-levels at Dungeness", *Geogr. J.*, 1940, **96**, pp. 258–77.

24. V.P. Zenkovitch, "On the Genesis of Cuspate Spits along Lagoon Shores", *J. Geol.*, 1959, **67**, pp. 269–77.

25. E.D. McKee and J.J. Bigarella, "Deformational Structures in Brazilian Coastal Dunes", *J. Sed. Petrol.*, 1972, **42**, pp. 670–81.

26. See for instance S.Y. Landsberg, "The Orientation of Dunes in Britain and Denmark in Relation to Wind",

Geogr. J., 1956, **122**, pp. 176–89; J.N. JENNINGS, "On the Orientation of Parabolic or U-dunes", *Geogr. J.*, 1957, **123**, pp. 474–81.

27. J.N. JENNINGS, "Coastal Dune Lakes as exemplified from King Island, Tasmania", *Geogr. J.*, 1957, **123**, pp. 59–70.

28. S.Y. LANDSBERG, "The Orientation of Dunes", pp. 176–89.

29. S.Y. LANDSBERG, "The Orientation of Dunes", pp. 176–89.

30. J.N. JENNINGS, "On the Orientation of Parabolic or U-dunes", pp. 474–81.

31. F.W. WHITEHOUSE, "Sandhills of Queensland — Coastal and Desert", *Qld Naturalist*, 1963, **17**, pp. 1–10.

32. See P.S. HOSSFELD, "Late Cainozoic History of the South-east of South Australia", *Trans. R. Soc. S. Aust.*, 1950, **73**, pp. 232–79; R.C. SPRIGG, "The Geology of the South-East Province of South Australia; with reference to Quaternary Coastline Migrations and Modern Beach Developments", *Bull. Geol. Surv. S. Aust.*, 1952, **29**; G. BLACKBURN, R.D. BOND and A.R.P. CLARKE, "Soil Development associated with Stranded Beach Ridges in South East of South Australia", *C.S.I.R.O. Soil Pub.*, 1965, **22**.

33. J.N. JENNINGS, "The Question of Coastal Dunes in Tropical Humid Climates", *Z. Geomorph.*, 1964, **8** NS, pp. 150–54.

34. E.C.F. BIRD, "The Formation of Coastal Dunes in the Humid Tropics. Some Evidence from north Queensland", *Aust. J. Sci.*, 1964, **27**, pp. 258–9.

35. J.N. JENNINGS, "Further Discussion of Factors affecting Coastal Dune Formation of the Tropics", *Aust. J. Sci.*, 1965, **28**, pp. 166–7.

36. M.G. WOLMAN and J.P. MILLER, "Magnitude and Frequency of Forces in Geomorphic Processes", *J. Geol.*, 1960, **68**, pp. 54–74.

37. J.N. JENNINGS, "Cliff-top Dunes", *Aust. Geogr. Stud.*, 1967, **5**, pp. 40–9.

38. J.L. DAVIES, "A Morphogenetic Approach to World Shorelines", *Z. Geomorph.*, 1964, **8** NS, pp. 127–44.

General and Additional References

D.W. JOHNSON, *Shore Processes and Shoreline Development*, Wiley, New York, 1919.

A. GUILCHER, *Coastal and Submarine Morphology*, Methuen, London, 1958. Translated by B.W. Sparks and R.H.W. Kneese.

E.C.F. BIRD, *Coasts*, Australian National University Press, Canberra, 1968.

Part 4

CHANGES IN TIME —

HISTORICAL GEOMORPHOLOGY

"The first and most natural application of geomorphic study is the history of the earth"

Kirk Bryan, 1950

CHAPTER 19

Models of
Landscape Evolution

A. GENERAL

The idea of change in time is implicit in much of what has been written in the foregoing chapters. The shape of Mt Lamington was altered to a pronounced degree early in 1951. The Anchorage area of Alaska changed on Good Friday, 1964, and the Meckering area of Western Australia on 14 October, 1968, as a result of earthquakes. The local topography of the Frank region, Alberta, changed quite dramatically on 29 April, 1903 as a result of a major landslide, and many other catastrophic changes in the form of the land surface caused by earthquakes, volcanicity, heavy rains and other storm effects come to mind.

Less obviously, but in a long-term sense continuously, various agents are gnawing away at the land surface and in total they achieve more widespread and important changes than the various catastrophic events that could be mentioned. Rivers wear away their beds and banks, not all the time[1] but regularly, and in some areas frequently. The sand dunes of the tropical deserts and coastal regions vary in detail

almost daily according to the strength and direction of the winds that mould them (see for example, Pl. 15.9). Glaciers are moving downslope, eroding their beds in some areas and depositing debris elsewhere. Weathering is proceeding at various and varying rates at or beneath most parts of the earth's surface.

Thus changes in the shape of the land surface in time are indisputable though only those due to cataclysmic events are readily discernible. Some workers have conceived that these changes over time take place in evolutionary sequence, and have claimed to have recognised evidence consistent with this interpretation. Others, however, consider that the entire land surface is self-regulating, in keeping with what takes place on slopes and in river channels, and in keeping with physical principles. Two major types or models of landscape development have been suggested, one involving sequential or evolutionary changes in the landform assemblages, the other involving a condition of dynamic equilibrium or steady state.

Plate 19.1   Area of subdued relief developed on gently-dipping strata near Amman, Jordon. The slopes are essentially graded (though there are ribbed slopes in several places) and there has been rejuvenation of tributaries close to the main river valley which has a flat alluviated floor. *(Hunting Surveys.)*

B. EVOLUTIONARY MODELS

1. Geomorphic Cycles: the Peneplain Concept

In several parts of the world (see for instance, Pls 13.4a-b and 19.1) the landform assemblage consists of a large number of rather shallow, broad valleys separated by low, rounded interfluves. In some river valleys, though by no means in all and possibly not even in most, it appears that the angle of inclination of the valley side slopes increases upstream (Fig. 13.10—see also Chapter 13).

W.M. Davis[2] interpreted the latter situation as implying that slopes both decrease in inclination and are lowered in time, a suggestion corroborated by statistical analysis of slopes in such areas,[3] and conceived the rolling plains of low relief amplitude to be the result of long-continued downwasting and downwearing of the land mass. He called the low relief plain, dominated by river valleys and by convex interfluves, a *peneplain* (that is, almost a plain). He explained its development in terms of an idealised though admittedly unrealistic situation. For the sake of simplicity of understanding and exposition he assumed that the uplift of an area of low relief (perhaps a former continental shelf—see Fig. 2.20) was virtually instantaneous so that no significant erosion took place during the uplift, that the block was structurally homogeneous and that over a very long period of time there was both tectonic and climatic stability. Given these conditions Davis deduced that a certain sequence of landforms would develop (Fig. 19.1).

At first, in what Davis called the *youthful* stage, streams cut down rapidly and erode deep valleys which are V-shaped

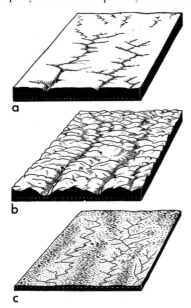

Fig. 19.1 Stages in the development of a peneplain. (*After A. Holmes,* Principles of Physical Geology, *Nelson, Edinburgh, 1944, p. 187.*)

in cross-section. Areas of low relief persist high in the landscape (Fig. 19.1a). As the rivers approach baselevel, they cut down more slowly and the various agencies at work on the valley side slopes, always effective, become even more dominant so that valleys are widened. The river systems extend until the whole land mass is dissected. At this stage (*maturity*) there are no significant areas of flat land either in valley floors or high in the relief: the region is one of *all-slopes*, and the relief amplitude is at its greatest (see Pl. 3.5). The rivers continue to erode their beds but henceforth weathering, mass movements, wash and streamwork cause the divides to be lowered more rapidly than the river beds.

Thus the relief amplitude decreases in this stage of *old age* or senility. As the rivers migrate laterally to a greater and greater degree, flood plains are developed, and eventually these, together with the gently undulating or broadly rolling interfluves worn down by weathering and wash, come to form an extensive area of low relief—the *peneplain* (Pl. 19.2). The flood plains may continue to extend laterally, gradually destroying the interfluves which separate them (Fig. 19.2), so that they eventually merge to form an extensive plain of lateral corrasion or *panplain*[4] as an integral part of the larger peneplain.

Unlike plains of deposition (Pls 12.9, 19.3), peneplains are not flat, for there may be 50–75 m of relief per km[2] (Pl. 19.4). Moreover, isolated ranges or hills, which are essentially remnants of circumdenudation and which are called *monadnocks* (after Mt Monadnock in New Hampshire, USA), stand above the general plain level. But slopes are graded, even on the upland remnants. Weathering and erosion continue but at a decreasing rate as streams and hillslopes approach closer and closer to regional baselevel.

Because the end-product of this long continued erosion, the peneplain, is morphologically similar to the postulated original surface of low relief, the whole sequence of events is referred to as a *cycle of erosion* or, because more than erosion is involved, as a *geomorphic cycle*.

In the peneplain concept, the convex interfluves are gradually lowered and the inclination of slopes is gradually reduced. The concept endeavours to explain the sequence of forms developed in the landscape over time. To understand more clearly the one factor, namely changes in time, Davis isolated that factor by making various assumptions as to the stability of exogenetic and endogenetic forces. Davis and his followers (notably D.W. Johnson in the USA, C.A. Cotton in New Zealand, H. Baulig in France and S.W. Wooldridge in Britain) considered both the method and the hypothesis valid, in some circumstances at least. They thought that they could discern in the landscape assemblages of forms which they believed represented stages in this evolutionary sequence, and they, and many other workers, thought they could detect in the present landscape remnants of what appeared to be former peneplains which are now dissected and partially destroyed.

Because of the realities of the landscape, because of Davis' persuasive logic, and because the idea was promulgated in the aftermath of Darwin when evolution was very much in the air, the peneplain concept attained wide acceptance; indeed, too wide an acceptance, for peneplains were recognised in many parts of the world on insufficient evidence. Initially Davis applied the peneplain concept to

Plate 19.2 Carpentaria plains near the Gulf coast and on the Northern Territory/Queensland border. Note the vast area of subdued relief, the oxbow and numerous abandoned stream channels. The darker patches are burns caused by grass fires. (*RAAF.*)

Plate 19.3 The flat Wondoola plains (part of the Carpentaria plains), northwest Queensland, built of Pleistocene riverine and shallow swamp (paludal) deposits. (*C.R. Twidale/C.S.I.R.O.*)

Plate 19.4   Broadly rolling relief of a high peneplain just north of the Barossa Valley, South Australia. *(C.R. Twidale.)*

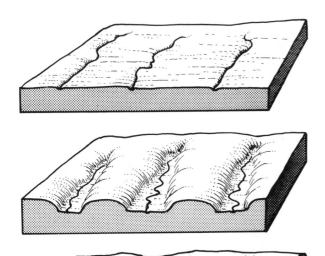

Fig. 19.2 Development of a panplain. *(After C.H. Crickmay, "Later Stages of the Cycle", pp. 337–47, by permission of* Geol. Mag.)

temperate (what were to him "normal") climates, though he also conceived of cycles of landscape development in arid and glaciated areas,[5] and Johnson[6] applied the cyclic concept to coastal forms (Fig. 19.3a-d).

However, toward the end of his life, showing remarkable flexibility and lack of dogma for one of such advanced years and almost unchallenged eminence in his profession, Davis allowed that slope, and hence landform, development in the arid tropics could follow a different course from that described for the temperate areas of the globe.[7]

2. Geomorphic Cycles: Scarp Retreat and Pedimentation

In several parts of the world, especially in tropical arid regions, the slopes which border valleys and uplands and which are developed on similar bedrock display only minor variations of form and inclination. For example, the slopes which delimit the plateau, mesa and butte shown in Pl. 5.2 are nearly identical. Even in areas of graded slopes (for instance, Pls 13.4a and b, 19.5), the slopes continue at the same inclination from the mouths to the heads of the valleys. Such situations have been interpreted as implying that the slopes do not decline, but retreat, maintaining a similar form and inclination as they do so. Parallel slope retreat was conceived by Fisher[8] and Penck,[9] but has been developed into a thoroughgoing hypothesis of landscape development by King,[10] who has argued that the world's plainlands are not peneplains but *pediplains* formed by stream incision,

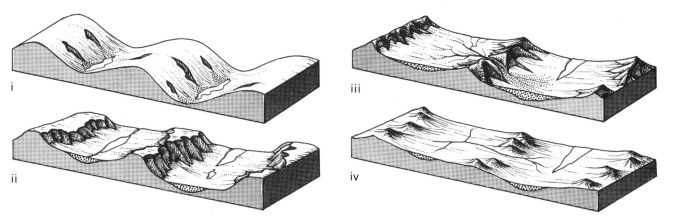

Fig. 19.3a Cyclic development of landscapes in the arid tropics.

Fig. 19.3b Cyclic development of landscapes in glaciated country.

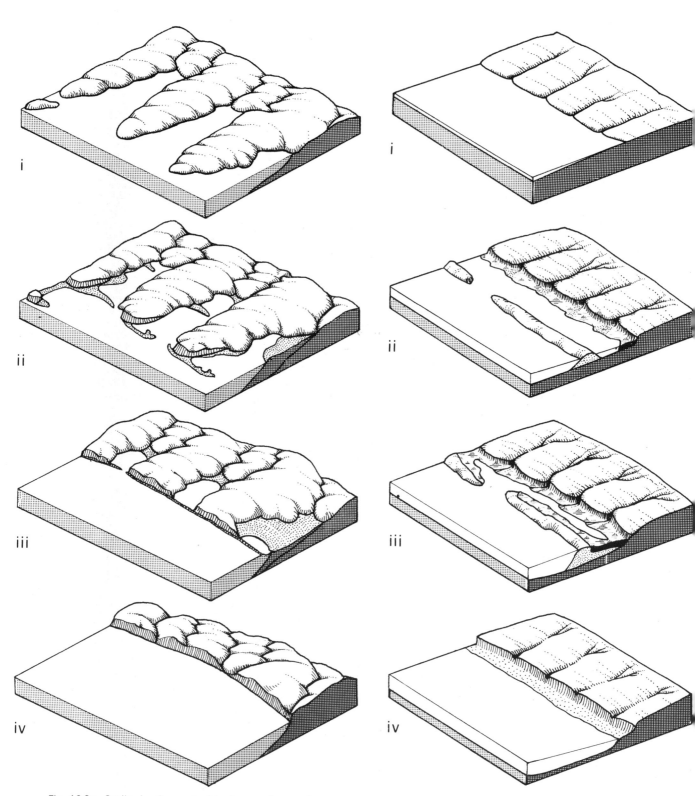

Fig. 19.3c Cyclic development of landscapes in coastline suffering erosion.

Fig. 19.3d Cyclic development of landscapes in constructional coastline.

Plate 19.5 The low Pertnjara Hills, N.T., developed on argillaceous sediments and with uniformly graded slopes of similar inclination and morphology virtually throughout. The shape of some of the upper shoulders varies, but on the other hand they are quite constant from valley mouth to valley head in many areas, for example, in valley X. *(Dept Nat. Dev., Canberra.)*

scarp retreat and pedimentation, and that they consist of coalesced *pediments* (See Chapter 14).

Such pediplains differ from Davisian peneplains in that slopes are predominantly concave or rectilinear rather than convex; in that streams are widely spaced and valleys narrow; in that there is little relief on the pediment surface in profile parallel to the upland front; and in that the piedmont angle (see Chapter 10) is widely and on the whole well developed. Thus there is typically a sharp break of slope between the plains and the uplands. There is also strong suggestion that pediplains are best, if not exclusively, developed in arid and semiarid lands, and particularly in tropical and subtropical regions. Finally it may be mentioned that absence of marked surface lowering, plus the activity of scarp recession has in many areas permitted the preservation in the present landscape of remnants of ancient surfaces of low relief.

How significant and real are these differences? In view of the uncertainties attaching to the origin of pediments (see Chapter 14) and to the nature of the processes responsible for the shaping of slopes, it is not possible to base any differentiation between peneplain and pediplain on genetic speculations. Morphologically there is a continuum between the planate undissected surfaces of the north African deserts and the undulating plains of low relief found in many areas. Moreover pediments which are convex in profile normal to the mountain front are not unknown and the lower sectors of

graded hillslopes are commonly concave. Any classification or identification based on either drainage network density or on slope profile would be both arbitrary and illusory. Surely, and despite the contrary avowals of some writers [11] the only significant contrast between planate surfaces in arid and in humid climates is the piedmont angle? It was the sharp break of slope between hill and plain which so caught the attention of the early scientific explorers of the American West and of central and southern Africa. It was this characteristic which led the German explorers of the latter regions to give to the isolated hills and ranges which stand so dramatically above the great plainlands the name by which they are now known the world over: *inselberg* or island-mountain. It is this feature, due to the concentration of water, weathering and hence erosion and surface lowering on the lower slopes of the uplands which can logically be explained as being best developed in arid or semiarid contexts where pediments are most characteristically displayed. It is aridity primarily which causes drainage networks to be coarse, and causes the erosional surfaces to be so little dissected. The arid conditions also permit the preservation of coarse regolithic veneers which protect and preserve the planate surfaces and also, in detail, are the reason for the quite remarkable smoothness of the pediment surfaces. [12]

Thus it is suggested that the peculiar conditions of weathering and erosion developed in arid and semiarid lands are responsible for the formation of the piedmont angle which is the prime distinction between plains in arid and humid regions. But if this is the case, what if no uplands survive? For slow as degradational forces may be in arid lands, no surface is quite immune to the forces of weathering and erosion, and eventually the uplands are destroyed. In these conditions it is difficult indeed to determine the character of a surface of low relief. Such problems have led some workers to suggest that instead of labelling plains of low relief as peneplains or pediplains according to their morphology, junction with residual remnants and a specific implied mode of development, the term *oldland* should be used.[13] An oldland is a surface of low relief which is a result of long-continued erosion, but the term carries no suggestion as to the processes active, the behaviour of slopes and the ancestral forms developed during the evolution of the feature.

But whatever label is put on a given surface of low relief, whether peneplain, pediplain or oldland, the concept still implies a cyclic or evolutionary development of the land surface.

C. RIVER REJUVENATION AND LANDSCAPE REVIVAL

1. General

Remnants of what appear to be old valley-floors and river beds, now located well above even the highest known flood level, occur in many river valleys. In several small valleys which are cut in the alluvial fan deposits forming the Willunga scarp south of Adelaide, for example, terraces and valley side facets (Pl. 19.6, Fig. 19.4) are commonplace. The terrace remnants are vegetated and are never touched by floodwaters, yet on them are remnants of the old stream channels (Pl. 19.7), with rounded pebbles and boulders still *in situ*. Clearly the river formerly flowed at this level but has since incised its bed. The valleys of New Zealand abound in examples of such river terraces (see Chapter 13). In all cases, the rivers clearly have been endowed with new energy and are said to have been *rejuvenated*, for reasons which vary from

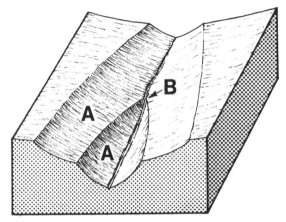

Fig. 19.4 Valley side facets (A) and nick point (B).

location to location (tectonism, climatic change, variation in lithologic environment, human settlement).

On a broader scale, many landform assemblages include surfaces of subdued relief located high in the landscape. They transgress the structure of the bedrock and are of erosional type. Thus in the Mt Lofty Ranges (Fig. 4.17) there is a distinct high plain or plateau. In southern Africa there are extensive stepped erosional surfaces.[14]

Such flights of erosion surfaces have been interpreted as due to the development of surfaces of low relief, close to regional baselevel and to subsequent river rejuvenation, the inception of new cycles of erosion, and the development, extension and coalescence of new valleys to form new surfaces of low relief. Such landscapes which bear the imprint of more than one cycle of erosion are known as *multicyclic* landscapes.

Plate 19.6 Gully eroded in the alluvial fanglomerates which front the Willunga scarp, south of Adelaide, South Australia (see Figs 4.17 and 20.1). Two alluvia are represented, and two phases of cut and fill. Constructional terrace remnants are preserved on the younger alluvium. (*C.R. Twidale.*)

Plate 19.7 Remnant of former stream channel preserved on a terrace remnant, Willunga escarpment, S.A. (*C.R. Twidale.*)

2. Rejuvenation

Though they differed radically in their interpretation of landscapes, both Davis and King accepted that major movements of baselevel, which were assumed to be caused by positive movements of the land, have caused interruptions of geomorphic cycles.

Evidence which clearly indicates variations in stream behaviour is observed in many river valleys. The longitudinal profiles of streams display marked breaks of slope, manifested in rapids and waterfalls (Pls 3.1b, 19.8); valley-floors and side slopes also display marked breaks of slope separating distinct facets (Fig. 19.5); old valley-floors, now dissected and standing above the present flood plain and bounded by river bluffs (Fig. 19.6a-d), form river terraces (Pls 19.4, 19.9a and b).

In terms of cyclic landscape development these features are commonly interpreted as being a result of negative movements (that is, a lowering) of baselevel.[15] A fall of sealevel allows a stream to erode its bed to the new lower baselevel. It erodes a new, steeper profile and is said to be rejuvenated (cf. the stages of youth, etc., used to describe the evolution of the peneplain). The break of slope in the river profile indicating the junction of the older and newer stream stages is called the *nick point*. The nick point gradually migrates upstream so that the new valley extends at the expense of the old. However, a resistant stratum may retard such upstream migration and indeed several nick points may be held up at such an obstacle. Simultaneously, because the river forms the local baselevel for its tributaries and for the adjacent slopes, the latter are regraded to the new lower baselevel so that valley side facets form (Fig. 19.4). Alternatively, if the river had developed a flood plain before rejuvenation, remnants of this plain may be left high and dry as paired or matching terraces (Figs 19.5, 19.6).

Plate 19.8 The Barron Falls, a nick point in coastal Queensland. (*Qld Govt Tour. Bur.*)

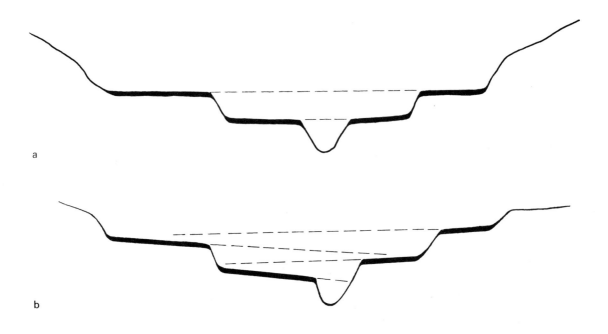

a

b

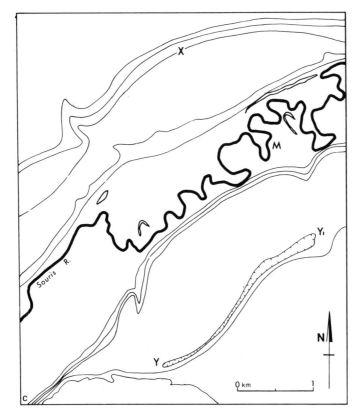

c

Fig. 19.5 (a) Paired terraces; (b) unpaired terraces (in both the black indicates a layer of alluvium); (c) terraces in the valley of the Souris River, North Dakota, with X indicating a meander bluff separating the terrace treads, M a meander loop and Y a linear swampy depression.

Fig. 19.6 Terrace standing high above the present channel of the Shotover River, Otago, South Island, New Zealand.

But do such features necessarily indicate rejuvenation in the sense of negative movements of baselevel? The field evidence suggests not. As mentioned earlier (Chapter 13), below the Aswan High Dam the Nile is incising its bed because of changes in its load characteristics.

Below the Willunga fault scarp, South Australia, narrow valleys cut in unconsolidated alluvial fan debris all display terrace remnants which are themselves constructed of riverine deposits (see for example, Pls 19.6, 19.9). As the streams which cut the valleys drain to the coast nearby, it is reasonable to suggest that the oscillations of river level evidenced are related to fluctuations in sealevel. If this were so, the assemblage of forms in these valleys should differ from valleys located in other inland situations. It does not. Could it be then that the location of the valleys near an active fault zone is relevant? Possibly, but if the upfaulted side of the zone has continued to rise, rejuvenation effects would be primarily on the up-faulted side of the fault. They are not. There should be a fault scarp affecting the alluvium. There is not. A further and crucial argument against the terraces and related features being due to faulting is that similar assemblages occur in valleys which (as far as is known) are remote from faults.

Land clearance in the interests of agricultural settlement could be cited as a possible cause of some of the features and there is little doubt that accelerated erosion, particularly the deepening and extension of gullies, has followed European settlement in this South Australian area during the past century or so (see Chapter 25). But to attribute all the phases of cut and fill evidenced in these valleys to this argues a rate of erosion difficult to sustain by reason and by comparison with known recent erosion. In any case

there have been phases of alternating cut and fill, not merely erosion, in these valleys; has man come and gone, has the vegetation been cleared and then been regenerated? Clearly not; at least not in the century which has elapsed since European settlement. The one factor which explains the field evidence is climatic change, resulting in changed volume:load ratios in the streams and hence in phases of cut and fill. This explains the occurrence of such features in river valleys in many parts of southern South Australia, and also the special effects of man, which have merely added to a pattern imposed by climate. If, however, climate is responsible for the cut and fill evidenced, then the comparable landforms in each valley should be of approximately the same age, though some allowance should possibly be made for migration of climatic belts.[16] The original deposition of the alluvial fans, in which many of the valleys are cut, occurred in the late Pleistocene some 34,000 years ago, in both the southern Flinders Ranges and the Willunga scarp,[17] so that the cut and fill must have taken place quite recently.

Rapids and waterfalls may be, and commonly are, caused not by rejuvenation but by resistant strata on the stream course: the obstacle may take the form of a distinct stratum or a massive compartment of crystalline rock, or merely a variation in mineralogy, lithology or joint spacing. Therefore, as with valley side facets and terraces, waterfalls and rapids do not of necessity imply rejuvenation; rapids may be a nick point, they may form for structural reasons, or they may combine elements of both modes of formation. Each area must be examined on its merits and the distribution of the landform assemblages in particular must be examined before the causes of variations in stream behaviour can be identified.

Plate 19.9a Terrace remnants (X) of constructional origin in a river valley cut into the alluvial
fanglomerates fronting the Willunga scarp, S.A. (*C.R. Twidale.*)

Plate 19.9b Constructional terrace
remnants (X) in unconsolidated fill, some
350 m thick, in the Fraser Valley, British
Columbia. Note the alluvial fan (A) and
the sparse vegetation which reflects the
arid climate. (*Dept Lands, Forests and
Water Resources, B.C.*)

3. Multicyclic Landscapes

In broader view, many landscapes present a stepped appearance, with surfaces of subdued relief separated by distinct escarpments or standing clearly above present stream channels (Pl. 19.10). What appear to be remnants of oldlands occur high in the relief, though their limits are indistinct. Such surfaces of low relief located above present regional baselevel have been interpreted as the result of widespread stream rejuvenation, the development of new valleys and the merging of these to form new surfaces of low relief. In such landscapes there are, in the terms outlined, surfaces related to more than one cycle of erosion; the landscape is said to be *multicyclic*. The new peneplain or pediplain extends inland at the expense of the old, with remnants of the older, higher surface or surfaces being preserved in the interiors and high in the relief. A few regional examples are described below.

SOUTHERN AFRICA

Multicyclic landscapes have been described from many parts of the world, though evidence for some oldlands and former cycles is uncertain, and in almost all cases there are complications due to tectonism and undue reliance on purely morphological evidence. One of the most convincing examples of a multicyclic landscape is southern Africa, where King[18] has recognised several surfaces of low relief which are described as pediplains (Fig. 19.7 and Pl. 19.10d) and which are listed below:

Gondwana	—	Mesozoic
African	—	Early Tertiary
Post African (two phases)	—	Late Tertiary-Quaternary
Congo	—	Quaternary

Though developed in a region dominated by strata which are essentially flat-lying, there is no doubt that the surfaces cut across various beds are of erosional, and not structural, type. Admittedly some of the surfaces may have complex histories (some for instance may be locally of etch character —see below, p. 420) and there are many minor facets and phases as King himself recognises, but there can be no doubt as to the reality of these extensive and magnificently displayed and described erosional surfaces of low relief.

This raises the question of what type of interruptions separate cycles and which events can, as it were, be accommodated within a cycle and treated as intracyclic events? No part of the crust is stable and, as will be described later (Chapter 21), there is reason to believe that climates have changed and are changing constantly. Furthermore, sealevel is rarely stable for long (Chapter 22), so that waves of cutting and filling due to various factors in theory can be expected to affect rivers. The matter is obviously subjective but in practice most cycles and major surfaces seem to be separated by major tectonic events, commonly in the form of widespread warping or gentle folding. Thus, in southern Africa, the various major cycles were interrupted essentially by phases of uplift, as well as by disturbances at the margins of the continent, such as in the Natal Monocline.

As described below, major (though complex) surfaces of similar age ranges occur in several parts of the world, for instance Australia, South America and southern Africa.[20] In the case of the older surfaces, similar age of interruption of cycles in what are now distant land masses has been taken as evidence supporting continental drift (see Chapter 2).[21]

AUSTRALIA

Three cycles and related surfaces of low relief have been identified in northwestern Queensland.[22] The first two have been described in other contexts in previous chapters. The first and youngest is the late Tertiary-Quaternary surface represented in areas of Cretaceous outcrop by the rolling Julia plain (Pl. 13.8 and Fig. 19.8a), and by pedimented surfaces on the areas of crystalline outcrop which border the Mesozoic basin to both east and west. Standing above this late Cainozoic surface are remnants, such as the Kynuna Plateau (Pl. 13.8), of an older, lateritised surface of low relief which seems to have developed in middle or late Tertiary times. It is certainly post Mesozoic, for Cretaceous sediments were affected by the weathering processes which resulted in the formation of laterite in many areas.

The third surface is not so obvious at first inspection of the landscape. The Cretaceous sediments for the most part are restricted to the Carpentaria plains, between the

Plate 19.10a Multicyclic landscape in the southern Arcoona Plateau, S.A. Developed on a sequence of flat-lying Precambrian sediments. A is the Arcoona Surface, represented by domed plateaux, B is the duricrusted and pedimented Beda Surface, and C is the Torrens Surface, located in the present valley floors. (*C.R. Twidale.*)

Plate 19.10b This planate surface, now high in the topography, is cut across a monoclinal flexure involving sandstones. The plateau borders the Wittenoom Gorge, W.A.
(*Geol. Surv. W.A.*)

Plate 19.10c The Kaikoura Surface north of Wellington, New Zealand, takes the
form of a high plain and flat-topped crests. (*W.D. Mackenzie.*)

Plate 19.10d The Drakensberg escarpment in Natal is eroded in flat-lying Mesozoic lavas and is almost 1·5 km high. The upper surface is a facet of the Gondwana Surface. Note the structural benches and planate surfaces in the foreground. (*C.R. Twidale.*)

Fig. 19.7 Distribution of major erosion surfaces in southern Africa. (*From L.C. King,* Morphology of the Earth, *Oliver and Boyd, Edinburgh, 1960, opposite p. 300, by permission of Professor King.*)

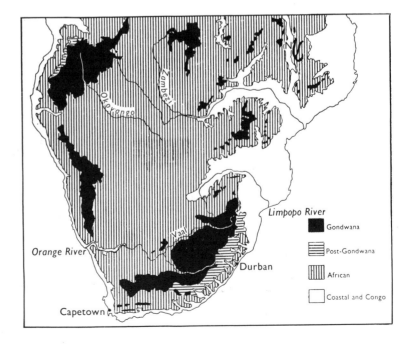

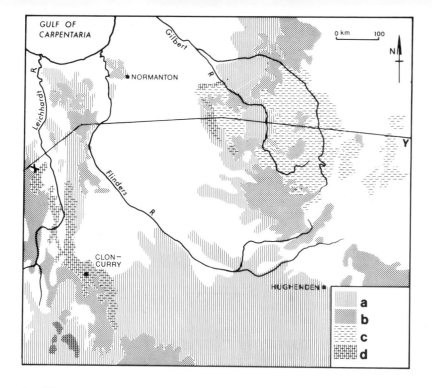

Fig. 19.8a Distribution of erosion surfaces in northwest Queensland: a—late Cainozoic surface of erosion; b—early-mid Cainozoic surface (lateritised); c—pre Cretaceous surface (at present high in the relief); d—pre Cretaceous surface, low in the relief and concordant with the late Cainozoic surface. Both c and d are exhumed.

Isa Highlands and Einasleigh uplands. They rest unconformably on folded or crystalline strata (Fig. 19.8b). This unconformity emerges both east and west of the Carpentaria plains without break of slope. Thus, in Pl. 19.11, A–B is a surface which was initially cut in Precambrian rocks and upon which the Cretaceous sediments were deposited; it was then the unconformity between the two rock groups; and it is now part of the present land surface, being the unconformity from which the Cretaceous strata have been eroded. It was a surface in pre Cretaceous time and it has recently been re-exposed, resurrected or exhumed. This *exhumed* land surface can be traced into the Einasleigh and Isa uplands where it is identified by the remnants of Cretaceous rocks which overlie it in a few, widespread localities. In the southwest of the region, near the Queensland–Northern Territory border, a still older exhumed surface, of pre middle Cambrian age,[23] is locally represented.

The exhumed pre Cretaceous surface is widely represented all along the western margin of the Great Artesian Basin. It has been described from the uplands of central Australia[24] the north of South Australia[25] and the northern Flinders Ranges,[26] whence it can be traced to the adjacent Frome embayment where it disappears beneath the Cretaceous sediments (Fig. 19.9). It is represented in the Flinders Ranges by a prominent summit surface, including some distinct flats high in the relief, cut across contorted Precambrian cystalline and sedimentary rocks. Standing above this summit surface,

however, is Mt Babbage—a remnant or outlier (see Appendix) of flat-lying Cretaceous strata; the level of the unconformity between the Mt Babbage Cretaceous and the underlying Precambrian can be projected without break into the summit surface.

In the central and southern Flinders this high and ancient surface is probably represented by flats high in the relief.[27] There are, however, no Cretaceous remnants to confirm this interpretation. The most prominent remnants of ancient landscapes here are relics of old valley-floors, which occur in many areas within and marginal to the ranges. They take the form of small, flat-topped spurs and mesa remnants, in each case protected by a mantle of quartzite gibber, and standing 10–30 m above the present valley or plain level.

The Flinders landscape is an example of Appalachian-type relief (Chapter 5, p. 122), developed on a fairly simple fold belt. There is every reason to believe that recurrent faulting has affected the Flinders block and that such movements are responsible for the interruptions of cycles evidenced in the landscape. This area demonstrates two important features of cyclic development. The first is that a cycle need not be complete before it is interrupted. The pre Cretaceous surface may well have been an oldland, but the old valley-floor cycle, which was developed throughout the ranges, clearly did not extend to peneplanation or pediplanation, that is, it was not of regional extent, before rivers

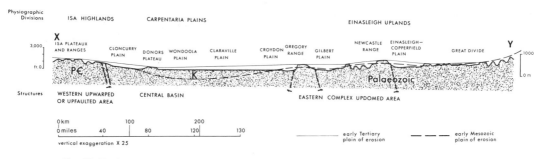

Fig. 19.8b Section along the line X–Y in Fig. 19.8a showing behaviour and relationships of the principal surfaces. K—Cretaceous sediments. (*After C.R. Twidale, "Chronology of Denudation", pp. 867–82, by permission of the Royal Society of South Australia.*)

Plate 19.11 Three surfaces and cycles are represented in this small section exposed south of Cloncurry, northwest Queensland. A–B–C is the unconformity between the folded and eroded Precambrian rocks below and the Mesozoic marine sediments above. The latter have been partly eroded, exposing the plain remnant A–B, and exhumed surface. D is a remnant of a lateritised middle Tertiary surface developed on the Mesozoic strata, and E is the present pediment surface which is extending up river valleys at the expense of the two older surfaces. (*C.R. Twidale/C.S.I.R.O.*)

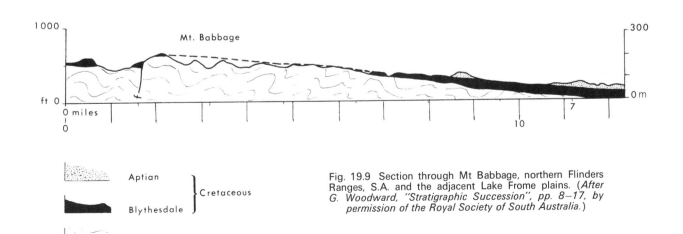

Fig. 19.9 Section through Mt Babbage, northern Flinders Ranges, S.A. and the adjacent Lake Frome plains. (*After G. Woodward, "Stratigraphic Succession", pp. 8–17, by permission of the Royal Society of South Australia.*)

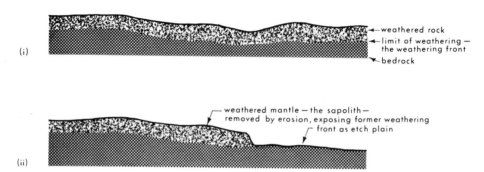

Fig. 19.10a Stages in the development of an etch plain.

again incised their beds to erode the present valleys and plains. The second feature is that, though in the Davisian cycle relief amplitude decreases in time, there are strong grounds for suggesting that relief amplitude in the Flinders has continued to increase since the Cretaceous. The landscape is a structural one, with prominent sandstone ridges; run-off from the ridges flows to the valleys underlain by siltstone and shale. Thus, a lithological weakness tends to be emphasised because moisture is the principal weathering agent. Whereas the valley-floors are deeply weathered and eroded, the sandstone ridges, though not untouched, must surely be lowered at a slow rate than the valley-floors, so long as baselevel conditions allow.

The bedrock beneath many of the old valley-floor remnants in the Flinders Ranges is weathered and in many areas the level of the weathering front which is preserved beneath the mesa remnants is coincident with that of the surrounding plain. It is reasonable to suggest that the incising streams have cut down rapidly to the weathering front and readily removed the weathered debris, but that erosion of the still fresh rock has proved a slower process. Such plains which are etched from the regolith and which in essence represent the former weathering front are called *etch plains* (Fig. 19.10a). Many extensive etch plains have been described from central and eastern Africa[28] and from many other areas. In Western Australia, for instance, Mabbutt attributes the new plateau of Jutson[29] to the stripping of weathered granite and the exposure of the weathering front. There is a close coincidence between the elevation of the front, preserved beneath old plateau remnants, and that of the new plateau (Fig. 19.10b and c).

Similarly a high, partly dissected plain surface cut in and underlain directly by virtually unweathered gneisses and schists in the eastern Mt Lofty Ranges, South Australia, is interpreted as an etch plain because scattered, small remnants of a lateritic regolith[30] are standing above it. It is argued that as the lateritic remnants are scattered, they could have formerly extended over the whole area; and that as the base of the regolith is at the same elevation as the high plain surface (which is conspicuously lacking a weathering mantle) the regolith has been removed from the greater part of the region, leaving the old weathering front exposed as the high plain surface (Fig. 19.11).

In the Australian areas discussed, some of the erosion surfaces and particularly the pre Cretaceous surface marginal

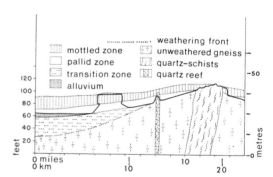

Fig. 19.10b The newer and older plateaux in Western Australia. Note the essential etch character of the former. (*After J.A. Mabbutt, "A Stripped Land Surface", pp. 104, Fig. 2, by permission of Trans. Inst. Br. Geogr.*)

to the Great Artesian Basin are dated within rather narrow limits by virtue of the minimum age of rocks across which the surfaces were eroded, and the maximum age of the sediments which overlie them. But the erosion of a land surface inevitably gives rise to deposition, and in theory the dating of related deposits should greatly assist the dating of land surfaces (Fig. 19.12).

WESTERN U.S.A.

In Wyoming, the Sherman Surface truncates Precambrian crystalline rocks, including granite weathered to a depth of some 15 m.[31] The surface which slopes eastwards (Fig. 19.13) to merge without break of slope with the Pliocene depositional surface of the Great Plains in the Gangplank, west of Cheyenne,[32] is remarkably smooth though in places it is surmounted by residual remnants. Though called a peneplain,[33] it is so smooth and so little dissected that it would seem rather to be a pedimented surface. Possibly because the Sherman Surface was earlier called a peneplain, residual remnants rising from it have been called monadnocks.[34] though again they are so steep-sided that they would be called inselbergs by many workers. It is of interest, however,

Fig. 19.10c The old plateau in W.A. represented here by a laterite-capped mesa near Cue, W.A.

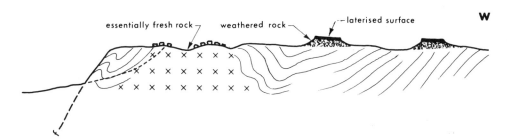

Fig. 19.11 The etch surface of the high plain in the Mt Lofty Ranges west of Palmer, S.A.

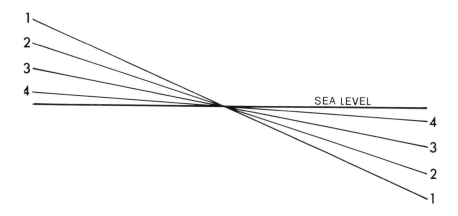

Fig. 19.12 Relationship of erosional surfaces and resultant deposits, in section.

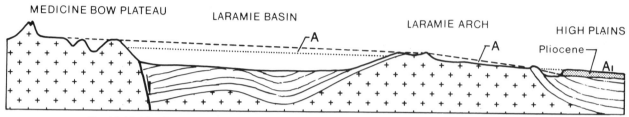

Fig. 19.13 The Sherman Surface (A), its easterly extensions, and possible associated deposits near Laramie, Wyoming. The surface above the Sherman is the Medicine Bow Surface and the valley floors below it are related to a Lewis Surface. (*After E. Blackwelder, "Cenozoic History", pp. 429–44, by permission of J. Geol. and the University of Chicago Press.*)

that the more prominent residuals standing above the granite plain are of a different rock type. Whatever it is called the Sherman Surface is undoubtedly of erosional type. It apparently continues downslope (to the east) across sediments of middle Tertiary age and its age may be adduced as Pliocene. On the other hand if the Sherman Surface continued beneath these flat-lying sediments then the latter could be derived from the erosion of the former, in which case the Sherman is of middle Tertiary (Miocene) age.

THE APPALACHIANS

Another area from which erosion surfaces have long been recognised is the Appalachians in the eastern USA. The surfaces are generally considered to be of peneplain type though there are some [35] who attribute them to marine processes. Three surfaces, in order of decreasing elevation the Schooley, Harrisburg and Somerville peneplains (and their local equivalents) have been identified, and are represented by distinct flats atop the sandstone ridges of the area, by accordant crests and spurs, and by valley floors. Johnson[36] regarded the Schooley Surface as of early Tertiary age, and Harrisburg and Somerville as late Cainozoic. The Schooley is represented by accordant ridge crests which stand at about 600 m above sealevel in the north, rising to 1300 m in the south. The lower surfaces are represented by valley-floors and basins within and marginal to the uplands. The Harrisburg Surface may however, be of greater antiquity, for it may be related to the Fall Line Surface which dips beneath the Cretaceous sediments of the coastal plain and continental shelf (Fig. 19.14). In these terms the Harrisburg Surface is pre Cretaceous and post Triassic in age for

it cuts across Triassic sediments and in places is overlain by Cretaceous rocks. It is also of exhumed type.

The relationship between the Harrisburg and Schooley surfaces is rendered difficult to interpret by faulting and also by the latter being represented only by accordant ridges.[37] Are these ridges merely standing above the Harrisburg Surface, or do they in reality represent a dissected, older surface? There is additional uncertainty because of the complications introduced by structure and because benches and flats have been recognised at levels intermediate between those thought to be related to the major cycles and surfaces.[38] For instance, a pre-Schooley Surface has been suggested.

WALES

Erosion surfaces have long been recognised in Wales.[39] Four major surfaces have been identified:

1. a *summit plain* (900 m plus) along the crests of the highest mountains which are regarded as monadnocks;
2. a *high plateau* (500–600 m) consisting of a series of flat-topped summits;
3. a *middle erosion plain* (400–450 m); and
4. a *low erosion plain* or *coastal surface* (200–350 m).

However, both in Wales and in other parts of Britain, numerous other levels and plains have been located, which are thought to represent relatively minor erosional stages. Thus in North Cardiganshire, although the three lower major surfaces are represented, nine surfaces and stages are recognised altogether.[40] The Cardiganshire erosion surfaces are, following Ramsay,[41] regarded as basically of marine

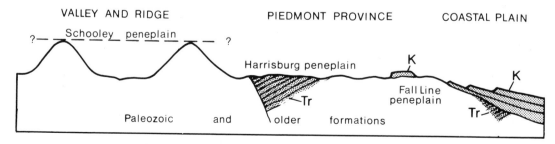

Fig. 19.14 Diagrammatic representation of the major erosion surfaces in the southern Appalachians. (*From* Physiography of the United States *by Charles B. Hunt. W.H. Freeman and Company. Copyright 1967.*)

origin and as having been cut during irregular uplift of the land. But some writers[42] have emphasised the significance of Triassic planation under arid conditions in the evolution of the main upland surface in what has been called Highland Britain.[43] In any case these stepped surfaces have been significantly modified by subsequent events: the High Plateau in Cardiganshire, for instance, presents the aspect of a peneplain moulded by subaerial agencies, and the Coastal Surface appears to have suffered glaciation. All the surfaces in Cardiganshire are regarded as of later Cainozoic age, though the higher one is presumably older and may represent one of the Mesozoic surfaces considered to exist in other parts of the uplands of northern and western Britain.

One of the remarkable features of this flight of erosion surfaces is that none of them have apparently suffered severe distortion since their formation: all remain only very gently sloping and undeformed by earth movements. This situation has of course encouraged their interpretation as of fundamentally marine origin. It has also encouraged their treatment and analysis from map data. In this connection it is of interest to note that the occurrence and separation of the surfaces in North Cardiganshire as described by Brown[44] has recently been corroborated by trend surface analysis,[45] though the statistical techniques and methods used both there and in northern England[46] have been severely criticised as "completely inappropriate".[47]

DISCUSSION

Thus surfaces of low relief attributed to ancient cycles of erosion and deposition have been recognised from many parts of the world. But increasingly it is appreciated that many have complex histories, being of etch or exhumed type. For instance a mature topography around Bari in central India is now interpreted as an exhumed surface: as a late Cretaceous topography which was buried and preserved by the lava flows of the Deccan Traps and which has subsequently been exposed to form part of the modern land surface.[48]

The best preserved erosion surfaces are undoubtedly those which carry a protective duricrust and those which are located in arid and semiarid regions. Hence the prominence of surfaces of low relief, or their remnants, in these areas. The best studies of denudation chronology are undoubtedly those which relate erosional surfaces to sedimentary deposits or regolithic mantles. Many of the planate surfaces recognised in Africa and Australia, for example, are preserved by virtue of duricrusts and are dated (in approximate terms anyway) by their correlation with stratigraphic sequences in adjacent sedimentary basins. Not only are the sediments proof of the reality of erosion but they may allow the surface and cycles to be dated, and may shed some light on the conditions under which the plains were moulded. Regoliths also betoken conditions of relatively slow erosion, of low gradients, and again allow insights as to the climatic and other conditions prevailing when the regolith was formed: lateritic veneers imply humid tropical or monsoonal conditions, calcretes arid or semiarid climates.

Unfortunately many peneplains have been postulated on insufficient evidence. Davis' vigorous and persuasive exposition of the cyclic concept is often blamed for this, but the fault surely lies in the uncritical acceptance of these ideas, and with the failure to examine alternative explanations for the features cited as evidence of former peneplains. For example, the accordant height (or limited height range) of peaks and crests in many upland areas has been taken to imply that the area in question was formerly a peneplain and that it has suffered dissection; the peaks and crests supposedly representing the last fragmentary remnants of the surface of low relief. The *gipfelflur* of the Alps and elsewhere has already been cited as an example of this sort of interpretation. In Pls 17.12 and 17.13, the accordant peaks would be seen in this light. Such interpretations are not necessarily incorrect. Neither sediments, nor a regolith, are preserved beneath the alleged former surface, but on the other hand this is scarcely surprising in view of the intense dissection and steep slopes. It is possible to suggest that weakly consolidated or unconsolidated sediments were removed by erosion and that the accordant crests represent elements of an etch plain. But is there no other mechanism which could account for the undoubted evenness of the cumulative horizon, any mechanism which would tend to reduce crests and peaks to a common or similar elevation?

For instance, in uplands subject to frost action (provided that the area is not so high and cold that oscillations of temperature around freezing point are not excluded), the higher peaks will surely tend to be reduced more rapidly than the lower ones because at higher elevations frost action would be frequent. In areas subjected to strong dissection such frost-shattered material is readily evacuated so that there is rapid renewal of weathering.

Another possible mechanism involves stream spacing and slope development and applies not only to areas such as that depicted in Pl. 3.5 but also to river valleys characterised by intrenched meanders and flat-topped or only gently sloping spurs (Fig. 19.15). Taking the minor example first, because crests of spurs are essentially of the same elevation, it has been assumed that spur tops are remnants of a surface formed during a phase of baselevel stability, when there was little or no stream incision. As the baselevel involved is sealevel in many cases, such terrace or spur remnants have been used to reconstruct former stands of the sea, particularly if the spur crests stand at the same or similar elevations above the river bed. The question is, do such even-crested spurs in fact represent former river flood plains? If they carry remnants of river sediments on their crests then this interpretation is valid: they represent marked lateral shifts of the river course, the sediments and point bar deposits being laid down on the inside of the river curves during the lateral migration. But in many, perhaps most, cases no such evidence exists. Either it has been destroyed by erosion, which is a reasonable supposition, or it never existed. But why, then, are the spurs even-crested? In such situations the river runs at the base of the slopes which bound the spur, or it is not far distant and has not, in geological terms, been long departed from the slope foot. Hence, although the precise shape of the slope—faceted or graded, steep or gentle—depends on the local bedrock, on climate and vegetation, and on the slope budget which results from their interplay, the footslope is either being actively attacked by the river or it has been in the recent past. Thus slope recession is implied (see Chapter 13) and the height of the slope crest is determined by the width of the spur (Fig. 19.15c) and not by baselevel controls. And as many of the elongate

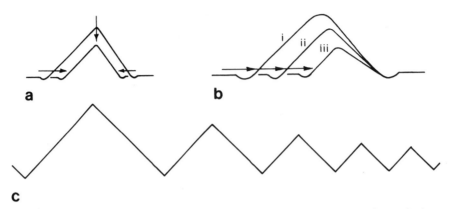

Fig. 19.15 Variation of height of spur crest with spur width. In (a) the spur is attacked from both sides, in (b) from only one. In (c), which represents a hilly or mountainous upland, there is a direct correlation between crest height and stream spacing.

arms of intrenched meanders are structurally controlled and hence parallel or subparallel, there is strong tendency for even-crested spurs to evolve.

D. STEADY STATE OR DYNAMIC EQUILIBRIUM

1.   General Statement

"Lines in the sky" is a description commonly and reasonably given to some alleged peneplains or other oldland surfaces of low relief. However, lack of sound evidence is not the only reason that doubts have been cast on the cyclic concept of landscape development. Just as fundamental is the assertion that the cyclic concept is at odds with physical realities expressed in well-established laws and relationships.[49]

Davis made several assumptions in his deductive outline of the normal geomorphic cycle, commonly known as the peneplain concept. One of the most important is that the cycle is initiated by an instantaneous uplift of the land mass. Modern rates of uplift appear to be of the order of eight times as great as modern maximum denudation,[50] and thus seemingly corroborate Davis' supposition. Nevertheless it seems to be unreasonable, on several counts. For example modern faulting, though uncomfortably effective at certain times and places, does not accomplish instant uplift of the order required in Davis' exposition of the cycle. Furthermore, what is envisaged? Block faulting of a huge continental mass? This is not borne out by field evidence. Davis was aware of these problems, but presented the case as he did in order to simplify the model. Indeed (as is mentioned later), he was well aware that uplift could be related to and caused by erosion, a point developed by Penck,[51] who endeavoured to relate slope morphology to rates of uplift of the land.

Another implication of this instantaneous uplift of the land mass is that the whole system is viewed as what is called a closed system.[52] After the initial input of energy through uplift of the land mass, there was no renewal of energy from any internal source. Gravitationally induced forces cause a rundown of free energy; as relief decreases in the Davisian cycle so does the amount of energy available for work. The peneplain itself may be regarded as an equilibrium condition of minimum free energy, that is, of what is called maximum entropy.[53] Given a certain uplift, the end result—the peneplain—is inevitable in a closed system model. Both

Hack and Chorley[54] take strong exception to a geomorphic system being regarded as closed. Chorley especially regards an open system view of landforms as being more realistic and as directing attention to the processes at work on the land surface; and as a view of landscape which is not restricted by what he regards as evolutionary blinkers.

An important aspect of the open system is that there is a continuous exchange, import and export, of energy and material. A change in one part of the environmental complex induces a response or reaction opposite in sense but equal in amount (Le Chatelier's Principle).

2.   Dynamic Equilibrium

In terms of landscape development in humid temperate regions, Hack, and Hack and Goodlett[55] envisaged that rivers would cut into a rising land mass. The rate at which they could incise their beds would depend partly on climatic factors (amount, duration and intensity of rainfall, temperature regime, etc.) but if these were regarded as constant in the long term then the only limitation on river incision would be that of baselevel; and in broad view this in turn is determined by the rate of uplift. Thus the greater the uplift the more rapid the incision by rivers. The rivers themselves, however, are the local baselevels for the valley side slopes, and the more rapid the incision of the stream, the steeper the adjacent slopes. The steeper the slopes, the more rapid the removal of waste[56] and the greater the renewal of weathering. If uplift is increased a chain reaction starts: more stream incision, more rapid transport and weathering of waste: adjustment of slopes to incision and uplift. Decrease uplift and rivers cut down more slowly, slopes become more gentle, and again there is adjustment.

In addition to the notion of reaction and adjustment or self-regulation of all facets of the landform assemblage, both to each other and to external and internal forces, a vital aspect of Hack's scheme is that given constant climatic and tectonic conditions, the whole landscape will in time become adjusted. Though undergoing constant loss of material through erosion, it nevertheless remains morphologically constant. It attains a *steady-state,* and is in *dynamic equilibrium,* in contrast with the evolutionary changes of landforms inherent to the cyclic schemes of landscape development.

Though Hack's is a most thorough exposition of the steady-state concept, it is not new. Nikiferoff[57] made similar proposals with respect to pedology and Davis himself[58] was well aware of the possibility of dynamic interaction between the active crust and external agencies on the one hand, and the likelihood of self-regulatory processes on the other. His discussion of grade in streams bears witness to the latter type of thinking and as evidence of the former, he conceived the idea of an "old-from-birth" peneplain. This was a surface of low relief which, subjected to slow uplift, was eroded at a rate adequate to balance the uplift and thus maintain its morphology despite erosional losses. The peneplain was thus not the result of the downwasting and downwearing of a once-high land mass, but had never been above baselevel. A similar distinction was recognised by Penck[59] who differentiated between *endrumpf*—the end product of long-continued erosion—and *primärrumpf*, or primary peneplain which was never above baselevel.

The significance of crustal response to erosion is implicit in the concept of isostasy. Estimates of the time required for peneplanation to run its course also make allowance for isostatic response to erosion. Ten million years would be required, without isostatic adjustments, for the base-levelling of the continental United States, but making allowance for the increased amount of rock that has to be eroded because of isostatic response, a period of some thirty-three million years would be necessary. Others have suggested that continental baselevelling could take up to 110–115 million years.

However, despite its apparent applicability to such humid and tectonically active areas as the Andes, parts of New Zealand and New Guinea (where uniform erosion over the entire slope is consistent with the idea of steady-state[60] and where late Cainozoic uplift has been dramatic), the concept has not received universal acceptance. In part this is because it cannot apply in regions where slope retreat prevails. It is also due partly to the difficulty of defining critical criteria by which the cyclic and noncyclic conditions may be distinguished, so that a philosophical element enters into interpretations (but see below). It is partly that in some comparatively stable areas of the globe, cyclic interpretations fit the field evidence as well as, or better than, the steady-state concept.[61]

There are logical difficulties with the steady-state hypotheses related to baselevel control. No matter how rapid the uplift of, say, a fault block, the effects of that uplift, transmitted to the landscape by the rivers and streams, are not felt for some time in the central area of the block. In the Mt Lofty Ranges (Figs 4.17, 19.16) for instance, dissection of the horst has not so far affected the major part of the upland: the larger rivers are deeply incised and large areas of the Tertiary laterite remain only at the margins. Hack would, with some justification, interpret this as a passing phase, a period of 'maladjustment''. But in areas of recurrent faulting like the Mt Lofty Ranges, such phases are a consistent if "maladjusted" feature of the landscape.

Another problem is that the rock rising from depth to

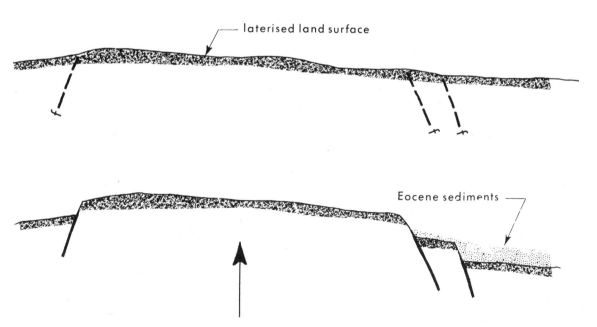

Fig. 19.16 Orthodox interpretation of recent evolution of the Mt Lofty Ranges with planation and deep weathering (lateritisation) followed by block faulting.

replace that lost through erosion is unlikely to be lithologically similar so that adjustments may again be anticipated on that account alone. The supply of strata from depth (and this is a problem with the concept of isostasy as a whole) also presents certain difficulties.

But the concept of dynamic equilibrium is one that should be borne in mind when landscapes are subjected to analysis. Some of Chorley's critical essays on Davis' geomorphological work are so devastating that one is left with the impression that nothing of the cyclic concept either survives or deserves to survive.[62] But more balanced and realistic views prevail elsewhere[63] and it is allowed that erosion surfaces are preserved in the arid and semiarid tropics, and the validity and value of some studies of denudation chronology linked with stratigraphy are recognised. The idea of dynamic equilibrium is presented as an alternative to the cyclic scheme. The cyclic theories are allowed as a special low energy case of open systems (for as Chisholm[64] points out, all natural systems are open and vary only in the amount of available energy and the relative rates of activity of the processes moulding the land surface), but it is argued that it is better to see them as open systems because this focuses attention on the processes at work on the landscape.

3. Interrelation of Baselevel, Streams and Wasting

Although the occurrence of surfaces of low relief high in the landscape, even those which carry a mantle of duricrust, may suggest former oldlands which have since been uplifted, it is important to appreciate that such features need not necessarily develop close to regional baselevel.

The Mt Lofty Ranges have already been mentioned in several contexts. The upland is a horst block (Figs 4.17, 19.16) which has suffered recurrent upfaulting.[65] The upland is dominated by remnants of a surface of low relief, carrying weathered profiles, including laterite, in many areas (Fig. 19.17) and located at elevations of 400–500 m near Adelaide and 350 m in the Fleurieu Peninsula to the south. The deep weathering of this surface occurred, in part at least, in the Mesozoic[66] though later phases of pedogenic ferruginisation are not ruled out.

The orthodox or conventional interpretation[67] of this situation is that an oldland or peneplain developed close to regional baselevel during the late Tertiary or even the early Pleistocene. The surface was carried to its present elevation by faulting, and has since suffered considerable dissection (Fig. 19.17). Such an interpretation may be valid; but certain deducible consequences stem from it. In particular, evidence of the former land surface, carrying laterite and disrupted by faulting, should be found not only on the raised blocks but on those which subsided and were covered with alluvium and other debris (Fig. 19.16). Such evidence has not been found despite numerous bores put down in the critical areas, in connection with underground water supplies. It may be argued that the laterite has suffered solution by circulating groundwaters. It may be that the laterite was never developed on these particular blocks, not because they were lithologically unsuitable (the strata are similar to those on the raised blocks) or because of climatic differences

(which are not possible since the raised and depressed blocks are juxtaposed), but because the laterite and the surface of low relief on which it occurs did not develop on a stable block close to regional baselevel.

A theory which could account for this situation has been proposed by Kennedy.[68] Though principally concerned with the volume and type of sediments contributed to depositional basins by continental areas undergoing varied modes of attack, Kennedy draws attention to the interrelations between baselevel, stream erosion and wasting in order to interpret the sedimentological history better.

In theory, landscapes may be analysed in terms of three factors: uplift, river incision and wasting (by which is meant the complex of processes responsible for the reduction of interfluves). Uplift and river incision are related because wasting cannot proceed beyond the level of river incision, which thus provides a local baselevel of wasting. But there is no connection between uplift and wasting. Thus in combination there are nine possible ways in which landform assemblages may develop, according to the relative rates of uplift, incision and wasting (see Table 19.1).

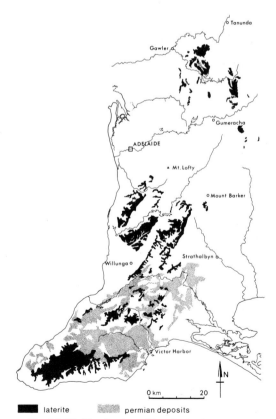

laterite   permian deposits

Fig. 19.17 Distribution of laterite in the southern Mt Lofty Ranges, S.A. Also shown are the glacigene deposits of Permian age. (*After* S.A. Geol. Surv. Map Sheets Barker, Adelaide, Gawler, and Mannum.)

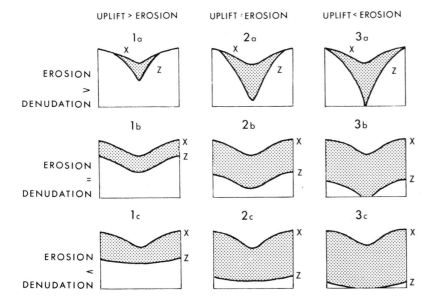

UPLIFT > EROSION   UPLIFT = EROSION   UPLIFT < EROSION

Fig. 19.18 Theoretical results of interplay between uplift, wasting and stream erosion according to Kennedy. (*After W.Q. Kennedy, "Some Theoretical Factors", pp. 305–12, by permission of* Geol. Mag.)

Typical profiles for the nine cases are shown in Fig. 19.18. Cases 1–3 result in various types of "perched" relief well above baselevel; cases 7–9 in relief at regional baselevel; and 4–6 in relief located at an intermediate level. An important conclusion resulting from this theoretical analysis is that surfaces of low relief can develop, in theory at least, high in the landscape as well as close to baselevel.

Some aspects of Kennedy's theoretical outline have been further explored by Schumm and Lichty[69] who have shown that, on the data available, it appears that modern orogeny and uplift are proceeding at about eight times the rate of average maximum erosion (8 m per 1000 years uplift as opposed to 1 m from denudation). The conclusion is drawn that with such disparity between uplift and erosion, baselevel control is all important. Local, climatically controlled variations in weathering and erosion will guide slope (thence landscape) development; and time-independent forms (that is forms in dynamic equilibrium) are unlikely to develop. In these terms the high surface of the Mt Lofty Ranges could have evolved as a perched peneplain, the level of which was controlled by local, slowly evolving baselevel beneath which a deep weathering profile was formed (Fig. 19.19).

In theory it may, with increasing knowledge of the stratigraphy of depositional basins, be possible to suggest reasonable tests. For instance it should be possible to estimate the volume of rock eroded from a particular region and to compare this with the volume of sediment deposited in adjacent basins. Due allowance would have to be made for such factors as compaction, or lack of it, and for material taken out of the local system in solution or in suspension, but in broad terms the volume of sediment contributed as a result of the dissection of an uplifted oldland should be less than that derived from the development of a perched peneplain. In the latter case there is not only the dissection of the high plain to be taken into account but also the lowering of relief during the formation of the perched surface of low relief. Such a test is particularly apposite for areas of limited extent where the effects of possible isostatic adjustment are likely to be slight.

Table 19.1
Landform assemblage development

Uplift > Incision	Incision > Wasting	1	Increasing relief
	Incision = Wasting	2	Static relief
	Incision < Wasting	3	Decreasing relief — P
Uplift = Incision	Incision > Wasting	4	Increasing relief
	Incision = Wasting	5	Static relief
	Incision < Wasting	6	Decreasing relief — P
Uplift < Incision	Incision > Wasting	7	Increasing relief
	Incision = Wasting	8	Static relief
	Incision < Wasting	9	Decreasing relief — P

(P: peneplain or other surface of low relief.)

4. Stable Landforms

Those forms which are *stable* (using the term in a relative sense) are not to be confused with forms in dynamic equilibrium. This is not to suggest that landforms remain absolutely static and permanent. The earth's surface is constantly changing. But it is clear that whereas some areas are eroded

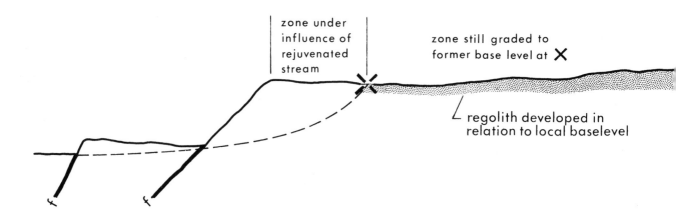

Fig. 19.19 Development of lateritised high plain as a perched peneplain.

rapidly, others change only slowly in time, and have been as they are for very long periods of earth history. Thus granite has the property of being susceptible to prolonged attack by moisture and a subsurface mass of granite may be altered; but once exposed at the surface the granite is weathered and eroded comparatively slowly. This is the basis of the interpretation of the stepped topography (see Chapter 10) developed on granite in the Sierra Nevada.[70]

The contrast in rates of weathering between various members of geosynclinal sequences leads again to relative perpetuation of forms. In the Flinders Ranges, the principal ridges are underlain by various sandstones which are interbedded with other sediments. Though the dimensions and shapes of the ridge and valley patterns have changed with the erosion of the structures in depth, tongues of early Tertiary lacustrine sediments extend into valleys eroded between sandstone ridges, showing that both ridges and valleys, which remain prominent features of the landscape, were already in existence some 70 million years ago, and that they have not changed in essence since that time.[71]

In this same area since the close of the Mesozoic there have been close to 6000 m of rock eroded from the crestal areas of the major, complex anticline, which underlies the central and southern Flinders Ranges. But there is reason to suggest that the erosion of the upland, though it has by no means ceased, is proceeding at a decreased rate. This is probably due first to the ease of weathering and erosion in the stretched and shattered crestal rocks, and second to the fact that thus far erosion has proceeded in the upper zone of tension, whereas at present in many areas the rivers have attained the neutral plane in the structure (see Chapter 5) and are now attacking the strata which are in compression.*

* A similar situation is reported in the Sierra Nevada, where the Mesozoic granite batholiths were very rapidly exposed by deep stream dissection, but where subsequent uplift and slow erosion have permitted the development of high relief. Again the ease of erosion in the shattered crestal areas may be the reason for this change in rate of erosion.

The various landforms already developed on duricrusted surfaces in several contexts may be cited as examples of features which persist over long periods—several scores of millions of years—particularly in arid conditions where the high energy processes which are necessary for the dismemberment of the duricrusts do not act with sufficient frequency to affect the surfaces materially. Duricrusts are attacked and undermined at their margins, but the surfaces are weathered and eroded only very slowly.

Such stable land surfaces are still developing. For instance, around the granite domes of northwestern Eyre Peninsula (Pl. 1.2), the soils display thick accumulations of calcrete which, under the prevailing arid conditions, are resistant to erosion. This duricrust caps the plains which may be expected to persist until some major change of climate occurs at baselevel.[73]

Similarly, frost-riven blocks deriving from a former periglacial regime survive and, in view of the seeming ineffectiveness of modern processes, appear likely to survive for a long time in such areas as the Pennsylvanian Appalachians. In a broader context Crickmay[74] has challenged the cyclic and the dynamic concepts of landscape development by suggesting that the field evidence indicates unequal activity in valleys and on interfluves respectively. He allows that rivers are actively eroding but cites evidence which indicates either that interfluves are at a standstill or that they are changing only so slowly that they may be considered stable. Certainly there is evidence that some scarps have not retreated to any important extent for many millions of years. Crickmay goes so far as to claim that all scarps which do not have an active agent of erosion such as a river at their base, are effectively static. Hence the principle of what Crickmay calls Unequal Activity.

E. CONCLUDING STATEMENT

No one model of landscape development satisfactorily explains all landform assemblages. There are several theoretical possibilities, each of which may best fit the field evidence in particular regions at particular times.

References Cited

1. C.R. Twidale, "Erosion of an Alluvial Bank at Birdwood, South Australia", *Z. Geomorph.*, 1964, **8** NS, pp. 189–211.
2. W.M. Davis, *Geographical Essays*, Dover, Boston, 1909, pp. 249–78, 350–412.
3. A.N. Strahler, "Davis' Concepts of Slope Development viewed in Light of Recent Quantitative Investigations", *Ass. Am. Geogr. Ann.*, 1950, **40**, pp. 209–13.
4. C.H. Crickmay, "The Later Stages of the Cycle of Erosion", *Geol. Mag.*, 1933, **20**, pp. 337–47.
5. W.M. Davis, *Geographical Essays*, pp. 296–322, 617–89.
6. D.W. Johnson, *Shore Processes and Shoreline Development*, Wiley, New York, 1919.
7. W.M. Davis, "Rock Floors in Arid and Humid Climates", *J. Geol.*, 1930, **38**, pp. 1–27.
8. O. Fisher, "On Cirques and Taluses", *Geol. Mag.*, 1872, **9**, pp. 10–12.
9. W. Penck, *Die Morphologische Analyse*, Engelhorns, Stuttgart, 1924; *idem*, *Morphological Analysis of Landforms*, translated by H. Czeck and K.C. Boswell, Macmillan, London, 1954.
10. L.C. King, "A Study of the World's Plainlands", *Q. J. geol. Soc. London*, 1950, **106**, pp. 101–31; *idem*, "Canons of Landscape Evolution", *Bull. geol. Soc. Am.*, 1953, **64**, pp. 721–52; *idem*, "The Uniformitarian Nature of Hillslopes", *Trans. geol. Soc. Edinburgh*, 1957, **17**, pp. 81–102; *idem*, *Morphology of the Earth*, Oliver and Boyd, Edinburgh, 1960.
11. See for instance G.H. Dury, "A Partial Definition of the Term *Pediment* with Field Tests in Humid Climate Areas of Southern England", *Trans. Inst. Br. Geogr.*, 1972, **57**, pp. 139–52.
12. C.R. Twidale, "Hillslopes and Pediments in the Flinders Ranges, South Australia", Ch. 6 in J.N. Jennings and J.A. Mabbutt (Eds), *Landform Studies from Australia and New Guinea*, Australia National University Press, Canberra, 1967, pp. 95–117; *idem*, "The Origin of the Piedmont Angle as evidenced in South Australia", *J. Geol.*, 1967, **75**, pp. 393–411.
13. E.S. Hills, "Die Landoberfläche Australiens", *Die Erde*, 1955, **7**, pp. 195–205; A.A. Öpik, "Geology and Palaeontology of the Headwaters of the Burke River, Queensland", *Bull. Bur. Miner. Resour. Geol. Geophys. Aust.*, 1956, **53**.
14. See L.C. King, *Morphology of the Earth*, Oliver and Boyd, Edinburgh, 1960, and several references contained therein to King's earlier papers on this aspect of South African and world landscapes.
15. H. Baulig, "The Changing Sea Level", *Inst. Br. Geogr. Pub.*, 1935, **3**.
16. C.R. Twidale, "A Possible Late Quaternary Change in Climate in South Australia", in *Quaternary Geology and Climate (Proc. VII INQUA Congress)*, National Academy of Science, Washington, D.C., 1969, **16**, pp. 43–8.
17. G. E. Williams, "Glacial Age of Piedmont Alluvial Deposits in the Adelaide Area, South Australia", *Aust. J. Sci.*, 1969, **32**, p. 257.
18. L.C. King, *Morphology of the Earth*, p. 235 *et seq.*
19. L.C. King, *The Natal Monocline*, University of Natal Press, Durban, 1972.
20. L.C. King, "The Cyclic Land-surface of Australia", *Proc. R. Soc. Vic.*, 1950, **62**, pp. 79–95; *idem*, "A Geomorphological Comparison between eastern Brazil and Africa (central and southern)", *Q. J. geol. Soc. London.*, 1957, **112**, pp. 445–74.
21. L.C. King, "Basic Palaeography of Gondwanaland during the late Palaeozoic and Mesozoic Eras", *Q. J. geol. Soc. London.*, 1958, **114**, pp. 47–70.
22. C.R. Twidale, "Chronology of Denudation in northwest Queensland", *Bull. geol. Soc. Am.*, 1956, **67**, pp. 867–82; *idem*, "Geomorphology of the Leichhardt-Gilbert Area, northwest Queensland", *Land Res. Ser.*, (C.S.I.R.O.) 1966, **16**.
23. A.A. Öpik, "Geology and Palaeotology of the Headwaters of the Burke River, Queensland", *Bull. Bur. Miner. Resour. Geol. Geophys. Aust.*, 1956, **53**. E.K. Carter and A.A. Öpik, "Explanatory Notes on the Lawn Hill 4-mile Geological Sheet", *Bur. Min. Resour. Geol., Geophys. Expl. Notes*, 1961, **21**.
24. J.A. Mabbutt, "The Weathered Land Surface in Central Australia", *Z. Geomorph.*, 1965, **9** NS, pp. 82–114.
25. H. Wopfner, "Permian-Jurassic History of the western Great Artesian Basin", *Trans. R. Soc. S.Aust.*, 1964, **87**, pp. 117–28.
26. G.D. Woodard, "The Stratigraphic Succession in the Vicinity of Mt Babbage Station, South Australia", *Trans. R. Soc. S.Aust.*, 1955, **78**, pp. 8–17; C.R. Twidale, "A Possible Late Quaternary Change", pp. 43–8.
27. C.R. Twidale, "Chronology of Denudation in the southern Flinders Ranges, South Australia", *Trans. R. Soc. S.Aust.*, 1966, **90**, pp. 3–28; *idem*, "Geomorphology of the Flinders Ranges", in D.W.P. Corbett (Ed.), *National History of the Flinders Ranges*, Public Library Adelaide, 1969, pp. 57–137.
28. E.J. Wayland, "Peneplains and some Erosional Landforms", *Geol. Surv. Uganda Ann. Rep. Bull.*, 1934, **1**, pp. 77–9.
29. J.A. Mabbutt, "A Stripped Land Surface in Western Australia", *Trans. & Papers Inst. Br. Geogr.*, 1961, **29**, pp. 101–14; J.T. Jutson, "The Physiography (Geomorphology) of Western Australia", *Bull. Geol. Surv. W.A.*, 1934, **95**.
30. C.R. Twidale, *Geomorphology, with Special Reference to Australia*, Nelson, Melbourne, 1968, pp. 315–16.
31. E. Blackwelder, "Cenozoic History of the Laramie Region, Wyoming", *J. Geol.*, 1909, **17**, pp. 429–44.
32. D.H. Eggler, E.E. Larson and W.C. Bradley, "Granites, Grusses, and the Sherman Erosion Surface, southern Laramie Range, Colorado-Wyoming", *Am. J. Sci.*, 1969, **267**, pp. 510–22.
33. E. Blackwelder, "Cenozoic History", pp. 429–44.
34. D.H. Eggler *et al.*, "Granites, Grusses", pp. 510–22.
35. J. Barrell, "The Piedmont Terraces of the northern Appalachians", *Am. J. Sci.*, 1920, **49**, pp. 227–58, 327–62, 407–28; Florence Bascom, "Cycles of Erosion in the Piedmont Province of Pennsylvania", *J. Geol.*, 1921, **29**, pp. 540–59.
36. D.W. Johnson, *Stream Sculpture on the Atlantic Slope*, Columbia University Press, New York, 1931.

37. C.B. Hunt, *Physiography of the United States*, Freeman, San Francisco, 1967.

38. See, for example, P. Macar, "Appalachian and Ardennes Levels of Erosion Compared", *J. Geol.*, 1955, **63**, pp. 253–67.

39. A.C. Ramsay, "The Denudation of South Wales and Adjacent English Counties", *Mem. geol. Surv. Gt Brit.*, 1846, **328**; E.H. Brown, "Erosion Surfaces in north Cardiganshire", *Trans. Papers Inst. Br. Geogr.*, 1950, **16**, pp. 51–66; *idem*, "The Physique of Wales", *Geogr. J.*, 1957, **123**, pp. 208–21.

40. E.H. Brown, "Erosion Surfaces in north Cardiganshire", pp. 51–66.

41. A.C. Ramsay, "The Denudation of South Wales".

42. See, for example, O.T. Jones, "Some Episodes in the Geological History of the Bristol Channel Region", in *Rep. 95th Meeting, Brit. Assoc. Adv. Sci.* 1930, Brit. Assoc., London, 1931, pp. 57–82.

43. C. Fox, *The Personality of Britain*, National Museum of Wales, Cardiff, 1932.

44. E.H. Brown, "Erosion Surfaces in north Cardiganshire", pp. 51–66.

45. J.C. Rodda, "A Trend-surface Analysis Trial for the Planation Surfaces of North Cardiganshire", *Trans. Inst. Br. Geogr.*, 1970, **50**, pp. 107–14.

46. C.A.M. King, "Trend-surface Analysis of Central Pennine Erosion Surfaces", *Trans. Inst. Br. Geogr.*, 1969, **47**, pp. 47–59.

47. J.R. Tarrant, "Comments on the Use of Trend-surface Analysis in the Study of Erosion Surfaces", *Trans. Inst. Br. Geogr.*, 1970, **51**, pp. 221–2.

48. V.D. Choubey, "Pre-Deccan Trap Topography in central India and Crustal Warping in Relation to Narmada Rift Structure and Volcanic Activity", *Bull. Volc.*, 1972, **35**, pp. 660–85. (Part I, *Proc. Inter. Symp.* on "Deccan Trap and Flood Eruptions", Sagar, M.P., 1969.)

49. J.T. Hack, "Interpretation of Erosional Topography in Humid Temperate Regions", *Am. J. Sci.*, 1960, **238A**, pp. 80–97; R.J. Chorley, "Geomorphology and General Systems Theory", *U.S. Geol. Surv. Prof. Paper*, 1962, **500-B**.

50. S.A. Schumm, "Disparity between Present Rates of Denudation and Orogeny", *U.S. Geol. Surv. Prof. Paper*, 1963, **454**.

51. W. Penck, *Die Morphologische Analyse*, Engelhorns, Stuttgart, 1924; *idem, Morphological Analysis of Landforms*, translated by H. Czeck and K.C. Boswell, Macmillan, London, 1953.

52. L. Von Bertalanffy, "The Theory of Open Systems in Physics and Biology", *Science*, 1950, **3**, pp. 23–9; *idem*, "An Outline of General System Theory", *J. Br. Phil. Sci.*, 1951, **1**, pp. 134–65.

53. See, for example L.B. Leopold and W.B. Langbein, "The Concept of Entropy in Landscape Evolution", *U.S. Geol. Surv. Prof. Paper*, 1962, **500-A**.

54. J.T. Hack, "Interpretation of Erosional Topography", pp. 80–97; R.J. Chorley, "Geomorphology and General Systems Theory".

55. J.T. Hack, "Interpretation of Erosional Topography", pp. 80–97; J.T. Hack and J.C. Goodlett, "Geomorphology and Forest Ecology of a Mountain Region of the central Appalachians", *U.S. Geol. Surv. Prof. Paper*, 1960, **347**.

56. Cf. F. Ahnert, "The Role of the Equilibrium Concept in the Interpretation of Landforms of Plural Erosion and Deposition", *Proc. Symp. L'Évolution des Versants*, (Liège), 1967, **40**, pp. 23–41.

57. C.C. Nikiferoff, "Fundamental Formula in Soil Formation", *Am. J. Sci.*, 1942, **240**, pp. 847–66; *idem*, "Weathering and Soil Evolution", *Soil Sci.*, 1949, **67**, pp. 219–30; *idem*, "Hardpan Soils of the Coastal Plain of southern Maryland", *U.S. Geol. Surv. Prof. Paper*, 1955, **267B**; *idem*, "Re-appraisal of the Soil", *Science*, 1959, **129** (3343), pp. 186–96.

58. W.M. Davis, "Peneplains and the Geographical Cycle", *Bull. geol. Soc. Am.*, 1922, **23**, pp. 587–98.

59. W. Penck, *Morphological Analysis*, pp. 144 *et seq.*, 215 *et seq.*

60. B.P. Ruxton, "Slopewash under Mature Primary Rainforest in northern Papua", in J.N. Jennings and J.A. Mabbutt (Eds), *Landform Studies from Australia and New Guinea*, Australian National University Press, Canberra, 1965, Ch. 5, pp. 85–94.

61. See for example J.H. Bretz, "Dynamic Equilibrium and the Ozark Landforms", *Am. J. Sci.*, 1963, **260**, pp. 427–38.

62. R.J. Chorley, "A Re-evaluation of the Geomorphic System of W.M. Davis", in R.J. Chorley and P. Haggett (Eds), *Frontiers in Geographical Teaching*, Methuen, London, 1965, Ch. 2.

63. R.J. Chorley, "A Re-evaluation".

64. M. Chisholm, "General Systems Theory and Geography", *Trans. Inst. Br. Geogr.*, 1967, **42**, pp. 45–52.

65. M.F. Glaessner, "Some Problems of Tertiary Geology in southern Australia", *J. & Proc. R. Soc. N.S.W.*, 1953, **87**, pp. 31–45.

66. B. Daily, C.R. Twidale, and A.R. Milnes, "The Age of the Lateritised Summit Surface in Kangaroo Island and Adjacent Areas of South Australia", *J. geol. Soc. Aust.*, 1974, **21**, pp. 387–92.

67. C. Fenner, "The Major Structural and Physiographic Features of South Australia", *Trans. R. Soc. S. Aust.*, 1930, **54**, pp. 1–36; *idem, South Australia: A Geographic Study*, Whitcombe and Tombs, Melbourne, 1931, pp. 29–49.

68. W.Q. Kennedy, "Some Theoretical Factors in Geomorphological Analysis", *Geol. Mag.*, 1962, **99**, pp. 305–12.

69. S.A. Schumm, "Disparity between Present Rates"; S.A. Schumm and R. Lichty, "Time, Space and Causability in Geomorphology", *Am. J. Sci.*, 1965, **263**, pp. 110–19.

70. C. Wahrhaftig, "Stepped Topography of the southern Sierra Nevada, California", *Bull. geol. Soc. Am.*, 1965, **76**, pp. 1165–90.

71. C.R. Twidale, "The Neglected Third Dimension", *Z. Geomorph.*, 1972, **16** NS, pp. 283–300; *idem*, "Chronology of Denudation in the southern Flinders Ranges".

72. G.H. Curtis, J.F. Evernden and J. Lipson, "Age Determination of Some Granitic Rocks in California by the Potassium-Argon Method", *Calif. Div. Mines Spec. Rep.*, 1958, **54**.

73. See C.R. Twidale, Jennifer A. Bourne and Dianne M. Smith, "Reinforcement and Stabilisation Mechanisms in Landform Development", *Rev. Géomorph. Dyn.*, 1974, **23**, pp. 115–125.

74. C.H. Crickmay, *Some Central Aspects of the Scientific Study of Scenery*, Crickmay, Calgary, 1968; *The Role of the River*, Crickmay, Calgary, 1971.

CHAPTER 20

Drainage Patterns

A. STRUCTURAL CONTROL OF DRAINAGE PATTERNS

An examination of drainage patterns shows that both in gross and in detail streams have adapted to structure. That is to say, they flow and have excavated valleys along lines of weakness in the underlying bedrock. On a local scale, the shape of river courses seems, even in some upland regions, to be dominated by hydraulic shapes (river curves). Close examination, however, shows that the relatively straight sectors between curves, as well as the arms of some curves, are to a greater or lesser extent aligned parallel to structural trends of various types. On a regional scale, many of the major rivers of eastern and central Australia run along major fault zones and this is also true in local detail (Pl. 4.3c). In fold mountain belts, rivers run parallel to the regional strike

for long distances (Pls 5.12, 5.13, 5.14; Fig. 20.1). Significantly, however, these long sectors, controlled by structure, are linked by short sections which flow across the structural grain (see below).

Some drainage patterns considered as entities are closely adjusted to joints (Pl. 3.1b), cleavage, bedding planes, faults and outcrops of weaker strata. Zernitz[1] to whom we owe the best analysis of structurally defined stream patterns, has identified several types. *Trellis drainage* develops on outcrops of alternate resistant and weak strata. Long strike streams are linked by short *cataclinal* or *anaclinal*[2] streams, that is, by dip or antidip elements. This pattern is typical of fold mountain belts (Fig. 20.2). *Annular drainage* develops on dome and basin structures which have been breached, exposing concentric outcrops of rocks with contrasted resistance to weathering and

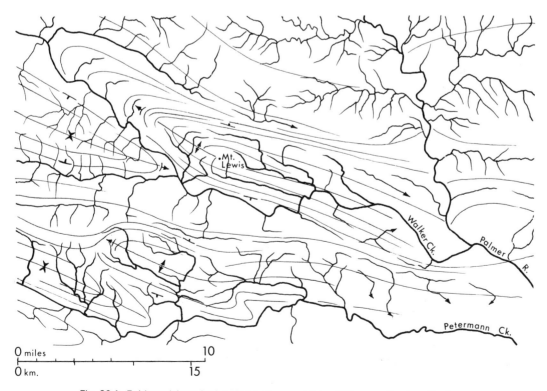

O miles 10

O km. 15

Fig. 20.1 Folds and rivers in the Mt Lewis area, Krichauff Ranges, Northern Territory.

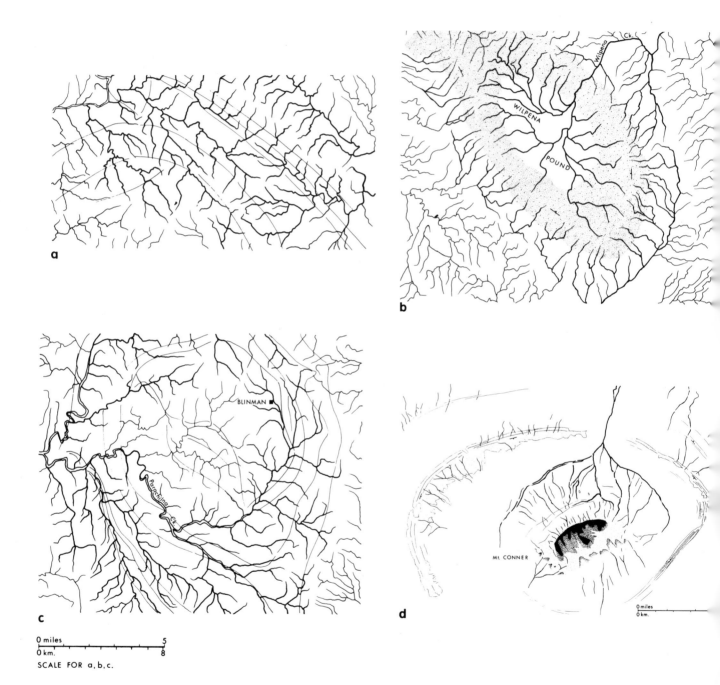

Fig. 20.2 (a) Trellis drainage pattern—central Flinders Ranges, S.A. The hair lines indicate lithological boundaries. (b) Annular pattern around Wilpena Pound, central Flinders Ranges, with centripetal drainage inside the sandstone basin (dotted) and radial drainage on the outer slopes. (c) Annular pattern in the Blinman Dome, central Flinders Ranges. (d) Annular pattern around Mt Conner, N.T.

erosion. As in the trellis pattern, the main streams develop along the strike of the weaker outcrops and are linked by dip or antidip streams (Figs 20.2b-d) which breach the intervening hard rock ridges. In *rectangular drainage*, all streams are subsequent or strike streams, that is, they follow along lines of weakness. The main streams and tributaries display right-angled junctions and bends due to joint or fault control. This pattern is characteristic of many granite outcrops (Pl. 3.1a), of heavily faulted exposures, and of outcrops of massively jointed and bedded sedimentary sequences.

Some streams follow the regional slope. In *radial drainage*, which is typical of dome structures (Fig. 20.2b), streams radiate from a central area. *Centripetal drainage* is the opposite of the radial type; all streams flow towards a central area and the several elements converge (Fig. 20.2b). *Parallel drainage* develops on long, straight slopes, and simply reflects the regional slope of the land while *dendritic* (like the branches of a tree) *drainage* implies uniform resistance of the local bedrock and minimal slope of land.

Most drainage networks conform to one of these patterns and their several minor variants. That streams have adjusted to lines of structural weakness does not imply that they can in any way adjudicate or choose or know which course will involve them in the least work. Rather it is based on natural selection. For, if a number of streams in a drainage network flow over bedrock which is resistant to weathering and erosion, and others flow over weak strata, the latter should incise and extend their channels and valleys more rapidly than the former. Those streams draining outcrops of weak rock should (and usually do) enlarge their valleys, extend their catchments, capture rival streams and in general become the dominant elements of the network. Furthermore, during incision, rivers become remarkably well adjusted to the structures of the bedrock over which they flow. This was noted in the American West by Marvine, who advocated what is now called *superimposition* (see below) in explanation of the eccentric course of some of the major rivers of Colorado, but who also noted that "... the structure of the lower rocks has begun to affect the courses of the streams, and in places to a considerable extent." Similar conclusions have been reached by many writers. For instance both Strahan[4] and Marr,[5] though invoking drainage superimposition in South Wales and the English Lake District respectively, nevertheless describe many examples of streams which are concordant with respect to the structures over which they presently flow.

In the Torrens Gorge, east of Adelaide, the river displays what at first appear to be purely hydraulic forms in its well-developed intrenched meanders. But many of the forms are closely accommodated to the local bedrock. For instance, for long sectors between Gumeracha and Cudlee Creek the river runs parallel to the variable strike of the Proterozoic sediments, and near the Kangaroo Creek junction the elongate and nearly straight arms of several of the prominent river loops run parallel to the foliation and lineation of the older Precambrian schists and gneisses.

Local structural trends should certainly be examined before a stream sector is judged to be deviant. It is not merely a matter of matching stream direction against the dominant structural grain, but of plotting all structural trends. For example in the Torrens Gorge, joints, bedding planes, linea-tion and schistosity together provide exploitable lines of weakness which display four main trends but which, because of their local variability, range through a considerable part of the circle. Some streams or stream sectors are guided by each of these factors.

Nevertheless many streams or stream sectors flow across the structural grain or follow other courses which appear to be abnormal in terms of natural selection. These rivers which are *anomalous* in their courses are considered to form *transverse* drainage. On a regional scale some rivers run directly across the structural grain, running in *diaclinal*[6] valleys—valleys which pass through folds, strictly speaking at right angles to the fold axes, but in any event at an obtuse angle to them. Thus, in the Appalachians the Susquehanna and several of its tributaries like the Juniata, North Branch and West Branch, flow essentially southeastwards across the NE–SW fold axes of the Ridge and Valley Province, though for long stretches they run parallel to the strike and fold axes before plunging through gorges excavated in sandstone ridges (Figs 5.16, 20.3). In the Zagros Mountains the fold axes trend WNW–ESE, but rivers like the Kashgar, Sehzar and Bakhtiari run southeastwards, though here again there are long subsequent sectors (Fig. 20.4). Examples of transverse drainage abound in the Cape Fold Belt of South Africa. The Tradour River near Barrydale, and many minor streams in the Riebeek Oos area cut across fold axes, though here, as elsewhere, the major stream sectors follow the strike. In central Australia, the Finke (Fig. 20.5) and other streams which flow southwards across the James Range toward Lake Eyre, show "the paradox of cutting through the great one thousand foot ridges at right angles and ploughing their way across the anticlines."[7] On a local scale the short dip and antidip streams which link strike streams in trellis and annular patterns also require explanation.

On analysis, and ignoring hydraulic forms, transverse patterns (at whatever scale they are considered) are found to consist of two elements. The first comprise the considerable strike sectors which are adjusted to structure. These are linked by the second element: the anomalous sectors, of which there are three types.

1. Streams which have breached ridges built of comparatively resistant rocks and usually by way of gorges (Pl. 20.1a and b). Many of these are the links between long subsequent elements and form an integral part of the trellis and annular patterns already referred to.

2. Less commonplace but of crucial importance to the understanding of river pattern evolution in fold mountain belts are the rivers which flow across and breach the hard strata exposed in the snouts of plunging anticlines. In many instances there are low plains and valleys eroded in weaker strata only a short distance from the breach (Pls 20.2, 20.3, 20.4; Fig. 20.3 and 20.5).

3. Most strike streams flow along the more-or-less broad plains they themselves have eroded in weak outcrops, but in a few instances the rivers twist into and flow for a short distance within the bordering ridge formed by resistant rocks (Fig. 20.6). These short sectors commonly run in sinuous gorges, and in some cases the river weaves back and forth from lowland to ridge. For this reason

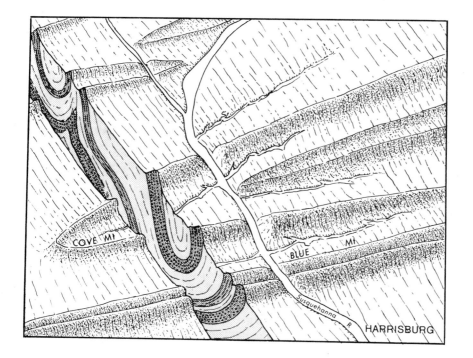

Fig. 20.3 Section through breached snout in the Appalachians. (*After A.K. Lobeck,* Geomorphology, *McGraw-Hill Book Co., New York, 1939, pp. 581– 612.*)

they may be called in-and-out gorges* or disconnected strike gorges.

Thus the three types of features eroded by streams which require explanation are the breached ridge, the breached snout, and the in-and-out-gorge.** Together with structurally adjusted sectors, these make up the transverse drainage. How are such illogical and uneconomic (in terms of energy expenditure) drainage patterns evolved?

B. POSSIBLE EXPLANATIONS OF DRAINAGE ANOMALY

Five types of explanation have been offered for the stream courses which run across the structural grain: diversion, antecedence, superimposition, inheritance and headward erosion and stream persistence.

---

* "In-and-out channel" was used for glacial spillways by Kendall.[8] The valleys under discussion are of similar morphology, but of course of different origin and for this reason the slightly different terminology is desirable.

** According to some writers,[9] gorges are anomalous features *per se*. But it will be clear from what has been written in Chapter 13 that gorges merely reflect a rapid rate of stream bed incision compared to the rate of valley side slope development. In particular structural conditions, gorges are a normal and expectable feature of river valleys.

1. Stream Diversion

Stream diversion can be caused by several agencies, and gives rise to rivers which are aberrant with respect to the regional pattern.

Lava flows have caused stream diversions on both regional and local scales.[10] The advance of glaciers against or across regional slope during the Pleistocene caused many modifications of drainage. Most were ephemeral but some have been maintained and remain part of the regional pattern (see Chapter 21). Major examples include the *urstromtäler* of the north European plain, but many cases have been described in the literature.[11] The course of the Grand Canyon of the Yellowstone, Wyoming, is attributed in some measure to diversion at different times by lava and ice.[12]

Earth movements have produced similar results. A classic case of a rising fault block diverting drainage during the late Pleistocene has been described from the Echuca district of Victoria and the adjacent parts of N.S.W (see p. 97).

Regional warping is less spectacular and less easily detected but its results are no less significant. The curious "palm-tree" pattern of the headwaters of the Diamantina in northwest Queensland, as well as apparent changes in the behaviour of such rivers as the Flinders and Leichhardt (which run to the Gulf of Carpentaria) appear to be due to recent activity of the Selwyn Upwarp.[14]

2. Antecedence

Tectonic movements are also invoked in antecedence. It is

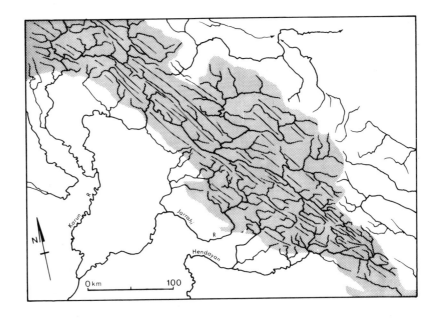

Fig. 20.4 Part of the Zagros Mountains, Iran, showing main structural and drainage lines. (*After T. Oberlander, "The Zagros Streams"*, Syracuse Geogr. Ser., *1965,* **1**.)

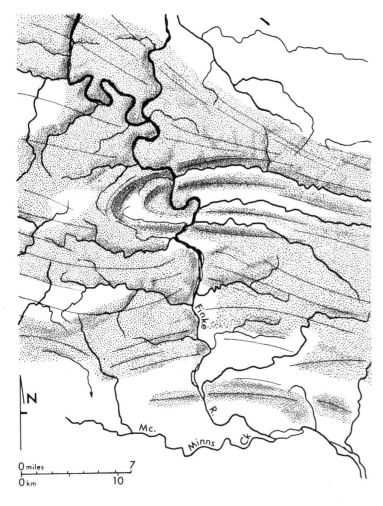

Fig. 20.5 Sector of the Finke River crossing a complex ridge in the James Range, N.T.

Plate 20.1a The Finke crossing the James Range, Northern Territory. The river course is virtually normal to the rock strike. (*Dept Interior, Canberra.*)

Plate 20.1b Simpson Gap, a narrow gorge through a sandstone ridge near Alice Springs, N.T. (*C.R. Twidale.*)

Plate 20.2 Plunging limestone anticline in the Zagros Mts. The ridge has been breached by two streams which have carved gorges or *tangs* through the ridge. (*Hunting Surveys.*)

Plate 20.3 Breached dome in the James Range, N.T., showing several strike streams (for example, X), but also the major river (Y) which flows across the structure. (*C. Wahrhaftig.*)

Plate 20.4 The Susquehanna River cutting through the sandstone snouts of plunging folds in the northern Appalachians near Harrisburg, Pennsylvania. (*J. S. Shelton.*)

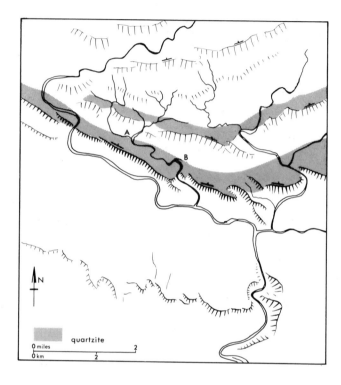

Fig. 20.6 In-and-out gorge on the northern dip slope of a sandstone ridge near Arkaroola, northern Flinders Ranges, S.A.

quartzite

0 miles   2
0 km   2

visualised that a river maintains its course by eroding its bed through a rising fold or fault block.

The concept is of some antiquity. Medlicott[15] considered the rivers of the Himalayas to be of this origin, and Wager[16] corroborated this general view with respect to the Arun. Hayden used the same idea to account for some of the Rocky Mountain canyons. But the term antecedence, as distinct from the idea, was first used by Powell[18] of the course of the Colorado River: "All the facts... lead to the inevitable conclusion that the system of drainage was determined antecedent to the faulting and folding."

Subsequently the concept was widely applied and Dutton[19] went so far as to suggest that wherever examples are found of rivers flowing through mountain chains, antecedence must be assumed to be the cause (though he used the term "persistence", see below p. 440 *et seq.*).

Matthes[20] attributed the Grand Canyon to antecedence. He considered that the crust had gradually arched after the Colorado had become established in its course, and that the river was able to excavate the gorge through the rising block. Subsequent work[21] suggests that the origin of the Grand Canyon is complex, involving both antecedence and superimposition, with a strong element of broad structural guidance.

In order to demonstrate antecedence it is necessary to prove that regional or local tectonism has occurred and that the river in question predates the earth movements. This is difficult to do. One clear case of an antecedent drainage channel is reported from Mesopotamia, where an irrigation canal constructed some 1700 years ago has maintained its course through the rising Shaur Anticline, deepening its bed by almost four metres in the process. Adjacent channels are warped and abandoned.[22] The course of the Salzach, between Werfen and Ofenau in Austria, is also cited as a likely example of an antecedent stream.[23] The lower limit of glaciation as marked by such features as striae and *roches moutonnées*, appears to be arched, falling towards both ends of the gorge from a point near the centre. There is no suggestion that the glacial limit was structurally determined, that is, there was no resistant bar in the central sector which affected the profile of the preglacial valley. There are, for instance, no lake deposits such as should accumulate behind such a rock barrier. The gorge is of preglacial origin for the glacier clearly penetrated along it and it is suggested that the arched shape of the glacial limit results from postglacial distortion.

In most cases in which antecedence is invoked, however, the evidence is not so persuasive. Invalid criteria may be cited, as in the cases mentioned earlier; or tectonism, for which there is no proof, is invoked; or the patterns can be explained adequately, and possibly more reasonably, in terms of another mechanism.

Thus, Fenner[24] suggested that the rivers which drain the western Mt Lofty Ranges in South Australia are antecedent, maintaining their courses through rising fault blocks. It is doubtful whether there has been such marked differential faulting within the Ranges during the Cainozoic—the period in question—and in any case, as is outlined below, the courses of rivers like the Myponga, Sturt and Onkaparinga appear understandable in terms of headward erosion and capture of rivers flowing along lines of structural weakness.

3. Superimposition

The most commonly advanced explanation of anomalous drainage patterns is superimposition. Superimposed stream patterns develop in accordance with a superior geological formation (*overmass* or *cover* deposit). The latter rests with marked angular unconformity on an *undermass*, and as the rivers penetrate through the base of the overmass, their patterns are incongruous with respect to the structures of the undermass in which they now flow.

The idea of superimposition was used by Jukes[26] to explain the courses of rivers in southern Ireland, but the technical term was first used by Maw in 1866: "...a series of valleys superimposed in part unconformably over an ancient buried series."[27] Superimposition was cited by both Marvine[28] and Powell[29] in the Rocky Mountains, on a local scale in the Appalachians[30] and on a regional scale in the the same area,[31] where the pattern is allegedly derived from a Permian overmass.

Strahan[32] considered the rivers of South Wales to have been superimposed from a cover of Upper Cretaceous rocks on to the basined Upper Palaeozoic strata which they are now eroding. Marr[33] advocated a double superimposition in explanation of the radial drainage pattern of the Lake District in northwestern England. The stream pattern, which

While Lowl's[25] suggestion that under no circumstances can erosion keep pace with mountain building and that transverse valleys are never formed by antecedence is in some instances manifestly incorrect, antecedent streams are nevertheless rare.

had been in some measure perpetuated by glacial erosion, is discordant with respect to the folded Palaeozoic strata exposed in the core of the dome structure (Fig. 20.7a and b). But these strata were formerly buried beneath a cover of Carboniferous Limestone, on the domed form of which the radial pattern may have developed and which certainly assisted in its maintenance. However, Marr considers it likely that the dome may have been capped by Cretaceous or even Tertiary strata and that the radial pattern may initially have formed on that.

The lower course of the Russian River, north of San Francisco, is explained by its gradual extension westward as the Pliocene sea withdrew, and its superimposition from the Pliocene Merced Formation into the folded Mesozoic strata below.[34] Rivers such as the Binar, Kair and Sohar, in Madhya Pradesh, central India, are anomalous with respect to the structures of the Precambrian Vindhyan rocks over which they now flow and are regarded as superimposed from the Deccan lavas, remnants of which are abundant high in the local topography and which formerly extended over the entire region.[35]

Thus the concept of superimposition is sound enough, and is commonly cited, but it is not always wisely applied. In the English Lake District, the presence of a Carboniferous cover is clearly implied by the occurrence of limestone displaying quaquaversal dips at the margins of the Lake District; in view of the proven former extent of the Cretaceous seas, the former presence of a Cretaceous cover is a reasonable suggestion both there and in South Wales. But the concept is commonly applied in areas where there is less reason for suggesting the former occurrence of an overmass. For instance, there is little evidence for the Permian cover from which the transverse streams of the Appalachians are alleged to have been superimposed.

In South Australia, superimposition has been suggested to account for the courses of major rivers of the Mt Lofty Ranges such as the Torrens and the Onkaparinga. Fenner envisaged that the Precambrian and Palaeozoic undermass had been covered by a Miocene overmass on which the roughly radial pattern of major streams developed:

> Deep valleys were cut into the overmass of limestones, and ultimately these continued into the ancient and more resistant undermass (superimposed streams).[36]

There is no evidence of such a Miocene (or any other) overmass. On the contrary, there is evidence that throughout Tertiary times, the upland now known as the Mt Lofty

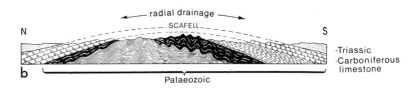

Fig. 20.7 Superimposed drainage in the Lake District of northwestern England: (a) plan; (b) section through the region showing reconstruction of earlier geological situation. (*After A. Holmes,* Principles of Physical Geology, *Nelson, Edinburgh, (2nd ed.), 1965, p. 564.*)

Ranges existed, and that the seas lapped up against the foot-hills,[37] penetrating only into low-lying fault angle depressions such as the Willunga basin[38] and into such depressions as the Myponga basin, the Miocene floor of which lies below present sealevel.[39] Despite this long established evidence, the idea of superimposition from an imaginary cover of Tertiary marine sediments died hard. It was reiterated as recently as 1969.[40]

Ward has suggested that the courses of the Sturt Creek and other minor streams, like the Field River, are superimposed[41] from a cover of Pleistocene marine sediments. He believes that there were high stands of the sea during the Pleistocene,[42] but both the lack of local evidence and the absence of regional support render the suggestion dubious.[43] Certainly no remnants of Pleistocene marine strata remain on the fault block in question. And, like other nearby rivers of similar behaviour, the Sturt, can be explained in terms of stream piracy (see below).

The drainage of the central Flinders is said to be partly superimposed,[44] but one might reasonably ask from what. Remnants of the old lateritised surface "uplifted in the Tertiary" are mentioned, but ignoring the facts that inheritance and not superimposition should have been cited where is the lateritised surface in the central Flinders Ranges? It does not exist.

In the southern Flinders Ranges the courses of the Kanyaka and Willochra creeks for one short sector could be superimposed from local lacustrine deposits of early Tertiary age, though in the main the abnormal courses of the rivers in this section (which are further discussed below) are due to what has been called "a rather unusual type of superimposition".[45] The only other possible overmass deposit is the marine Cretaceous, of which there are remnants in the northern Flinders,[46] but the southern limits of which remain obscure. There has been some suggestion that surfaces of possible Cretaceous age, which were cut by streams graded to the baselevel of the Cretaceous seas, survive in the southern and central Flinders,[47] but any streams which developed on these surfaces and which maintained their courses during incision are not superimposed but inherited.

4. Inheritance

An inherited stream is one which derives from a former surface of low relief, where structural control is slight. Rather than originating on and being derived from a distinct overmass deposit, the pattern is inherited from alluvial deposits and the regolith formed on a surface of low relief.

Intrenched meanders are commonly cited as evidence of a former surface of low relief[48] and subsequent stream incision, but it has been shown that such meanders are autogenic forms[49] which evolve as the stream migrates laterally during incision.

Discussing the braided pattern eroded in Precambrian quartzite near the junction of Lawn Hill and Widallion creeks, in northwestern Queensland, Whitehouse, after noting that Ball[50] had suggested Recent or sub-Recent capture in explanation of the intertwined channels, goes on:

...in the rugged country of the Pre-Cambrian quartzites this is not likely to have happened so effusively as the tangle of braids would require. Rather the features suggest that the present stream system, as an entity,

has been impressed on this topography. That is to say, it is the characteristic system of a very flat surface; and it is suggested accordingly that it developed on a plain and that by gradual denudation the same system has been superimposed on the harder rocks below.[51]

Though Whitehouse uses the word superimposed, he mentions no overmass deposits and clearly had in mind what Cotton[52] called inheritance. To confuse matters further it may be mentioned that superimposition from a Cretaceous cover, remnants of which occur at high elevations within the nearby Isa Highlands,[53] is distinctly possible on general grounds, though an examination of air photographs of the area suggests that headward erosion along prominent joints and resultant piracy accounts for much of the complex braiding in the vicinity of the stream junctions.

Possibly because of the adjustment to structure during incision, good examples of inherited streams are not plentiful though minor examples can be identified. In any case the distinction between superimposition (involving a stratigraphically distinct overmass) and inheritance (involving a combination of regolith and local alluvial deposits) is a fine one, though of course the geological implications of the two concepts stand in strong contrast.

5. Stream Persistence and Valley Impression*

In his exhaustive analysis of the streams of the Zagros Mountains, Oberlander[55] makes the point that no one mechanism or factor can account for the patterns of a major region: several mechanisms need to be invoked. One, involving the breaching of anticlines at their structural culminations, has already been described; many more sectors of the Zagros streams are adjusted to structures, and some are locally superimposed from synorogenic strata. But some are due to what Oberlander called *autosuperimposition*, that is, derivation from weak but conformable members of the fold sequence. In principle this is comparable to the idea of superimposition from asymmetric folds advocated by Meyerhoff and Olmstead[56] in explanation of the transverse drainage of the Appalachians, though headward erosion was also implied.

Independent work in the Flinders Ranges led to a similar conclusion.[57] At the northern margin of the Willochra plain, an intermontane basin in the southern part of the Ranges, there is a peculiar double breach of a quartzite ridge by the Willochra and Kanyaka creeks (Fig. 20.8a). This may be explained by suggesting that the two converging streams developed initially (or at least at some earlier stage) in weaker members of the tilted sedimentary sequence, and that through erosion they lowered themselves on to—or were impressed upon—the hitherto buried, stratigraphically lower, and resistant formation which was gradually exposed and through which the twin gorges were eroded (Fig. 20.8b-d). As both rivers were powerful enough to maintain or persist in their courses despite flowing over resistant strata, this anomalous pattern remains as an integral part of the regional drainage pattern.

---

* The term persistence was used by Dutton[54] in the sense of antecedence, but was never established; likewise reference to valleys being impressed was made by Whitehouse, but again the term has not achieved wide acceptance in this context.

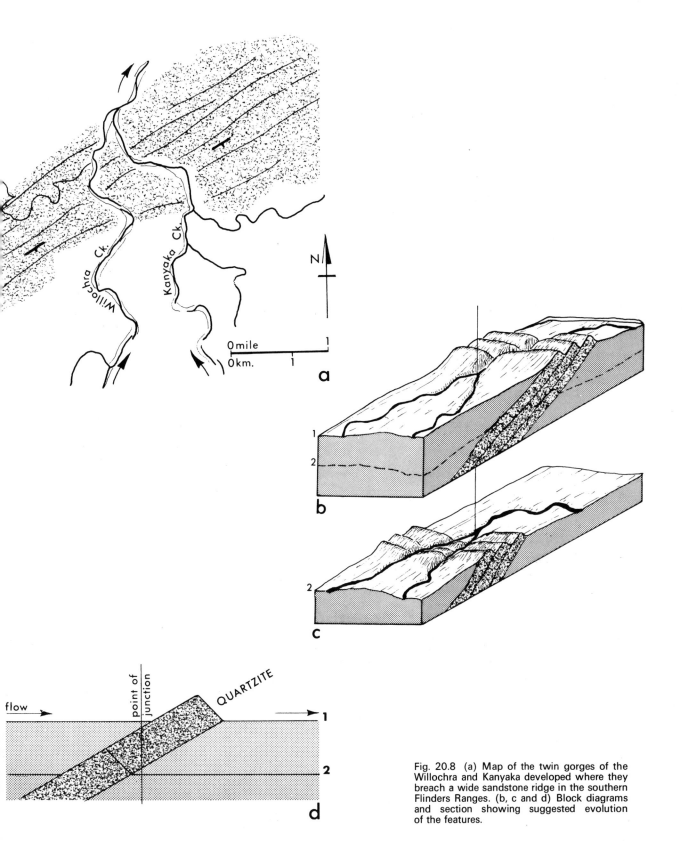

Fig. 20.8 (a) Map of the twin gorges of the Willochra and Kanyaka developed where they breach a wide sandstone ridge in the southern Flinders Ranges. (b, c and d) Block diagrams and section showing suggested evolution of the features.

The term autosuperimposition tends to mislead, but the phrases "persistence of streams" and "impression of valleys" are evocative, and these terms are therefore preferred.

C. TRELLIS AND ANNULAR PATTERNS AND THE PROBLEM OF BREACHED RIDGES

1. General

Trellis and annular patterns are characteristic of fold mountain belts. Long subsequent streams dominate the pattern but they are linked by comparatively short streams which have breached the ridges or other divides separating the adjacent subsequent valleys. It is these breaches which have proved controversial.

Almost a century ago, Lowl[58] considered that ridges could be effectively breached by headward eroding streams, and there is no doubt that he was right. Thompson[59] attributed many elements of the transverse drainage of the Appalachians to progressive stream piracy caused by headward erosion. In the Flinders Ranges the mechanism has been invoked in connection with the development of trellis patterns.[60] There is so much evidence for the breaching of hard rock ridges by regressively eroding streams that it is difficult to believe the mechanism was ever doubted. But Davis[61] took issue with Lowl[62] on the point. Davis claimed that rivers in the Appalachians exploited the lines of weakness caused by the occurrence of cross faults; however, he conceded that if there were a strong elevational contrast from one side of a ridge to the other, then headward breaching could occur. Nevertheless, he denied the general application of the process and asserted that he could observe no later stages of headward erosion in the region.

Similar argument was brought to bear by Strahler in denying Thompson's thesis: "Unfortunately, critical intermediate stages are omitted and doubtful stages of the process left undiscussed."[63] He also questioned the assertions made by Ashley,[64] for instance, concerning the presence of transverse faults which influence the course of streams, and points out that there are:

> repeated instances of long, roundabout subsequent streams flowing many miles along weak rock belts to drain lowlands only a few miles back of resistant barriers...[65]

These, he says, testify to the insignificance of headward erosion in breaching divides and, all in all, he questions the "assumption that the small stream...will push the divide entirely across a high and broad resistant rock barrier and into the lower weak rock lowland beyond."[66]

Cotton also considered headward erosion a mechanism of doubtful effectiveness.

> Though headward development of subsequents on crushed zones is quite probably the correct explanation of some minor transverse gorges, there is no evidence or probability of such guidance in the development of great gorges through mountain ranges, and without it there is little to be said in favour of the hypothesis of headward erosion.[67]

2. Stages in the Headward Breaching of Ridges

Every stage in the headward extension of streams through ridges, both simple and narrow, and broad and complex, can be observed in the field in the Flinders Ranges alone. The earliest stage in the development of a gorge or valley cutting across a hard rock ridge is the notch, such as those observed near Buckaringa Gorge (Pl. 20.5), where they are developed in steeply dipping quartzite. The notches are enlarged (Pl. 20.5) as is seen at Bellaratta, and then evolve into minor valleys such as are developed in many parts of the Ranges, for instance, adjacent to Aroona Valley, near Wilpena (Pl. 20.5). At Yarrah Vale, north of Quorn, the stage penultimate to breaching is displayed (Fig. 20.9a). A few miles to the north, Skeleroo Creek has extended back through a complex sandstone ridge to capture the headwaters of a stream system which drains the anticlinal valley beyond (Fig. 20.9b). Many other examples could be cited (see for instance, Fig. 20.10).

3. Processes Involved

As the stream incises its bed, the gradients in the headwater region become so steep that slopes become unstable and collapse. In this way run-off which also "eats back" into the upland (Fig. 20.11) is gradually funnelled more and more into the stream. In broad complex ridges, run-off from within the ridge is also tapped and the original notch gradually becomes a cleft, and the cleft a valley. In porous or permeable strata, such as limestone and sandstone, subsurface flushing and piracy contribute to the extension of streams.[68]

4. Structural Influences

Transverse faults are few in the Flinders and some other fold mountain belts, and they play an insignificant part in guiding stream courses. There are, however, other more important and widespread structural weaknesses. Joints, for instance, are abundantly developed in the sandstones which commonly form the ridges and these joints are quite clearly exploited by streams. The courses of several rivers in gorge sections in the MacDonnell Ranges of central Australia, for instance, are clearly controlled by such major joints and the notches at Buckaringa (Pl. 20.5a) are also evidently joint-controlled.

Another structural factor involved concerns the calibre and density of debris available for stream transport. Stream gradients adjust to the calibre of load available, being of gentler inclination where the bedrock disintegrates to fine debris, and steeper where large blocks are to be carried. Thus strike streams and their tributaries vary in elevation according to the bedrock in which they develop; and local variations and captures develop on this account.

5. Application to Transverse Drainage

Though of only local significance in one respect, headward erosion effectively links structurally adjusted elements and, in this way (and as has been previously stated) it is an important part of many transverse drainage patterns. The question of scale, therefore, does not arise.

Plate 20.5a Cleft in sandstone ridge near Buckaringa Gorge, southern Flinders Ranges, South Australia. Note numerous joints. (*C.R. Twidale.*)

Plate 20.5b Small breach in low sandstone ridge near Bellaratta, southern Flinders Ranges, S.A. (*C.R. Twidale.*)

Plate 20.5c Transverse valley crossing sandstone ridge, Aroona Valley, central Flinders Ranges, S.A. (*C.R. Twidale.*)

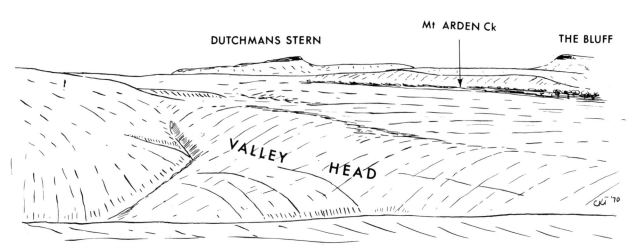

DUTCHMANS STERN   Mt ARDEN Ck   THE BLUFF

VALLEY   HEAD

Fig. 20.9a Head of valley near Yarrah Vale, southern Flinders Ranges. The ridge is almost breached and the Mt Arden Creek is in danger of capture.

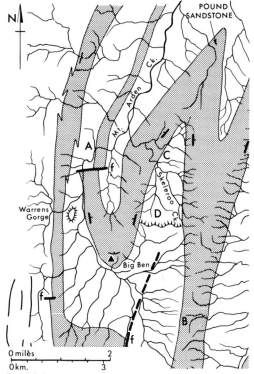

POUND SANDSTONE

Fig. 20.9b Pattern of relief and drainage around Big Ben, north of Quorn, S.A. At A, Mt Arden Creek crosses a sandstone ridge. At B (see Fig. 20.9a), headward breaching is, in geological terms, imminent; at C, the Skeleroo Creek has breached the complex sandstone ridge and is extending in the weaker sediments beyond, eroding a plain at a lower level than the original southward sloping valley, from which it is separated by a distinct escarpment up to 12 m high.

Headward erosion, it is emphasised, is an integral mechanism in the development of trellis drainage patterns. In limestone bedrock, similar subterranean capture may be responsible for similar patterns.

The effectiveness of headward erosion is not restricted to local drainage development. Sprigg,[69] for example, has suggested that the courses of rivers such as the Onkaparinga, Torrens and Sturt (and, it may be added, the Myponga, to the south) to which earlier reference has been made, are due to capture of long subsequent streams occupying fault-angle depressions and fault zones by rivers heading back from the steep west-facing scarps of the fault blocks dominating the relief of the Mt Lofty Ranges (Fig. 20.12). Close examination shows that for considerable sections the breaching rivers run along fault zones within the fault blocks, and it is not difficult to visualise antidip streams heading back and gradually capturing the consequent and strike drainage of the dip slopes (Figs 20.12a-c).

The rate at which the breaching of ridges takes place is not clear and the mechanism is not only still in train, but it has been effective over long geological periods and some of the breaches are of great antiquity.[70] However, similar features may result from other, different processes: all breaches of ridges are not due to headward erosion, and some may be caused by stream persistence and valley impression.

D. STREAM PERSISTENCE AND VALLEY IMPRESSION

In the Ridge and Valley Province of the Appalachians, in the James Range (Northern Territory), in the Zagros Mountains, in the Flinders Ranges and elsewhere, some of the most puzzling features of the drainage patterns are the breached snouts of plunging anticlines. In some localities, rivers which for the most part follow along the arcuate or U-shaped outcrops encircling the ridges of hard rock, plunge into gorges they have eroded through the ridges (Pls 20.2, 20.3, 20.4, 20.6a and b). Many of these gorges are located only a short distance from the lowland or valley, though elsewhere they

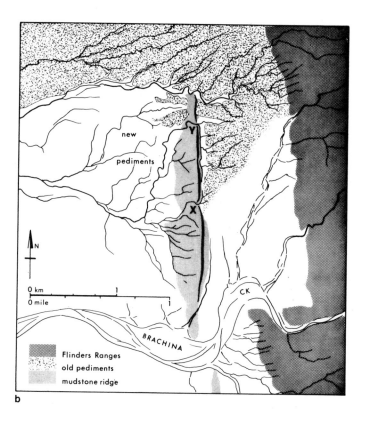

b

Flinders Ranges
old pediments
mudstone ridge

Fig. 20.10 Stages of headward breaching of a siltstone ridge near Brachina Gorge, S.A. At Y the breaching is complete; at X it has only recently occurred and the headwaters have not extended far into the plain east of the ridge; and many other streams are actively attacking the western slope of the ridge.

Fig. 20.11 Headward eroding gully in siltstone ridge, Willunga Scarp, S.A. The most obvious gully developed as a result of deforestation by European settlers in the second half of the nineteenth century, but there was clearly a wider valley extending into the ridge before this.

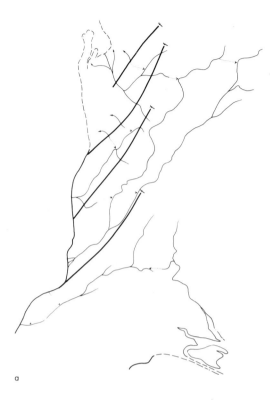

a

are a considerable distance from the termination of the resistant outcrop. In the Mern Merna Dome, west of Hawker in the Flinders Ranges, there are two breaches of the same snout of quartzite, as well as a centrally located transverse stream (Pl. 20.6a and b).

When considering the origin of such breached snouts, the three-dimensional geometry of the structures has to be considered.[71] The fact that the dome structures involved in these features are breached implies considerable erosion, and at least in many of the uplands in which such breached snouts are represented, the occurrence of surfaces of low relief high in the landscape[72] argues deep erosion. This is important as the geometry and dimensions of the folds change vertically and an attempt should be made to reconstruct the location of resistant and weak strata, and of ridge and valley, during earlier stages in the lowering of the land surface.

Thus, considering Fig. 20.13a-d, the subsequent stream follows the law of natural selection and follows around the least resistant outcrop. But as it incises its bed and lowers its valley floor, it approaches closer and closer to the divergent resistant strata included in the limbs of the fold structure. Eventually it encounters them. Because the river is confined between banks it does not necessarily merely slip off down

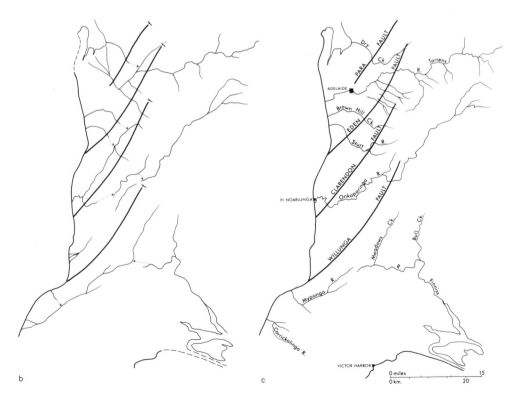

b

c

Fig. 20.12 Stages in the development of drainage patterns in the southern Mt Lofty Ranges. (*After R.C. Sprigg, "Some Aspects of the Geomorphology",* Trans. R. Soc. S. Aust., *1945, 69, pp. 277–303, by permission of the Royal Society of South Australia.*)

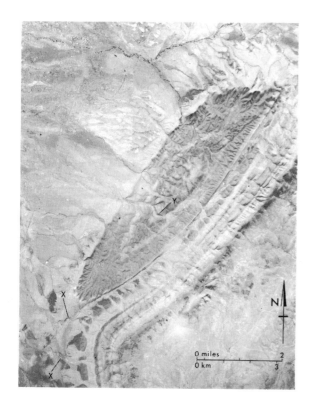

Plate 20.6a Vertical air photograph of the Mern Merna Dome, west of Hawker, central Flinders Ranges. The dome is outlined by a sandstone formation, and shales are exposed in the core. Note the transverse stream (Y) and the double breach of the southern snout (X). (*S.A. Lands Dept.*)

Plate 20.6b Mern Merna Dome from the south showing transverse stream (Y) and double breach of snout (X). (*C.R. Twidale.*)

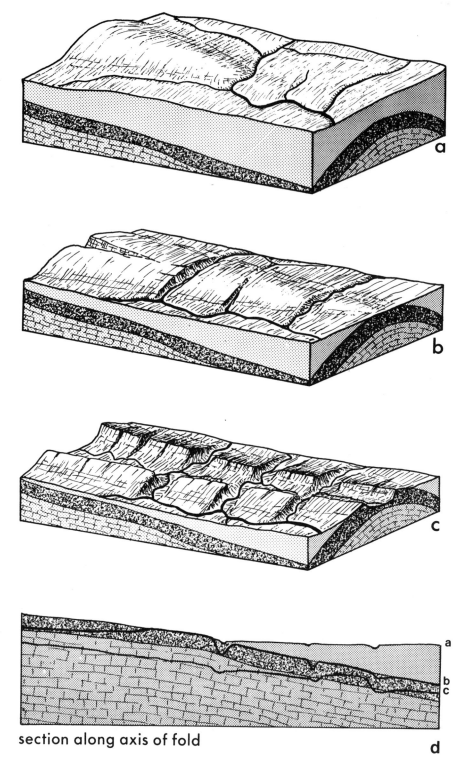

section along axis of fold

Fig. 20.13a–c Stages in the development of breached snout through persistence of streams and
impression from conformable strata.
(d) Stages a–c in section along the fold axis.

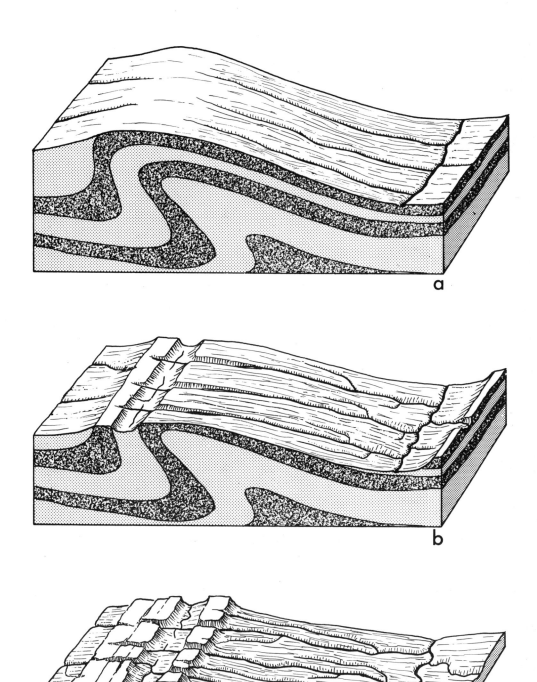

Fig. 20.14 Development of transverse drainage from conformable strata. (*After H.A. Meyerhoff and E.W. Olmstead, "Origins of Appalachian Drainage",* Am. J. Sci., *1936,* **36,** *pp. 21–42, by permission of* Am. J. Sci. *and H.A. Meyerhoff.*)

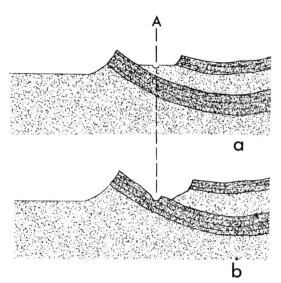

Fig. 20.15 Development of in-and-out gorges by stream impression.

the bedding plane; if the stream energy is sufficient, the stream may *persist* in its course and *impress* its bed in the resistant strata where, although it may develop intrenched meanders, it becomes essentially fixed. In this way the snout is breached. Successive tributaries may behave in similar ways (Fig. 20.13) and thus multiple breaches develop (Pl. 20.5a-c). As erosion proceeds to lower and lower levels, the limbs and snouts of the plunging anticlines "expand" and so the erstwhile breached snout becomes more and more centrally located, forming a transverse valley. If the anticlinal core is breached, the transverse stream may come to flow through two gorges separated by a lowland (Pl. 20.4).

In similar fashion transverse drainage can be impressed on resistant strata of the same fold sequence (Fig. 20.14), and strike streams eroding downwards in weak strata may erode gorges in underlying resistant rocks, forming strike gorges or, if the river was winding, in-and-out gorges (Figs 20.6, 20.15).

Thus wherever there is evidence of deep erosion of folded sedimentary sequences, and where in consequence the geometry and character of the folded strata subject to stream erosion have changed in time, persistence and impression of streams and valleys comprise a normal and expected phase in the evolution of stream patterns.

References Cited

1. EMILIE R. ZERNITZ, "Drainage Patterns and their Significance", *J. Geol.*, 1931, **40**, pp. 498–521.

2. J.W. POWELL, *Exploration of the Colorado River of the West and Its Tributaries Explored in* 1869, 1870, 1871 *and* 1872, Government Printing Office, Washington D.C., 1875, p. 160.

3. A.R. MARVINE, "The Stratigraphy of the east Slope of the Front Range", *U.S. Geol. & Geogr. Surv.* (Hayden Survey), Annual report, 1873, 1874.

4. A. STRAHAN, "On the Origin of the River System of South Wales and its Connection with that of the Severn and the Thames", *Q. J. geol. Soc. London*, 1902, **58**, pp. 207–22.

5. J.E. MARR, "The Influence of the Geological Structure of English Lakeland upon its Present Features—a Study in Physiography", *Q. J. geol. Soc. London*, 1906, **62**, pp. lxvi–cxxviii.

6. J.W. POWELL, *Exploration of the Colorado*, p. 160.

7. C.T. MADIGAN, "The Physiography of the western MacDonnell Ranges, central Australia", *Geogr. J.*, 1931, **78**, pp. 417–33, at p. 428.

8. P.F. KENDALL, "A System of Glacier-lakes in the Cleveland Hills", *Q. J. geol. Soc. London*, 1902, **58**, pp. 471–571.

9. See for example A.S. FRASER, "Wilpena Pound", Section 2.3, in M. Williams (Ed.), *South Australia from the Air*, ANZAAS/ Melbourne University Press, 1969, p. 22.

10. See for example T.G. TAYLOR, "Physiography of Eastern Australia", *Comm. Met. Bur. Bull.*, 1911, **8**; C.R. TWIDALE, "Physiographic Reconnaissance of some Volcanic Provinces in north Queensland, Australia", *Bull. Volc.*, 1956, **2**, No. 18, pp. 3–23; A. CUNDARI and C.D. OLLIER, "Australian Landform Examples No. 17. Inverted Relief due to Lava Flows along Valleys", *Aust. Geogr.*, 1970, **11**, pp. 291–3.

11. For example, P.F. KENDALL, "A System of Glacier-Lakes", pp. 471–571; F. LEVERETT and F.B. TAYLOR, "The Pleistocene of Indiana and Michigan and the History of the Great Lakes", *U.S. Geol. Surv. Monogr.*, 1915, **53**; J.K. CHARLESWORTH, *The Quaternary Era*, Arnold, London, 1957, 2 vols.

12. A.D. HOWARD, "History of the Grand Canyon of the Yellowstone", *Geol. Soc. Am. Spec. Paper*, 1937, **6**.

13. W.J. HARRIS, "Physiography of the Echuca District", *Proc. R. Soc. Vic.*, 1939, **51**, pp. 45–60, and Chapter 4.

14. See T.G. TAYLOR, "Physiography of Eastern Australia"; A.A. ÖPIK, "Geology and Palaeontology of the Headwaters of the Burke River, Queensland", *Bull. Bur. Miner. Resour. Geol. Geophys. Aust.*, 1956, **53**; C.R. TWIDALE, "Late Cainozoic Activity of the Selwyn Upwarp", *J. geol. Soc. Aust.*, 1966, **13**, pp. 491–4.

15. H.B. MEDLICOTT, "On the Geological Structure and Relations of the southern portion of the Himalayan Range between the River Ganges and Ravee", *Mem. geol. Surv. India*, 1860, **3**.

16. L.R. WAGER, "The Arun River Drainage Pattern and the Rise of the Himalaya", *Geogr. J.*, 1937, **89**, pp. 239–50.

17. F.V. HAYDEN, "Some Remarks in regard to the Period of Elevation of those Ranges of the Rocky Mountains near the Sources of the Missouri River and its Tributaries", *Am. J. Sci.*, 1862, **33**, pp. 305–13.

18. J.W. POWELL, *Exploration of the Colorado*, p. 198.

19. C.E. DUTTON, "Physical Geology of the Grand Canyon District", *U.S. Geol. Surv.*, 2nd annual report, 1880–1, pp. 60–1.

20. F.E. MATTHES, *The Grand Canyon of the Colorado River.* Commentary on "Bright Angel Quadrangle, Arizona", *U.S. Geol. Surv. Topo. Map*, Series 1903.

21. C.B. HUNT, "Geology and Geography of the Henry Mountains Region, Utah", *U.S. Geol. Surv. Prof. Paper*, 1953, **228**.

22. G.M. LEES, "Recent Earth Movements in the Middle East", *Geol. Rdsch.*, 1955, **42**, pp. 221–6.

23. E. SEEFELDNER, "Die Entstehung der Salzachofen", *Mitt. Gesellschaft Salzburger Landskunde*, 1951, **91**, pp. 153–69; A. COLEMAN, "The Terraces and Antecedence of a Part of the River-Salzach", *Trans. & Papers Inst. Br. Geogr.*, 1958, **25**, pp. 119–34.

24. C. FENNER, *South Australia, A Geographical Study*, Whitcombe and Tombs, Melbourne, 1931, p. 254.

25. F. LOWL, "Die Entstehung der Durchbruchsthaler", *Pet. Mitt.*, pp. 405–16.

26. J.B. JUKES, "On the Mode of Formation of some of the Early River Valleys in the south of Ireland", *Q. J. geol. Soc. London*, 1862, **18**, pp. 378–403.

27. G. MAW, "Notes on the Comparative Structure of Surfaces produced by Subaerial and Marine Denudation", *Geol. Mag.*, 1866, **3**, pp. 439–51.

28. A.R. MARVINE, "The Stratigraphy of the east Slope".

29. J.W. POWELL, *Exploration of the Colorado*, p. 116.

30. W.M. DAVIS, "The Rivers and Valleys of Pennsylvania", in *Geographical Essays*, Dover, Boston, 1909, Chapter 19. pp. 413–84.

31. D.W. JOHNSON, *Stream Sculpture on the Atlantic Slope*, Columbia University Press, New York, 1931; A.N. STRAHLER, "Hypotheses of Stream Development in the Folded Appalachians of Pennsylvania", *Bull. geol. Soc. Am.*, 1945, **56**, pp. 45–88.

32. A. STRAHAN, "On the Origin of the River System", pp. 207–22.

33. J.E. MARR, "The Influence of the Geological Structure", pp. lvxi–cxxviii.

34. C.G. HIGGINS, "Lower Course of the Russian River, California", *Univ. Calif. Pub. Geol.*, 1952, **29**, pp. 181–264.

35. W.D. West and V.D. Choubey, "The Geomorphology of the Country around Sagar and Katangi, M.P.; an example of Superimposed Drainage", *J. geol. Soc. India*, 1964, **5**, pp. 41–55.

36. C. Fenner, *South Australia*, p. 253; see also pp. 43–8, 78–81, 253–4 and *idem*, "The Major Structural and Physiographic Features of South Australia", *Trans. R. Soc. S. Aust.*, 1930, **54**, pp. 1–36.

37. R. Tate, "Leading Physical Features of South Australia", *Trans. Phil. Soc. Adel.*, 1879, p. lix; M.F. Glaessner, in B. Campana, "Geology of the Gawler Military Sheet", *S.A. Geol. Surv. Rep. Invest.*, 1955, **4**.

38. M.F. Glaessner and M. Wade, "The St Vincent Basin", in M.F. Glaessner and L.W. Parkin (Eds), *The Geology of South Australia*, Melbourne University Press, 1958, Chapter 9, pp. 115–26.

39. R.C. Horwitz, "Geologie de la Région de Mt Compass (feuille Milang), Australie Meridionale", *Eclog. géol. Helv.*, 1960, **53**, pp. 211–63.

40. A.S. Fraser, "Torrens Gorge" Section 2.4 in *South Australia from the Air*, p. 24.

41. W.T. Ward, "Field Excursion Guide", *Proceedings of Symposium on Geochronology and Land Surfaces in Relation to Soils in Australia*, Adelaide, 1961.

42. W.T. Ward, "Eustatic and Climatic History of the Adelaide Area, South Australia", *J. Geol.*, 1965, **73**, pp. 592–602.

43. C.R. Twidale, B. Daily and J.B. Firman, "Eustatic and Climatic History of the Adelaide Area, South Australia, Discussion", *J. Geol.* 1967, **75**, pp. 237–42.

44. A.S. Fraser, "Wilpena Pound", *South Australia from the Air*.

45. C.R. Twidale, "Chronology of Denudation in the southern Flinders Ranges, South Australia", *Trans. R. Soc. S. Aust.*, 1966, **90**, pp. 3–28, at pp. 24–5.

46. G.D. Woodard, "The Stratigraphic Succession in the Vicinity of Mount Babbage Station, South Australia", *Trans. R. Soc. S. Aust.*, 1955, **78**, pp. 8–17; B. Campana, R.P. Coats, R.C. Horwitz and D. Thatcher, "Mooloowatana Map Sheet", *S.A. Geol. Surv. Geol. Atlas 1-mile Series*, **612**, p. 6.

47. C.R. Twidale, "Chronology of Denudation", pp. 3–28; *idem*, "Geomorphology of the Flinders Ranges" in D.W.P. Corbett (Ed.), *Natural History of the Flinders Ranges*, Public Library of South Australia, Adelaide, 1969, pp. 57–137.

48. C.T. Madigan, "The Geology of the western Mac-Donnell Ranges, central Australia", *Q. J. geol. Soc. London*, 1932, **88**, pp. 672–711; B. Campana, "The Mt Lofty-Olary Region and Kangaroo Island" and "The Flinders Ranges" in *The Geology of South Australia*, pp. 3–27 and 28–45 respectively.

49. R.H. Mahard, "The Origin and Significance of Intrenched Meanders", *J. Geomorph.*, 1942, **5**, pp. 32–44; C.R. Twidale, "Interpretation of High-level Meander Cut-offs", *Aust. J. Sci.*, 1955, **17**, pp. 157–63; *idem*, "Australian Landform Example No. 3: Abandoned Ingrown Meander", *Aust. Geogr.*, 1964, **9**, pp. 246–7.

50. L.C. Ball, "The Burketown Mineral Field" *Qld Geol. Surv. Pub.*, 1911, **232**, p. 13.

51. F.W. Whitehouse, "Studies in the Late Geological History of Queensland", *Univ. Qld Geol. Papers*, 1940, **2**, No. 1, p. 50.

52. C.A. Cotton, *Landscape*, Whitcombe and Tombs, Christchurch, (2nd ed.) 1948, p. 56.

53. E.K. Carter and A.A. Öpik, "Geological Map of north-western Queensland. 4-mile Series QG 34", *Bur. Miner. Resour. Geol. & Geophys.*, 1959.

54. C.E. Dutton, "Physical Geography of the Grand Canyon District".

55. T. Oberlander, "The Zagros Streams", *Syracuse Geogr. Ser.*, 1965, **1**.

56. H.A. Meyerhoff and E.W. Olmstead, "The Origins of Appalachian Drainage", *Am. J. Sci.*, 1936, **36**, pp. 21–42.

57. C.R. Twidale, "Chronology of Denudation", pp. 3–28.

58. F. Lowl, "Die Enstehung der Durchbruchsthaler", pp. 405–16.

59. H.D. Thompson, "Hudson Gorge in the Highlands", *Bull. geol. Soc. Am.*, 1936, **47**, pp. 1831–48; *idem*, "Drainage Evolution in the southern Appalachians", *Bull. geol. Soc. Am.*, 1939, **50**, pp. 1323–55; *idem*, "Drainage Evolution in the Appalachians of Pennsylvania", *N.Y. Acad. Sci. Ann.*, 1949, **52**, pp. 33–62.

60. C.R. Twidale, "Chronology of Denudation", pp. 3–28.

61. W.M. Davis, "The Origin of Cross Valleys", *Science (N.Y.)*, 1883, **1**, pp. 325–7, 356–7.

62. F. Lowl, "Die Enstehung der Durchbruchsthaler", pp. 405–16.

63. A.N. Strahler, "Hypotheses of Stream Development", p. 80.

64. G.H. Ashley, "Studies in Appalachian Mountain Sculpture", *Bull. geol. Soc. Am.*, 1935, **46**, pp. 1395–1436.

65. A.N. Strahler, "Hypotheses of Stream Development", p. 80.

66. Ibid.

67. C.A. Cotton, *Landscapes*, p. 147.

68. I.B. Crosby, "Methods of Stream Piracy", *J. Geol.*, 1937, **45**, pp. 465–86; J.N. Jennings, "Australian Landform Example No. 18. Ingrown Meander and Meander Cave", *Aust. Geogr.*, 1970, **11**, pp. 401–2.

69. R.C. Sprigg, "Some Aspects of the Geomorphology of Portion of the Mt Lofty Ranges", *Trans. R. Soc. S.Aust.*, 1945, **69**, pp. 277–303.

70. C.R. Twidale, "Chronology of Denudation", pp. 3–28.

71. H.A. Meyerhoff and E.W. Olmstead, "The Origins of Appalachian Drainage, pp. 21–42; H.A. Meyerhoff, "Postorogenic Development of the Appalachians", *Bull. geol. Soc. Am.*, 1972, **83**, pp. 1709–27; C.R. Twidale, "The Neglected Third Dimension", *Z. Geomorph.*, 1973, **16** NS, pp. 283–300.

72. See D.W. Johnson, *Stream Sculpture*; J.A. Mabbutt, "The Weathered Land Surface in central Australia", *Z. Geomorph.*, 1965, **9** NS, pp. 82–114; C.R. Twidale, "Chronology of Denudation", pp. 3–28.

CHAPTER 21

Climatic Change and Direct Effects of Late Cainozoic Glaciation

A. GENERAL STATEMENT

If the complex, climatically induced geomorphological processes described in Part 3 were solely responsible for the landform assemblages we see about us, geomorphological interpretation would be much more simple than in fact it is, though it would also be much less challenging. But many lines of evidence show that the various climatic zones and associated processes with which we are familiar have not always been located where they are now found on the earth's surface. The range of activity of geomorphological processes has changed as the pattern of climatic zones has changed. Although the glacial zones, for instance, have expanded and contracted in time, it is not yet clear whether other zones, such as the arid deserts, have simply migrated while maintaining a more-or-less constant area, or whether they too have varied in extent. Furthermore, the present climates and hence the range of geomorphological processes at work shaping the earth's surface are rather unusual in terms of the climate of geological time, and some of the agents at work at present have not always (or even at most times) been active in the past.

Most parts of the earth's surface have evolved under the influence of more than one set of climatically induced geomorphological processes. Even in the humid tropics, signs of previous arid land processes have been detected.[1] Areas displaying landforms which have evolved under more than one climatic regime are described as *polygenetic*, that is, they are of complex genetic origin. Land surfaces of this type are also referred to as *palimpsest surfaces* and landforms which evolved under climatic conditions different from those which now obtain in the area in which they occur are said to have been *inherited*.

B. CLIMATIC CHANGES AND FLUCTUATIONS

Although a broad outline of world climatic history is now discernible, many details remain obscure. Many different, often contradictory, interpretations have been offered for the same data, though the fundamental problem in this field, as in so many others, is the shortage of facts and a surplus of speculations. But no grand design has emerged; no simple pattern of climatic change has been established; rather the picture is one of complexity and local variation.

Several types of evidence have contributed to our identification of climatic changes, and to interpretations of their sense, degree and chronology. Stratigraphic, morphological and biological evidence shows that ice ages have developed at intervals of about 240–250 million years. Glacial deposits are interbedded with nonglacial sediments, glacial features formed on nonglacial strata are buried in the geological sequence, and for brief periods cold water and life forms indicative of a cold climate appear in the stratigraphic record. All these factors attest to these brief periods of climatic aberration. At least four major periods of glaciation have been recognised—Precambrian, Siluro–Devonian, Permian and Quaternary—but together they amount to a mere 3 per cent of geological time.

The "normal", or more frequently experienced, climate of geological time was apparently warmer, drier and less differentiated zonally than it now is. The atmospheric circulation was less vigorous. Average world temperature was about 22°C (72°F) and what would today be called savanna type climate prevailed over wide areas. The polar regions were inevitably cooler than the tropics but there were no icecaps. Even Antarctica was ice-free during the early Tertiary and there is no sign of even local glaciation there until well into the Cainozoic—at the earliest in the late Oligocene or early Miocene, with the main glacial phase beginning abruptly in the late Miocene or early Pliocene.[2] But after long eons when such warm semiarid conditions prevailed over much of the land surface, the world climate changed for brief periods. The atmospheric circulation took on a new vigour, precipitation levels increased, and the average world temperature dropped to around 7°C (45°F). Zonation between the polar and equatorial regions became pronounced. Vast sheets of ice developed and spread out from regions of high latitude. Smaller glaciers evolved at high altitudes in the low latitude regions, even near the equator in New Guinea, central Africa and South America. During these same periods of refrigeration in high latitude regions, the subtropics became very dry, and deserts appeared on the earth's surface which were more arid than those known during periods of weaker atmospheric circulation. This is not to say that glaciation and aridity developed in precise contemporaneity, but rather that extremes of cold and/or aridity, which were unusual in a global and geological context, were in evidence during the complex periods we call the ice ages.

Thus, at present, in what is either part or the aftermath of the Pleistocene ice age (*sensu lato*), the tropical deserts cover approximately 15 per cent of the continents. Two large ice-sheets survive. Similarly the Permian ice age is well documented in the geological record, as is evidence of Permo-Triassic tropical deserts. It may also be noted that both these periods of climatic extremes follow close on the heels of major phases of widespread orogenic activity.

The ice ages were of brief duration, but their geomorphological effects were widespread, profound, and in some cases, enduring. For, as is described later (in Chapter 23), glacial landforms dating from Permian and Precambrian times constitute minor elements of the landscape in several parts of the world, and the geomorphological effects of the Pleistocene glaciation dominate landform assemblages over very large areas of the terrestrial world.

The climates of the major nonglacial and glacial periods were not, however, stable. In addition to the evidence provided by sediments, landforms, and flora and fauna, which occur in environments in which they could not have lived or developed, many lines of evidence point to climatic changes of great complexity within this major framework.

For example, there are three isotopes of oxygen with atomic weights of 16, 17 and 18 respectively. Of these O_{16} and O_{18} are the most common. When water which consists of hydrogen and oxygen evaporates, the water molecules which contain the lighter oxygen isotope are removed more rapidly than those with the heavier isotope. Thus the concentration of these two isotopes of oxygen in water is in considerable measure a function of temperature, for the higher the temperature, the greater the amount of evaporation and of selective evaporation.[3]

The external skeletons of many marine organisms are made of minerals which contain oxygen: many marine creatures have calcareous shells ($CaCO_3$) and others, such as some radiolaria and the extinct belemnites, have tests made of silica (SiO_2). In some there are periodic growth rings. By analysing the ratio of O_{16} to O_{18} in the shells and comparing these ratios with those modern shells from waters of known temperature, it is possible to deduce the temperature of the water in which these marine organisms lived. The shells and tests survive in fossil form in sedimentary sequences, the order of deposition of which is known, and so it is possible to reconstruct temperature curves for the period of time represented by the sedimentary sequence. The fact that calcareous shells contain carbon as well as oxygen is an added advantage, for it enables younger sediments at any rate to be dated by the C_{14} method. In the case of creatures with annual growth rings, even seasonal and annual temperatures can be detected: for instance belemnite tests enable the temperature of some Cretaceous summers—seasons dating from 70–100 million years ago—to be deduced.

By taking long cores from the sea-floor it has proved possible to reconstruct temperature graphs for quite long periods of geological time. The major temperature fluctuations of the Quaternary, for instance, emerge quite clearly (Fig. 21.1), with six cold periods during the Pleistocene.[4]

Ice is only frozen water and cores from the Greenland ice-sheet have enabled scientists to reconstruct the climatic record for a period possibly as long as 100,000 years, using

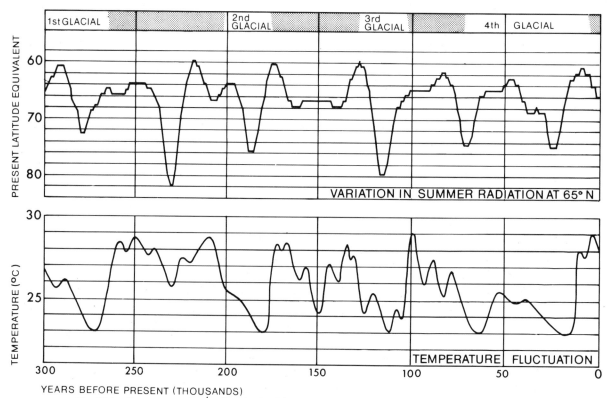

Fig. 21.1 Temperature fluctuations in the Pleistocene as revealed by $O_{16}:O_{18}$ analysis of shells in deep sea cores. *(From C. Emiliani, "Ancient Temperatures", p. 61. Copyright   1958 by Scientific American, Inc. All rights reserved.)*

normal stratigraphic methods and with oxygen isotopes as indicators of temperature. By use of very refined statistical analysis, cycles of temperature change of varied duration—2400, 400, 181, 78 and 40 years—are discernible (Fig. 21.2).[5]

Such short period cyclic changes have been identified or corroborated by other evidence. Various cycles determined by tree ring analysis or dendrochronology, which are apparently related to similar sunspot cycles, have long been known. Trees usually produce one growth ring each year, and though the thickness of the ring decreases with age, there are irregular or annual variations due to a particular seasonal condition superimposed on this regular variation. In cold areas such as Labrador, a thick annual growth ring implies a warm summer, but in a dry region such as Arizona a thick growth ring is a result of a particularly wet year. Thus, though tree rings apparently reflect climate, different factors are involved in different areas. Nevertheless, long climatic records have been constructed by means of tree ring analysis, particularly in the American Southwest where there are very old trees. An important feature is that a climatic chronology should be available from a simple count back since the growth rings are annual, though necessarily the record extends back for only the last few thousand years at most. However, many factors influence tree growth and dendrochronology provides a general guide rather than an infallible record of climate.

Some events of recorded history at the same time are explained by and corroborate these short-term and small-scale climatic fluctuations. Vines were grown in the open in several parts of southern and central England during the twelfth century A.D., indicating a climate much warmer than at present. The Swedish army crossed the frozen Baltic Sea during the 1420s; the Thames froze several times during the sixteenth century, allowing Henry VIII to ride in a carriage on the river in 1537 and Elizabeth I to hold a frost fair with oxen roasted on fires built on the ice about thirty years later: these events of recorded history indicate more severe winters than those of the present day. Literature also provides clues. In Dickens' novels, descriptions of harsh winters stem from a series of heavy snowfalls in his youth, in the early years of the nineteenth century. Similarly, the balmy summer days of P.G. Wodehouse's youth in Edwardian England form the background for many a country house party in his delightful world of make-believe.

These impressions are confirmed by direct observations. For instance, most glaciers were in retreat for most of this century up to about 1960, and particularly during the period 1900—1960. (There were exceptions, but these are explicable in terms of local heavy snowfalls, earthquakes and so on.)

Some of the evidence is anthropological. About five thousand years ago, the elephant, the giraffe and the hippopotamus, as well as large herds of domesticated cattle, lived in what is now the Sahara, indicating a moister climate. Again there is confirmation from a study of pollen spectra obtained from old soils and sediments, the stratigraphic position of which is known. Pollen analysis or palynology has been more widely used in temperate and cool regions where pollens are more readily preserved than in the oxidising conditions prevalent in the arid tropics. Pollen blown from trees and grasses settles on the land surface, or on lakes or marshes whence it sinks to the bottom and is incorporated in sedimentary layers. Microscopic in size, pollens are exceedingly tough and it is possible to identify the trees or grasses from which they derive. By examining the pollen of a given layer, the flora of the area at the time of the deposition of the layer can be established, and by comparing the pollens of successive sedimentary layers (each of which represents a certain time period), floristic and hence climatic changes can be reconstructed. Despite many possible pitfalls, palynology seems to work; indeed it is an invaluable tool by means of which the postglacial climatic sequence has been firmly established in many parts of the world (Table 21.1).

Several grades or scales of climatic change can be distinguished. First there are the changes between glacial and nonglacial climates (variation of temperature from 22° to 7°C); next between glacial periods and interglacial periods such as the present when the average world temperature is of the order of 10°C; then the various moderate variations typified by the contrast between the so-called Boreal and Atlantic type climates which affected the

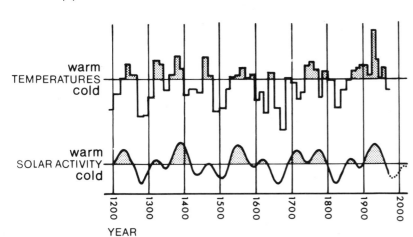

Fig. 21.2 Recent temperature fluctuations as indicated by oxygen isotope analysis of layers in core, some 1500 m long, from the Greenland ice. (*From The Observer, 2 August 1970.*)

Table 21.1
The Postglacial Climatic Sequence in western Europe and northeastern USA
as established by Pollen Analysis.*

		Maine (Deevey)		Ireland (Jessen)		Denmark (Jessen)		Swedish zones	European dates	Swedish recurrence surfaces	Climate
Postglacial	Sub-Atlantic	C3	Hemlock, Spruce return; Beech maximum	VIII	Alder-Birch-Oak	IX	Birch-Spruce-Pine	I / II	A.D. −1000 / −0 B.C.	I / II / III	Cool, Moist
Postglacial	Sub-Boreal	C2	Oak maximum; Hemlock, Beech at minimum	VIIB	Alder-Oak	VIII	Oak-Birch	III / IV	−1000 / −2000 −3000	IV / V	Warm, Dry
Postglacial	Atlantic	C1	Hemlock max.; Oak, Beech	VIIA	Alder-Oak-Pine	VII	Mixed Oak; Lime	V / VI	−4000 −5000		Moist; Thermal Maximum
Postglacial	Boreal	B	Pine maximum; Birch falls	VI	Hazel-Pine	VI	Alder-Elm	VII	−6000		Warm, Dry
Postglacial	Boreal	A3	Fir-Birch	V	Hazel-Birch	V	Pine-Birch-Hazel	VIII			
Postglacial	Pre-Boreal	A2 / A1	Spruce maximum / Birch; Nap falls	IV	Birch	IV	Birch	IX	−7000		Cool
Late-Glacial		L3	Nap higher	III	Younger Salix Herbacea	III	Younger Dryas	X	−8000		Cold; Glacial Advance
Late-Glacial		L2	Nap lower; Spruce, Birch	II	Birch	II	Birch	XI	−9000		Warmer; Allerod= Two Creeks Interval
Late-Glacial		L1	Nap high	I	Older Salix Herbacea	I	Older Dryas	XII	−10,000 B.C.		Cold; Glacial Advance

*E.S. Deevey, "Palaeolimnology and Climate" in H. Shapley (Ed.), *Climatic Change*, Harvard University Press, Cambridge, 1953, p. 275.

Atlantic coasts during the postglacial periods (Table 22.1) and the shorter term fluctuations of the last millenium.

Though for a philosophical reason which seems obscure, it was at one time considered desirable, or even essential, to explain all types and scales of climatic change in terms of one factor,[7] it is far more reasonable to suggest that many factors are involved and that they have caused the changes acting either singly or in combination. Many theories have been advanced to explain climatic change. Some call upon variations in extraterrestrial factors. These include variations in the sum total of solar radiation, either an increase or a decrease; variations in the output of part of the solar spectrum such as ultraviolet radiation; and variations in the effects due to elements contained in the earth's atmosphere, such as ozone and water vapour. These external variations may be manifested as changes of the atmospheric circulation. Some hypotheses call upon variations in terrestrial factors such as the astronomic properties of the earth (obliquity of the ecliptic, eccentricity of orbit, precession of equinoxes); variations in the latitude of a given zone, either relative, involving polar wandering, or absolute, involving continental drift, or both; variations in the distribution and elevation of mountains; variations of sealevel and hence of the distribution of land and sea and of the activity of ocean currents; and variations in the composition of the atmosphere caused by terrestrial (including human) activity, including volcanic dust and carbon dioxide.

Two examples will suffice to expose the fallacy of attempting to explain all magnitudes and types of climatic change in terms of variation in one factor. First, the theory of continental drift was originally conceived in an attempt to explain major climatic changes demonstrated in the ancient geological record. Though the distribution of Permian glaciation may in part be explained in such terms, continental drift clearly cannot be responsible for the Pleistocene glaciations, for the palaeomagnetic evidence shows that the relative positions of the continents and the zones changed but little during the Cainozoic (see Chapter 2). Second, Plass[8] has urged that the global rise of temperature evidenced for the first part of this century is due to the increased content of carbon dioxide in the atmosphere. This resulted in less shortwave radiation escaping into space, thus causing an overall heating. The increase in carbon dioxide is attributed to the use of organic fuels by man: but clearly such a factor cannot explain any climatic change which predates man, or indeed the Industrial Revolution.

Whatever their causes, however, there have been climatic changes and few areas, if any, have escaped their effects. Though climatic changes are probably typical of geological time, the Pleistocene period is of great importance from a geomorphological point of view, if only because the climatic changes characteristic of it were so pronounced, widespread and recent.

C. DURATION OF THE PLEISTOCENE

The Pleistocene comprises all but the last 10,000 years of the Quaternary Era, the last period being called the Recent or Holocene. The distinction between the Pleistocene and the Recent is purely and necessarily arbitrary. Antarctica and Greenland may be regarded as being still in the Pleistocene, because they retain their massive ice-sheets. Most tropical and mid-latitude regions were never directly affected by glaciers, and those regions which were reached by the spreading ice-sheets were deglaciated at widely varying times. The last glacier did not retreat from central Labrador until about 7000 years ago. Eastern England emerged from beneath the ice some 15,000 years ago, while Kansas has not been touched by glaciers for several hundreds of thousands of years. Thus the time boundary between the Pleistocene and the Recent or between postglacial and present interglacial (depending on whether the ice age is regarded as being over, or whether the present is viewed as merely an interlude between glacial advances) in reality varies from place to place. In order to avoid confusion, the beginning of postglacial time in northwestern Europe is regarded as marking the end of the Pleistocene. As this period in northwestern Europe began some 10,000 years ago, this is considered the duration of the Recent.

It is even more difficult to establish a date for the beginning of the Pleistocene. For a long time the Pleistocene was regarded as the last million years; it was also regarded as the age of man. Then, about ten to fifteen years ago, deep sea cores revealing their characteristic temperature curves showed the repeated cooling experienced in the oceans during the Pleistocene. The upper parts of these sections were dated by means of C_{14}; in the absence of any evidence of erosion in the cores studied, and assuming constant rates of deposition in the oceanic environments, it was thought possible to deduce the date of early Pleistocene events by comparing the thickness of sediments and extrapolating from the radiometrically dated section. On this basis,[9] the Pleistocene was reduced in length, with the maximum of the first glaciation dated about 600,000 years ago. But the radiometric dating of lavas, either interbedded with glacial moraines or overlying glaciated features such as striated pavements, has shown that, as indicated by the development and spread of glaciers, the Pleistocene began at least two million years ago in many areas, and that in some areas it may have begun more than twenty-two million years ago.[10] The problem is of course to determine the extent and duration of these earliest glaciations, which can be identified in such areas as Alaska, Antarctica and the Sierra Nevada of California. If they really mark the beginning of the Quaternary ice age and the older dates are correct, then much of the conventional later Tertiary in fact disappears and becomes part of the Pleistocene. There are possibly large errors in the datings, though many which have been duplicated check out; but in any case the Pleistocene is very much longer than was envisaged a decade ago.

D. PLEISTOCENE GLACIAL CHRONOLOGY

Sediments deposited by glaciers and by meltwaters from the ice provide a complex, yet in many instances comprehensive, history of Quaternary events and climatic changes. In many parts of the world, ground moraines have been separated and identified as being related to different periods of glacier advance. The drifts are distinguished in various ways. Some are separated by nonglacial sediments (Fig. 21.3), some of which include fossils: examples of such interglacial deposits are peat, windblown loess, riverine gravels and

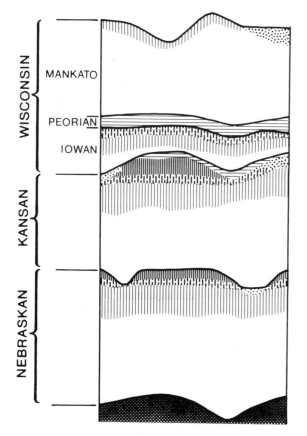

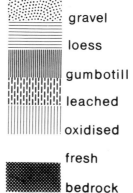

Fig. 21.3a Diagrammatic stratigraphic section in north-central Ohio showing several glacial and interglacial deposits. (*After R.F. Flint,* Glacial Geology and the Pleistocene Epoch, *Wiley, New York, 1947, p. 191.*)

gravel

loess

gumbotill

leached

oxidised

fresh

bedrock

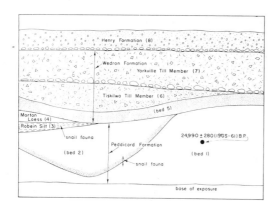

Fig. 21.3b Section exposed in roadcut in Illinois, USA, showing glacial beds and interbedded interglacial layer with fossils and windblown loess. (*After H.B. Willman, A. Byron Leonard, John C. Frye, "Farmdalian Lake Deposits and Faunas in northern Illinois",* Ill. State Geol. Surv. Circ., *1971,* **467**.)

sands deposited in valleys cut into the drift. Some morainic deposits were weathered and eroded before the arrival of the next glacier. Successive glaciers followed slightly different paths, and their deposits are not only superimposed one on the other, but also display different patterns in plan.[11] The suites of pebbles and boulders (erratics) vary in composition from one moraine to the next, for they reflect the source and path (the provenance) of the glacier in which they occur. Thus, treated statistically and making due allowance for contamination from earlier glaciers and moraines, the erratics provide an invaluable clue to the source area and also serve to distinguish drifts and associated ice-sheets.

The use of such evidence, connected by the investigation of hundreds of sites and by correlating and linking the sequences exposed in them, has served to distinguish four major periods of ice advance in the northeastern United States (Table 21.2). Various minor periods of advance and retreat have been recognised in relation to the last major glacial phase there (Table 21.3). In the European Alps also, four phases of glacial advance and recession have been recognised (Table 21.2), but in northern Germany, Denmark and Britain only three have been identified with certainty

Table 21.2
Pleistocene Chronology in North America, northern Europe, the European Alps, northwestern Europe and European USSR.*

N.America	N. Europe	Alps	N.W. Europe	European USSR
Wisconsin	Mecklenburgian	Würm	Weichselian	Waldai
Sangamon	Neudeckian	R/W	Eemian	Mikulino, Mgi
Illinoian	Polandian	Riss	Warthe	Moscow
Yarmouth	Helvetian	M/R	?	Odintzovo
Kansan	Saxonian	Mindel	Saale	Dnieper
Aftonian	Norfolkian	G/M	Holstein	Lichwin
Nebraskan	Scanian	Günz	Elster	Oka
			Cromerian	Belovezhsky
		Donau	Menapian	(Apsheron and
			Waalian	Akchagyl
		Biber	Eburonian	formations)
			Tiglian	
			Praetiglian	

*R.G. West, *Pleistocene Geology and Biology*, Longmans, London, 1968, p. 219.

(Table 21.2). As in North America, it has proved possible to subdivide the latest major glaciation into various substages (Table 21.3).

In mountain regions the glacial sequences tend to be more complex. This is possibly because these areas are more prone to tectonic movements which disturbed the glaciers and related deposits as well as the rate of glacier flow. More complex chronologies with more phases of glacier advance and retreat are evidenced in such areas. Possibly because the valley glaciers were less massive than the ice sheets, they were more sensitive to climatic changes: unlike the ice-sheets which, as it were, enveloped themselves in a protective blanket of cold of their own making, the smaller valley glaciers were vulnerable to all climatic fluctuations. Moreover, as valley glaciers occurred in lower latitudes, radiation would have been more intense than in areas closer to the poles.

Table 21.3
Detailed Chronology of the later Pleistocene in northeastern North America.*

Mankato Advance	11,400 years BP
Cary Advance	15,200 years BP
Tazewell Advance	18,000 years BP
Late Wisconsin	25,000 years BP
Early Wisconsin	60,000 years BP

*After H.T. Stearns, "Eustatic Shorelines on Pacific Islands", pp. 3–16.

Whatever the reasons, upland chronologies tend to be more complex. Thus, four major periods of glaciation have been recognised in New Zealand from the late Pleistocene (Table 21.4), with one more known from the earlier Pleistocene. Several stages of the later glaciations have been recognised.[12] In the Western Cordillera, also, many separate glacial phases have been recognised. In the Great Basin, for instance, some ten glaciations have been postulated (Fig. 22.4a and b), though they have been correlated with the classic sequences of the Great Plains and northeastern USA.[13] A complex local chronology has also been established in the Sierra Nevada of California (Table 21.5).

Although there are disputes and controversies concerned with local sequences and with the events of the earlier Pleistocene times in glaciated areas, Pleistocene glacial chronology is crystal clear compared to that of the nonglaciated tropical regions. For example, there is general agreement that glaciations were synchronous on both sides of the Atlantic. The climatic changes of the later Quaternary, including periods of glacial advance during the later Pleistocene, were broadly synchronous in northern and southern hemispheres,[14] though the position with regard to the earlier Pleistocene in

the two hemispheres remains obscure because of dating problems. But in the tropics it is not yet possible to match chronologically comparable climatic and geomorphological events on the polar and equatorial margins of the Sahara; and thus it is not possible to say with any confidence whether, during say the glacial periods of the Pleistocene, the subtropical deserts were drastically reduced in area or whether they remained constant in area but migrated toward the equator.

The reality of climatic changes of Quaternary date in the tropical regions has long been recognised. Evidence of pluvial periods (that is, periods of higher rainfall of sufficient duration to have significant geological effects) was identified early in this century. These pluvial phases were suggested by evidence of high shorelines surrounding the present enclosed lakes of central Africa, such as Lake Victoria,[15] and by geological and archaeological findings in the Fertile Crescent, that is, in northeastern Africa and southwestern Asia. More recently it has been recognised that many present semiarid tropical areas were once truly arid, and that even the humid tropics, the selvas, have not been immune to the effects of climatic change.

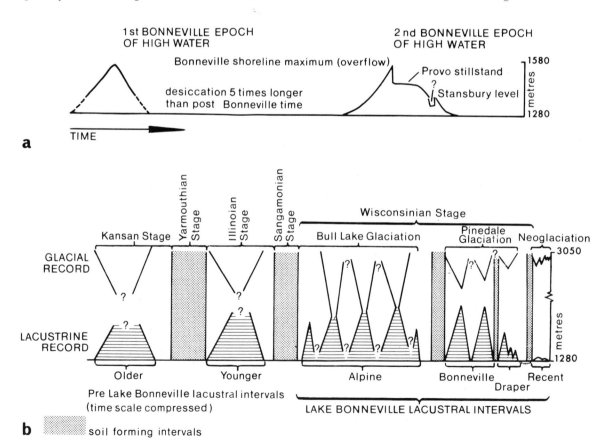

Fig. 21.4 Interpretations of the Lake Bonneville history: according to Gilbert (a); according to Morrison (b).

Table 21.4
Pleistocene Sequence in the New Zealand Alps.*

West Coast	Sequence of Glaciations	Canterbury Glacial Advances
Kumara–3 Advance		Poulter Advance
2 ⎫ ⎬ Kumara–2 Advance 1 ⎭	Otira Glaciation	Blackwater Advance Otarama Advance
Kumara–1 Advance	Waimea Glaciation	Woodstock Advance
Hohonu Advance	Waimaunga Glaciation	? ⎫ Avoca
Porika Glaciation	Porika Glaciation	? ⎭ Glaciation

*After R.P. Suggate, "Late Pleistocene Geology".

There has always been a strong tendency to relate the pluvials to periods of glacial advance in higher latitudes. Possibly this assumption was because of work in North America and Europe which had rightly been widely admired and was well known. Gilbert,[16] in his classic monograph on Lake Bonneville, unravelled the Pleistocene climatic chronology of that part of the Great Basin, recognised various high stands of the ancient lake, and correlated them with periods of glaciation (Fig. 21.4a). In similar fashion, Penck and Brückner[17] recognised the classical four-fold division of the Alpine glaciation. Part of their evidence derived from periods of cut and fill in the river valleys which radiate from the Alps, and which are represented by series of terraces (Fig. 21.5a). By tracing the terrace gravels into the glacial moraines, they were able to show that periods of high river flow and gravel deposition coincided with periods of glacial advance, and conversely, that periods of glacial recession were represented in the valleys below by dissection of the flood plains and deposition of, for example, loess on the land surface. Similar sequences occur elsewhere in Europe (Fig. 21.5b). Thus there seemed to be a rational basis for correlating pluvial and glacial periods, especially as such a chronology provided a simple and easily identifiable basis for archaeological studies.

In the Great Basin of North America, Gilbert's classic studies of the Bonneville basin have been generally confirmed, though greatly refined and detailed by later work. The early work of Russell in the adjacent Lahontan basin has also been extended.[18] The major feature of this work (some of which is referred to below) is that high stands of the lakes are firmly correlated with glacial phases in the Rocky, Wasatch and other mountain ranges, for some of the tills are claimed to show signs of having been deposited in lake waters.

Table 21.5
Pleistocene Chronology in the southern Sierra Nevada.*

Age	Correlation	Glacial Deposits
Recent...............	Neoglaciation	⎰ Matthes till ⎱ Recess Peak till
Pleistocene	⎧ Wisconsin ⎨ ⎩ Correlation with mid-continental sequence uncertain	⎧ Hilgard till ⎪ Tioga till ⎨ Tenaya till ⎪ Tahoe till ⎩ (Sawmill Canyon basalt, 90,000+90,000 years) ⎧ Mono Basin till ⎪ Till (?) on Mammoth Mountain (less than 370,000 years) ⎪ At least two possible glaciations as yet not fully described in the ⎪ literature (Bishop tuff, 700,000 years) ⎨ Sherwin till ⎪ McGee till ⎪ (2.6×10^6-year-old basalt) ⎪ (2.7×10^6-year-old quartz latite) ⎪ Deadman Pass till ⎩ (3.1×10^6-year-old andesite)

*After R.P. Sharp, "Semiquantitative Differentiation", p. 73, *by permission of* J. Geol. *and the University of Chicago Press.*

O.D. — Older Deckenschotter
Y.D.   Younger   „
H.   High Terrace gravel
L.   Low   „   „

Fig. 21.5a Section across the Inn Valley near Scharding, Austria, showing terrace sequence. (*After W.B. Wright,* The Quaternary Ice Age, *Macmillan, London, 1914, p. 132.*)

Thus correlation of glacial and pluvial periods is upheld by many workers though there are complications.[19] Indeed there is a suspicion, as Wayland[20] has said, that correlation of pluvials with periods of glacial advance was so attractive because of its convenience and simplicity that the wish became father to the thought. And there are considerable grounds for suggesting that this relationship does not hold everywhere. For instance, the glacials were times of low sealevel and the interglacials of higher sealevel (though as is discussed below, how high is controversial), so that ancient Pleistocene strandlines should provide morphostratigraphic markers with which climatically diagnostic deposits and forms are correlated. If the glacial/pluvial and interglacial/arid phase relationship is valid, then in all cases desert sand dunes should be related to higher stands of the sea. Yet Fairbridge[21] has claimed that sand ridges extend beneath sealevel in the northwest of both Western Australia and West Africa, which areas are assumed to have been tectonically stable during the relevant time period.

Some of the difficulties and contradictions of Quaternary climatic chronology in the low latitude areas are real: the climatic changes were not simple in their pattern, and evidence for both cool and warm pluvials has been found, for instance. The problems arise partly from the lack of a clear-cut, extensive and continuous sedimentary record. In glaciated regions, the moraines and associated fluvioglacial deposits form an essentially continuous and extensive blanket over the landscape. Correlation is comparatively straightforward. In the tropics, erosion and deposition occur simultaneously on different parts of the land surface, and erosion and deposition take place at different times at the same site. The sedimentary record is fragmentary. Moreover, under the oxidising conditions which prevail, the preservation of fossils is rare.

Furthermore, it is difficult to be sure of the causes of the observed effects because of what might be called the oasis influence. Does a sedimentary sequence in the Nile or in the Murray valley record the climatic conditions of the area of deposition, including the adjacent (now arid) areas, or does it reflect the prevailing climates of the source region in central Africa or eastern Australia respectively, or worse still from an interpretational point of view, is it an amalgam

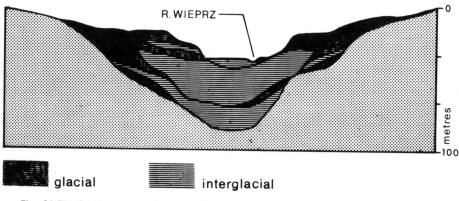

Fig. 21.5b Section across tributary of Vistula near Lublin, Poland, showing glacial and interglacial sequence in a Pliocene valley. (*After A. Holmes,* Principles of Physical Geology, *Nelson, Edinburgh, (2nd edition) 1965, p. 681.*)

of both? Likewise desert uplands receive higher rainfall than the surrounding plains, but from time to time run-off spreads out from the upland on to the plains, possibly giving quite an erroneous impression of the climate of those particular plains if the vegetation of the time is later reconstructed.

Another difficulty is that climatic chronologies have not been constructed on the basis of field evidence, but on what the climates should have been, or what a particular worker thought they should have been, assuming certain theories concerning the climatic pattern of time. Thus Zeuner[22] has applied the Milankovitch temperature curves to the problem. Others have assumed a simple migration of all climatic zones toward the equator. Others have assumed an expanded arid zone during the glacials.[23] Still others[24] believe that the pluvials of the poleward fringe of the deserts were contemporaneous with the glacials of higher latitudes, but that pluvials of the equatorial fringe were coincident with the interglacials (migration of deserts rather than expansion and contraction). Fairbridge considers the sequence to have been very complex. The monsoons were more extensive during the interglacials, bringing warm pluvials to the mid-latitudes, where two wet and two dry cycles accompanied every glacial/interglacial oscillation.

Thus, although in many areas there is a broad correlation between glacial and pluvial and between interglacial and dry phases, there are exceptions and many complications. But there is no doubt that climatic changes affected the tropics, nor that these changes left their imprint on the landscape. Both in glacial and in extraglacial regions, the interpretation of landforms rests more and more on palaeoclimatology, and throughout the world, landforms developed during the Pleistocene are of great significance.

E. DIRECT EFFECTS OF THE PLEISTOCENE GLACIATIONS

1. General Statement

The Pleistocene is distinguished as a geological period by marked changes of climate, and most obviously by the development and spread of glaciers at high latitudes and altitudes. It is above all an ice age. But it would be wrong to think of the Pleistocene or the Quaternary, of which the Pleistocene forms the major part, as a time period when only exogenetic forces were at work shaping the earth's surface, for the crust remained at least as active as in many previous geological periods. For instance, though the volume of volcanic rocks extruded does not compare with that which poured out of the earth's interior during the Tertiary, volcanoes were active during the Quaternary. Volcanic lavas are interbedded or associated with glacial moraines in the Sierra Nevada, in the Canadian Rockies and in Alaska. The Banks Peninsula craters in New Zealand are of Pleistocene age, and the glacial age of Mt Garabaldi has already been mentioned (see Chapter 6). Mt Schank and the various craters in the Mt Gambier area of the southeast of South Australia were formed only a few thousand years ago and Mt Gambier was active only some 1400 years ago.

Faulting has also been active in the Pleistocene. The Cadell Fault Block (Chapter 4, Fig. 4.25) rose and diverted the then River Murray during late Pleistocene times. In New Zealand, the Alpine Fault in the South Island has been active, and indeed the major uplift of the Southern Alps, the Kaikouran orogeny, occurred during the middle Pleistocene and predated four major glacial phases in that island.[1]

Thus the Pleistocene is not merely a climatic event or series of events. Nevertheless climatically induced geomorphological processes and events (particularly the spread of huge ice-sheets) distinguish the period from the rest of the Cainozoic.

The two major areas of glacier development were around the north and south poles. Since it was engulfed during the middle or late Tertiary, Antarctica has remained occupied by ice.[25] There is no evidence that the Antarctic ice-sheet was significantly smaller during the Quaternary. The ice-sheet has at times been substantially thicker, and it has been marginally more extensive in the past; but it has never melted away. In the northern hemisphere there was no single ice-sheet as there was in the south. Ice-sheets developed on various land masses located in the high northern latitudes, and they spread south from these centres. Except for various small glaciers in North America and northern Europe, the only major body of ice remaining in the northern hemisphere is in Greenland. Yet during the Pleistocene, ice-sheets extended to about 37°N in the Mississippi valley and almost as far south as the Thames and the Mendip Hills in England (Figs 21.6, 21.7). Moreover, such advances occurred not once but several times: multiglaciation is well evidenced in many areas.

Though there is argument as to the precise duration of the Pleistocene, even on the most generous interpretation of the data it is a very brief moment of geological time. But because of its recency and the dramatic climatic changes which took

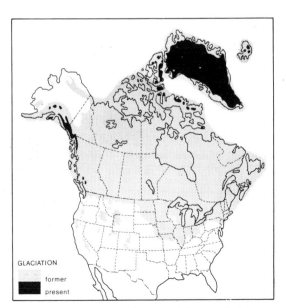

GLACIATION

■ former

☐ present

Fig. 21.6 Limits of present and former glaciation in North America.

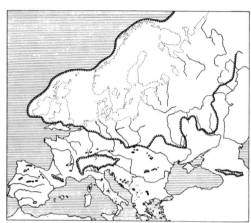

Fig. 21.7 Limit of maximum glaciation in Europe.
(*After R,G, West*, Pleistocene Geology and Biology,
Longmans, London, 1968, p. 220.)

place, Pleistocene chronology looms large in any interpretation of landforms. As Gilbert has said, "When the work of the geologist is finished and his final comprehensive report is written, the largest and most important chapter will be on the latest and shortest of the geological periods."

2. Glaciation and Deglaciation

Antarctica and Greenland remain buried under vast sheets of ice. Our knowledge of their subglacial morphology is very general, and is inferred from seismic investigations. But deglaciated landscapes are exposed to view in many parts of the northern hemisphere and to a lesser extent in places around the margin of Antarctica. Rather more than half of the total area of North America was formerly blanketed by ice (Fig. 21.6), as were large areas of northwestern Europe (Fig. 21.7). Smaller valley glaciers affected parts of the Himalayas, the European Alps, the Andes, New Zealand and Tasmania, New Guinea and the Kosciusko region within Australia, as well as upland areas of central Africa.

The changes wrought by glaciation in some of these areas were dramatic, though nowhere total. Thus, while in north-western Europe and northeastern North America the various erosional and depositional landforms described in Chapter 17 are much in evidence and at first glance dominate the landscape, some major features of the relief predate the Quaternary. Over much of the Labrador Peninsula and the North American Arctic generally, the landscape is dominated by a surface of low relief: it is a high plain out of which glaciers have gouged depressions and valleys, and above which stand residuals scoured by the ice, but it is nevertheless a surface of preglacial origin which is comparable with the Palaeozoic–Tertiary peneplain of Fennoscandia[27] and also parts of arctic North America.[28] Throughout the European Alps, the Alaskan uplands and the Himalayas, the sharp-edged crests and pointed summits of the glaciated mountains display a notable accordance of summit level. Linked together in the imagination they form not a horizontal

surface, but rather a broadly arched surface. This surface is called the *gipfelflur*[29] in the European Alps. It is said to represent not so much an ancient peneplain related to a single baselevel and now uplifted and greatly dissected, but a surface of low relief resulting from the coalescence of several minor surfaces, each of which is related to different baselevels according to the catchment involved, and was upwarped in late Tertiary and Quaternary times.

In glaciated uplands, the vast majority of glaciated troughs are modified river valleys, and their pattern reflects structural trends exploited initially by river erosion. The major outlines of the chalk uplands in eastern England remain as they were in the late Tertiary. They are modified here and there by the deepening of pre-existing valleys or by the creation of new waterways (see below), with boulder clay masking their rounded outlines, filling the valleys and obliterating other features in places, but they remain intrinsically similar to chalk uplands in southern England and northern France which were not affected by glaciers.

In still other areas, the cover of glacial ice prevented weathering and stream erosion; the ice-sheets had a protective influence. Thus the Greenland basin seems at present to be subject to little or no glacial scouring. This may appear anomalous in view of the proven thicknesses of ice present above the rock surface, but the evidence[30] suggests that glacial erosion occurs only in the coastal regions where the ice pours out through passes in the coastal ranges to reach the sea.

Nevertheless glacial landforms are prominent in many areas. Glacial features not only give a distinctive character to many regions of northwestern Europe, northern North America and the European Alps, but the deposition of glacial debris over considerable areas of eastern England, northwestern Europe and northeastern USA is responsible for the existing build-up above present sealevel in those areas.

3. Proglacial Lakes and Drainage Modifications

One of the most interesting and widespread direct effects of glaciation was the blocking and diversion of drainage by ice (see also Chapter 20). Most of these diversions were ephemeral but some have persisted.

Modern glaciers in Alaska, western Canada, Greenland, Iceland and Switzerland have impounded lakes between their margins and the declivity of the land (Pl. 21.1) Such lakes (*ice-dammed marginal lakes* or *proglacial lakes*) are deeper and larger than the meltwater ponds found in summer on the glacier surface, or the ponds impounded between glaciers and recently abandoned moraines.

Several such proglacial lakes have been described from Iceland.[31] For instance, Graenalon is 20 km^2 in area today (Fig. 21.8a). The Britannia So or lake (Fig. 21.8b) is an example from Greenland and others have been described from the Himalayas, the Caucasus and the Rocky Mountain regions. Perhaps the best known example is or was the Märjeelen See, dammed by the Aletsch Glacier in the Bernese Oberland of Switzerland (Fig. 21.9). However, this lake which attained a maximum length of 1600 m and a depth of 78.5 m was reported to be drained in 1965.[32]

No fewer than sixty proglacial lakes which either exist now or are known to have been impounded in historical times, have been described from Alaska.[33] The largest, Lake George, is 20 km long and 6 km wide. Several, such as Lake George and Tazlitwo Lake, display terracettes cut in unconsolidated debris which mark higher stands of the shore. Several are known to drain regularly. One, North Fork Lake (Fig. 22.10a), located between the North Fork and Wolf Creek glaciers, no longer exists but has left behind clear evidence of its former presence.[34] During its advance some 170 years ago, the North Fork Glacier failed by about 4 km to reach confluence with the Wolf Creek Glacier, which occupies the trunk valley (Fig. 21.10a). In the lower sector of the ice-free tributary valley, a lake was impounded by the Wolf Creek Glacier. It was 2·5 km long and 0·6 km wide. Its site is clearly marked by the well-bedded silts laid down in a lake, and by a cliff and terrace which mark its highest stand. The terrace is about 22 m wide where it is cut in till, and 5 m where cut in bedrock. The North Fork River poured sediment into the lake, and the coarser fraction was built out in a delta which eventually occupied about half the area of the lake.

Such observations of modern proglacial lakes are significant, because, as Sharp[35] has shown, when such water bodies drain, lake floor deposits, deltas and shoreline features are left behind as clear indications of former lacustrine conditions. Such sedimentary and morphological features have been located in many areas which formerly were marginal to ice-sheets, where the ice has flowed across or against the declivity of the land and the regional drainage direction, thus impounding large bodies of water (see for instance Fig. 21.10b).

Many of the best known proglacial features are found in North America, the northeastern part of which was depressed beneath the weight of ice so that the slope of the land (and the rivers) was then the reverse of the present slope. For instance, in late Wisconsin times, the Hudson–Lake George–Champlain corridor was occupied by a tongue of ice which at one stage terminated just north of the present Tri-cities complex of Albany–Troy–Schenectady. Lake Albany was impounded in this broad valley (Fig. 21.11) and various small streams carried into it debris which was deposited partly as deltaic deposits and partly as bottom sediments. When the ice withdrew northwards it eventually opened the valley to the rising sea—the Champlain Sea—extending up the St Lawrence lowland. The present Lake George which occupies a graben (Fig. 4.8b), is a remnant of the Champlain Sea, which flooded into the northern part of the corridor about 11,000 years ago.[36]

The Pleistocene ice deranged the drainage of much of northeastern USA. Formerly, much of Ohio and Virginia was drained by the Teays River, which ran west–northwest before swinging south toward the Gulf of Mexico. The

Plate 21.1 The Tiedemann Glacier, British Columbia. Note cirque glaciers on adjacent uplands on south side of main valley, hanging valleys, medial moraines and two small proglacial lakes in foreground. (*Dept of Lands, Forests and Water Resources, B.C. Photo No. B.C. 1414:64.*)

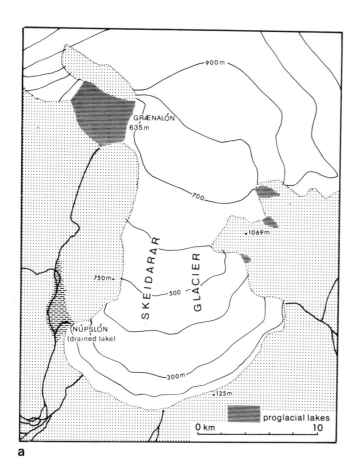

a

Fig. 21.8a Graenalon, a proglacial lake in Iceland. (*After S. Thorarinsson, "The Ice-dammed Lakes", pp. 216–42.*)

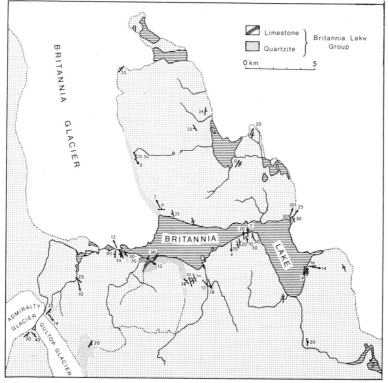

b

Fig. 21.8b Britannia Sø, a proglacial lake in North Greenland. (*After R.H. Hamilton et al., "British North Greenland Expedition 1952–4; Scientific Results", Geogr. J., 1956, 122, pp. 203–37.*)

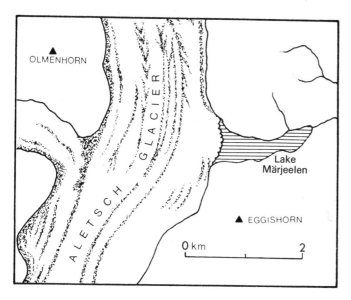

Fig. 21.9 The Märjeelen See (Lake Märjeelen)
and the Aletsch Glacier, Switzerland.

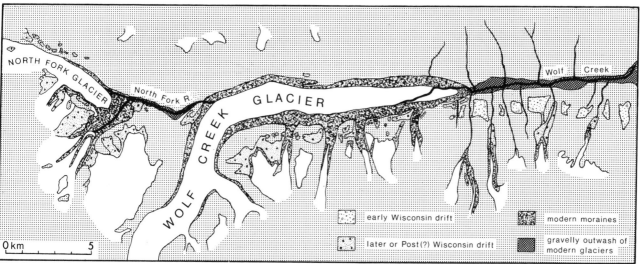

Fig. 21.10a The Wolf Creek Glacier, Alaska, and related glacial features. (*After R.P. Sharp, "The Wolf Creek Glaciers", pp. 26–52. Reprinted from the* Geographical Review (*Vol. 37, 1947*), *copyrighted by the American Geographical Society of New York.*)

Fig. 21.10b Field sketch of beach berms (foreground) and other shoreline features now standing well above the present level of Lake Wakatipu, South Island, N.Z. The berms which stand about 1 m above the intervening swales, are located about 15–20 metres above lake level near the mouth of the valley of Collins Creek. On the far or eastern side of the lake, at the foot of the Remarkables, there are benches indicating other high stands of the lake.

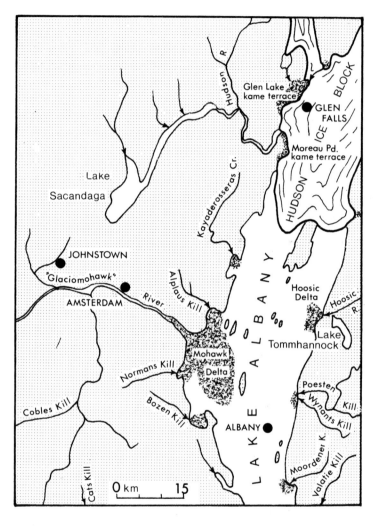

Fig. 21.11 Proglacial Lake Albany and associated features. (*After R. La Fleur, "Glacial Lake Albany", in Fairbridge (Ed.),* Encyclopaedia of Geomorphology, *Van Nostrand Reinhold Co., copyright 1968 by Litton Educational Publishing Inc.*)

ancestral Mississippi was a tributary of the Teays. During the ice ages, the old valley was blocked and the drainage was diverted to form what is now the Ohio system, but the course of the old Teays is traceable over long distances.[37]

One of the most complex histories belongs to the present Great Lakes and their precursors. During late Wisconsin times, huge volumes of meltwater gathered around the ice margin, trapped between the ice front and the land which sloped down towards Hudson Bay because of the weight of the ice-sheet. One large water body, known as Lake Agassiz, at various times extended over some 500,000 km² (200,000 square miles) of Manitoba, Saskatchewan, Western Ontario, North Dakota and Minnesota. Others accumulated on the southern margin of the ice-sheet in large preglacial valleys which were greatly deepened and broadened by glacial scour, and which became the basins now occupied by the Great Lakes. Though they probably have histories which extend back through the Pleistocene, the earliest known stages

of development are thought to have taken place no earlier than about 15,000 years B.P. Even their recent history is complex,[38] as indicated in Fig. 21.12, but in general the lakes overflowed southward into the Mississippi and Hudson systems until just under 12,000 years ago when ice retreat opened a northeastern outlet to the St Lawrence.

Proglacial lakes also developed in what are now the Great Plains (Fig. 21.13) and in the Western Cordillera of the USA, where lakes Missoula and Coeur d'Alene formed through glaciers blocking old river valleys. Overflows from there and from the Great Lakes system formed the ancestor of the present Mississippi drainage system (Fig. 21.14). In all, over fifty such proglacial lakes have been described from North America alone.

A number of the classic accounts of proglacial lakes and related features have come from Britain. For instance the so-called parallel roads of Glen Roy (Pl. 21.2) constitute some of the clearest examples of shorelines associated with pro-

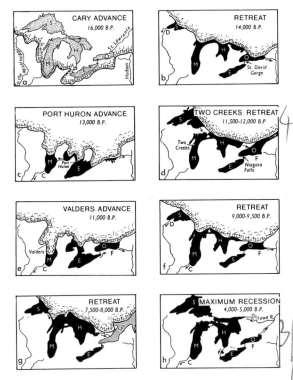

Fig. 21.12 Stages in the late Pleistocene development of the Great Lakes. Dates in years B.P. (*After J.L. Hough,* Geology of the Great Lakes, *pp. 284–96.*)

glacial lakes. Described by Jamieson[39] in 1863, the roads are in fact narrow shore platforms which stand at elevations of 351 m, 327 m and 258 m and represent successive stands of a lake which occupied the valley in late glacial times. P.F. Kendall's[40] account of the proglacial lakes and associated spillways or overflow channels in north Yorkshire provided essential foundations for the recognition of such lakes (Fig. 21.15). For instance, the direct spillway which carried meltwaters across the North York Moors was Newton Dale. It emptied into a lake occupying the Vale of Pickering which was blocked at its eastern end by the Hessle (late Wisconsin) ice. Where Newton Dale emptied into this lake, it deposited a delta on which the town of Pickering now stands. The gravels and sand of the delta pass distally into finer silts and clays which gradually settle in such proglacial lakes and which can be seen in suspension in modern examples. Any free sand settles first, followed by the slower-settling platey silt and clay particles. With the next (annual) influx of sediment the process is repeated so that the lake floor sediments come to consist of very fine lamellae, each of which is composed of relatively coarse sand below fine silt. Such fine strata, which in many cases constitute annual layers and thus offer possibilities of dating by direct counting,[41] are called *varves*.

In Europe, the most extensive system of glacial spillways is that of the North German *urstromtäler* (Fig. 21.16) which extend from the North Sea across the north European plain into the USSR. They were formed by meltwaters from the Baltic ice-sheet which were impounded between the ice and the northern slope of the north European plain. The meltwaters flowed around the ice margin and the highly concentrated pattern of the *urstromtäler* reflects successive retreat stages of the ice flow. They provide ideal channels for the canals which form part of the great commercial waterway

Fig. 21.13 Proglacial lakes in Montana and North Dakota: 1. Lake Cutbank; 2. Lake Great Falls; 3. Lake Musselshell; 4. Lake Jordan; 5. Lake Circle; 6. Lake Glendive. (*After W.D. Thornbury,* Regional Georphology of the United States, *Wiley, New York, 1965, p. 294.*)

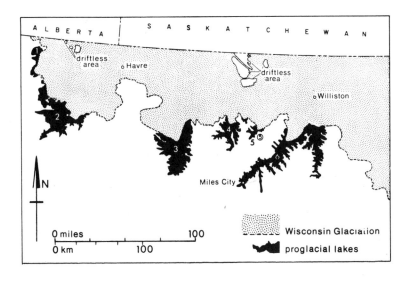

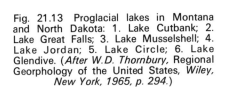

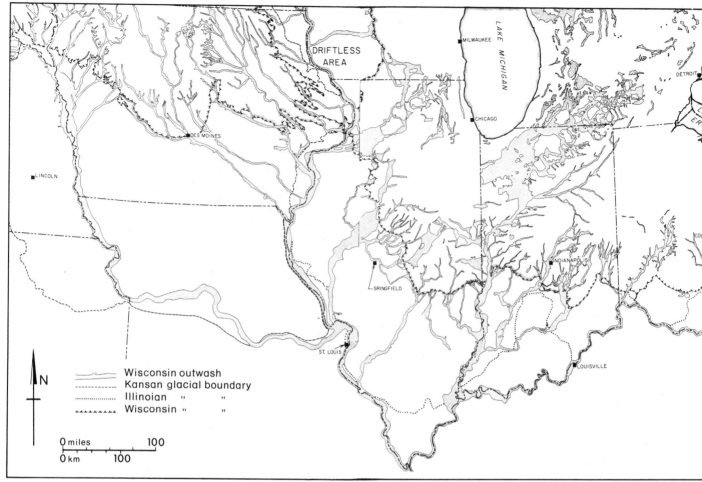

Fig. 21.14 Glacial overflows or sluiceways and associated valley and outwash plains in north central USA. (*After W.D. Thornbury,* Regional Geomorphology, *pp. 224–5.*)

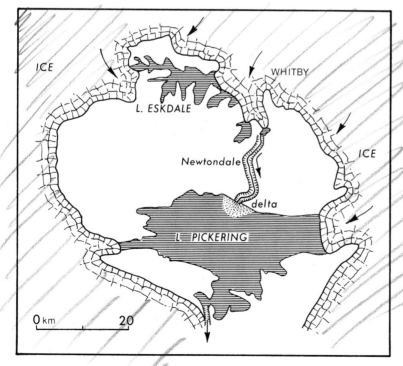

Fig. 21.15 Proglacial lakes and overflow channels in North Yorkshire. (*After P.F. Kendall, "A System of Glacier Lakes", pp. 471–571.*)

Plate 21.2 The parallel roads of Glen Roy, Inverness-shire are narrow benches, 12–15 m wide and following round the contours at elevations of 351 m, 327 m and 258 m. They mark separate shorelines and beaches of a proglacial lake which occupied the valley during late Pleistocene times. (*Inst. Geol. Sciences, Crown copyright reserved.*)

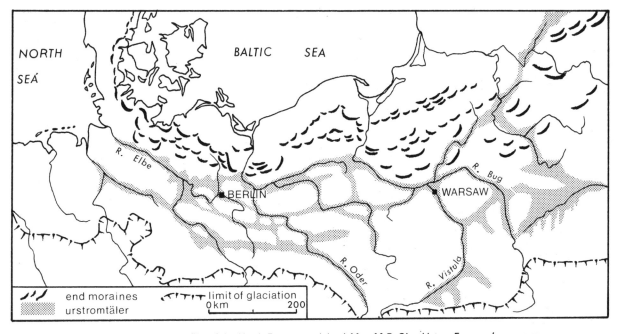

Fig. 21.16 The *urstromtäler* of the North European plain. (*After M.R. Shackleton,* Europe, *Longmans, London, (6th edition) 1955, p. 245.*)

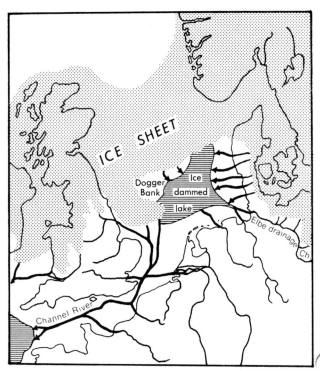

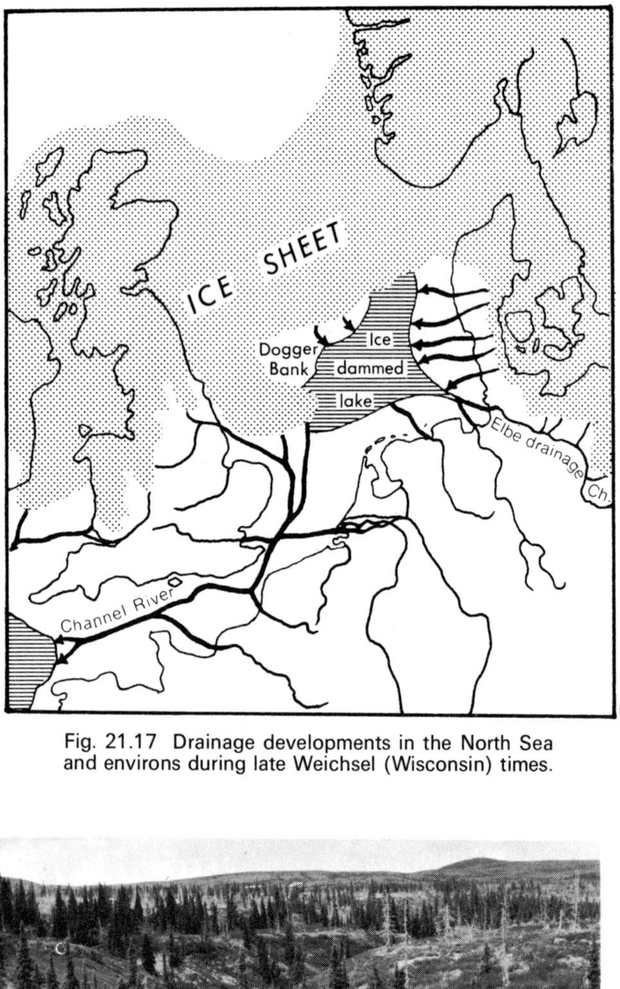

Fig. 21.17 Drainage developments in the North Sea and environs during late Weichsel (Wisconsin) times.

running across Europe from west to east. The *urstromtäler* flowed westward into a great lake which occupied much of the area of the present North Sea (Fig. 21.17), and then via the so-called "Channel River" into the Atlantic.

From Siberia, Kartashov[42] reports that geomorphological evidence indicates that Lake Krasnoye is a relic of a proglacial lake formed when the Anadyr River was forced southward by a glacier from the Pekulney Range and was impounded behind a barrier of moraines and fluvioglacial deposits.

Though in many cases the evidence for the former presence of proglacial lakes is incontrovertible, in some cases the nature and origin of the spillways or overflow channels are open to dispute. Some are certainly of curious morphology. They are generally trough-shaped in section (Pls 21.3a, b), and though some are sinuous and winding, they have no tributaries. Their characteristic feature is that they do not fit into any "normal" drainage pattern. Many display humped profiles and have large depressions in their floors, which are now filled with debris[43] (Fig. 21.18). The origin of these profiles, plus a scarcity of evidence for the lakes from which they allegedly drained, has led Sissons[44] to suggest that many of these channels are not due to overflow but rather to the flow of meltwaters under stagnant ice.

Some channels are unquestionably of this character. In central Labrador, for instance, some have no intake but were simply initiated on a hillside where the meltwater reached the subglacial floor.[45] Moreover, observations and historical records of modern proglacial lakes in Iceland[46] show that in that country they drain frequently, but always under the ice, never around or over adjacent rock spurs. However, in Northern Greenland,[47] meltwaters from the icecap run along its margin, eroding channels not only in the

Plate 21.3a A small spillway in central Labrador, showing characteristic trough-shape in cross-section. (*C.R. Twidale.*)

Plate 21.3b A large meltwater channel (left side of picture runs south from Ekwan Lake, Fort Nelson Lowland, B.C. Note the beautifully developed river meanders and oxbow lakes. (*Dept of Lands, Forests and Water Resources, B.C. Photo No. B.C. 1198:71.*)

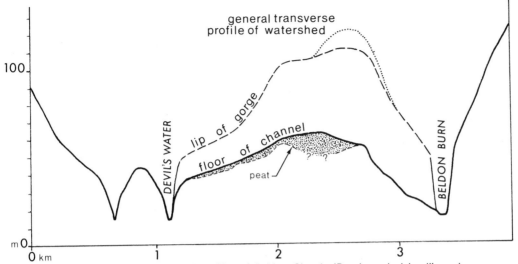

Fig. 21.18 Longitudinal profiles of Beldon Cleugh (Burn), a glacial spillway in Northumberland. (*After R.F. Peel, "A Study of Two Northumbrian Spillways", p. 82, Fig. 9, by permission of the Institute of British Geographers.*)

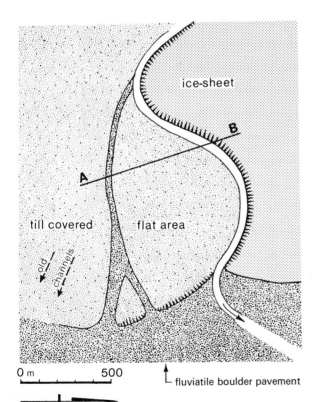

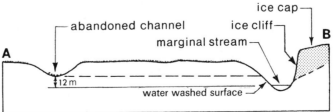

Fig. 21.19 Abandoned meltwater channel and modern marginal drainage channel on Inglefield Land, North Greenland. (*After R.L. Nichols, "Geomorphology of Inglefield Land", p. 188, by permission of* Med. Grønland *and Robert L. Nichols.*)

ice and snow banks, but also in glacial drift and in the local crystalline bedrock (Fig. 21.19). These meltwater channels are of considerable size: 4 m deep, 60 m wide, 0·4 km long. Some are active, some abandoned. Similar large meltwater channels have been identified in northwestern Canada.

4. Isostatic Adjustments

In several parts of the previously glaciated zones in Scandinavia and Canada, what are clearly marine shorelines of postglacial age are now found high above sealevel and are no longer level. Elsewhere, as in the Great Lakes region of North America, there is evidence that rivers have reversed their flow. What has caused these features to develop?

Many parts of the subglacial floor in Antarctica and Greenland are now below sealevel. Thus at present Greenland is not one island but two or three, with the uncertainty arising from the paucity of data on the elevation of the subglacial form. However, as the thickness and average density of the ice are known, it is possible to estimate the rebound of the floor which would result from the melting of the Greenland icecap (Fig. 17.4).

Following periods of glaciation in the Pleistocene, the melting of the ice-sheets led to the return of water to the ocean basins and to a world-wide rise of sealevel; a *eustatic* rise of sealevel. But the melting of glaciers also led to isostatic adjustment of the areas which had suffered glaciation and they rose in response to the lighter loading. In the high and middle latitudes, the structure of many landforms is related to the interplay of two (in a sense opposing) factors—a tendency of the land to submerge due to a rise in sealevel, and to emerge due to isostatic rise. The rise of sealevel followed immediately upon deglaciation in many areas, with the result that shoreline features were widely developed. Isostatic uplift lagged so that the shoreline features and sediments related to the marine transgressions of immediate postglacial date had developed and had been deposited before the land was uplifted. Some cover extensive areas (see Fig. 21.20). These postglacial marine sediments and shorelines now occur above sealevel. They display warping, which indicates that the isostatic uplift consequent upon deglaciation has not been uniform.

Although such events have been detected in many areas, the details remain controversial. In Scandinavia, the broad outlines of the postglacial history of the Baltic area are well established. The history involves the retreat of the icecap, a rising sealevel and the isostatic recovery of the Scandinavian peninsula (Fig. 21.20b). As the ice front retreated northwards, meltwaters were impounded in the still-depressed area to the south and particularly in what is now the shallow basin of the Baltic Sea. This early lake was the Baltic Ice Lake. It drained to the sea by way of spillways, perhaps three in number, to the north of Denmark in the present position of the Skagerrak (Fig. 21.21). Several strandlines related to this stage have been found at various levels above sealevel due to progressive uplift of the land.

With the further retreat of the ice, a broad connection with the sea was uncovered and the lake became a great eastward extension of the Yoldia Sea, named after a characteristic mollusc (*Yoldia arctica*) which lived in its waters and which today inhabits only the colder parts of the Arctic Ocean. This sea persisted for a long time and many strandlines were devel-

oped at its margin, each to be raised as the isostatic recovery of the Scandinavian land mass continued. Eventually, as a result of the continued uplift, the connection with the sea was narrowed and a freshwater lake, the Ancylus lake (after the mollusc *Ancyclus fluviatalis*) was formed. This existed for at least a thousand years and with the continued retreat of the ice, the lake waters no longer lapped up against the ice front. This lake discharged to the sea through the Oresund, the channel which today separates Sweden and Denmark. But a general rise of sealevel due to continued deglaciation submerged the Oresund and once again converted the lake into an arm of the sea, the Littorina or Tapes Sea. The fauna indicated a warmer climate than heretofore, and with the continued isostatic recovery of Scandinavia, the Littorina Sea was gradually converted to the present Baltic Sea.

This is a very broad outline of a complex history, and in detail there are differences of interpretation. For instance, some workers have considered that a hinge zone from which uplift took place was in operation through late Pleistocene time; others have considered that the hinge zone has migrated in time or that differential block uplifts have occurred.[48] But the uplift continues, and at measurable rates (Fig. 5.5c).

It is generally agreed that Scotland and northern Ireland display evidence of postglacial uplift due to deglaciation, but almost the only other point upon which workers concur is that the isobases representing rates of uplift drawn by Valentin[49] are wrong. Various arguments have been advanced,[50] but an uplift of some 25 m seems well established around the Firth of Clyde, with lesser amounts of recovery evidenced in adjacent areas (Fig. 21.22).

Much the same sort of picture has emerged in northeastern Canada where the evidence for postglacial marine submergence and isostatic recovery is widespread, though in some cases rather confusing. Postglacial strandlines at elevations of up to 250 m in northwestern Labrador have been reported, but in general estimates are rather more conservative (Fig. 21.23). Indeed the higher isobases of uplift published by Farrand and Gadja[51] have been challenged by Ives.[52]

A postglacial marine submergence over rather more than 200 m is clearly attested in central Arctic Canada,[53] an area which is near the limit or edge of isostatic deformation. The greatest recovery of the land has taken place where the ice was thickest, that is, around Hudson Bay (Fig. 21.23). Considerable uplift is in evidence to the east, west and north of this; there is, for instance, a raised beach at 53·3 m dated at 8275 ± 320 B.P., and others up to 70–71·5 m above sealevel dated 9075 ± 275 B.P. on Melville Island in the North West Territories.[54] But some of the features taken as evidence of raised shorelines have not been accurately levelled and so the heights attributed to them are erroneous; others are ice margins and not marine features. Thus the details remain to be ascertained although the broad picture is clear.

Isostatic movements are not only due to glaciation and deglaciation: as noted in Chapter 2, vertical movements can be caused by erosion and deposition. In many parts of European USSR which were never glaciated, movements of a similar magnitude to those attributed to deglaciation in Scandinavia have been detected. On what grounds then, can these vertical movements in Scandinavia and northern North America be attributed to glaciation and deglaciation?

Several lines of argument can be brought to bear.[55] First,

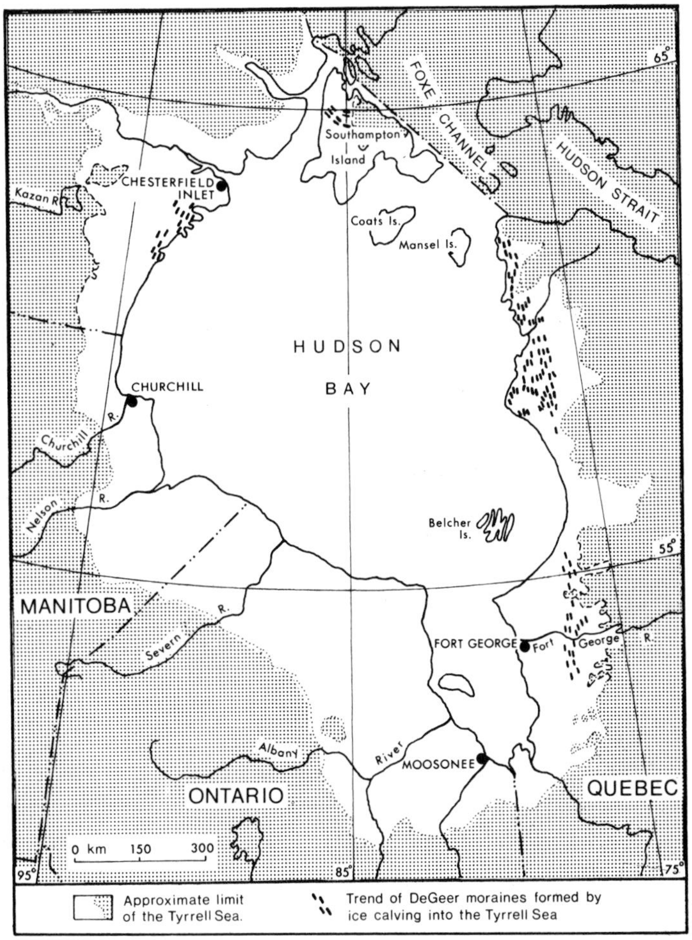

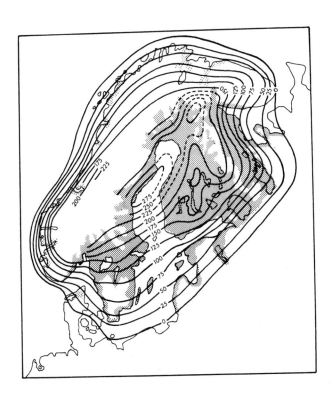

Fig. 21.20a Extent of the Tyrrell Sea, which attained its maximum some 7–8000 years ago as a result of retreat and melting of the ice sheets. Subsequently there has been isostatic rebound.

Approximate limit of the Tyrrell Sea.

Trend of DeGeer moraines formed by ice calving into the Tyrrell Sea

Fig. 21.20b Postglacial isostatic uplift of Fennoscandia (in metres). Postglacial marine transgression shown stippled. (*After R.A. Daly, The Changing World of the Ice Age, Yale University Press, New Haven, 1934, p. 66.*)

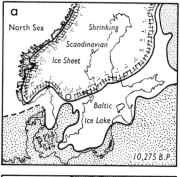

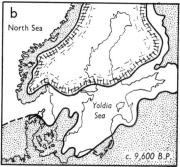

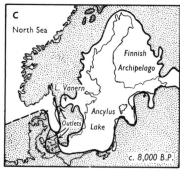

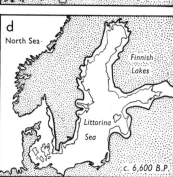

Fig. 21.21 Evolution of the Baltic Sea and Lake. *(After A. Holmes,* Principles of Physical Geology, *Nelson, Edinburgh, (2nd edition) 1965, p. 689.)*

Fig. 21.22 Postglacial emergence in east-central Scotland (isobases in feet). *(After F.M. Synge and N. Stephens, "Late- and Post-glacial Shorelines, and Ice Limits in Argyll and northeast Ulster",* Trans. Inst. Br. Geogr., *1966,* **39***, p.114, Fig. 5, by permission of the Institute of British Geographers.)*

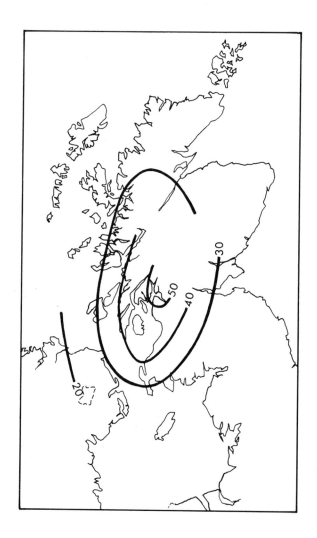

the outer limit of the areas affected by rebound are parallel to (though a little toward the centre of glaciation) the known position of glaciated areas. Second, the isobases of movement are arranged concentrically around areas which on other evidence are known to have borne the greatest thicknesses of ice, such as the Hudson Bay area of North America and the northern part of the Gulf of Bothnia in Fennoscandia. Third, every major area which was covered by an ice-sheet during the Quaternary also displays signs of isostatic rebound. Fourth, the rates of rebound are similar in the various glaciated and deglaciated areas; and fifth, some areas which have suffered rebound are also areas of negative gravity

anomaly, suggesting that rebound is not yet complete and that the depressions in the geoid remain to be adjusted.

These isostatic adjustments due to deglaciation have occurred on a regional, subcontinental scale. Some evidence has been advanced to suggest that on a local scale as well, crustal movements may be caused by glaciation and deglaciation. For example, in the Swiss Alps near Andermatt, Jackli[56] has described a number of small parallel reverse scarps which decrease in amplitude upslope. He considers that they are due to the isostatic reactivation of old faults caused by deglaciation, the maximum recovery being in the centre of the valley where the glacier was thickest.

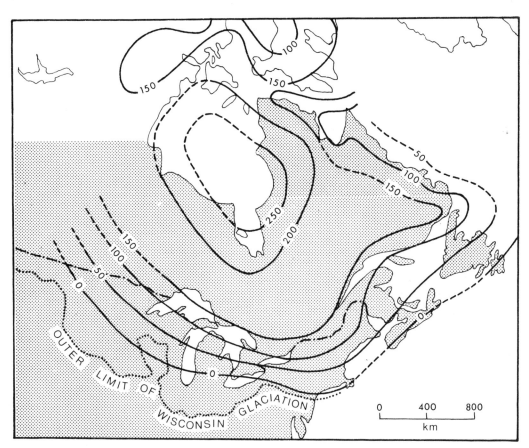

Fig. 21.23 Postglacial isostatic uplift of northeastern North America (in metres). *(After W.R. Farrand and R.T. Gadga, "Isobases", pp. 5—22 and others, as shown in Phillip B. King, "Quaternary Tectonics in middle North America" in H.E. Wright, Jr and David G. Frey (Eds),* The Quaternary of the United States, *(copyright 1965 by Princeton University Press): Fig. 4, p. 836. Reprinted by permission of Princeton University Press.)*

References Cited

1. H.F. GARNER, "Stratigraphic-sedimentary Significance of Contemporary Climate and Relief in Four Regions of the Andes Mountains", *Bull. geol. Soc. Am.*, 1959, **70**, pp. 1327–68; *idem*, "Tropical Weathering and Relief" in R.W. Fairbridge (Ed.), *Encyclopaedia of Geomorphology*, Reinhold, New York, 1969, pp. 1161–72; J.J. BIGARELLA and G.O. DE ANDRADE, "Contribution to the Study of the Brazilian Quaternary", *Geol. Soc. Am. Spec. Paper*, 1965, **84**, pp. 433–51.

2. J.T. HOLLIN, "The Antarctic Ice Sheet and the Quaternary History of Antarctica", *Palaeoecol. Afr.*, 1969, **5**, pp. 109–38; D.E. HAYES and L.A. FRAKES *et al*, "Leg 28 Deep-Sea Drilling in the Southern Ocean", *Geotimes*, 1973, **18** (6), pp. 19–24.

3. C. EMILIANI, "Pleistocene Temperatures", *J. Geol.*, 1955, **63**, pp. 538–78; *idem*, "Ancient Temperatures", *Sci. Am.*, 1958, **198**, pp. 54–63.

4. C. EMILIANI, "Palaeotemperatures Analysis of Caribbean Cores P 6304-8 and P 6304-9 and a Generalized Temperature Curve for the past 425,000 Years", *J. Geol.*, 1966, **74**, pp. 109–24.

5. See for instance C.C. LANGWAY Jr, "Stratigraphic Analysis of a Deep Ice Core from Greenland", *Geol. Soc. Am. Spec. Paper*, 1970, **125**.

6. See K.W. BUTZER, *Environment and Archaeology*, Methuen, London, 1964.

7. H.C. WILLETT, "Solar Variability as a Factor in the Fluctuations of Climate during Geological Time", *Geogr. Ann.*, 1949, **31**, pp. 295–315.

8. G.N. PLASS, "The Carbon-dioxide Theory of Climatic Change", *Tellus*, 1956, **8**, pp. 140–54.

9. H.T. STEARNS, "Eustatic Shorelines on Pacific Islands", *Z. Geomorph. Suppl.*, 1961, **3**, pp. 3–16.

10. J.T. HOLLIN, "The Antarctic Ice Sheet and the Quaternary History of Antarctica", *Palaeoecol. Afr.*, **5**, pp. 109–38.

11. R.P. SHARP and J.H. BIRMAN, "Additions to the Classical Sequence of Pleistocene Glaciations, Sierra Nevada, California", *Bull. geol. Soc. Am.*, 1963, **77**, pp. 1079–86; R.P. SHARP, "Semiquantitative Differentiation of Glacial Moraines near Convict Lake, Sierra Nevada, California", *J. Geol.*, 1969, **77**, pp. 68–91.

12. M. GAGE, "New Zealand Glaciations and the Duration of the Pleistocene", *J. Glaciol.*, 1961, (3) pp. 940–3; M. GAGE and R.P. SUGGATE, "Glacial Chronology of the New Zealand Pleistocene", *Bull. geol. Soc. Am.*, 1958, **69**, pp. 589–98; R.P. SUGGATE, "Late Pleistocene Geology of the Northern Part of the South Island, New Zealand", *Bull. Geol. Surv. N.Z.*, 1965, **77**.

13. R.B. MORRISON, "The Quaternary Geology of the Great Basin" in H.E. Wright and D.G. Frey (Eds), *The Quaternary of the United States*, Princeton University Press, Princeton, 1965, pp. 265–85; R.B. MORRISON and J.C. FRYE, "Correlation of the Middle and Late

Quaternary Successions of the Lake Lahontan, Lake Bonneville, Rocky Mountain (Wasatch Range), southern Great Plains, and eastern Midwest Areas", *Nevada Bur. Mines Rep.*, 1965, **9**.

14. See for instance, R.P. SUGGATE, "Late Pleistocene Geology".

15. C.E.P. BROOKS, "Variations in the Levels of the central African Lakes Victoria and Albert", *Met. Office Geophys. Mem.*, 1923, **20**.

16. G.K. GILBERT, "Lake Bonneville", *U.S. Geol. Surv. Monogr.*, 1890, **1**.

17. A. PENCK and E. BRÜCKNER, *Die Alpen im Eiszeitalter*, Tauchnitz, Leipzig, 1909.

18. I.C. RUSSELL, "Geological History of Lake Lahontan, a Quaternary Lake of northwestern Nevada", *U.S. Geol. Surv. Monogr.*, 1891, **11**; R.B. MORRISON, "The Quaternary Geology", pp. 265–85; R.B. MORRISON and J.C. FRYE, "Correlation of the Middle and Late Quaternary".

19. W.W. BISHOP, "Pleistocene Chronology in East Africa", *Adv. Sci.*, 1961, **18**, pp. 491–4.

20. E.J. WAYLAND, "The Geological History of the Great Lakes", *Geol. Surv. Uganda Ann. Rep.* 1928, 1929, pp. 35–8.

21. R.W. FAIRBRIDGE, "World Palaeoclimatology of the Quaternary", *Rev. Géogr. Phys. Géol. Dyn.*, 1970, **12** (2), pp. 97–104.

22. See for instance, F.E. ZEUNER, *Dating the Past*, Methuen, London, 1950, pp. 138–42.

23. K.W. BUTZER, "Climatic Change in Arid Regions since the Pliocene", in L.D. Stamp (Ed.), *History of Land Use in the Arid Regions*, UNESCO, Paris, 1961, pp. 31–56; *idem*, *Environment and Archaeology*, Methuen, London, 1964; K.W. BUTZER and C.R. TWIDALE, "Deserts in the Past" in E.S. Hills (Ed.), *Arid Lands*, Methuen/UNESCO, London, 1967, Chapter 7, pp. 127–44.

24. R.W. FAIRBRIDGE, "World Palaeoclimatology", pp. 97–104.

25. J.T. HOLLIN, "The Antarctic Ice Sheet and the Quaternary History of Antarctica", *Palaeoecol. Afr.*, 1969, **5**, pp. 109–38.

26. G.K. GILBERT, "Lake Bonneville", *U.S. Geol. Surv. Monogr.*, 1890, **1**.

27. V. TANNER, *Outlines of the Geography, Life and Customs of Newfoundland-Labrador*, Acta Geographica, Helsinki, 1944, Vol. 8.

28. J.W. AMBROSE, "Exhumed Paleoplains of the Precambrian Shield of North America", *Am. J. Sci.*, 1963, **262**, pp. 817–57.

29. A. PENCK, "Die Gipfelflur der Alpen", *Sitzber Preuss Akad. Wiss.* (Berlin), 1919, **17**, p. 256.

30. See also M. BOYÉ, *Glaciaire et Périglaciaire de L'Ata Sund, Nord Oriental Groenland*, Hermann, Paris, 1950.

31. S. THORARINSSON, "The Ice-dammed Lakes of Iceland,

with particular reference to their Values as Indicators of Glacier Oscillations", *Geogr. Ann.*, 1939, **21**, pp. 216–42.

32. N.E. ODELL, "The Märjeelen Sea and its Fluctuations", *Ice*, 1966, **20**, p. 27.

33. K.H. STONE, "Alaskan Ice-dammed Lakes", *Ass. Am. Geogr. Ann.*, 1963, **53**, pp. 332–49.

34. R.P. SHARP, "The Wolf Creek Glaciers, St Elias Range, Yukon Territory", *Geogr. Rev.*, 1947, **37**, pp. 26–52.

35. Ibid.

36. R.G. LA FLEUR, "Glacial Geology of the Troy N.Y. Quadrangle", *N.Y. State Mus. & Sci. Serv. Map*, No. 7, 1965; *idem*, "Glacial Lake Albany" in R.W. Fairbridge (Ed.), *Encyclopaedia of Geomorphology*, Reinhold, New York, 1969, p. 1295.

37. R.E. JANSSEN, "The History of a River", *Sci. Am.*, 1952, **186**, pp. 74–80; *idem*, "The Teays River, Ancient Precursor of the East", *Sci. Monthly*, 1953, **77**, pp. 306–14; E.C. RHODEHAMEL and C.W. CARLSTON, "Geologic History of the Teays Valley in West Virginia", *Bull. geol. Soc. Am.*, 1963, **74**, pp. 251–74.

38. J.L. HOUGH, *Geology of the Great Lakes*, University of Illinois Press, Urbana, 1958, p. 313.

39. T.F. JAMIESON, "On the Parallel Roads of Glen Roy and their Place in the History of the Glacial Period", *Q. J. geol. Soc. London*, 1863, **19**, pp. 235–59.

40. P.F. KENDALL, "A System of Glacier Lakes in the Cleveland Hills", *Q. J. geol. Soc. London*, 1902, **58**, pp. 471–571.

41. G. DE GEER, "A Thermographical Record of the late Quaternary Climate" in *Die Veranderungen des Klimas*, International Geological Congress, Stockholm, 1910.

42. I.P. KARTASHOV, "The Origin of Lake Krasnoye", *Akad. Nauk. SSSR, Doklady, Earth Sci. Secs.*, 1962, **142**, pp. 8–10.

43. See R.F. PEEL, "A Study of Two Northumbrian Spillways", *Trans. & Papers Inst. Br. Geogr.*, 1949, **15**, pp. 75–89; also C.R. TWIDALE, "Longitudinal Profiles of Certain Glacial Overflow Channels", *Geogr. J.*, 1956, **122**, pp. 88–92.

44. J.B. SISSONS, "Some Aspects of Glacial Drainage Channels in Britain", *Scot. Geogr. Mag.*, 1960, **76**, pp. 131–46.

45. E. DERBYSHIRE, "Fluvioglacial Erosion near Knob Lake, central Quebec-Labrador, Canada", *Bull. geol. Soc. Am.*, 1962, **73**, pp. 1111–26.

46. S. THORARINSSON, "The Ice-dammed Lakes of Iceland", pp. 216–42.

47. R.L. NICHOLS, "Geomorphology of Inglefield Land, North Greenland", *Med. Grønland*, 1969, p. 188.

48. For review, see for instance R.F. FLINT, *Glacial and Pleistocene Geology*, Wiley, New York, 1957, pp. 242–7. But see also M. OKKO, "The Relation between Raised Shore and Present Land Uplift in Finland during the past 8000 Years", *Ann. Acad. Scient. Fennicae, Series A* (III Geol. and Geogr.), 1967, p. 83.

49. H. VALENTIN, "Present Vertical Movements of the British Isles", *Geogr. J.*, 1953, **119**, pp. 299–305.

50. K. WALTON, "Vertical Movements of Shorelines in Highland Britain", *Trans. Inst. Br. Geogr.*, 1966, **39**, pp. 1–8, and other papers in this volume.

51. W.R. FARRAND and R.T. GADGA, "Isobases on the Wisconsin Marine Limit in Canada", *Geogr. Bull.*, 1962, **17**, pp. 5–22.

52. J.D. IVES, "Determination of the Marine Limit in eastern Arctic Canada", *Geogr. Bull.*, 1963, **19**, pp. 117–22.

53. J.B. BIRD, "Postglacial Marine Submergence in central Arctic Canada", *Bull. geol. Soc. Am.*, 1954, **65**, pp. 457–64.

54. W.E.S. HENOCH, "Postglacial Marine Submergence and Emergence of Melville Island, N.W.T.", *Geogr. Bull.*, 1964, **22**, pp. 105–26.

55. W.R. FARRAND, "Postglacial Isostatic Rebound", in R.W. Fairbridge (Ed.), *Encyclopaedia of Geomorphology*, p. 1295.

56. H.C.A. JACKLI, "The Pleistocene Glaciation of the Swiss Alps and Signs of Postglacial Differential Uplift", paper presented to VII Congress INQUA, Boulder, Colorado, 1965.

General and Additional References

C.E.P. BROOKS, *Climate through the Ages*, Yale University Press, New Haven, 1938.

M. SCHWARTZBACH, *Das Klimat der Vorzeit*, Enke, Stuttgart, 1950.

F.E. ZEUNER, *Dating the Past*, Methuen, London, 1950.

H. SHAPLEY, *Climatic Change. Evidence, Causes, Effects*, Harvard University Press, Cambridge, 1953.

P. WOLDSTADT, *Die Eiszeitalter*, Enke, Stuttgart, 1958.

R.G. WEST, *Pleistocene Geology and Biology*, Longmans, London, 1968.

CHAPTER 22

Indirect Effects of Late Cainozoic Climatic Changes

A. GENERAL

In many areas which escaped glaciation, there is abundant evidence that different climatic conditions prevailed from those which now obtain; and moreover that they prevailed for periods of sufficient length to allow the development of landform assemblages which are alien to the present environment. Such *inherited* forms are widely distributed in the mid- and low-latitude regions of the world.

B. INHERITED LANDFORMS

Periglacial or nival landforms and deposits are not only found marginal to former glacial limits (Pl. 22.1), as for instance in Alaska,[1] southern Finland,[2] southern Quebec,[3] and Ontario;[4] they also exist at high altitudes in many tropical regions, and at moderate altitudes in subtropical and mid-latitude areas.[5] Evidence of ground ice activity was reported from southern Africa as early as 1944,[6] and a considerable range of forms and features related to frost and ice action has recently been identified there: solifluction (geliflual) deposits, aprons and blocks, and patterned ground.[7] Frost-heaved soils have been found in the English Midlands.[8] Soliflual lobes, aprons and deposits recognised in the high country around Mt Kosciusko in N.S.W., as well as in central and southern Tasmania are in both areas below present levels of significant nival action;[9] and a whole range of glacial and periglacial features have been described from New Guinea.[10] Similar forms and deposits occur in southern South America.

Relict pingos have been identified in central Alaska[12] and it is alleged that similar relict features occur in north-western Europe, though doubt has been cast on this interpretation.[13]

Evidence of a once high and effective rainfall is widespread in the mid-latitude and arid tropical and subtropical regions. Spreads of lacustrine sediments, abandoned shoreline deposits and landforms all attest the previous existence of extensive pluvial lakes which were distant from the ice front, although they left evidence similar to that associated with the proglacial lakes described in Chapter 21 (Pl. 22.2, Fig. 22.1a).

Some of the best known examples are to be found in the Great Basin and in the southwest of the United States. Lake Bonneville[14] was one of the earliest recognised and it remains one of the best known. Gilbert identified four major Pleistocene lacustrine stages separated by arid phases occurring either when the lake level was very low or when the bed was quite desiccated (Fig. 21.4a); and though various substages have since been distinguished, this scheme remains essentially valid in a relative sense though many complexities have since been revealed, and Gilbert's simplistic assumptions concerning dating have been shown to be invalid. At its earliest stage, the Bonneville lake had an

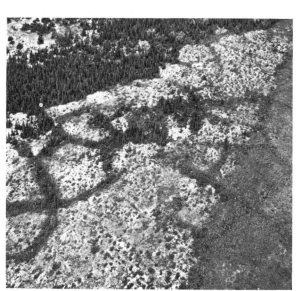

Plate 22.1 Relict ice-wedge polygons, Donnelly Dome Area, Alaska. *(T.L. Pewe.)*

Plate 22.2 Old beaches and cliffs related to higher Pleistocene stands of Pyramid Lake, north of Reno, Nevada. *(C.R. Twidale.)*

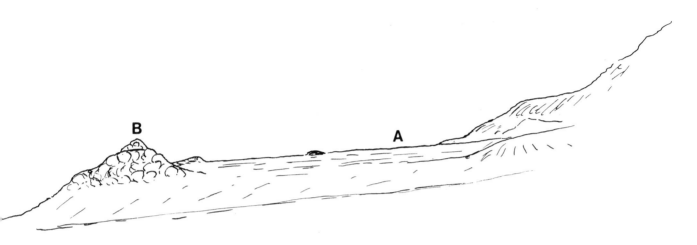

Fig. 22.1a Sketch of Travertine or Tufa Point (B), southern California, showing a raised shore platform (A) related to a former high stand of Pleistocene Lake Cahuilla, the predecessor of the Salton
Sea. The platform stands at about sealevel, and some 72 m above the present level of the Salton Sea.
The Point is an old stack built of gneiss with a veneer of calcareous tufa. It stands 24 m above the
old platform and lake level.

area of 51,700 km and a depth of 335 m, 305 m above
the level of the present Great Salt Lake which is the last
remnant of this ancient lake. Subsequently the lake
rose 27 m (Bonneville stage), and at this level waters overflowed northward through the Red Rock Pass, and thence
into the Snake River system and the Pacific. The erosion
of the outlet col caused the lowering of the lake level by
142 m from the Bonneville to what was called the Provo
level, 190 m above the Great Salt Lake. Several lower
stages of the lake have been recognised (Fig. 21.4a and b),
prominent among which is the Stansbury level, 100 m above
the present lake bed.

At each stable level of the lake, a characteristic suite of
shoreline features was developed by wave action (see
Chapter 18): shore platforms, cliffs, beaches, cuspate forelands (Fig. 22.1b) and tombolos, as well as lake floor
sediments. The great mass of water depressed the crust, and
when it disappeared through overflow and aridity, there was
isostatic rebound so that the originally horizontal shoreline
features are now distorted[15] (Fig. 22.2).

All these developments are of Pleistocene age, and it
seemed reasonable to assume that the lakes resulted from
periods of higher rainfall in the catchment area, and that
such pluvials coincided with glacial phases of the Pleistocene.
Both assumptions have been corroborated by subsequent
work, though the amount of precipitation increase required
to produce the run-off represented by Lake Bonneville is
quite small when cooler conditions prevail (thus causing a
more effective rainfall). The coincidence of glaciation and
high stands of the lake is suggested by the apparent deposition
of moraines from the local mountain glaciers in the lake
waters, and in this case it can be reasonably argued that
glacial and pluvial phases were synchronous. However,
this is not necessarily true everywhere.

Many other major Pleistocene lakes have been recognised

in the American West. Lake Lahontan was a major water
body with a maximum area of 22,442 km[2] and a depth of
213 m, and evidence of many smaller lakes has been detected
in southern California (Fig. 22.3). Some, like Searles Lake,[16]
have clearly suffered climatic change, for beneath their beds
are alternating saline deposits and muds, representing arid
and pluvial phases respectively (Fig. 22.4). In west Texas,
Reeves[17] reports lake and playa depressions of several origins,
but some are due to deflation in late Pleistocene and mid
Recent times (indicating arid conditions) and subsequent
infilling by water and sediment during pluvial periods. In
Asia, the former Aral-Caspian lake stood some 76 m above
the level of the present Caspian Sea and overflowed into
the Black Sea. Around Lob Nor, in central Asia, Berkey
and Morris[18] detected at least seven old shorelines which
are related to the former lake and which are up to 9 m above
the 1922 level of the playa floor.

Lake Chad, in central Africa, occupies a broad downwarp
in the Precambrian crystalline basement. The lake at present
stands some 282 m above sealevel, is joined to the lower
Bodele depression by the Bahr al Ghazal and is underlain
by some 800 m of Pleistocene lacustrine sediments. There
are Pleistocene beaches at 287 and 300 m and evidence of
alternating arid and pluvial phases:[19]

Arid phase: sand dunes deposited on the pre-Pleistocene
                  land surface.
Pluvial phase: Aterian (late Palaeolithic) lake some 180 m
                  deep and 1 million km[2] in extent.
Arid phase: lake deposits moulded into dunes during
                  early–mid Recent; dunes trended NNE-
                  SSW (the ancient erg of Hausaland; see
                  below).
Pluvial phase: lake at 320–332 m and Bama ridge formed.
Arid phase: NNW–SSE dunes formed.

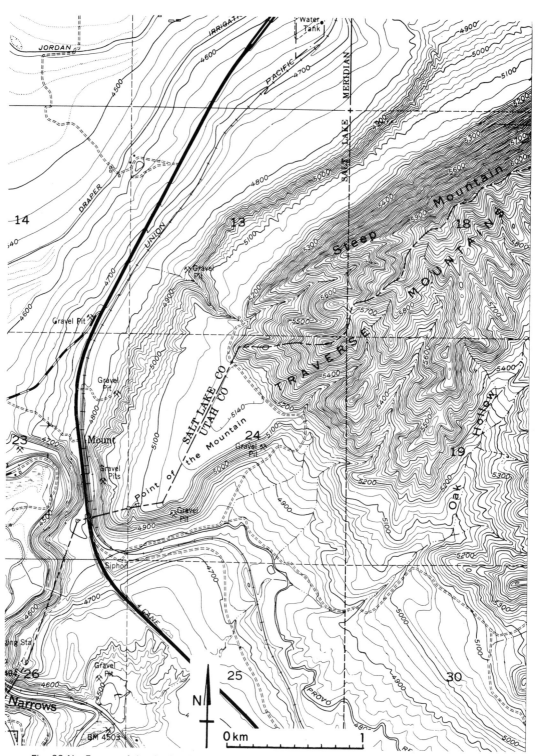

Fig. 22.1b Extract of the Jordan Narrows Quadrangle, Utah, showing former lake shore platforms at elevations of approximately 1425 m (4600 feet) and 1580 m (5100 feet), a cuspate foreland, and old lake cliffs related to a former stand of Lake Bonneville, the precursor of the present Great Salt Lake. (U.S. Topographic Map Series.)

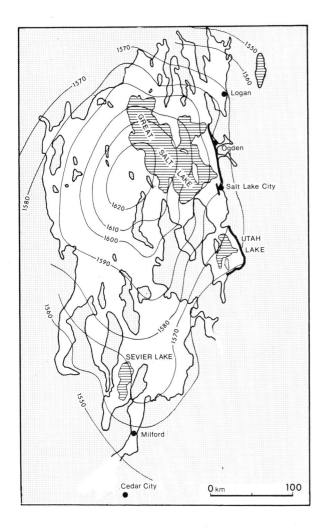

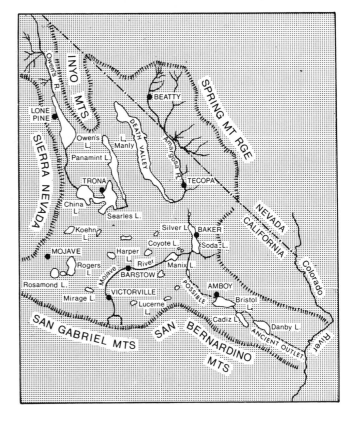

Fig. 22.2   Deformation of Bonneville shoreline, bordering the Pleistocene lake of the same name, by isostatic rebound consequent on the drying of the old lake. *(After M.D. Crittenden, U.S. Geol. Surv. Prof. Paper, 1963, **454-E**.)*

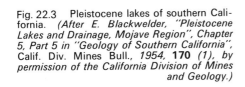

Fig. 22.3   Pleistocene lakes of southern California. *(After E. Blackwelder, "Pleistocene Lakes and Drainage, Mojave Region", Chapter 5, Part 5 in "Geology of Southern California", Calif. Div. Mines Bull., 1954, **170** (1), by permission of the California Division of Mines and Geology.)*

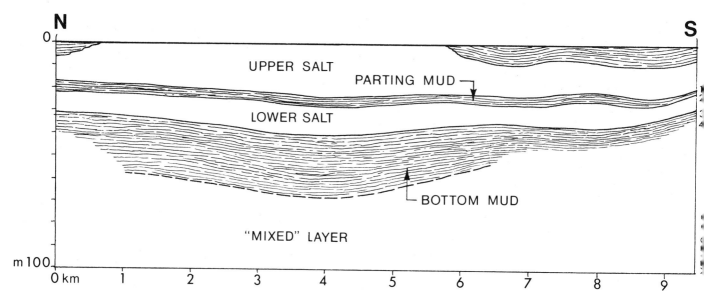

Fig. 22.4 Section through Searles Lake, southern California, showing alternating layers of salt and mud which reflect changing climatic conditions in the catchment. *(After R.F. Flint and W.A. Gale, "Stratigraphy and Radiocarbon Dates", pp. 689–714, by permission of Am. J. Sci.)*

In Australia, Lake Woods near Newcastle Waters, N.T, formerly covered some 2850 km[2] though the playa is now only 180 km[2]. The older, larger lake existed before the present phase of dune building, for dunes encroach upon the old shoreline features.[20] Lake Eyre was much more extensive some 40,000 years ago when considerable thicknesses of sediments were deposited in a body of water, the depth of which remains unknown but which possibly extended far to the northeast of the present lake margins (Fig. 22.5).[21]

Alluvial fans, though at present forming in many areas, have evidently developed under different conditions in some areas in the past. For instance alluvial fan development in southern Arabia is attributed to a pluvial period some five to six thousand years ago.[22] The fans which front the Mt Lofty and southern Flinders Ranges on their western sides are not forming now; on the contrary they are suffering dissection. They accumulated some 34,000 years ago.[23] Whether they represent a pluvial phase or an arid period is arguable: they could be the result of increased run-off related to increased rainfall; but on the other hand it is possible that increased rainfall resulted in a more profuse vegetation cover and hence in less run-off and lower sediment yield.

Remnants of old drainage systems which survive in areas where rivers no longer exist have been identified in the Nullarbor Plain and the Amadeus and Officer basins,[24] all in arid central and southern Australia (see Fig. 22.6a). None of the ancient rivers has been dated and they do not necessarily indicate the same pluvial period. More profuse drainage systems in the Riverina of western N.S.W.[25] (Fig. 22.6b) are also open to varied interpretation. According to Langford-Smith,[26] they developed during glacial pluvial phases, whereas Butler[27] believes the old channels were formed in arid phases and the soil development in evidence

took place in the pluvial periods.

Another aspect of river patterns with implications for climatic change is the geometrical relationship between present stream meanders and river curves. Observations of river meanders (see Chapter 12) have disclosed an empirical relationship between the bankfull discharge, channel width and meander geometry. It has been assumed that the geometry of winding or meandering valleys reflects that of the rivers which formerly occupied them. But the geometry of the valley shape in plan is anomalous with that of the present river channels. The streams are described as *misfit*. If the radius of the meander bluffs preserved in some of the winding valleys is related to channel width, and hence to discharge, then it is clear that these streams carried huge volumes of water in the recent past.

River capture cannot provide a general explanation for the recent decrease in discharge, for misfit streams are widely distributed in many parts of Britain, the USA (Fig. 22.7a and b) and around Sydney, N.S.W.[28] (Fig. 22.7). According to Dury, they can only be explained in terms of climatic change.

The question is this: are the meander bluffs related to gigantic stream meanders of former times? It is certainly difficult to conceive of climatic changes which could bring about the quite enormous former discharges indicated by the empirical comparison.[29]

One possible explanation is that the meander bluffs do not represent the outlines of former meanders but rather outlines of meander belts (see Chapter 12)—an altogether larger scale feature. The difficulties inherent in the use of river valley morphology as a climatic indicator may be illustrated with reference to the valley shown in Fig. 19.5. Here the Souris River in North Dakota is meandering and

flows in a narrow flat-floored valley incised into glacial
drifts. River terraces are preserved above the present flood
plain, and the meander bluffs which back these remnants
are clearly of a different geometry than the present meanders.
Their radius of curvature is much greater. Hence Dury
and his followers would regard the Souris as *underfit*, that is,
as a river which has been reduced in discharge. But are the
river bluffs really a reflection of meander geometry, and
hence of discharge, or are they formed by the meander
belt? Are they scoured essentially during floods when the
river runs high and fast and straight, and erodes straight
channels such as displayed at Y on Fig. 19.5? Are such
elongate lagoons true riverine forms, or are they in part
due to spring sapping and flushing at the base of the bluffs?

Another problem concerns the relationship between
these straight sections and the meanders. If they are the
remains of old river channels, are they a manifestation of
higher volume and/or velocity streams (Fig. 12.20)? Climatic
change is not the only factor which can cause alterations in
channel pattern, for as has been pointed out with respect to
some of the major rivers of northwest Queensland,[30] increased
gradients (in that case due to the tectonic uplift in the
headwater region of the Flinders and Leichhardt rivers) can
give rise to a change from meandering to braided, from
winding to comparatively straight.

Thus the nature and origin of these misfit streams remains
controversial.

Former aridity is clearly indicated by the occurrence of
dune fields which are stabilised by vegetation and are no
longer active. Such forms are in disequilibrium with their
present environment. Fields of relict dunes occur along the
southern edge of the Sahara (Fig. 22.8) from Nigeria to the
Sudan[31] and in many parts of southern Australia, from the
Western Australian Wheat Belt to Eyre Peninsula (Pl. 22.3),
in the Yorke Peninsula, on the Adelaide plains (where they
extend to within 32 km of Adelaide), from the Murray plains
in South Australia to the Wimmera of western Victoria,
and in what are now the western suburbs of Melbourne.

The western Great Plains of the USA, in Nebraska and
Arkansas, also display relict dunes, but these are related to
cold periglacial conditions rather than to hot, dry environments.

The occurrence of pediment forms in mid-latitude
regions, as for instance around the Sea of Japan and in
Korea,[32] is a less reliable climatic indicator, but is nevertheless worthy of consideration (see comments in Chapter
14).

River terraces are often considered reliable indicators
of climatic change: and they may well originate as a result
of changes in discharge or regime consequent upon climatic
change. But great care is necessary in such interpretations
with regard to terraces, for there are serious general difficulties.

River terraces are either paired or unpaired, but in either
case they form as a result of stream dissection. The latter
may result from tectonic uplift, changes of sealevel, changes
in land use (for instance, European settlement and land
clearance—see Chapter 24) or climatic change. Thus it
is necessary to take a regional view and to consider the
distribution of the terraces with respect to the coast, evidence
of sealevel movements, known faults, and the dates of
terrace cutting on the one hand and settlement on the other.

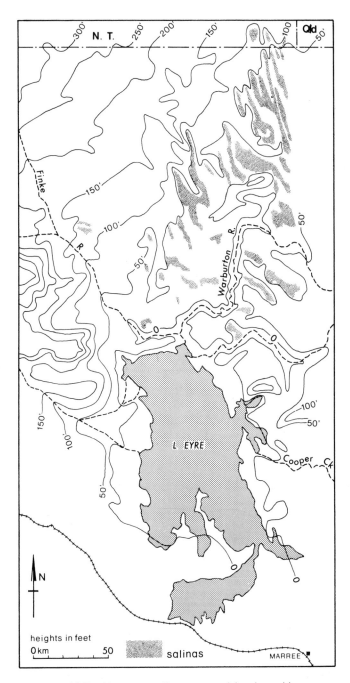

Fig. 22.5 Numerous salinas separated by dune ridges
which have migrated across the floor of the former extensive and continuous salt pan of the late Pleistocene
Lake Eyre, subdividing it into many small linear salt-
encrusted playas. *(From unpublished F.C.P.A. map.)*

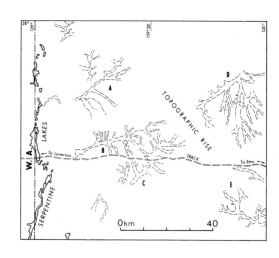

Fig. 22.6a Old stream courses (A–E) on the Nullarbor Plain in South Australia close to the Western Australian border. *(After G.W. Krieg, "Relic Drainage", pp. 1–2, by permission S.A. Dept Mines.)*

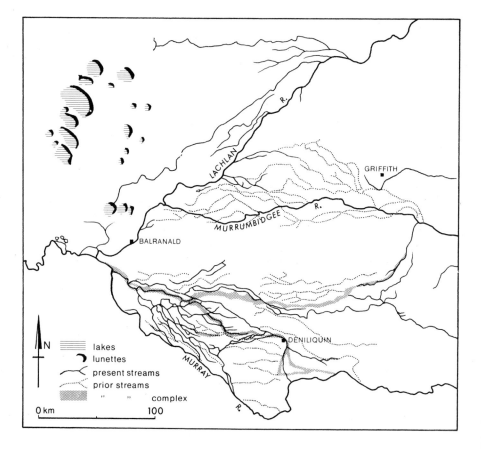

Fig. 22.6b Former drainage lines in the Riverina of western N.S.W.: present streams are represented by continuous lines, earlier channels which are now abandoned are shown with dotted lines. *(After S. Pels, "Late Quaternary Chronology of the Riverina Plain", pp. 27–40, by permission of the Geological Society of Australia.)*

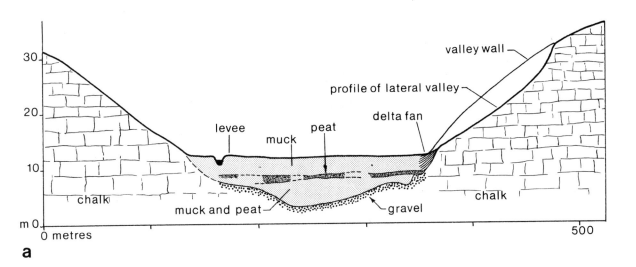

a

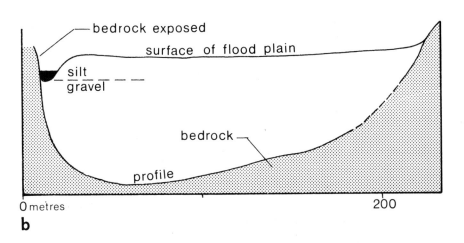

b

Fig. 22.7 Sections through stream valleys which are said to be of misfit type: (a) River Rib, Hert-fordshire, England; (b) Mineral Point Branch, East Pecatonica River, Wisconsin. *(After G.H. Dury, "Principles of Underfit Streams".)*

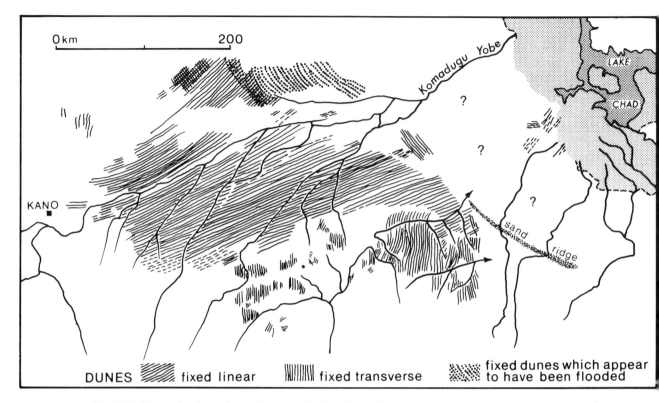

Fig. 22.8 Fixed relict dunes in northeastern Nigeria. *(After A.T. Grove, "The Ancient Era", pp. 528–33.)*

Plate 22.3 Relict Pleistocene dunes fixed by vegetation and left in their virgin state to prevent sand drifting, central Eyre Peninsula, South Australia. The dunes trend NW–SE and the interdune areas are used for cereal cultivation. *(C.R. Twidale.)*

In addition, it must be remembered that even within a limited area all terraces need not necessarily be of the same origin.

Even if the evidence suggests that terraces are of climatic origin, it is frequently difficult to judge the trend and degree of climatic change responsible, as the interplay of rainfall, temperature, vegetation and run-off is complex. Moreover, precipitation and temperature commonly vary simultaneously. Again, climatic changes in the headwater regions generally have far-reaching effects downstream, and finally terraces in tributary valleys may form wholly as a result of the lateral migration of the main stream which cause fluctuations of local baselevel within the minor valleys (Fig. 13.5).

Finally, features indicative of climatic change also occur in the humid tropics. For instance, Garner[33] has reported that in parts of the South American Andes, tropic soils give way at depth to coarse mineral soils derived from arid weathering. Bigarella and de Andrade[34] describe pediments, presumably eroded under semiarid conditions, which are now clothed by a tropical rain forest in Brazil. And Garner attributes the braided and anastomosing patterns of the Rio Caroni, Venezuela, to a variation in river regime consequent on climatic change.

C. EUSTATIC MOVEMENTS OF SEALEVEL
1. General

The Pleistocene period saw the repeated growth and recession of ice-sheets in the northern hemisphere. As water was trapped in the glaciers during the glacial periods, and released when the ice-sheets melted, sealevel fluctuated repeatedly during the Pleistocene, falling during glacial periods and rising during the interglacials.

World-wide movements of sealevel are called *eustatic* movements. In theory, two types can be distinguished with respect to Quaternary sealevel changes. During glacial periods, water was withheld from the ocean basins, resulting first in a fall in sealevel. On the other hand due to the unloading of the ocean floors and continental margins, the former are believed to have risen in readjustment, while the latter (including the continental shelves) may have suffered broad-scale warping or flexing. In these ways the capacity of the ocean basins was reduced, compensating in some measure for the lowering of sealevel caused by the growth of glaciers. When the ice-sheets melted, sealevel rose and the continental margins were depressed beneath the weight of water, causing an enlargement of the basins, but it was insufficient to compensate for the rise in level brought about by the massive accumulation of water.

Thus two separate and unequal effects are involved. First, those shifts of sealevel which were a direct response to the withdrawal and release of water consequent upon the waxing and waning of the ice masses were *glacio-eustatic* changes of sealevel. With very rare and trivial exceptions, modern glaciers—the Antarctic and Greenland ice-sheets in particular—are shrinking. As a result, sealevel·is still rising. Statistical analysis of tide gauge data from all over the world suggest that the level of the oceans is rising by 0.12 to 0.5 cm per annum, and that it has risen by 4.5 to 9.5 cm since the middle of the last century.

Second, there are changes occasioned by tectonic movements of the sea-floors and adjacent continental areas: these are *tectono-eustatic* movements.

In theory, the two types of movements can be distinguished by their relationship in time with dated glacier fluctuations, but in practice such correlations are complicated by several factors concerning the inherent strength and rigidity of the crust.

In addition, variations in the volume of ocean water were probably occasioned by the heating and cooling of those waters in sympathy with the same factors which caused the more dramatic development and disintegration of the ice-sheets. If the temperature of all the water in the oceans were raised by 1 °C, there would be an increase in volume which would cause sealevel to rise by some 60 cm, but the circulation of the deep ocean basins is slow, the cycle being of the order of several thousands of years, by which time atmospheric and surface temperature regions may have changed. Furthermore, an increase in atmospheric temperatures would result in greater melting of ice and calving of icebergs which would tend to lower oceanic temperatures. Thus the significance of this factor also is uncertain.

Until relatively recently, most abandoned sealevels were considered to be of glacio-eustatic origin. The best known exception is the Daly level, named after a geologist, R.A. Daly, who first recognised its world-wide significance. It stands some 5-6 m above present sealevel, and is considered to date from 3000–3500 years ago. On the southern shore of the Gulf of Carpentaria, for instance, shell beds standing a little more than 6 m above sealevel at Karumba have been dated as being 3300 years old and probably represent the Daly level.

However, over the past decade it has become clear that synchronous shoreline features occur at varied elevations. A number of writers, notably Wellman, Bloom and Higgins,[36] have endeavoured to explain this in terms of isostatic depression of the continental shelves under the weight of superincumbent water. The shelves were submerged (the amount of depression has been estimated at $\frac{1}{3}-\frac{1}{4}$ of the depth of water), and then, as a result of the melting of glaciers, the shelves rebounded by various amounts and at various rates.

2. Criteria for the Recognition of Old Sealevels

Present shoreline features have been described in Chapter 18. When any of these forms or deposits occur beyond (either above or below) the reach of modern shore processes, a change in the relative positions of land and sea is indicated: the level of the sea has either risen or fallen (eustatism), or the land has moved (tectonism). The two cases are distinguished by the behaviour of the old shore features along the coast: if sealevel has changed relics of the same former strand-line occur at similar (though not identical) elevations on all coasts, whereas deformation results in unequal dislocation or tilted shoreline features (Pl. 22.4), so that the coastal features dating from the same period in this instance occur at different levels along the coast.

Plate 22.4 Raised shore platforms near Wellington, North Island, New Zealand. As these are tilted, they have clearly been affected by, and may have been raised by, tectonism. *(D.W. Mackenzie.)*

Plate 22.5a Raised beach some 10 m above present high tide level, southeastern South Australia. *(C.R. Twidale.)*

Plate 22.5b Shore platform cut in granite gneiss, southern Yorke Peninsula, S.A. The granite bench is covered by several metres of aeolianite, and is therefore related to a previous period of marine planation. *(C.R. Twidale.)*

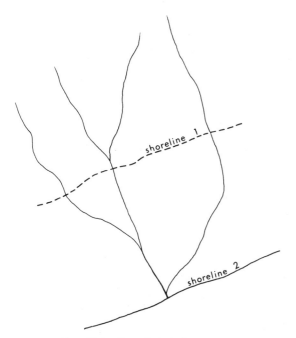

Fig. 22.9 Extended drainage on an emerged coastal plain.

The types of evidence whereby old shorelines are recognised will be briefly listed and discussed later in relation to specific areas. Morphological evidence includes shore platforms (Pl. 22.5a and b) raised stacks, notches and cliffs (Pl. 22.6) which have been either raised or depressed relative to sealevel. The general configuration of the coast is also useful in this regard. Rias, fjords and truncated drainage systems result from submergence (see for example Pls 3.1a, 17.4, 17.13), while extended drainage (Fig. 22.9) and smooth, flat coasts are in some areas indicative of emergence. The slope of river terraces may suggest their grading to a sealevel (baselevel) different from that existing now, but care is necessary in using these forms and it is difficult to ascertain the level to which they were graded unless the then shoreline can be located (Pl. 22.7). Submerged forests or man-made features, such as harbours and buildings which have been wholly or partially drowned,[37] may also be cited; though again care in application is necessary so as not to confuse silting up with submergence in this case. Finally, sediments or soils in alien environments are suggestive of elevational changes.

Plate 22.6 Raised stack standing some 8 m above present high water mark near Cape Willoughby, Kangaroo Island, South Australia. In (a) the basal notch marking the former high water mark is facing the camera and in (b) the general situation and the backing cliff are seen. *(Nicholas Twidale.)*

Plate 22.7 Terrace deposit grading to a level of some 6–8 m above present high tide level, Pearson Island, Great Australian Bight. Note that the interpretation of this form in terms of former sealevel rests on its projection downslope, and how far this is carried; hence the level of the old sealevel depends on the amount of coastal recession since its foundation. In this case, however, there is reason to believe that only minimum retreat of the coast has taken place in the late Pleistocene. *(Keith P. Phillips.)*

3. The Traditional Glacio-eustatic Scheme

For obvious reasons, the interglacial high stands of the sea are more easily recognised than the glacial low sealevels. Nevertheless, Quaternary low sealevels are widely recognised. Off the east coast of North America, the Franklin shore at between −75 and −100 m can be traced for 290 km.[38] On the Sahul Shelf, off northwestern Australia, there is a consistent and well marked step at −115 to −120 m which probably denotes the lowest Pleistocene sealevel.[39] Remnants of an intricate major drainage system have been identified on the Sunda Shelf between Borneo and Sumatra, at a depth of 90 m below sealevel (Fig. 22.10), again indicating drainage to the sealevel of a former glacial maximum. A lowering of sealevel during glacial maxima of at least 100 m is widely accepted; off the eastern and southern shores of North America evidence suggests a fall of as much as 140 m.

The classic area for Pleistocene high sealevels is the central Mediterranean, where it has been claimed the ancient shorelines and intervening breaks of slope form a flight of terraces extending several hundreds of metres above sealevel.[40] Since this thesis was first advanced in 1918, many such flights of shorelines have been recognised in many parts of the world. They apparently formed at successively lower levels, which suggests a gradual lowering of sealevel through the Pleistocene (Fig. 22.11), with

glacio-eustatic rises and falls superimposed, as it were, upon this long-term decline.[41] The several shorelines, the type areas for which are indicated by the names, stand at the elevations indicated in Table 22.1, and are interpreted as being related to preglacial and interglacial phases of higher sealevels.

If these are glacio-eustatic sealevels, it has correctly been reasoned that they should have affected the continental margins the world over. However, many remnants associated with such shorelines have been identified not because local evidence proves their association, but primarily because they ought to be there. Even where features can be cited as indicating a former stand of the sea at the level— and much evidence susceptible of alternative interpretation has been used—there is no evidence of their age, either relative or absolute. Caution should be exercised before suggesting that features are related to Quaternary high stands of the sea for several reasons. First, it has been shown conclusively that the types of sections in the central Mediterranean have been subjected to severe earth movements during the Quaternary, a fact which renders elevation itself invalid as a determinant of the age of the coastal features.[42] The same is true of several other areas where the traditional scheme has been applied uncritically.[43]

Second, the validity of the whole scheme has been called to question by a study of the behaviour of the Antarctic

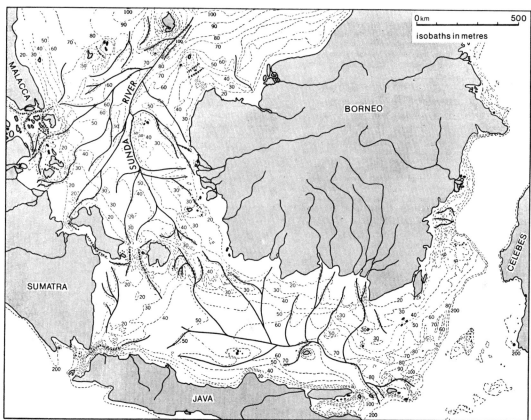

Fig. 22.10 The Sunda River in Pleistocene glacial times. *(After P.H.Kuenen,* Marine Geology, *Wiley, New York, 1950, p. 482, by permission of P.H. Kuenen.)*

Table 22.1
The Traditional Scheme of Quaternary High Sealevels

	Name	Average elevation		Age
		metres	feet	
Present				
	Monastirian (Main)	7.5	24	
	(Late)			Postglacial
	Tyrrhenian	18	58	Interglacial 3
	Milazzian	32	104	Interglacial 2
	Sicilian	60	195	Interglacial 1
Pliocene	Calabrian	100	330	Preglacial

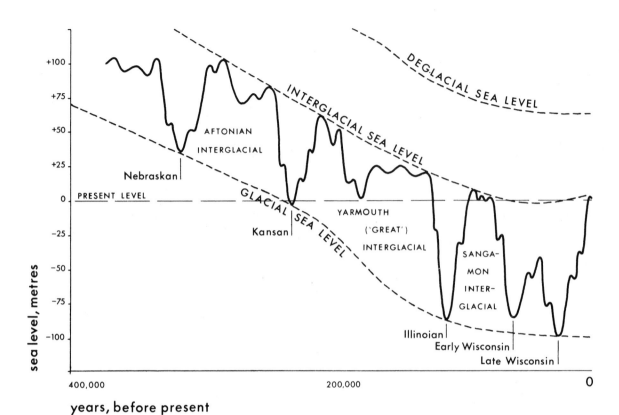

Fig. 22.11 Fluctuations of sealevel during the Pleistocene. *(After R.W. Fairbridge, "The Changing Level", p. 76. Copyright 1960 by Scientific American, Inc. All rights reserved.)*

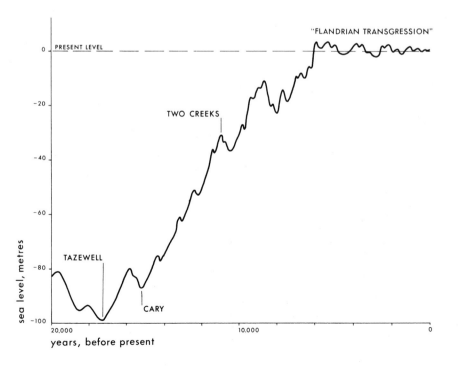

Fig. 22.12 Fluctuations of sealevel during the late Pleistocene and the Recent. *(After R.W. Fairbridge, "The Changing Level", p. 77. Copyright 1960 by Scientific American, Inc. All rights reserved.)*

ice-sheet.[44] The Antarctic ice mass is so huge that its size has an important bearing on sealevel. If it and all the other ice in the world melted, sealevel would rise by about 100–115 m. This approximates the preglacial level, and it has been tacitly assumed that considerable melting took place during each interglacial, returning the sealevel to elevations well above that of today. In Western Australia, however, ancient strandlines standing only 3 m above sealevel have been dated as 35,000 and 37,000 years old.[45] In other words (see Table 22.1), these date from the minor interglacial between the Early and Late Wisconsin glaciations, and should therefore be well above sealevel instead of a mere three metres above datum. Was the Antarctic ice mass essentially static during this interval? To what extent did it melt during the interglacials, and to what extent therefore can high Pleistocene sealevels be anticipated?

Third, several late and postglacial sealevels have been postulated quite close to modern datum (see Fig. 22.12) on the basis of platforms and other evidence; however, as suggested earlier (Chapter 18), at present several benches are forming simultaneously, close to modern sealevel.

There can be little doubt that several high Pleistocene stands of the sea have been recognised and substantiated. For example, in North Carolina, the Surry Scarp which is related to a Pleistocene interglacial high sealevel and is at an elevation of some 25–30 m, has been traced over considerable distances (Fig. 22.13). But in many instances the interpretation of raised beaches (using the term in the loose, though widely accepted sense, implying any form or deposit of a former strand line now located above sealevel) is difficult and complicated.

In the South Shetland Islands, for instance flights of surfaces and benches occur between 275 m above and 110 m below sealevel. Those above about 120 m are considered to be of subaerial origin, but phases of marine planation are responsible for benches at 185–102, 27–50 and 11–17 m. However the islands have suffered recent isostatic recovery and only the major benches at 18.5 and 6 m represent true high stands of sealevel.[47]

Again in the South East of South Australia, a magnificent series of old coastal foredunes is preserved (Fig. 22.14) but their interpretation is rendered impossible by undoubted tectonic deformation, by our present inability to identify any sort of shoreline datum relevant to the Pleistocene in the area, and by evidence that the sequence does not represent a simple emergence.[48]

Even the history of sealevel changes over the past few thousands of years is controversial. On the one hand Fairbridge[49] argues for a postglacial rise of the sea to about its present level some 6000 years ago, since which it has been essentially steady though with minor fluctuations (Fig. 22.12). But others have suggested rather different rates and chronologies of postglacial rise of sealevel.[50] Some evidence points to a continuous rise of sealevel from 7000 to 2000 years B.P. But other features in other parts of the world are suggestive of a diminishing rate of rise of sealevel over the past 6000 years. In both cases there is however disagreement with Fairbridge's curve. Thus doubts as to tectonic stability, the behaviour of the Antarctic ice mass, and the precise effects of present processes suggest that all possible Pleistocene shorelines should be carefully examined

and judged on the available evidence, and not "identified" and dated according to the tenets of a scheme which is surely illusory, despite its widespread acceptance.

D. **SUBMARINE CANYONS**

Some apparent submarine canyons located by geophysical methods are cut in Precambrian rocks and buried beneath Tertiary sediments at the northern margin of the Eucla basin in South Australia. Thus these canyons are not restricted in their development to the Pleistocene. Nevertheless submarine canyons were widely initiated during the Pleistocene because conditions were particularly suitable to their formation then.

Submarine canyons are deep gorges scoured in the continental slope which head back into the shelf. They are not the only channels found on the sea-floor, though they are the most widespread and the most spectacular. Other channels developed on the shelf include drowned river

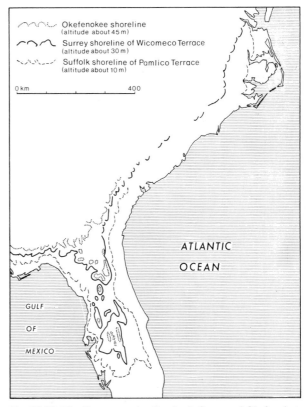

Fig. 22.13 The Sangamon (or Pamlico), Surry and Okefenokee shorelines, all of Quaternary age, in the eastern USA. *(From Phillip B. King, "Quaternary Tectonics in middle North America" in H.E. Wright and David G. Frey (Eds), The Quaternary of the United States, (copyright 1965 by Princeton University Press), Fig. 5, p. 838. Reprinted by permission of Princeton University Press.)*

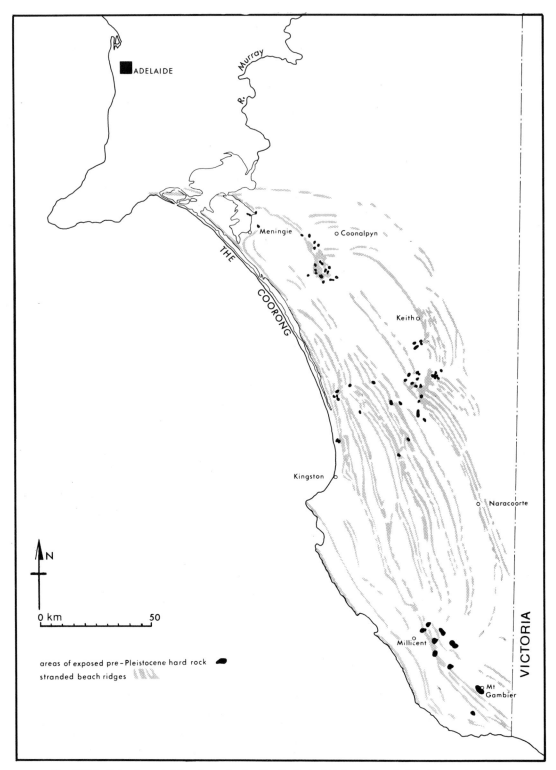

Fig. 22.14 Raised coastal foredunes in the South East of S.A. *(After G. Blackburn et al., "Soil Development".)*

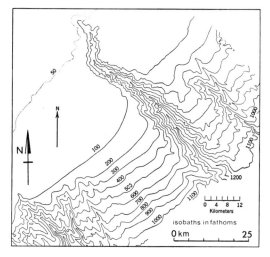

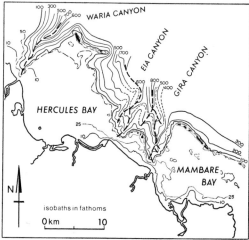

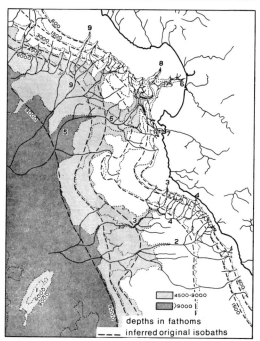

valleys, such as those of the Hudson, Elbe and Rhine, (see Fig. 21.17) and notably the "Sunda River", which consists of an intricate branching system of channels at a depth of about 90 m on the Sunda Shelf, (Fig. 22.10).

A few channels are due to tidal scouring or more probably to non-deposition in narrow passages between land masses, and others are glacial troughs which are morphologically similar to fjords—of which they are believed to be continuations.

But submarine canyons are essentially features of the continental slope. Some extend across the slope for a short distance,' but they then plunge down the slope to great depths (Fig. 22.15). Their walls are steep, and they are usually V-shaped in cross-section, though detailed surveys of some canyon heads near La Jolla, southern California, have shown that they are flask shaped, and that they widen with depth. Canyons pursue sinuous courses, are steepest in the headwater regions and flatten near their termini. They are usually accordant, though one or two hanging valleys have been detected. Some canyons extend to about 6000 m below present sealevel, and some extend to greater depths in rather trough-shaped depressions.

Dredging has revealed that the canyons are excavated in rocks of various types and ages. For instance the Carmel Canyon, off the coast of southern California, is cut wholly in granite, but others are eroded in sediments which range in age from old to young, but which in some cases include strata of Pliocene age. Submarine canyons have been reported from all parts of the world: the greatest number have been described from the coastal margins of the USA, but this reflects the extensive submarine surveys carried out there rather than any genetic factor; there are enough from other parts of the world—Congo, New Guinea, Tagus, Indus, Rhone—to suggest they have a world-wide distribution. Many canyons occur off the southern coast of Australia (Fig. 22.16a-d) including the Murray Canyon, (Fig. 22.16a and b) which with a depth of 2 km is one of the deepest so far discovered anywhere.[51]

Many hypotheses have been advanced in explanation of these features. They have been attributed to faulting; but their sinuous courses and their discordance with known tectonic trends are irreconcilable with this suggestion, as is the accordance of the shelves on both sides of the canyons. This is is not to imply that tectonism plays no part in the development of any canyon, for parts of some canyons off southern California trend parallel to the regional tectonic trend. It is not, however, a general explanation.

Fig. 22.15 Contour plans of submarine canyons: (a) Hudson Canyon, east coast USA; (b) several canyons off the north coast of New Guinea; (c) off the coast of California—1) Davison Seamount; 2) Lucia Canyon; 3) Sur Canyon; 4) Partington Canyon; 5) Monterey Trough; 6) Monterey Canyon; 7) Carmel Canyon; 8) Soquel Canyon; 9) Ascension Canyon. *(After P.H. Kuenen,* Marine Geology, *by permission of P.H. Kuenen.)*

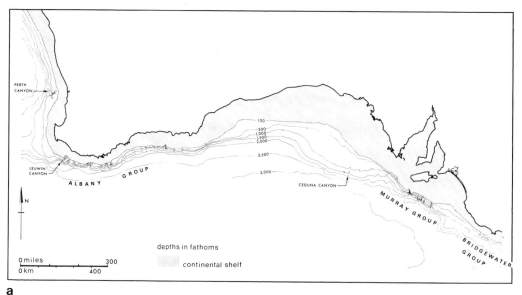

a

Fig. 22.16a Distribution of submarine canyons off the coast
of southern Australia.

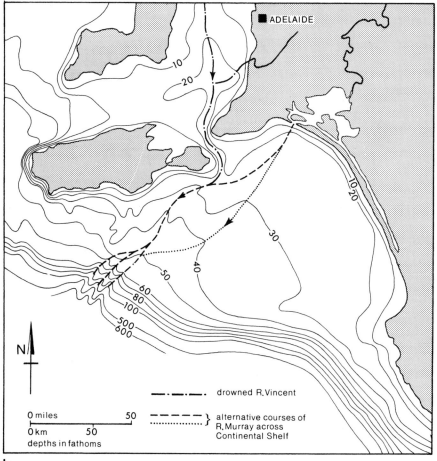

Fig. 22.16b The Murray canyons and the truncated drainage
related to them. (*After R.C.
Sprigg, "Submarine Canyons
of the New Guinea and Australian Coasts",* Trans. R. Soc.
S. Aust., *1947,* **71***, pp. 296–
310.*)

b

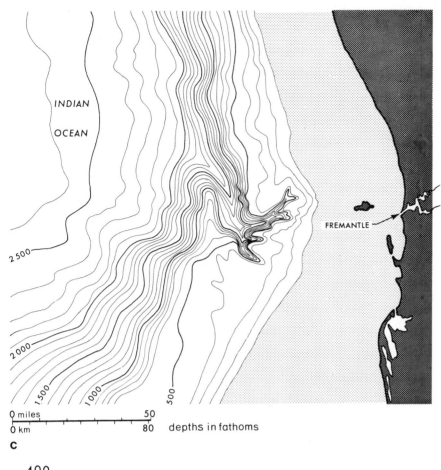

Fig. 22.16c Plan of the Perth Canyon.

INDIAN

OCEAN

FREMANTLE

2 500

2 000

1 500

1 000

500

0 miles   50

0 km   80   depths in fathoms

c

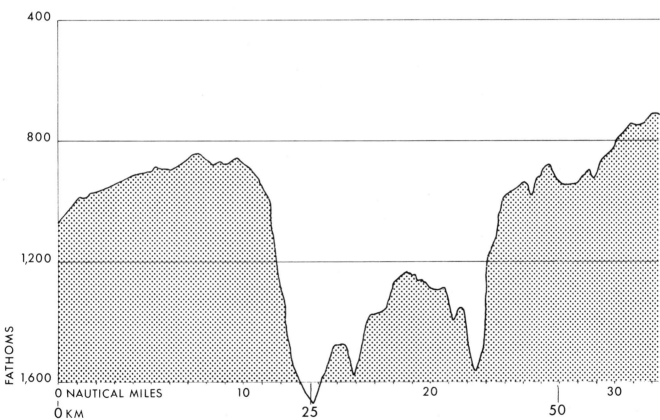

400

800

1,200

FATHOMS

1,600

0 NAUTICAL MILES   10   20   30

0 KM   25   50

d

Fig. 22.16d Cross-section of the Leeuwin Canyon, W.A. (Figs 22.16a, c, d after C. von der Borch, "Southern Australian Submarine Canyons", pp. 395–404 by permission of Mar. Geol.)

It has been suggested[52] that the canyons are riverine gorges cut in the continental margins when these areas were tectonically raised in response to tensional forces. However, there is no evidence of such activities and none of the consequences expected of such deformation exist; for instance, if the continental edges were raised, the drainage should have been diverted inland, but this is not evidenced by accumulated fluviatile deposits. The same comment applies to the theory of submarine or spring sapping[53] though in addition this idea is inherently unlikely. Tsunamis, or tidal waves, caused by earthquake shocks may contribute to canyon development in some areas by rendering masses of debris unstable, as may hydraulic currents when set upon by strong onshore winds along the shallow sea-floor,[54] but neither explanation can account for all features of the canyons. Two mutually exclusive theories have emerged as possible explanations of submarine canyons.

The first is that they are of subaerial origin, that is, normal river gorges, which have subsequently been submerged.[55] The similarities between submarine canyons and terrestrial gorges are indeed striking, and Shepard considered that they had been eroded by rivers during the Pleistocene, at times of low sealevel, and then drowned with the return of water to the ocean basins. The great difficulty facing this hypothesis is that, whereas a lowering of sealevel of the order of 100 m during glacial periods of the Pleistocene is accepted, the canyons commonly extend to depths of 200 m and deeper in some instances. Shepard suggested that the ice-sheets were thicker and more extensive than usually believed and that more water was extracted from the oceans, and further that with the lowering of sealevel the continental margins were upwarped; when water returned to the oceans on the melting of the ice-sheets, the continental margins were again depressed under its weight. However, although there is some evidence for these events (especially the isostatic marginal warping of the shelves) they do not either singly or together bridge the enormous gap between the depth of the ocean during the glacials and the lower limits of the canyons.

Thus the drowned river-eroded gorge hypothesis is difficult to support as a general explanation of submarine canyon formation. But there is one region where because of exceptional circumstances it seems applicable. Through much of the early Tertiary the present Mediterranean basin was not occupied by the sea but formed a depression over $1\frac{1}{2}$ km deep into which poured rivers from the European and African continents and the floor of which was a great salina. The incoming rivers eroded great gorges as they plunged into the deep basin and when the sea eventually entered the latter to form the present Mediterranean Sea the gorges were drowned to form submarine canyons.[56]

The second major theory is more probable as a general explanation. It states that submarine canyons are eroded by turbidity currents—rivers of water heavily laden with

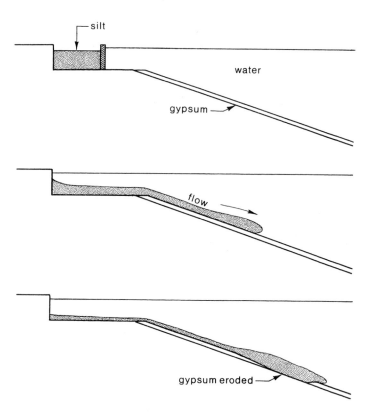

Fig. 22.17 Experimental induction of turbidity currents. (*After P.H. Kuenen, "Density Currents", pp. 241–9, by permission Geol. Mag.*)

silt and other debris which run down the continental shelf and slope, and which can erode some rocks, though there remains a question as to whether they are capable of excavating channels in rock as hard as granite. The suggestion that turbidity currents are responsible for canyons is due to Daly and Kuenen,[57] the latter having demonstrated experimentally that such currents can erode soft materials such as gypsum (Fig. 22.17). It is argued that during the Pleistocene glacials, unconsolidated sediments were churned up on the exposed continental shelves and were carried out to sea by hydraulic currents, which were more dense than normal seawater (with one per cent silt added, the specific gravity of seawater is increased from 1·026 to 1·040). This heavier water tended to concentrate in and flow down chance depressions in the shelf, forming a suspension or turbidity current, which was capable of erosion and thus of cutting a canyon in the outer part of the shelf, especially on the steep gradients of the continental slope.

Such a hypothesis accounts for the similarity between canyons and the gorges cut by terrestrial rivers: in the latter the water transports sediment, but in the submarine current it is more a case of the sediment carrying along water. The world-wide distribution and Pleistocene age of the canyons is also satisfactorily explained.

Suspension currents are known to occur in both natural and artificial lakes; slumping has been witnessed in the Redondo Canyon, off southern California; the occurrence of canyons off the mouth of main rivers (Tagus, Indus, Hudson) is also explained, for the rivers bring sediment to the sea and thus provide debris for the currents. The role of tsunamis is taken into account, for they provide debris for the currents. The hypothesis also accounts for the common origin of canyons off the Pacific coast of USA on the northward side of projecting headlands (Fig. 22.18) where, because of the prevalent northwestward wind, consequent southerly longshore drift, and accumulation of debris in the embayments on the northern sides of headlands, there is an abundance of material with which to initiate turbidity currents.

Finally, impressive evidence that turbidity currents exist comes from studies of telegraphic cables. On 18 November, 1929, an earthquake occurred off Newfoundland. The transatlantic telegraph cables were broken, and the disruption was attributed to the direct effects of the earthquake. However Heezen and Ewing,[58] many years later, took the trouble to examine the precise schedule of service breaks. They found that the cables broke at different times, and not simultaneously as might be expected had their rupture been due to faulting. It was discovered that the

cable breaks were delayed progressively away from the earthquake epicentre, and away from the land margin: the further downslope the break, the later it occurred. Heezen and Ewing suggested that the earthquake initiated a turbidity current and that it was this current which, flowing downslope, had broken the cables and disrupted the telegraph service. Similar conclusions were reached from a study of cable breaks between France and Algeria during an earthquake centred near Orleansville on 9 September, 1954.

Thus, although this hypothesis is not universally accepted as the sole cause of submarine canyons, the case for it is very strong.

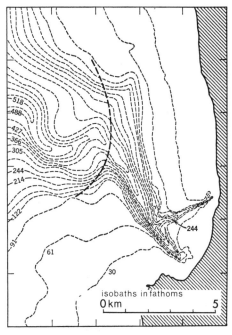

Fig. 22.18 Submarine canyons located on the northern side of a headland on the California coast: the La Jolla and Scripps canyons. (*After P.H. Kuenen,* Marine Geology, *p. 515, by permission of P.H. Kuenen.*)

References Cited

1. T.L. Péwé, R.E. Church and M.J. Andersen, "Origin and Palaeoclimatic Significance of Large-scale Patterned Ground in the Donnelly Dome area, Alaska", *Geol. Soc. Am. Spec. Paper*, 1969, p. 103.

2. T. Aartolahti, "On Patterned Ground in southern Finland", *Ann. Acad. Scient. Fennicae, Series A* (III—Geol. and Geogr.), 1969, **99**, p. 104.

3. J-C. Dionne, "Formes de Cryoturbation Fossilés dans le sud-est du Québec", *Geogr. Bull.*, 1967, **9**, pp. 13–24.

4. A. Straw, "Periglacial Mass Movement on the Niagara Escarpment near Meaford, Grey County, Ontario", *Geogr. Bull.*, 1968, **8**, pp. 369–76.

5. M. Boyé, "Gélivation et Cryoturbation dans le Massif du Mont-Perau (Pyrenées Centrales)", *Inst. Estud. Piren. (Saragossa)*, 1952.

6. C. Troll, "Strukturboden, Solifluktion und Frostklimat der Erde", *Geol. Rdsch.*, 1944, **34**, pp. 546–672.

7. D.L. Linton, "Evidences of Pleistocene Cryonival Phenomena in South Africa", *Paleoecol. Afr.*, 1969, **5**, pp. 71–87.

8. L.F. Curtis and J.H. James, "Frost-heaved Soils of Barrow Rutland", *Proc. geol. Soc. London*, 1959, **20**, pp. 310–14; E.M. Bridges, "Examples of Periglacial Phenomena in Derbyshire", *East. Midl. Geogr.*, 1964, **3**, pp. 262–6.

9. A.B. Costin, "Mass Movements of the Soil Surface with Special Reference to the Monaro Region of New South Wales", *J. Soil Conserv. Serv. Soc. Tas.*, 1958, pp. 151–4; J.L. Davies, "Tasmanian Landforms and Quaternary Climates" in J.N. Jennings and J.A. Mabbutt (Eds), *Londform Studies from Australia and New Guinea*, Australian National University Press, Canberra, 1967, pp. 1–25; R.W. Galloway, "Glaciation in the Snowy Mountains—a Reappraisal", *Proc. Linn. Soc. N.S.W.*, 1963, **88**, pp. 180–98.

10. E. Reiner, "The Glaciation of Mount Wilhelm, Australian New Guinea", *Geogr. Rev.*, 1960, **50**, pp. 491–503; E. Loeffler, "The Pleistocene Glaciation of the Saruwaged Range, Territory of New Guinea", *Aust. Geogr.*, 1920, **11**, pp. 463–72.

11. V. Auer, "The Pleistocene of Fuego-Patagonia", *Ann. Acad. Scient. Fennicae, Series A* (III—Geol. and Geogr.), 1970, **100**, p. 194.

12. C. Wahrhaftig, "Physiographic Divisions of Alaska", *U.S. Geol. Surv. Prof. Paper*, 1965, p. 482.

13. G.C. Maarleveld and J.C. van den Toorn, "Pseudosolle in Noord-Nederland", *Tidjschr. Kon. Ned. Aar. Genootsch.*, 1955, **74**; A. Pissart, "Les Traces de 'Pingos' du Pays de Galles (Grande Bretagne) et du Plateau des Hautes Fagnes (Belgique)", *Z. Geomorph.*, 1963, **7** NS, pp. 147–65; R.T. Slotbloom, "Comparative Geomorphological and Palynological Investigation of the Pingos (Viviers) in the Hautes Fagnes (Belgium) and the Mardellen in the Gutland (Luxembourg)", *Z. Geomorph.*, 1963, **7** NS, pp. 193–231.

14. G.K. Gilbert, "Lake Bonneville", *U.S. Geol. Surv. Monogr.*, 1890, **1**; R.B. Morrison, "Quaternary Geology of the Great Basins", in H.E. Wright and D.G. Frey (Eds), *The Quaternary of the United States*, Princeton University Press, Princeton, 1965, pp. 265–85; R.B. Morrison and J.C. Frye, "Correlation of the Middle and Late Quaternary Successions of the Lake Lahontan, Lake Bonneville, Rocky Mountain (Wasatch Range), southern Great Plains and eastern Midwest Areas", *Nevada Bur. Mines Rep.* **9**, 1965.

15. M.D. Crittenden, "New Data on the Isostatic Deformation of Lake Bonneville", *U.S. Geol. Surv. Prof. Paper*, 1963, **454-E**.

16. R.F. Flint and W.A. Gale, "Stratigraphy and Radiocarbon Dates at Searles Lake, California", *Am. J. Sci.*, 1958, **258**, pp. 689–714.

17. C.C. Reeves, "Pluvial Lake Basins in west Texas", *J. Geol.*, 1966, **74**, pp. 269–91.

18. C.P. Berkey and F.K. Morris, *Geology of Mongolia*, American Museum of Natural History, New York, 1927.

19. A.T. Grove and R.A. Pullan, "Some Aspects of Pleistocene Palaeogeography of the Chad Basin", in F.C. Howell and F. Bouliere (Eds), *African Ecology and Human Evolution*, Methuen, London, 1964, pp. 230–45.

20. See E.S. Hills, "Die Landoberfläche Australiens", *Die Erde*, 1955, **7**, pp. 195–205.

21. See D. King, "The Quaternary Stratigraphic Record on Lake Eyre North and the Evolution of Existing Topographic Forms", *Trans. R. Soc. S.Aust.*, 1956, **79**, pp. 93–103; H. Wopfner, "Environment and Age of the Lake Eyre Basin", *Trans R. Soc. S.Aust.*, in press; C.R. Twidale, "Evolution of Sand Dunes in the Simpson Desert, central Australia", *Trans. Inst. Br. Geogr.*, 1972, **56**, in press.

22. G.F. Brown, "Geomorphology of western and central Saudi Arabia", *Rep. 21st Int. Geol. Cong. Copenhagen*, 1960, **21**, pp. 150–9.

23. G.E. Williams, "Glacial Age of Piedmont Alluvial Deposits in the Adelaide Area, South Australia", *Aust. J. Sci.*, 1969, **32**, p. 257.

24. J.N. Jennings, "Some Karst Areas of Australia", in J.N. Jennings and J.A. Mabbutt (Eds), *Landform Studies*, pp. 256–92; G.W. Krieg, "Relic Drainage in the Serpentine Lakes Area", *Q. Geol. Notes Geol. Surv. S.Aust.*, 1971, **37**, pp. 1–2; J.A. Mabbutt, "Aeolian Landforms in central Australia", *Aust. Geogr. Stud.*, 1968, **6**, pp. 139–50; H. Wopfner, "Lithology and Distribution of the Observatory Hill Beds, eastern Officer Basin", *Trans. R. Soc. S.Aust.*, 1969, **93**, pp. 169–85.

25. S. Pels, "Lake Quaternary Chronology of the Riverina Plain of southeastern Australia". *J. geol. Soc. Aust.*, 1966, **13**, pp. 27–40.

26. T. Langford-Smith, "The Dead River Systems of the Murrumbidgee", *Geogr. Rev.*, 1960, **50**, pp. 368–89; see also S.A. Schumm, "River Adjustment to Altered Hydrologic Regimen—Murrumbidgee River and Palaeochannels, Australia", *U.S. Geol. Surv. Prof. Paper*, 1968, **598**.

27. B.E. Butler, "Theory of Prior Streams as a Causal Factor in Soil Occurrence in the Riverine Plain of

southeastern Australia", *Aust. J. Agric. Res.*, 1950, **1**, pp. 231–52.

28. G.H. DURY, "Test of a General Theory of Misfit Streams", *Trans, & Papers Inst. Br. Geogr.*, 1958, **25**, pp. 105–18; *idem*, "Misfit Streams: Problems in Interpretation, Discharge and Distribution", *Geogr. Rev.*, 1960, **50**, pp. 230–3; *idem*, "Results of Seismic Exploration of Meandering Valleys", *Am. J. Sci.*, 1962, **260**, pp. 691–706; *idem*, "Principles of Underfit Streams", *U.S. Geol. Surv. Prof. Paper*, 1964, **452A**; *idem*, "Incised Valley Meanders on the lower Colo River, New South Wales", *Aust. Geogr.*, 1966, **10**, pp. 17–25.

29. W.F. GEYL, "Tidal Stream Action and Sea Level Changes as One Cause of Valley Meander and Underfit Streams", *Aust. Geogr. Stud.*, 1968, **6**, pp. 24–42.

30. C.R. TWIDALE, "Late Cainozoic Activity of the Selwyn Upwarp", *J. geol. Soc. Aust.*, 1966, **13**, pp. 491–4.

31. A.T. GROVE, "The Ancient Erg of Hausaland", *Geogr. J.*, 1958, **124**, pp. 528–33; A. WARREN, "Dune Trends and their Implications in the central Sudan", *Z. Geomorph. Suppl.*, 1920, **10**, pp. 154–80.

32. Y. AKAGI, "Pediment Morphology in Korea", *Geogr. Rev. Jap.*, 1965, **38**, pp. 20–36.

33. H.F. GARNER, "Stratigraphic-sedimentary Significance of Contemporary Climate and Relief in Form Regions of the Andes Mountains", *Bull. geol. Soc. Am.*, 1959, **70**, pp. 1327–68; *idem*, "Tropical Weathering and Relief" in R.W. Fairbridge (Ed.), *Encyclopaedia of Geomorphology*, Reinhold, New York, 1969, pp. 1161–72.

34. J.J. BIGARELLA and G.O. DE ANDRADE, "Contribution to the Study of the Brazilian Quaternary", *Geol. Soc. Am. Spec. Paper*, 1965, **84**, pp. 433–51.

35. H.F. GARNER, "Derangement of the Rio Caroni, Venezuela", *Rev. Geomorph. Dyn.*, 1966, **16**, pp. 54–83.

36. H.W. WELLMAN, "Delayed Isostatic Response and High Sea-levels", *Nature (Lond.)*, 1964, **202**, pp. 1322–23; A.L. BLOOM, "Pleistocene Shorelines: a New Test of Isostasy", *Bull. geol. Soc. Am.*, 1967, **76**, pp. 537–66; C.G. HIGGINS, "Isostatic Effects of Sea-level Changes" in H.E. Wright (Ed.), *Quaternary Geology and Climate*, Proc. VII Congress INQUA, pp. 141–5.

37. N.C. FLEMMING, "Archaeological Evidence for Eustatic Change of Sea-level and Earth Movements on the western Mediterranean during the last 2000 Years", *Geol. Soc. Am. Spec. Paper.*, 1969, **109**.

38. A.C. VEATCH and P.A. SMITH, "Atlantic Submarine Valleys of the United States and the Congo Submarine Valley", *Geol. Soc. Am. Spec. Paper*, 1939, **7**.

39. R.W. FAIRBRIDGE, "The Sahul Shelf, northern Australia: its Structure and Geological Relationships", *J.R. Soc. W.A.*, 1953, **27**, pp. 1–33.

40. DE LAMOTHE, "Les Anciennes Lignes de Rivage du Bassin de la Somme et leur Concordance avec celles de la Mediterranée occidentale", *C.R. Acad. Sci. Paris*, 1916, **162**, pp. 948–51; C. DEPERET, "Essai de Co-ordination Chronologique general des Temps Quaternaires", *C.R. Acad. Sci. Paris*, 1918, **167**, pp. 418–22.

41. See R.W. FAIRBRIDGE, "The Changing Level of the Sea", *Sci. Am.*, 1960, **202** (5), pp. 70–9; *idem*, "Eustatic Changes in Sea-level" in *Physics and Chemistry of the Earth*, 1961, **4**, pp. 99–185.

42. G. CASTANY and F. OTTMANN, "Le Quaternaire Marin de la Mediterranée occidentale", *Rev. Géogr. Phys. Géol. Dyn.*, 1957, **2**, pp. 46–55.

43. W.T. WARD, "Eustatic and Climatic History of the Adelaide Area, South Australia", *J. Geol.*, 1965, **73**, pp. 592–602.; *idem*, "Geology, Geomorphology, and Soils of the South-Western Part of County Adelaide, South Australia", *C.S.I.R.O. Soil Pub.*, 1966, **23**.

44. J.T. HOLLIN, "On the Glacial History of Antarctica", *J. Glaciol.*, 1962, **4**, pp. 173–95.

45. R.J. RUSSELL, "Recent Recession of Tropical Cliffy Coasts", *Science*, 1963, **139** (3549), pp. 9–15.

46. See for instance W.T. WARD and R.W. JESSUP, "Changes in Sealevel in Southern Australia", *Nature (Lond.)* 1965, **205** (4973), pp. 791–2.

47. B.S. JOHN and D.E. SUGDEN, "Raised Marine Features and Phases of Glaciation on the South Shetland Islands", *Br. Antarct. Surv. Bull.*, 1971, **24**, pp. 45–111.

48. P.S. HOSSFELD, "Lake Cainozoic History of the southeast of South Australia", *Trans. R. Soc. S.Aust.*, 1950, **73**, pp. 232–79; R.C. SPRIGG, "Geology of the Southeast Province, South Australia", *Bull. Geol. Surv. S.Aust.* 1952, **29**; G. BLACKBURN, R.D. BOND and A.R.P. CLARKE, "Soil Development associated with Stranded Beach Ridges in southeast South Australia", *C.S.I.R.O. Soil Pub.*, 1965, **22**.

49. R.W. FAIRBRIDGE, "Eustatic Changes of Sea-level".

50. F.P. SHEPARD and J.R. CURRAY, "Carbon-14 Determination of Sealevel Changes in Stable Areas", *Progr. Ocean.*, **4**; *idem*, *The Quaternary History of the Ocean Basins*, Pergamon, New York, 1967.

51. J.R. CONNOLLY, "Submarine Canyons of the Continental Margin, East of Bass Strait (Australia)", *Mar. Geol.*, 1968, **6**, pp. 449–61; C. VON DER BORCH, "Southern Australian Submarine Canyons: their Distribution and Ages", *Mar. Geol.*, 1968, **6**, pp. 267–79.

52. A. DU TOIT, "An Hypothesis of Submarine Canyons", *Geol. Mag.*, 1940, **77**, pp. 395–404.

53. D.W. JOHNSON, *Origin of Submarine Canyons*, Columbia University Press, New York, 1939.

54. W.M. DAVIS, "Submarine Mock Valleys", *Geogr. Rev.*, 1934, **24**, pp. 297–308.

55. F.P. SHEPARD, *Submarine Geology*, Harper, New York, 1948.

56. K.J. HSU, "When the Mediterranean Dried Up", *Sci. Am.*, 1972, **227**(6), pp. 27–36.

57. R.A. DALY, "Origin of Submarine Canyons", *Am. J. Sci.*, 1936, **31**, pp. 401–20; P.H. KUENEN, "Density Currents in Connection with the Problem of Submarine Canyons", *Geol. Mag.*, 1938, **75**, pp. 241–9.

58. B. HEEZEN and M. EWING, "Turbidity Currents and Submarine Slumps and the 1929 Grand Banks Earthquake", *Am. J. Sci.*, 1952, **250**, pp. 849–73; B. HEEZEN, "The Origin of Submarine Canyons", *Sci. Am.*, 1956, **195** (2), pp. 36–41.

CHAPTER 23

Older Inherited Forms

A. GENERAL

Most higher latitude regions are dominated by erosional and depositional landforms dating from the Pleistocene, which are superimposed on features which originated in earlier times under diverse climatic conditions. Only a few landforms of greater antiquity survive in areas affected by Pleistocene glaciations.

In lower latitudes, however, and especially in the arid and semiarid tropics, pre Quaternary features are widespread and are dominant landform assemblages in some areas. For instance, in many tropical regions duricrusts of various mineralogical types form caprocks to ancient erosional surfaces of low relief. Where they are dissected they form plateau, mesa and butte assemblages (see Chapters 5 and 10).

Silcrete-capped plateau forms dominate the landscape in many parts of central and southern Australia (Pl. 23.1), and laterite-capped residuals are common in many parts of the continent.[1] These duricrusts were formed in the Mesozoic and Tertiary, probably under climatic conditions which no longer obtain in the particular regions where the residuals are preserved; they survive partly by virtue of their inherent toughness, and partly by virtue of the arid climates in which some of them now exist, and the consequent slow rate of weathering and erosion there. Some survive because of the resistance of the strata in which they are cut. Ancient landforms more commonly have survived because they were buried after their development and were thus protected by sediments or possibly volcanic lavas, and have only recently been re-exposed to the elements.

Plate 23.1 Silcrete dissected to form plateaux, Rumbalara, Northern Territory. (*C.S.I.R.O.*)

B. PLIOCENE MARINE BENCH IN SOUTHEASTERN ENGLAND

The major landforms of southeastern England are related to gentle folds in the Chalk (see Pl. 3.33). But the Chalk has been eroded, and a rather complex history of denudation has been elicited from the field evidence. Early Tertiary strata were deposited in the plunging syncline which underlies the lower Thames valley, and in places the Chalk surface on which they were laid down has been re-exposed (by subsequent erosion) as a pre Eocene exhumed surface (Fig. 23.1). There is a distinct surface of low relief high in the Chalk uplands in the Chilterns and in the North and South Downs, which has been interpreted as a peneplain of Miocene–Pliocene age.[2] Below this, at an elevation of 180–210 m and cut into the dip slope of the Chalk in the London Basin, is a narrow bench with which marine sediments of Pliocene age are associated (Fig. 23.1). The bench, which appears to be represented in the Hampshire Basin and the Weald, is clearly a shore platform of Late Tertiary age which has been well preserved in the London Basin by virtue of the relative immunity of the Chalk to stream erosion. In adjacent areas, however, dissection has taken effect and in the Weald, for instance, the feature is represented only by accordant hill summits on the flanks of the chalk slopes. A platform of similar age and origin is preserved in southwest Cornwall, for instance on The Lizard.[3]

C. PERMO-TRIASSIC ARID LANDFORMS

Permo-Triassic desert sandstones are found in many parts of the world. Associated salt deposits are of economic significance in such areas as Stassfurt, north Germany, and the Tees and Cheshire areas of England. The sandstones display all the characteristics of modern desert sands, with rounded, frosted grains of quartz, often a patina of iron oxide, an absence of mica grains, and good aeolian cross-bedding (see Chapter 15 and Appendix I). The bedding in these sandstones, and in those of Tertiary (Uinta and Chuska sandstones) or Jurassic (for example, Navajo—Pl. 23.2, Coconino, Tensleep, Casper) ages which abound in the western United States, enables the direction of the prevailing strong wind at the time of deposition to be determined. For example, the Lower Permian sandstones of Britain were deposited by winds which were easterly with reference to the modern poles. Though the probability of continental drift complicates the usefulness of such deductions, former circulations can be reconstructed by the reorientation of the continents on the basis of palaeomagnetic determination.[4]

However, in the Parana Basin of southern South America (Uruguay, Paraguay, northern Argentina, southern Brazil), dune forms of Triassic age are preserved beneath lava flows (Fig. 23.2), and these are clearly discernible where they have been sectioned for railway cuttings and other excavations.[5]

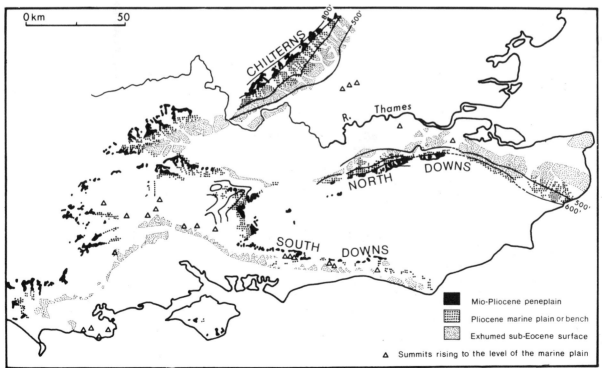

Fig. 23.1 Map of southeastern England, showing the 200 m marine bench eroded in the dip slope of the Chiltern Hills and on the flanks of the Downs. *(After S.W. Wooldridge and D.L. Linton,* Structure, Surface and Drainage, *p. 59.)*

Plate 23.2 Aeolian cross-bedding in Navajo Sandstone, Zion National Park, Utah. (*Nat. Parks Service, U.S. Dept Int.*)

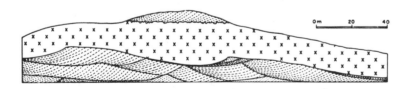

Fig. 23.2 Barchans preserved beneath basalt lava flow and exposed in railway cutting, Parana Basin, Paraguay. (*After F.F.M. de Almeida, "Botacatu", pp. 9—24.*)

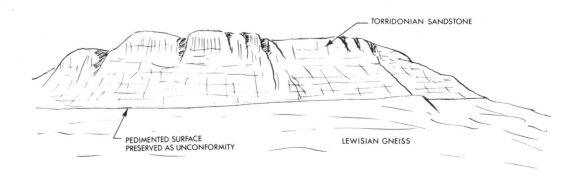

Fig. 23.3 An ancient pediment in northwest Scotland represented by the unconformity between the Lewisian gneisses below and Torridonian Sandstone above. (*Sketched from a photograph in G.E. Williams, "Characteristics and Origin", pp. 183—207.*)

During the later Triassic, the Parana Basin was a broad shallow depression. At one stage it was occupied by a field of dunes no less than 1·3 million km² in area. The sands of which these dunes were built constitute the Botacatu Sandstone, a typical aeolian deposit. The cross-bedding indicates that the deposit was laid down by winds from the "north", that is, the north in terms of the present orientation of the continent. The nature of the bedding (curvilinear strikes), plus the morphology of the dunes preserved beneath the basalts, show that the dunes were crescentic in shape, with distinct avalanche slopes on the lee side: barchans.

Wind-faceted pebbles (*dreikanter*) occur in the Permo-Triassic deposits of the English Midlands. This and other characteristics of the formation indicate that they formed under tropical desert conditions. At Charnwood in Leicestershire, Precambrian rocks overlain by Triassic strata display fretting and other signs of sand-blasting.[6]

D. PERMIAN GLACIAL FEATURES

Glacial or fluvioglacial sediments of Late Palaeozoic age, principally Permian, occur quite widely in the southern continents.[7] The age of ancient glacial sediments can be inferred from the ages of the underlying and overlying sediments; or established by tracing them laterally into fossiliferous marine sediments. They are associated with polished and striated pavements, and with erratic blocks, though most of these remain buried and do not form part of the present landform assemblage. Evidence of Permian glaciation has been reported from several parts of South America (Argentina, Uruguay, southern Brazil), the Falkland Islands, southern Africa (South Africa, with the well-known Dwyka Tillite, South West Africa and Angola), Madagascar, and Australia (N.S.W., Victoria, South Australia, Northern Territory and Western Australia). Evidence is also reported from India (the Talchin Tillite of the Central Provinces, Bihar and Orissa), and from east Turkestan. In a few areas relics of these ancient glacial landscapes, buried beneath sediments either of glacial or nonglacial origin, have been exhumed, and they form minor but interesting elements of the present landform assemblage. Thus, near Bacchus Marsh and adjacent areas of Victoria[8] and in several parts of the southern Mt Lofty Ranges,[9] well-preserved Permian glaciated pavements are exposed (Pl. 23.3). Naturally the inferred direction of ice movement varies from locality to locality, but in South Australia, for instance, the general movement is northwesterly.

Resting on the Permian pavement at Halletts Cove, a few kilometres south of Adelaide (a site of great historical significance since it was here Walter Howchin established the pre Quaternary age of the feature) is an erratic block of Sturtian (Precambrian) tillite—an unusual example of multi-glaciation insofar as the two ice ages are rather more than 500 million years apart in time. Erratics related to the Permian glaciation are common and are important indicators of the direction of ice movement. Granite boulders from Victor Harbour, for instance, are scattered through the southern Mt Lofty Ranges, though some erratics in this area apparently derived from the South East of South Australia. Features around Victor Harbour, and in the Inman Valley generally, have been interpreted in terms of Permian glaciation. For example, The Bluff and Crozier Hill have been called large *roches moutonnées*, This remains to be proven however: the asymmetry of The Bluff is the reverse of what would be expected if it had been plucked by a northward-moving glacier, and the form of Crozier Hill is consistent with

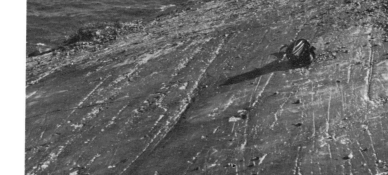

Plate 23.3 Permian glaciated pavements at Halletts Cove, south of Adelaide, South Australia (*C.R. Twidale.*)

structural control. But the presence of minor glacial forms of Permian age in this area is undeniable.

E. PRECAMBRIAN GLACIATED PAVEMENTS

Precambrian glacial or glacigene sediments have been described from no fewer than fifty-eight localities throughout the world according to Harland.[10] In a few, glaciated pavements are associated with the deposits: in India, Norway, North America, Africa, South Australia (though here some of the alleged glacial features appear to be of tectonic origin,[11]) and northwestern Western Australia. Seven such pavements have been described from the latter region[12] (see also Pl. 1.1). Overlain by a tillite, the pavement is eroded in rocks which are 1400–1800 million years old. Sediments overlying the tillite have been dated between 600 and 700 million years, so that the Precambrian age of the pavement cannot be doubted.

F. PRECAMBRIAN EROSION SURFACES

One of the oldest land surfaces yet identified has been described from the northwestern extremity of Scotland, near Cape Wrath.[13] It is a planate surface eroded in weathered Lewisian Gneiss, largely overlain and protected by late Precambrian Torridonian sediments, the bedding of which parallels the inclination of the underlying unconformity (Fig. 23.3). Where still buried, the surface is gently sloping and remarkably smooth, being disturbed by only one hill of

gneiss. Its character suggests a pediment surface eroded in the weathered crystalline rock and subsequently buried beneath a series of coalescing Torridonian alluvial fans. The nature of the weathering associated with the gneiss suggests a warm, moderately humid climate with dry periods or a dry season.[14] It may be that the pedimented surface evolved in an environment which was desert or semidesert as much in a biological (that is, lacking vegetation, and especially ground cover) as a climatic sense.

Another far more extensive Precambrian land surface has been described from northern and eastern Greenland[15] and the Canadian Shield.[16] Cut in crystalline rocks and displaying considerable relief in places,[17] it is of low relief over wide areas and is preserved and partially exhumed from beneath Proterozoic sediments. To a minor degree it is part of the present land surface (Fig. 23.4). No regolith is preserved, which suggests that this palaeoplain is of etch character.

Similar exhumed palaeosurfaces, of later Proterozoic age form minor but nevertheless significant elements of the present land surface on Eyre Peninsula north of Whyalla and in the Gawler Ranges east of Lake Gairdner.[18]

G. CONCLUDING COMMENT

Though not common, inherited landforms which predate the Quaternary have been recognised in several parts of the world. They are especially important in the arid tropics, but in several localities in the high latitudes they are emerging from beneath the sedimentary covers which have protected them, for hundreds of millions of years.

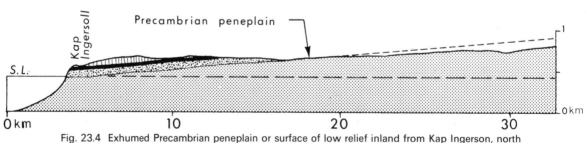

Fig. 23.4 Exhumed Precambrian peneplain or surface of low relief inland from Kap Ingerson, north Greenland. *(After J.W. Cowie, "Contributions to the Geology", by permission Med. Grønland and J.W. Cowie.)*

References Cited

1. W.G. WOOLNOUGH, "The Influence of Climate and Topography in the Formation and Distribution of the Products of Weathering", *Geol. Mag.*, 1930, **67**, pp. 123–32; C.R. TWIDALE, "Chronology of Denudation in northwest Queensland", *Bull. geol. Soc. Am.*, 1956, **67**, pp. 867–82; C.G. STEPHENS, "The Soil Landscapes of Australia", *C.S.I.R.O. Soil Pub.*, 1961, **18**; R.L. WRIGHT, "Deep Weathering and Erosion Surfaces in the Daly River Basin, Northern Territory", *J. geol. Soc. Aust.*, 1963, **10**, pp. 151–64; J.A. MABBUTT, "The Weathered Land Surface in central Australia", *Z. Geomorph.*, 1965, **9** NS, pp. 82–114.

2. S.W. WOOLDRIDGE and D.L. LINTON, *Structure, Surface and Drainage in southern England*, Philip, London, 2nd ed., 1955.

3. See for example E.M. LIND HENDRIKS, "The Physiography of south-west Cornwall, the Distribution of Chalk Flints, and the Origin of the Gravels of Crousa Common", *Geol. Mag.*, 1923, **60**, pp. 21–31.

4. A.E.M. NAIRNE, "Evidence of Latitude from Permian Palaeomagnetic Data" in A.E.M. Nairne (Ed.), *Problems in Palaeoclimatology*, Interscience, London, 1964, pp. 550–8.

5. F.F.M. DE ALMEIDA, "Botacatu, a Triassic Desert of South America", *19th Int. Geol. Cong., Algiers*, 1952, **VII**, pp. 9–24.

6. W.W. WATTS, "Charnwood Forest, a Buried Triassic Landscape", *Geogr. J.*, 1903, **21**, pp. 623–33.

7. A.L. DU TOIT, *Our Wandering Continents*, Oliver and Boyd, Edinburgh, 1937; J.C. CROWELL and L.A. FRAKES, "Lake Palaeozoic Glaciation of Australia", *J. geol. Soc. Aust.*, 1971, **17**, pp. 115–55.

8. E.S. HILLS, *Physiography of Victoria*, Whitcombe and Tombs, Melbourne, 1940, pp. 150–3; for an account of very early investigations in this area, see E.J. DUNN, "Glaciation in South Africa, Australia and Tasmania", *Geol. Mag.*, 1923, **60**, pp. 11–20.

9. A.R.C. SELWYN, "Geological Notes of a Journey in South Australia from Cape Jervis to Mount Serle", *Parl. Paper Adelaide*, 1859, **20**, p. 4; R. TATE, "Glacial Phenomena in South Australia", *Aust. Ass. Adv. Sci.*, 1887, pp. 232–3; W. HOWCHIN, "Evidence of Glaciation in Hindmarsh Valley and Kangaroo Island", *Trans. R. Soc. S. Aust.*, 1901, **27**, pp. 75–90; R.P. BOURMAN, "Landform Studies near Victor Harbour", unpublished B.A. (Hons) thesis, University of Adelaide, Adelaide, 1969.

10. W.B. HARLAND, "Evidence of Late Precambrian Glaciation and its Significance" in A.E.M. Nairne (Ed.), *Problems in Palaeoclimatology*, pp. 119–49.

11. B. DAILY, V.A. GOSTIN and C.A. NELSON, "Tectonic Origin for an Assumed Glacial Pavement of Late Proterozoic Age, South Australia", *J. geol. Soc. Aust.*, 1973, **20**, pp. 75–8.

12. W.J. PERRY and H.G. ROBERTS, "Late Precambrian Glaciated Pavements in the Kimberley Region, Western Australia", *J. geol. Soc. Aust.*, 1968, **15**, pp. 51–6.

13. G.E. WILLIAMS, "Characteristics and Origin of a Precambrian Pediment", *J. Geol.*, 1969, **77**, pp. 183—207.

14. G.E. WILLIAMS, "Palaeogeography of the Torridonian Applecross Group", *Nature (Lond.)*, 1966, **209** (5030), pp. 1303–06.

15. J.W. COWIE, "Contributions to the Geology of North Greenland", *Med. Grøn.*, 1961, **164** (3).

16. J.W. AMBROSE, "Exhumed Palaeoplains of the Precambrian Shield of North America", *Am. J. Sci.*, 1964, **262**. pp. 817–57.

17. E.R. ROSE, "Manicouagan Lake–Mushalagen Lake Area, Quebec", *Can. Geol. Surv. Paper*, 1955, **55-2**.

18. C.R. TWIDALE, JENNIFER A. BOURNE and DIANNE M. SMITH, "Age and Origin of Palaeosurfaces on Eyre Peninsula and in the southern Gawler Ranges, South Australia", *Z. Geomorph.* In press.

CHAPTER 24

Landforms Related to the Activities of Organisms, Including Man

A. GENERAL

The influence exerted by organisms on landform development is widespread and in some cases profound. Organisms contribute to the formation of many features and are directly and largely responsible for several features, including some of great magnitude. The varied roles played by organisms in several weathering processes have been mentioned before: lichens and plant roots cause both chemical and physical weathering, and earthworms and termites aerate the upper layers of the soil. The fundamental importance of vegetation in protecting and binding the land is brought out by the comparison of the effectiveness of water and wind in the arid tropics and in humid temperate regions. Comparative rates of erosion on ploughed and grassed fields[1] highlight the same point.

Table 24.1
Water and Soil Loss with different Ground Covers*

Cover	Water loss	Soil loss
Forest	0·5%	0
Grass	0·2%	0
Potatoes	88·0%	480 kg (1055 lbs)

*U.S. Department of Agriculture, 1934

The marked erosion on the ploughed field is achieved partly through lack of root binding and partly because raindrop impact during an intense storm on an unprotected soil surface can cause up to 41·13 tonnes per ha (100 tons per acre) to be exploded into the air and hence put into motion. The effects of kelp, seagrapes and seaweed in general, of boring creatures and of algae in coastal weathering have been mentioned, though their precise quantitative effects are not known; however, marine borers can efficiently destroy the fabric of a rock with astonishing rapidity. Some biological effects are suspected rather than proven. For example pitted and striped patterned ground in parts of lowland New Guinea has been attributed not to the presence of cracking or hydrophilic clays, of which there is no evidence in the areas concerned, but rather to the work of large earthworms over the past few centuries.[2]

B. MOUNDS

Over wide areas of northern Australia the plains are distinguished by the presence of rather turret-like mounds, of the same colour as the local soil, but of widely varying shapes and sizes (Pl. 24.1). On inspection they are found to be riddled with interlaced tunnels, which are inhabited by termites and littered with grass fragments and seeds. The mounds are the homes of colonies of these tiny creatures and are called *termitaria*. Some stand 4 m above ground level, and although generally of rather irregular shape in detail, they are roughly cylindrical with a crude conical peak (Pl. 24.1). In some areas, however, the termitaria are blade-like, and are preferentially oriented, the long axis consistently running north–south. This is supposed to be a heat-avoidance arrangement, for oriented as they are these termitaria present the least surface area to the direct

Plate 24.1 Cylindrical termitaria, Barkly Tableland, Northern Territory, near Queensland border. (*C.S.I.R.O.*)

rays of the northern sun. Termitaria occur in groups, and from afar look rather like micro-cities; clearly very large volumes of debris are involved in their construction, and some settling of the surface beneath which the detritus originated can be expected.

There is no doubt about the origin of these micro-landforms, but this cannot be said of the so-called Mima mounds of the northwestern USA (Pl. 24.2) and the prairie mounds common on the Great Plains of the USA. Though some regard them as of periglacial or nival origin, that is, as a form of patterned ground, the gravels on which the silt mounds rest are quite undisturbed and there are no signs of intense frost action; the mounds are clearly not due to gelifraction. A case has been made for attributing them to the activities of the pocket gopher, *Thomomys talpoides*.[3]

The mounds, between 30 cm and 2 m high average 6 m in diameter, though they range between 2 and 12 m and they are round to slightly oval in plan. Like termitaria they occur in large groups (Pl. 24.2). Some of the prairie mounds of Arkansas are undoubtedly old dunes now fixed by vegetation. But many in northwestern USA are thought, by Scheffer in particular, to be due to gophers. Each mound is thought to be the home of an individual gopher; it is built of a lenticular mass of black silt and gravel which overlies stratified gravel and other bedrock. In digging its burrow, the gopher deposits the sediment from the tunnel at the surface as a mound. It is estimated that a colony of these animals brings to the surface between 200 and 300 kg of soil per hectare (5–8 tons per acre). In support of his interpretation of the Mima mounds, Scheffer points out that the mound sediment is of a size that the gophers are capable of moving; that the "roots" typical of many mounds could be merely abandoned and infilled tunnels; and that there is a strong correlation, both positive and negative, between the distribution of gophers and of mounds. On the other hand, if they are responsible for the construction of the features, the pocket gopher must be one of the tidiest creatures known, for he left neither litter nor food debris nor skeletal remains behind.

Mounds with morphology similar to the Mima mounds, but of quite different origin, have been described by Mackay from an island in the MacKenzie Delta of northwestern Canada.[4] Standing about 80 cm above the level of the alluvial plains, and with a span of about 10 m, the mounds were at first thought to be incipient pingos (see Chapter 16) or pingos in course of development. However, when the crest was drilled in the expectation of finding an ice core, Mackay graphically describes how, when the drill rods had reached some 135 cm below the surface, they suddenly dropped by about a metre and gas, which proved to be inflammable and which is thought to consist primarily of methane, spurted from the vent. Mackay regards these mounds as gas blisters and attributes them to the doming of turf and sods by gases which are produced beneath the surface by the anaerobic decay of plant material.

C. FLAT-FLOORED VALLEYS

Many glaciated valleys are not perfectly U-shaped in cross-section but are flat-floored (see Chapter 17). This is commonly attributed to aggradation by braided streams which originated either as meltwater streams emanating from glacier snouts in glacial times and which deposited valley trains, or as heavily laden rivers formed from summer meltwaters derived from snow and ice in postglacial times. In either case the aggraded floors of the catenary valleys are attributed to riverine deposition, though lake deposits have been recognised in some areas such as the Yosemite Valley of California (Fig. 17.9).

These lakes have been regarded as having been impounded by moraines and other obstacles of inorganic origin. But Ruedemann and Schoonmaker, and Rutten,[5] have

Plate 24.2 Mima mounds on Mima Prairie, Thurston County, Washington, the type locality of Mima microrelief. (*V.B. Scheffer.*)

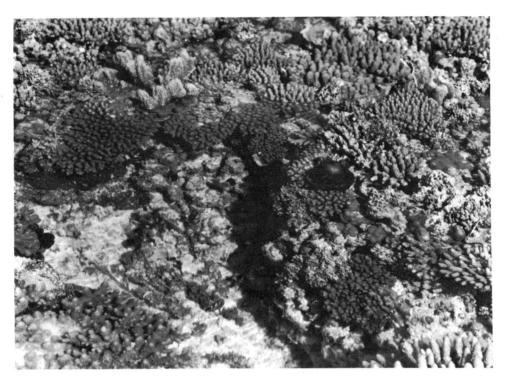

Plate 24.3a Coral on Heron Island, Great Barrier
Reef. (*Qld Railways.*)

Plate 24.3b Aerial view of Northwest Island, a coral island in the Capricorn Group off Rockhampton,
Queensland. Only a small part of the coral mass appears above the water surface. Note the sandy
beach (B) and the channel (C) which has been eroded, probably at times of low sealevel, on the flank
of the submerged coralline growth. *(B.A. Rudd.)*

argued that such valley-floor deposition has been materially assisted by the erection of beaver dams, which are well known in some parts of the North American mountain areas today, but which in postglacial times were more widespread both on that continent and in such European alpine areas as Scandinavia and Switzerland.

Rutten points out that though flat-floored valleys are associated with braided streams in the Alpine valleys of Switzerland, they are not formed by the braided rivers which occur in the circum-Mediterranean region. For this reason it is suggested that an additional, contributory factor is required in order to explain satisfactorily the flat-floored valleys in cold lands. Rutten argues that the additional factor is the beaver, which built large, intricate and numerous dams, thus forming shallow lakes in the rivers which occupied the recently deglaciated valleys. Sediments normally carried downstream became impounded in these beaver-made lakes, and formed a spread over the valley floors. With the extermination of the beaver in some areas and a depletion of its numbers elsewhere, the dams fell into disrepair, crumbled and were washed away; the shallow lakes drained; but the alluvial spreads remain in the floors of these flat-bottomed and steep-sided valleys.

D. **COASTAL AND OCEANIC REEFS**

Reefs and chains of reefs are common in coastal and oceanic settings. Some are of mineral rock, but many are found to be of organic origin. Along the coast of the Gulf of Mexico and the east coast of the USA, for example, there are reefs and chains of reefs up to 40 km long which are built of oyster shells. The tops of the reefs extend from the intertidal zone, and the shells continue in some areas to a depth of 24 m below sealevel. C datings of the shells indicate that the reefs started growing upward in response to a rising sealevel (see Chapter 22) almost 10,000 years ago. Large as they are, however, oyster reefs do not compare in size with coralline reefs, which are the largest landforms of organic origin.

Coral reefs take numerous forms, the most important of which are illustrated in Fig. 24.1a-d. But whatever their morphology, each is composed of the external calcareous skeletons of coral polyps. Only the uppermost layer in the intertidal zone is built of live, actively growing coral (Pl. 24.3a and b). This extends to a variable but shallow distance below sealevel. The shells of other marine animals such as foraminifera, gastropods and lamellibranchs contribute to the bulk of the reefs but most of the reef mass is made of myriads of polyps and their skeletons. The reefs are formed of colonial corals which have grown *in situ*: they are not built of the transported and deposited detritus of coralline materials.

Because there are a number of specific requirements necessary for its survival and growth, the colonial reef-forming coral has only a limited environmental or geographical range. These environmental conditions are concerned with salinity, light and temperature. The water in which coral lives must be saline, and it must be warm. Because the temperature of the sea must be between 20° and 30°C coral growth is restricted to the tropics, and usually within 30° of the equator. Coral growth is especially profuse on the western sides of oceans where there are no upwellings of cold water.

Temperature requirements also restrict coral growth to the surface layers of the seas, for even in the tropics tempera-

ture decreases with depth. In practice, and depending on local conditions, living corals extend, at most, to depths of 90–100 m below the surface, but more commonly the lower limit of growth is 50 m. There is another reason for corals being restricted to the near-surface layers of the tropical seas. Green algae, which live symbiotically with the coral polyps and which provide them with oxygen and possibly carbohydrates, require light for photosynthesis, and light does not penetrate far below the surface of the sea; though again local conditions vary greatly. Thus near the mouths of major rivers where large volumes of silt are poured into the ocean (the sea is discoloured for a few hundred kilometres from the coast by the waters and sediments brought to it by the Amazon), the silt in suspension causes dispersion of light to such an extent that coral growth is severely inhibited. There are other factors involved, however; the fresh water prevents growth, and most colonial corals prefer a clean non-silty basement on which to build.

Thus, actively growing coral reefs are restricted to the shallow zones of the tropical seas. But fossil reefs extend far beyond this limited area. They occur in the geological column in many parts of the world, and indicate either climatic change through geological time, or continental drift of some description, or an environmental tolerance different from that of modern reef-forming corals, or a combination of these variables. Even where reefs continue to grow, bores and seismic evidence indicate that the dead coral extends to depths far greater than 100 m. For example, coralline rock has been reported from depths of 777 m at Bikini Atoll, in the Marshall Islands, and the bore did not reach the basement (that is, in this context, non-coralline) rocks; from −319 m at Oahu in the Hawaiian Islands; and from −429 m at Maratoa, northeast of Borneo.

This is the crux of the problem concerning the genesis of coral reefs. Why is coral found at such great depths, far beyond its growth limits? Some workers have placed great emphasis on the implications of Pleistocene lowerings of sealevel and the related climatic changes generally.[7] The Bermudas, which are themselves coral islands, stand on a coral platform some 75 m below sealevel which extends beneath the entire archipelago, supporting the suggestion that the reefs have grown upon a foundation planed off during a low Pleistocene stand of the sea. It is likely that an overall lowering of seawater temperatures during the Pleistocene killed off those reefs near the latitudinal limits of growth, thus rendering them susceptible to wave attack and planation.

But although such factors are important, this is clearly not the whole story. If a maximum lowering of sealevel during the Pleistocene of 100 m is allowed, then, adding 100 m maximum as the lower limit of coral growth, the greatest depth to which reefs should extend is of the order of 200 m, which is not in keeping with the observed facts. The same objection stands in the way of general acceptance of the hypothesis involving stable platforms (that is, platforms which conveniently and fortuitously occur at the right elevation for corals to be able to colonise them), proposed over a century ago by Sir John Murray.[8]

Some parts of the crust are still active; folds, particularly rising anticlines, are still developing (see Chapter 5) in Melanesia and the western Pacific.[9] This has been

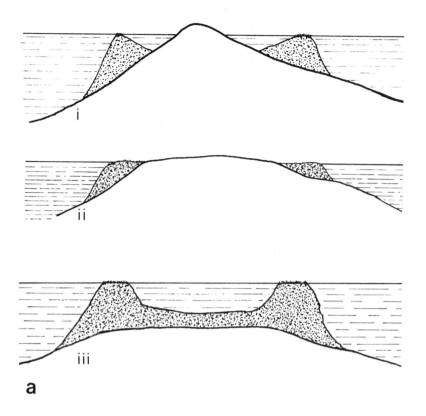

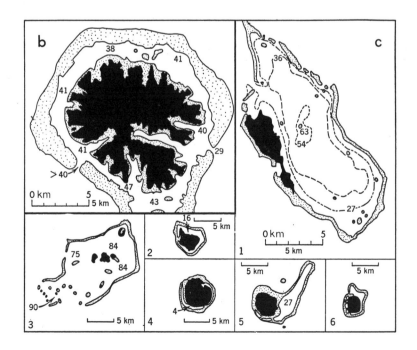

Fig. 24.1 Various types of coral reef: (a) in section *i.* fringing reefs; *ii.* barrier reefs; *iii.* atoll; : (b) and (c) several types in plan—depths in metres: (b) Tahaa, Society Islands, barrier reef with strongly embayed island and fringing reef; (c) Fiji Islands: 1) Wakaya, strongly asymmetrical barrier reef with wide lagoon, passing into fringing reef; 2) Naitomba, 3) Budd Reef, an almost-atoll with incomplete barrier, deep lagoon and half a dozen lagoon reefs, 4) Mango, narrow lagoon, partly fringing reef, 5) Kanathea, asymmetrical barrier, partly fringing reef, 6) Vanua Vatu, barrier joined to island by several patches of reef. *(After P.H. Kuenen,* Marine Geology, *Wiley, New York, 1950, by permission of P.H. Kuenen.)*

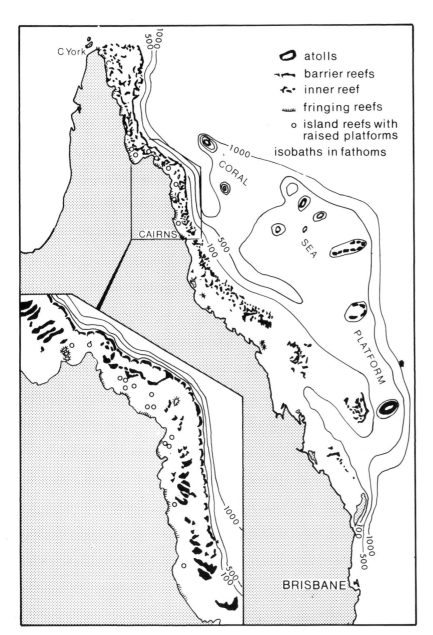

atolls
barrier reefs
inner reef
fringing reefs
o island reefs with
  raised platforms
isobaths in fathoms

C York

CORAL

SEA

PLATFORM

CAIRNS

BRISBANE

Fig. 24.1d The Great Barrier Reef.

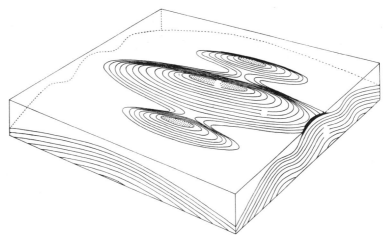

Fig. 24.1e Rising island with coral cap-rock. (*After C.A. Cotton, "Growing Mountains", pp. 209–11.*)

demonstrated in several parts of the continental areas and is probably true of the Indonesian archipelago and the Antilles, for example. New Guinea is a tectonically unstable region which has risen rapidly through the Pleistocene. It is possible that earlier, pre Wisconsin glaciations are not in evidence in the New Guinea mountains because until the later Pleistocene they had not risen to an elevation sufficient for snowfields and glaciers to form and persist. Similarly Pliocene coral reefs are reported from elevations of over 5000 m in the Carstenz Mountains. In such positive tectonic regions, reefs could become established on rising platforms while they were still in the critical zone between sealevel and −100 m. In that case, bores or seismic investigations should reveal the reef base at depths of less than 100 m and such evidence has indeed been cited by Verstappen,[10] who considers that all "reef caps"—carapaces of coral which cover islands of considerable extent—are evidence of such recent and continuing tectonism (Fig. 24.1e). Eighty-five per cent of the island of Barbados is covered by a layer of coral of Quaternary age, the thickness of which varies between 50 and 80 m[11] but which extends high above sealevel as a result of anticlinal uplift since the beginning of the Quaternary.

The only explanation of the great depths to which coral reefs extend is that advanced by Charles Darwin[12] during the voyage of the *Beagle* in 1831–36, namely that the reefs were established on and around islands, many of which were volcanic, which have since subsided, carrying the reefs with them, but sinking at such a rate that upward growth of the coral can keep pace with the negative movement and thus allow the reef to persist. As the island becomes submerged, the barrier or fringing reef formations continue their upward growth but a lagoon develops, not because of solution or erosion, but because of slower coral growth in this sheltered zone of less food and oxygen.

The subsidence hypothesis receives corroboration from the ria coasts common in reef areas, from the great depths to which coastal cliffs extend in reef areas, from the discovery of dolomite and other shallow water facies in deep bores and from the occurrence of *guyots* or flat-topped sea mounts at great depths in the Pacific and Atlantic (Fig. 24.2). These guyots are interpreted as sinking volcanoes which subsided too rapidly for reef-building to be maintained, even where conditions were conducive to coral growth.

Superimposed on this picture of gradually subsiding islands and their associated reefs are the effects of Pleistocene sealevel changes. These changes caused phases of planation and renewed growth evidenced in some areas by the occurrence of distinct generations of coral, one superimposed on the other (Fig. 24.3).

E. MAN AS A GEOMORPHOLOGICAL AGENT

In many parts of the world, and especially in those areas recently settled by Europeans, gullying is commonplace; though measurements in the American Southwest indicate that this form of erosion accounts for only 1·4 per cent of all soil removal, sheet erosion being of overwhelming importance in this respect. Nevertheless gullies are the most obvious and in some ways the most troublesome manifestations of a worldwide problem. This recent erosion, which because it has

clearly taken place far more rapidly than is usual for the area in which it occurs, is called *accelerated* erosion.

In the southwestern states of the USA the gullies or arroyos are incised in alluvial sediments only a few thousand years old at most. The same is true of gullies in many parts of Australia (Pl. 24.4a and b). The gullies themselves are of recent date, indeed it can be shown that many have formed since European settlement. It is of course possible that a climatic fluctuation has occurred during the same period, and this argument has been espoused by some workers; but several lines of evidence suggest that the gullies are the result of increased run-off, or the localisation and concentration of run-off through man's activities. Man has not actually excavated the gullies, but in various ways he has rendered the land surface more vulnerable to stream erosion.

The gullies eroded in the alluvial fans which front the Willunga scarp have already been referred to in connection with their river terraces (Chapter 19). Early plans indicate that most of the present stream lines existed before 1850, though the occurrence of a well alongside a very deep gully (Pl. 24.5) strongly indicates that the valleys were not deep at that time. It is possible still to find eye-witnesses who attest the recent initiation of some minor gullies and the rapid extension and broadening of others. Erosion implies deposition, and the recency of many of the gullies, in whole or in part, is indicated by the burial of fence posts, horse bones, barbed wire and corrugated iron sheeting by fills and fans downstream from the valley (Pl. 24.5b).

These all occur in areas which have been cleared of vegetation for farming, and it is still arguable, on the evidence presented, that a minor climatic change or even a particular heavy rainstorm could have initiated this valley deepening or development. There is no evidence of climatic fluctuation, that is, a secular change in climate, even of small magnitude, in the years since the colonisation of South Australia;[13] but there is field evidence which suggests that clearing of the land has been a crucial factor in the development of gullies. Deforestation has been marked on the Willunga scarp and on the plateau above in the areas southwest of Willunga township, but it is not nearly as extensive to the northeast. Though there are numerous streams in the latter area, they run in shallow valleys which stand in strong contrast to the deep, steep-sided chasms of the area to the southwest. The bedrock is the same and the two areas are juxtaposed: the two points of contrast are the degree of clearance and the intensity or depth of stream erosion.

In New Mexico though there has been no discernible change in total rainfall over the past century, the area "experienced a relatively low frequency of small rains, both in summer and winter".[14] On the other hand, there were more frequent heavy storms. On this basis, though he admits that the introduction of grazing was a contributory factor, Leopold[15] attributes the arroyo cutting of the region primarily to the depletion of vegetation by low rainfalls and to the incidence of heavy rains. Bull,[16] in his investigation of gully erosion in the central valley of California, attributes two major phases of accelerated erosion in 1875–95 and 1935–45 to particularly heavy rainfall.

Similar findings, though with possibly more emphasis on the role of man than is allowed by the American workers, have been reported from many parts of the "New" World

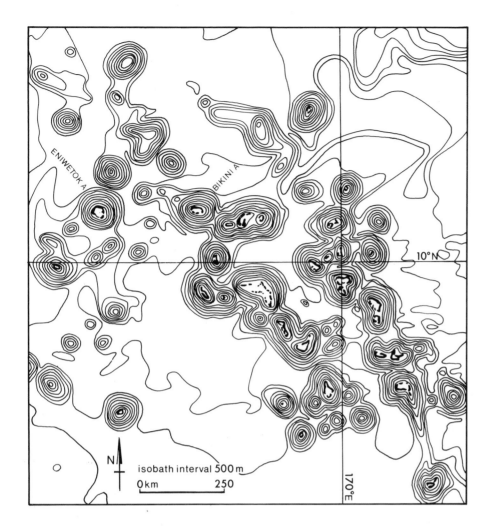

Fig. 24.2 Guyots in the north Pacific (the Marshall Islands.)

Fig. 24.3 Development of coral reefs through interplay of glacioeustasy and subsidence: 1) growth of barrier in pre-glacial times by subsidence as swift as post-glacial rise of sealevel—A, barrier; B, fringing reef; C, deep lagoon; D, very deep passages; E, very steep outer slope; 2) glacial control of a barrier formed by slow subsidence without a deep passage—A, abrasion behind a growing reef; B, infilling of deep lagoon; C, denudation of elevated fringing reef; 3) postglacial growth of reefs —A and B, faros—small reefs shaped like atolls but forming part of a larger reef (barrier or atoll); C, fringing reef with boat channel; D, sediments veneering platform and forming a basin with greatest depth in centre; no passages deeper than about glacial sealevel. *(After P.H. Kuenen,* Marine Geology, *p. 461, by permission of P.H. Kuenen.)*

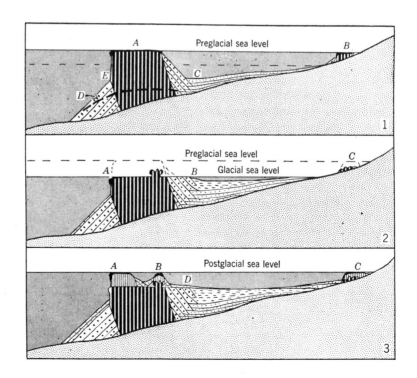

Plate 24.4a Gully erosion in weathered shales, Chace Range, central Flinders Ranges, South Australia. Note the sheet erosion involving the removal of the A-horizon of the soil as for example, at X. *(C.R. Twidale.)*

Plate 24.4b Gully erosion in Permian glacigene sediments near Yankalilla, Mt Lofty Ranges, S.A. (*C.R. Twidale.*)

(see for example Cumberland,[17] working in New Zealand, and Holmes,[18] summarising work up to the end of World War II in Australia). Europe has not escaped such erosion, though hopefully for the newer agricultural areas, equilibrium appears to have been attained once more. Gully erosion was widespread in Western Europe in the eighteenth century, [19] and where there has been human interference, as in parts of the heathlands of the New Forest in southern England, pronounced gully erosion continues to be initiated.[20] Accelerated soil erosion continues in many parts of the world besides those mentioned, including such long-settled areas as Spain [21] and many parts of the underdeveloped world. So widespread and devastating is this erosion of soil resulting from man's interference that it has been called the "Rape of the Earth".[22] From a geomorphological view, a new anthropogenic minor cycle or *epicycle* of erosion and deposition is in train.

Destruction of the vegetation cover and consequent increased run-off has caused higher flood-peaks and larger sediment loads in the rivers of the Murrumbidgee system in the Riverina of western N.S.W. As a result, there have been major changes in the dimensions, shape and pattern of the modern stream channels.[23]

Land clearance for agriculture is the most obvious means of inducing accelerated erosion, but grazing (particularly overgrazing) and bushfires achieve similar results; not that all grazing and all bushfires are caused by man or his animals, but man has certainly introduced many creatures in great numbers into new areas, and he certainly either directly or indirectly initiates many fires. In addition to depleting vegetation, grazing animals such as cattle tend to follow certain paths: the soil beneath becomes compacted and rendered vulnerable to erosion. Much of the recent erosion of the Great Plains has been attributed to a combination of vegetation depletion and soil compaction by cattle.[24]

In addition to these obvious and important effects, there are rather more subtle effects and causes, all stemming either from man's misuse of the soil, or from his interference with the vegetation. Removal of timber from an area can cause a rise in the water-table. In some areas in South Australia where the groundwater is saline, the salts have killed off the natural vegetation and left the soil surface bare. Depletion of plant litter causes loss of soil structure, and particularly of soil porosity, so that run-off and hence potential erosion are increased. Intensive farming and especially monocropping in some areas of England and Wales have caused a serious depletion in the organic content of soils, and soil structure and fertility have declined alarmingly, despite frequent dressings with artificial manures. Intensive and continuous cereal growing on the clay soils of the English Midlands has reduced the organic content of the soils to about 3 per cent compared with the recommended minimum of 8 per cent. Soil structure has fallen away and erosion is likely unless there is a reversion to the traditional mixed rotation of crops, including some periods under pasture. In such clay areas, poor drainage and compaction through the use of heavy machinery have contributed to the breakdown of the soil structure.

In certain areas of the Mt Lofty Ranges, many of the gullies are remarkably straight and are parallel to each other. Some occur in areas of arable farming and are caused by modern ploughing. Others, however, are located in fields

Plate 24.5a Deep gully in alluvial fan deposits fronting the Willunga scarp, Mt Lofty Ranges, S.A. The early settlers sank a well at X, and this suggests that the gully was not then as deep as it is now. (*C.R. Twidale.*)

Plate 24.5b Fence posts partially buried by valley fill, Yankalilla, Mt Lofty Ranges. (*C.R. Twidale.*)

a

b

Plate 24.6 Furrows exploited by run-off to form gullies in the Yankalilla area, Mt Lofty Ranges. In (a) the furrows are clearly visible, despite the sward. Some of the lower furrows are eroded. In (b) there has been more erosion, though in both cases the erosion is discontinuous, suggesting that subsurface flushing and subsidence is more significant than erosion by surface streams. *(C.R. Twidale.)*

Plate 24.7 Washed-out track in the Gawler Ranges, S.A. *(C.R. Twidale.)*

which are under pasture. Many of the furrows in these areas are narrow and shallow but quite distinct. Investigations have shown[25] that these grooves mark the boundaries of field units called *lands*, which were introduced by early European settlers in the area (Pl. 24.6). The English settlers especially brought with them the agricultural habits of their homelands, and when they ploughed the fields for cereal cultivation, they ploughed with the single share plough and used the field pattern they knew at home. Subsequently, with the improvement of communications and the opening up of more suitable wheatlands, these fields in the Mt Lofty Ranges were returned to pasture, thus preserving the shallow furrows which delineated the lands. Shallow as they are, however, they are sufficient to have concentrated run-off with the result that in many areas they have been greatly enlarged to form gullies.

Other straight gullies (Pl. 24.7) occur in the semiarid and arid areas of South Australia where wheat has never been grown. These prove to be former tracks used by carts and cars. The soil was compacted by the passage of wheels, run-off increased and erosion occurred after particularly heavy rains. In the moister climate of England, the passage of carts over the centuries has caused the lowering of the surface by the adhesion of mud to the wheels of the vehicles. Some of the sunken lanes stand 2–3 m below the level of the adjacent fields. Even where the road is surfaced and protected by bitumen, and the porosity of the surface reduced to virtually nil, the camber commonly built into the roads causes water to run to the roadside areas which are vulnerable to gullying because they are often depleted of vegetation as a result of construction and maintenance work, and because they are exposed to the fumes from exhausts (Pl. 24.8). This problem is so serious in the eastern States of Australia that a research

station has been established at Wagga Wagga, N.S.W., to study the problem of roadside erosion.

The introduction of exotic plants and animals can also lead to increased erosion. In the Flinders Ranges of South Australia, some of the gullying and sheet erosion is due to overgrazing (vast flocks of sheep were kept during the early years of settlement); but some is due to the depletion of the native grasses which were perennials, and their replacement

Plate 24.8 Roadside gullying in the southern Flinders Ranges, S.A. *(C.R. Twidale.)*

Plate 24.9 Numerous gullies created on the slopes of a conical shale hill during a summer storm in 1955 in the central Flinders Ranges. *(C.R. Twidale.)*

by exotic species which were annuals. Thus in summer the ground is quite unprotected and summer storms create havoc (Pl. 24.9). Again the introduction of the rabbit into Australia not only caused severe depletion of pasture but in some areas their warrens clearly form collecting areas for run-off, and the concentrated flows which emerge from them cause collapse of the burrows as well as pronounced erosion downslope.

Vast areas of once fertile land in the Levant and North Africa are now bare rock slopes. The hills behind Aleppo were well-forested in Biblical times but are now outcrops of limestone, with virtually no soil. In considerable measure this erosion results from the destruction of the vegetation, which in turn relates back to religion. For instance, the natural browser of many of the oak forests of Tunisia is the pig, but this animal is considered "unclean" and large flocks of goats forage in these areas instead. The goat is a voracious nibbler with an insensitive digestive-system. It is responsible for the devastation of the vegetation not only on the southern fringes of the Mediterranean, but also on the southern margin of the Sahara where the goats of the nomadic herdsman have so depleted the vegetation that the wind is again able to move the naturally fixed dunes of the area. Because he keeps herds of goats it has been said that in some parts of Africa and the Levant the Arab is not so much the son of the desert as its father.[26] But the goat is not the only offender: overgrazing by sheep and cattle and the unwise clearance of vegetation for agricultural purposes have allowed the dunes to be reactivated on the southern fringe of the Sahara and in the relict deserts of Eyre Peninsula, the Adelaide plains and the Murray valley both in South Australia and Victoria. Indeed, the removal of vegetation from large dunes is now controlled by government regulation in South Australia.

Technology must also be considered as a contributor to erosion and landform modification. Pushing roads through the boreal forest in parts of Alaska and northern USSR has entailed the felling of trees. This has altered thermal conditions near the land surface and has caused the swelling of ground ice: as a result, thermokarst has formed (Pl. 16.6). Merely by his presence, man disturbs the environment. Over one million trees have been asphyxiated by smog around Los Angeles, thus increasing the erosion on the hillslopes they inhabited. Aboriginal man has influenced the detailed morphology of bedrock slopes in the Murray valley, South Australia. It can be shown [27] that alternate occupation and abandonment of Aboriginal shelters has caused variations in the rates of weathering and sedimentation on the bedrock wall in the shelters or caverns and has in effect caused the evolution of a stepped rock surface (Fig. 13.9). But modern man, armed with his machines and know-how (or alleged know-how), operates on an altogether bigger scale. In building reservoirs, man has caused local depression of the crust because of the weight of water and sediment collected in them. Thus the Hoover Dam and Lake Mead, in Nevada, concentrate a mass of $42,68910^6$ tonnes (42 billion tons— 40 billion of which are water) in an area of 600 km² (232 square miles). The pressure exerted by this man-induced mass is 103.34 tonnes/m² (147 lbs/sq. in) and subsidence is taking place at a rate of 10 m per millennium, though total subsidence is expected to be only about 25 cm because of compression of the underlying rocks.[28]

Recent subsidence of the land surface has been noted in many areas. In Los Angeles, parts of the surface have sunk by as much as 10 m and the subsidence continues at a rate of 30 cm per annum.[29] In parts of the Central Valley of California (Fig. 24.4) west of Fresno, the surface has subsided by over 7 m over the past fifty years.[30] In parts of Mexico City (Fig. 24.5), the surface of the old city has subsided by 7 m over the past century,[31] due to pumping of water from the lake sediments which underlie the city. The lake,

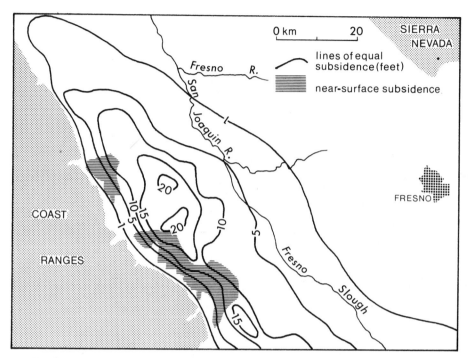

Fig. 24.4 Surface subsidence over the past fifty years in the Fresno area, Central Valley of California. *(After B.E. Lofgren, "Land Subsidence", pp. 140–2.)*

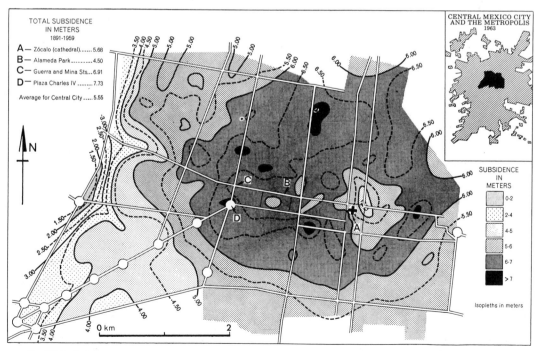

Fig. 24.5 Subsidence in the older part of Mexico City during the past century. *(After D.J. Fox, "Man–Water Relationship", pp. 523–45. Reprinted from the* Geographical Review *(Vol. 55, 1965), copyrighted by the American Geographical Society of New York.)*

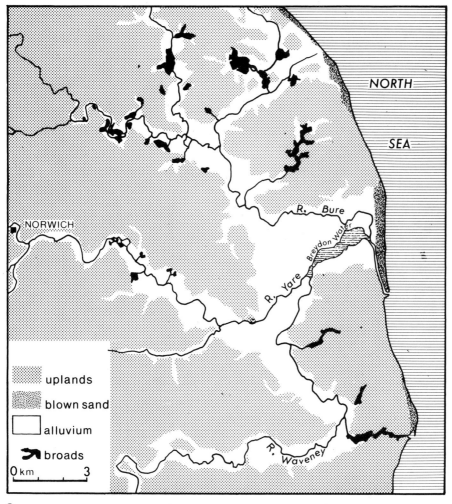

Fig. 24.6 (a) Map of the
Norfolk Broads and (b) section
through one of the old peat
quarries. (*After J.M. Lambert
et al., "The Making of the
Broads".*)

a

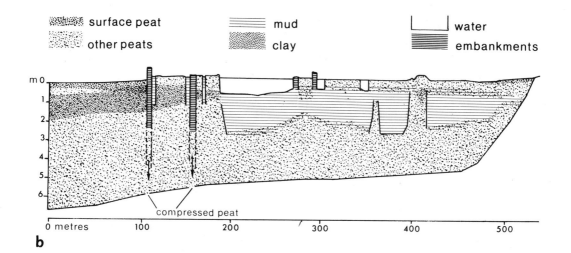

b

Texcoco, has shrunk (Fig. 24.5). In all cases, the timing, location and distribution of the disturbances points to their having a common origin. The mining of underground waters and oils has induced a decrease in total volume and a consequent settling and compaction of the rock, which is reflected in surface subsidence. In Los Angeles it is oil, in the Fresno region water for irrigation, and in Mexico City water for domestic use, but the result is the same.

In Cheshire, England, the mining of salt and brine has caused a similar local subsidence, which is also widespread and well known in many coal mining areas. Some of the depressions due to salt extraction in Cheshire have filled with water to form small lakes known as meres or *flashes*.[32]

Shallow lakes of particular interest occur in East Anglia. These are the Norfolk Broads. They were originally thought to be natural features caused by the ponding back of waters in peaty hollows located between low clay ridges or in tributary valleys blocked by clay which was deposited at the margin of a marine transgression dating from approximately Romano–British times. But detailed investigations have shown that the peat deposits beneath the lake depressions are separated from the adjacent older sediments by vertical unconformities (Fig. 24.6a and b); that residual walls of older sediments lie within the peat; and that the boundaries between older and newer sediments are often rectilinear.[33] Historical evidence shows that the Broads existed as lakes by Elizabethan times. There is no evidence that they existed before the beginning of the fourteenth century. Turf or peat cutting was a major industry in the area from the twelfth through to the middle of the sixteenth century, and was especially pronounced in the late thirteenth and early fourteenth centuries. This industry is considered to be responsible for the Broads; they are in fact great peat quarries which were flooded, possibly as a result of storms and high tides, mostly during the fourteenth century. This explains their regular outlines (though these have been softened by time) and their recent date. It has been calculated that the foundation of the Broads in this way involved the removal of some 836 million m³ of peat.

To the northwest of the Broads, but still in eastern England, are the Fens, now one of the richest agricultural areas in Britain. This fertile land is also a product of man's activities.[34] Originally a vast expanse of marsh and shallow swamp, the Fens were drained largely by Dutch engineers in a number of stages: beginning in the sixteenth century, principally during the seventeenth century and continuing to the present day. The main rivers were straightened, increasing their gradient and thus increasing the flow of water and sediment to the sea. The deposition of sediment in the Wash has given rise to outfall problems. The draining of the land also introduced unexpected difficulties. In addition to the subsidence which resulted from drying, bacterial action further reduced the volume of the newly-drained soils so that the land surface was gradually lowered. Hence the task of the pumps—lifting water from drains into dykes which ran in embayments above the level of the land—became greater and greater. Windmills were unequal to the task and only the invention of the steam engine saved the Fens from reverting to an area of poor drainage. Subsidence continues apace: at Holme Fen the surface was lowered about 4 m between 1848 and 1970, over 1·5 m of this occurring in the last ten

years of this period.[35]

Of course the greatest feat of the Dutch engineers is undoubtedly the draining of the Zuider Zee (Fig. 24.7) and the gradual conversion of this former embayment of the sea to agricultural land. It should not be overlooked that both in Holland and in eastern England very large and agriculturally important areas are kept dry—at least at most times—because of the construction of artificial seawalls and river embayments (Fig. 24.8).

Finally, in connection with lakes and coasts, mention may be made of the Aswan High Dam and Lake Nasser. Since the dam was built, there has been much wave erosion of the northern margin of the Nile Delta. What is the connection between these two features 900 km apart? The High Dam, as mentioned in Chapter 13, acts as a silt trap, and not only the Nile Valley but also the coast of the Nile Delta is deprived of silt which in normal circumstances would be carried by the river. Before man interfered with the system, equilibrium had been attained between wave erosion and replacement.

Thus although man's effects on the land surface are numerous and varied, his main influence has been a result of his depletion of vegetation. The effect in fact has been simulation, to a greater or lesser degree, of the conditions

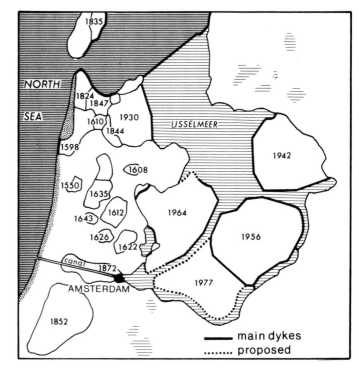

Fig. 24.7 General plan of Zuider Zee reclamations. (*After P. Wagret,* Polderlands, *Methuen, London, 1968, p. 114.*)

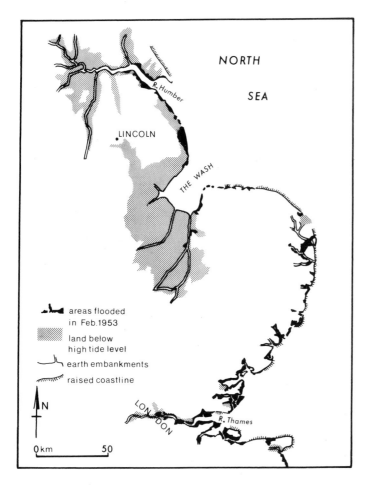

Fig. 24.8 The east coast of England, showing areas vulnerable to flooding, and coastal defences. (*After P. Wagret,* Polderlands, *p. 247.*)

which obtain in areas more arid than the areas concerned. A "desert effect" is imposed by the creation of biological deserts, and the results are commonly disastrous.

It is of course true that modern man is not responsible for all soil erosion. Aboriginal man undoubtedly made fires as an aid to hunting (fires can be and still are generated naturally; modern man has simply added to the problem). Natural droughts also caused vegetation to be depleted and the land surface to be made vulnerable to wind and water erosion; again man has extended and augmented this natural effect. [36] In many areas it is demonstrable that the erosion postdates modern settlement. Man inhabits all the continents but Antarctica and even there he has now established small but quasi-permanent settlements in the shape of research stations. The duration of man's habitation of the planet is controversial, and rests in considerable measure on the definition of *Homo sapiens*. In Africa, Asia and Europe there is evidence of very ancient man, but man's entry to the Americas and Australia was much more recent—a few thousands of years ago at most. However, during the past three or four millennia man has become sufficiently numerous, widespread and technologically advanced to modify and even drastically change his environment. During the past century especially he has become so numerous that the environment has been seriously affected. The geomorphological works of man are so important that even in the nineteenth century — before the population and technological explosion — a Russian geographer wrote that he "could not conceive Physiography [geomorphology] from which Man is excluded." [37]

F. PLANTS, MAN AND UNIFORMITARIANISM

It is clear that the spread of organisms over the continents has brought about significant and dramatic short-term changes in the shaping of the land surface. Plants, especially ground cover, are a particularly important factor in the environmental complex: they control the rate at which erosion takes place to a considerable degree. Plants did not emerge from the oceans until quite late in geological history, and probably did not achieve a significant cover until about mid Tertiary. Before that time even the humid lands were biological deserts; from a geomorphological point of view, conditions in many areas were similar to those which presently obtain in the semiarid savanna regions; slopes would tend to be faceted and to retreat, and pediplanation was the most common model of landscape development. With the spread of plants, the protective cover of grasses and soils induced a tendency, as Russell[38] has pointed out, toward a more even distribution of weathering and erosion, toward the rounding of landscapes, toward slope decline and peneplanation. Man has contributed a little towards reversing this trend by clearing forests and destroying grasslands and soil.

Thus the law of uniformitarianism is in one sense right, but in another sense wrong. The same processes which shaped the land surface in the past are still at work now; but their effective extent and their rate have changed as a consequence of biological evolution and plant dispersal over the continental areas. For how long will man reverse this evolutionary trend?

References Cited

1. See "Influence of Vegetation and Watershed Treatments on Runoff, Silting and Streamflow", *U.S. Dep. Agric. Misc. Pub.*, 1940, **397**. Also W.D. ELLISON, "Some Effects of Raindrops and Surface Flow on Soil Erosion and Infiltration", *Trans. Am. geophys. Un.*, 1945, **26**, pp. 415–29; *idem*, "Studies of Raindrop Erosion", *Agric. Eng. St Joseph Mich.*, 1944, **25**, pp. 131–6 and 181–2; P.C. EKERN, "Raindrop Impact as a Force Initiating Soil Erosion", *Proc. Soil Sci. Soc. Am.*, 1950, pp. 7–10; Y. MICHARA, "Raindrop and Soil Erosion", *Bull. Nat. Inst. Agric. Sci.* (Tokyo), 1951, (Ser. A), **1**, pp. 48–51.

2. H.A. HAANTJENS, "Morphology and Origin of Patterned Ground in a Humid Tropical Lowland Area, New Guinea", *Aust. J. Soil. Res.*, 1965, **3**, pp. 111–29.

3. W.W. DALQUEST and V.B. SCHEFFER, "The Origin of the Mima Mounds of western Washington", *J. Geol.*, 1942, **50**, pp. 68–84; V.B. SCHEFFER, "Mima Mounds; their Mysterious Origin", *Pac. Search*, **3** (5); *idem*, "Do Fossorial Rodents Originate Mima Type Microrelief?", *Am. Midl. Nat.*, 1958, **59**, pp. 505–10.

4. J.R. MACKAY, "Gas-domed Mounds in Permafrost, Kendall Island, NWT", *Geogr. Bull.*, 1965, **7**, pp. 105–15.

5. R. RUEDEMANN and W.J. SCHOONMAKER, "Beaver Dams as Geologic Agents", *Science (N.Y.)*, 1938, **88**, pp. 523–4; M.G. RUTTEN, "Flat-bottom Glacial Valleys, Braided Rivers and the Beaver", *Geol. Mijnbouw*, 1967, **46**, pp. 536–60.

6. W.A. PRICE, "Oyster Reefs" in R.W. Fairbridge (Ed.), *Encyclopaedia of Geomorphology*, Reinhold, New York, 1969, pp. 799–803.

7. R.A. DALY, *The Changing World of the Ice Age*, Yale University Press, New Haven, 1934; W.M. DAVIS, "The Coral Reef Problem", *Am. Geogr. Soc. Spec. Pub.*, 1928, p. 9.

8. J. MURRAY, "On the Structure and Origin of Coral Reefs and Islands", *Proc. R. Soc. Edin.*, 1880, **10**, pp. 505–18; *idem*, "Coral Reefs", *Nature (London.)*, 1889, **40**, p. 222.

9. J.E. HOFFMEISTER and H.S. LADD, "The Antecedent Platform Theory", *J. Geol.*, 1944, **52**, pp. 388–402; H. VERSTAPPEN, "On the Geomorphology of Raised Coral Reefs and its Tectonic Significance", *Z. Geomorph.*, 1960, **4** NS, pp. 1–28; C.A. COTTON, "Growing Mountains and Infantile Islands on the western Pacific Rim", *Geogr. J.*, 1961, **127**, pp. 209–11.

10. H. VERSTAPPEN, "Geomorphology of Raised Coral Reefs", pp. 1–28.

11. R.J. RUSSELL, "Coral Cap of Barbados", *Tijdschr. Kon. Ned. Aar. Genootsch.*, 1966, **8**, pp. 298–302.

12. C. DARWIN, *The Structure and Distribution of Coral Reefs*, Appleton, London, 1842; see also A. MOOREHEAD, *Darwin and the Beagle*, Penguin, London, 1971.

13. E.A. CORNISH, "On the Secular Variation of Rainfall at Adelaide", *Aust. J. Phys.*, 1954, **7**, pp. 334–46.

14. L.B. LEOPOLD, "Areal Extent of Intense Rainfalls, New Mexico and Arizona", *Trans. Am. geophys. Un.*, 1942, **23**, pp. 558–63.

15. L.B. LEOPOLD, "Areal Extent", pp. 558–63.

16. W.B. BULL, "History and Causes of Channel Trenching in western Fresno County, California", *Am. J. Sci.*, 1964, **262**, pp. 249–58; see also S.A. SCHUMM and R.F. HADLEY, "Arroyos and the Semiarid Cycle of Erosion", *Am. J. Sci.*, 1957, **255**, pp. 161–74.

17. K.B. CUMBERLAND, "A Geographical Approach to Soil Erosion in New Zealand", *Aust. Geogr.*, 1943, **4**, pp. 120–31.

18. J. McD. HOLMES, *Soil Erosion in Australia and New Zealand*, Angus and Robertson, Sydney, 1946.

19. J. VOGT, "Erosion des Sols et Techniques de Culture en Climat Tempéré Maritime de Transition (France et Allemagne)", *Rev. Géomorph. Dyn.*, 1953, **4**, pp. 157–83.

20. C.G. TUCKFIELD, "Gully Erosion in the New Forest, Hampshire", *Am. J. Sci.*, 1964, **262**, pp. 795–807.

21. H.H. BENNETT, "Soil Erosion in Spain", *Geogr. Rev.*, 1960, **50**, pp. 59–72.

22. G.V. JACKS and R.O. WHYTE, *The Rape of the Earth*, Faber & Faber, London, 1939.

23. S.A. SCHUMM, "River Adjustment to Altered Hydrologic Regimen—Murrumbidgee River and Palaeochannels, Australia", *U.S. Geol. Surv. Prof. Paper*, 1968, **598**.

24. J.T. DUCE, "The Effect of Cattle on the Erosion of Canon Bottoms", *Science*, 1918, **47**, pp. 450–2.

25. C.R. TWIDALE, G.A. FORREST and JENNIFER A. SHEPHERD, "The Imprint of the Plough: 'Lands' in the Mt Lofty Ranges, South Australia", *Aust. Geogr.*, 1971, **11**, pp. 492–503; C.R. TWIDALE, "Farming by the Early Settlers and the Making of Ridges and Furrows in South Australia", *Tools and Tillage*, 1971, **1**, pp. 205–23; *idem*, " 'Lands', or Relict Strip Fields, in South Australia", *Agric. Hist. Rev.*, 1972, **20**, pp. 46–60.

26. A. REIFENBERG, "The Struggle between the Desert and the Sown", *Desert Research*, Res. Counc. Israel Spec. Publ., 2; for an earlier expression of the same sentiment see E.H. PALMER, *The Desert of the Exodus*, Cambridge University Press, London, 1871.

27. C.R. Twidale, "Effects of Variations on the Rate of Sediment Accumulation on a Bedrock Slope at Fromm's Landing, South Australia", *Z. Geomorph. Suppl.*, 1964, **5**, pp. 177–91.

28. H.R. Gould, "Amount of Sediment" in W.O. Smith, C.P. Vetter, G.B. Cummings, *et al.*, "Comprehensive Survey of Sedimentation on Lake Mead, 1948–9", *U.S. Geol. Surv. Prof. Paper*, 1960, **295**; C.R. Longwell, "Interpretation of Leveling Data", *U.S. Geol. Surv. Prof. Paper*, 1960, **295**.

29. H.J. Nelson, "The Spread of an Artificial Landscape over southern California", *Ass. Am. Geogr. Ann.*, 1959, **49**, pp. 80–99.

30. B.E. Lofgren, "Land Subsidence due to Artesian Head Decline in the San Joaquin Valley, California" in *Northern Great Basin and California*, Guidbook Field Conference I, VII INQUA Congress, Boulder, Colorado, 1965, pp. 140–2; R.H. Meade, "Compaction of Sediments underlying Areas of Land Subsidence in central California", *U.S. Geol. Surv. Prof. Paper*, 1968, **497-D**; B.E. Lofgren and R.L. Klausing, "Land Subsidence due to Ground-water Withdrawal, Tulare-Wasco Area, California", *U.S. Geol. Surv. Prof. Paper*, 1969, **437-B**: F.S. Riley, "Land-surface Tilting near Wheeler Ridge, southern San Joaquin Valley, California", *U.S. Geol. Surv. Prof. Paper*, 1970, **497-G**.

31. D.J. Fox, "Man–Water Relationships in Metropolitan Mexico", *Geogr. Rev.*, 1965, **55**, pp. 523–45.

32. K.L. Wallwork, "Subsidence in the mid-Cheshire Industrial Area", *Geogr. J.*, 1956, **122**, pp. 40–53.

33. J.M. Lambert, J.N. Jennings, C.T. Smith, C. Green and J.N. Hutchinson, "The Making of the Broads", *R. Geogr. Soc. Res. Ser.*, 1960, **3**.

34. H.C. Darby, *The Drainage of the Fens*, Cambridge University Press, London, 1940.

35. Bruce F. Curtis, personal communication.

36. F.N. Ratcliffe, "Soil Drift in the Arid Pastoral Areas of South Australia", *C.S.I.R. Pamphlet*, 1936, **63**; *idem*, "Further Observations on Soil Erosion and Sand Drift, with special reference to southwest Queensland", *C.S.I.R. Pamphlet*, 1937, **70**.

37. P. Krapotkin, "On the Teaching of Physiography", *Geogr. J.*, 1893, **2**, pp. 350–9.

38. R.J. Russell, "Geological Geomorphology", *Bull. geol. Soc. Am.*, 1958, **69**, pp. 1–22.

Additional Reference

Dianne M. Smith, C.R. Twidale and Jennifer A. Bourne, "Kappakoola Dunes — Aeolian Landforms Induced by Man", *Austr. Geogr.*, 1975, **13**, pp. 90–6.

Part 5

Has some vast imbecility
Mighty to build and blend
But impotent to tend
Framed us in jest and
Left us now to hazardry?

Thomas Hardy, "Nature's Questioning"

CHAPTER 25

Factors in the Analysis of Landforms

The foregoing analysis of landforms has demonstrated that landforms are genetically complex, and that although it is possible to suggest how and when *some* features have evolved, the origin of many forms and facets of the land surface remains indeterminate. So complex are some landforms that at best one can only arrive at the most likely explanation either statistically — through the application of probability theory to determine the most likely origin of the forms[1] — or in qualitative terms. Long before statistical techniques acquired their present vogue in geomorphology, Baulig remarked that "Truth in geomorphology...is seldom more than increasing probability".[2]

The complexity of landforms is due to the number and variability of factors which influence and have influenced their development: both exogenetic and endogenetic forces have changed in time, and landforms are clearly multivariate features. Nevertheless, certain general impressions or principles emerge from the analysis — though they are not sufficiently specific to be elevated to the status of laws.

The first is that both in detail and in gross, *structure* exerts a significant and all-pervasive influence on landform development. Not only is the nature of the crust expressed directly in a wide range of tectonic and structural landforms, but properties such as the porosity, permeability, texture and mineralogy of its constituent rocks determine to a large degree the effectiveness of various climatically-induced processes operating at and near the earth's surface. Moreover, though depositional forms are in large measure independent of structural considerations, in many areas their broad distribution is delineated by structural lines or determined by structural factors. Thus, in northwest Queensland, parts of the Carpentaria plains are of depositional origin, the sediments having been laid down in either riverine, paludal (that is, shallow swamp), or marine conditions.[3] They occupy part of a relatively narrow corridor which separates the Isa Highlands to the west and the Einasleigh uplands to the east, and which is delineated by major faults or warps. The playas and coalesced alluvial fans which are such prominent features of the American Southwest occupy downfaulted bolsons. The great alluvial plains of central Australia are related to Lake Eyre which is in detail a down-faulted block, and which in a regional context occupies a broad basin of subsidence; the depositional sand ridges or dunes of the Simpson Desert are located to windward of the supply of sand, namely the subsiding depositional basin centred on Lake Eyre.

Second, it is difficult not to be impressed by the over-whelming significance of *water* in the shaping of the earth's surface. This is remarkable when it is recalled that over 97 per cent of the earth's total water supply is contained in the oceans and that only 2.87 per cent is located on land (rivers and lakes plus glaciers and groundwater), with a mere 0.01 per cent in the atmosphere.[4] But the cycling and recycling of the small percentage of the total moisture present on land is vital to many widespread geomorphological processes. It is this small percentage which causes solutional effects and is thus the precursor of much weathering of rocks. In addition hydration and hydrolysis are dominant forms of weathering in many areas. The significant role of subsurface weathering by moisture in initiating landforms is being increasingly appreciated.

Water is also the medium responsible for the transport and deposition of weathered debris, either through the agency of rivers themselves, or through wash, through some forms of mass movement, through glaciers or through wind-driven waves.

Next, the importance of *storms* — brief periods of high-energy impact on the land surface — cannot be overlooked. Although it is necessary to temper the undoubted ferocity of storms with consideration of their infrequent occurrence,[5] river floods, heavy rains, wind storms, coastal storms and internally-generated events such as earthquakes and volcanic eruptions all have left their imprint on the landscape long after the event occurred. It is true that the more usual levels of process activity eventually modify and mask the effects of these catastrophes, but specific landforms related to such brief periods of intense activity are probably more widespread and significant than is generally conceded.

Fourth, cognisance should be taken of an effect which pervades many aspects of geomorphology: in many situations, a small initial variation may be responsible for the ultimate development of major relief contrast — small differences are emphasised or *reinforced* in time. For instance, weathering effects are in this category, with the effect of one mechanism or process opening the way for the operation of others. Chemical weathering, for example, may result in a loss of volume, increased porosity and permeability, and vulnerability not only to further chemical attack but also perhaps to salt crystallisation or to frost action, depending on local conditions. The structure of a fissile rock renders it vulnerable to frost shattering, which in turn leads to further susceptibility as the fractures are opened up. The widening of one fracture by weathering can reduce pressure on an immediately adjacent rock enough to release stress and allow the development of a swarm of minor fissures parallel with the original fracture. Even a slight contrast in jointing results in one compartment of granite standing slightly above the adjacent areas. But this slightly higher elevation means that run-off from the residual runs down to the plains, which therefore receive more than their proportional share of water, and which are therefore weathered more intensely than the rocks which underlie the inselberg. Thus in time, quite a considerable relief contrast is developed from a minor textural or joint contrast. This contrast persists until some major change occurs in the system—tectonism, change in stress conditions in depth (see Chapter 5), or climatic change. Likewise, sandstones in a folded sedimentary sequence are tougher and chemically less susceptible than are most other sediments. Again they form ridges, and again they are less exposed to chemical (moisture) attack on this account, whereas the valley floors are rapidly and intensely weathered and hence are prone to erosion. Thus in the Flinders Ranges, the ridges have, since the end of the Mesozoic, become more and more prominent because of the comparative rapidity with which the valleys have been worn down. Relief amplitude appears to have increased in time. The implications of localised scarpfoot weathering have been discussed in several earlier chapters. Because of volume reduction consequent on weathering, or as a result of stream erosion in the scarpfoot zone, depressions or valleys are commonly developed at the hill foot in arid and semiarid regions. The valleys, and especially the depressions, gather run-off from the hillslopes above, thus inducing further weathering of the scarpfoot zone.

Depositional effects, too, are reinforcing. Sand grains laid down by the wind themselves roughen the surface, decrease near-surface wind velocity and induce further deposition. River deposition, by increasing the wetted perimeter, is also a cumulative process as discussed in relation to alluvial fans.[6]

A fifth conclusion, which stems in part from the previous two, is that many forms commonly attributed to processes which were thought to be characteristic of and restricted to a particular climatic region, are now known to occur in more than one climatic zone. In many instances this is not a matter of inheritance, but due to similar forms developing in different ways: they are *convergent* forms. Thus braided streams occur in both periglacial regions and in arid and semiarid lands: though the hydraulic conditions are similar in both cases, the causative factors are not. Meandering forms occur in a wide variety of environments, including tidal flats; and although the basic physical conditions are similar in all areas where meanders develop, the precise contexts vary. Patterned ground (*sensu lato*) occurs in arctic and in seasonally dry tropical lands.

Though inselbergs are by definition residuals standing above planate surfaces in tropical arid regions, forms similar to typical inselbergs occur both in temperate and cold lands.[7] On Dartmoor, both castellated and domed inselbergs (locally called tors) occur within a few kilometres of each other; they compare morphologically with forms called inselbergs from other parts of the world (see Chapter 3). Though the break of slope between hill and plain is more pronounced for climatic reasons in the arid and semiarid tropics (see Chapter 10), structural factors are of primary importance in inducing the development of the piedmont angle. These obtain as much in Lapland[8] or on Dartmoor as in Angola[9] or central Australia. In this case structural

factors are of over-riding importance, and have produced convergent forms which are morphologically similar but which have evolved through different processes.

Gnammas or weather pits are another form which evolves in varied circumstances: they have been described from many climatic contexts,[10] and different processes are responsible for their development in different parts of the world; moreover they occur in coastal regions as well as in areas far distant from the sea.

Planate surfaces are another source of confusion. Plains which conform morphologically to the peneplain type are not restricted to temperate regions; indeed, they appear to be more widespread in arid and semiarid lands at present, though this may be more a matter of preservation than genesis. Again, there are similarities, both in morphology and (as far as it is understood) in genesis, between the pediments of tropical arid regions and similar surfaces of cooler lands.

But the problem arises whether one could or should distinguish between forms which are morphologically similar— either on the basis of real or alleged contrasts in their precise mode of origin, or on an age basis. The inselberg tors of Dartmoor have been attributed to a two-stage development involving differential subsurface weathering of the granite under tropical conditions during the Tertiary, followed by differential erosion under periglacial conditions during the Quaternary. Should one differentiate between these and similar forms which have been brought into relief by river erosion rather than frost-induced processes? Possibly if the interpretations were certain, such distinctions would be justified, but alternative views have been advanced in explanation of the Dartmoor tors.[12] As with the interpretation of most landforms, there are considerable areas of doubt, and primary classifications and terminologies based on morphology are best at this stage.

There is no doubt that to a considerable extent geomorphology is, and has been, captive to its own past. Climatic associations of landforms are established (possibly on inadequate grounds) and become entrenched both in the literature and in our thinking. There is all the more reason then to recognise that some of these associations are of dubious validity, and that many features evolve in more than one way and are in reality convergent forms.

A sixth and final general observation on practical problems of analysis is that *climatic changes* have enormously complicated interpretations of landforms. Although "the interpretation of landforms rests more and more on palaeoclimatology", it is desirable to have the climatic chronology established independently of morphology (except in certain well-established cases), before it is used as a basis for geomorphological thinking; otherwise cause and effect can easily become confused.

At a more general, philosophical level the doctrine of uniformitarianism appears to form the only reasonable foundation on which to base the interpretation of landforms at present. Further detailed studies and greater understanding of the processes at work at the earth's surface may, however, reveal more discrepancies between the forms we see and those that should have developed under the conditions we know existed in the past, and thus may cause the doctrine to be modified in part. In any case, the present period of the earth's history, following on a major ice age, is not an ideal yardstick by which to evaluate the past. This is not to suggest that the laws of physics have altered, but that the intensity of activity of various processes could have changed; in the past, critical values could have been exceeded and new forms of activity could have developed. This is particularly relevant to the present age, for one external variable has not been constant but has evolved in time: the biological factor. As Russell has said: "Never before has the earth been so well armoured [by soils and vegetation] against processes of weathering and erosion".[13] In the past, certainly prior to the Cainozoic, the earth's surface was a biological, if not a climatic, desert. Thus, to take one example, it is arguable that the forms most readily developed in arid or semiarid regions today are likely to be those typical of geological time. Pedimentation rather than peneplanation may, as King has suggested, have been dominant in the past.[14] On the other hand, the human population explosion and the technological revolution have together brought about at least temporary reversion to biological deserts over wide areas. These have critical if ephemeral implications for measured rates of erosion and deposition.

Nevertheless, uniformitarianism forms the backdrop to the first stage in investigation. It should reveal those aspects of a land surface which are evidently alien to the present condition and which therefore warrant even closer study; it "should be looked on only as supplying a beginning for investigation".[15]

In view of the undoubted complexity of landform developments, and the manifest uncertainties which at present attach to their evolution, are valid and logical interpretations of landforms possible? It is too much to expect that all facets of a landscape are susceptible to rational analysis, if only because of their antiquity and the fragmentary survival of evidence. The number and variability of the factors which influence the land surface present another kind of difficulty; as Melton has stated: "The variability in any natural environment is the product of happenings in many geological periods... To argue that this variability could ever be entirely explained is absurd".[16]

Nevertheless, though the nature of the basic processes at work is obscure in many areas and the possibility of rational analysis appears remote, even in "the absence of hope it is still necessary to strive".[17] Complexity is not synonymous with chaos, any more than uniformity is to be equated with simplicity.[18] Despite its complexity, the geomorphological landscape has already proved susceptible to reasoned analysis and interpretation. The analysis must obviously be incomplete, but the potentialities nevertheless compare satisfactorily with the possibilities of the other earth and natural sciences. Faced with considerable philosophical problems, and large gaps in the web of evidence, we may wryly console ourselves that there is little likelihood of geomorphologists ever being so dogmatic or confident of their findings that complacency will stultify further investigations.[19] On the contrary, far from being allayed or satiated, scientific curiosity tends rather to be quickened by the ever-increasing radius of the sphere of ignorance revealed by the ever more powerful illumination provided by man's increasing knowledge of the earth.

References Cited

1. A. Scheidegger and W.B. Langbein, "Probability Concepts in Geomorphology", *U.S. Geol. Surv. Prof. Paper*, 1966, **500-C**.

2. H. Baulig, "The Changing Sea Level", *Inst. Br. Geogr. Publ.*, 1935, **3**, p. 44.

3. C.R. Twidale, "Geomorphology of the Leichhardt-Gilbert Area of northwest Queensland", *C.S.I.R.O. Land. Res. Ser.*, 1966, **16**.

4. R.L. Nace, "Water Management, Agriculture and Groundwater Supplies", *U.S. Geol. Surv. Circ.*, 1960, **415**.

5. M.G. Wolman and J.P. Miller, "Magnitude and Frequency of Forces in Geomorphic Processes", *J. Geol.*, 1960, **68**, pp. 54–74.

6. See for example C.R. Twidale, Jennifer A. Bourne and Dianne M. Smith, "Reinforcement and Stabilisation Mechanisms in Landform Development", *Rev. Geomorph. Dyn.*, 1974, **23**, pp. 115–25.

7. H. Wilhelmy, *Klimamorphologie der Massengesteine*, Westermann, Brunswick, 1958; C.R. Twidale, *Structural Landforms*, Australian National University Press, Canberra, 1971.

8. H. Schrepfer, "Inselbergs in Lappland und Newfoundland", *Geol. Rdsch.*, 1933, **24**, pp. 137–43.

9. O. Jessen, *Reisen und Forschungen in Angola*, Reimer, Andrews and Steiner, Berlin, 1936.

10. C.R. Twidale and Elizabeth M. Corbin, "Gnammas", *Rev. Geomorph. Dyn.*, 1964, **14**, pp. 1–20.

11. D.L. Linton, "The Problem of Tors", *Geogr. J.*, 1955, **121**, pp. 470–87.

12. L.C. King, "The Problem of Tors", *Geogr. J.*, 1958, **124**, pp. 289–91; J. Palmer and R.A. Nielson, "The Origin of Granite Tors on Dartmoor, Devonshire", *Proc. York. geol. Soc.*, 1962, **33**, pp. 315–40.

13. R.J. Russell, "Geological Geomorphology", *Bull. geol. Soc. Am.*, 1958, **69**, pp. 1–22, at p. 5.

14. L.C. King, "Canons of Landscape Evolution", *Bull. geol. Soc. Am.*, 1953, **64**, pp. 721–52; *idem*, "The Uniformitarian Nature of Hillslopes", *Trans. geol. Soc. Edin.*, 1957, **17**, pp. 81–102.

15. J. Barrell, "Rhythms and the Measurements of Geologic Time", *Bull. geol. Soc. Am.*, 1917, **28**, pp. 745–904.

16. M.A. Melton, "Correlation Structure of Morphometric Properties of Drainage Systems and their Controlling Agents", *J. Geol.*, 1958, **66**, pp. 442–60.

17. Generally attributed to William of Orange.

18. F.H. Lahee, "Theory and Hypothesis in Geology", *Science*, 1909, **30**, pp. 562–3.

19. P. Medawar, "The Effecting of All Things Possible", *Adv. Sci.*, 1969, **26**, pp. 1–7.

Appendix: Geological Background

A. GENERAL

Many important geomorphological features, major and minor, owe their origin either directly or indirectly to the nature of the underlying crust. In this appendix, the origin and classification of rocks, their texture and structure, and the means of determining their ages are briefly described.

B. ROCK TYPES

The rocks which form the earth's crust are of three types. *Igneous* rocks are either emplaced in molten form intrusively, at depth, or they are poured out or extruded at the surface in the form of lavas. All *volcanic* rocks, of which basalt is a common example, are lavas. The rocks intruded deep in the crust are known as *plutonic* rocks. Because they cool slowly, there is time for considerable aggregates of the various rock-forming minerals to come together, and so the average size of crystals is large; these rocks are coarse-grained. Volcanic rocks, on the other hand, cool rapidly as the lavas come in contact with the atmosphere, and so these rocks are fine-textured, featuring crystals so small that in many instances they can be discerned only under the microscope. Igneous rocks intruded at moderate depths, in dykes and sills for example (Pl. A.1), are of *hypabyssal* type, and exhibit crystals of a size intermediate between these two extremes. Sills are injected along pre-existing structures such as joints, whereas dykes are injected across them.

Igneous rocks are the primary material of the earth's crust and igneous rocks crop out over very large areas of the earth's surface. They are classified partly on the basis of their origin (plutonic, hypabyssal, volcanic) and partly on their mineralogy: the principal types are shown in Table A.1. The weathering and erosion of these crystalline rocks provides materials which, after transportation by water, wind, waves or ice, are laid down either on the land or (more commonly) in water, to form *sedimentary* rocks. These may be regarded as forming a veneer or blanket overlying a crystalline basement in many areas. They are characteristically layered, bedded or stratified (Pl. A.2), commonly display sorting or grading, and usually contain fossils. Sandstone, shale and siltstone are common examples of *clastic* sediments, built of particles derived from other rocks (Table A.2). Other sediments, such as limestone, bog iron ore and coal, originate as *chemical* precipitates or through *organic* agencies.

Both igneous and sedimentary rocks may be subjected to heat or pressure and thereby changed to a rock of quite different structure and mineralogical character. Some such *metamorphic* rocks are associated with intrusions or extrusions of igneous rocks. The rocks in the immediate vicinity of an intrusion of granite or immediately beneath a lava flow are baked, and for obvious reasons this type of alteration is often known as contact metamorphism. The zone of alteration is called a *metamorphic aureole*. Regional metamorphism, which is much more important, occurs when large areas suffer compression and associated heating. The more intense the compression, the more marked are the changes brought about. It is common for all traces of the original bedding of a sediment, for instance, to be lost and replaced by a new set of fractures—the cleavage—which is disposed normal to the direction of greatest stress, and for extensive re-crystallisation of the rock to occur. The type of rock produced

Plate A.1 Sill of fine-grained aplite in granite bedrock, Yarwondutta Rocks, Eyre Peninsula, S.A. (*C.R. Twidale.*)

Table A.1
Classification and Setting of Igneous rocks*

		Oversaturated			Saturated			Undersaturated		
		I Quartz	II Quartz+Felspars		III Felspars			IV Felspars + Felspathoids	V Felspathoids	VI Mafic Minerals Predominant
			Predominant Orthoclase	Predominant Plagioclase	Predominant Alkali Felspars (Or, Ab).	Predominant Soda-Lime Plagioclase	Predominant Lime-Soda Soda Plagioclase			
PLUTONIC	Felsic	Igneous Quartz Veins (Arizonite; Silexite)	Granite	Granodiorite (Tonalite)	Syenite	Diorite	Anorthosite	Nepheline-Syenite	×	
	Mafelsic	×		×.	×	×	Gabbro	Theralite and Teschenite	Ijolite	
	Mafic				×	×	×	×	×	Peridotite Picrite
HYPABYSSAL			←————————————————Aplites————————————————→							
			←————————————————Porphyries————————————————→							
				←————————————Lamprophyres————————————→						
			Granophyre Felsite				Dolerite	Tinguaite		
			←————————Pitchstone————————→				Tachlyte			
VOLCANIC		Rhyolite	Dacite	Trachyte	Andesite	Basalt	Phonolite	Leucitophyre Nepheline-Basalt Leucite-Basalt	Olivine-rich Basalts	
				←————————Pitchstone————————→			Tachylyte			Limburgite
		←————Obsidian————→								
Average Silica Percentage		90	72	66	59	57	48	54·5	43	41

*After G.W. Tyrrell, *The Principles of Petrology*, Methuen, London, 1949, p. 108.

Table A.2
Generalised Classification of Sediments*

Composition			Quartz ± Chert	Quartz + Chert + Micas + Chlorite		Quartz + Feldspar ± Clay
				− Feldspar	+ Feldspar	
	Texture		Quartz conglomerate	Graywacke conglomerate		Arkosic conglomerate
Detrital Rocks (Clastic Texture)	Coarse Conglomerates			Low rank	High rank	
	Medium Sandstones		Quartzite and quartzose sandstone	Graywacke		Arkose
				Low rank	High rank	
	Fine Shales		Quartzose shale	Micaceous shale	Chloritic shale	Kaolinitic and feldspathic shale
Chemical Rocks	Sandy (clastic)		Limestone, Dolomite, Chert, Salt, Gypsum, etc.			
	Pure (crystalline)					

*After P.D. Krynine, "Sediments of the Search for Oil", *Prod. Monthly*, 1945, **9**, pp. 12—22.

Plate A.2   Thinly-bedded, horizontally disposed shales exposed in the gorge of Joffre Creek, Hamersley Range, W.A. *(Geol. Surv. W.A.)*

Table A.3
Classification and Setting of Metamorphic Rocks*

Zone	Temperature	Uniform Pressure	Directed Pressure	Kind of Metamorphism	Minerals Formed	Rocks
Epizone Near-surface	Low to moderate.	Small.	Often strong; occasionally absent.	Cataclastic to Dynamo-thermal.	Stress minerals (p. 266). Hydroxyl-bearing minerals common.	Phyllites, sericite-, talc-, epidote-, chlorite-, glaucophane-schists. Quartz-schist; schistose-grit, etc.
Mesozone Intermediate depth	Considerable.	Considerable.	Mostly strong.	Dynamo-thermal. Load metamorphism.	Stress minerals predominant.	Mica-schist; garnet-mica-schist, staurolite-schist, hornblende-schist. Mica-, and hornblende-gneiss.
Katazone Deep	High.	Very great.	Feeble or absent.	Load or static metamorphism. Plutonic metamorphism.	Anti-stress minerals (p. 263) predominant.	Coarse biotite-, pyroxene-, sillimanite-, and cordierite-gneiss. Granulites, eclogites, etc

*After G.W. Tyrrell, *Principles of Petrology*, p. 258.

by metamorphism depends on the intensity of the pressure; various zones have been distinguished (Table A.3) according to their depth of occurrence.

C. **STRUCTURES**

Rocks of any origin may be subjected to tectonism or earth movements (indeed metamorphism is commonly associated with earth movements) and as a result these rocks are either *folded* or *faulted*.

It is clear from detailed studies of rocks, both in the field and in the laboratory, that some bend more easily than others, while others fracture more readily than they bend. Some rocks are inherently stronger than others; some are brittle, some plastic. In addition, the environment and thickness of the stratum and the time factor are important in determining whether a bed will bend (fold) or break (fracture). Limestone and sandstone, for instance, tend to fracture or to form only broad folds because they are strong and rather brittle; but if either of these is interbedded in a thick sequence of shales (which tend to be soft and plastic and hence to fold easily), the flow form of the shale is imposed upon the interbedded sandstone. Again, where the overlying thickness of rocks is great, limestones and sandstones tend to fold rather than fracture; and with any rock type, a pronounced long-term stress results in fracture. Rocks which have a high inherent strength and possess the ability to lift overlying strata, like massive limestones and sandstones, are said to be *competent*, while rocks which flow easily and are weak are *incompetent*.

It will be recalled that sedimentary rocks are laid down in layers or strata. These beds may have a slight initial inclination, but usually most are horizontally disposed, or nearly so. The outcrop of a stratum which is precisely

horizontal traces the contours of landforms and a whole succession of beds is exposed in a valley or hillside.

An outcrop of younger rocks preserved on a hill top and surrounded by older strata is called an *outlier* (Fig. A.1); an exposure of older rocks (say on an isolated hill or on a valley floor) surrounded by younger rocks is called an *inlier*. Mt Babbage, in the northern Flinders Ranges, is an example of an outlier of Cretaceous rocks surrounded by Precambrian strata (Fig. A.1 and Pl. A.3). Many of the Permian outcrops of the southern Mt Lofty Ranges occur in basins bounded by Precambrian or early Palaeozoic strata, and thus also form outliers.

Strata which have been compressed, however slightly, are inclined and the bedding planes are said to dip. The *dip*, or *true dip*, of a stratum (Fig. A.2) is the maximum angle between the bedding plane and the horizontal. On geological maps the direction of dip of a stratum is indicated by an arrow, and the amount of dip in degrees. The direction at right angles to true dip, measured in the horizontal plane, is the *strike*. Just as the form of a hill can be indicated by the use of contours, so the behaviour or form of a stratum can be shown by *structure* or *stratum contours* and (as is the case with topographical contours) the distance of the contours from one another indicates the dip of the stratum, the amount of which can be obtained precisely by the use of simple trigonometry. The dip of a stratum is rarely constant, for beds usually follow irregular or curved planes, but their attitude and behaviour can be reconstructed very easily from a study of the stratum contours. If there is a monocline or simple flexure or bend in a stratum, for instance, it is expressed in the stratum contours as shown in Fig. A.3a and b.

Folds are generated by several mechanisms. Stresses from two opposed directions give rise to *anticlines* and *synclines*

in section   in plan

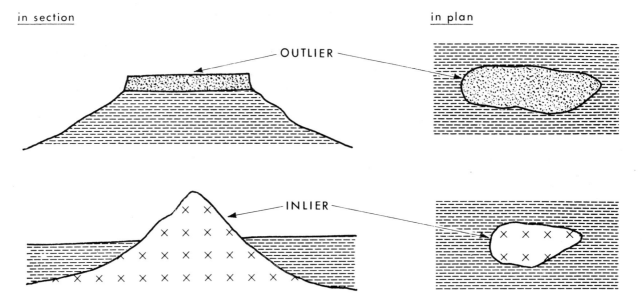

Fig. A.1 Outlier and inlier.

Plate A.3 Mt Babbage, an outlier of Cretaceous rocks (K) resting on Precambrian crystallines (P€), in the northern Flinders Ranges, S.A. The unconformity between the sediments and crystallines is indicated by the dashed line. The crest ridge, X, is an exhumed surface remnant of pre Cretaceous age. (*C.R. Twidale.*)

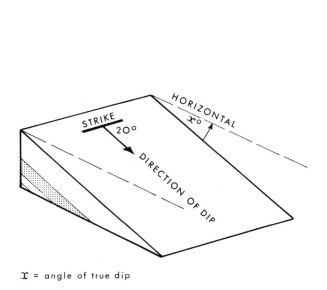

x = angle of true dip

Fig. A.2   Dip and strike.

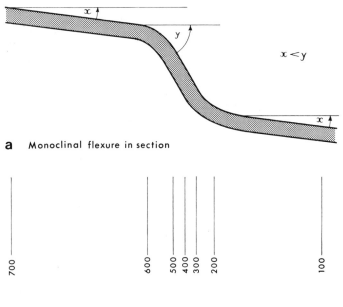

a   Monoclinal flexure in section

b   on map, as expressed in Stratum Contours

Fig. A.3   A monocline (a) in section and (b) as expressed in stratum contours.

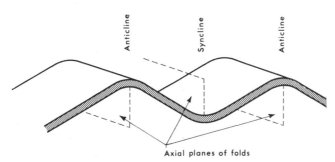

a   Anticline & Syncline

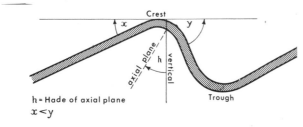

h = Hade of axial plane

x < y

b   Asymmetrical folds

Fig. A.4   Anticline, syncline and asymmetrical fold, showing axial plane, crest and trough.

(Fig. A.4). The former is a fold which is convex upward, the latter one which is convex downward. Each comprises two limbs which meet either in the *crestal* or *hinge line* or in the *trough line*. The crestal (hinge) line is the imaginary line which joins all the highest points on any bedding surface of an anticline or *antiform* (an apparent anticline in which, however, the correct stratigraphic sequence is not determined. The trough line is an imaginary line which joins all the lowest points on any bedding plane in a syncline or *synform*. The plane which connects the hinge line between the two limbs of a fold is the axial plane, *axial surface*, or axis of the fold. The trace of the axial plane on any stratum is the *axial line* (Fig. A.4). Symmetrical and upright folds have axes which bisect the structures.

Inequality in opposed stresses or failure of parts of the sedimentary sequence causes the development of asymmetrical and inclined folds, with limbs unequally inclined and axes which are not vertical. (Figs A.4, A.5). Complex *anticlinoria* and *synclinoria* may be developed with minor anticlinal and synclinal structures within the broader forms. The angle between the sloping axial line and the horizontal, as measured in a vertical plane, is the *plunge inclination*. Such *plunging folds*, are very common and are geomorphically significant by virtue of their convergent and divergent limbs and outcrops (see Pls 5.13 and 5.14). The inequality of opposed stresses may be so great that the dip of the steeper limb exceeds the vertical (*overturned folds*), and even the axial plane may become horizontal or nearly so. These are called *recumbent folds*. On the other hand, in *isoclinal folds* the lateral compression has been so intense that the limbs and axial planes are parallel.

Two sets of opposed stresses cause *crossfolding* and the development of domes and basins. This occurs in the Wilpena Pound (Pl. 5.16) of the central Flinders Ranges and in the Maverick Domes of Wyoming (Fig. A.6 and Pl. A.4). A *dome* is a double plunging anticline, the length of which in ground plan is not more than three times its width; a *basin* is a double plunging syncline, the length of which in ground plan is not more than three times its width. In a dome (also known as a periclinal structure or pericline, though the terms seem superfluous), the strata dip outwards from the centre of the structure in all directions: this is called a a *quaquaversal* dip.

A *fault* is a fracture in the earth's crust along which differential movement has taken place. Usually a fault does not consist of a single fracture, but rather is a narrow zone containing several fractures. Furthermore, once developed as a result of stresses in the earth's crust, a fault forms a zone of weakness along which further dislocation readily takes place, so that most faults are of recurrent character.

Faults which run parallel to the dip of the country rock are called *dip* faults; those which trend parallel to the local strike are called *strike* faults; while those which cut across both dip and strike are termed *oblique* (Fig. A.7). In Fig. A.8 various geometrical values of faults are shown. The *hade* of the fault is the angle between the fault plane and the vertical; the *dip* is the angle between the fault plane and the horizontal. Hade and dip thus total 90°. The *slip* of the fault (A–C in Fig. A.8) is the amount of movement that has taken place along the fault plane; the amount of vertical dislocation is the *throw* (A–B); the *heave* (B–C) is the measure of the horizontal displacement brought about by faulting; and the amount of dislocation of outcrop in plan by faulting is called *displacement* (O–P).

The nature of the stresses which cause faults provides a convenient basis for their classification. *Tensional* faults (also known as *normal* and *gravity* faults) are associated with extension of the crust. In such faults, the plane hades towards the upthrow side (Fig. A.9). *Compressional* faults with a dip of more than 30° (a hade of less than 60°) are known as *reverse faults* (Fig. A.9); but those displaying lower dips (or more gently inclined planes of dislocation) are called *thrust* faults. In a *normal thrust*, the upper block rides over the lower (Fig. A.9) but in a *lag thrust*, the lower block is thrust forward and upwards beneath the upper (Fig. A.9).

In both normal and reverse faults the amounts of slip and throw commonly vary not only from fault to fault but also along any given fault plane. This variation is commonly irregular, but in some instances an overall regular increase or decrease in these dimensions along the plane results from a strong tensional component in the fault dislocation, and causes the development of *rotational* faults. These are of two principal types. A *hinge* fault is one in which the throw increases away from the hinge point (Fig. A.10a). In a *pivotal* fault, one block rotates with respect to another along a fault plane and about a point (Fig. A.10b) causing one block to be raised with respect to the other on one side of the hinge, and depressed on the other side.

There is a distinct horizontal component to the movement in rotational faults, and the same is true in some small measure of normal and reverse faults. In *wrench, tear* or *transcurrent* faults, however, the horizontal dislocation is

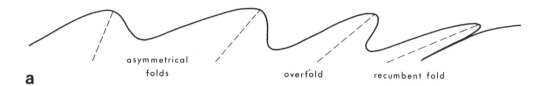

a   asymmetrical folds   overfold   recumbent fold

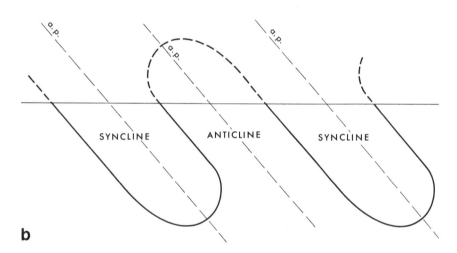

b

SYNCLINE   ANTICLINE   SYNCLINE

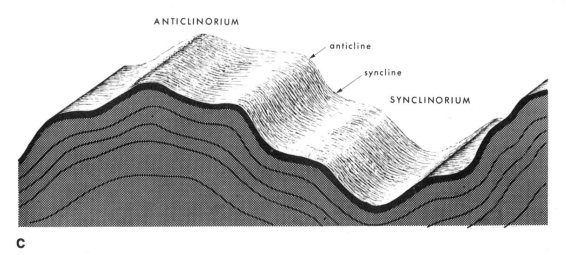

c

ANTICLINORIUM

anticline

syncline

SYNCLINORIUM

Fig. A.5   (a) Overfolds, (b) isoclinal folds, and (c) anticlinorium and synclinorium.

Plate A.4 Maverick Domes, Wyoming, from the north, showing three domes *en echelon* developed in upper Palaeozoic and Mesozoic strata (see Fig. A.6). In the foreground is Circle Ridge Dome, beyond that is Big Dome, and just visible in the distance is Little Dome. *(John S. Shelton.)*

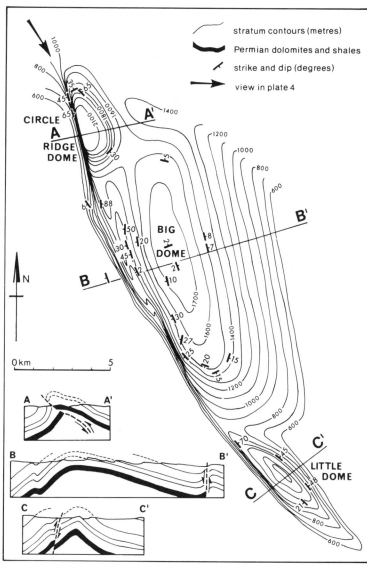

Fig. A.6 Stratum or structure contours of a stratigraphic horizon and sections across major features in the Maverick Domes, Wyoming (see Pl. A.4). *(After A.J. Collier, "Anticlines near Maverick Springs, Fremont Country, Wyoming", U.S. Geol. Surv. Bull., 1920, 711, pp. 149–66; and Bruce F. Curtis, personal communication.)*

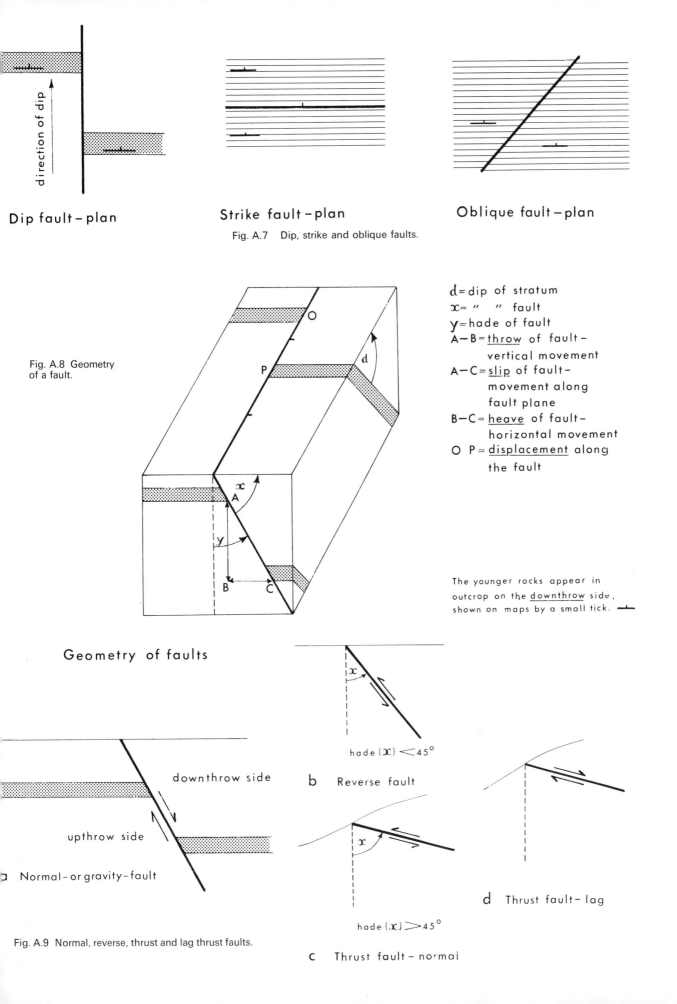

Dip fault – plan

Strike fault – plan

Oblique fault – plan

Fig. A.7 Dip, strike and oblique faults.

direction of dip

Fig. A.8 Geometry of a fault.

d = dip of stratum
x = " " fault
y = hade of fault
A–B = <u>throw</u> of fault –
vertical movement
A–C = <u>slip</u> of fault –
movement along
fault plane
B–C = <u>heave</u> of fault –
horizontal movement
O P = <u>displacement</u> along
the fault

The younger rocks appear in outcrop on the <u>downthrow</u> side, shown on maps by a small tick. ⊥

Geometry of faults

downthrow side

upthrow side

a Normal – or gravity – fault

hade (x) < 45°

b Reverse fault

hade (x) > 45°

c Thrust fault – normal

d Thrust fault – lag

Fig. A.9 Normal, reverse, thrust and lag thrust faults.

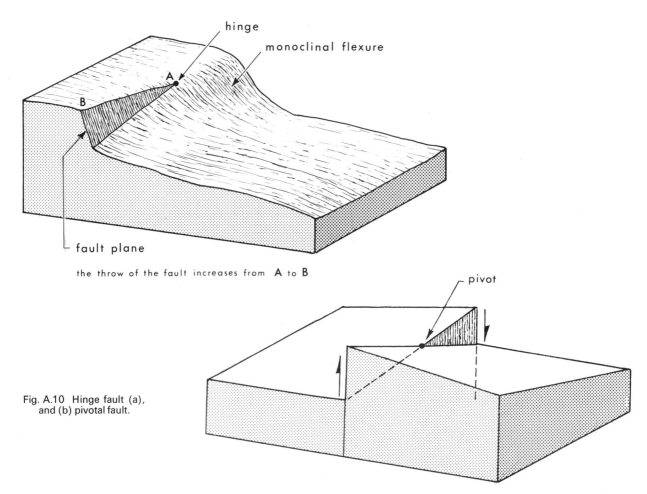

the throw of the fault increases from **A** to **B**

Fig. A.10 Hinge fault (a), and (b) pivotal fault.

dominant and any vertical movement is so small as to be incidental. Wrench faults are described as either dextral (called right lateral by U.S. workers) or sinistral (left lateral). If it is imagined that the observer is facing a fault block with the wrench fault running at right angles to the line of view, the fault is described as dextral or right lateral if the block on the opposite side of the fault trace has moved to the observer's right (Fig. A.11). The fault is called left lateral or sinistral if this block has been offset to the observer's left.

The evidence for faulting takes several forms. When rocks adjacent to each other or to a fault plane are moved relative to each other, the plane surfaces are gouged and scratched, resulting in the formation of scratches or striations called *slickensides* (Pl. A.5), or distinct ridges and furrows called *mullion structures*. The opposed surfaces may also be polished and plucked, causing the development of *fault steps* (Pl. A.5). Movement along the fault also generates frictional heat, so that a skin of rocks immediately adjacent to the fault is cooked or recrystallised; that is, it is metamorphosed. There may also be intense shattering of rocks along fault planes, and aggregates of angular fragments of the country

rock form *fault breccias* when consolidated. Some rocks do not, however, disintegrate into sufficiently large fragments to form a breccia; they break up into a powder of flour-like consistency which forms a fine-grained *fault gouge* when consolidated. Such a powder, if silicified, is called *mylonite*. The presence of any of these indicates that a fault is nearby. Strata are abruptly displaced and are repeated or disappear. They are commonly distorted in the immediate vicinity of fault planes, and such *fault drag* can be used to detect the sense of movement (Pl. A.6). Faulting is also indicated by topographic features, as described in Chapter 4.

D.   DATING

Rocks can be dated in two ways: relatively and absolutely (or at least within narrow limits on the absolute scale). The *Law of Superposition* is obvious, but fundamental. It states that in a sequence of rocks which is not overturned by tectonism, the age of the rock is directly related to its position in the sequence: the lower one goes in the sequence, the older

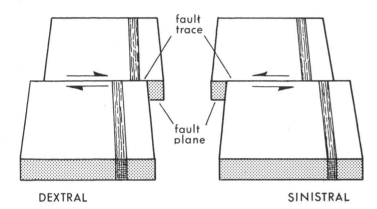

fault
trace

fault
plane

DEXTRAL   SINISTRAL

Fig. A.11   Dextral and
sinistral wrench faults.

Plate A.5 Slickensides and fault steps formed
on a sandstone block near Quorn, southern
Flinders Ranges, S.A. Clearly the relative
movement was vertical and common sense
indicates that the block nearest the camera—
the block which has been eroded away—
moved downwards relative to the opposite side
of the fault plane. The fault steps would seem
to indicate this, but some experimental
evidence (*see M.S. Paterson,* Bull. geol. Soc.
Am., *1958,* **69**, *pp. 465–76*) suggests that the
near block may have moved from bottom to
top (possibly rucking up flakes of rock and
forming fault steps). (*B. Daily.*)

Plate A.6 Flat-lying early Tertiary lake sediments dragged up by movements
on the Eden Fault, Mt Lofty Ranges, S.A. *(C.R. Twidale.)*

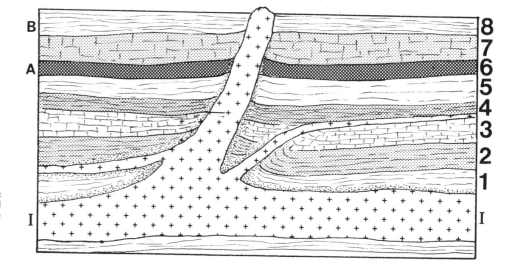

Fig. A.12   Diagrammatic
section through geological
sequence to illustrate the
Law of Superposition.

the rocks are; and, conversely, the higher the rocks, the younger they are. Thus in Fig. A.12, A is clearly older than B because the latter overlies the former—similarly the granite and related sills and dykes are obviously younger than both A and B, because they intrude into the sediments which must have already been there when the vulcanicity took place.

As has been mentioned, sediments commonly contain fossils: these are the skeletons, imprints, or any vestiges of any type of organism which lived in the past.

Generally speaking, both fossilised flora and fauna which are preserved in the geological column have evolved closer and closer to modern forms as time has passed. Many modern organisms do not appear until late in the geological record; others, abundant in the past, are now extinct. Some fossil forms have a restricted time range; others are widespread and continued almost unchanged for many millions of years. By considering groups or assemblages of fossils as characteristic of and peculiar to limited time spans, and bearing in mind the Law of Superposition, a relative time scale can be arbitrarily constructed for sediments (Column 2 of Table A.4); a time scale for igneous and metamorphic rocks can also be constructed on the basis of their relations with assemblages of fossils.

Table A.4
Time Scales

A. Geological Time-scale (M. YEARS)

Era (1)	Period (2)	Duration (3)	M. of years ago (4)
Cainozoic......	Quaternary	1	
	Tertiary	68	70+2
Mesozoic ·......	Cretaceous	65	135+5
	Jurassic	45	180+5
	Triassic	45	225+5
Palaeozoic......	Permian	45	270+5
	Carboniferous	80	350+10
	Devonian	50	400+10
	Silurian	40	440+10
	Ordovician	60	500+15
	Cambrian	100	600+20
Precambrian ...	Proterozoic		
	Archaean		

B. Cainozoic Time-scale (M. YEARS)

Quaternary......	Recent	0.01 M. (10,000 years)	
	Pleistocene*	1	
Tertiary·.........	Pliocene	10	11
	Miocene	14	25
	Oligocene	15	40
	Eocene	20	60
	Palaeocene	10	70+2

*The Pleistocene may be as little as 600,000 years, or as long as 3–3·5 M. years duration.

Many attempts were made to give absolute values to range this widely used and very useful relative scale, but this was not satisfactorily accomplished until the discovery of radio-activity. In the last years of the nineteenth century and the early years of this century, it was recognised that many elements are unstable and lose particles in the course of time. Through this decomposition, called radioactivity, they are transmuted to new substances. Thus uranium, with atomic weight 238, gradually loses four particles of helium (atomic weight 8) changing in the process to isotopes of uranium, then omnium, radium, polonium and finally uranium-lead (atomic weight 206). The rate of decay is constant and can be determined. Thus by measuring the proportional amounts of uranium 238 to uranium-lead, the time during which decay has been going on, which is the same as the age of the minerals in which the elements genetically occur, can be calculated.

Other common and useful pairs of elements linked by chains of disintegration are potassium-argon, strontium-rubidium, and thorium-actinium. Great care has to be taken with the application of the principle, but consistent results have been obtained which enable absolute limits to be given to the various periods and epochs listed in Table A.4 The methods mentioned can be used on materials about one million years old or more.

A similar and especially interesting method, since it can be used to date recent geological events which are beyond the limits of accuracy of the radioactive decay methods already described, concerns carbon-14. Ordinary carbon has an atomic weight of 12, but carbon with an atomic weight of 14 forms from nitrogen in the upper atmosphere. It diffuses into the lower atmosphere and some is taken up by and incorporated in the cells of plants and animals. There is a fixed ratio of C_{14} to C_{12} in the air, and in organisms the two are also in equilibrium, for the intake of new C_{14} is balanced by the loss of electrons from the C_{14} already there, which is thus reconverted to nitrogen of the same atomic weight. When an organism dies, however, the accession of new C_{14} stops, while the C_{14} already in it continues to disintegrate. Thus, the proportion of C_{14} to C_{12} decreases. The rate of decay of C_{14} has been carefully measured and is found to be constant, so that a measure is provided of the age of the organic material in which the C_{14} occurs. Half of the C_{14} disappears in about 5568 ± 30 years (a value known as the half-life of C_{14}) in another 5568 years, half of the remainder will have disintegrated, and so on.

Thus, by measuring the radioactivity of living matter, and the radioactivity of the sample of fossil material, the ratio between the two can be determined. Bearing in mind the half-life of C_{14}, the material from which the sample was taken can be dated within close limits. Thus let us imagine that a sample when measured yields 3.825 disintegrations per minute compared with the 15.3 ± 0.1 which is usual for living matter. The sample is only one-fourth as active as modern material, which means that the C_{14} in the sample has run through two half-lives: in the first the radioactivity was reduced by half, and in the second the remaining half would be halved, to a quarter of the original; the material in question is $2 \times 5568 \pm 30$ or roughly 11,136 years old. C_{14} dates are expressed in years before present (B.P.) or years before Christ (B.C.).

For technical reasons, at present the method is only valid for material up to 50,000–60,000 years old, and for certain types of sample the validity span is even shorter. Thus there is a vital and frustrating gap in our absolute dating technique between, say 50,000 years C_{14} and about one million years (beyond which the other radioactive methods can be used).

It should also be appreciated that the $\bar{C}_{14}$ and fossil methods of dating can only be applied to sedimentary rocks, and that the other techniques based on radioactive decay can be used only with certain rocks. The potassium-argon method is applied extensively to determine ages of glauconites in sedimentary rocks, but usually the radiometric methods (apart from C_{14} determinations) are most commonly applied to igneous rocks. Great care must be taken in evaluating the results of such determinations, for there are many pitfalls. For example an age so obtained may not be that of the rock, but merely of the last phase of equilibrium of the minerals in the rock. In many respects ages derived from fossil evidence are still the most reliable, and in every case the radiometric date should conform to the limits imposed by stratigraphic (including fossil) evidence.

Absolute methods only provide milestones, and traditional methods are still essential for their application and extrapolation as well as to test their validity. In Fig. A.12, for instance, the relative ages of the units can be readily determined. The geological age of the sediments can be determined from fossils, and the absolute age of the igneous unit by radioactive methods. These should be consistent: for instance in Fig. A.12 the determined age of I cannot be less than 1–8, for I intrudes into the sedimentary sequence.

General References

J. Verhoogen, F.J. Turner, L.E. Weiss, C. Wahrhaftig and W.S. Fyfe, *The Earth*, Holt, Rinehart and Winston, New York, 1970.

H.H. Read and Janet Watson, *Introduction to Geology. I. Principles*, Macmillan, London, 1963.

E.S. Hills, *Elements of Structural Geology*, Methuen, London, 1963.

P. Badgley, *Structural and Tectonic Principles,* Harper and Row, New York, 1965.

Index

aa, 133
Aartolahti T., 502
abandoned channel, 247
  meander loop, 248
ABC Quartzite, 89
ablation, 355
  moraine, 355, 356, 357
aboriginal man—and soil erosion, 526
abrasion
  by waves, 371
  in river, 233
Abu Musa, Persian Gulf—river erosion, 256
accelerated erosion, 516
acid volcano, 138
Ackermann, E., 77, 80
active volcano, 125
  distribution, 126
Adelaide Geosyncline, 107
adventitious cone, 131, 132
aeolian action, 282
  deposition, 292
  in cold lands, 317, 318
aeolianite, 201, 374, 490
Africa, central—etch plain, 420
Africa, southern
  erosion surfaces, 417
  relict nival forms, 480
African Rift and volcanoes, 125
  shield, 87
  Surface, 415, 417
Agnew, A. F., 215
Ahmed, F., 41
Ahlmann, H. W., 398
Ahnert, F., 272, 430
air fluidisation, 223
Akagi, Y., 503
Akaroa Harbour, N.Z.—drowned volcanic
  crater, 132
Alaska
  pingo, 323
  proglacial lake, 464, 645
  relict nival forms, 480
  relict pingos, 480
Alderman, A. R., 169
Aldrich Station, Yerington, Nevada—
  pediments, 277
Aletsch Glacier, Switzerland, 464, 467
Alexander, L. T., 195, 215
Alexandria, Egypt, 21
algae—and weathering, 188
Algodones Dunes, 282, 285
Alison, I. S., 365
Allerød interval, 456

allogenic river, 77, 232
all-slopes topography, 42, 49, 51, 404
alluvial fans, 243 et seq., 414
  characteristics, 243
  early Recent, 274
  late Quaternary, 484
  origin, 243, 244
alluvial plain, 250
alluvial storage in valleys, 244
Almannagja Fault, Iceland, 86
Alpine Fault, N.Z., 97, 463
Alps (European)—Pleistocene chronology,
  459, 461
altiplanation bench
  description, 329, 331
  occurrence, 331
altiplanation level, 331
alveolar relief, 292
alveolar weathering, 65, 205, 206
Amadeus Basin, N.T.—relict drainage, 484
Amazon River—fault control, 81, 85
Ambrose, J. W., 478, 509
American Southwest—yardangs, 289
Amman, Jordan—peneplain, 277
amphitheatre, natural, 116
anaclinal stream, 431
anastomose streams, 252
Anchorage, Alaska—Good Friday
  earthquake, 1964, 94, 95, 403
ancient inherited forms, 504
Ancyclus Lake, 474
Andersson, J. G., 228, 332
Andermatt, Switzerland—possible isostatic
  rebound, 477
Andersen, M. J., 502
Andes, South America—landslides, 221
Angepena Syncline, S.A., 83
annular drainage pattern, 431, 432, 442
anomalous drainage patterns, 433
  types, 433
  explanations, 434
Antarctic ice sheet, 338
  Cainozoic history, 463
  description, 339
  dimensions, 333
  ice shelf, 339
  plan, 336
  relict ice, 339
  sections, 337
  significance for sealevel, 492
antecedence—concept, 434, 438
antecedent river, 112
anthropogenic effects, 167

anticlinal valley, 121
anticline, 107, 538, 540
anticlinorium, 540, 541
antidip stream, 431, 444
antidune, 235
Antilles
  karst, 74
  volcanoes, 125
antipodal arrangement of land and sea, 24
Anvil Cave, Alabama, 71
Appalachian type of relief, 122
Appalachians, U.S.A.
  drainage pattern, 433, 434, 437, 447
  erosion surfaces, 422
  fold mountains, 121
  fold pattern (map), 120
  ridge and valley, 61
Aquitainian Depression, 30
Arabia
  basalt flows, 132
  complex barchan, 296
  early Recent alluvial fans, 274, 484
Arabian Sea, 29
Aral Sea—Pleistocene condition, 481
Arauco Peninsula, Chile—uplift, 97
Arcoona Plateau, S.A.
  weathering, 66
  plateau forms, 103 et seq.
  erosion surface, 415
Arctic Canada
  post glacial marine submergence, 474
  isostatic recovery, 474
Arellano, A. R. V., 215
arête, 86, 338, 347, 348
arid tropics, 273
  wind action, 282
  structural forms, 282
  types of desert, 282
aridity—in past, 485
aridity and glaciation, 453
arid zone—slope behaviour, 406
Arizona Meteor Crater, 161
Arizona—volcanoes, 133
Arkansas River, U.S.A.—fault control, 81
Arm Creek, N.T.,—tidal meanders, 246
armchair-shaped hollow, 203
Arroyo de los Frijoles, California—
  honeycomb weathering, 377
artificial satellites and earth's shape, 22
ash (volcanic), 147
  cone, 138, 152
  crater, 136
  layers, 137

Ashanti Crater, Ghana, 159
Ashaye, T. I., 289, 315
Ashitaka Volcano, Japan, 152
Ashley, G. H., 451
Ashuanipi Lake, Labrador—esker, 363
Asia, central—yardang, 289
associative method, 6
Aswan, Egypt and Eratosthenes'
    calculation, 21
Aswan High Dam, Egypt, 237, 256, 413,
    525
asymmetrical fold, 540, 541
Atera Fault, Japan, 99
Atlantic Ocean and sea floor spreading
    south, 25, 35
    north, 29, 30
Atlantic type coast, 367, 368, 369
Atlantic (climatic) phase, 455, 456
attrition, 233
avalanche, 7, 318
A-tent, 58, 59
Auer, V., 100, 502
aureole, metamorphic, 535
Aurès Mts., Algeria—graben, 91, 93
Australia
    central, floods, 273
    soil erosion, 519
    volcanic provinces, 132, 133
australite, 161
    distribution in Australia, 160
autigenic meander loop development, 248
autosuperimposition, 440, 442
Awatere Fault, N.Z., 89, 91
Axelrod, D. I., 281
axial plane, 540
axial surface, 540
Ayers Rock, N.T.
    cliff-foot cave, 206
    flared slopes, 7, 13, 66
    form in plan, 56, 57
    scarp foot depression, 203
    sheet structure, 59
    surrounding alluvial plain, 207

Bacchus Marsh, Vic—Permian glaciation,
    507
backwash, 370
Badgley, P., 548
Bagnold, R. A., 6, 18, 255, 292, 301, 315,
    316
Bain, A. D. N., 42, 80, 211, 216
Baker, H. B., 18
Bakhtiari River, Iran, 433
Bakker, J. P., 347, 365
Balchin, W. A. V., 398
ball, 387
Ball, L. C., 451
Baltic Sea
    late and postglacial history, 474, 476
    spits and lagoons, 388
Banks Peninsula, N.Z.—breached volcanic
    crater, 132
Barbados—coral cap, 516
Barbeau, J., 216
Barbier, R., 194
barchan, 295, 296
    distribution, 307

barchan (*continued*)
    rate of movement, 297, 298
    structure, 297, 299
    Triassic, 506
Barlow, B. C., 169
Barre, Vermont—sheet structure, 61
Barrell, J., 17, 429, 534
barrier bar, 387
    orientation, 389
Barron Falls, Qld.
    nick point, 411
Barron Gorge, 257
Barrungen Crater, Mexico, 161
Barrydale, S. Africa, 433
Barton, D. C., 193
Bartrum, J. A., 383, 398
basal notch, 491
basal slipping (ice), 343
basalt
    caprock, 107
    crater, 131
    distribution, 36
    flows, distribution, 25, 36
    layer, 28
    plain, 135
    plateau, 134
    volcano, 128
    weathering, 188
Bascom, Florence, 429
Bascom, W., 398
baselevel
    local, 256, 257
    movements and cycles, 415
    regional, 256, 257
    streams—wasting relationship, 426
    ultimate, 256, 257
Bassett, H., 215
batholith, 122
Bathurst, N.S.W., 188
Battle of Tit, 273
Battle, W. R. B., 365
Baulig, H., 272, 281, 404, 429, 531, 534
bauxite, 188, 189, 195
    profile, 196
Bay of Biscay soil, 168
Bayliss, P., 193
The "Bayonets", N.Z.—schist towers, 79
beach drifting, 385
    forms, 381, 387
    profile—changes in time, 386
    ridge, 393
Beadnell, H. J. L., 287, 315
The Beagle, 516
Beals, C. S., 169
Bean Gulch, California—honeycomb
    weathering, 377
Beardsmore Glacier, Antarctica, 348
Beaufort Sea—pingos, 323
beaver (and flood plain development), 248
    dam, 513
    former distribution, 513
Beda Surface, 415
bed forms
    and grain size, 237
    and stream velocity, 237
    riverine, 236
bed load, 235

"beehive" in sandstone, 65
Beldon Cleugh, Northumberland—glacial
    spillway, 473
Bellaratta, S.A.—breaching of ridge, 442,
    443
Beloussov, V. V., 37, 41, 122, 124
Bendefy, L., 124
Bennett, H. H., 527
Benson, W. N., 100
Berkey, C. P., 481, 502
Berkeley Hills, Calif.—relief inversion, 121
bergschrund hypothesis (of cirque
    development), 345
berm, 381, 387
Bernouilli effect, 231
Berridale, N.S.W.—faulted granite outcrop,
    81, 84
Betoota Anticline, 112
Bigarella, J. J., 398, 478, 484, 503
Bik, M. J., 80
Bikini Atoll—thickness of coral, 513
"billiard table" rock surface, 183
Bird, E. C. F., 178, 398, 399
Bird, J. B., 429
birdsfoot delta, 237
Birkeland, P. W., 194, 216
Birman, J. H., 478
Birot, P., 100, 124, 193, 272, 280, 281
Biscay Rift, 30
Bishop Tuff, 152
Bishop, W. W., 478
Black, H. P., 18, 332
Black, P. M., 132, 158
Black, R. F., 228, 332
Black Forest, Germany—horst, 92, 94
Black Hill, S.A.—boulders in norite, 42
Blackburn, G., 399, 503
Blackhawk Landslide, Calif., 223, 225
Blackwelder, E., 193, 225, 228, 255, 315,
    429
Blanche Point, S.A.—erosional coastal
    forms, 382
Blench, T., 232, 255
blind valley, 70, 71
Blinman Dome, S.A., 432
Blissenbach, E., 255
blockschollen flow, 341
block stream, 226, 329
blocky disintegration, 182
Bloom, A. L., 489, 503
"Blue" granite, Dartmoor, 202
bluff, river, 229
Bobek, H., 315
Boderbreen, Norway—esker, 364
Boer, G. de, 398
Bogardi, J., 236, 255
bogburst, 223
boiling mud, 156, 157
Bolinas Lagoon, Cal.—spit and lagoon, 392
bolson, 91, 282, 295, 531
Bond, R. D., 399, 503
Bonneville Lake, 481
Bool Lagoon, S.A.—lunette, 312
Boreal (climatic) phase, 455, 456
boreal region, 173
bornhardt, 42, 46, 47, 57
    distribution, 174

Bostrom, R. C., 327, 332
Bosworth, T. O., 273, 281
Bolt, M. H. P., 124
bottomset beds, 238
Boulaine, J., 316
Boulder, Colorado—pediments, 277
boulder clay, 355
boulder, granite, 3, 45, 49, 55, 205
  erratic, 4
  perched, 45
  sandstone, 62, 65
  shape, 54
  two stage development, 42, 50
boulder tongue, 318
bouldery disintegration, 46, 47
Bourman, R. P., 509
Bourne, Jennifer A., 430, 509, 528, 534
bourrelet, 311
Bowen, Qld—granite cliff, 378
Bowermans Nose, Dartmoor, England, 56
Bowler, J. M., 100, 316
box pattern of dunes, 294
Boyé, M., 318, 332, 343, 365, 478, 502
Brachina, S.A.—alluvial fan, 244
Brachina Gorge, S.A., 84, 445
Bradley, W. C., 80, 398, 429
braided channel, 253, 254, 362
  meandering stream relationship, 252
  stream discharge, 252
  stream distribution, 252
  stream gradient, 252
  stream morphology, 252
"The Brain", Ayers Rock, N.T., 191
Branner, J. C., 17
Brazil
  depth of weathering, 190
  pediments and climatic change, 489
breached anticline, 118
breached dome, 115, 437, 447
breached ridge, 433, 434, 445
  discussion, 442
  stages in development, 442
  structural influence, 442
breached snout, 433, 434, 444, 446
  stages in development, 448
"Breadknife", Warrumbungles, N.S.W.—
  vertical dyke, 155
breakpoint bay, 387
breaking of wave, 370
Brett, R., 169
Bretz, J. H., 365, 430
Bridges, E. M., 502
Bridgman, P. W., 5, 18
Brigalow country, 166
Britannia Sφ, Greenland—proglacial lake,
  464, 466
British Columbia
  fjord coast, 353
  proglacial lakes, 465
Brook, G. H., 80
Brookfield, Muriel, 315
Brooks, C. E. P., 478
Brown, A. R., 169
Brown, E. H., 422, 430
Brown, G. F., 281, 502
Brown, R. G., 169
Brückner, E., 461, 478

Brückner, W., 80, 215
Brunnschweiler, R. O., 101
Brunsden, D., 228
Bryan, K., 272, 401
Bucher, W. H., 161, 169
buckling
  of salt plates, 165, 167
  of calcrete, 167
Buchanan, F., 215
Buckaringa Gorge, S.A.—breached ridge,
  442, 443
Büdel, J., 80, 175, 178
  landform regions, 174
  world map of landform assemblages, 176
buffalo wallow, 5
Bukken Fjord, Norway—plan, 353
bulge (valley), 163, 164
Bull, W. B., 516, 527
Bullard, E., 41
bulldozing (by ice), 343, 347, 350
buried barchans
  Parana Basin, 505, 506, 507
  fence posts, 519
  soil, 192
Burke, R. O'H., 3, 180
bush fire—and weathering, 179
Bushire, Iran—salt dome, 123
Butler, B. E., 484, 502
buttressing, 162
Butzer, K. W., 478
Büssersteine, 77
butte, 101, 103, 106, 107, 201, 206, 504
  definition, 105
  sandstone, 105
Buwalda, J. P., 100
by-passing (in stream), 236

Cadell Fault Block, Vic., 97, 403
Cady, J. T., 195, 215
Cailleux, A., 124, 178, 365
Cairo, Egypt—granite weathering, 181
calcite—reaction with acidulated water, 185
calcium bicarbonate, 68
calcrete, 167, 195
  composition, 201
  hardpan (sheet), 200, 201
  honeycomb, 201
  origin, 201
  nodular, 200, 201
caldera, 137, 138, 143
caliche, 201
California
  earthquake, 1857, 81
  earthquake, 1906, 81
  Pleistocene lakes, 481, 483
  relict water, 230
  shore profile, 378
  soil erosion (Central Valley), 515, 516
  submarine canyons, 497
  subsidence (Central Valley), 522
  vertical movements, 109
Callen, R., 215
camber, 163, 164
Campana, B., 451
Campaspe River, Vic., 138
Campbell, Elizabeth M., 313, 316
Camsell Range, Canada, 118

Canada
  postglacial isostatic recovery, 474
  Precambrian erosion surfaces, 508
Canadian Shield—preglacial forms, 343
Canterbury Plains, N.Z.—depositional,
  132, 250
capacity (river), 235
Cape Fold Belt, S. Africa—drainage
  anomaly, 433
Cape Hatteras, U.S.A.—cuspate foreland,
  391
Cape Northumberland, S.A.—"raised
  beach", 374
Cape Paterson, Vic.—coastal pits, basins,
  hollows, 375
Cape Willoughby, S.A.—emerged stack, 491
caprock, 101, 102, 105, 107, 201
  coral, 515
carbon 14 dating, 547, 548
carbon dioxide flux and climatic change,
  457
carbonic acid, 68
carbonisation, 188
Cardiganshire, Wales—erosion surfaces,
  422
Carey, S. W., 29, 41
Carlston, C. W., 429
Carol, H., 343, 344, 365
Carmel, Calif.—beach profile, 386
Carpentaria plains, Qld., 189, 267, 277,
  405, 415, 531
Carroll, Dorothy, 316
Carruthers, R. G., 357, 365
Carstenz Mts., West Irian—raised coral
  reef, 516
Carter, E. K., 429, 451
cascade, 162, 163
Cascade Inlet, B.C.—fjord, 353
Caspian Sea—Pleistocene condition, 481
Castany, G., 503
castle koppie, 42, 48, 51
castellated inselberg, 42, 48, 57
cataclinal stream, 431
catena—soil, 192
Cathedral Rocks, Yosemite Valley—
  granitic turrets, 55, 352
causes of climatic change, 457
  extraterrestrial, 457
  terrestrial, 457
causes of soil erosion, 519, 521, 522
cave pool, 73
caves, 77, 267
  cross section, 71
  origin, 71
  pattern in plan, 71
  vertical zonation, 71
cavernous weathering, 204, 205
caverns—and slope development, 101
cavitation, 13, 233
Cedar Creek alluvial fan, 244
centripetal drainage, 432, 433
Chagnon, J. Y., 228
chain dune, 305, 309
chalk escarpment, 76
  landforms, 76
  uplands, 464
Chamberlin, T. C., 18

Chambers Pillar, N.T.—butte in sandstone, 105
Champlain Clay, 217
Champlain Sea, 465
Chang Houng-Yi, 216
channel characteristics, 230
"Channel River", 472
channelled scablands, 362
    origin, 363
Chapel-le-Dales, Derbyshire, clints, 70
Chapman, C. A., 80
Charlesworth, J. K., 365, 451
chemical weathering, 179
    definition, 185
    optimum conditions, 190
chenier, 387
Cheshire, England
    salt deposits, 505
    subsidence, 525
Chesil Bank or Beach, England
    shingle spit, 386
    orientation, 389
Chezy equation, 231
Chief Mt., Montana—*klippe*, 120, 121
Chile
    graben, 91
    earthquake, 94
    faults, 96
China, southern—karst, 74
Chisholm, M., 426, 430
Chorley, R. J., 365, 424, 426, 430
Choubey, V. D., 430, 451
Christian, C. S., 17
chronology of Pleistocene, 457, 458
Church, R. E., 502
chute, 229, 245
circum Pacific "Ring of Fire", 125
circumdenudation—remnants of, 403
cirque
    glacier flow, 346
    origin, 345
    section, 346
Clarke, A. R. P., 399, 503
clay desert, 282
clay dune, 311
clay formation and climate, 187
claypan, 291
Clayton, R. W., 216
cleavage, 77, 535
cleft, 63
Clements, T., 215
cliff, 378, 382, 387
    aeolianite, 374
    granite, 378
    related to Lake Bonneville, 482
cliff-foot notch, 373
cliff-top dune—origin, 396, 397
climate
    and clay formation, 187
    and weathering, 190
    and process, 175, 177
climatic change, 453
    arid tropics, 485
    causation, 457
    evidence (anthropological, historical,
       ice cores, tree rings, palynological, etc.),
       454, 455

humid tropics, 489
    and landform distribution, 174, 175
    nature, 463
    scales, 455
    significance, 531
climatic control of landform development,
    397 et seq.
    coastal, 177, 385, 396, 397
climatic fluctuation, 453
climbing dune, 396
clint, 68, 69, 74
    joint control, 70
closed hypotheses—pingo formation, 323,
    326
closed system
    weathering, 186
    landscape evolution, 424
cluse, 121
Coanda effect, 13
coastal landforms, 366
    and climatic control, 177, 385, 396, 397
    modern foredunes, 369, 393, 394, 395,
       396
    morphogenetic classification, 397
    Pleistocene foredunes, 369, 395
    processes, 366
    and structure, 369, 370
Coats, R. P., 451
Cobbing, E. J., 37, 41
cockpit country, 74
coesite, 161
Coffey, G. N., 316
cold loess, 286, 364
Coleman, Alice, 451
Coleman, J. M., 398
collapse, 159, 226
    debris, 217
    doline, 70, 71
Colorado Plateau, U.S.A., 59, 62
Colorado River, U.S.A., 106, 177, 438
Columbia Plateau, U.S.A.—basalt flows,
    132, 133, 134
Columbia River, U.S.A., 107
Columbus, Christopher, 21
columnar jointing, 153
    volcanic, 152
combe, 121
comminution of shale, 184
comparative method, 6
competence (river), 235
competent rock, 538
complexity in geomorphology, 5, 531
composite volcano, 144
compressive flow (in glacier), 341
compressive stress, 58
    and sheet structure, 59
compressional fault, 540
concentric dunes, 314
conchoidal fracture, 66
Condolias, E., 255
cone karst, 74
Congo Surface, 415
Connolly, J. R., 503
Conrad discontinuity, 28
conservation, 4
constant slope, 259
constructional terrace, 414

consumption of crust, 35
continents—tectonic regions, 39
continental drift, 33
    and climatic change, 457
    difficulties, 32
    evidence, 32
continental shelf, eastern U.S.A., 39
continents
    area, 23
    distribution, 24
contortions due to ice wedging, 319
contraction, 167
contraction of earth, 28
convection currents, 34
    in crust, 33
convergence of plates, 36
convergent evolution of landforms, 6, 174,
    531
convergent ridges, 114
Cook, K. L., 100
Cook, P. J., 161, 169
Cooke, R. U., 193, 315, 316
cooling and contracting earth, 28, 29
Coorong Lagoon, S.A., 394
    folds in Recent muds, 164, 165
coral reefs, 512
    distribution, 513
    environmental tolerance, 513
    evolution, 517
    forms, 514
    theories of origin, 513–16
    thickness, 513
    types, 514
Corbel, J., 80, 177, 178, 228, 281
Corbin, Elizabeth M., 216, 534
corduroy soil, 168
core of earth, 27
corestone, 49, 50, 52
cornice, 317
Cornish, E. A., 527
Cornish, V., 315, 332
corrasion, 233
corrosion, 233
corrie, 345
Costin, A. B., 18, 332, 502
Cotton, C. A., 100, 158, 178, 272, 281, 332,
    365, 383, 398, 404, 451, 527
County Kerry, Ireland, 223
course of weathering, 188
cover deposit, 439
Cowie, J. W., 509
Cox, A., 41, 158, 332
crab hole, 168
Craig, D. C., 193
Crary, A. P., 365
Crater Bluff, Warrumbungles, N.S.W.—
    volcanic plug, 155
Crater Lake, Oregon, 144, 145, 146, 147
craters—and steam upthrust, 161
craton, 87
creep
    debris, 217
    rates, 222
    sand grains, 288
    slope, 217
    soil and rock, 222
crenulated fault scarp, 91

crescentic coastal dune, 299
crescentic snow drifts, 317
crest (of fold), 540
crestal line, 540
crevasse, 341, 349
    pattern, 342
    ridge, 357, 359
Crickmay, C. H., 428, 429, 430
criteria for recognising former sea stands, 489, 491
critical drag velocity, 286
critical tractive force, 235
Crittenden, M. D., 502
Crocker, R. L., 311, 316
Crook, K. A. W., 169
Crosby, I. B., 451
cross-bedding in sandstone, 63, 500
Cross, C. I., 169
crossfold, 540
cross-section area of channel, 231
Crowell, J. C., 509
crust of earth, 28
    consumption of, 35
    destruction of, 35
    making of, 35
croûtes désertiques, 274, 288
cryostatic pressure, 323
cryoexplosion structure, 159
crystal surfaces—and water, 186
crystalline limestone, 68
Cuba—karst, 75
Cue, W.A.—laterite mesa, 421
cuesta, 115, 201
    definition, 116
    on pingo, 323
Cuillen Hills, Skye—forms of glacial erosion, 344
Cumberland, K. B., 519, 527
Cundari, A., 150, 451
Curalle Anticline, 112
Curray, J. R., 503
Currie, K. L., 161, 169
Curtis, B. F., 528
Curtis, G. H., 430
Curtis, L. F., 216, 502
cusp, 387
cuspate foreland, 389, 391
    of Lake Bonneville, 482
cut off, incipient, 247
cwm, 345
cycle (geomorphic)—duration, 425
cycle of erosion, 404
cyclic concept, 404
    arid, 406, 407
    glaciated, 406, 407
    coastal, 406, 408

Dahran, Arabia—relief inversion and travertine, 201
Daily, B., 430, 451, 509
Dalmatia, Yugoslavia, 368
Dalmatian coast, 367
Dalquist, W. W., 527
Dalrymple, G. B., 41, 158
Daly, R. A., 489, 503, 527
Daly level, 489
Danel, P., 255

Darby, H. C., 17, 528
Darling Downs, Qld—gilgai, 167, 168
Darling River—fault control, 81
Dartmoor, England
    alteration of granite, 50
    granite turret, 56
    inselberg tor, 49
    loganstone, 45
    pseudobedding, 58
Darton, N. H., 17
Darwin, Charles, 185, 193, 516, 527
dating of rocks, 547
dating of surfaces, 420, 421
Davenport, California
    coastal pits, basins, hollows, 375
    structural bench, 379
Davies, J. L., 319, 332, 397, 399, 502
Davis, S. N., 281
Davis, W. M., 17, 18, 100, 171, 178, 404, 411, 424, 425, 426, 429, 430, 451, 503, 527
Dead Sea
    delta, 240
    intersection of graben, 87
De Almeida, F. F. M., 509
De Andrade, G. O., 489, 503
Death Valley, Calif.
    alleged wind erosion, 292
    ash cone, 152
    pediment (?) remnants, 277, 279
    recent faulting, 90
debris
    accumulation and slope development, 266
    flow, 217
    lobe, 319
    mantle (on pediments), 279
    mound (and linear dunes), 305-07
    tongue, 319
    trail, 318
    slide, 223
    slope, 259
Deccan, India
    laterite, 132, 196
    traps, 133, 134
Deccan lavas, 439
deep erosion and drainage problems, 446, 450
deepening of valleys by glaciers, 347
defile, 262
deflation, 286
deforestation and gullying, 445
De Geer, G., 479
deglaciation, 464
De Lamothe, 503
delta, 237
deltaic bedding, 238
    deposition, 237
de Martonne, E., 80
Dence, M. R., 169
dendritic drainage, 433
dendritic glacier, 333
dendrochronology, 455
denudation—definition, 173
Dent, O. F., 398
Deperet, C., 503
depositional forms—coastal, 385

depositional plain, 405
depression, 166
depth of weathering, 190
Derbyshire, E., 479
Derruau, M., 318, 332
desert
    floods, 273
    landforms, 282, et seq.
    of past, 485
    varnish, 205
destruction of crust, 35
destructive work of waves, 371
Devils Marbles, N.T.—granite boulders, 45
Devils Postpile, Calif.—volcanic plug, columnar jointing, 7, 8, 152, 154
Devils Tower, Wyoming—volcanic plug, 152, 154
dextral wrench fault, 544, 545
diaclinal valley, 433
Diamantina flood plain, 306, 307, 308
Diamantina River, Qld—"palm tree" pattern, 434
diapir, 7, 122
diatreme, 152
Dickens, Charles, 455
Dickinson, W. R., 41
Dickson, H. N., 158
Dietz, R. C., 169, 398
differential
    erosion, 203
    flow in glacier, 341
    weathering, 203
differentiation of types of isostatic movement, 474, 477
Dionne J.-C., 502
dip (of strata), 538, 539
    fault, 540, 543
    of fault, 540
    stream, 431
direct effects of Pleistocene glaciation, 453, 463
dirt band, 334
disaggregation of shale, 184
discharge
    braided streams, 252
    meandering streams, 252
discontinuity
    Conrad, 28
    M, 28
    Mohorovicic, 28
displacement of fault, 540
Disraeli Glacier, Ellesmere Island, 338
Disraeli Fjord, Ellesmere Island, 338
dissected fault scarp, 87
distant earthquake, 27
distribution of
    active volcanoes, 126
    basalt, 25
    continents and oceans, 24
    earthquakes, 25
    laterite in eastern Australia, 199
    mountain ranges, 25
    permafrost, 322
    plates, 28
    silcrete in eastern Australia, 199
    submarine ridges, 25
Dirac, P. A. V., 35, 41
divergence of plates, 36

divergent ridges, 114
diversion (stream), 434
Dixey, F., 100
D.L. Bliss State Park, Calif., 49
Doell, R. R., 158
doline, 67, 74, 76, 77
   solution, 70, 71
   collapse, 70, 71
Dombashawa, Rhodesia—granite dome, 8
dome dune, 294, 309
dome—granite, 46, 47
domed inselberg, 42, 51, 52, 57, 660
   distribution, 174
   sandstone, 62
domed plateau, 101, 107, 415
Douglas, I., 281
Dorman, F. H., 18
dormant volcano, 125
downs—chalk, 76
downstream sweep of meanders, 248
draa, 296, 297
drainage diversion—volcanic, 138
drainage modifications—glacial, 464
drainage patterns, 431
   adjustment to structure, 433
   and deep erosion, 446, 450
drainage systems—relict, 484
Drakensberg Escarpment, Natal, 107, 417
dreikanter, 289, 507
drift, 358
drowned coast, 367
drumlin, 355, 359
   distribution, 362
   drowned, 361
   plan form, 360
   N. Ireland, 360
   origin, 362
drumlinised surface, 344
dry valley—chalk, 75
   origin, 76
Duce, J. T., 527
Dumanowski, B., 216
dunes
   barchan, 295
   characteristics, 292
   cold lands, 317, 318
   depositional forms, 295
   en rateau, 314
   longitudinal, 295
   massif, 292
   nonaligned, 295
   relict, 485
   riverine, 236
   sand source, 295
   snow, 317
   transverse, 295
   tropical and subtropical, 292 et seq.
   wind tunnel experiments, 309
Dungeness, England—cuspate foreland
   orientation, 389
   stages in development, 391
Dunn, E. J., 509
Durand, R., 255
duration of Pleistocene, 457
duricrust, 195, 415, 428
   caprock, 107
   origin, 195

duricrust (*continued*)
   morphostratigraphic marker, 195
Dury, G. H., 17, 215, 255, 274, 281, 429,
   503
dust devil, 286
duststorm
   examples, 286
   Greenland, 318
du Toit, A., 32, 33, 41, 503, 509
Dutton, C. E., 87, 438, 451
dyke, 154, 155
dynamic equilibrium and landscape
   change, 403, 424, 425
Dyson, J. L., 18, 332

Eakin, H. M., 332
earth
   core, 27
   cross sections, 22
   frequency distribution curve, elevations,
   24
   hypsographic curve, 24
   magnetic field, 23, 33
      reversals, 35
   major relief features, 21
      explanations, 28 et seq.
   polar flattening, 22
   shape, 21, 22, 23
   shear patterns, 33
   sphericity, 21
   structure, 24 et seq.
earthflow, 217, 218, 222, 223
earthslip, 217, 223
earthworms, 510
Eastwood, T., 398
Echuca, Vic.—diversion of Murray, 434
Eden Fault, S.A., 546
Eden, W. J., 228
Eggler, D. H., 429
Egyptian desert—wind erosion, 289
Einasleigh River, Qld—braided, 253
Ekern, P. C., 527
El Capitan, Yosemite Valley, 352
Ellef Rignes Is., Canada—salt domes, 122
Ellison, W. B., 228, 527
El Salvador—rate of weathering, 192
Embleton, C., 332, 365
emerged coastal plain, 491
Emery, K. O., 398
Emiliani, C., 478
end moraine, 355, 359, 388
   Long Island, 388
   Ohio, 360
   Sweden, 360
endogenetic forces, 5
endrümpf, 425
England, eastern
   coastal defences, 526
   areas vulnerable to flooding, 526
England, southern—pediments, 277
Ennis, Montana—alluvial fan, 244
entropy, 424
ephemeral stream, 232
epicycle of erosion, 519
equifinality, 6
Eratosthenes—calculation of earth's
   circumference, 21

Ericksen, G. E., 228
erosion, 173
   by boulders, 226
   cycle, 404
   deposition and weathering, 192
   effect of ground cover, 510
   roadside, 521
   stream, 233
   surface, 410, 423
   velocity (in stream), 235
erosional landforms
   coastal, 328
   glacial, 344
erosional work of waves, 371
erosionist school, 355
erratic, 4, 355, 458
eruptions (volcanic)—classification, 125
esker, 357, 363
   distribution, 363
   origin, 363, 364
etch plain
   development, 420
   Western Australia, 420
etch surface, 420
Et-then Series, 81, 82
Europe—soil erosion, 519
European U.S.S.R.
   Pleistocene chronology, 459
   vertical movements, 109
eustatic movements of sealevel, 474, 489
eustatism, 489
Evans, I., 193
Everard Ranges, S.A.
   joint control, granite domes, 42, 44, 54
   alluvial plains, 207
Evernden, J. F., 430
evidence of multiglaciation, 457, 458
evolutionary concepts of landscape change,
   403, 404
evorsion, 233
Ewing, M., 503
exhumed surface, 418
Exmoor, Devon—storm of 1952 and
   landslides, 219
exogenetic forces, 5
Exon, N. F., 215
exotic river, 232
expanding earth, 35, 36
expansion and contraction, 168
expansion of surface materials, 167
exposed dune, 119, 121
exsudation, 183
extended drainage pattern, 491
extending flow (glacier), 341
extinct volcano, 125
extraterrestrial bodies, 159
extraterrestrial factors, climatic change, 457
extrusion flow (glacier), 341
Eyre Peninsula
   deep weathering, 52
   granite forms, 7 et seq., 42 et seq.
   pediments, 277, 279
   Precambrian erosion surfaces, 508
   relict dunes, 485, 488

faceted slope, 101, 259
   complex, 105, 106

diagram, 261
in granite, 260
simple, 105
Fagg, C. C., 80
Fairbridge, R. W., 385, 398, 462, 478, 495, 503
Fairweather Ranges, B.C.—glaciated, 355
Falcon, N. L., 169, 228
Fall Line Surface, 422
fan-shaped delta, 237
Farmin, R., 184, 193
Farrand, W. R., 474, 478
fault
    angle depression, 94, 96
    angle valley, 97
    breccia, 544
    creep, 81
    crenulated scarp, 91
    dip, 540
    displacement, 540
    dissected scarp, 87
    drag, 544, 546
    fissure, 86, 91
    geometry, 543
    gouge, 555
    gravity, 542
    hade, 540
    heave, 540
    normal, 542
    oblique, 543
    plane, 87
    reverse, 543
    scarp, 86, 87, 88, 90, 91, 94, 253
    and sheet structure, 59
    slip, 540
    splinter, 97
    step, 544, 545
        interpretation, 545
    strike, 543
    structural forms, 81
    tectonic forms, 81
    throw, 540
fault and fault-line scarp—controversy, 87, 91
fault-line—scarp, 81, 87, 357
    resequent, 81
    obsequent, 81
    valley, 81, 84, 85
faulting
    and piedmont angle, 21
    and stream diversion, 434
Féderovitch, B. A., 301, 315, 316
Fennemann, N. F., 272
Fenner, C., 430, 438, 451
feldspar—reaction with water, 186
Felsenmeer, 182, 328
Fennoscandia
    glacial depression, 475
    postglacial isostatic rebound, 475
    postglacial marine transgression, 475
    vertical movement, 109
Fens, England
    landslides in cuts, 221
    draining, 525
    subsidence, 525
    modern subsidence, 525
Ferns Glacier, B.C., 355

Ferpecle Glacier, Switzerland, 349
ferricrete, 196, 197
fetch, 370
Field River, S.A., 440
Finger Lakes, N.Y. State, 347, 350, 352
"Finger of God", 105
"fingerprints" on limestone, 68, 71
Finke River, N.T.—transverse drainage, 433, 435, 436
Finkel, H. J., 297, 315
Finland—relict nival forms, 480
Finlayson Channel, B.C.—depth of fjord, 350
fire—and weathering, 179, 181
Firman, J. B., 216, 451
firn, 333
Firth of Clyde, Scotland, 474
Fisher, O., 272, 406, 429
Fisher, R. L., 41
Fisk, H. N., 100
fissure eruption, 129
Fitzroy Valley, W.A.—pseudoanticlines, 167
fixed dune, 309, 310
fjord, 491
    Antarctica, 350, 354
    British Columbia, 353
    common features, 354
    depth, 350
    development, 350
    Norway, 353
flaking—due to fire, 181
flap, 162, 163
flared slope, 7–12, 203
    characteristics, 10 et seq.
    comparative rarity outside Australia, 16
    in arkose, 59
    origins, 13–15
    two-stage development, 14
flash, 525
flatiron, 61, 118
    definition, 116
The Fleet, England—lagoon, 386
Flemming, N. C., 503
Fletcher, E. B., 228
Fleurieu Peninsula, S.A.—lateritic high plain, 426
Flinders Ranges, S.A.
    drainage patterns, 431 et seq.
    erosion surfaces, 428
    fold pattern, 108
    late Pleistocene alluvial fans, 243, 274
    pediments, 277
    piedmont angle, 209, 210
    pre Cretaceous surface, 418
    structural forms, 107 et seq.
Flinders River, Qld
    fault control, 81
    steepening of gradient, 434, 485
Flint, R. F., 365, 479, 502
flood
    deposits, 245
    desert, 273
    plain, 245, 248, 411
    significance, 232, 233
flood basalt, 132
flow
    debris, 217

flow (*continued*)
    earth, 217
flowstone, 72
fluting
    glacial, 344
    limestone, 72
fluvioglacial deposition, 354, 355
    forms, 362
fold
    asymmetrical, 541
    isoclinal, 541
    mountain belt, 61
    pattern, Flinders Ranges, 108
    recumbent, 541
    silcrete surface, 111, 112, 198
    stress, 112, 113
    surface, 107
foliation, 77
Folk, R. L., 316
Folkestone, Kent—landslide, 221
footslope, 259
forced wave, 370
foreset beds, 238
former proglacial lakes—relict features, 465
former sealevels—criteria, 489
formkreisen, 173
Forrest, G. J., 527
Fox, C., 430
Fox, D. J., 528
fractures in crust, 42
Frakes, L. A., 478, 509
France
    chalk, 76
    salt domes, 122
    vertical movements, 109
Frank, Alberta—landslide, 223, 224
Frankel, J. J., 215
Franklin Shore, 492
Franz Josef Glacier, N.Z., 362
Fraser, A. S., 451
Fraser River, B.C., 86
Fraser Valley, B.C.—riverine forms, 414
Frederickshaab Glacier, Greenland, 333
Frederickson, A. F., 187, 193
Freeburg, Jaquelyn H., 169
free face, 259
free wave, 370
freeze-thaw activity—principle, 182
Fresno, Calif.—subsidence, 522, 523
friction crack, 7
Friedkin, J. F., 6, 18, 255
Friese, F. W., 281
Frog Pool, Rotorua, N.Z., 157
frost
    action, principle, 182
        optimal areas, 182, 183
    heaved soils, 480
    polygon, 322
    riving, 182
    shattering, 182
    splitting, 182
frozen ground features, 217, 222
Fry, E. Jennie, 193
Frye, J. C., 478, 502
Fuji Yama, Japan, 144 et seq.
"Fuji Yama", South Australia—gullying, 522

fulje, 292
    examples, 292
Fulking, Sussex—chalk scarp, 76
fumarole, 156
furrows and soil erosion, 520, 521
Fyfe, W. S., 548

gable, 163
Gadja, R. T., 474, 479
Gage, M., 478
Gagliano, S. M., 298
Gaillard Cut, Panama Canal—landslides,
    221
Gale, W. A., 502
Galloway, R. W., 502
The "Gangplank", Wyoming, 420
Garber, L. W., 228
Garlock Fault, Calif., 98
Garner, H. F., 478, 489, 503
gas mound, 511
Gawler Ranges, S.A.
    joint control, 42, 44
    sheet structure, 59
    washed out track, 521
Gefügerelief, 77, 79
gel—and silcrete development, 198
genetic geomorphology, 3, 4
geo, 369
geoid, 22, 23
geological background to geomorphology,
    535
geological time
    problems of simulating, 6
    scale, 547
geometry of faults, 543
geomorphic cycle, 404
    duration, 425
    interpretation, 415
    intracycle events, 415
    termination, 415
geomorphology
    aims, 3
    and climatic change, 531
    complexity, 5
    definition, 3
    genetic, 3, 4
    indeterminacy in, 6
    laboratory work, 6
    methods, 4, 5
    "principles", 531
    scope, 3
    and storms, 531
    and structure, 531
    and water, 531
Georgina River, Qld and N.T., 250
    fault control, 81
geosyncline, 29, 39
Germany—salt domes, 122
Geyl, W. F., 503
geyser, 156, 157
Gèze, B., 216
Ghana—superficial disturbance structures,
    163
"Giant" granite, Dartmoor, 202
gibber, 288
    modes of formation, 288
Gibbons, F. R., 169

Gifford, Joyce, 228
Gilbert, C. M., 281
Gilbert, G. K., 6, 18, 57, 80, 255, 272, 461,
    464, 478, 480, 502
Gilbert, O. L., 194
Gilberton Plateau, Qld—sandstone, 61, 65
gilgai, 166, 220
    development, 168
    linear, 167
    network, 160
    patterns, 168
Gill, E. D., 158
Gilluly, J., 37, 41, 124, 281
gipfelflur, 464
glacial chronology
    European Alps, 458
    Great Basin, U.S.A., 460
    Great Plains, U.S.A., 460
    New Zealand, 460
    northeastern U.S.A., 458
    Sierra Nevada, 460, 461
    Western Cordillera, 460
glacial
    climate, 453
    deposition, 354, 355
    drainage modifications, 434, 464
    drift origin, 359
    erosion, 344, 345
    erratics, 4
    limit, Europe, 464
    limit, N. America, 463
    plucking, 343, 344
    spillways, 469
    spillways, north central U.S.A., 470
    striae, 7
    valley, 344–52
glacial/pluvial, interglacial/arid
    relationship, 462
glaciation
    duration, 5
    Permian, 508
    Pleistocene, 253 et seq.
    Precambrian, 7
glaciation and aridity, 453
glaciated pavement, 7
    Permian, 507
    Precambrian, 506
glacier
    motion
        differential, 339
        rates, 339
        variation in time and space, 339
    movement
        intergranular movement, 343
        slippage, 343
        basal slipping, 343
        rotational flow, 343
    types
        cirque
        ice cap, 333
        piedmont, 333
        valley, 333
glacio-eustatic changes of sealevel, 489
Glaessner, M. F., 100, 430, 451
Glasshouse Mts., Qld—volcanic plugs, 156
Glennie, K. W., 301, 316
Glen Roy—parallel "roads", 468, 469, 471

Glen Tarbert, Scotland, 349
gliding surface, 219
Glikson, A. Y., 169
globule, 163
gnamma, 66, 202
    distribution, 203
    origin, 203
gneiss
    landforms, 77
    weathering, 189
gneiss dome, 122, 123
goats—and soil erosion, 522
Goguel, J., 161, 169
goletz terrace, 331
Gondwanaland, 32
Gondwana Surface, 415
Good Friday earthquake, Alaska, 1964, 94
    95, 403
Goodchild, J. G., 365
Goodlett, J. C., 424, 430
gopher, 511
gorge—definition, 257
Gorge Creek valley, S.A., 200
Gosses Bluff, N.T., 159, 160, 161
Goudie, A., 193, 215, 216
Gould, H. R., 528
Gostin, V. A., 509
graben, 35, 86, 92
    origin, 94, 95
grade, 256
graded slope, 259
    diagram, 261, 269
gradient—meandering streams—braided
    streams, 252
Graenalon, Iceland—proglacial lake, 464,
    466
Grampians, Vic.—sandstone, upland,
    ridges, 61, 65, 66, 121
Grand Canyon, Grampians, Vic., 63, 65
Grand Canyon of Yellowstone, Wyoming,
    434
Grand Canyon, U.S.A.—origin, 63, 65,
    105, 106, 257, 262, 438
granite batholith, 122
granite
    boulder, 3, 42 et seq.
    composition, 42
    distribution, 42
    dome, 46, 47, 356
    "flats", 189
    landforms, 42 et seq.
    massif-faulted, 81, 84
    minor landforms, 7 et seq., 203 et seq.
    nubbin, 208
    spire, 55
    stepped topography, 211
    turret, 55
    weathering, 180 et seq., 211 et seq.
Granite Mts., Calif., 203
granitic layer, 28
Granitrillen, 203, 204
Grantz, A., 100
granular disintegration, 181
Gravenor, C. P., 359, 365
gravity, 23
    collapse structures, 161–62
    fault, 542

gravity (*continued*)
 flow, 341
 induced folds, 162
gravitational hypothesis—pingo formation, 326
grazing—and soil erosion, 519
Great Artesian Basin—exhumed pre Cretaceous surface, 418
Great Australian Bight—cliffed coast, 366
Great Barrier Reef—coral reef, 515
Great Basin, U.S.A.—Pleistocene lakes, 480
 glacial chronology, 460
Great Escarpment, Cape Province, S. Africa, 211
Great Glen, Scotland—fault-line valley, 84
Great Lakes—Pleistocene evolution, 468, 469
Great Plains, U.S.A.
 glacial chronology, 460
 relict dunes, 485
Great Salt Lake, Utah, 481, 482, 483
Green, C., 17, 528
Green, L., 29, 41
Greenland
 basalt flows, 132
 description, 339
 dust storms, 318
 glacial protection, 464
 ice stagnation, 343
 isostatic rebound (theoretical), 341
 modern meltwater channels, 372, 373
 patterned ground, 320, 321
 pingo, 323, 325
 plan, 341
 Precambrian erosion surface, 508
 relict ice, 339
 section, 341
 wind erosion, 318
Greenwood, G., 230
Greenwood, J. E. G. W., 281
Gregory, J. W., 41
Greyhound Rock, Calif.—tombolo, 390
Griggs, D. T., 6, 18, 193
grike, 68, 69, 71, 74
gritstone tors, 3
groove
 glacial, 344
 granite, 203
 riverine, 233, 235
ground cover and erosion, 510
ground ice, 319
 and mass movement, 219
ground moraine, 355, 359
Grove, A. T., 215, 503
grus, 49
Guilcher, A., 332, 399
Guiter, Victoria, 332
Gulf of Akaba—drowned graben, 87
Gulf of Bothnia—isostatic rebound, 477
Gulf of Carpentaria coast
 saline flats and beach ridges, 393
 tidal meanders, 246
gullies and gully erosion, 516, 518, 519, 520, 521, 522
gully gravure, 268, 269, 280
gullying, 516–22
 coastal, 373

gullying (*continued*)
 of cliff, 372
gumbotil, 458
Gutenberg, B., 27
gutter, 52, 203
 beach, 387
guyot, 516, 517
Gym Beach, Yorke Peninsula—solutional forms on coast, 375, 376

Haantjens, H. A., 527
Hack, J. T., 315, 316, 424, 425, 430
Haddon Syncline, 112
hade—of fault, 540
Hadley, R. F., 527
Hager, D., 169
Haggett, P., 17
Hakone Volcano, Japan, 152
Halemaumau, Hawaii, 129
 contour plan, 130
 firepit, 129
Half Dome, Yosemite, Calif.—glacial erosion, 347, 352
Halletts Cove, S.A.—Permian glaciation, 507
 serrated platform, 384
 shore platform, 371, 372
Hallsworth, E. G., 169
hamada, 288
Hamersley Ranges, W.A.
 butte, 105
 plateau, mesa, 101
Handreck, K. A., 215
hanging valley, 345, 347, 351
hardpan calcrete, 201
Hardy, Thomas, 529
Harford, L. B., 100, 316
Harker, A., 344, 365
Harland, W. B., 509
Harris, G. F., 80
Harris, W. J., 100, 451
Harris, W. K., 18, 215
Harrison, J. V., 169, 228
Hartshorne, R., 17
Hartt, C. F., 17
Hawaii
 mass movements, 223, 225
 volcanoes, 127 et seq.
Hawaiian eruption—characteristics, 125, 127, 128
Hay, R. L., 158
Hayden, F. V., 438, 451
Hayes, D. E., 478
haystack hill, 74
Haytor, Dartmoor, England—tor or castle koppie, 48, 202
Hayward Fault, 81
Hayward-Calaveras Fault, 81, 87
head, 328
headward erosion, 434, 442, 444, 445
heave of fault, 540
Hedin, S., 315
Heezen, B., 503
Hegben Lake earthquake, 1959, 223
Hehuwat, F., 158
helictite, 74
Hellstrom, B., 10, 14, 18

Henbury Craters, N.T., 161
Henderson, E. P., 332
Hendriks, E. M. Lind., 509
Henin, S., 6, 18
Henoch, H. E. S., 478
Herbert River, Qld—granite boulders, 50
Herculaneum, 152, 156
Hess, H. H., 41
Hessle ice, 469
Heymaey Island, Iceland—recent volcanic eruption, 36, 128
Higgins, C. G., 451, 489, 503
high plain, 101
High Plains, U.S.A., 101
higher platform, 382
hill-plain junction, 52
Hillock Point, Yorke Peninsula
 shore platform, 373
 platform, pool, rainfall, 376
Hills, E. S., 19, 41, 100, 124, 311, 316, 385, 398, 429, 502, 509, 548
Himalayas—and isostasy, 28
hinge fault, 96, 97, 540, 544
 line, 540
Hirayama, K., 193
historical evidence for climatic change, 455
historical geomorphology, 401 et seq.
Hjülstrom, F., 232, 255
Hobbs, P. V., 332
Hoddy Well, W.A.—laterite, 196
Hodgkin, E. P., 398
Hodgson, J. H., 41
Hoffmeister, J. E., 527
hogback, 117, 119, 201
 definition, 116
Hoggar Mts., Algeria
 granite inselbergs, 47–55
 volcanic form, 139
Holderness, Yorks.—wave erosion, 371
Hollin, J. T., 478, 503
Hollingworth, S. E., 124, 158, 169, 365
Hollister, Calif.—movement of fault, 87
Holm, D. A., 316
Holmes, A., 41, 80, 100
Holmes, J. McD., 519, 527
Holocene—definition, 457
homoclinal ridge, 118, 201
 definition, 116
honeycomb calcrete, 201
honeycomb weathering, 377
Honshu, Japan, 26
Hooke, R. LeB., 216
Hoover Dam, U.S.A.—isostatic depression, 522
Hope-Simpson, J. F., 194
Hornby Bay Fault, N.W.T., 82
horst, 91
 origin, 94
Horwitz, R. C., 451
Hossfeld, P. S., 398, 503
hot loess, 286
hot springs, 138
Hough, J. L., 478
Howard, A. D., 281, 451
Howchin, W., 509
Hoyle, A. C., 194
Hsu, K. J., 503

Hudson Bay, Canada—isostatic rebound,
477
Hudson Bay lowland—frost polygon, 322
Hudson Canyon, U.S.A., 497
Hudson River and submarine canyon, 497
hum, 74
Humber Estuary, 390
Hume, W. F., 281, 289, 315
humid tropics
    climatic change, 489
    river work, 273
hummocky microrelief, 319
humped profile of spillway, 472, 473
Hungary—vertical movements, 109
Hunt, C. B., 429, 451
Hurst Castle spit, England—orientation,
389
Hutchinson, J. N., 17, 228, 528
Hutton, James, l, 230
Hutton, J. T., 124, 193, 215
Huttons Lagoon, S.A.—lunette, 312
Hwang Ho, China, 286
    turbidity, 245
Hyden Rock, W.A.—granite residual, 10,
15, 56
hydration, 186
hydraulic force, 233
    lift, 234
hydrologic cycle, 230
hydrolysis, 186
hydrophilic clay, 168, 220
hypabyssal, 535

ice
    ages, 453
    cores and climatic change, 454, 455
    dammed marginal lake, 464
    formation, 317
    masses, 317
    shelf, 339
    stagnation, 359
    wedge, 319
    wedge polygons (relict), 480
Iceland
    graben, 86, 91, 92
    and mid Atlantic ridge, 36
    modern proglacial lakes, 472
    proglacial lakes, 464
    sandur, 362, 363
    subglacial drainage, 472
    wind erosion, 318
Icelandic type of eruption, 129
igneous rock, 535
    classification (table), 536
Ikebe, N., 100
Illinois, U.S.A.—glacial and interglacial
    deposits, 458
imbricate structure—stream deposits, 234
immobile dune, 309
in-&-out channel, 434
    gorge, 433, 434, 438, 450
incipient (meander) cut off, 247
incompetent rock, 538
indeterminacy in geomorphology, 531
Indian Mt., Hughes, Alaska—altiplanation
    terraces, 331

Indo-Gangetic plain—isostatic subsidence,
28
Inglefield Land, Greenland—modern
    meltwater channel, 473
Ing Soen Oen, 80
inheritance, 434, 440
inherited forms, 175, 453, 480 et seq.
    dunes, 485, 488
    lakes, 480, 485
    nival, 480
    periglacial, 480
    pingo, 480
    pre Quaternary, 504 et seq.
    river channels, 484, 486
Indonesia, 188
    and volcanoes, 125
induration, 179
Ingleborough, Yorks.—karst, 71
inlier, 538
Inman Valley, S.A.—Permian glaciation,
507
inner core of earth, 27
Inn Valley, Austria—terraces, 462
inselberg, 9, 50, 274, 409, 420
    origin, 42
    multicyclic landscapes, 54
    sandstone, 66
    scarp retreat and pedimentation, 53, 54
    shape, 54–56
    two stage development, 52, 53
insolation weathering, 179
    arguments for, 180
    doubts, 181
    theory, 181
interdune corridor, 302
interglacial/arid, glacial/pluvial
    relationship, 462
intergranular movement (ice), 343
interlocking spurs, 248, 250, 266
intermittent stream, 232
intertidal platforms, 383
interpretation of fault step, 545
interruption of geomorphic cycle, 415
intrenched meander, 248, 250, 251, 440
inversion of relief, 121
Investigator Group, S.A.—literal island
    mountains (inselberge), 383
Ireland, southern—antecedent rivers,
439
Isa Highlands, Qld
    exhumed surface, 418
    possible superimposition, 440
Isafjordur, Iceland—Tertiary plateau
    basalt, 134
Ishikawa, K., 100
Isla Mocha, Chile—recent fault
    displacement, 97
isochron, 25
isoclinal fold, 540, 541
isostasy, 28
isostatic adjustments to deglaciation
    Canada, 474
    Great Lakes, 474
    Scandinavia, 474
isostatic depression
    Antarctica, 474
    Greenland, 474

isostatic rebound
    Fennoscandia, 475
    Greenland (theoretical), 341
isostatic subsidence, 242
Ives, J. D., 474, 478

Jabel Ram, Jordan—sandstone tower, 104
Jackli, H. C. A., 477, 479
Jack, R. L., 124
Jacks, G. V., 527
Jaggar, T. A., 158
James, J. H., 502
James Range, N.T.
    drainage pattern, 435 et seq.
    flatiron, 61
    folds, ridge and valley, 114, 115
    lateral corrasion and pediment
        development, 212, 279
    ridge and valley form, 61, 114, 115
Jamieson, T. F., 469, 479
Janssen, R. E., 479
Japan
    mass movements, 221, 222
    thixotropic clays, 220
    vertical movement, 109, 110
    vertical movement 1920–1928, 110
"Jaws of Death", Grampians, Vic., 65, 66
Jeffreys, H., 27
Jennings, J. N., 17, 18, 80, 101, 124, 169,
    178, 255, 332, 396, 398, 399, 451, 502,
    528
Jenny, H., 186, 193
Jessen, O., 534
Jessup, R. W., 18, 398, 503
Jewell Cave, W.A., 73
Joffre Creek, W.A.—flat lying sediments,
537
John, B. S., 503
John F. Kennedy Rock, Rhodesia, 45
Johnson, D. W., 100, 209, 216, 272, 399,
    404, 422, 429, 451, 503
Johnson, W. D., 17, 124, 345, 365
joint
    block, 48
    control and caves, 71
    control and clints, 70
    control in sandstone, 63
    definition, 42
    marginal weathering of blocks, 50
    orthogonal, 57
    pattern, 42, 44
    pattern, rectangular, 64
    pattern, sandstone, 64
jokhulhaup, 252
Joly, J., 398
Jones, L. H. P., 215
Jones, O. T., 430
Jordan Narrows Quadrangle, Utah, 482
"Judges Wig", Jewell Cave, W.A., 73
Judson, S., 18, 178
Jukes, J. B., 451
Julia Plain, Qld—peneplain, 415
Junction Valley Graben, Utah, 95
Juniata River, U.S.A.—transverse, 433
Jura Mts., 122
Jura type of relief, 121, 122
Jutson, J. T., 183, 193, 281, 315, 398, 420, 429

Kaamasjoki-Kieltajoke river basin, Finland
—parabolic dunes, 318
Kachadoorian, R., 100
Kaikouran orogeny, 463
Kaikoura Surface, N.Z., 416
Kair River, Madhya Pradesh, India, 439
Kalahari Desert, 292, 295
Kalinske, A. A., 236, 255
Kalk, E., 193
kame terrace, 363
kamenitzas, 68
Kangaroo Tail, Ayers Rock, N.T., 59
kankar, 201
Kanyaka Ck., S.A.—anomalous, 440, 441
Keraskov, N. P., 194
Karlsruhe, Germany—meanders on Rhine,
248
karren, 68
karrenfeld, 74
karst, 66 et seq., 77
karst window, 74
Kartashov, I. P., 472, 478
Kashgar River, Iran—anomalous, 433
Katherine River Gorge, N.T., 64
Katili, J. A., 158
Kaye, C. A., 398
Kayser, E., 124
kegelkarst, 74
Keilambete, Vic.—maar, 153
Kellaway, G. A., 169
Keller, W. D., 187, 193
Kelso Dunes, Calif., 299
Kendall, P. F., 451, 478
Kennedy, W. Q., 426, 430
Kent, L. E., 215
Kent, P. E., 228
Kermadec Trench, 3, 35
kettle hole, 358, 359
    lake, 363
    pond, 354, 359
Keyes, C. R., 281, 315
Kharga Depression, Libya, 292
    oasis, 297
Kiel Canal—landslides, 221
Kilauea volcano
    1967 eruption, 128
    crater, 130
Kimberley Plateau, W.A., 7, 61, 72, 105, 106
kinetic energy of stream, 230
King, C. A. M., 18, 318, 332, 365, 430
King, D., 18, 315, 316, 502
King-Hele, D. G., 41
King Island, Bass Strait—coastal dunes, 396
King, L. C., 54, 80, 209, 216, 278, 279, 281,
    406, 411, 415, 429, 533, 534
Kingscote, S.A.—rock creep in basalt, 222
Kingston, S.A.—aggradational coast, 369
Kinki depression, Japan, 94, 95
Kirk, L., 215
Kiso River, Japan—faulted terraces, 99
Kitasu Hill (volcano), B.C., 43
Klamath Lake, Oregon—graben, 91
Klausing, R. L., 528
klippe, 120, 121
kneefold, 162, 163
Knox, J. C., 215
Kojima, G., 193

Kokerbin Rock, W.A.—flare and tafoni,
204
Komitake volcano, Japan, 152
Koonalda Cave, S.A., 71
Korea—pediments, 485
Krakatoa, 143, 152, 156
    explosion, 144
    rate of weathering, 192
    wave erosion, 371
Kranck, E. H., 124
Krapotkin, P., 528
Krenkel, E., 100
Krichauff Ranges, N.T.
    drainage pattern, 431
    folds, 431
    pediments and lateral corrasion, 279
Krieg, G. W., 502
Krinitzsky, E. L., 315
Krinon, E. L., 169
Kuenen, P. H., 503
Kuh-i-Hamak, Iran—salt dome, 123
kunkar, 201
Kupsch, W. O., 359, 368
Kwanto earthquake, 26
Kynuna plateau, Qld
    laterite, 267
    lateritised erosion surface, 269, 415

laboratory work
    dunes, 309
    freeze thaw, 182
    geomorphology, 6
    insolation weathering, 182
    meanders, 252
    turbidity currents, 501
    wind erosion, 289
Labrador
    patterned ground, 319
    postglacial isostatic recovery, 474
Labrador Trough, Canada, 28, 39
    volcanic rocks, 125
Lacroix, A., 125, 158
    classification of volcanic eruption and
    form, 127
Ladd, H. S., 527
La Fleur, R. G., 479
lag thrust fault, 540, 543
lagoon, 386, 393
lahar, 152, 223
Lahontan Basin, U.S.A., 461
Lahee, F. H., 534
La Jolla Canyon, Calif., 501
Lake Agassiz, 468
Lake Albany—proglacial lake, 468
Lake Baikal, U.S.S.R.—graben, 91
Lake Bonneville, 461
    deformation of shorelines, 481–83
    isostatic rebound, 483
    Pleistocene history, 428, 460, 480, 481
    relict forms, 481
Lake Bosumtwi—crater, 159
Lake Cahuilla, Calif.—(Pleistocene lake),
    282, 285, 295, 481
Lake Chad, central Africa, 4
    Pleistocene history, 481

Lake Coeur d'Alène, 468
Lake Cooper, Vic.—fault blocked, 97
Lake District, England
    superimposed drainage, 433, 439
Lake Eyre, S.A., 295, 531
    catchment, 283
    fault control, 284
    late Pleistocene conditions, 484, 485
    regional baselevel, 257
    windrift dunes, 290
Lake Eyre Basin, 198
    silcrete, 197
Lake Eyre plains, 253
Lake Gairdner, S.A.
    salt plates, 165
    salt weathering, 183
Lake George, N.Y. State—graben, 91
Lake George Rift, N.Y., 88
Lake George, Alaska—proglacial lake, 465
Lake Greenly, Eyre Peninsula
    isolated lagoon, 376
    shore platform, 384
Lake Krasnoye, U.S.S.R.—relict of
    Pleistocene lake, 472
Lake Lahontan—Pleistocene lake, 481
Lake Mead, U.S.A.—and isostatic
    adjustment, 522
Lake Missoula—Pleistocene lake, 468
Lake Phillipi, Qld—salina with lee mound,
    305
Lake Plateau, Labrador, 323
Lake Texcoco, Mexico—Pleistocene lake,
    525
Lake Thingvallavatur, Iceland, 86
Lake Toba, Sumatra—crater lake, 144
Lake Torrens plains
    oscillating dunes, 308
    fixed dunes, 310
Lake Wakatipu, N.Z.—abandoned
    shorelines, 467
Lake Woods, N.T.—late Pleistocene
    history, 454
lakes
    as result of glaciation, 364
    impounded by glaciers, 464
    impounded by landslide, 220
    Pleistocene, 480
Laki fissure, Iceland, 156
Lambert, J. M., 17, 528
laminar flow, 231
lamination, 57, 58
Lamplugh, G. W., 2, 195, 196, 215
land clearance—and soil erosion, 519
"lands"—and soil erosion, 521
Landsberg, H., 315
Landsberg, S. Y., 395, 398, 399
landscape development—according to
    Davis, 404–06
    Hack, 424–26
    Kennedy, 426–27
    King, 406–10
landscape revival, 410
landslide—deformation, 222
landslip, 218
Langbein, W. B., 18, 255, 430, 534
Langford-Smith, T., 215, 484, 502
Langway, C. C., 478

lapiés, 68, 74
lapilla, 138
Laramie Arch, Wyoming, 422
Laramie Basin, Wyoming, 422
La Rochelle, P., 228
Larsen, E. S., 17, 80
Larson, E. E., 429
Larsson, W., 158
late Pleistocene drainage—North Sea, 472
late Pleistocene glacial chronology—North
    America, 459
late Quaternary sealevel changes, 494
lateral (spit), 389
lateral corrasion, plain of, 248
lateral moraine, 354–62
laterite, 107, 415, 419, 421
    description, 6
    distribution, 195
    distribution in eastern Australia, 199
    profile description, 195–96
    surface, 419, 426
    theories of origin, 196
Lauterbrunnen valley, Switzerland—
    glaciated valley, 351
lava, 125
    cave or cavern, 133
    flows and stream diversion, 434
    formation, 128
    plain, 133
    plateau, 133
    stalactite, 133, 135
Law of Parsimony, 5, 6
Law of Superposition, 544, 546, 547
Law of Unequal Activity, 428
Lawn Hill Ck., Qld, 440
Lawson, A. C., 216, 272
Lawton, G. H., 272
layered structure of earth, 27, 28
Le Chatelier's Principle, 424
Leda Clay, 219
Leech, G. B., 100
Lees, G. M., 124, 451
Leeuwin Canyon, W.A.—cross-section, 499
Leffingwell, E. de K., 332
left lateral wrench fault, 544
Legget, R. F., 228
Lehmann, H., 80
Lehmann, O., 272
Leighly, J., 171
Leichhardt River, Qld—steepening of
    gradient, 434, 485
Lena River, U.S.S.R.—thermokarst, 327
Lenk-Chevitch, P., 255
Leopold, L. B., 18, 232, 255, 430, 527
Le Pichon, X., 41
Lester, J. G., 80
Levant, 42, 44
    soil erosion, 522
    faults, 91
Leverett, F., 451
Levi Fault, S.A., 88
Lewis, W. V., 345, 364, 365, 398
Lewis Surface, 422
Lewis Terrace, Nevada—pediment, 277
Lewisian gneiss, 506
Libya
    dunes, 294

Libya (*continued*)
    dune structure, 300
lichens—and weathering, 184, 187, 188,
    510
Lichty, R., 427, 430
limb of fold, 540
limestone, 7, 66 et seq.
    crystalline, 68
limestone pavement, 68, 69
Limpsfield, Surrey, England—chalk scarp,
    76
Lindsay, J. M., 18
lineaments, 32, 87
linear depression, 203
    escarpment, 81
    gilgai, 167
    outcrops, 190
    relief, 78
lineation, 77
Linton, D. L., 17, 80, 124, 178, 319, 332,
    502, 509, 534
Lipson, J., 430
Lisbon Scarp, 30
literary evidence for climatic change, 455
Little Dome, Wyoming, 542
Little Shuteye Pass, Calif.—sheet
    structure, 60
Livingstone, D. A., 255
The Lizard, Cornwall—Pliocene bench, 505
The Lizard, N. T.—scarp foot
    depression, 211
Lliboutry, L., 365
load transference, 164
loading of crust, 163
lobe, 222
Lob Nor, central Asia, 481
local baselevel, 256, 257
Loch Lochy, Scotland—glacial deepening
    of fault line valley, 84
Loch Oich, Scotland—glacial deepening of
    fault line valley, 84
Lochaber Swamp lunette, S. A., 311
Loeffler, E., 502
loess
    cold, 286, 364
    hot, 286
Lofgren, B. E., 528
loganstone, 45
Long, J. T., 315
Long Beach, Calif., 107
Long Island, N. Y. State—end moraines as
    spits, 388
longitudinal dune, 292, 294, 295, 302
    form, 301
    form, change in time, 303
    structure, 300
    origin, 301
longitudinal wave, 27
longshore drifting, 385
    wave induced, 387
Los Angeles, Calif.
    smog, trees and erosion, 522
    subsidence, 522
Loughnan, F. C., 193
Lovering, T. S., 17, 193, 215
low, 387
low tide cliff, 374

Lowe, F., 439, 442, 451
Lublin, Poland—glacial and interglacial
    sequence, 462
Lugeon, M., 255
lunette—description, 311, 313
    theories of origin, 311, 313
lunettes—and coastal foredunes, 313
Louisiana, U.S.A.—salt domes, 122
Luxor, Egypt—granite weathering, 181
Lyell, C., 230
Lynmouth, Devon, 201
    flood, 1952, 203
Lyttleton Harbour, N.Z., 132

maar, 152, 153
Maarleveld, G. C., 502
Mabbutt, J. A., 80, 193, 215, 279, 281, 316,
    420, 429, 502, 509, 451
Macar, P., 430
MacDonnell Ranges, N. T., 442
MacFadyan, W. A., 281
Machida, H., 158, 228
Mackay, J. R., 323, 332, 511, 527
Mackenzie, D. P., 41
Mackenzie delta, Canada, 511
    pingos, 323, 324
    section, 327
Mackenzie River, Canada, 229
Mackenzie valley, Canada, 118, 229
Mackin, J. H., 255
Macumber, D., 316
Maddock, T., 232, 255
Madigan, C. T., 18, 301, 316, 317, 332, 451
Madison, Wisconsin—rate of erosion due
    to mass movement, 225
Madison Canyon landslip, Montana, 223
Maggie Springs, Ayers Rock, N. T.—
    flared slopes, 66, 204
magma—and development of volcanoes, 125
magnetic field of earth, 33
magnetic reversal, 25
Mahard, R. H., 451
Maignien, R., 18, 215
Mainguet, Monique, 80
Mainwaring, E. A., 169
major features of earth's relief, 21
major plates—distribution, 25
Makaopuhi crater, Kilauea, Hawaii, 128
making of crust, 35
Malaspina Glacier, Alaska, 333, 335
mamillation, 344
man
    geomorphological agent, 516
    and weathering, 192
Manicouagan Lake, Que., 39
Mannerfeldt, C. M., 368
Manning equation, 231
mantle of earth, 27, 28
Maratoa, thickness of coral, 513
Mariana Trench, 3
maritime region, 173
Marjeelen Sea, Switzerland, 464, 467
Marker, Margaret E., 80
Marr, J. E., 433, 439, 451
Marshall Islands, 517
    coral reefs, 513
Marvine, A. R., 433, 439, 451

mass movement, 217
　classification, 222
　conditions, 219
　and earth tremors, 219
　and ground ice, 219
　and structure, 219
　and water, 219
　and undermining, 221
matched river terraces, 411, 412
Mathews, D. H., 41
Mathews, W. H., 158
Matterhorn, Switzerland, 349
Matthes, F. E., 4, 158, 255, 451
Mattox, R. B., 124
mature stage of landscape development, 404
Mauna Loa, Hawaii, 129, 130, 132
　recent lava flows, 130
Maverick Domes, Wyoming—stratum
　contours, 540, 542
Maw G., 451
Mawson, D., 317, 332
Maya, U.S.S.R.—thermokarst, 328
McAllister, J. F., 215, 281
McCall, G. J. H., 169
McCall, J. G., 365
McCallien, W. J., 169
McColl, D. H., 161, 169
McDonald Scarp, N.W.T., Canada, 81, 82
McDougall, I., 215
McGee, W. J., 281
McGowran, B., 18
McGregor Plateau, Canadian Plateau—
　glaciated surface, 344
McKee, E. D., 315, 398
McKinley, D. G., 232, 255
McNab, R. F., 332
M discontinuity, 28
Meade, R. H., 528
meander, 244, 252, 340, 472
　belt, 248, 258
　bluff, 412
　cone, 248, 251
　Coriolis force, 252
　development, 248
　discharge and gradient, 252
　geometry, 252
　laboratory experiments, 252
　loop, 412
　origin, 252
　pools and shoals, 252
　tidal zone, 445
meandering-braided streams—relationship,
　252
mechanical weathering, 179
Meckering earthquake 1968, 91
Medawar, P. B., 18, 534
medial moraine, 350, 354, 355
Mediterranean Sea, 87
　drowned gorges, 500
　Pleistocene sealevels, 492
　Quaternary tectonism, 492
　submarine canyons, 500
　Tertiary history, 500
　volcanoes, 125
Medlicott, H. B., 438, 451
Meinesz, F. A. V., 41, 100
Melanesia—coral reefs, 513

mélange deposits, 36, 78, 79, 220
melon hole, 168
Melton, M. A., 534
meltwater channel—British Columbia, 472
　modern, 472, 473
meltwater hypothesis—cirque development,
Melville Island, N.W.T., Canada—raised
　beach, 474
Menan Buttes, Idaho—volcanic craters, 136
Menard, H. W., 41
Mencl, V., 228
Meramac River, Minnesota—intrenched
　meander, meander core, 251
Merrill, G. P., 80
Mern Merna Dome, S.A., 446, 447
Merscherikov, Y. A., 124
mesa, 101, 105, 107, 197, 201, 406, 504
Meseta, Spain, 29
Messia Channel, Chile—depth of fjord, 350
metamorphic rocks, 353
　aureole, 535
　classification (table), 537
meteor shower, 161
meteorite craters, 159
　distribution, 161
　impact, 161
methods in geomorphology, 4, 5
Mexico—pseudoanticlines, 167
Mexico City—subsidence, 522, 523
Meyer, B., 193
Meyerhoff, A. A., 37, 41
Meyerhoff, H. A., 451
Michara, Y., 527
microyardang, 318
Mid Atlantic Ridge, 35, 36
　earthquakes and volcanoes, 125
Milankovitch, M., 463
Milbank strandflat, B. C., 43
Milford Sound, N.Z.—fjord, 348
Miller, J. P., 232, 255, 396, 399, 534
Miller, R. P., 216
Milne, A. A., 215
Milne, G., 194
Milnes, A. R., 124, 215, 430
Milton, D. J., 169
Mima mound, 511
Mineral Point Branch, East Pecatoma
　River, Wis.—section and misfitness, 487
misfit streams
　definition, 484
　in Britain, U.S.A., N.S.W., 484
　origin, 484, 485
misfit valleys—sections, 487
Mississippi delta, 237
　growth, 243
　recent variation, 239
　variation, 1885–1935, 240
Mississippi River
　ancestral, 465
　bed forms, 236
　deposition, 244
　fault control, 81
　loess, 286
　turbidity, 245
Mississippi valley
　fault control, 81

Mississippi valley (*continued*)
　warping, 242
Mitre Peak, N.Z.—pyramidal peak, 348
Miyabe, N., 124
Miyamura, S., 124
Mizoue, M., 124
moat, 203
mobile dunes
　conditions, 292
　types, 292
model of landscape development
　Davis, 404–06
　Hack, 424–26
　Kennedy, 426–27
　King, 406–10
models of landscape evolution, 403
moderate region, 173
modern meltwater channel
　Canada, 474
　Greenland, 472, 473
modern proglacial lakes, 465
mofettes, 156
mogote, 74
Mohorovicic—discontinuity, 28
Mohr, E. C. G., 188, 193
Mojave Desert, Calif., 203, 289, 290
　barchan movement, 298
　warping, 107
monadnock, 404, 420
Monaro district, N.S.W.—block stream,
　329
monkstone, 77
monocline, 416, 539, 544
Mono Craters, Calif., 152
Mono Rock, Calif.—bornhardt, 46
Montana—proglacial lakes, 469
montmorillonite, 168
monument, 345, 347
monzonite, 55
Moore, G. W., 80
Moore, J. G., 41
Moorehead, A., 17, 180, 193, 527
moraine, 355
　ablation,
　end,
　ground,
　lateral,
　medial,
　recessional,
　terminal,
Morisawa, Marie, 255
morphogenetic regions, 173
morphogenetic system, 173
Morison, G. C. T., 194
Morris, F., 481
Morris, F. K., 502
Morrison, R. B., 478, 502
Mortensen, H., 175, 178
Morteratsch Glacier, Switzerland, 334
Morton, D. M., 228
Moss, F. J., 169
Motpena, S.A.—oscillatory dune, 308
moulin, 350
mountains—distribution, 25
mountains—submarine, 24
Mt Arenal, Costa Rica—recent eruption,
　142, 156

Mt Babbage, S.A.
  exhumed pre Cretaceous surface, 418
  unconformity, 538, 539
Mt Bandai, Japan—composite volcano, 148
Mt Brown, Qld—residual remnant, 189
Mt Buffalo, Vic.—flared boulders, 7, 8,
  10, 14
Mt Conner, N.T.
  annular drainage pattern, 432
  blocky disintegration, 182
  scarp foot depression, 211
Mt Egmont, N.Z.—volcano, 144, 148
Mt Etna, Sicily, 132
Mt Everest, 4
  alleged uplift, 28
Mt Fort Bowen, Qld.—residual remnant,
  189
Mt Fuji, Japan, 144, 149
  geological map, 150
  Pleistocene history, 152
  section, 150
  tephra, 151
Mt Gambier, S.A.—Quaternary
  volcanicity, 463
Mt Garibaldi, B.C.—Pleistocene volcano,
  152
Mt Kosciusko—glacial and nival forms, 480
Mt Lamington, New Guinea, 139, 142, 144,
  403
  1951 eruption, 142, 143
Mt Lofty Ranges, S.A.
  drainage evolution, 444
  etch plain, 420
  etch surface, 421
  fault angle valleys, 94
  orthodox geomorphological evolution, 425
  antecedent streams, 438
Mt Margaret Fault, S.A., 88
Mt Mazama, Oregon—old volcano, 144,
  152
Mt Monadnock, New Hampshire, 404
Mt Pelée, Martinique, 138, 139
Mt Ruwenzori—horst, 91, 93
Mt Schank, S.A.—ash cone, 152, 463
Mt Somma, 140, 156
Mt Tabletop, Qld.—volcano, 131
Mt Vesuvius, 129, 140, 156
mudflow, 7, 222, 223, 225, 226
  volcanic, 152
Mueller Plateau, N.T.—flat lying
  sandstone, 63, 100
Muir Glacier, Alaska, 5
Mukorob, 105
Müller, F., 323, 332
multicyclic landscape, 410
  Appalachians, 422
  Australia, 415
  Queensland, 415
    map, 418
  South Australia, 415
  southern Africa, 415
    map, 417
  Wales, 422
multicyclic landscape—and sheet structure,
  60
multiglaciation—evidence, 457
multiple working hypotheses, 6, 16, 17

mullion structure, 544
Mulvaney, D. J., 272
muren, 217
Murray, G. W., 258, 315
Murray, J., 513, 527
Murray Basin, S.A., 62
Murray Canyons, S.A., 497, 498
Murray Plains—relict dunes, 485
Murray mallee, 314
Murray Valley, S.A.—slope variations,
  262 et seq.
Murrumbidgee River, Australia, 245, 519
mushroom, 163
Mutitjilda Waterhole, Ayers Rock, N.T.—
  flared slopes, 66
mylonite, 544

Nace, R. L., 534
Nagooka, S., 193
Nairne, A. E. M., 509
naled, 323
Namib Desert, South West Africa—
  barchan, 296
Nansen F., 382, 398
Napier Range, W.A.—karst, 72
nappe, 121
nari, 201
Natal—Pleistocene coastal dunes, 395
Natal Monocline, 16, 415
natural amphitheatre, 116
  bridge, 103
Navajo region, Arizona—dune forms, 314
Navajo Sandstone, 63
near earthquakes, 28
The Needles, Isle of Wight—chalk stacks,
  381
Neff, G. E., 365
negative exfoliation, 66
Nelson, C. A., 509
Nelson, H. J., 528
Nerdi Syncline, Algeria, 93
Netterberg, F., 215
network of gilgai, 166
névé, 333
Newbiggin Crags, Westmoreland—karst
  forms, 69
New England, Australia
  flared forms, 7
New England, U.S.A.—sheet structure, 59
New Forest, England—soil erosion, 519
New Fuji, Japan, 152
New Guinea
  glacial and nival forms, 480
  karst, 74
  no evidence pre Wisconsin glaciations,
    516
  patterned ground, 510
  slopes, 273
  submarine canyons, 497
New Mexico, U.S.A.
  causes of gullying, 516
  volcanic forms, 133
new plateau, W.A., 420
New Quebec Crater, Canada, 159
New South Wales
  Blue Mts, 64, 105
  coastal forms, 367, 375, 383

New South Wales (*continued*)
  depth of weathering, 190
  fault control, 81, 84
  gilgai, 166
  granite forms, 7, 13, 49, 56, 180
  inselbergs, 275
  Pleistocene glacial and nival action, 464
  relict drainage, 484
  volcanic forms, 155, 156
Newton, Isaac, 22
Newton Dale, Yorkshire—overflow
  channel, 469
New Zealand
  alluvial fans, 243
  alluvial plain, 132, 250
  coastal forms, 13, 348, 375, 377
  erosional comparison with central
    Europe, 175
  faults, 89, 91, 97, 463
  glacial chronology, 460, 461
  glacial forms, 348, 362, 467
  mass movements, 221
  soil erosion, 519
  terraces, 410
  vertical movements, 109
    (Recent), 111
  volcanic forms, 132, 152, 156, 157
Nichols, R. L., 318, 332, 478
nickel iron, 161
nick point, 410, 411
Nielson, R. A., 178, 534
Nigeria, 185
  depth of weathering, 190
Nigeria—relict dunes, 487
Nikiferoff, C. C., 425, 430
Nile Delta, 237, 241, 525
nival—terminology, 319
nivation hollow, 318, 347
Nobles, L. H., 228
nodular calcrete, 200, 201
Nome Creek, Tolovana, Yukon—
  solifluction lobes, 329
nonaligned dune, 295
non-faceted slope, 259
nonsorted slopes, 321
Norfolk Broads, England—origin, 524, 525
normal climate of geological time, 453
  fault, 542
  morphogenetic region, 173
  thrust fault, 540, 543
Norrish, K., 18, 193, 215
Norris, K. S., 315
Norris, R. M., 315
North Africa—soil erosion, 522
north African desert
  geological map, 40
  relict water, 230
North America
  late Pleistocene glacial chronology, 459
  Pleistocene chronology, 459
  Pleistocene proglacial lakes, 465
  proglacial lakes, 468
Northamptonshire ironstone field—
  superficial structures and forms, 163
North Branch, U.S.A.—transverse
  drainage, 433
north central U.S.A.—glacial spillways, 470

North Dakota—proglacial spillways, 469
North Downs, England—late Tertiary
  peneplain remnants, 505
North European plain—*urstromtäler*, 471
North Fork Lake, Alaska—recent proglacial
  lake, 465
North Sea—late Pleistocene drainage, 472
Northern Ireland
  drumlin, 361
  post glacial isostatic recovery, 474
Northern Territory
  boulders, 45
  butte, 105
  fold mountains, 61, 114, 115, 212, 279,
    431, 435 et seq.
  graded slopes, 261, 409
  inselbergs and related forms, 7, 13, 56,
    59, 66, 182, 191, 204, 206, 207, 211, 432
  meteorite craters, 159–61
  relict drainage, 484
    lake, 454
  sandstone plateau, 63, 100
  silcrete, 197, 504
  tidal meanders, 246
northwestern Europe—Pleistocene
  chronology, 459
North York Moors—proglacial lakes, 469
North Yorkshire—proglacial lakes and
  overflow channels, 470
Norway—fjord coast, 353
Norwest Island, Capricorn Group, Qld,—
  coral island, 512
notch, 382
Nova Scotia—drowned drumlin, 361
nuclear explosions—weathering effects, 180
nueés ardentes, 7, 125, 138, 139, 142, 147, 152
Nullarbor plain
  karst, 67 et seq.
  relict drainage, 484–86
  sediplain, 101
nunatak, 338, 339, 348
Nye, J. F., 365
Nye, P. H., 185, 193

Oahu, Hawaii—thickness of coral, 513
Oberlander, T., 124, 216, 228, 451
oblate spheroid, 22
oblique fault, 540, 543
obsequent fault line scarp, 81
Obst, E., 80
ocean basin, 23
oceans
  area, 23
  distribution, 24
  pattern, 24
Odell, N. E., 479
offloading hypothesis, 57, 184
offset ridge, 81, 82
  stream, 97, 98
  topography, 84
  valley, 81
Ohio—glacial and interglacial deposits, 458
Okefenokee shoreline, U.S.A., 495
old age stage of landscape development, 404
"old-from-birth" peneplain, 425
Old Fuji, Japan, 152
Old Hat platform, 383, 384

oldland, 410, 415
  and continental drift, 415
old plateau, W.A., 420, 421
Olgas, N.T.
  flared slope, 7, 13
  plan form, 56, 57
  sheet structure, 59
  split boulder, 182
Ollier, C. D., 18, 158, 193, 194, 216, 451
Olmstead, E. W., 440, 451
onion weathering, 180, 181, 205, 207
Ontario—relict nival forms, 480
Oodomari, Japan—landslide, 221
open hypothesis—pingo formation, 323
opaline silcrete, 197
open system—weathering, 186
Öpik, A. A., 215, 267, 272, 429, 451, 530
organic sediments, 535
Ordos Plateau—loess, 286
organisms
  and coastal processes, 378
  and landform development, 510
Orkney Islands, Scotland—stack, 380
orocline, 29
  concept, 29, 30
orogen, 121, 125
Orinoco Basin, S. America, 61
orthogonal joint system, 57
Otago, N.Z., 78, 79
Ottman, F., 503
Ouichita River, U.S.A.—fault control, 81
outlier, 538, 539
outwash
  apron, 354
  fan, 354
  plain, 357, 362, 363
overbank deposits, 245
overflow channels—North Yorkshire, 470
overfold, 121, 541
overloading—local, 159
overloading of crust, 159
overmass deposit, 439
overturned fold, 540
Owens Range, Calif., 8
Owens Valley, Calif.—mudflows, 226
Owikeno lineament, 85
oxbow, 229, 340
Oxbow Lake, 245, 248, 472
oxidation, 188
oxygen isotopes, 454
oyster reef, 513

Paarl Mt., Cape Province—flared slopes,
  7, 16
Pacific Basin, 24, 221
Pacific "Ring of Fire", 125
Pacific Ocean—floor of, 35
Pacific type coast, 367, 368, 369
pahoehoe, 133
paired river terraces, 411, 412
Pakefield, eastern England—wave erosion,
  371
palaeomagnetism, 33, 34, 35
Pallister, J. W., 215
palimpsest surface, 453
Palmer, E. H., 527
Palmer, J., 178, 534

Palmer, S.A.
  boulders, 49, 50
  boulder split by tree, 184
  etch surface, 421
"palm-tree" drainage pattern, 434
paludal plain, 405
palynology, 455
Pamlico shoreline, U.S.A., 495
pan
  coastal, 375, 376
  granite, 202, 203
  limestone, 375, 376
Panamint Valley, Calif.—bolson and dunes,
  295
Pandasteco, E. B., 289, 315
Pangaea, 32
panplain, 404, 406
parabolic dune, 313, 314, 394, 395
  example, 395
  in cold lands, 317, 318
Paraiba Valley, Brazil—graben, 91
parallel drainage, 433
parallel roads of Glen Roy, 468, 469, 471
parallel slope retreat, 406
Parana Basin—Triassic desert, 505, 506, 507
Paris Basin—ridge and valley forms, 121,
  122
Parramore Island, Virginia—mounds due
  to gravitational pressure, 164
particles in motion—turbulent eddies, 285
Passarge, S., 315
Patagonia—basalt flows, 132
paternoster lakes, 347
Paterson, M. S., 545
patterned ground
  desert gilgai, 288, 289
  gilgai, 166–68
  nival or periglacial, 168, 222, 319, 323
  relict, 480
pavement
  desert, 284, 288
  limestone, 68
Peake and Denison Ranges, S.A.—faulted,
  88
Peake River, S.A.—distributory system, 253
Pearson Island, S.A.
  flared slope, 9
  joint pattern, 56
  raised beach, 492
  sheet structure on coast, 379
peat quarry—section, 524
pebble shape, 233
pediment, 103, 260, 409
  definition, 274
  dissected, 210, 211, 275
  distribution, 277
  lithological control, 277
  map (New Mexico), 276
  pass, 278
  processes, 278, 279
  relict, 485
  remnants, 212, 277
  and scarp foot weathering and erosion,
    278
  and slope behaviour, 278
  theories of origin, 278, 279
pedimentation, 91, 406, 409, 533

pediplain, 101, 406
  contrasts with peneplain, 409
  importance of piedmont angle, 409
Pedro, G., 6, 18
Peel, R. F., 429
Pekulney Range, U.S.S.R., 472
Peléean eruption—characteristics, 125, 127, 128, 138
Pelletier, B. R., 332
Pels, S., 502
Peltier, L. C., 173, 175, 177, 178
Penck, A., 461, 478
Penck, W., 272, 406, 424, 429, 430
Pendleton, R. L., 215
peneplain, 405, 406, 420
peneplain concept
  assumptions, 404
  description, 404
penitent rock, 77
Penny, L. F., 17
pepino hill, 74
perception, 189
perched
  boulder, 45
  peneplain, 428
  sandstone boulder, 65
  syncline, 119, 121
perennial stream, 232
periglacial—terminology, 319
permafrost, 323
  distribution, 322
Permian glaciation, 508
  distribution, 507
Permian ice age, 453
Permo-Triassic desert, 453, 505
Perry, W. J., 18, 509
persistence—stream, 438, 444, 448, 450
Perth (submarine) Canyon, W.A., 499
Pertnjara Hills, N.T.
  graded slopes, 261
  uniform slopes, 409
Peru
  barchan movement, 298
  desert flood, 273
  landslides, 1970, 219
Péwé, T. L., 502
Philippine Trench, 3
phreatic water, 72
physical weathering, 179
  optimum conditions, 190, 192
Pickering, Yorkshire—delta, 469
Pic Parana, Brazil—faulted inselberg, 60
piedmont angle
  definition, 207
  description, 209
  due to change of process, 209
  faulting and scarp retreat, 209, 211
  origin, 209
  scarp foot weathering and erosion, 209
  stream erosion, 209
  structural control, 209
Piketberg, Cape Province—relief inversion, 121
"Pile of Logs", Qld—horizontal joint columns in granite, 55
Pinchemel, P., 124

pingo, 167
  description, 323
  distribution, 323
  ice cone, 324
  plan, 323
  relict, 480
  ruptured, 325
  theories of origin, 323 et seq.
"The Pinnacles", N.S.W.—inselbergs, 275
pipkrake, 319
pisolitic structure in laterite, 196
Pissart, A., 502
pit
  coastal, 375
  granite, 203
Pitcher, W. S., 37, 41
pitted end moraine, 358
Pitty, A. F., 315
pivotal fault, 540, 544
Piz Bernina, Switzerland—pyramidal peak, 317
Plafker, G., 100, 228
plain of lateral corrasion, 248, 258
"Plain of the Parliament", Iceland, 86
planated relief, 121
planeze, 156
plant roots—and weathering, 187
plants and animals—in weathering, 184
Plass, G. N., 478
plastic clay, 159, 164
plateau, 101, 105, 106, 197, 201, 406, 504
  basalt, 134
  domed, 101, 103
Plate Theory, 36, 37
  problems, 37
plates
  distribution, 25
  map, 37
platey gibber, 284
platform (shore), 382
  structural, 39
playa
  deflational, 292
  tectonic, 292
  South Australian, 292, 293
playa stone track, 195
Pliocene marine bench—S.E. England, 505
ploughing—and soil erosion, 519
Playfair, J., 229, 230, 255
Pleistocene
  chronology, glacial, 457–61
    European Alps, 459–61
    European U.S.S.R., 459
    North America, 459
    northern Europe, 459
    northwest Europe, 459–61
    Sierra Nevada, U.S.A., 461
  chronology, tropical regions, 460
  definition, 457
  direct effects, 463 et seq.
  duration, 457
  effects in nonglaciated areas, 480 et seq.
  extent of ice sheets, 463, 464
  lakes, 480 et seq.
  nonglacial geological events, 463
  temperature fluctuations, 454

plucking, 343
plug, volcanic, 152–56
plunging anticline, 83, 118
  syncline, 115
plutonic rocks, 535
pluvial periods, 460
Pohutu Geyser, Rotorua, N.Z., 156
point bar, 229, 245, 247
  deposit, 244, 245, 259
point of inflexion, 259, 270
polar flattening of earth, 22
polar wandering, 35
polished surface, 7
polje, 74
polygenetic landform, 453
Pompeii, 156
Pondalowie Bay, S.A.—Pleistocene foredunes, 396
pool weathering, 373, 375
Popper, K. R., 18
Port Clinton, Qld—parabolic dunes, 395
Port Hedland, W.A.—coastal dune, 299
Port Phillip Bay, Vic.—fault control, 94
Porlock, Devon—infilled lagoon and shingle bar, 392
Porsild, A. E., 332
Portsea, Vic.—shore platform and cliff, 374
postglacial isostatic recovery
  Arctic Canada, 474
  Canada, 474
  Labrador, 474
  North Ireland, 474
  Scotland, 474
potential energy of stream, 230
pot-hole, 233
Potts, A. S., 193
Powell, J. W., 451
prairie mounds
  as old dunes, 511
  due to nival action, 511
  as gopher spoil heaps, 511
Prairie Terrace, U.S.A.—warping, 242
Precambrian erosion surfaces
  Canada, 508
  distribution, 508
  Eyre Peninsula, 508
  Greenland, 508
Precambrian glaciation, 7
  distribution, 508
  glaciated pavement, 508
Precambrian pediment, 508
pre Cretaceous surface, 418
preglacial survivals, 464
pre Middle Cambrian surface, 418
Prescott, J. A., 215
present rise of sealevel, 489
"present is the key to the past", 5
pressure release hypothesis, 60, 184
  difficulties, 57
  outline, 57
pressure wave, 27
  pattern, 27
Price, W. A., 169, 527
Prider, R., 215
primärrumpf, 425
primary wave, 27

proglacial lake
  British Columbia, 465
  modern, 464–65
  Montana, 469
  North Dakota, 469
  north Yorkshire, 469, 470
  problems and disputes, 472
protalus rampart, 318
protectionist school, 355
provenance of erratics, 458
Provo level, Lake Bonneville, 481
pseudo-anticline, 167
  in calcrete, 201
pseudo bedding, 58, 184
pseudo structural landforms, 159
Puerto Rico—karst, 74
puff, 166, 168
Pullan, R. A., 502
push ridge, 359
push wave, 27
puy, 152
P wave, 27
Pyramid Lake, Nevada, 480
pyramidal peak, 347, 349, 351, 355
Pyrenees Mountains—orocline concept,
  29, 30
pyroclastic, 125
Pythagorean School, 21

qanat, 112
Qishin Island, Persian Gulf—diapir, 123
quaquavasal dip, 540
Quaternary sealevel changes, 494
Quaternary vertical movements, Japan, 110
Quebec—relict nival forms, 480
Queensland
  coastal
    parabolic dunes, 395
    Pleistocene foredunes, 395
  depositional plain, 405
  gilgai, 166–68
  granite forms, 50, 55, 378
  north
    karst, 72
    volcanic areas, 135
  northwest
    erosion surfaces, 415, 418
    map, 418
    section, 419
    exhumed pre Cretaceous surface, 419
  volcanic forms, 131, 135, 156
Quisapu, Chile
  1932 eruption, 152
  tephra, 152

rabbits—and weathering, 184
Racetrack Playa, Calif., 288
radial drainage, 131, 432
radial rifts (on volcano), 130
Radiant Glacier, B.C., 334
radioactive decay, 29
  and dating, 547
rainfall effectiveness in deserts, 273, 274
raised beach, 490
  shore platform, 374, 490
  stack, 491
Raistrick, A., 194

ramp, 383
Ramsay, A. C., 422, 430
Rapp, A., 228
Rassirs Syncline, Algeria, 93
Ratcliffe, F. N., 315, 528
rate of creep, 222
  geomorphological work, 177
  glacier motion, 339
  lava flow, 128
  weathering, 192
Raussell-Colom, J. A., 6, 18
Razumova, V. N., 194
reaction series—feldspar weathering, 186
Read, H. H., 548
reasons for climatic change, 457
Reba, I., 18
Recent—definition, 457
recession of fault scarp, 90
recessional moraine, 354, 359
  Ohio, 360
  Sweden, 360
rectangular drainage pattern, 433
recumbent fold, 163, 540, 541
Red River, U.S.A.—fault control, 81
Red Rock Pass—overflow from Lake
  Bonneville, 481
Red Sea—graben, complex origin, 87, 91
Red Sea Hills—granitic, all slopes, 87
reduction, 188
Reeves, C. C., 502
reg, 288
regional baselevel, 256, 257
regolith, 54, 179
Reiche, P., 194
Reifenberg, A., 527
reinforcement effect, 531
Reiner, E., 502
rejuvenation—river, 410, 411
rejuvenation of streams by faulting, 87
relative effectiveness of glaciers and rivers,
  355
relict drainage systems, 484
  Amadeus Basin, 484
  Nullarbor Plain, 484
  Officer Basin, 484
  Riverina, 484
relict dune field, 485
  Adelaide Plain, S.A. 488
  Eyre Peninsula, S.A., 485, 488
  Great Plains, U.S.A., 485
  Murray Plains, S.A., 488
  southern edge of Sahara, 485, 488
  Wimmera, Vic., 485
  Yorke Peninsula, S.A., 488
relict dunes—soil erosion, 522
relict ice—Antarctica and Greenland, 339
relict ice wedge polygons, 480
relict nival forms
  Alaska, 480
  Finland, 480
  Ontario, 480
  Quebec, 480
  southern Africa, 480
relict water, 230
relief inversion, 106, 121, 201
  coastal, 375
remanent magnetism, 33

Remarkable Rocks, Kangaroo Island,
  S.A.—tafoni, 205
remnants of circumdenudation, 404
renewal of weathering, 192
reptation, 288
resequent fault line scarp, 81
retreat of fault scarp, 90
retreat of slope, 105
reticulate dune pattern, 292, 309
Revelle, R., 41, 398
reversals of earth's magnetic field, 35
reverse fault, 543
revival, landscape, 410
Reynolds, M. W., 281
Reynolds number, 231
Rhine graben, 91, 92
Rhodehamel, E. C., 479
ria coast, 367, 368, 491
ribbed surface, 114, 117
Rich, J. L., 281
Ridge and Valley Province, Appalachians,
  433, 444
ridge, shape in cross-section, 113
ridge forms—distribution, 121
ridge and valley, 107, 115, 121
  Appalachians, 120, 121
Ridlon, J. B., 158
Riebeck Oos, S. Africa—transverse
  drainage, 433
riegel, 347
riffle, 252
rift valley, 42, 91
Rift Valley, Africa, 44
rift zone—and volcanoes, 125
right lateral wrench fault, 544
Riley, F. S., 528
Rillen, 52, 203
  subsurface initiation, 204
Rillenkarren, 71, 72
"Ring of Fire", 125
Rio Carona, Venezuela—braided and
  anastomose, 489
Rio de Janeiro, Brazil
  faulted bornhardt, 59
  rate of weathering, 192
  sugar loaf, 47
Rio Grande River, turbidity, 245
rip channel, 387
ripple
  snow, 317
  water, 236
ripple mark, 387
rise of sealevel—contemporary, 489
rising anticlines, and coral reefs, 513
rising island, 515
Ritter, D. F., 178
Rittman, A., 158
River Add, Argyll, Scotland—meandering
  stream, 247
River Amazon—braided, 252
River Arun, Himalayas—antecedent, 438
river bluff, 229, 258, 259
River Carcajou, N.W.T., Canada—riverine
  forms, 229
River Indus—exotic stream, 232
River Jordan—delta, 240
River Karun, Iran—transverse, 112

River Murray, S.A.—ancestral, 97
River Nile
    exotic stream, 232
    recent incision, 413
River Ouse, England—bank failure, 221
river rejuvenation, 410
River Rhine—meanders at Karlsruhe, 249
River Rib, Hertfordshire—section and
    misfitness, 487
River Salzach, Austria—antecedent, 438
river terrace, 258, 410, 411, 412, 414, 491
    anthropogenic, 413
    causation, 413
    climatic, 413
    interpretation, 413
    tectonic, 413
river terraces
    displaced by faulting, 97, 99
    indicators of climatic change, 485, 489
River Vistula, Poland, 462
River Wieprz, Poland, 462
Riverina, N.S.W., 519
    relict drainage, 486
rivers, 229, 273
    depth-width-discharge relationship, 232
    discharge-width-depth relationship, 232
    width-depth-discharge relationship, 232
    see also streams
Rivière aux Vases, Quebec—landslide, 217
road—washed out, 521
roadside erosion, 521
Roberts, H. G., 18, 509
Robertson, G. K., 169
Robinson, A. H. W., 398
roches moutonnées, 355, 438
    development, 343, 344
    Permian, 507
rock composition—and weathering, 189
rock fall, 226
rock glacier, 328, 331
rock meal, 183, 185
Rock of Ages Quarry, Barre, Vermont, 61
rock platform, 11
rock types, 535
rockslide, 217, 223
Rocky Mt. National Park, Colorado, 182
Rocky Mt. Trench, Canada, 86, 87
Rocky Mts, Colorado—glaciated, 356
Rodda, J. C., 430
Rohleder, H. P. T., 169
Romanes, J., 17
Ronai, A., 124
roof and wall structure, 162, 163
ropy lava, 133
Rose, E. R., 509
Ross, D. A., 41
Rosser, H., 124, 215
rotational fault, 540
    slip, 221, 223
rotational flow
    in cirque glaciers, 345
Rotorua, N.Z.—solfataric activity, 152, 156
Royston Head, S.A.—climbing dune, 396
Rubey, W. W., 230, 255
Ruby Range, Nevada
    hinge fault, 96
    horst, 94

triangular facets, 88
Ruedemann, R., 511, 527
Rumbalara, N.T.—silcrete, 504
Rundkarren, 69, 171
runnel, 381, 387
running water, 229
    extent of activity, 256
ruptured pingo, 325
Russell, I. C., 365, 478
Russell, R. J., 17, 100, 281, 289, 213, 503,
    527, 533, 534
Russian Platform, 39
Russian River, Calif.—superimposed, 79,
    439
Rutland, R. W. R., 124, 158
Rutten, M. G., 511, 513, 527
Ruxton, B. P., 169, 193, 281, 430
Ryder, J. M., 255

S wave, 27
sable mammelonné, 309
Sagami Bay, Japan—topographic changes
    due to earthquake, 26
sag pond, 97
Saguenay River, Quebec—landslide, 217
Sahara desert
    dune forms, 294
    geology, 40
    relict dunes (southern margin), 485, 488
Sahul Shelf, 492
Saidmarreh Landslip, Iran, 220, 221
Saidmarreh River, Iran—dammed by
    landslide, 221
Saint-Amand, P., 100
Saint Jean-Vianney landslide, Quebec, 217
St Lawrence lowland, 465
St Pierre, Martinique—destroyed by
    volcanic explosion, 138, 156
    1971 landslide, 219
salina, 291 et seq.
salt
    crystallisation, 375
        difficulties, 183
        principle, 183
    deposits in Permo-Triassic deserts, 505
    dome, 122, 123
    fretting, 183
    plate, 165
    ridges, 167
    and soil erosion, 519
    weathering and tafoni, 205
saltation, 235, 236, 288
Salton Sea, Calif., 282, 285
    barchan, 296
Salzsprengung, 183
San Andreas Fault, Calif., 87, 97, 98, 99, 225
    zone, drowned, 368
San Fernando, Calif.—landslides, 1971, 219
sand
    blasting, 289
    dunes, 4, 13, 293 et seq.
    ridge, 292, 294
    ripple, 297
    sheet, 292
    storm, 286
sandstone
    bluff, 62

Sandstone (*continued*)
    butte, 105
    definition, 61
    domed inselberg, 62
    joint pattern, 64
    landforms, 61–63
    minor landforms, 62
    plateau, 62
    tower, 102, 104
    turrett, 64
    sheet structure, 59
    weathering, 62
sandur, 362
Sangamon shoreline, U.S.A., 495
San Francisco—earthquake, 1906, 81
saprolith, 179
sastrugi, 317
scales of climatic change, 455
scallops—riverine, 233, 235
Scandia Fault, New Mexico, 90
scarp foot depression, 210
scarp foot valley, 210
scarp foot weathering, 14, 15, 209–211
scarp retreat, 91, 406, 409
    and inselberg development, 53, 54
Scharding, Austria, 462
Scheffer, F., 193
Scheffer, V. B., 511, 527
Scheidegger, A., 534
schist
    landform, 77
    tor, 79
    tower, 79
Schlee, J., 41
Schmidt, M., 124
Scholl, D. W., 158
Schooley Surface, 422
Schoonmaker, W. J., 511, 527
Schrepfer, H., 534
Schreve, R. L., 228
Schumm, S. A., 215, 228, 255, 420, 427
    502, 527
Schwartzbach, M., 479
scoria, 125
scoria cone, 138, 152
Scotland
    post glacial emergence, 476
    post glacial isostatic recovery, 474
    unconformity Lewisian and Torridonian,
        506
scree, 226, 330
Scripps Canyon, Calif., 501
scroll pattern, 229
scroll plain, 136
scrub fire—and weathering, 179
sea ice, 340
sea floor spreading, 35
sealevel changes
    Quaternary, 494
    late Quaternary, 494
sealevel and temperature changes, 489
Searles Lake, Calif., 481
    section, 484
Seddon. G., 169
sedimentary rocks, 535
    classification (table), 536
    landforms, 101 et seq.

sediplain, 101
Sedmik, E. C. E., 169
Seefeldner, E., 451
Segilman, G., 365
seif dune, 301
    development, 304
seismic shock wave, 27
seismograph, 27
seismology, 24
Selima sandsheet, Sudan, 292
selvas, 273
Selwyn, A. R. C., 509
Selwyn upwarp, 434
senile stage of landscape development, 404
Seppala, M., 332
serir, 288
serrated shore platform, 383, 384
settling velocity, 236
shadow zone, 27
Shag Point, Palmerston, N.Z.
    basin and relief inversion, 377
    coastal flares, 13
    coastal pits, basins, hollows, 375
shake wave, 27
shale—landforms, 79
Shaler, N., 365
shape of
    beach pebbles, 234
    boulders, 54
    earth, 21
    inselbergs, 54–6
    stream pebbles, 234
Shapley, H., 478
Sharp, R. P., 100, 228, 315, 465, 478
Sharpe, C. F. S., 228
shatter cone, 161
Shaur Anticline, 112, 438
shaved surface, 328, 330
Shearer, J. M., 332
shear patterns of earth, 33
shear plane, 222
shear wave, 27
Sheep Mt., Greybull, Wyoming—breached
    dome, 115
sheet
    calcrete, 200, 201
    erosion, 518
    silcrete, 197
    structure, 57, 58, 60, 61, 181
        faulting, 59
        granite, 57 et seq., 205, 379
        porphyry, 59
        sandstone, 59
sheetwash, 229
shelf ice, 339
shelter, 103, 267
Shepard, F. P., 503
Shepherd, Jennifer A., 124, 527
Sherman Surface, 420, 422
shield, 39
shield volcano, 129
shingle beach, 381, 386
Shiprock, New Mexico, 152, 154
shoal, 252
Shoemaker, E. A., 169

shore platform, 371–85, 491
    interpretation, 373, 385
    lacustrine, 384
    raised, 490
shore profile
    smooth, 378
    stepped, 378
Shotover River, N.Z., 413
shower
    tectite, 160, 161
    meteor, 161
shute, 229, 245
shutteridge, 96, 97
sial, 28, 39
Siberia
    basalt flows, 132
    meteorite fall, 161
    pingos, 323
    proglacial lake, 472
    thermokarst, 327, 328
Sidmouth, Devon—cliff gullying, 372
Sierra Nevada, Calif.
    flared slopes, 7, 13
    granite "flats", 189
    granite forms, 46 et seq.
    Pleistocene glacial chronology, 460, 461
    sheet structure, 60
    stepped topography, 213, 214, 428
Sikhole Alin, Siberia—meteorites, 161
silcrete, 5, 112, 185, 195
    age, 197
    caprock, 197, 504
    central Australia, 199
    composition, 197
    distribution, 197, 199
    folded, 111, 198
    gibber, 284
    opaline, 197
    origin, 198 et seq.
    profile, 197
    scarp foot, 197
    sheet, 197
    skin, 198
silica flower, 157
silicate structures, 187
Silikatrillen, 203
sill, 535
sima, 28
Simpson Desert, 295, 304, 531
    dune change in time, 303
    longitudinal dunes, 295–302
    origin of dunes, 301–09
Simpson Gap, N.T., 436
Sinai Peninsula
    granite hills, 51
    fault control, 87, 88
sinistral wrench fault, 544, 545
sinkhole, 67
Sissons, J. B., 472, 478
Sivarajasingham, S., 215
skaargard, 382
Skeidarsandur, Iceland—braided channel,
    363
Skeleroo Ck., S.A.—breached ridge, stream
    piracy, 442, 444
Skelton, J. S., 215, 313

Skelton Fjord, Antarctica, 350
    section, 354
skin—silcrete, 198
slabslide, 223
slickensides
    faulting, 7, 87, 544, 545
    associated with landslides, 222
slipe, 217
    of fault, 540
    off slope, 248, 250
    sheet, 162, 163
slippage in glaciers, 343
slope
    behaviour in arid lands, 406
    budget, 270
    decline in piedmont zone, 211, 278
    disequilibrium, 270, 271
    equilibrium, 270, 271
    faceted, 259 et seq.
    failure, 217
    form, and relation to structure, 262
    graded, 259 et seq.
    humid tropics, 273
    instability, 217
    lowering in piedmont zone, 211, 278
    variations in time and space, 262–65
    wash, 222
    and weathering, 192
Slotbloom, R. T., 502
sluff, 318
slumping (river bank), 247
Smiggins Holes, N.S.W.—flared granite
    boulder, 13
Smith, C. T., 17, 526
Smith, Dianne M., 430, 509, 528, 534
Smith, D. L., 17
Smith, H. T. U., 365
Smith, J., 332
Smith, P. A., 503
Smith, T. B., 332
Smith, W. G., 398
snow, 317
    dune, 317
    "mushroom", 317
snow patch sliding, 318
Snowy Mts., N.S.W.
    corestone, 49
    split boulder, 56
So, C. L., 228
Sogne Fjord, 350
    section, 354
soil catena, 192
soil erosion, 519, 520, 524
    causes of, 519, 521
    goats, 522
    grazing, 519
    land clearance, 519
    ploughing, 519
    salt, 519
solfatara, 156
solfataric activity, 156
solifluction, 219, 328
    deposits, 480
    lobe, 329
    stream features, 328
solution, 185, 186

solution (*continued*)
coastal, 373, 375
cup, 68, 71, 375
depression, 68
doline, 70, 71
pan, 69
river, 233, 235
Somerville Surface, 422
somma, 138, 140
Sorrento, Vic.—shore platform and cliff, 374
sorted polygon, 319
sorted stripes, 321
sorting action of wind, 285, 287, 288
Soufriere (volcano), St Vincent, 138, 142, 144
late Pleistocene activity, 138, 139
rate of weathering, 192
Souris River, N. Dakota
misfitness, 484, 485
terraces, 412
South America—glacial and nival forms, 480
South Australia
alluvial fan, 243, 244, 484
coastal forms, 369, 371–76, 382, 384, 491, 495, 496
desert forms, 165, 183, 257, 282–84, 290–95, 301 et seq., 484, 485, 531
drainage anomalies, 441 et seq.
erosion surfaces, 415, 421, 426
etch surface, 421
faults and related forms, 88, 94, 293
Flinders Ranges, 107 et seq., 441 et seq.
fold mountains, 107 et seq., 441 et seq.
Gawler Ranges, 42, 44, 59, 508, 521
granite forms, 9 et seq., 42–50, 54–7, 77, 184, 205, 379
karst, 67 et seq., 71
Lake Eyre, 197, 198, 257, 283, 284, 295, 484, 485, 531
laterite, 426
lunette, 311, 312
Mt Lofty Ranges, 94, 162, 163, 259, 420 et seq., 433
Murray Valley and River, 97, 262 et seq., 314, 485
Permian glaciation, 507
recent folds, 164, 165
relict forms, 484 et seq.
soil erosion, 516 et seq.
submarine canyons, 497, 498
volcanic forms, 152, 463
South Downs, England—late Tertiary erosion surface, 505
South East of South Australia—stranded coastal foredunes, 495, 496
South Nation River, Ontario—1971 landslide, 219
South Shetland Islands—isostatic recovery, former strandlines, 495
South Wales—superimposed drainage, 433, 439
Sparks, B. W., 215
Speight, J. G., 365
speleothem, 72
Spencer-Jones, D., 124
spheroidal weathering, 180, 181, 205

spillway (glacial)
characteristics, 472
disputed, 472
Labrador, 472
spires, granite, 55
spit, 386, 387, 388, 391
orientation, 389
Spitzbergen, karst, 74
splitting of rocks—by heat, 180
spray jets, 371
Sprigg, R. C., 100, 124, 399, 444, 451, 503
Springer, M. E., 168, 169
Spurn Head, Yorks—growth, 390
stable landforms, 427, 428
Stace, H. C. T., 193
stack
Blanche Point, S.A., 382
Isle of Wight, 381
Orkney Islands, 380
Port Campbell, Vic., 380
Whakapohai Beach, N.Z., 381
raised, Kangaroo Island, 491
Stager, J. K., 323, 332
stages in breaching of ridge, 442, 445
stagnation of ice, 359
stalactite, 66, 72, 201
stalagmite, 72
Stanley, G. M., 215
Stansbury level, Lake Bonneville, 481
Stapledon, D. H., 169
Stassfurt, Germany—salt deposits, 505
static threshold, 288
steady state landscape change, 403, 424
difficulties, 425, 426
steam upthrust—and craters, 161
Stearns, H. T., 478
Steers, J. A., 41
step (valley), 347
Stephens, C. G., 215, 311, 316, 509
stepped topography, 211, 213
origin, 214
Sternberg, H. O'R., 100
Stettin Gulf, East Germany—cuspate spit and lagoon, 391
Stirling Ranges, W.A., 116
Stone, K. H., 479
stone
banked terrace, 328
circle, 168
mantle, 168
polygon, 168
step, 168
stream, 328
stripe, 168
Stone Mountain, Georgia—granite dome, 52, 54
stony desert, 3, 282
stony rise—basalt, 135
storms—significance, 531
Strahan, A., 433, 439, 451
Strahler, A. N., 429, 442, 451
strain in folds, 112, 113
strandflat, 378, 382
stratovolcano, 144
stratum contour, 538, 539
Straw, A., 502

stream
capture, 442, 444
antiquity of, 444
deposition and grain size, 233
deposition and stream velocity, 233
deposition, velocity and grain size, 232
energy, 230
erosion, 233
erosion and grain size, 233
erosion and stream velocity, 233
erosion, velocity and grain size, 232
floods, velocity and grain size, 232
flow, 230
meander, 244–52
persistence, 434 et seq.
piracy, 442
rejuvenation, 87
sorting, 236
transport and grain size, 233
transport and stream velocity, 232
transport, velocity and grain size, 232
velocity, 231
velocity and erosion, 233
velocity and deposition, 233
velocity and transportation, 233
volume, 231
streams
baselevel, wasting—relationship, 426
—*see also* rivers
striae, 341, 438
strike, 538, 539
fault, 540, 543
stream, 431, 437
Strombolian eruption—characteristics, 125 127
structural
bench, 106, 262,
coastal, 379
control of drainage patterns, 431
effect on coastal forms, 369, 370, 379
landform, 42 et seq.
definition, 40
plain, 101
plateau, 106
structure
contour, 538
of crust, 27, 28
of earth, 24 et seq.
and mass movements, 219
and weathering, 190
Strzelecki Ck., S.A.—fault control, 81
Sturt, C., 3, 17, 295, 315
Sub Atlantic phase, 456
Sub Boreal phase, 456
subdebris convex slope, 266
subduction zone, 36
subglacial channel, 472
submarine canyon
description, 495–97
drowned river gorges, 500
origin, 497 et seq.
southern Australia, 498
spatial distribution, 497–98
time distribution, 495
turbidity currents, 500
submarine mountain ranges, 24
and volcanoes, 125

submarine ridges
  distribution, 25
  geophysical characteristics, 35
submarine trench, 35
submerged forest, 491
subsidence, 159, 217
substitution, 187
subsurface flushing, 280
  planation, 279
  weathering, 279
  weathering and flared slopes, 14, 15
subterranean stream capture (limestone),
  444
sugar loaf, 47, 60
  and faulting
Sugden, D. E., 503
Suggate, R. P., 478
Sullivan, Margaret E., 316
sulphur springs, 156
Sumatra—acid lava, 152
Sumatra Fault, 98
  and volcanoes, 127
Sunda River (Pleistocene river), 493, 497
Sunda Shelf, 492, 497
superimposition, 433, 434, 439
surface creep, 288
Surry Scarp, N. Carolina, 495
Surtsey, Iceland—wave erosion, 371
Susitna Lowland, Alaska, 357
suspension, 235
Susquehanna River, U.S.A.—transverse
  drainage, 433, 437
Suswa Volcano, Mountains of the Moon,
  Kenya, 135
Swallow Cliff, S.A.—gully gravure, 269
swash, 370
Sweatman, T. R., 18
Sweeting, M. M., 80, 169
swell, 370
Swindle Sound, B.C., 43
Swithinbank, C., 365
sword-shaped dune, 301
Syene, Egypt—and Eratosthenes'
  calculation, 21
syncline, 107, 540
  perched, 119
synclinorium, 540, 541
synform, 60
Szentes, F., 124

Taber, S., 332
tabular iceberg, 340
Taft, Calif.—offset stream, 99
talus apron, 226
  cone, 351
  creep, 222
tang, 161, 437
Tangle Lakes, Alaska—rock glacier, 331
Tanner, V., 428
Tanner, W. F., 255
Tardy, Y., 187, 193
Tarling, D. H., 41
Tarling, M. P., 41
tarn, 344, 346
Tarrant, J. R., 430
Tasman Glacier, N.Z., 356, 359
Tasmania—glacial and nival forms, 480

Tassili Mts, Algeria—sandstone plateau,
  mesa, 61, 62, 102, 104
Tate, R., 451
Tator, B. A., 281
Tavernas, F., 228
Taylor Dry Valley, Antarctica, 206
Taylor, F. B., 451
Taylor, G. A., 18, 158
Taylor, J. H., 169
Taylor, T. G., 158, 451
Tazlitwo Lake, Alaska—modern proglacial
  lake, 465
tear fault, 540
Teays River, U.S.A.—Pleistocene river,
  465, 468
tectite, 161
  distribution in Australia, 160
  shower, 161
tectonic landform, 42 et seq.
  definition, 40
  instability and mass movements, 219
  regions of continents, 39
tectono-eustatic changes of sealevel, 489
Tees Valley, England—salt deposits, 505
temperate region, 173
temperature fluctuations—Pleistocene, 454
temperature and sealevel changes, 489
Tenaya Lake, Calif.—inverse sheet
  structure, 59, 60
tension scar, 222
tensional fault, 540
tephra, 152
tephrochronology, 152
Te Punga, M., 332
terminal moraine, 355, 359, 362
termination of geomorphic cycle, 415
termitaria, 510, 511
termites, 510
  and weathering, 184, 185
termite mound, 208
terrace, riverine, 410, 411
  as indicators of climatic change, 485
terrestrial factors—climatic change, 457
Tethys, 32
tetrahedral theory, 29
tetrahedron, 29
Texas
  clay dune, 311
  late Pleistocene-early Recent lakes, 481
  relict water, 230
  salt domes, 122
texture (rock)—and weathering, 190
textured relief, 79
thalweg, 231
Thatcher, D., 451
thermokarst, 326, 327, 328
  man-induced, 522
Thesiger, W., 193
Thingvellir, Iceland—faulted region, 86
thixotropic clay, 220
thixotropy, 220
Thom, B. G., 18, 332
Thomamys talpoides, 511
Thomas, M. F., 193
Thompson, C., 398
Thompson, H. D., 124, 451
Thomson, Robyn M., 124

Thorarinsson, S., 158, 255, 365, 479
Three Sisters, Blue Mts, N.S.W., 64
throw of fault, 540
thrust fault
  lag, 543, 540
  normal, 543, 540
Tibbitts, G. C., 315
tidal lagoon, 393
Tiedemann Glacier, B.C., 465
tierra caliente, 74
tierra fria, 76
tierra templada, 74
Tigris-Euphrates delta, 237
  growth, 238
till, 355
till plain, 357
time—landscape changes, 401, 403
Tintina Trench, Yukon Territory, Canada,
  225
tjale, 323
Tokachioki earthquake, 26
Tomales Bay, Calif.—drowned fault zone,
  368
tombolo, 361, 389, 390
tombstone, 77
Tonga Trench, 35
tongue structure, 163
Tooma Dam Site, N.S.W.—onion
  weathering, 180
Toowoomba, Qld—gilgai, 166
topography—and weathering, 192
topset beds, 238
tor, 42
  gritstone, 3
Torrens Gorge, S.A.
  gravity induced bulges, 162, 163
  regrading of tributaries, 259
  accommodation of river curves to
  structure, 433
Torrens Surface, 415
tower—sandstone, 102
Tower Hill, Vic.—caldera and ash layers,
  137, 138, 149, 152
towerkarst, 74
  Sierra de los Organos, Cuba, 75
track—washed out, 521
tracks of tectite shower, 160
traction load, 235
traditional glacio-eustatic scheme of
  Pleistocene high sealevels, 492
  difficulties, 492, 495
  table, 493
Tradour River, Cape Province—transverse
  stream, 433
transcurrent fault, 96, 97, 99, 540
transection glacier, 333
transference of load, 164
transportation, 173, 235
  and weathering, 195
transverse drainage, 433 et seq.
  components, 433
transverse dune, 292, 295
  distribution, 297
  structure, 297
transverse ripple, 294
transverse wave, 27
travel paths of saltating sand grains, 288

travertine, 200, 201
 and relief inversion, 201
Travertine Point, Calif.—Pleistocene
 shoreline, 481
Treak Cliff Cave, Castleton, Derbyshire, 73
tree ring analysis, 455
tree roots—and weathering, 184
trellis drainage, 431, 432, 442
Trendall, A. F., 193
triangular facet
 erosional, not associated with faults, 89
 related to faults, 87, 88, 89
Tricart, J., 18, 124, 173, 177, 178, 272, 281,
 316, 365, 398
 world map of morphogenetic regions, 175
Troll, C., 502
tropical lands—Pleistocene chronology, 460
trough (of fold), 540
trough line, 540
true dip, 538
truncated drainage, 491
truncated spur, 345
Tsuya, H., 158
Tuckfield, C. G., 527
Tufa Point, Calif.—Pleistocene shoreline,
 481
tundra, 159
Tungkillo, S.A.—tombstones, 77
turbidity
 Hwang Ho, 245
 Mississippi, 245
 Rio Grande, 245
turbidity currents
 laboratory experiments, 501
 and submarine canyons, 500
turbulent eddies, 285
 updraft, 236, 286
turbulent flow, 231
turf-banked terrace, 328
turmkarst, 74
Turnbull, W. J., 315
Turner, F. J., 80, 548
turrets—granite, 55
Turtle Mt., Canada—landslide, 223
Twain, Mark, 3
Twelve Apostles, Vic.—stacks, 380
Twidale, C. R., 17, 18, 80, 100, 124, 158,
 178, 193, 215, 216, 279, 281, 313, 315, 316,
 429, 430, 451, 502, 503, 509, 527, 528, 534
Twistleton Scars, Ingleborough, Yorks—
 karst, 69
Two Creeks interval, 456
two stage development
 of boulders, 50
 of inselbergs, 52, 53
types of mass movement, 221
type of weathering—factors, 189
Tyrrell Sea, Canada, 475

U-dune, 313, 395
ultimate baselevel, 256, 257
 platform, 385
Umbgrove, J. H. F., 41
Umgeni River, Natal
 pot hole, 233
 scallops, grooves, 235
 scarp retreat, 278

unbuttressing, 163, 226
unconformity, 506, 539
undercut bank or bluff, 227, 247, 248, 250
underfit stream, 485
undermass deposit, 439
undermining, 226
 and mass movement, 221
Unequal Activity—law of, 428
uniclinal shifting, 262
uniformitarianism, 5, 533
 and evolution, 526
unglaciated areas—Pleistocene effects, 480
unloading of crust, 163
unpaired terraces, 258, 259
upper slope, 259
urstromtäler—North European plain, 434,
 469, 471, 472
U.S.A.—deserts
 drainage development, 433 et seq.
 glacial chronologies, 458 et seq.
 Great Lakes, 468, 469
 modern proglacial lakes, 465
 pediments, 276 et seq.
 Pleistocene lakes and drainage
 modifications, 465 et seq.
U-shaped valley, 344, 345, 347, 349
 and beaver, 511
 development, 350
U.S.S.R.
 depth of weathering, 190
 thermokarst, 327, 328
 vertical movements—contemporary, 109,
 474
Utah—pediments, 277
uvula, 74

vadose water, 72
Valentin, H., 474, 478
Vale of Pickering—Pleistocene proglacial
 lake, 469
Valery, Paul, 6
valley
 deepening, 256
 glacier, 344 et seq.
 impression, 440, 442, 444
 shoulder, 349
 step, 244, 347
 train, 363
 widening—processes at work, 258
valley-in-valley
 form, 445
 related to faults, 88
Valley of the Thousand Hills, Natal—scarp
 retreat, 54, 278
valleys
 origin, 229, 230
 side facet, 410, 411
Van Baren, H., 188, 193
Van den Toorn, J. C., 502
Van Rhynesdorp, Cape Province, S. Africa
 —scarp foot depression, 211
Van Son, J., 169
variation in velocity in glaciers, 342
Varvau, Tonga—sealevel notch, 373
varve clays, 469
Veatch, A. C., 503
velocity—stream activity, 232

velocity gradients (wind)above ground, 287
velocity variation (river)
 along thalweg, 231, 232
 in time, 232
 in cross-section, 231
ventifact
 cold lands, 318
 hot desert, 289
 Permian desert, 507
Venturi effect, 231
Verhoogen, J., 548
Verstappen, H., 516, 527
vertical movements
 in Canada (eastern), 477
 in crust, 37
 in France, 109
 in Japan, 109
 in New Zealand, 109, 111
 in Scandinavia, 109, 475
 in Scotland, 476
 in U.S.S.R. (European), 109
vertical zonation of caves, 71, 72
vesicular structure
 in laterite, 196
 silcrete, 197
Vesuvius, 129, 140
Victoria
 coastal forms, 94, 374, 380
 gilgai, 166
 Grampians, 61–6, 121
 lunette, 311
 Mt Buffalo, 7–14
 Permian glaciation, 507
 Port Phillip Bay, 94
 recent faulting, 97, 403, 434
 relict dunes, 485
 volcanic features, 133, 137, 138, 149, 152,
 153
 Werribee Gorge, 257
Victoria Falls—joint control, 43
Victorville, Calif.—warping, 107
Vine, F. J., 41
visor, 206
Vogt, J., 527
volcanic
 ash, 147
 crater, flooded, 132
 drainage diversion, 138
 landforms, 125 et seq.
 lava—extent, 125
 magma, 125
 plug, 152–56
 rocks, 535
volcano
 active, 125
 characteristics, 125–27
 classification, 125–27
 composite, 144 et seq.
 distribution, 125
 dormant, 125
 extinct, 125
 former distribution, 125
 general significance, 156
 Hawaiian, 128 et seq.
 Peléan, 138 et seq.
 strato, 144 et seq.
 variations, 128

Volga Delta, 241
volume change, 159
volume increase—water, 167
volume of stream formula, 231
volume variation in time (river), 232
Von Bertalanffy, L., 430
Von der Borch, C. C., 503
Von Heune, R., 158
Vosges, France—horst, 92, 94
Vredeford Dome, S. Africa, 161
V-shaped valley—in relation to faults, 87
Vulcanean eruption—characteristics, 125, 127

Wadham, S. M., 215
wadi, 102
Wager, L. R., 438, 451
Wahrhaftig, C., 216, 228, 332, 430, 502, 548
Wales—erosion surfaces, 422, 423
Walkers Flat, S.A.—slope profiles and variations, 264, 265, 267
Wallwork, K. L., 528
Walton, B. J., 169
Walton, K., 479
Warbrick Terrace, Rotorua, 157
Ward, W. H., 228
Ward, W. T., 18, 398, 451, 503
warped surface, 107
    California, southern, 109
    European U.S.S.R., 109
    Fennoscandia, 109
    France, 109
warping due to deltaic subsidence
    Mississippi valley, 242
    shorelines, 474
    stream diversion, 434
Warren, A., 316, 503
Warrens Gorge, Flinders Ranges, S.A.
    clefts, 62
    breached ridge, 444
Warrumbungle Mts, N.S.W.—volcanic, 155, 156
Warth, H., 193
Wasatch Mts, Utah
    faulted terrace, 91, 94, 95
    glaciation, 461
Wase Rock, Nigeria—volcanic plug, 153
Washburn, A. L., 169, 332
Wast Water, England—scree, 330
wasting, baselevel stream—relationship, 426
Watanabe, T., 193
water
    and crystal surface, 186
    and mass movements, 219
    effect with feldspar, 186
    relict, 230
    role in chemical weathering, 185
water and wind action compared, 282
Watern Tor, Dartmoor, England, 58
Waters, R. S., 193
Watson, Janet, 548
Watson, R. A., 228
Watts, W. W., 509
wave, 366 et seq.
    breaking, 370
    diffraction, 371
    erosion, 371 et seq.

wave (*continued*)
    forced, 370
    free, 370
    height, 370
    length, 370
    period, 370
    reflection, 370
    refraction, 370, 371, 385
    snow, 317
Wave Rock, W.A.—flared slope, 10, 15
Wayland, E. J., 429, 478
Wealden Anticline, Kent, 113
weathering, 179 et seq.
    alveolar, 65
    cold lands, 317
    definition, 179
    economic significance, 195
    front, 54, 179
    geomorphological significance, 195
    mechanical, 179 et seq.
    and pediment development, 280
    physical, 179 et seq.
    spheroidal, 180
    tropical lands, 188 et seq.
weathering pit, 66, 202
Weertman, H., 365
Wegener, A., 32, 33, 41
Weiblen, P. W., 216
Weipa, Qld—bauxite, 188, 189, 196
Weiss, L. E., 548
Wellington, N.Z.—shaved surface, 328
Wellington Fault, N.Z., 87
Wellman, H. W., 124, 193, 216, 489, 503
Wells, C. B., 18
Wells, J. F., 281
Wentworth, C. K., 18, 228, 398
Werribee Gorge, Vic., 257
West, R. G., 478
West, W. D., 452
West Africa
    crater, 159
    laterite, 196
    superficial disturbances, 163
West Barnes River, W.A.—fault controlled, 85
West Branch, U.S.A.—transverse stream, 433
Western Australia
    coastal forms, 299, 367
      evidence of late Quaternary sealevel change, 495
    erosion surfaces, 416, 420
    etch plain, 420
    faulting and fault control, 85, 91
    granite forms, 10, 15, 56, 204, 420
    karst, 67 et seq., 72, 73
    laterite, 196, 421
    new plateau, 420
    old plateau, 420, 421
    plateau, mesa, butte, 7, 16, 72, 101, 105, 106, 421, 537
    pseudo-anticline, 167
    Stirling Range, 116
    submarine canyon, 499
Western Cordillera, U.S.A.—glacial chronology, 460
Western Districts, Victoria—volcanic, 107

Westralian Shield, 107, 122
Westward Ho!, Devon, 392
wetted perimeter, 231
Weye, R., 192, 194
Whakapohai Beach, N.Z.—stack, 381
"Whale's Jaw", Grampians, Vic., 65, 66
"white horses", 370
White Island, N.Z.—volcanic, 132, 133
White, P., 17
White, W.A., 365
Whiteman, A.J., 100
Whitehouse, F.W., 451
Whitsunday Island, Qld, 55
whorled structure—silcrete, 197
Whyte, R.O., 527
Wick, Scotland—work of waves exemplified, 371
Widallion Ck., Qld,—anomalous, 440
Wilhelmy, H., 178, 216, 534
Willett, H.C., 478
William of Orange, 534
Williams, G. E., 161, 169, 281, 429, 502, 509
Williams, H., 158
Williams, M. A. J., 185, 193, 228
Williams, P. J., 332
Williams, R. S., 41
Williams, W. W., 398
Willochra Ck., S.A.—anomalous, 440, 441
Willouran Range, S.A.—folds, 115
Wills, W. J., 3, 180
Willunga Scarp, S.A.—gullying, 516
Wilpena Pound, S.A.—natural amphitheatre, 116, 181, 432, 540
Wilson, A. T., 193, 216
Wilson, I., 297, 315
Wilson, J. T., 41
Wimmera, Vic.—relict dunes, 485
wind
    action, 282 et seq.
    deposition, 292 et seq.
    erosion, 291, 318
    faceted stones (cold areas), 318
    fluting (gneiss), 291
    sorting, 285 et seq.
    transportation, 285 et seq.
wind-rift dunes, 289, 290
wind and water action compared, 282
Windy Point, Calif.—wind erosion, 289, 291
wineglass valley, 87, 88, 90
Wissman, H. Von, 80
Wittenoom Gorge, W.A.—erosion surfaces, 416
Wodehouse, P. G., 455
Woldstadt, P., 479
Wolf Creek Crater, W.A., 159
Wolf Creek Glacier, Alaska, 465, 467
Wolff, R. G., 216
Wolman, M. G., 232, 255, 396, 397, 534
Wondoola Plains, Qld—depositional, 405
Woodard, G. D., 429, 451
Woods, J. E. T., 313, 316
Wooldridge, S. W., 17, 124, 404, 509
Woolnough, W. G., 195, 215, 509
Wopfner, H., 124, 215, 272, 315, 316, 429, 502
worms—and weathering, 184, 185
Wray, D. A., 80

wrench fault, 97, 540
Wright, H. E., 228
Wright, R. L., 509
Wright, W. B., 365
Wrightwood, Calif.—mudflow, 225, 226, 227

Yamasaki, M., 193
Yampi Sound, W.A.—irregular coast, 367
yardang, 289, 290
Yarrah Vale, S.A.—imminent drainage capture, 442, 444
Y-junction, 301, 304
Yoldia Sea, 474

Yolnir, Iceland—volcanic eruption 1966, 152, 371
Yorke Peninsula—relict dunes, 485
Yorkshire, northwestern—karst, 69
Yosemite National Park, Calif., 59
Yosemite Valley, Calif.—glaciated with granite domes, 347, 352, 511
Yoshikawa, T., 124
Young, G. A., 169
youthful stage of landscape development, 404
Yukon coastal plain and sea ice, 340

Zabriskie Point, Death Valley, Calif., 211

Zagros Mts., Iran
    structures, 116, 117, 118, 119, 121, 161, 250
    alluvial fans, 243, 435, 437
    drainage patterns, 440, 443, 444
Zambezi River—joint controlled course, 43
Zaruba, Q., 228
zastrugi—cf. with sand dunes, 317
Zenkovitch, V. P., 398
Zernitz, Emilie R., 431, 451
zeugen, 289
Zeuner, F. E., 18, 463, 478
Zion National Park, Utah—sandstone, 63, 506
Zuider Zee, Holland—reclamation, 525